HANDBOOK ON THE PHYSICS AND CHEMISTRY OF RARE EARTHS

VOLUME 15

HANDBOOK ON THE PHYSICS AND CHEMISTRY
OF RARE EARTHS

HANDBOOK ON THE PHYSICS AND CHEMISTRY OF

RARE EARTHS

VOLUME 15

EDITORS:

Karl A. GSCHNEIDNER, Jr.

Ames Laboratory–US DOE, and
Dept. of Materials Science and Engineering
Iowa State University
Ames, Iowa 50011
USA

LeRoy EYRING

Department of Chemistry, and
Center for Solid State Science
Arizona State University
Tempe, Arizona 85287
USA

NORTH-HOLLAND

1991 AMSTERDAM, LONDON, NEW YORK, TOKYO

ISBN: 0 444 88966 3

Published by:

North-Holland

Elsevier Science Publishers B.V.

P.O. Box 211
1000 AE Amsterdam
The Netherlands

Sole distributors for the USA and Canada:

Elsevier Science Publishing Company, Inc.

655 Avenue of the Americas
New York, NY 10010
USA

Library of Congress Cataloging in Publication Data

(Revised for volume 15)

Handbook on the physics and chemistry of rare earths.

Vols. 5– distributed by: Elsevier Science Pub Co.
Errata [for volumes 1–6]: v. 8, p. 375–378
Includes bibliographies and indexes.
Contents: v. 1. Metals – v. 2. Alloys and intermetallics. – v. 3–4. Non-metallic compounds.– [etc.] – v. 12, 15 [without special title].
1. Earths, Rare. I. Gschneidner, Karl A. II. Eyring, Leroy. III. Rare Eaths.
QC173. R2H26 546'.4 78-12371
ISBN 0-444-85022-8 (set)

Printed on acid-free paper

Transferred to digital printing 2006

Handbook on the Physics and Chemistry of Rare Earths, Vol. 15
edited by K.A. Gschneidner, Jr. and L. Eyring

PREFACE

Karl A. GSCHNEIDNER, Jr., and LeRoy EYRING

These elements perplex us in our rearches [sic], *baffle us in our speculations, and haunt us in our very dreams. They stretch like an unknown sea before us – mocking, mystifying, and murmuring strange revelations and possibilities.*

Sir William Crookes (February 16, 1887)

About a decade ago, volume 5 of this serial publication signaled a decision by the Editors and Publishers to continue to provide topical and authoritative reviews of the entire scientific and technical literature of the rare earths. This decision followed the encouraging reception of the first four volumes presented to the world as an advanced, comprehensive introduction to the field. Now, ten years later, continuing this voyage of discovery, we are pleased to launch volume 15 containing chapters approaching and beyond 100.

In chapter 98, Julian Sereni adds significantly to an evaluation of systematic, experimental low-temperature studies of the ambivalent behaviors of cerium (ferromagnetism, antiferromagnetism, spin glass, superconductivity, valence fluctuations, heavy Fermion, Kondo and spin fluctuations) which depend upon its environment in materials. The systematic conclusions arrived at should provide new data against which the theory can be advanced.

Drs. G.-y. Adachi, N. Imanaka, and Z. Fuzhong review the rare earth carbides (chapter 99) emphasizing the thermodynamics, phase diagrams, crystal structures, and physical properties. The binary rare earth carbides present an exceptionally wide range of compositions and structures both as solids and gas-phase molecules. More complex carbides with additional metal and non-metal components also receive attention.

Metal-rich halides (i.e. compounds with an X/R ratio ⟨2) are systematized and reviewed in chapter 100 by A. Simon, Hj. Mattausch, G.J. Miller, W. Bauhofer, and R.K. Kremer. These compounds are based upon octahedral M_6 cores either with X atoms above the eight faces or above the 12 edges, forming M_6X_8 or M_6X_{12} clusters, respectively. Many of these metal-rich phases are stabilized by an interstitial atom (generally oxygen, carbon and hydrogen) which occupies the center of the R_6X_{12} cluster. The compounds are classified according to their structure and chemical bonding characteristics. Their electrical and magnetic properties are also reviewed.

Heavy-metal fluoride glasses, into which large amounts of rare earths can be added, possess a great potential for optical applications in the mid-infrared range as fiber optic glasses for communication and transmission of information, optical wave guides, fiber lasers, and sensors. In chapter 101, Professor R.M. Almeida discusses the preparation of these glasses, their structure and chemical and physical properties. He pays particularly close attention to those properties that are critical for their practical use.

In chapter 102, Drs. K.L. Nash and J.C. Sullivan explore the chemical kinetics of solvent and ligand exchange in aqueous lanthanide solutions. These authors deal with redox reactions readily available only from the Ce(IV)/Ce(III) and the Eu(II)/Eu(III) couples among the lanthanides. A wealth of tabulated information on rate and equilibrium constants is provided in textual and tabular form.

The fundamentally important reactions of the lanthanide ions with water are considered by Drs. E.N. Rizkalla and G.R. Choppin in chapter 103. These interactions are discussed for both solids and solutions. In aqueous solution the hydrated species are considered in detail, revealing the consequences of the lanthanide series sequence.

Professor L.M. Vallarino's review of macrocyclic complexes formed by rare earth and dioxouranium(VI) ions as templates concludes this Volume as chapter 104. Synthetic trends and reactivity are considered as well as potential uses of these intriguing wrap-around structures.

CONTENTS

CONTENTS OF VOLUMES 1–14

VOLUME 1: Metals

VOLUME 2: Alloys and intermetallics

VOLUME 3: Non-metallic compounds – I

VOLUME 4: Non-metallic compounds – II

VOLUME 5

VOLUME 6

VOLUME 7

VOLUME 8

VOLUME 9

VOLUME 10: High energy spectroscopy

VOLUME 11: Two-hundred-year impact of rare earths on science

VOLUME 12

VOLUME 13

VOLUME 14

Handbook on the Physics and Chemistry of Rare Earths, Vol. 15
edited by K.A. Gschneidner, Jr. and L. Eyring

Chapter 98

LOW-TEMPERATURE BEHAVIOUR OF CERIUM COMPOUNDS

Specific Heat of Binary and Related Intermetallics

JULIÁN G. SERENI
División Bajas Temperaturas, Centro Atómico Bariloche, and CONICET,
8400 S.C. de Bariloche, Argentina

Contents

Symbols

A	coefficient of the T^2 term of $\rho(T)$
B	coefficient of C_M in the magnetically ordered phase
C	Curie constant
C_E	free-electron specific heat
C_L	lattice specific heat
C_M	magnetic specific heat
C_N	nuclear specific heat
C_p	constant-pressure specific heat

C_{Sch}	Schottky specific heat
C_v	constant-volume specific heat
CF	crystalline field
d	spatial dimensionality
dT	$T_f - T_i$
E_g	energy gap in the magnon spectrum
F	Helmholtz free energy
g	degeneracy ratio between the excited and ground CF states
H	applied magnetic field
HF	heavy fermion
I	intensity of the neutron diffraction lines
I_0	I at $T \to 0$
IV	intermediate valence
J	total angular momentum
$\mathcal{J}$	exchange interaction
k	Boltzmann constant
Ln	lanthanides
M	magnetization
M_0	M at $T \to 0$
m	exponent of the magnon dispersion relation
N	Avogadro's number
$N(\varepsilon_0)$	density of the electronic states at the Fermi energy
NM	noble metals
n	exponent of the term BT^n
P	external pressure
R	gas constant
S	entropy
$\mathcal{S}$	spin
T	temperature
T_C	Curie temperature
T_f	final temperature
T_i	initial temperature
T_K	Kondo temperature
T_m	temperature of the maximum of $\chi(T)$
T_{max}	temperature of the maximum of $C_{Sch}(T)$
T_N	Néel temperature
T_S	spin fluctuation temperature
T_{SC}	superconducting temperature
TM	transition metals
$\mathcal{U}_M$	internal magnetic energy
V	unit cell volume of a hypothetical trivalent Ce compound
Z	partition function
β	phononic coefficient
Γ_7	doublet CF state
Γ_8	quartet CF state
γ	Sommerfeld electronic factor
γ_{HT}	C_M/T at $T > 2T_0$ ($T_0 = T_C$, T_N or T_m)
γ_{LT}	C_M/T at $T < T_0/2$ ($T_0 = T_C$, T_N or T_m)
γ_0	bare electronic contribution to the specific heat
γ_p	γ_{HT} factor in the paramagnetic phase
Δ	crystalline field energy splitting
ΔC_M	specific heat jump at the magnetic transition
ΔS	entropy gain
ΔV	volume contraction with respect to V

θ_D	Debye temperature
θ_P	Curie–Weiss temperature
λ_{em}	electron–magnon interaction
λ_{eph}	electron–phonon interaction
μ_B	Bohr magneton
ρ	electrical resistivity
χ	magnetic susceptibility
χ_0	χ at $T \to 0$

1. Introduction

"In its elemental form Ce is the most fascinating member of the Periodic Table" (Koskenmaki and Gschneidner 1978); such a statement can be also applied to the Ce intermetallic alloys and compounds due to the large variety of physical phenomena they present. The exceptional behaviour of this lanthanide is due to the vicinity of the energy of the "inner" $4f^1$ shell level to the "outer" 5d and 6s levels.

As a normal trivalent lanthanide, Ce forms a large number of ferromagnetic, ferrimagnetic and antiferromagnetic compounds with ordering temperatures predictable from its De Gennes factor (see, e.g., Buschow 1977, 1979). The exceptions, however, are easily found as, e.g., in the body-centred cubic (bcc) CeX compounds (X = Zn, Tl, Mg, Cd) or in the face-centred cubic (fcc) ones (X = Sb, Bi) with ordering temperatures ten times higher than those expected, or in the ternary compound $CeRh_3B_2$ where the Curie temperature reaches $T_C = 115$ K (Dhar et al. 1981a). On the other hand, some compounds behave as enhanced Pauli paramagnets due to the strong hybridization of the 4f shell with the conduction states, as e.g. in the $Ce(TM)_2$ Laves compounds (TM = transition metals) (Weidner 1984).

The ability of Ce for sharing electrons is also capable of reducing the magnetism of its partner element. In the case of $CeFe_2$ the ferromagnetic temperature is reduced by a factor of three with respect to that of $ZrFe_2$ (while the change in the Fe–Fe spacing is only 5% (Muraoka et al. 1976)), and $CeCo_2$ does not order magnetically at all. With Ni, there is no magnetic order reported in the Ce–Ni system up to 70% of Ni concentration (Buschow 1977). When diluted in Pd, 10% of Ce reduces the susceptibility of the matrix by a factor of 13 (Kappler et al. 1984).

Among the highly correlated electron systems, the Ce compounds are again found as the limit examples. The heavy-fermion compounds $CeAl_3$ and $CeCu_6$ were reported to have electronic coefficients of the specific heat of about 1.6 J mol^{-1} K^{-2} (see, e.g., Stewart et al. 1984), which is more than a hundred times larger than that of CeRh: 12 mJ mol^{-1} K^{-2} (Sereni and Kappler 1990). Within these systems, $CeCu_2Si_2$ was the first superconductor to be found (see, e.g., Steglich 1985). Superconductivity is also present in some $Ce(TM)_2$-Laves binary compounds. Among them $CeRu_2$ shows the highest superconducting transition at $T_{sc} = 6.2$ K (Wilhelm and Hillebrand 1970). This compound was recently claimed to be an unconventional type of superconductor by Sereni et al. (1989a).

A considerable number of lanthanides show valence instabilities. Among them Sm, Eu, Tm and Yb are able to fluctuate between the trivalent and divalent electronic configurations, depending on the environmental conditions. On the tetravalent side, although Pr and Tb are suspected to have some hybridization of localized-conduction states, only Ce clearly shows such an effect. Some of its compounds are unique for the study of the electronic fluctuation between the trivalent and tetravalent configurations. The $4f^1$ electron hybridization is the origin of the Ce demagnetization, volume reduction, anomalous temperature dependence of the transport properties, and a number of other abnormal behaviours.

Due to the 11% reduction of its molar volume, CeN was the first compound to be recognized more than fifty years ago by Iandelli and Botti (1937), as having valence instabilities. A similar volume reduction was also found in several $Ce(TM)_2$ Laves compounds (Olcese 1966a). In particular, the Ce ionic radius in $CeCo_2$ was evaluated to be 0.89 Å, which is smaller than the value reported for ionic Ce^{4+} (0.94 Å) (Iandelli and Palenzola 1979). Furthermore, Ce is the only lanthanide to form a $LnRh_3$ compound with the $AuCu_3$ structure, for which an extremely small atomic volume of the Ln element is required.

In this brief overview of the striking properties of the Ce compounds, we have to include the still unsolved disagreement between the thermodynamical and spectroscopical determination of the Ce valence. Many structural and magnetic properties of the Ce compounds were earlier explained by including the energy of the tetravalent configuration within the thermal energy range (i.e., smaller than the cohesive energy) (see, e.g., Wohlleben 1976, Sales 1977). However, from physicochemical considerations it was shown that the tetravalent configuration ($6s^2 5d^2$) of the Ce ion should stand almost 2 eV above the trivalent one ($6s^2 5d4f$) (Johansson 1978). Nevertheless, other values of interconfigurational energy difference are obtained if electronic configurations with p components and the metallization energy are taken into account (Sereni 1985). The spectroscopic techniques, mainly the X-ray absorption spectroscopy (XAS), have drastically reduced the upper limit of the Ce valence. At present there are no intermediate-valence (IV) compounds reported by XAS when the Ce partner is a p element (see, e.g., Röhler 1987), regardless of the fact that their volume contraction, magnetic susceptibility and specific heat are characteristic of group-IV systems, the limiting case being CeN with the maximum volume contraction, lowest magnetic susceptibility and a valence of 3.00 (Sereni et al. 1989b). On the contrary, Ce ferromagnetic compounds formed with d elements without any volume contraction are not spectroscopically recognized as trivalent (see again Röhler 1987).

These and many other effects such as crystalline electric field, spin glass, spin fluctuations, coherence, quadrupolar and martensitic transitions, etc., showed by Ce compounds allow us to say that "Ce compounds are the most fascinating family among the lanthanide compounds".

In this review we shall concentrate our analysis on the specific heat measurements of Ce binary and related compounds, giving examples of each characteristic behaviour. We shall at first discuss briefly the information that can be extracted from the specific heat itself and its derived thermodynamical parameters: internal energy and entropy. In the following section, we shall propose a general classification of the Ce binary compounds based on their magnetic properties. This procedure will help us to define some parameters upon which to organize this work, to define sets of similar behaviour, to relate their respective physical properties and to show the evidence that there has been no systematic study of some sets of compounds, suggesting new topics for research (sections 4–7). The use of the specific heat data to recognize the magnetic behaviour of Ce in some binary related (or pseudoternaries) compounds will be considered in sect. 8. Finally, we shall discuss the crystal-field effects and the relationships between the specific heat and both magnetic and transport measurements in sections 9 and 10.

2. Specific heats

Thermodynamically the specific heat of a substance is defined as the amount of heat required to raise the temperature of a unit mass by a unit degree of temperature, and it is expressed by the equation $C_x(T) = (\mathrm{d}Q/\mathrm{d}T)_x$, where x represents the thermodynamic parameter kept constant during the measurement (x = pressure, volume, magnetic field, number of particles, etc.). Thermal homogeneity and equilibrium requires that $\mathrm{d}T \ll T$. Although this definition is powerful in specifying the general laws governing a phenomenon, the definition provided by statistical mechanics allows one to extract information concerning the microscopic behaviour of a system. Here $C_x = -T(\partial^2 F/\partial T^2)_x$, where F is the free energy. When x represents the volume, F is the Helmholtz free energy, which is related to the partition function Z by $F = -kT \ln Z$. We can now write the general expression

$$C_v = T(\partial^2(kT\ln Z)/\partial T^2)_v. \quad (1)$$

In a solid the number of particles is implicitly constant. Usually, the theoretical models provide a description of Z for the system under study, from which the specific heat is deduced by using eq. (1). Nevertheless, one should remark that experimental measurements (C_p) are ordinarily done at constant pressure because the pressures required to keep a constant volume in a solid are not practicable. Because the condition $C_p = C_v$ depends on temperature, thermal expansion α and compressibility, in the low-temperature region it is valid to take $C_p - C_v$, if also $\alpha(T \to 0) = 0$. Such a condition is not always observed in real systems and in cases like $CeAl_3$, where $\alpha(T < 10\ \mathrm{K})$ shows an anomalous behaviour (see, e.g., Andres et al. 1975, Ribault et al. 1979), the equality is not absolutely valid and the theoretical interpretation of the experimental results has to be made with care.

2.1. *Contributions to the specific heat*

The total free energy of a system is the sum of the free energies of its components so that the total specific heat is the sum of these contributions. From the correct choice of the range of temperature where each of these contributions is dominant emerges the possibility of extracting a single contribution from the total specific heat. The main contributions to C_p in a solid at low temperatures are due to lattice vibrations (or phonons), C_L, conduction electrons, C_E, and magnetic (localized electrons), C_M. The nuclear contribution, C_N, is present only in a few cases, usually at temperatures below 1 K. The phase transitions (mainly magnetic in this study) are usually related to only one of the mentioned contributions. Nevertheless, in some cases the phase transition involves magnetic, structural and/or electronic effects simultaneously. For example, the charge density waves at martensitic transformations also affects the electronic density of states. A complete treatment of the specific heat of simple systems at low temperatures can be found in the book by Gopal (1966).

The conduction electron and phonon contributions in a metal at low temperature are given by

$$C_p = C_E + C_L = \gamma T + \beta T^3 \qquad (\text{or } C_p/T = \gamma + \beta T^2), \tag{2}$$

where $\gamma = \frac{2}{3}\pi^2 k^2 N(\varepsilon_0)$ is the Sommerfeld electronic factor, $N(\varepsilon_0)$ is the density of electronic states at the Fermi energy and $\beta = \frac{12}{5}\pi^4 R(1/\theta_D)^3$, θ_D being the Debye temperature (this expression is valid at $T < \theta_D/24$). In a metallic system some corrections have to be taken into account for the γ-factor. They arise from the electron–phonon and electron–magnon interactions: λ_{eph} and λ_{em}, respectively. As a first approximation they only renormalize the bare γ_0 value as $\gamma = \gamma_0(1 + \lambda_{eph} + \lambda_{em})$. Such a renormalization factor ranges between 0.5 and 2 in most cases.

In a magnetic system, the total specific heat can be described as $C_p = C_E + C_L + C_M$, where the first and second terms represent the nonmagnetic contributions as in eq. (2). Then, the term C_M accounts for the magnetic contribution to the specific heat. In practice, the free-electron (C_E) and lattice (C_L) contributions are extracted from a reference (nonmagnetic) compound, usually formed with La or eventually Y or Lu. The term C_M is therefore defined as $C_M = C_p - (C_E + C_L)$.

Most of the Ce compounds show a linear contribution at some range of temperature. Although this has an electronic origin, it does not necessarily represent a density of states as γ does. It is, however, illustrative to compare the C_M/T ratio of related compounds, then we define γ_{LT} and γ_{HT} as C_M/T at $T < T_0/2$ and $T > 2T_0$, respectively; T_0 being a characteristic temperature of the system (T_C, T_N, T_m, etc.). In particular, γ_{HT} is also known as γ_p, the paramagnetic γ.

The magnetic contribution, ΔC_M, which originates in the localized 4f electrons in the case of Ce is responsible for the magnetic order that presents a discontinuity at the transition temperature T_C. The mean-field theory predicts that $\Delta C_M = 1.5R$ for a spin $\frac{1}{2}$, which is the typical spin for the magnetic Ce ground state. However, magnetic fluctuations giving rise to short-range order above T_C (T_N), are usually present in real systems making such a ΔC_M value only a theoretical upper limit. At $T \ll T_C$ (T_N), $C_M(T)$ is described by magnetic spin waves, with the general form $C_M = BT^{d/m}$, where d is the dimensionality and m is defined as the exponent in the dispersion relation $\omega \sim k^m$ (for antiferromagnetic magnons $m = 1$ and for ferromagnetic magnons $m = 2$). The coefficient B is $\sim NkT/|\mathcal{J}|\mathcal{S}$, where $\mathcal{J}$ is the exchange interaction and $\mathcal{S}$ the spin. Magnetic anisotropies arising from crystal-field effects (CF) introduce a gap in the magnon dispersion becoming $C_M \sim BT^n \exp(E_g/kT)$, where $n = d/m$ and E_g is the energy gap. For a more detailed discussion of this topic see, e.g., Sundström (1978).

TABLE 1

Temperature (T_{max}) of the maximum of the Schottky anomaly (C_{Sch}) and entropy as a function of the degeneracy ratio (g) and splitting (Δ), for the three possible CF excitations of Ce^{3+} ($J = \frac{5}{2}$).

Excitation	g	C_{Sch}(max)*	T_{max}/Δ	Entropy
quartet → doublet	0.5	2.00	0.448	$R\ln\frac{3}{2}$
doublet → doublet	1	3.64	0.417	$R\ln 2$
doublet → quartet	2	6.31	0.377	$R\ln 3$

* In units of $J\,K^{-1}\,mol^{-1}$.

The CF plays an important role in the formation of the ground states of the lanthanides. From the Hund rules the total angular mometum of Ce^{3+} is $J = \frac{5}{2}$, with a degeneracy of $2J + 1 = 6$. The CF splits this degeneracy into a doublet (Γ_7) and a quartet (Γ_8) in a cubic symmetry and into three Kramer doublets in a lower symmetry. Except in a few cases, most of the Ce compounds have a doublet as a CF ground state, thus an entropy gain of $\Delta S = R\ln 2$ is expected to be associated with their magnetic phase transition. Then thermal promotion of the localized electrons gives rise to the so-called Schottky specific heat contribution which is given by

$$C_{Sch} = R(\Delta/T)^2 g\exp(\Delta/T)\,[1 + g\exp(\Delta/T)]^{-2}, \qquad (3)$$

where Δ is the energy splitting and g is the degeneracy ratio between the excited and the ground state. The quite simple CF level scheme of Ce^{3+} allows us to predict all the possible g-values and to evaluate all the possible maximum values of C_{Sch} with their respective temperatures T_{max} and corresponding entropy gain. See table 1.

2.2. *The internal magnetic energy*

Some thermodynamic functions can be deduced from the knowledge of the specific heat. They are the enthalpy, the internal energy and the entropy. The first function appears as the latent heat in first-order transitions like the ferromagnetic–antiferromagnetic transition or where the magnetic transition is related to a structural transition.

The internal magnetic energy, $\mathcal{U}_M$, of the magnetically ordered phase can be evaluated as

$$\mathcal{U}_M(T) = \int_0^T C_M\,dT. \qquad (4)$$

From simple thermodynamic relations we know that $C_M \simeq -T\partial M^2/\partial T$ (where $M(T)$ is the magnetization of the system). This allows one to evaluate the spontaneous magnetization and the order parameter in a ferromagnet (the same concept applies to the magnetic sublattices in an antiferromagnet). Thus, C_M can be compared with the intensity of the neutron diffraction lines, $I(T)$, by means of eq. (4) and

$$\mathcal{U}_M(T) = -[T_M M_0^2/2C][I(T)/I_0], \tag{5}$$

where C denotes the Curie constant, and M_0 and I_0 are the values of the magnetization and intensity of neutron diffraction lines at $T = 0$.

In addition, the so-called "energetic susceptibility", $\chi T/C$, is related to $\mathcal{U}_M$ by

$$\chi T/C \simeq 1 - |\mathcal{U}_M|, \tag{6}$$

(see de Jongh and Miedema 1974). When compared with $C_M(T)$ one can extract information about the presence of short-range order above T_N. Equation (6) implies that the maximum derivative of χT occurs at T_N (where $C_M(T)$ is a maximum) and then the maximum of χT will occur somewhat above T_N (at T_{max}). The difference between T_N and T_{max} reflects the presence of short-range order above T_N, which is enhanced by lowering the dimensionality. Basic models for magnetic ordered systems, i.e., the Ising, X–Y and Heisenberg models allow us to evaluate $\mathcal{U}_M$ at T_N for simple magnetic structures.

2.3. *The entropy*

The entropy gain as a function of temperature is given by

$$\Delta S(T) = \int_0^T (C_p/T)\,\mathrm{d}T. \tag{7}$$

Since the entropy is defined in statistical mechanics as $S = k \ln Z$, where the partition function Z represents the accessible states at a given temperature, the calorimetric technique can be considered as a type of "thermal spectrometry". In this, the heat ($\mathrm{d}Q$) is the excitation energy absorbed by the system, the initial state is characterized by the temperature T_i and the final one by $T_f = T_i + \mathrm{d}T$. The system remains in its final state because it is always at thermal equilibrium.

3. Classification of the cerium binary compounds

About 150 Ce binary intermetallic compounds, Ce_jX_k, are reported with a recognized crystal structure, see *Pearson's Handbook of Crystallographic Data for Intermetallic Phases* (1985). About eighty of these can be included in our study because of the knowledge of their magnetic and structural properties. We exclude those Ce–transition-metal compounds where the magnetic behaviour is governed by the partner, such as $CeFe_2$, $CeCo_5$ or Ce_2Ni_{17}. In any case, the large variety of properties – magnetic order, valence instabilities, electronic correlations, electric crystal-field effects, etc. – makes it difficult to develop a simple classification. The use of theoretical models gives fundamental information on the microscopic mechanisms, but they are useful only within their range of validity. When the models are extended out of their specific limits, one finds that the characteristic energies (e.g., the Kondo temperature T_K) reach unphysical values, as, e.g., in the case of Ce impurities diluted in the Th matrix (Luengo et al. 1975). Usually, different models applied to the same experimental result give different values for the same characteristic parameter (see, e.g. Fillion et al. 1987).

In the search for an ample but unique criterion for this classification, we can profit from the fact that all the magnetic interactions are governed by an exchange parameter, $\mathscr{J}$. Although such a parameter is not directly measurable (it can be evaluated only qualitatively through the use of different models), it relates two macroscopic or phenomenological properties: magnetic properties and interatomic distances. A first attempt at this criterion was made by Hill (1970), where the order temperature (only antiferromagnetic or superconducting) was plotted against the Ce–Ce spacing. This early work was applied to the small number of Ce compounds known at that time. An up-to-date generalization of such a classification can be made by taking as the Ce–Ce spacing the minimum Ce–Ce distance. We have to note that the shortest distance is not necessarily directly related to the dominant interaction. This point becomes relevant in the cases where there are inequivalent positions in the lattice for the Ce atoms. The measurable magnetic parameter is the ordering temperature, whether ferromagnetic (T_C) or antiferrimagnetic (T_N). The systems with magnetic structures that are not simple are taken as ferromagnetic or antiferromagnetic, depending on the appearance or absence of spontaneous magnetization, respectively. The compounds that do not order magnetically are included in this systematic classification by taking as their characteristic temperature the temperature of the maximum (T_m) of their specific heat (heavy-fermion compounds) or their magnetic susceptibility (intermediate-valence compounds). In terms of energy, T_m is related to the hybridization energy or the binding energy of the Kondo singlet (see, e.g., Doniach 1977).

In fig. 1 we plot these magnetic temperatures against the minimum Ce–Ce distance (D). Because the ordering temperatures (T_C or T_N) do not have the same physical meaning as T_m, they are presented in different sectors. The squares represent the ferromagnets (T_C) and the circles the antiferromagnets (T_N). The diamonds identify those compounds with more than one magnetic transition observed in their specific heat, impeding a clear identification of the nature of their magnetic phase. Such is the case when frustration effects do not allow the development of a long-range order, or in crystalline structures with inequivalent Ce sites. In this case, we shall take the highest-order temperature. A more detailed discussion will be made in the analysis of each particular case.

The compounds that do not order magnetically are represented by a triangle (T_m) on a logarithmic scale because of their wide range of values (from a few degrees for the heavy-fermion compounds to a thousand for the intermediate-valence compounds). The Greek letters α, β and γ in fig. 1 represent the respective allotropic phases of Ce metal for comparison.

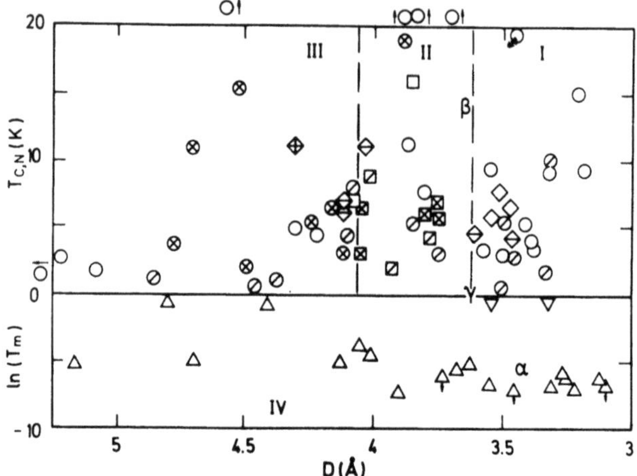

Fig. 1. Characteristic temperatures plotted against Ce–Ce spacing (D) of eighty Ce binary compounds. Squares indicate T_C, circles T_N, upright triangles T_m, diamonds and inverted triangles identify compounds with more than one transition. Those crossed once (twice) have fractional (full) entropy gain at the transition, see the text. The Roman numbers (I–IV) identify different models.

From the large number of Ce compounds included in fig. 1 it becomes evident that the ferromagnetic and the antiferromagnetic compounds are not randomly distributed as a function of D. On the contrary, the ferromagnetic compounds are concentrated in the region $4.1 > D > 3.7$ Å. A recent analysis proposes the limits: $3.83 > D > 3.65$ Å for the compounds with a TM partner (Huber 1989, De Long et al. 1990a). The upper limit becomes 4.1 Å if the ferromagnetic compounds with "p" partners are included. Considering that some physical reason underlies this distribution (a detailed discussion of this point has been given by De Long et al. 1990b), we shall subdivide fig. 1 into the following regions:

(I) With antiferromagnetic compounds and $D < 3.7$ Å. Such a boundary value coincides with the diameter of a trivalent Ce atom (see e.g. Koskenmaki and Gschneidner 1978).

(II) With ferromagnetic compounds and D ranging between 4.1 and 3.7 Å. The upper D-boundary is given by the limit where the ferromagnetic compounds appear.

(III) With antiferromagnetic compounds and $D > 4.1$ Å.

(IV) Within this region are included all the Ce–X compounds that do not order magnetically – heavy-fermion compounds and intermediate-valence compounds.

As fig. 1 is intended to show all the compounds within a unique pattern, we have subordinated the exact coordinates of some of them to the clarity of the figure itself. Figures 2, 9 and 12 show in detail the Ce_jX_k compounds, where j is the number of Ce atoms and k the number of partners, X, per formula unit. There the compounds will be labeled as ${}_jX_k$.

4. Region I: antiferromagnets with small Ce–Ce spacing

The Ce_jX_k compounds belonging to this region are shown in fig. 2 as $_jX_k$. Their structural, magnetic and specific heat data are summarized in table 2. The variety of the crystalline structures of these compounds renders it difficult to make a strict comparison of their properties. Nevertheless, by taking a "general rule" that applies to 90% of such cases, some distinctive characteristics can be attributed to these compounds. They are: (i) all these compounds order antiferromagnetically; (ii) where measured, the entropy gain at T_N is significantly smaller than the expected value of $\Delta S = R\ln 2$. In the best of the cases, ΔS reaches 90% of the value at $T \approx 3T_N$; (iii) the parameters γ_{LT} and γ_{HT} are about one order of magnitude larger than those observed in the nonmagnetic reference compounds, see $\gamma(LaX_j)$ in table 2; (iv) the Curie–Weiss temperature, θ_P, is negative and $|\theta_P|$ is significantly larger than T_N, almost in one of the crystallographic axes; (v) no appreciable volume contraction with respect to the neighbouring lanthanide (La and Pr) isocompounds is observed.

As already mentioned, these compounds show a variety of crystalline structures. Among them only three are cubic, but most of them have much lower symmetry, making their physical properties strongly anisotropic. In those cases single crystals are required for microscopic studies. Another characteristic of this group are the crystalline structures where Ce occupies inequivalent sites in the lattice, which induce

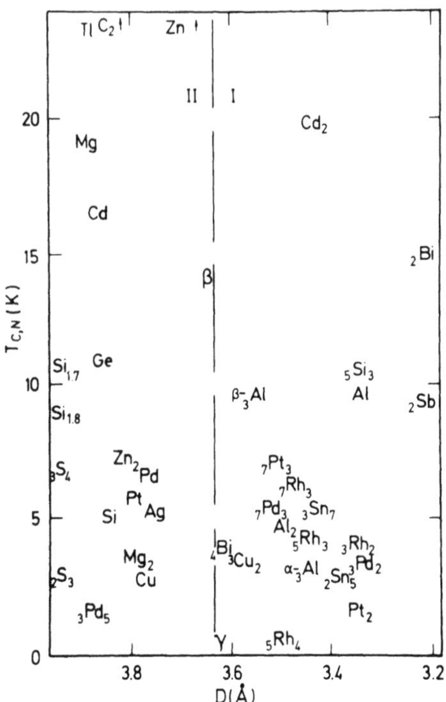

Fig. 2. Ce_jX_k compounds belonging to regions I and II labelled as $_jX_k$. β and γ are the Ce allotropic phases.

TABLE 2
Antiferromagnetic compounds belonging to region I. The parameters are defined in the list of symbols.

Compound	Structural data		Magnetic data			Specific heat data								
	Type	D (Å)	T_N (K)	θ_P (K)	Ref.*	T_N (K)	C_{max} ($J K^{-1}$/Ce atom)	B ($J K^{-4}$/Ce atom)	ΔS ($R \ln 2$)	H (T)	γ_{LT} (mJ K^{-2}/Ce atom)	γ_{HT} (mJ K^{-2}/Ce atom)	$\gamma(LaX_j)$ (mJ K^{-2}/Ce atom)	Ref.*
CeAl	CeAl	3.32	10	−19	[1]									
$CeAl_2$	$MgCu_2$	3.49	3.85		[2a]	3.9	[a]	0.14	0.83[k]	0	135			[4]
			3.45		[2b]	2.6				5				
				−33	[2c]	3.8	7.8	[b]		0	120	150	3.63	[5a]
						4.6	5.0				180	150		[5b]
								0.09		0	187			[6]
										2	184			[6]
										4	172			[6]
										8	136			[6]
α-Ce_3Al	Ni_3Sn	3.52	2.6	−50[c]	[7]	2.5	11		0.4[k]			85		[27]
β-Ce_3Al	Cu_3Au	3.54	9.0	−44[c]	[8]									
Ce_4Bi_3	Th_3P_4	3.60				3.8	12		0.72[i]			200		[3]
						3.5	[q]		0.2					[3]
Ce_2Bi	La_2Sd[d]	3.25	15	−52	[9]									
$CeCd_2$	$CeCd_2$	3.44	20	−56	[10]	22/18.5								[10]
$CeCu_2$	$CeCu_2$	3.57	3.0	270[m]	[11]	3.0	4[l]		0.60	0		180		[11]
			3.5	−45	[12]	3.3	6	1.4[n]	0.66	0	82			[12]
										4	74			[13]
										8	56			[13]
Ce_7Ni_3	Th_7Fe_3	3.44		−38	[28]	1.7	1.74		0.41[k]		1310	900		[28]
Ce_3Pd_4	Pu_3Pd_4	3.45	3.1	−27	[14]									
Ce_3Pd_2	Er_3Ni_2[e]	3.42	3.3	−26	[14]									
Ce_7Pd_3	Th_7Fe_3[e]	3.51	5.4	−33	[14]	5.4	4.2		0.60			160		[28]
$CePt_2$	$MgCu_2$	3.35	1.5	−25	[15]	1.6	5.7		0.81	0		< 15[f]	1.77	[16]
				5	[16]	1.5	3.8			0.9				

Ce_7Pt_3	Th_7Fe_3[e]	3.49	7.0[g]	−12	[17]									
Ce_5Rh_4	Sm_5Si_4[e]	3.49	0.8	−200	[18]	0.75	2.8	[h]	0.40[k]	0		90	3.77[r]	[19]
Ce_5Rh_3	Pu_5Rh_3	3.47	4.4/2.5[j]	−45	[18]	4.2/2.4[j]	3.6/6.4	0.3	0.79	0	130	44		[20]
						3.9/2.0	3.6/3.0	0.46	0.76	1	150			[20]
						3.6/1.5	3.6/1.5	0.64	0.70	2	170	≈ 50		[20]
Ce_3Rh_2	Er_3Ni_2[e]	3.42	3.6	−37	[18]									
Ce_7Rh_3	Th_7Fe_3[e]	3.47	7.2[g]		[18]	6.5	5.4	0.012	0.42			160		[25]
						1.3	1	0.6	0.11					
Ce_2Sb	Ti_2B[d]	3.22	9.5	−57	[9]									
Ce_4Sb_3	Th_3P_4	3.54				3.9[p]	12	0.3	0.75[k]	0		160		[26]
						5.2[l]	15	0.05	0.73[k]	1	–			[26]
Ce_5Si_3	Cr_5B_3[d]	3.33	10	−50	[21]	10	3.7		0.90	0	250			[21]
Ce_2Sn_5	Ce_2Sn_5[d]	3.45	2.9	−43	[22]	2.9	6.7	0.3	0.55[k]	0	190	38		[22]
				−160	[24]	3.0	7		0.42[k]	0		22		[23]
						2.75				2.5				[23]
						≈ 5	[l]			5.3				[23]
Ce_3Sn_7	Ce_3Sn_7[d]	3.42	5.3	−140	[24]	5.07	6.7		0.55[k]		103	≈ 0		[29]

[a] λ-type.
[b] $C_M = AT^{1/2}\exp(-E_g/T)$.
[c] Estimated from ref. [8].
[d] Two different sites of Ce in the cell.
[e] Three different sites of Ce in the cell.
[f] Estimated from ref. [16].
[g] Spontaneous magnetization.
[h] $C_M = A\ln|(T_N - T)/T_N|$.
[i] At $3T_C$.
[j] Two magnetic transitions.
[k] At $2T_N$.
[l] Broad maximum.
[m] Direction parallel to c axis.
[n] $C_M = BT^3\exp(-E_g/T)$.
[p] Ferromagnetic ground state.
[q] First-order transition.
[r] Superconductor at 1.2 K.

*References: [1] Kissel and Wallace (1966); [2a] Croft et al. (1979); [2b] Ref. [2a], measured under $P = 8$ kbar; [2c] Swift and Wallace (1968); [3] Suzuki et al. (1987); [4] Bredl et al. (1978), Armbrüster and Steglich (1978); [5a] Peyrard (1980); [5b] Ref. [5a], measured under $P = 6$ kbar; [6] Breal (1987); [7] Sera et al. (1987); [8] Sakurai et al. (1988); [9] Isobe et al. (1987); [10] Tang and Gschneidner (1989); [11] Ōnuki et al. (1985b); [12] Gratz et al. (1985); [13] Bredl (1985, 1987); [14] Kappler et.al. (1985); [15] Vijayaraghavan et al. (1968); [16] Joseph et al. (1972a); [17] Jamari-Saele (1980); [18] Kappler et al. (1988a, b); [19] Sereni et al. (1990a); [20] Sereni and Kappler (1989); [21] Kontani et al. (1987); [22] Boucherle et al. (1987a, b); [23] Dhar et al. (1987a, b); [24] Bonnet et al. (1988); [25] Kappler and Sereni (1990); [26] Suzuki et al. (1990); [27] Chen et al. (1989); [28] Sereni et al. (1991); [29] Stunault (1988).

different types of behaviour for each Ce sublattice. We remind the reader here that the minimum Ce–Ce spacing taken for fig. 1 does not necessarily correspond to the magnetically ordered sublattice. Noteworthy is the fact that all the intermediate-valence (IV) compounds with equivalent D-values to those of region I are cubic close packed; we shall discuss this point later (cf. sect. 7) in connection with the role of the symmetry in the orbital hybridization.

The absence of a dominant effect of the hybridization (valence instability) over the magnetic interactions (magnetic order) in the compounds belonging to region I forces a reconsideration of the concept of "chemical pressure", which is usually taken as the driving force that induces valence instability in Ce. Taking the limiting cases of Ce_2Sb and Ce_2Bi, where $D \approx 3.2$ Å, one sees that D is 20% smaller than that for γ-Ce (Isobe et al. 1987), but only 2% smaller than the La–La spacing in La_2Sb (Stassen et al. 1970). Because 2% is expected to be within the normal lanthanide contraction, one concludes that in this case the "chemical pressure" only induces a strong overlap of the band states. We should, therefore, distinguish between the "chemical pressure" that acts on the whole electron system and that which acts on the f orbitals. In the aforementioned examples the available Ce volume is so small that one should also expect strong crystal field effects and, as quoted by Isobe et al. (1987), strong anisotropy.

Although the f orbitals retain their magnetic moment, some evidence of the states mixing is present in the compounds belonging to region I: the lack of the expected entropy gain and the large values of γ_{LT}, γ_{HT} and θ_P. Such apparent ambiguity can be understood in the compounds with inequivalent atomic sites, where the different environments of each sublattice induce different Ce behaviours. In that case the entropy gain as a function of temperature may provide the fraction of atoms involved in each sublattice. Within these compounds we can therefore find magnetic order superimposed on the Kondo or intermediate-valence behaviour, or eventually two independent magnetic transitions. Among the compounds whose specific heat is known we can quote Ce_5Rh_4 as an example, with $T_N = 0.85$ K, $\Delta S = 0.4 R\ln 2$ at $T = 4T_N$ (see fig. 3) and a saturation moment of 0.34 μ_B/Ce atom at 1.4 K (Sereni et al. 1990a). These results indicate that not all the Ce atoms are involved in the magnetic order. In this compound the Ce atom has three inequivalent sites in the lattice: Ce_I, Ce_{II} and Ce_{III}. The atoms identified crystallographically as Ce_{II} are located at the corners of connected canted squares, they have a larger available volume than the Ce_I and Ce_{III} atoms, and represent 40% of the Ce atoms per formula unit. From these characteristics the Ce_{II} atoms can be proposed as those responsible for the magnetic order in Ce_5Rh_4. The rest of the atoms (Ce_I and Ce_{III}) should be involved in the Kondo-like behaviour of this compound, which is recognized by a maximum in C_M at 18 K (see fig. 4), a $\gamma_{HT} = 90$ mJ K^{-2}/Ce atom, $\theta_P = -200$ K and anomalous $\chi(T)$ and $\rho(T)$ behaviours (Sereni et al. 1990a). Similar experimental evidence was found in Ce_2Sn_5, which orders magnetically at $T_N = 3$ K, has an entropy gain of $\Delta S = 0.42 R\ln 2$ at $T = 3T_N$, $\theta_P = -43$ K and $\gamma_{HT} = 22$ mJ K^{-2}/Ce atom (Dhar et al. 1987). The studies performed in a single crystal of this compound report: $T_N = 2.9$ K, $\Delta S = 0.55 R\ln 2$ at $T = 3T_N$, θ_P $(H \parallel c) = -160$ K and $\gamma_{HT} = 56$ mJ K^{-2}/Ce atom (Boucherle et al. 1987a, b, Bonnet et al. 1988). The compound Ce_5Si_3 was reported by

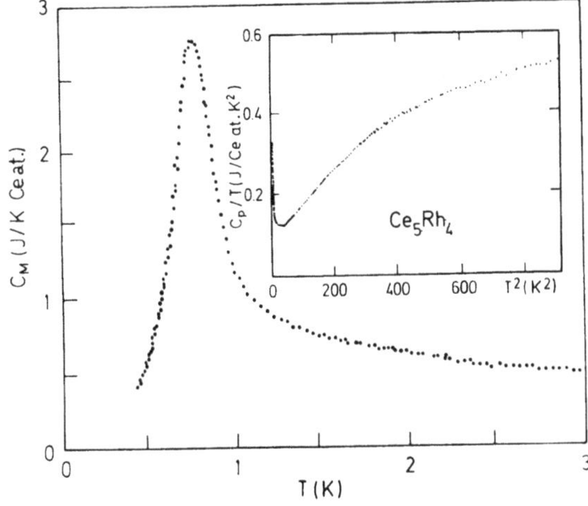

Fig. 3. Magnetic specific heat of Ce_5Rh_4 at the antiferromagnetic transition. Inset: the total specific heat in a C_p/T versus T^2 representation.

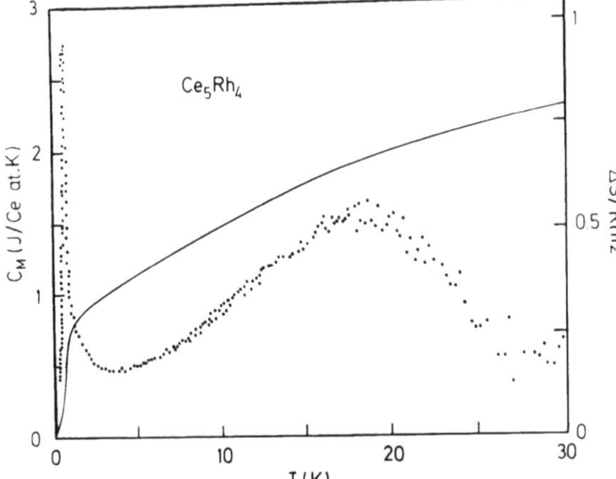

Fig. 4. Magnetic specific heat of Ce_5Rh_4 up to 30 K. The continuous curve is the entropy gain in that range of temperature.

Kontani et al. (1987) as showing two anomalies in the specific heat around 2.5 K and 10 K, a large γ_{LT} value of 250 mJ K^{-2}/Ce atom and $\theta_P = -50$ K. The total entropy gain reaches 0.9 $R\ln 2$ at 20 K. The magnetic origin of each specific heat anomaly is not conclusive and further studies on this compound are required because of some particular features: (i) a Ce_{II}–Ce_{II} dimer is claimed to form by the authors and (ii) the cubic symmetry seen by the Ce_I atoms gives the possibility of quadrupolar effects.

The final example for this group is given by Ce_7Rh_3. Here Ce atoms also have three inequivalent positions in the lattice. The specific heat measurements show two peaks, at 1.3 K and at 6.5 K. The C_M contribution is shown in fig. 5, after subtracting a linear term $\gamma_{HT} = 160$ mJ K^{-2}/Ce atom (Kappler and Sereni 1990a). The entropy gain is distributed as follows: 0.11 $R\ln 2$ for the 1.3 K anomaly, 0.42 $R\ln 2$ for the 6.5 K transition and the rest should be associated with the γ_{HT} term. By comparing these values with the number of Ce atoms in each sublattice (14%, 43% and 43% of the

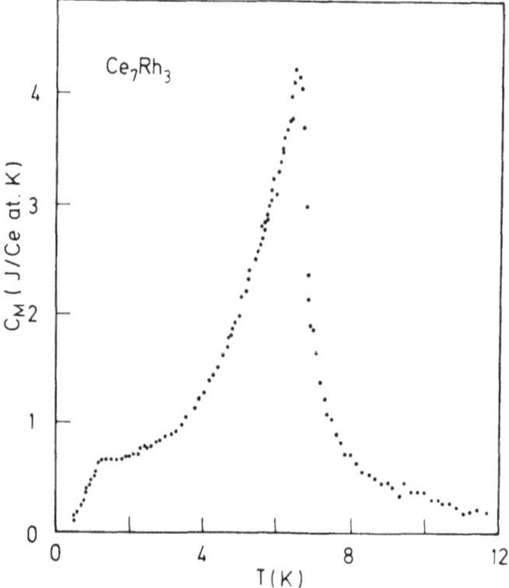

Fig. 5. Magnetic specific heat of Ce_7Rh_3 after the $\gamma_{HT}T + \beta T^3$ subtraction.

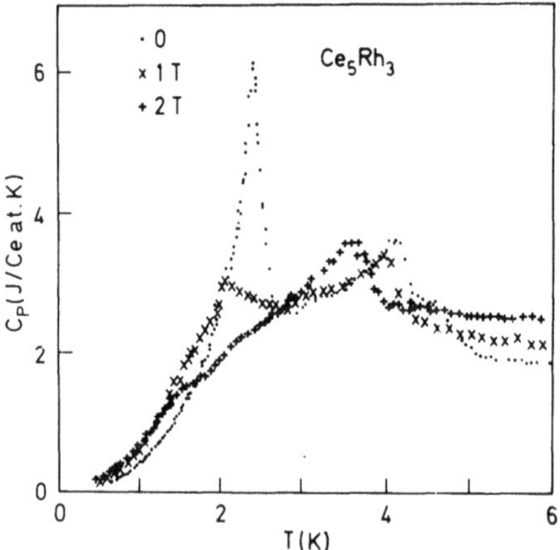

Fig. 6. Total specific heat of Ce_5Rh_3 at zero magnetic field and with applied magnetic fields of 1 and 2 T.

formula unit, respectively) one can conclude that each sublattice behaves independently in a first approximation. The nature of the magnetic transition is rather confusing because, although a spontaneous magnetization arises at $T \leqslant 7.2$ K, a T^3-dependence is observed in the specific heat between 3.5 and 6 K. The low-temperature anomaly (around 1.3 K) can be fitted by a D/T function (with $D = 0.74$ J/Ce atom) in the 1.5 to 3 K range as an indication of short-range (spin glass like) magnetic order.

The case of Ce_5Rh_3 represents a system where both sublattices order antiferromagnetically and at 4.2 K and 2.4 K, as shown in fig. 6. The low-temperature transition is strongly field-dependent, while the high-temperature one is one slighly shifted to lower temperatures. The entropy gain at the transitions up to 10 K is $0.79R\ln 2$ and $\gamma_{HT} = 130$ mJ K^{-2}/Ce atom, in qualitative agreement with the previously mentioned compounds. Under an external field of 2 T, these values do not change significantly: $0.70R\ln 2$ for the entropy and $170\,\mathrm{mJ\,K^{-2}}$/Ce atom for γ_{HT}. The $\theta_P = -45$ K value together with γ_{HT} suggests significant hybridization effects in almost all of the sublattices (Sereni and Kappler 1990).

Within the compounds with equivalent sites for Ce, the specific heat of Ce_4Bi_3 also shows two well-defined peaks (Suzuki et al. 1987). This compound lies on the borderline between regions I and II, it is actually ferromagnetic for $T < 3.5$ K but after undergoing a first-order transition. The entropy gain including the transition at 3.8 K is $0.8R\ln 2$ at 20 K. The field dependence of the specific heat in this compound was compared by the authors with that of CeB_6, where a quadrupolar transition was observed by neutron diffraction.

One of the most intensively studied Ce compounds is $CeAl_2$, which became the archetype of the magnetic Kondo lattice compounds (Bredl et al. 1978, Bredl and Steglich 1978, Armbrüster and Steglich 1978). It orders antiferromagnetically at 3.8 K, but the entropy gain at T_N is only $0.5R\ln 2$, reaching 80% of that value at $T = 2T_N$. The γ_{LT} and γ_{HT} coefficients are similar and within the typical values of $150\,\mathrm{mJ\,K^{-2}}$/Ce atom, see table 2. The similarity between the γ_{LT} and γ_{HT} values is remarkable in this compound, because different mechanisms can be the origin of these coefficients. The γ_{LT} coefficient may contain electronic and magnon contributions of the magnetic phase, together with an enhanced density of the quasiparticle state at the Fermi level resulting from contributions of the Kondo interaction or the spin fluctuations. On the paramagnetic side, γ_{HT} may contain CF contributions from the broadened levels, plus the same quasiparticle contribution present at low temperatures because $T_N < T_K$. However, if γ_{LT} and γ_{HT} have the same origin, the entropy gain has to be evaluated as the addition of two contributions, one from the magnetic order and the other from the quasiparticle density of states. The Schottky specific heat of $CeAl_2$ at high temperatures is clearly anomalous and will be discussed in sect. 9, in connection with the γ_{HT} value. The field dependence of the specific heat cusp at T_N depends on the direction of the applied field. While T_N is shifted towards lower temperatures, a hump appears around 6 K (Bredl et al. 1978, Bredl and Steglich 1978).

The specific heat of $CeCu_2$ falls within the already mentioned general trend of this region: low entropy gain (0.60–$0.66R\ln 2$), large γ_{LT} and γ_{HT} values (80 and $180\,\mathrm{mJ\,K^{-2}}$/Ce atom, respectively) and large θ_P values (-270 K in the c direction) (Ōnuki et al. 1985b, Gratz et al. 1985). Finally, $CePt_2$, which orders magnetically at $T_N = 1.6$ K with an entropy gain of $0.8R\ln 2$, has very low values of $\gamma_{HT} < 15\,\mathrm{mJ\,K^{-2}}$/Ce atom and $\theta_P = 5$ K (Joseph et al. 1972a). Indeed there is a disagreement in the literature concerning its θ_P value, which ranges between the mentioned value of $\theta_P = 5$ K and $\theta_P = -25$ K (Vijayaraghavan et al. 1968). These low values of γ_{HT} and θ_P are normal for a classical antiferromagnet, but they become abnormal in comparison with the values observed in the compounds of region I. The deficit of entropy and the low value of C_M at T_N admits the possibility of still unobserved anomalies at $T < 2$ K. Noteworthy is the fact that this compound, which has received practically no attention in the last few years because of its normal behaviour, becomes important for the understanding of twenty compounds that behave abnormally.

5. Region II: ferromagnets

A striking feature in fig. 1 is the relatively small range of D values where the Ce ferromagnets are located. Although in this region there are also some antiferromagnetic compounds, practically all of them are related in structure to the ferromagnets. In order to facilitate a comparison between these compounds and those of region I, we also give here the general characteristics of this group, see table 3. They are: (i) with only one exception (to be discussed in sect. 6), all the Ce ferromagnets are included in this region; (ii) the chemical composition of most of them is equiatomic and all the Ce atoms are in equivalent positions in the lattice; (iii) most of the ferromagnets reach the expected entropy value, $R\ln 2$, which is not the case for the antiferromagnets; (iv) the measured γ_{LT} and γ_{HT} values do not exceed the value 100 mJ K^{-2}/Ce atom (the only exception to this is $CeS_{1.7}$, and it also shows a deficit in the magnetic entropy); (v) the θ_p values are much smaller than those of region I.

The equiatomic compounds belong to the related structures CrB, CsCl and FeB, which can be predicted from the valence-electron concentration of the partner (Hohnke and Parthe 1966). Here this trend follows the d, s and p electronic character of the partners, as in the periodic table.

One of the most fascinating set of compounds in this region are the compounds with the CsCl structure, where the atoms are placed in the corner and body-centered positions. Within this crystalline structure the local Ce symmetry gives the possibility for a quartet (Γ_8) CF ground state, in coincidence with the highest ordering temperatures among the binary compounds. Concerning the nature of the magnetic order one sees that:

(i) CeZn, CeMg and CeTl order antiferromagnetically in a first-order transition. CeMg has a latent heat of $L = 5.6\ \mathrm{J\,mol^{-1}\,K^{-1}}$ at $T_N = 19.5$ K and $\Delta S(20\ \mathrm{K}) = 1.1 \ln 4$ (Pierre and Murani 1980).

(ii) CeAg and CeCd order ferromagnetically, having a structural (martensitic-like) transformation at $T > T_C$.

(iii) The link between both behaviours is given by the pressure-induced structural transition in CeZn (Kadomatsu et al. 1986), and in CeTl (Kurisu et al. 1985). The magnetic phase boundary (ferromagnetic–antiferromagnetic) is found at 10–12 kbar in both compounds. Further studies of the specific heat of these systems should provide basic knowledge about the nature of the transition, their entropy and the ground-state degeneracy.

Now that the tetragonality of the ferromagnets has been established, we can conclude that "with exception of $CeMg_2$, there are no Ce ferromagnets with a cubic structure". Concerning the compounds with the Γ_8-CF ground state, although other cubic compounds will be discussed in the next section, we can advance the proposition that "all the cubic Ce compounds having a quartet CF ground state undergo a quadrupolar transition at or above T_N" (the first-order transition temperature).

The second-order magnetic transition is usually preceded by local (short-range) magnetic order. As mentioned in sect. 3 the mean-field theory, which does not take into account the magnetic fluctuations, predicts a jump of $\Delta C_M = 12.5\ \mathrm{J\,mol^{-1}\,K^{-1}}$ at T_C for a spin 1/2. In this region there are some examples of a practically "pure" second-order transition, they are CePt (Holt et al. 1981) and CePd (Kappler et al. 1984), with the respective values of T_C at 5.8 K and at 6.4 K. Their specific heats near T_C are shown in fig. 7. Their respective entropies are $\Delta S = R \ln 2$ right above T_C.

Another interesting set of ferromagnets in this region, which define the upper D limit, are those formed with the p-type elements S and Si. The compounds are $Ce_{3-x}S_4$ and $CeSi_{2-x}$. Their common feature is to present vacancies x in the Ce or in the partner sublattice. The magnetic specific heat of the $Ce_{3-x}S_4$ ferromagnets (with $x = 0.255$ and 0.128) shows a broad maximum at $T_C = 2.9$ K and at 5.9 K, with an entropy gain of $R \ln 2$ and $0.92 R \ln 2$, respectively (Ho et al. 1982). The $CeS_{1.393}$ ($x = 0.128$) sample shows a value of $\gamma_{HT} = 71$ mJ K^{-2}/Ce atom that is attributed to a mixed-valence character. Although the magnetic susceptibility data show that Ce is trivalent (Ho et al. 1982), we recall the correlation between the entropy deficiency and the high γ_{HT} values observed in the compounds of region I. The other set of compounds, with the generic formula $CeSi_x$, forms in the tetragonal α-$ThSi_2$-type structure in the range $1.55 \leqslant x \leqslant 1.95$ (Dhar et al. 1987b). As the Si concentration is increased it undergoes a transition from magnetic to intermediate-valence behaviour, also showing spin fluctuation effects. The entropy gain associated with the magnetic transition is low enough to be significant (being $0.77 R \ln 2$ in $CeSi_{1.7}$, which has the highest value) and has large γ_{LT} coefficients, see table 3. A broad maximum of the specific heat at the magnetic transition is a characteristic of these compounds, nevertheless in a $x = 1.7$ single-crystal sample a double ferromagnetic transition was found by Sato et al. (1985, 1988b). This evidence suggests that the broad specific heat maximum at T_C in $x \geqslant 1.7$ compounds may be the result of their polycrystalline character that disguises a magnetic structure that is not simple.

Although we shall include the $CeSi_x$ compounds with $x \geqslant 1.9$ within the non-magnetic systems (region IV), the borderline as a function of x is not unequivocal. Specific heat measurements performed under a magnetic field in the $x = 1.85$ and 1.9 compounds shows the occurrence of a hump at about 3.5 K, with a very low associated entropy, which is mainly gained in the spin-fluctuation temperature region (Dhar et al. 1987). Similar behaviour was observed in $CeSn_3$ (Ikeda and Gschneidner

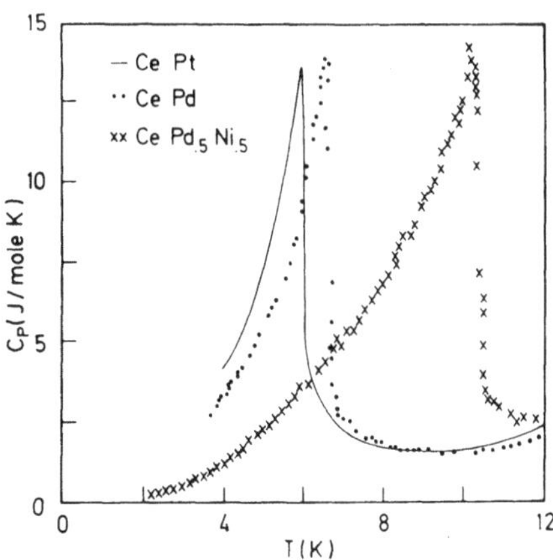

Fig. 7. Specific heat of the "second-order" ferromagnetic transitions. CePt from Holt et al. (1981), CePd from Kappler et al. (1984) and $CePd_{0.5}Ni_{0.5}$ from Nieva et al. (1988).

TABLE 3
Compounds belonging to region II. The parameters are defined in the list of symbols.

Compound	Structural data		Magnetic data					Specific heat data										
	Type	D (Å)	T_C (K)	T_N (K)	θ_p (K)	Ref.**	T_C (K)	T_N (K)	C_{max} ($J K^{-2}$/Ce atom)	B^*	n	ΔS ($R \ln 2$)	H (T)	γ_{LT} (mJ K^{-2}/Ce atom)	γ_{HT} (mJ K^{-2}/Ce atom)	$\gamma(LaX_j)$ (mJ K^{-2}/Ce atom)	Ref.**
CeAg	CsCl[a]	3.76	9	(15.6[c])	−26	[1,3]	5.5		18			1[b]	0	0		9.9	[4]
			5.3			[2]		(15.6[c])	26	0.11	1.5[d]		0				[5]
										0.07	1.5[d]		2				[5]
CeAu	CrB	3.83	4.9			[28]											
CeC_2	CaC_2	3.87		33		[6]											
				30	12	[7]											
CeCd	CsCl[a]	3.86	16.5	(216[c])		[8]											
CeCu	FeB	3.75		3.63	−1.6	[10]		3.2	4.9	0.9	3[e]	0.72	0	30			[10]
				2.7	0	[9]		2.9	4.3			0.63	1				[10]
							(13[f])		(2.8[f])			(0.23[f])	0				[10]
							(11.5[f])		(2.5[f])			(0.30[f])	1				[10]
CeGe	FeB	3.88		10.5	0	[11]											
$CeMg_2$	$MgCu_2$	3.78					3.6		8			0.8	0		30	12.7	[12]
CeMg	CsCl	3.91		20	−3	[1]		19.2	[g]			2[h]	0	40		6.8	[13]
Ce_3Pd_5	Th_3Pd_5	3.95	1.5		−35	[14]	1.45		5			0.8	0				[10]
CePd	CrB	3.77	6.5		−22	[14]	6.4		14			1	0		60		[15]
CePt	CrB	3.80	6.0		−16	[16]	5.8		13	0.057	3	1	0	100		6	[17]
Ce_2S_3[j]	Th_3P_4	4.04	2.8		−85	[18]	3		2.9[f]			1	0		11.3	3.6	[19]
							4.2		3.2[f]				2				[19]
Ce_3S_4[k]	Th_3P_4	4.03	6.6		−170	[18]	6.6		5.9			0.92	0		71	4.66	[19]

$CeSi_{1.85}$	α-ThSi	4.02			−110	[21]	1			-1.16^{p}	3^{p}		0	269^{p}		[20]
$CeSi_{1.83}$	α-$ThSi_2$	4.02	5.5		−70	[21]	5.5		1.6	-1.1^{p}	3^{p}	0.17	0	255^{p}		[20]
							≈8		≈ 1.3				5.3			
$CeSi_{1.8}$	α-$ThSi_2$	4.02					9		4.2		1.5	0.6	0	230		[22]
											1^{v}					[22]
$CeSi_{1.7}$	α-$ThSi_2$	4.01					13.7		3.7		1.5	0.77	0	78		[21]
							12.4^{m}		8.7		1^{v}					
CeSi	FeB	3.83		6.2	10	[27]		5.75	7.4	0.078	3	0.87	0	22		[27]
								4.6	3.2	0.063		2^{l}	5			[27]
CeTl	CsCl	3.89	(21.5^{q})	25.5		[23]										
$CeZn_2$	$CeCu_2$	3.81		7.5	-63^{r}	[24]										
					51^{s}	[24]										
CeZn	CsCl	3.71	(12^{q})	30	11^{t}	[25]		g					0	19	9	[26]
					18^{u}											

[a] Room temperature structure.
[b] At $1.5\,T_C$.
[c] Structural transition.
[d] In the range $0.4 < T < 1$ K.
[e] $C_M = BT^3 \exp(-E_g/T)$.
[f] Broad maximum.
[g] First-order transition.
[h] Quartet ground state.
[j] Actual composition $CeS_{1.457}$.
[k] Actual composition $CeS_{1.393}$.
[l] At 30 K.
[m] Double transition.
[p] Spin fluctuation behaviour, $C = GT + BT^3 + DT^3 \ln T$.
[q] Under pressure $P > 10$ kbar.
[r] Along the c axis.
[s] Along the b axis.
[t] From ref. [13].
[u] From ref. [1].
[v] At $T < 1.5$ K.
* In units of $J\,K^{-(n+1)}$/Ce atom.

** References: [1] Schmitt et al. (1978); [2] Takke et al. (1981); [3] Ray and Sivardiere (1976); [4] Morin et al. (1988); [5] Bredl (1985); [6] Wegener et al. (1981); [7] Sakai et al. (1981); [8] Nakazato et al. (1988); [9] Walline and Wallace (1965); [10] Sereni and Kappler (1989); [11] Buschow and Fast (1966); [12] Buschow et al. (1978); [13] Pierre and Murani (1980); [14] Kappler et al. (1985); [15] Kappler et al. (1984); [16] Huber and Luengo (1978); [17] Holt et al. (1981); [18] Taher et al. (1980); [19] Ho et al. (1982); [20] Dhar et al. (1987); [21] Sato et al. (1985, 1988b); [22] Yashima et al. (1982); [23] Kurisu et al. (1985); [24] Fujii et al. (1988); [25] Kadomatsu et al. (1986); [26] Pierre et al. (1985); [27] Dijkman (1982); [28] Kappler (1990).

1982b). The specific heat of these exchange-enhanced paramagnets will be discussed in sect. 7.

There are three antiferromagnetic equiatomic compounds showing the FeB crystalline structure: CeCu, CeSi and CeGe. In particular, CeCu does not follow the mentioned correlation between crystalline structures and valence electron concentration (an s-type partner should favour the CsCl-type structure). However, some unexpected similarities between the CeCu and CeSi specific heat results (see fig. 8 for the CeCu data) are very suggestive. The respective entropies associated with the magnetic transitions are 0.72 and $0.87R\ln 2$, the γ_{LT} coefficients are 30 and $22\,\text{mJ}\,\text{mol}^{-1}\,\text{K}^{-2}$, and both show an extra contribution to C_M with maximums at 13 K and at 30 K, see Sereni and Kappler (1989) for CeCu and Dijkman (1982) for CeSi. Under a magnetic field, C_M (at T_N) and T_N itself decrease by about 10% per tesla in both cases and the high-temperature contribution is enhanced proportionally to the entropy decrease of the antiferromagnetic transition (fig. 8).

As already mentioned, the difference between T_N and T_{max} is a sign of short-range order and low dimensionality. For both CeSi and CeCu, T_N is 10% lower than the maximum of the susceptibilities (T_{max}). From the analysis of the fraction of internal energy and entropy gain at T_N in CeCu, no conclusion can be extracted concerning the dimensionality. Nevertheless, the C_M contribution of the high-temperature (13 K) anomaly in CeCu has a specific heat maximum of $0.35R$ in coincidence with the Heisenberg model for a chain with a spin $\frac{1}{2}$, see fig. 8. Although under an external field of 1 T the temperature of that maximum is lowered by about 10%, no significant changes in its absolute value are detected, fig. 8. The total entropy gain, including the antiferromagnetic transition and the high-temperature anomaly reaches $0.95R\ln 2$. In conclusion we see that the specific heat measurements in CeCu allow us to recognize the presence of antiferromagnetic fluctuations of low dimensionality, which are related to the doublet ground state of CeCu. The canted chains formed by the Ce atoms provide the appropriate geometry for this behaviour.

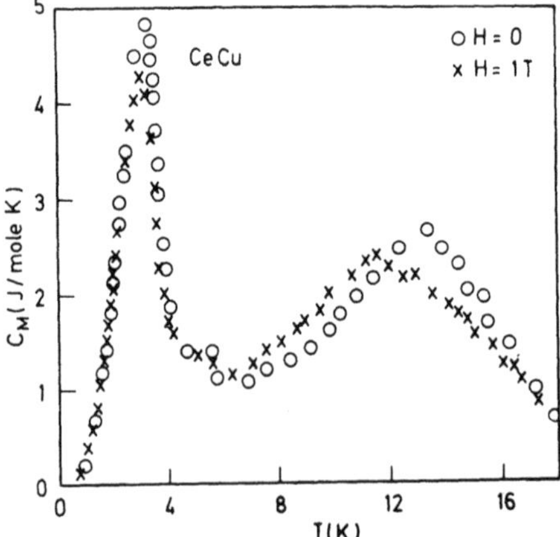

Fig. 8. Magnetic specific heat ($C_M = C_p - \beta T^3$) of CeCu at zero and an applied magnetic field of 1 T.

6. Region III: antiferromagnets with large Ce–Ce spacing

The compounds included in this region present a great variety of magnetic behaviours and crystalline structures. They are shown in fig. 9 and their properties are summarized in table 4. Although we use the term antiferromagnets as an identification of this group, ferromagnetism is present in some of their phase diagrams. In the limit, the only "pure" ferromagnet of this group, $CeGe_2$, can be used as the exception (see below) which confirms the trend that the direct magnetic exchange is not the dominant interaction for $D > 4.1$ Å. The large Ce–Ce spacing itself suggests as the most probable magnetic interaction one of RKKY-type, allowing the formation of a variety of magnetic structures together with complicated magnetic phase diagrams. Here, the entropy associated with the magnetic transition will be found: smaller, equal or larger than $R \ln 2$. The last case results from the low energy of the crystal-field levels. The γ_{LT} and γ_{HT} coefficients range between 0 and1.5 $J\,K^{-2}$/Ce atom and θ_P between 22 and -100 K.

There are three compounds with ferromagnetic behaviour: $CeGe_2$, $CeGa_2$ and Ce_3Al_{11}, whose specific heats are known. As already mentioned, $CeGe_2$ orders ferromagnetically at $T_C = 6.9$ K, undergoing a second-order transition, with the $R \ln 2$ value reported by Yashima et al. (1982) and 80% of that value by Mori et al. (1985). In the latter case, the authors argue that the entropy deficiency results from the Kondo effect. If this is so, then the remanent entropy should appear in the specific heat at $T > T_C$ as a γ_{HT} term, but here γ_{HT} is not larger than 30 $mJ\,K^{-2}$/Ce atom. Furthermore, γ_{LT} and θ_P values are found to be practically zero, which is not usual in Kondo-like systems. A similar entropic deficiency is observed in the Schottky anomaly, with the maximum at 25 K (Mori et al. 1985). There the authors note that the phonon contribution to the total specific heat was evaluated from the $LaGe_2$ specific heat,

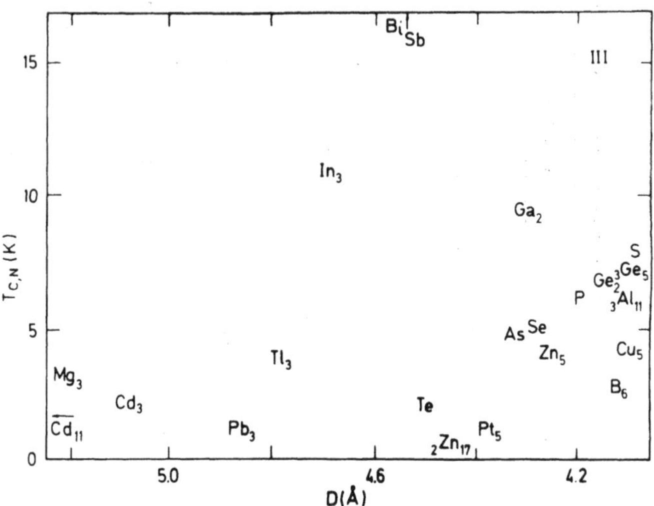

Fig. 9. Ce_jX_k compounds of region III, labelled as $_jX_k$.

TABLE 4
Antiferromagnetic compounds belonging to region III. The parameters are defined in the list of symbols.

	Structural data		Magnetic data				Specific heat data										
Compound	Type	D (Å)	T_C (K)	T_N (K)	θ_P (K)	Ref.**	T_C (K)	T_N (K)	C_{max} ($J K^{-1}$/Ce atom)	B^*	n	ΔS ($R \ln 2$)	H (T)	γ_{LT} (mJ K^{-2}/Ce atom)	γ_{HT} (mJ K^{-2}/Ce atom)	$\gamma(LaX_j)$ (mJ K^{-2}/Ce atom)	Ref.**
Ce_3Al_{11}	La_3Al_{11}	4.12	6.2		−25[a]	[1]	6.2	3.2[c]	7.6	b		0.42[m]		118	147		[3]
				3.2					5[n]			0.34[n]					
CeAs	NaCl	4.30		5.1	−21	[4]		7.2	10								[21]
				7.5	16	[21]											
CeB_6	CaB_6	4.14	(3.2[c])	2.4	−62	[6]	(3.4[c])	2.5	20(7[c])	0.95	3	1	0	300			[7, 8]
							(5[c])		(13)			2[d]	2				[8]
							(7[c])		(23)				8				[8]
CeBi	NaCl	4.58		25.2	12	[9]		25	[e]			2.6[f]					[10]
								12.5	[e]								[10]
$CeCd_3$	$BiLi_3$	5.11		2.0	−52	[12]		4	8			≈1		50			[13]
$CeCu_5$	$CaCu_5$	4.11		3.8		[13]		3.8	14								[13]
				3.6		[13]		11.3	10			1					[15]
$CeGa_2$	AlB_2	4.32		9.5	−107[g]	[14]			e			0.75					[15]
			8.5		22[k]	[14]	8.4	9.9	6.9				0	8.9		5.1	[32]
								≈12.5	7.5				3				[32]
$CeCd_{11}$	$BaHg_{11}$	6.59		<2	−55	[11]											
$CeGe_2$	$\alpha GdSi_2$	4.13	6.8		−3	[16]	7.0		8.3		3	1		0			[17a]
					1.3	[17b]	6.88		8.1	0.09	3	0.8		0		3.9	[17b]
Ce_3Ge_5	$\alpha ThSi_2$	4.11	7.2			[16]											
$CeIn_3$	$AuCu_3$	4.69		10.2	−64	[18]		10.1	13	0.012	3	1		140	140	6.3	[19]
$CeMg_3$	BiF_3	5.25		3.5	−44	[20]		8.2[h]	6.6[h]	0.019[h]	3	0.8[h]			290[h]		[19]
CeP	NaCl	4.18		7.1	6.5	[21]		7.0	7.2			≈1				0.8	[21]

$CePb_3$	$AuCu_3$	4.88	1.15	−10	[22]	1.1	3.8			0		200	10	[23]
						0.85	1.5			5.5				[23]
									0.64	0	1500			[24]
										10.7	1000			[24]
$CePt_5$	$CaCu_5$	4.38	1.03	−1.0	[25]	1.0	7	2.7	0.85	0	≈50		14	[25]
						1.5	2.5[j]	3		3	30			
CeS	NaCl	4.10	8.3	−20	[26]	8.35	10.6		≈1					[26]
CeSb	NaCl	4.53	18	5	[9]	16	[e]		>1					[10]
						8.5	[e]							[10]
											25			[27]
CeSe	NaCl	4.24	5.0	−7	[26]	5.0	9.3		≈1					[26]
								≈2	>1					[33]
CeTe	NaCl	4.50	2.2	−4	[26]	2.15	6.6	<3	≈1					[26]
$CeTl_3$	$AuCu_3$	4.77	3.8	−9	[28]	3.85	9.2		1	0		43	3.4	[28]
						2.9	6			3				[28]
						3.5	3.2			5				[28]
				−7	[29]	3.8	9.6		≈1	0				[29]
Ce_2Zn_{17}	Th_2Zn_{17}	4.42		−100[k]	[30]	1.6	15[e]	[l]	0.82			80		[30]
$CeZn_5$	$CaCu_5$	4.28	3.8	−38	[31]									

[a] From ref. [2].
[b] $C_M = BT^{-1/2} \exp(-E_g/T)$.
[c] Quadrupolar transition from ref. [5].
[d] At 50 K.
[e] First-order transition.
[f] Crystal field splitting $\Delta < T_N$.
[g] Parallel to the c axis.
[h] Under pressure, $P = 12$ kbar.
[j] Broad maximum.
[k] Perpendicular to the c axis.
[l] Logarithmic divergence at T_N.
[m] Between T_N and T_C.
[n] At T_N.
* In units of $J\,K^{-4}$/Ce atom.

** References: [1] Chouteau and Palleau (1980); [2] Van Daal and Buschow (1970); [3] Peyrard (1980), Berton et al. (1980); [4] Tsushida and Wallace (1965a), Ott et al. (1978); [5] Effantin et al. (1985); [6] Komatsubara et al. (1980); [7] Fujita et al. (1980); [8] Peysson et al. (1986); [9] Tsushida and Wallace (1965a); [10] Hulliger et al. (1975), Tsushida et al. (1973); [11] Tang and Gschneidner (1988); [12] Tang and Gschneidner (1989); [13] Willis et al. (1987); [14] Burlet et al. (1987); [15] Takahashi et al. (1988); [16] Lahiouel et al. (1986); [17a] Yashima et al. (1982); [17b] Mori et al. (1985); [18] Lawrence and Shapiro (1980); [19] Peyrard (1980); [20] Galera et al. (1981); [21] Hulliger and Ott (1978); [22] Sereni and Olcese (1979); [23] Lin et al. (1986); [24] Fortune et al. (1987); [25] Schroeder et al. (1988); [26] Hulliger et al. (1978); [27] Kitazawa et al. (1988); [28] Elenbaas et al. (1980); [29] Rahman et al. (1989); [30] Sato et al. (1987, 1988a); [31] Gignoux et al. (1987); [32] Dijkman (1982); [33] Busch et al. (1971).

whose Debye temperature ($\theta_D = 250$ K) may differ from that of $CeGe_2$. Concerning the presence of this ferromagnet in region III, we have to remark that in the α-$GdSi_2$ structure the Ce atoms are distributed in a triangular coordination and such a spatial distribution frustrates the antiferromagnetic interaction. Such frustration effects can also be present in the $CeSi_{2-x}$ compounds. The $CeGa_2$, where Ce lies in a honeycomb coordination, undergoes four magnetic transitions (between 11.3 and 8.4 K), becoming ferromagnetic in the lowest temperature phase (Takahashi et al. 1988). In this detailed study on a single-crystal sample, the specific heat clearly shows second-order transitions at 11.3, 10.3 and 9.9 K and a first-order one at 8.4 K (from the antiferromagnetic to the ferromagnetic phase), see fig. 10. The same measurement performed in a polycrystalline sample shows a unique broad anomaly (Dijkman 1982). Such a dense succession of magnetic phases can be understood by the fact that the order parameter cannot develop from short to long range due to frustration effects in the "honeycomb" lattice. At low enough temperature a simple magnetic configuration, such as the ferromagnetic one, is able to develop the long-range order, lowering the free energy of the system. There, the antiferromagnetic to ferromagnetic transition has to be of first order because of the already nonzero value of the order parameter. This quasi-bidimensional scheme is applicable for $CeGa_2$ because of its strong anisotropic exchange interactions, which result in a value of $\theta_P = -107$ K when the applied field is parallel to the c axis and $\theta_P = 22$ K when it is perpendicular. The entropy gain reaches the $R \ln 2$ value at the paramagnetic phase, being 75% of the value related to the ferromagnetic phase.

Another case of a ferromagnetic phase is shown by Ce_3Al_{11}. Between 6.2 and 3.2 K the magnetic order is found to be ferromagnetic and below 3.2 K it becomes

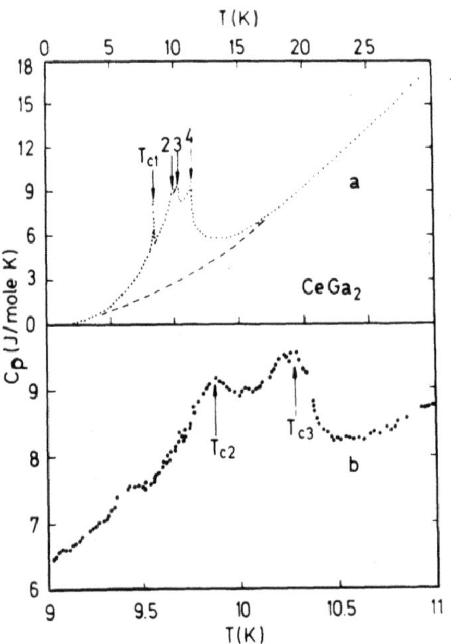

Fig. 10. (a) Total specific heat of $CeGa_2$ at the magnetic transition temperature. T_{c1} is ferromagnetic while 2, 3 and 4 denote the antiferromagnetic transitions. (b) Detailed structure at around 10 K, after Takahashi et al. (1988).

modulated (Benoit et al. 1979). The temperature dependence of C_M at $T < T_N$ follows a $C_M \sim T^{-1/2} \exp(-E_g/T)$ dependence because of the strong anisotropy of the system (Peyrard 1980). The entropy gain at $1.5T_C$ is about 80% of $R \ln 2$, which, together with the large γ_{LT} and γ_{HT} values (118 and 147 mJ K^{-2}/Ce atom, respectively), indicate the Kondo character of this compound, see also Peyrard (1980) and Berton et al. (1980). Under an external presure the specific heat measurements show the gradual disappearance of the modulated (antiferromagnetic) phase, without significant changes in the T_C transition, but with an increase of the γ_{LT} and γ_{HT} terms (250 mJ K^{-2}/Ce atom at 11 kbar) (Peyrard 1980), see fig. 11. This is one of the few cases where from the similarity in value and pressure dependence, γ_{LT} and γ_{HT} can be associated as having the same origin. Taking into account that the La_3Al_{11}-type structure has two different sites for Ce (one atom at position I and two atoms at position II) and that 60% of the entropy is associated with the magnetically ordered phase, one can conclude that each sublattice behaves independently as suggested in some compounds belonging to region I.

The following are compounds with prevailing antiferromagnetic characteristics. In the hexagonal CeX_5 (X = Cu, Pt, Zn), the Ce sites are equivalent but not the X ones. This structural characteristic permits some "ternaries", compounds of the form $Ce(X, Y)_5$ to be pseudobinaries. We shall discuss this kind of pseudobinary in sect. 8 as "related compounds". Similar concepts have to be taken into account in those compounds where Ce has inequivalent sites. For example, Ce_7Rh_3 should be considered as a $Ce^{I}Ce^{II}_3Ce^{III}_3Rh_3$ quaternary compound.

The $CeCu_5$ orders antiferromagnetically at $T_N = 3.8$ K, showing a structure in the specific heat jump at 4 K. This "double transition" is also seen in thermal expansion and in magnetic susceptibility measurements at 3.6 and 3.8 K, respectively. The entropy associated with both transitions is approximately $R \ln 2$ and $\gamma_{LT} = 50$ mJ K^{-2}/Ce atom (Willis et al. 1987). The temperature dependence of C_M at $T < T_N$ is not characteristic of a second-order transition, furthermore, the cusp observed at T_N and the strong anomaly in the thermal expansion suggest such a

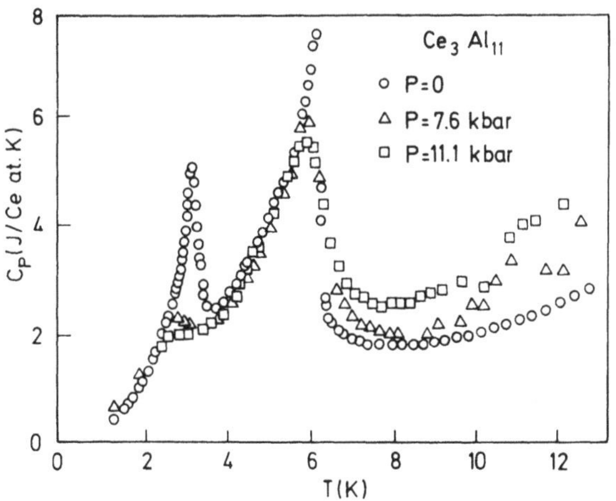

Fig. 11. Total specific heat of Ce_3Al_{11} under various applied pressures, after Peyrard (1980).

transition to be of first order. Unfortunately, to our knowledge, there are no specific heat measurements performed in $CeZn_5$, which would be the best compound for comparison. The $CePt_5$ shows a second-order type transition in the specific heat at $T_N = 1$ K, with a $T^{2.7}$ dependence of C_M at $T < T_N$ and a linear term of $\gamma_{LT} \approx$ 50 mJ K^{-2}/Ce atom. Under an applied field of up to 3 T, the γ_{LT} term does not change significantly, but the temperature dependence of C_M at $T < T_N$ takes the expected form of T^3 for an antiferromagnet, with an estimated entropy, up to $T \simeq 3T_N$, of $0.85R\ln 2$ (Schroeder et al. 1988). A long tail observed right above the specific heat jump suggests the presence of spin fluctuations at $T > T_N$. The above authors identify this compound as an XY-type antiferromagnet. As expected for an antiferromagnet the ordering temperature is shifted towards lower temperatures with increasing magnetic field and yields a Schottky-like anomaly for $H = 3$ T. In terms of a two-level scheme, the splitting depends on the magnetic field but not on its height; here C_M at T_{max} is closer to $0.35R$, the expected value for an XY antiferromagnet (de Jongh and Miedema 1974).

Region III is rich in cubic compounds and practically all of them are formed with p elements. Cerium monopictinides and monochalcogenides have the NaCl structure and CeX_3 (with X = In, Tl, Pb) the $AuCu_3$ structure, all of them face-centered cubic (fcc) type. CeB_6 has the CaB_6 structure, which is simple cubic and is closely related to the CsCl structure, whereby the B_6 octahedra occupy the Cl positions.

The smallest crystal field splitting, together with the highest entropy of the magnetic phase in Ce compounds is found in CeSb and CeBi, with values of $T_N = 18$ and 25 K (and $\Delta = 33$ and 10 K), respectively. In both cases the paramagnetic to antiferromagnetic transition coincides with a cubic to a tetragonal structural distortion, observed in the lattice parameters and in the thermal expansion (Hulliger et al. 1975). The specific heat of these transitions has a divergence according to a first-order character, which cannot be attributed simultaneously to the structural and to the magnetic transitions. Another first-order transition occurs at $\frac{1}{2}T_N = 8.5$ K (CeSb) and 12.5 K (CeBi), and a slight discontinuity is observed in the tetragonal lattice parameters together with an anomaly in the thermal expansion of CeSb. In the specific heat these transitions are reflected as a small hump in CeSb and a peak in CeBi; the latent heat there was evaluated to be 6 J mol^{-1}. These transitions are associated with a change in the periodicity in the antiferromagnetic arrangement (see also Hulliger et al. 1975). The low dimensionality of such an arrangement is seen in the T^2 dependence of C_M at $T < T_N$ in CeSb (Busch et al. 1971). The entropy evaluated by these authors is $\Delta S/R = 1.08$ at T_N and 1.71 at $T \simeq 50$ K, close to $\Delta S/R = \ln 6 = 1.79$, which is in agreement with the extremely low crystal field splitting Δ between the Γ_7 and Γ_8 levels, $\Delta = 33$ K from Furrer et al. (1979). A similar conclusion is extracted from the evaluation of the entropy gain of CeBi, excluding the latent heat at the transitions. There, one finds $\Delta s/R = 1.47$ at 26 K, again in agreement with the small value of $\Delta = 10$ K (Furrer et al. 1972).

The other two monopicnitides, CeP and CeAs, order antiferromagnetically at $T_N = 7.1$ and 7.2 K, respectively. No lattice distortions were observed, although there is a thermal expansion anomaly associated with the purely magnetic second-order transition. The absence of lattice distortion agrees with the zero quadrupolar matrix elements of the Γ_7 doublet ground state and a CF splitting of $\Delta \approx 150$ K (Hulliger and Ott 1978).

The three Ce monochalcogenides: CeX, with X = S, Se, Te also order antiferromagnetically at $T_N = 8.3$ K, 5.0 K and 2.15 K, respectively. The magnetic entropy is close to $R \ln 2$, but the maximum values of C_M at T_N are appreciably lower than the values predicted in a molecular-field approximation (Hulliger et al. 1978). From the high-temperature specific heat measurements the CF splitting can be obtained, being $\Delta = 155$, 120 and 45 K, respectively. The fact that none of these Schottky anomalies reach the expected maximum value will be discussed in sect. 9, together with other Ce compounds with similar characteristics.

As already mentioned, there are three Ce compounds in this region with the formula CeX_3 (with X = In, Tl, Pb) and crystalline structure of the $AuCu_3$ type. There is a fourth compound, $CeSn_3$, which is included by structure into this group, but from its intermediate-valence behaviour it belongs to the region-IV compounds (cf. sect. 7). $CeIn_3$ is considered to be a local-moment antiferromagnet with $T_N = 10.1$ K and the maximum value of C_M coincident with the mean-field prediction ($C_M(T_N) = \frac{3}{2}R$), and an entropy gain of $0.95 R \ln 2$ right above T_N (Peyrard 1980). Although the entropy reaches practically the maximum value, there is a significant contribution of the linear term included in this evaluation (γ_{LT} and $\gamma_{HT} = 140$ mJ mol^{-1} K^{-2}, Peyrard 1980), which is about 25% of $R \ln 2$ at T_N. Coincidentally, it shows a strong Kondo lattice resistivity (Maury et al. 1982), θ_P is large: -64 K (Lawrence and Shapiro 1980), and applied pressure shifts T_N towards lower temperatures, increasing γ_{LT} and γ_{HT} to the detriment of the entropy associated with the magnetic transition (Peyrard 1980). $CeTl_3$ is also a local-moment antiferromagnet, but with a normal resistivity at high temperatures ($T \geqslant T_N$). It orders at $T_N = 3.8$ K with the total magnetic entropy gain at 4.2 K (Rahman et al. 1989). The $\theta_P = -9$ K value (Elenbaas et al. 1980), is within that expected for a system with CF effects. Under applied fields, T_N is shifted towards lower temperatures as expected for an antiferromagnet and for fields stronger than 3 T the specific heat shows a Schottky anomaly described by two levels with $\Delta = 8.4$ K, with a maximum of 3.6 J mol^{-1} K^{-1}, which is independent of the applied field (Elenbaas et al. 1980). Finally, $CePb_3$ has been shown to be an itinerant antiferromagnet with an extremely small magnetic moment and large γ_{HT} value (Lin et al. 1985). It orders at $T_N = 1.1$ K, with an entropy associated with the magnetic transition of $0.64 R \ln 2$, a γ_{HT} contribution of 220 mJ mol^{-1} K^{-2} and a C_M/T value extrapolated to $T = 0$ of $\gamma_{LT}(0) = 1.5$ J mol^{-1} K^{-2}. Under a magnetic field of 11 T the transition is suppressed and the specific heat shows a broad maximum around 3 K, while the $\gamma_{LT}(0)$ value is reduced to 0.9 J mol^{-1} K^{-2} (Lin et al. 1986). The ordering temperatures of these three compounds follow an RKKY relationship, with a small "$k_F r$" argument (k_F is the Fermi wave vector and r is the distance between magnetic ions),

$$(r - r_0)^4 T_N = P, \qquad (8)$$

here the distance coincides with the lattice parameter: $r = a$; $r_0 = 4.43$ Å and $P = 0.045$ are the constants obtained from fitting eq. (8) with the CeX_3 magnetic compounds. The hypothesis that "$k_F r$" is small is supported by the fact that there is a power-four law and P is a constant within a 5% dispersion.

The specific heat of Ce_2Zn_{17} shows a sharp peak with a logarithmic divergence at $T_N = 1.6$ K, as a sign of low dimensionality, while the curves are also fitted by a power law with critical exponents $\alpha = \alpha' \simeq 0.1$, corresponding to a three-dimensional (3D) Ising magnet (Sato et al. 1987, 1988a). These authors' argument that the anisotropy of the magnetic interactions leads to a lower dimensionality is supported by the significant short-range order effect above T_N and the ratio of internal magnetic energy $\mathscr{U}(T > T_N)/\mathscr{U}(T < T_N \approx \frac{3}{4})$ (0.71 for the 2D Ising model for spin $\frac{1}{2}$), while the entropy $\Delta S(T_N)/R \ln 2 = 0.82$ is close to 0.84, the value for the 3D prediction (see, e.g., de Jongh and Miedema 1974). The entropy value may be an overestimate because of the lack of lower-temperature measurements ($T < 1.2$ K), where the C/T ratio becomes important and has to be compared with the reported linear term of 80 mJ K^{-2}/Ce atom extracted from $T > 5T_N$.

Finally, the CeB_6 specific heat has been intensively studied because of its double transition and the chance of it being one of the few cases having a guarded Γ_8-CF ground state. In the absence of an external magnetic field, CeB_6 shows a hump at 3.4 K and an antiferromagnetic transition at $T_N = 2.4$ K. The high-temperature anomaly was recognized as resulting from a quadrupolar transition (Effantin et al. 1985), which shifts towards higher temperatures and gains intensity under an applied magnetic field. On the contrary, the antiferromagnetic transition at 2.4 K is strongly depressed under a magnetic field (Fujita et al. 1980, Peysson et al. 1986). The value of C_M at T_N (20 J K^{-1}/Ce atom) largely exceeds the mean-field prediction. The entropy associated with this transition is $R \ln 2$, while the $R \ln 4$ value is reached at the paramagnetic phase (Fujita et al. 1980). The temperature dependence of the antiferromagnetic BT^3 (with $B = 950$ mJ K^{-4}/Ce atom) was observed between 0.5 and 1 K, while a linear term ($\gamma_{LT} = 300$ mJ K^{-2}/Ce atom) becomes dominant at $T < 0.5$ K as the antiferromagnetic excitation vanishes because of an anisotropy gap (Peysson et al. 1986).

7. Region IV: compounds with a nonmagnetic ground state

We shall include in this region the Ce compounds that, having equivalent sites for the Ce atoms, do not order magnetically. Indeed, some compounds where Ce lies in different sublattices could be included in this and other regions simultaneously. However, because the magnetic order usually eclipses weaker magnetic contributions, we chose the dominant one. Figure 12 shows the compounds that belong to this region. It is evident that there are two distinguishable groups: (i) the heavy-fermion compounds, with an energy scale of a few degrees Kelvin ($T_m \leqslant 10$ K) and (ii) the intermediate-valence compounds, with a much larger energy scale ($T_m > 100$ K) (see table 5).

7.1. *Heavy-fermion compounds*

Although the number of Ce compounds showing HF behaviour has increased in the last few years, most of them are ternary compounds. There is, however, a significant number of compounds included in the HF family because of their large γ_{LT} coefficient (of the order of 100 mJ K^{-2}/Ce atom), which actually have to be considered as spin glasses due to the varying electronic environment around the Ce ion. Such an effect is produced by nonmagnetic-atom disorder (NMAD), which leads to a large C_M/T peak near 1 K characteristic of the spin glass behaviour (Gschneidner et al. 1990). Some of these "false heavy-fermion compounds" will be discussed in sect. 8 within the pseudoternary group, where the Ce partner is partially substituted. Such atomic disorder is not possible in binary compounds and coincidentally there are only two well-recognized HF compounds ($CeCu_6$ and $CeAl_3$). We shall introduce another candidate to this group: Ce_3In, whose low-temperature specific heat has the characteristics of HF-compound behaviour.

One of the best examples of an HF compound is $CeCu_6$, its C_M/T ratio grows continuously by decreasing the temperature reaching a value of $\gamma_{LT}(0) =$ 1.67 J K^{-2}/Ce atom at $T < 700$ mK (Amato et al. 1987). Although the C_M/T ratio shows no maximum, the experimental results are fitted with a $C_M/T = \gamma_{LT}(0) - A'T$ equation. The presence of such a term $A'T$ indicates that some maximum should occur below the experimental temperature range (i.e., $T < 50$ mK). At temperatures $T > 10$ K, the γ_{HT} term is large: 250 mJ K^{-2}/Ce atom (Penney et al. 1987), but the total entropy gain of $R \ln 2$ is reached at 15 K (including the γ_{HT} contribution). This suggests that there is still a strong decoupling of the electrons involved in the interactions when the excited CF levels already contribute to the specific heat. The $\gamma_{LT}(0)$ term decreases quadratically with applied field (up to 4.5 T), with a ratio of 0.028 J K^{-2} T^{-2}/Ce atom. For $H > 5$ T a maximum in C_M/T appears as an effect of the strong polarization of a narrow band with Zeeman decoupling between the spin-up and spin-down bands (Amato et al. 1987). The $\gamma_{LT}(0)$ term also decreases under

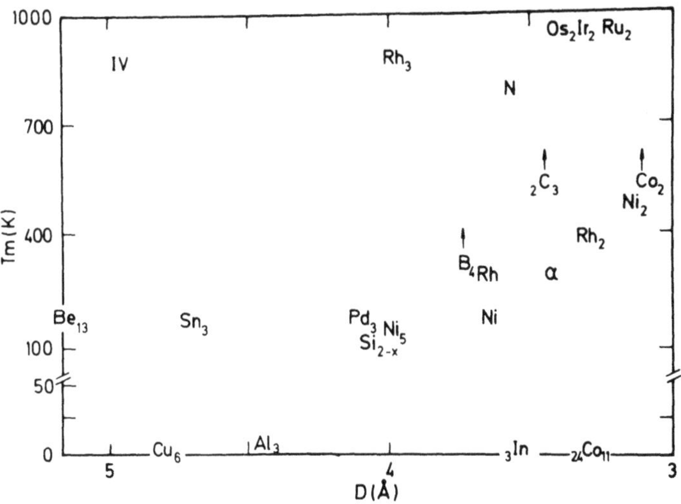

Fig. 12. Ce_jX_k compounds of region IV, labelled as $_jX_k$ and α-Ce.

TABLE 5

Compounds with nonmagnetic ground states, belonging to region IV. The parameters are defined in the list of symbols.

Compound	Structural data			Magnetic data			Specific heat data								
	Type	D (Å)	$\Delta V/V$ [a] (%)	T_m (K)	χ_0 (10^{-3} emu mol^{-1})	Ref.*	T_m (K)	C_M/T ($T=0$) (mJ K^{-2}/Ce atom)	γ_{LT} (mJ K^{-2}/Ce atom)	γ_{HT} (mJ K^{-2}/Ce atom)	H (T)	P (kbar)	$\gamma(LaX_j)$ (mJ K^{-1}/ Ce atom)	Ref.*	χ_0/γ (emu K^2 J^{-1})
$CeAl_3$	$SnNi_3$	4.43	1.1	0.6	36	[3]	0.35[b]	1400	1750[c]	250				[2]	
								1620						[3]	0.022
							0.36[b]	1400	1760[c]			0		[4]	
							0	1020				0.38		[4]	
							0	550				8.2		[4]	
							0.45	1200	2020[c]		0			[5]	
							0.4	≈ 1500	1820[c]		2			[5]	
							0.4		1730[c]		4			[5]	
										280	0			[6]	
										350	23			[6]	
CeB_4	ThB_4	3.71	3.5	> 500	1.1	[7]									
					0.55	[8]									
$CeBe_{13}$	$NaZn_{13}$	5.19	0.4	170	2.05	[9]				59			7.8	[9]	0.035
				140	2.17	[10]				115				[11]	
Ce_2C_3	Pu_2C_3	3.48	3.0	> 500	0.8	[7]									
$CeCo_2$	$MgCu_2$	3.1	7.9	> 500	1.23	[12]	(1.5)[t]		38	40			8.8[d]	[13]	0.032
$Ce_{24}Co_{11}$	$Ce_{24}Co_{11}$	3.37			14.0[r]	[51]	1.2		200[s]	87.5[s]				[52]	0.003[s]
$CeCu_6$	$CeCu_6$	4.8	0.0	− 12[f]	32[g]	[19]	2	1620		250[e]			8[h]	[15]	0.019[g]
								1670			0			[19]	
								800			5[i]			[19]	
								500			7.5[i]			[19]	
								1600				0		[20]	
								1370				2		[20]	
								1170				4.1		[20]	
								820				8.8		[20]	
Ce_3In	Cu_3Au	3.54	1.2	1.9	73	[21]	2.2	500	680	300				[21]	0.146
$CeIr_2$	$MgCu_2$	3.28	2.7	> 1000	0.55	[22]			24	24				[23]	0.023
CeN	$NaCl$	3.55	11	870	0.4	[24]			8.3				1.5	[25]	0.050
$CeNi_5$	$CaCu_5$	4.01	3.5	100	3.03	[26]			40				34	[27]	0.075
$CeNi_2$	$MgCu_2$	3.13	6.3	> 1000	0.57	[28]			29.4	30				[13]	0.019

CeNi	CrB	3.65	4.3	150	1.6	[29]		65	80		5	[30]	0.025
				230^j	0.8^j	[29]							
$CeOs_2$	$MgCu_2$	3.29	3.8	≈1000	0.45	[22]		24				[50]	0.019
	$MgZn_2$						$(1.1)^t$	22				[50]	
$CePd_3$	$AuCu_3$	4.16	0.4	145	1.28	[31]			38.2		0.28	[32]	0.033
$CePd_7$	$CuPt_7$	5.62			0.15	[54]		10				[55]	0.015
$CeRh_3$	$AuCu_3$	4.01	k	>800	0.31	[33]		14				[34]	0.022
$CeRh_2$	$MgCu_2$	3.27	2.4	380	0.77	[22]		26	23		6.4	[35]	0.033
CeRh	CrB	3.71	4.4	270	0.62	[33]		17	12.7			[36]	0.036
$CeRu_2$	$MgCu_2$	3.26	4.0	>1000	0.68	[22]	$(6.18)^t$	40.8	67		41.6	[37]	0.017
							$(6.15)^t$	6	64			[38]	0.010
$CeSi_2$	$\alpha ThSi_2$	4.01	k	≈100	4.2	[39]		104				[39]	0.040
Ce_3Sn	Cu_3Au	3.48			13	[53]	23	260				[53]	0.05
$CeSn_3$	$AuCu_3$	4.72	1.4	130	1.89	[40]		53	24		11.6^n	[43]	0.036
				140	1.98	[41]		75		0		[44, 45]	0.027
				130	2.38	[42]		63.4		10^l		[44]	
								53		10^m		[44]	
$CeSn_{3.05}$	$AuCu_3$			150	1.83	[47]		65.7				[47]	0.028
Ce-α	fcc	3.41	10.7^p	>300	0.42	[48]		12.8			11.5^q	[49]	0.033

[a] Volume of the unit cell, $V = \frac{1}{2}[V(LaX_j) + V(PrX_j)]$ and $\Delta V = V - V(CeX_j)$. Lattice parameters are taken from ref. [1].
[b] Temperature at the maximum of C/T.
[c] Maximum value of C/T.
[d] Value for $LaCo_2$ from ref. [14].
[e] See also ref. [16].
[f] θ_p value from ref. [17].
[g] Value of $\chi(T)$ at 2 K.
[h] From ref. [18].
[i] Magnetic field parallel to the c axis.
[j] Measured on the a axis.
[k] There are no reference compounds.
[l] Magnetic field in the direction [100].
[m] Magnetic field in the direction [110].
[n] From ref. [46].
[p] With respect to Ce-γ.
[q] For fcc La.
[r] Value of $\chi(T)$ at 1.5 K.
[s] Note that one mole $Ce_{24}Co_{11}$ contains 24 Ce atoms with 10 inequivalent positions in the lattice.
[t] superconducting transition.

* References: [1] *Pearson's Handbook* (1985); [2] Flouquet et al. (1982); [3] Andres et al. (1975); [4] Brodale et al. (1986a, b); [5] Bredl et al. (1984); [6] Andraka et al. (1989); [7] Olcese (1966a); [8] Sales (1974), Maple and Wohlleben (1973); [9] Besnus et al. (1983a); [10] Kappler and Meyer (1979); [11] Cooper et al. (1971); [12] Brochado Oliveira and Harris (1983); [13] Machado da Silva and Hill (1972); [14] Ikeda and Gschneidner (1980); [15] Penney et al. (1987); [16] Germann et al. (1988); [17] Zemirli and Barbara (1985); [18] Satoh et al. (1987, 1988); [19] Amato et al. (1987); [20] Phillips et al. (1987); [21] Sereni et al. (1989b), Sereni et al. (1990b), Chen et al. (1989); [22] Weidner (1984); [23] Sereni (1980b); [24] Olcese (1979); [25] Danan et al. (1969); [26] Gignoux et al. (1982); [27] Nasu et al. (1971); [28] Olcese (1980); [29] Gignoux and Voiron (1985); [30] Gignoux et al. (1983); [31] Besnus et al. (1982); [32] Besnus et al. (1983b); [33] Kappler et al. (1988); [34] Mihalisin et al. (1981); [35] Sereni (1981); [36] Sereni and Kappler (1989); [37] Joseph et al. (1972b); [38] Sereni et al. (1989a); [39] Yashima and Satoh (1982); [40] Sereni (1980a); [41] Shaheen et al. (1983); [42] Boucherle et al. (1986); [43] Costa et al. (1982); [44] Tsang et al. (1984); [45] Ikeda and Gschneidner (1982); [46] De Long et al. (1978); [47] Gschneidner et al. (1985); [48] Maple and Wohlleben (1973); [49] Koskenmaki and Gschneidner (1978); [50] Torikachvili and Maple (1984); [51] Canepa et al. (1989); [52] Kappler and Sereni (1990); [53] Chen et al. (1989); [54] Kappler et al. (1985); [55] Sereni et al. (1990c).

applied pressure (becoming $\gamma_{LT}(0) = 0.8\ J\ K^{-2}$/Ce atom under a pressure of 8.8 kbar) together with the γ_{HT} contribution (Phillips et al. 1987).

The $CeAl_3$ compound shows a $\gamma_{LT}(0)$ value of $1.2\ J\ K^{-2}$/Ce atom, but the C_M/T ratio increases up to $2\ J\ K^{-2}$/Ce atom at 0.5 K, which signals the transition to the coherent coupling of the Kondo resonances in a Kondo lattice (Bredl et al. 1984). Under an applied field the C_M/T maximum value is reduced to $1.7\ J\ K^{-2}$/Ce atom for $H = 4$ T and lightly shifted to lower temperatures (see also Bredl et al. 1984). The effect of pressure on $\gamma_{LT}(0)$ is similar to that in $CeCu_6$, reducing its value to $0.55\ J\ K^{-2}$/Ce atom under 8.2 kbar, but it is significant that the C_M/T maximum disappears under only 0.4 kbar of pressure.

The low-temperature specific heat of Ce_3In is presented in a C_M/T representation in fig. 13, after phonon subtraction (Sereni et al. 1989a, b). The extrapolated $\gamma_{LT}(0)$ value is about $0.5\ J\ K^{-2}$/Ce atom , with the maximum value of $C_M/T = 0.68\ J\ K^{-2}$/Ce atom at 2 K. The signature of the HF-compound behaviour is given by the ratio $(\chi T/C_M)(\pi k/\mu_{eff})^2 = 2.6$, which is constant for $T < 5$ K and close to the value predicted for a spin-$\frac{1}{2}$ Kondo system (from magnetization measurements, μ_{eff} was found to be about 1 μ_B for this compound) (Sereni et al. 1990b). At higher temperatures there is another contribution to C_M, which reaches its maximum at 18 K with a value of $C_M(18\ K) = 5.1\ J\ K^{-1}$/Ce atom. The total entropy gain at 30 K is $0.8 R \ln 4$, indicating that two doublets are involved in the low-temperature properties of this compound. In fact the cubic Cu_3Au-type structure gives the possibility of a quartet CF ground state. Desgranges and Rasul (1985) had studied the case of a Kondo system with CF splitting of the order of the characteristic temperature T_K; the experimental results from Ce_3In are compared with the theory in fig. 14. The fact that C_M shows a higher and sharper maximum at 18 K than any one of the values predicted by the theory may be due to the fact that Ce_3In actually has a Γ_8 quartet CF ground state that undergoes a quadrupolar transition at 18 K (as in the case of CeB_6) and therefore it becomes a real doublet ground-state system only at $T < 18$ K (Sereni et al. 1989b).

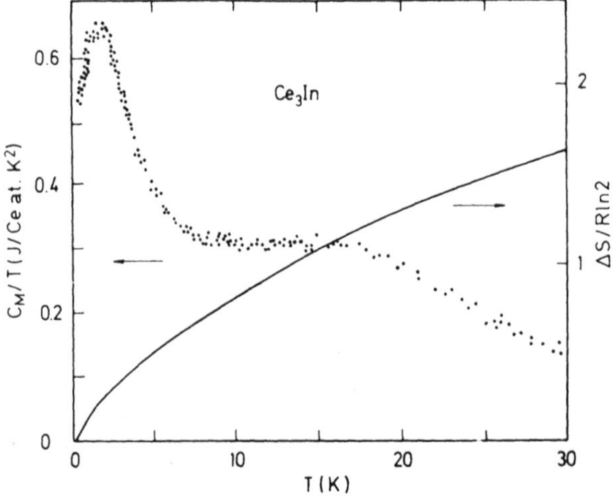

Fig. 13. Magnetic specific heat ($C_M = C_p - \beta T^3$) of Ce_3In in a C_M/T versus T representation. The continuous curve is the normalized entropy gain.

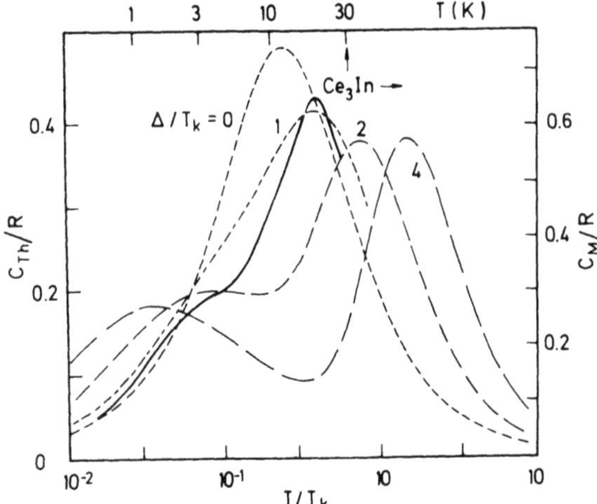

Fig. 14. Comparison of the Ce_3In specific heat (solid line) with the model of Desgranges and Rasul (1985) (dashed lines).

7.2. *Intermediate-valence compounds*

All the following compounds have a magnetic susceptibility that does not follow the Curie law, but shows a broad maximum (at T_m). The values of T_m are found between 100 K up to above the range of the measurements ($T > 10^3$ K) (see table 5). In all these compounds the Ce atoms have equivalent sites in the lattice, with a coordination number of 12. There are no s elements among the Ce partners. Coincidentally with the fact that large T_m values indicate larger energy scales than those of the HF compounds, the intermediate-valence compounds (IV) show much smaller γ terms, between 10 and 100 mJ mol^{-1} K^{-2}. The meaning of γ_{LT} and γ_{HT} will not be the same as for the compounds that order magnetically and they will be considered in each particular case.

The most important family of IV compounds is that formed with transition metals with the Laves structure. The volume reduction of these compounds (with respect to their related lanthanide compounds) is significant, reaching values up to 8%, see table 5. This leads to Ce–Ce spacings comparable to and smaller than those of the compounds of region I, see also table 5. The fact that the IV compounds are only cubic when $D < 3.5$ Å (more specifically fcc as α-Ce), suggests that the local symmetry of the Ce ion plays an important role in the f-conduction-electron hybridization (Sereni 1985). The low-temperature specific heat of these compounds is shown in figs. 15 and 16. $CeRu_2$ shows the highest temperature for the superconducting transition among the Ce compounds at $T_{sc} = 6.18$ K (Joseph et al. 1972b, Sereni et al. 1989a). The γ term extrapolated from $T > T_{sc}$ is $\gamma_{HT} \simeq 65$ mJ mol^{-1} K^{-2}. Because such a value of γ_{HT} is not consistent with the $\Delta C/T_{sc}\gamma = 1.43$ prediction of the BCS theory (where ΔC is the specific heat jump at T_{sc} nor with the thermodynamical condition that $\Delta S(T_{sc})$ for the superconducting phase and for the normal electronic contribution has to be equal, i.e. $\Delta S(T_{sc}) = \gamma T_{sc}$, the experimental results were interpreted in two different ways. Joseph et al. (1972b) evaluated a γ_{LT} term of 40.8 mJ mol^{-1} K^{-2}, which approaches both

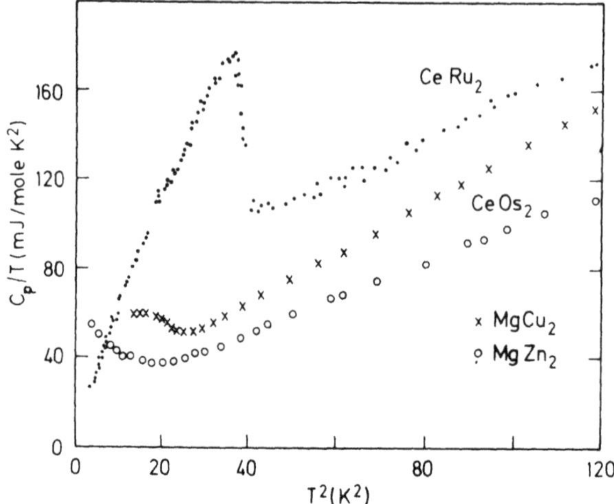

Fig. 15. Total specific heat of superconductive Ce Laves phases. $CeOs_2$ is shown for both allotropic forms, after Torikachvili et al. (1984). $CeRu_2$ was taken from Sereni et al. (1989).

conditions: $\Delta C/T_{sc}\gamma_{LT} = 1.33$ and $\Delta S(T_{sc}) = \gamma_{LT}T_{sc}$ (with $\Delta C = 336\ \text{mJ mol}^{-1}\ \text{K}^{-1}$). Sereni et al. (1989a) claim $CeRu_2$ to be an unconventional superconductor, with a measured value of $\gamma_{LT} = 6\ \text{mJ mol}^{-1}\ \text{K}^{-2}$ at $T < 0.1T_{sc}$, $\Delta C = 522\ \text{mJ mol}^{-1}\ \text{K}^{-1}$ and the ratio $\Delta C/T_{sc}\gamma_{HT} = 1.3$, close to the value predicted by Hirschfeld et al. (1986) for the "axial" superconductors. In the theory, a linear temperature dependence for $T \ll T_{sc}$ is also predicted. Note that the meaning of the γ_{LT} term is quite different in the two interpretations. In addition, in fig. 15 we show the specific heat of $CeOs_2$ in a C_p/T versus T^2 representation of its two allotropic forms, the cubic $MgCu_2$-type structure and the superconducting hexagonal $MgZn_2$-type structure. The γ_{HT} term is similar for both phases: 24 and 22 mJ mol^{-1} K^{-2}, respectively, and the jump $\Delta C(T_{sc})$ exhibited by the $MgZn_2$ phase is approximately one third of the value expected from the BCS theory (Torikachvili and Maple 1984).

In fig. 16, we also show the low-temperature specific heat of the Laves CeX_2 compounds (here X are TM belonging to the Co and Ni columns), in a C_p/T versus T^2 representation. The common feature of these compounds is an anomaly at $T \simeq 6.2$ K, which is usually attributed to the presence of Ce_2O_3 impurities in the sample. The hexagonal Ce_2O_3 compound, however, orders antiferromagnetically at $T_N = 8.5$ K (Westrum and Justice 1968). The observed anomaly may thus be the result of the allotropic cubic Ce_2O_3 phase, which is known to form preferentially at the surface of the sample, and is referred to by Besnus et al. (1983b). We should remark that such an anomaly appears in Ce IV compounds with partners belonging to the Co and Ni columns, including $CePd_3$ (Besnus et al. 1983b) and CeRh (Sereni and Kappler 1989), but not in those with partners belonging to the Fe column (Ru and Os) or in α Ce itself. Considering that this is an extrinsic effect, we find that the γ term extrapolated from $T > 6.5$ K (γ_{HT}) ranges between 38 and 23 mJ mol^{-1} K^{-2}, including $CePd_3$, which anyway is not far from the C/T values for $T \to 0$ (or γ_{LT}) which range between 40 and 24 mJ mol^{-1} K^{-2}. The linear terms of CeRh and $CeRh_3$ are somewhat lower than these values: $\gamma_{HT} = 12.7\ \text{mJ mol}^{-1}\ \text{K}^{-2}$ for CeRh (Sereni and Kappler 1989), and $\gamma_{HT} = 14\ \text{mJ mol}^{-1}\ \text{K}^{-2}$ for $CeRh_3$ quoted by Mihalisin et al. (1981).

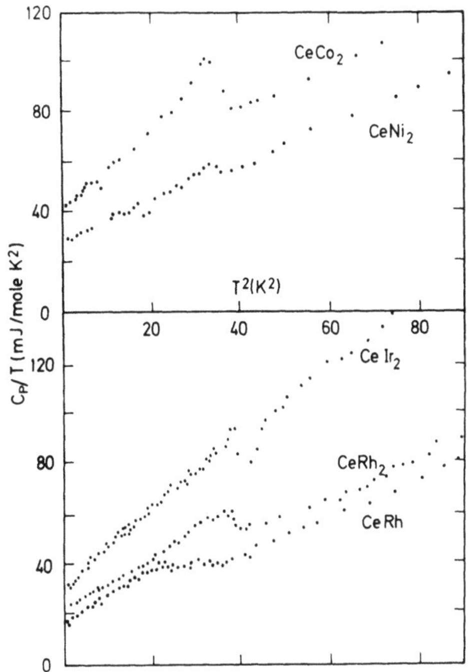

Fig. 16. Total specific heat of the intermediate valence Ce Laves phases and CeRh, in a C_p/T versus T^2 representation. $CeNi_2$ and $CeCo_2$ are taken from Machado da Silva and Hill (1972).

We have to mention that $CePd_3$ forms congruently, within a concentration of 24–26.5% of Ce. The γ_{HT} term depends on the composition and varies between $\gamma_{HT} = 34\ mJ\ mol^{-1}\ K^{-2}$ for 24.5% of Ce, has a maximum value of $\gamma_{HT} = 38.6\ mJ\ mol^{-1}\ K^{-2}$ at 25% of Ce and then decreases to $\gamma_{HT} = 36\ mJ\ mol^{-1}\ K^{-2}$ for 25.7% of Ce (Besnus et al. 1983b). Concerning the reference compound, it should be taken as YPd_3 (with $\gamma = 3.5\ mJ\ mol^{-1}\ K^{-2}$), because $LaPd_3$, which has an extremely low value ($\gamma = 0.28\ mJ\ mol^{-1}\ K^{-2}$), is not the appropriate reference since $CePd_3$ and $LaPd_3$ do not mix throughout the whole range of concentrations.

Another three IV compounds are formed with Ni, they are CeNi, $CeNi_2$ and $CeNi_5$. In the case of CeNi the specific heat was measured up to room temperature by Gignoux et al. (1983), from where a value of $\gamma_{LT} = 65\ mJ\ mol^{-1}\ K^{-2}$ and $\gamma_{HT} = 80\ mJ\ mol^{-1}\ K^{-2}$ were extracted. The γ_{HT} value will be discussed in sect. 9, in reference to CF effects in IV compounds. $CeNi_2$ has a γ_{HT} coefficient of $30\ mJ\ mol^{-1}\ K^{-2}$, with the mentioned anomaly at 6.2 K (Machado da Silva and Hill 1972). In the case of $CeNi_5$ the value of $\gamma_{LT} = 40\ mJ\ mol^{-1}\ K^{-2}$ has to be compared with that of $LaNi_5$ ($\gamma_{LT} = 34.3\ mJ\ mol^{-1}\ K^{-2}$) and $PrNi_5$($\gamma_{LT} = 37\ mJ\ mol^{-1}\ K^{-2}$) (Nasu et al. 1971). Here we see that the γ term has almost 80% of the "band" contribution and only 20% ($\sim 5\ mJ\ mol^{-1}\ K^{-2}$) of Ce IV contribution. If this is the correct interpretation, $CeNi_5$ shows the smallest γ contribution owing to the hybridized 4f orbital, i.e., $\sim 5\ mJ\ mol^{-1}\ K^{-2}$. Finally, $CeCo_2$ with a γ_{HT} coefficient of $40\ mJ\ mol^{-1}\ K^{-2}$ (Machado da Silva and Hill 1972) has a superconducting phase at $T_{sc} = 1.5$ K (Luo et al. 1968). Concerning the Ce superconductors, one finds that all of them are fcc-type with $D < 3.4$ Å, including α- and α′-Ce.

With the light p elements, Ce forms three IV compounds: CeB_4, Ce_2C_3 and CeN. Because of the difficulties in the sample preparation, to our knowledge, no specific heat results are available. Only a $\gamma = 8.3$ mJ mol^{-1} K^{-2} value for CeN was quoted by Danan et al. (1969).

Although the remaining Ce compounds that do not order magnetically: $CeSi_{2-x}$ ($0.05 < x < 0.14$), $CeSn_3$, $CeBe_{13}$ and $Ce_{24}Co_{11}$, are not related in structure or composition, they show relatively large γ values together with signs of spin fluctuations. The specific heats of $CeSi_{1.90}$ and $CeSi_{1.86}$ show the largest γ_{LT} terms, 184.6 mJ mol^{-1} K^{-2} (Dhar et al. 1987b) and 203 mJ mol^{-1} K^{-2} (Sato et al. 1988b), respectively. In both cases the C/T versus T^2 plot shows the characteristic minimum described by the $C/T = G + BT^2 + DT^2 \ln T$ equation. In a single-crystal sample, the cubic $CeSn_3$ was found to be an anisotropic spin fluctuator under an applied magnetic field (Tsang et al. 1984), with γ_{LT} values of 75 mJ mol^{-1} K^{-2} at zero field, and $\gamma_{LT} = 63.4$ and 59 mJ mol^{-1} K^{-2} when a field of 10 T was applied in the directions [100] and [110], respectively. In a polycrystalline sample, the C/T versus T^2 dependence is fitted with the spin fluctuation equation for the specific heat with a characteristic temperature of $T_s = 5.8$ K (Ikeda and Gschneidner 1982b). The γ_{LT} term decreases under an applied field, from 70 mJ mol^{-1} K^{-2} for $H = 0$ to 53 mJ mol^{-1} K^{-2} for $H \simeq 10$ T, where the spin fluctuations are practically quenched. Under fields stronger than 5 T a low entropic ($10^{-3} R \ln 2$) magnetic contribution appears at around 4.5 K, with similar characteristics in the T^3 dependence as observed in $CeSi_x$ ($x = 1.9$ and 1.85) (Dhar et al. 1987b). Together with the relatively large value of γ_{LT} ($\geqslant 50$ mJ K^{-2}/Ce atom), another common feature of the compounds that show spin fluctuation effects is their small volume contraction, $\Delta V/V \leqslant 1.5\%$, compared with other IV compounds where $\Delta V/V > 2.5\%$, see table 5.

The question why $CeSn_3$ does not order magnetically, while the isomorphic and isoelectronic neighbours $CeIn_3$, $CePb_3$ and $CeTl_3$ order antiferromagnetically, is still unanswered. Noteworthy is the fact that, for a Ce–Ce spacing in a $CeSn_3$ compound with trivalent Ce ($a = 4.74$ Å), the T_N value extracted from eq. (8) is 4.6 K, in coincidence with the observed anomaly under a magnetic field.

The compound $CeBe_{13}$, where the Ce–Ce spacing is the largest of this region ($D = 5.19$ Å), is considered an archetype of the Ce IV compounds from its magnetic behaviour. Although the γ_{HT} value fits within the expected values, 58.6 mJ mol^{-1} K^{-2} at around 5 K, there is a low entropic anomaly (Besnus et al. 1983a). The experimental results are fitted assuming that a Schottky contribution arises from magnetic clusters in the sample, nevertheless the possibility of spin fluctuations is not excluded by the authors.

Finally, $Ce_{24}Co_{11}$ shows a quite complicated hexagonal structure with 10 inequivalent sites for Ce and extremely short Ce–Ce spacings (Larson and Cromer 1962). This compound does not order magnetically down to 0.5 K, but shows a hump in the specific heat centered at $T_m \approx 1.2$ K. Below the hump C_M tends to be linear with temperature, with a value of $\gamma_{LT} \simeq 4.8$ J mol^{-1} K^{-2} (note that one mole contains 24 Ce atoms), see inset in fig. 17. Between 3 and 9 K the specific heat can be fitted with the function $C/T = G + BT^2 + DT^2 \ln T$, see fig. 17, with $G = 2.8$ J mol^{-1} K^{-2}, $B = 0.8$ J mol^{-1} K$^{-4} \ln K$ (Kappler and Sereni 1990). Because 10 different sites in the

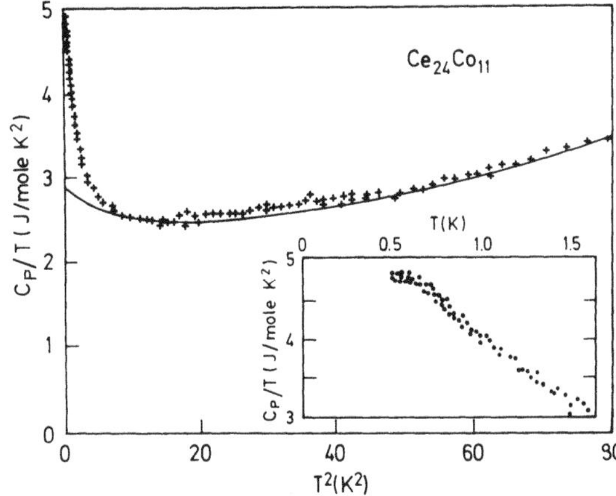

Fig. 17. Total specific heat of $Ce_{24}Co_{11}$ in a C_p/T versus T^2 representation in the range 0.5–9 K. The solid line is a fit to the relation $C_p/T = G + BT^2 + DT^2 \ln T$. The inset shows the lowest temperature data in a C_p/T versus T plot.

lattice may result in 10 different Ce environments, it is very speculative to attribute any particular C_M versus T dependence to certain Ce atoms. Qualitatively we can only say that some Ce atoms may behave as HF with a $T_m \simeq 1.2$ K and others exhibit a spin fluctuation regime between 3 and 9 K with a tentative value of $T_S \simeq 7$ K. From the $\chi(T)$ behaviour at higher temperature (Canepa et al. 1989), one can observe some similarities with $CeSi_{2-x}$.

A detailed phenomenological analysis of the specific heat and the susceptibility of the systems that show spin fluctuation behaviour was made by Gschneidner and Ikeda (1983).

8. Related compounds

In this section we shall include the Ce compounds based on binary compounds that are transformed into pseudoternaries by including interstitial atoms into the lattice or by a partial substitution of the partner. In any case the Ce ion is not changed, neither in its lattice position nor in its concentration, but there is an intrinsic disorder arising from the nonperiodic distribution of the nonmagnetic atom. The nonmagnetic atom disorder (NMAD) concept was recently introduced by Gschneidner et al. (1990) in connection with pseudoternary compounds that, showing large γ_{LT} and/or γ_{HT} values, were included within the heavy-fermion category.

8.1. *Interstitials*

The compounds with the antiperovskite $LaPd_3B$-type structure can be considered as a filled-up Cu_3Au-type. The boron atoms can be inserted in the T_6 octahedra of the binary compound $LaPd_3$ (or $CePd_3$ in our case), giving rise to a continuous solid solution of composition $CePd_3B_x$, with $0 \leqslant x \leqslant 1$ (Parthé and Chabot 1984). Although the atomic radius of B ($r = 0.98$ Å) is small, other light atoms with similar size, such as Be and Si ($r = 1.12$ Å and 1.32 Å, respectively) can be also inserted in that interstitial position (Kappler et al. 1983). The respective concentrations are limited by their size.

As quoted before (cf. sect. 7), the starting or matrix compound $CePd_3$ is a well-known IV compound. The addition of the interstitial atoms produces an expansion in the cell volume, which is much larger in $CePd_3B_x$ than in $LaPd_3B_x$ or $GdPd_3B_x$ compounds due to an induced change in the Ce valence (from mixed valence to trivalent) (Dhar et al. 1981b). The Ce ion becomes trivalent (i.e., magnetic) with more than two boron nearest neighbours, giving rise to clusters of magnetic Ce (Baurepaire et al. 1983). The larger the interstitial, the stronger the volume effect and, with only 20% Si per unit cell, an onset of antiferromagnetic order is detected. With 30% Si concentration a magnetic transition, with the full entropy and a jump of $\Delta C_M = 8\ J\,mol^{-1}\,K^{-1}$ is observed (Nieva 1988), see fig. 18. No long-range magnetic order is obtained by introducing Be as an interstitial for up to 45% of Be per formula unit. At that concentration, where the entropy is about $0.85R\ln 2$, the specific heat contribution of the magnetic Ce can be qualitatively described as that of a system with a random magnetic interaction. In fig. 19, we show the specific heat of the $CePd_3Be_{0.45}$ and the effect of an applied magnetic field, which shifts the temperature of the maximum of C_M and reduces the value of the maximum (Sereni et al. 1986).

As mentioned before, because of its small size, B can be included as an interstitial up to one atom per formula unit. At low concentrations ($x < 0.4$) the onset of some kind of cluster or spin glass configuration is shown by the specific heat. At higher concentrations a qualitative change in the specific heat anomaly is observed in the reduction of the temperature and the value of the maximum of the specific heat, see fig. 20 (Sereni et al. 1986). In addition, under an applied magnetic field there is a qualitative difference between the compounds with low and high concentrations of B. While the maximum of the magnetic contribution to the specific heat decreases with increasing magnetic field for $x = 0.3$ and 0.35, see Dhar et al. (1989) and Sereni et al.

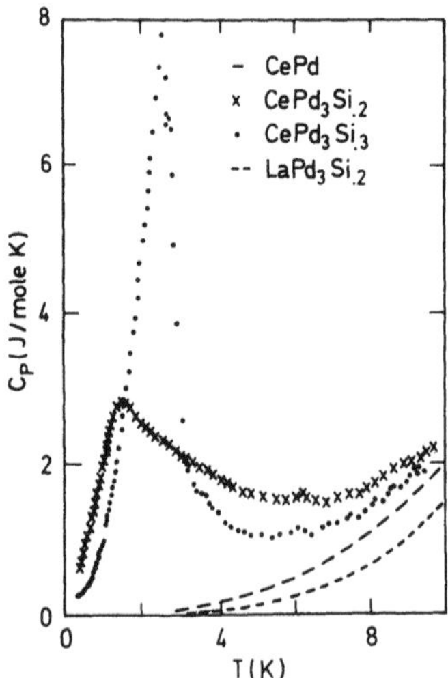

Fig. 18. Specific heat of $CePd_3$ with Si interstitials.

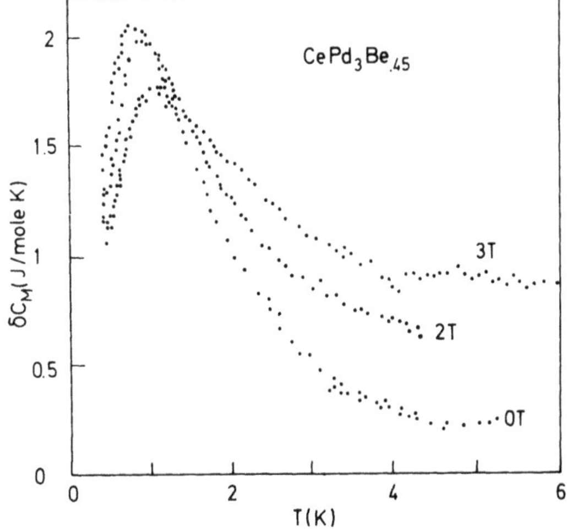

Fig. 19. Specific heat of $CePd_3Be_{0.45}$ in various magnetic fields after phonon subtraction.

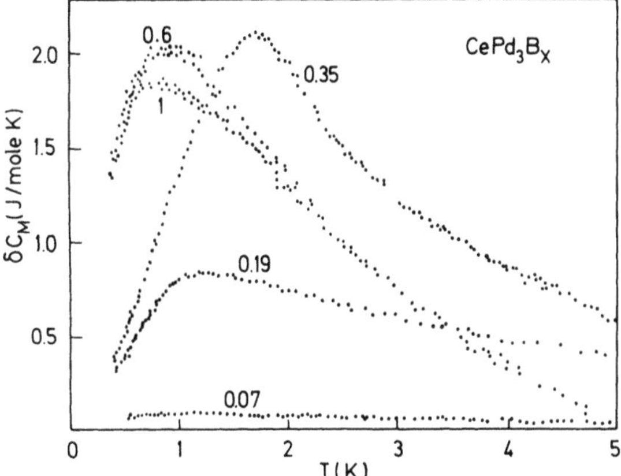

Fig. 20. Magnetic specific heat ($C_M = C_p - \gamma_{HT}T - \beta T^3$) of $CePd_3B_x$, after Sereni et al. (1986).

(1986), respectively (as also observed in $CePd_3Be_{0.45}$), the contrary occurs in $CePd_3B$ where C_M at T_m grows from 1.8 to 2.1 $J\,mol^{-1}\,K^{-1}$ between 0 and 2 T, see fig. 21 (Sereni et al. 1986). At the low-temperature experimental limit ($\simeq$0.4 K), it was found that the specific heat of $CePd_3B$ approaches a linear temperature dependence with a value of $\gamma_{LT} = 3.7\,J\,mol^{-1}\,K^{-2}$ in an "as-cast" sample ($3.3\,J\,mol^{-1}\,K^{-2}$ in an annealed one), which is the highest value reported for a γ_{LT} term (Sereni et al. 1986, Nieva 1988a). As remarked on by Dhar et al. (1989), measurements at $T < 0.4$ K are required to check if such a large γ_{LT} value arises from a Kondo interaction or decreases with $T \to 0$ as in $CePd_3B_{0.35}$.

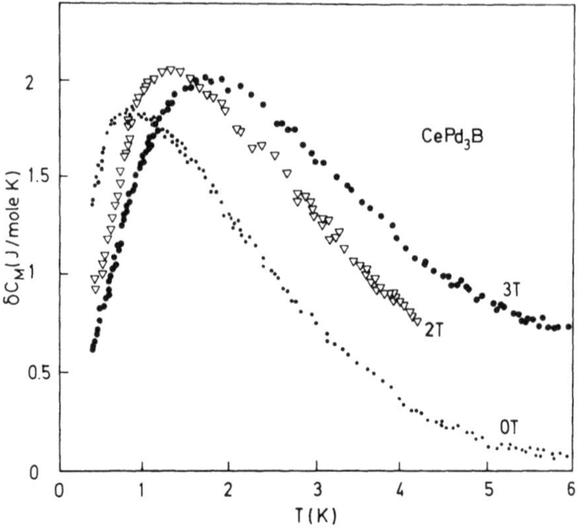

Fig. 21. Magnetic specific heat ($C_M = C_p - \gamma_{HT}T - \beta T^3$) of $CePd_3B$ in various magnetic fields, after Sereni et al. (1986).

8.2. *Partner substitution*

8.2.1. *Alloying*

As seen before, the change of the local Ce environment is able to drive the valence of Ce from an IV state to a Kondo-like or magnetic state. These changes can be studied from different and complementary properties, such as: magnetic, structural, transport and thermal. We intend to show here how the specific heat measurements are able to furnish information on the crossover from a magnetic behaviour to an IV one, induced by a change in the electronic structure of the band or by reducing the available volume of the Ce ion in the lattice. As shown in table 3 and fig. 9, CePd is a ferromagnet that forms with the CrB-type structure like CeRh or CeNi, which are IV compounds, see table 5 and fig. 16. These compounds can be mixed at any concentration without changing the structure. Because Pd and Rh are very close in size the Ce valence is continuously and homogeneously (in space) changed by the electronic concentration in the Ce(Pd, Rh) alloy (Kappler et al. 1988b, 1990b). The low-temperature specific heat of a sample containing 60% of Rh is shown in fig. 22a as $\delta C_M = C_M - \gamma_{HT}$. There one can see a ferromagnetic transition at $T_C = 2.7$ K, containing about 50% of the entropy. The rest of the entropy is associated with a broad anomaly, which is at the origin of the $\gamma_{HT} = 177\ \mathrm{mJ\,mol^{-1}\,K^{-2}}$ term and has a maximum around 17 K, see fig. 22b.

On the other hand, Pd and Ni have a similar electronic structure but different atomic size, which reduces the local available volume of Ce in the Ce(Pd, Ni) alloy. Up to 50% of Pd substitution, the reduction of the Ce–Ce spacing is reflected in the increase of T_C (from 6.6 K for CePd to 10.3 K in $CePd_{0.5}Ni_{0.5}$). The homogeneity of this effect can be verified by the specific heat, where the ferromagnetic transition shows a jump $\Delta C_N = 1.5R$ as predicted by the mean-field theory, with 90% of the expected entropy gain, see fig. 9 (Nieva et al. 1988). At higher Ni concentrations, the

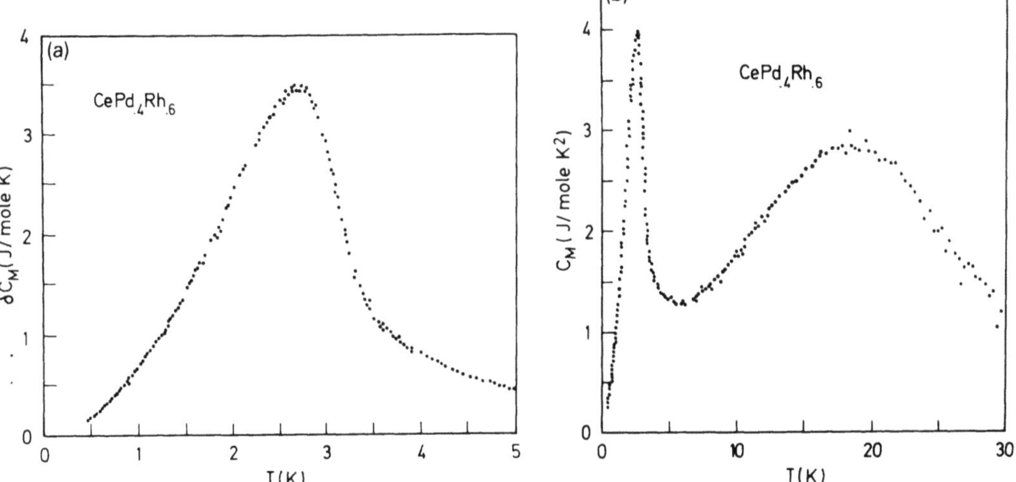

Fig. 22. (a) Low-temperature magnetic specific heat ($\delta C_M = C_p - \gamma_{HT}T - \beta T^3$) of $CePd_{0.4}Rh_{0.6}$. (b) Specific heat anomaly of $CePd_{0.4}Rh_{0.6}$ after phonon subtraction.

local pressure induces the IV state on Ce but in nonhomogeneous environments, as can be observed for 90% and 92% Ni concentrations (again see Nieva et al. 1988). The respective magnetic transitions at 6.3 K and at about 1.8 K are associated with broad maximums in the specific heat. The γ_{HT} terms (145 and 156 mJ mol^{-1} K, respectively) are related to the formation of mixed-valence states, which have a maximum in the specific heat around 25 K. The comparative entropy gain with temperature is shown in fig. 23 (Nieva et al. 1988). Similar effects were found by substituting Ni by Pt (Gignoux and Gomez-Sal 1984). The specific heat measurements on $CePt_{1-x}Ni_x$, with $x = 0.5$, 0.8, 0.95 were reported by Ferreira da Silva et al. (1989), with similar results to those of $CePd_{1-x}Ni_x$.

The effect of the Pd substitution on Ce strongly depends on the electronic configuration of the substituent. When Pd is replaced by a noble metal (NM = Cu, Ag, Au) instead of a transition metal (Rh, Ni), a few percent of NM are able to destroy the long-range ferromagnetic order, giving rise to a disordered intermediate phase (Kappler et al. 1988c). By applying a magnetic field in a sample with 10% of Ag, the disordered or incommensurate phase can be suppressed, while a second transition at lower temperature retains a ferromagnetic character, see fig. 24 (Sereni and Kappler 1989).

Strictly speaking, the compound $CeSi_{2-x}$ (cf. sect. 5) that has already been studied can be included in this group because Si is substituted by vacancies. It is also an example of a system that undergoes a continuous transition from ferromagnetic ($CeSi_{1.75}$) to IV behaviour. In this particular case, the spin fluctuations play an important role. Other systems to be included in this subsection are $CePd_3$, when Pd is substituted by Rh (displays an IV tendency) or Ag (trivalent tendency) and $CeSi_{2-x}$, when Si is substituted by Ge. The specific heats of $Ce(Pd, Rh)_3$ and $Ce(Pd, Ag)_3$ were

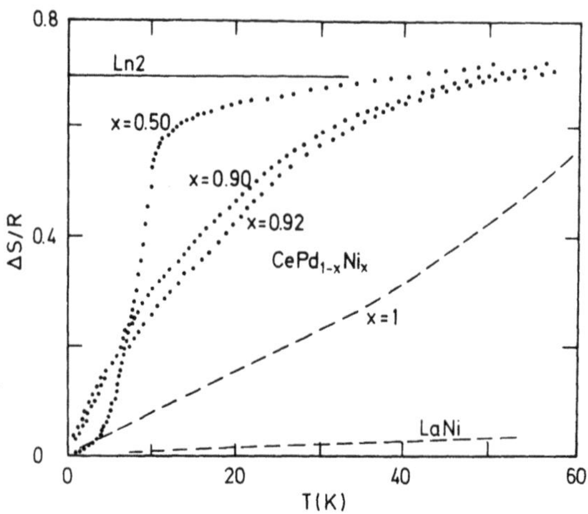

Fig. 23. Comparative entropy gain of the $CePd_{1-x}Ni_x$ pseudoternaries, after Nieva et al. (1988).

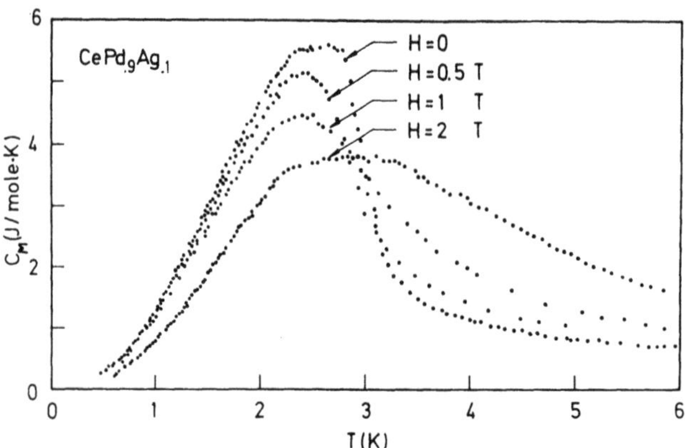

Fig. 24. Magnetic specific heat of $CePd_{0.9}Ag_{0.1}$ in various magnetic fields.

studied by Mihalisin et al. (1981), who found a continuous increase of γ_{HT} in going from $CeRh_3$ through $CePd_3$ to $CePd_{2.8}Ag_{0.2}$ as the valence of Ce goes from an IV state to a trivalent state. We find the comparison made by Pountney et al. (1978) between $CeRh_{3-x}Pd_x$ and $ZrRh_{3-x}Pd_x$ interesting. With the stable-valence Zr, the Cu_3Au-type structure, persists up to $x = 2.2$, becoming hexagonal ($TiNi_3$-type structure) for $x > 2.6$, but without deviation of the volume cell from Vegard's law. On the contrary, with the unstable-valence Ce, for $x > 2.2$ the alloys retain their cubic Cu_3Au-type structure, but deviate from Vegard's law. Such is a clear sign of a change in the electronic structure of Ce, at $x > 2.2$, from tetravalent (like Zr) to IV (as in $CePd_3$). Finally, the specific heat of $CeSi_{2-x}Ge_x$ was studied by Yashima et al. (1986), who found an extremely broadened transition attributed by the authors to the large fluctuation effect at the critical concentration.

8.2.2. *Stoichiometry*

Most of the ternary compounds have a binary-related crystal structure. This means that in the binary compound the partner has two (or more) inequivalent sites in the lattice which can be filled by different atoms. In the case that the substitution is made by completely filling one of the lattice sites, such a compound can be regarded as a ternary one because the periodicity is not affected and the Ce environment is well defined. However, if such is not the case the NMAD effect can occur. The specific heat is an appropriate tool to check the local disorder around the Ce ion because the eventual magnetic transition becomes a spin-glass-like anomaly. Such are the cases analyzed by Dhar and Gschneidner (1989) concerning the $CeCu_4Al$, $CeCu_3Al_2$ and $CeCu_4Ga$ compounds, which belong to the family with the $CeCu_5$ structure. The specific heats of these compounds were measured by Bauer et al. (1987, 1988), who found a temperature-dependent γ_{LT} term larger than $1\,J\,mol^{-1}\,K^{-2}$ (within the HF values). While the measurements reported by Willis et al. (1987) indicate that $CeCu_4Ga$ orders at 0.7 K, recent investigations performed by Kohlmann et al. (1989a) reveal that it neither shows typical signs of long-range magnetic order nor of spin glass behaviour.

In the absence of a well-defined maximum in C_M, the description of the dominant microscopic mechanism can become rather difficult. We can quote as an example the similar C/T versus T behaviours of $Ce_{24}Co_{11}$, $CeCu_3Al_2$ and $CeCu_3Ga_2$. The first compound was described in sect. 7 as a spin fluctuator, while the other two were recently described within the spin-$\frac{1}{2}$ single-impurity Kondo model (Kohlmann et al. 1990).

Similar values of $\gamma_{LT}(T)$ were observed in $CeCu_{6.5}Al_{6.5}$ (with the $NaZn_{13}$-type structure). Here the specific heat measurement was complemented by AC susceptibility measurements, which show that the "cusp" observed at zero field is efficiently smeared out by small values of the applied field (Rauchschwalbe et al. 1985). From this evidence the authors concluded that $CeCu_{6.5}Al_{6.5}$ should be considered as a "concentrated crystalline spin glass". Another example of spin glass behaviour is given by $CePtGa_3$ (Tang et al. 1990). This compound crystallizes in the $BaAl_4$-type structure, where the Ce partners randomly occupy the 4(e) and 4(d) lattice positions. After subtracting the Ga nuclear contribution, $C_M(T)$ shows a dominant T^2 term at low temperatures, as observed in the ThGd spin glass (Sereni et al. 1979). The broad maximum of C_M in the $CePtGa_3$ first grows (for $H = 2\,T$) and then decreases (for $H = 8\,T$) as expected in a disordered magnetic system.

The Schottky-like anomaly observed in the specific heat of the compounds discussed in this section can be derived phenomenologically using: (a) the resonance-level model, (b) the spin glass behaviour, (c) the crystal field (Schottky) contribution or even (d) low-dimensional magnetic fluctuations. The cases where an HF behaviour is deduced from a large γ_{HT} value will be discussed in sect. 9, in connection with the contribution to C_M of the excited crystal field levels. It is clear that complementary techniques, such as NMR, AC susceptibility and electrical resistivity, can easily reveal the magnetic character of the microscopic interactions. In some of the HF compounds the ratio between the γ term and the $\chi(T \to 0) = \chi_0$ value of the susceptibility, and between the γ term and the T^2 coefficient of the resistivity, A, have values predicted by the theory. These relationships will be discussed in sect. 10. Here, we shall summarize briefly some characteristics of this specific heat anomaly, which can be divided into three ranges of temperature: (i) $T \simeq T_m$, (ii) $T > T_m$ and (iii) $T < T_m$. The analysis of these three sectors provides independent conditions, which allow us to distinguish between the different types of behaviour.

(i) All the four mechanisms mentioned will result in a nearly symmetric shape around T_m on a ln T scale, unless large clusters are present. Theoretical models predict a C_M(at T_m) $\simeq 0.17R$ value for a spin-$\frac{1}{2}$ Kondo impurity (see, e.g., Rajan et al. 1982, Desgranges and Schotte 1982), while all the NMAD systems discussed before, including the $CeCu_{6.5}Al_{6.5}$ spin glass, show a value $C_M(T_m) > 0.25R$. For a crystal field anomaly a doublet–doublet excitation gives a value of $C_{Sch}(T_m) = 0.44R$ and only a quartet–doublet (Γ_8–Γ_7) excitation gives $C_{Sch}(T_m) = 0.24R$, see table 1. This case is unlikely because it requires a cubic symmetry and all the CF quartet ground states actually undergo a quadrupolar transition. Only in systems with a very small CF splitting that shows some Kondo interaction in the excited state, one may be able to reduce $C_M(T_m)$ to the experimentally observed values. This point will be discussed again in sect. 10. Finally, low-dimensional magnetic fluctuations are predicted to give $C_M(T_m) = 0.35R$ (de Jongh and Miedema 1974).

Under an applied magnetic field T_m is shifted to higher temperatures, while the value of C_M at T_m is found to decrease in $CePd_3Be_{0.45}$ and $CePd_3Be_{0.35}$ (Sereni et al. 1986) or to remain constant as in $CeCu_{6.5}Al_{6.5}$ (Rauchschwalbe et al. 1985). The resonance-level model for Kondo impurities predicts the narrowing and the increase of C_M (at T_m) with field, as verified by Bredl et al. (1978) for (La, Ce)Al_2; such a behaviour is also followed by $CePd_3B$.

(ii) More information can be extracted from the $T > T_m$ region. All the anomalies mentioned have a $C_M = QT^{-2}$ behaviour at $T \gg T_m$, but the difference is given by how close to T_m such a law remains valid. Qualitatively C_M deviates from QT^{-2} by decreasing the temperature ($T \to T_m$) in the following order: a Kondo system, a Schottky anomaly and finally a spin glass. A better analysis can be made by studying how a $C_M T^2 = Q(1 - T_0/T)^{-\alpha}$ law diverges (see, e.g., Souletie 1988). Such an analysis was applied to $CeCu_6$ and to $CePd_3B_{0.6}$ using a $\partial \ln T/\partial \ln(C_M T^2) = (T - T_0)/\alpha T_0$ versus T relationship. This static scaling gives a value of $T_0 = -5.8$ K for $CeCu_6$ and -0.96 K for $CePd_3B_{0.6}$, where the negative value $T_0 = -T_K$ corresponds to the Kondo regime (see Souletie 1988, Kappler et al. 1988d).

(iii) Finally, we can say that a direct indication of a heavy-fermion behaviour is given by very low temperature ($T \ll T_m$) specific heat measurements. One finds that only $CeCu_6$ tends to a constant γ_{LT} value as $T \to 0$. Taking advantage of the fact that theoy provides simple relationships for systems with spin 1/2, it is also possible to recognize a Kondo-like compound by the formula $\gamma_{LT} = \pi R/3aT_m$, which relates the linear term with the maximum of the specific heat. In this formula $a = 2.24$ or 1.29 for the Desgranges and Schotte (1982) or the Rajan et al. (1982) models, respectively. Noteworthy is the fact that only one binary Ce compound fulfils all the conditions for being an HF system, while many ternaries and pseudoternaries are claimed to be HF. Is this due to the fact that the NMAD effect is not possible in binary compounds?

9. High-temperature specific heats

The main electronic contribution to C_M at $T \gg T_0$ is related to the CF excitations, and is known as the Schottky anomaly, which has already been introduced in sect. 2. For well-localized f orbitals, the CF levels are described by delta functions centered at the corresponding energies and eq. (3) is applicable for describing such a contribution. Nevertheless, in many Ce compounds with Kondo interactions or interconfigurational mixing, the excited states are also hybridized and they have to be represented with an energy width (Wohlleben and Wittershagen 1985). In this case the Schottky anomaly is broadened and the maximum values for C_{Sch} (given in table 2) decrease. Simultaneously, the contribution of the broadened excited levels increases at low temperatures, smearing out the energy gap. In the IV limit the $2J + 1$ levels are mixed, giving a continuous spectrum, which should be observed as a linear contribution to C_M. Particular attention has to be paid to the term γ_{HT}, because it can be related not only to the remanent entropy of the ground state, but also to the contribution arising from the CF levels. In some cases both contributions are easily distinguished because $\Delta S(T)$ shows a kind of plateau at $T \ll \Delta$, see, e.g., fig. 23, in the low Ni concentration of the Ce(Pd, Ni) system. At the CeNi limit the levels are strongly mixed and ΔS increases continuously. At higher temperatures, however, the C_M/T versus T plot (something like a γ_{HT} versus T dependence) is not constant, see fig. 25 (Gignoux et al. 1983). This is a sign that CeNi is still not in the IV limit, as is confirmed by its $\chi(T)$ behaviour under pressure (Gignoux and Voiron 1985), which falls drastically under kbar. In this and the following figures, we take $\delta C = C_p - C_L$.

Because of the large phonon contribution to the heat capacity at high temperatures, the magnetic contribution determined experimentally has a significant error. Within

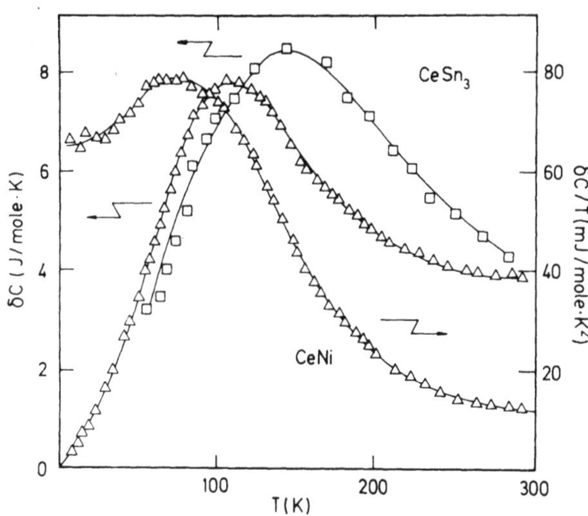

Fig. 25. High-temperature specific heat of CeNi (from Gignoux et al. 1983) and $CeSn_3$ (from Costa et al. 1982) after the respective phonon subtractions.

these limitations we also see in fig. 25 the high-temperature C_M of $CeSn_3$ (Costa et al. 1982), which is the only available example of an IV system. The entropy gain at 300 K is evaluated to be $\Delta S = 0.9R\ln 6$.

One can also observe that the compounds which show a Kondo-like anomaly at low temperatures do not have a "normal" Schottky anomaly at high temperatures. See, e.g., $CeAl_2$, $CeAl_3$ and $CeCu_6$ in fig. 26 (Felten 1987), and $CePd_3B_{0.6}$ and $CePd_3B$, in fig. 27 (Nieva et al. 1987). For simplicity, in the figures we have compared the Schottky contribution with that of a quartet (Γ_8) excited state, which is not the case for hexagonal compounds like $CeCu_6$ and $CeAl_3$. The hybridization effect of the excited levels in these compounds can be seen in detail in Onuki et al. (1985) and Mahoney et al. (1974), respectively. For comparison we show in fig. 27, the antiferromagnet $CePd_3Si_{0.3}$ (Nieva et al. 1987), where C_M at high temperatures is well described by the formula for C_{Sch} given by eq. (3).

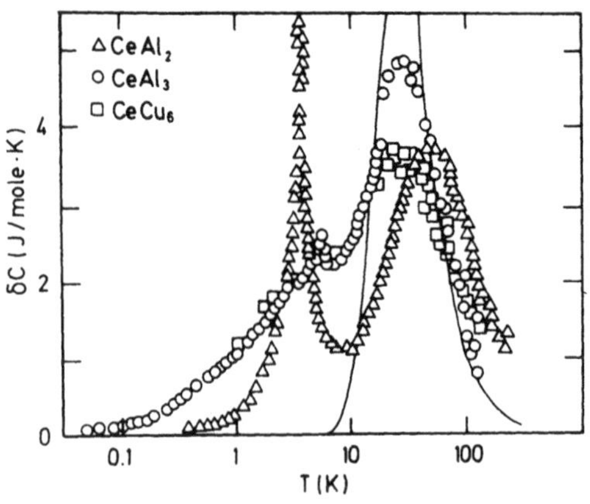

Fig. 26. High-temperature specific heat of three Kondo compounds with an abnormal Schottky anomaly, after de Boer et al. (1985) and Felten (1987). The continuous curve is a Schottky contribution for a Γ_7–Γ_8 thermal promotion.

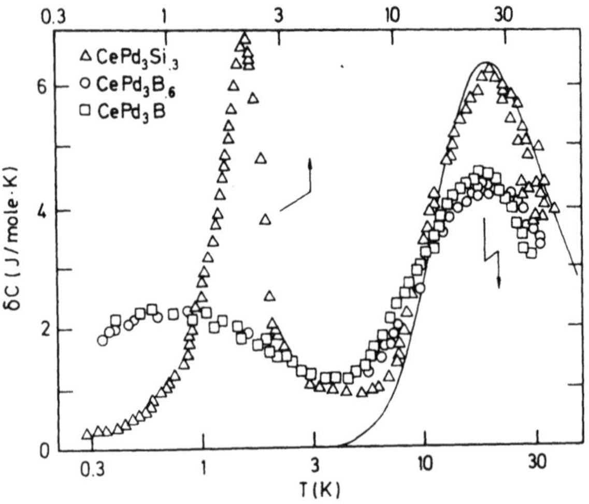

Fig. 27. High-temperature specific heat of two Kondo compounds ($CePd_3B_x$) and a normal antiferromagnet ($CePd_3Si_{0.3}$), after Nieva et al. (1987). The continuous curve is a Schottky contribution for a Γ_7–Γ_8 thermal promotion.

As pointed out by Gschneidner et al. (1990), most of the CF splittings in Ce compounds are larger than 100 K. However, there are some cases such as CeSb, CeBi (Hulliger et al. 1975), $CeCd_{11}$ (Tang and Gschneidner 1988) and $CeGa_2$ (Burlet et al. 1987) with $\Delta \leqslant 60$ K. In these cases a proper subtraction of the Schottky contribution drastically reduces the γ_{HT} term. See, e.g., Gschneidner et al. (1990) concerning $CeGa_2$. We have to note that in none of the NaCl-type Ce compounds, which order antiferromagnetically at low temperatures, can the high-temperature C_M be described properly by C_{Sch}: see Hulliger et al. (1978) for CeS, CeSe and CeTe, and Aeby et al. (1973) for CeP. Finally, we observe that the γ_{HT} values of the Kondo or heavy-fermion systems (including Ce ternaries) are quite similar, $\gamma_{HT} = 160 \pm 40$ mJ K^{-2}/Ce atom, independent of the degeneracy and the splitting of the excited levels.

10. Miscellanea

The low-temperature specific heat of the Ce compounds is dominated by magnetic contributions, which in most cases are related by a certain degree of hybridization between the 4f state and the conduction band. The magnetic properties of a system can obviously be studied by other techniques such as magnetization and electrical resistivity, or by more sophisticated techniques like neutron scattering. In this section we shall briefly discuss some relationships between the specific heat and the techniques mentioned.

10.1. *Magnetization and susceptibility*

In a ferromagnetic system the order parameter is given by the spontaneous magnetization, M_s, which in an ideal second-order transition rises from zero at T_C (see, e.g., Belov 1959). As mentioned in sect. 2, C_M and M_s are related by $C_M \simeq -T\partial M_s^2/\partial T$ and therefore, if the temperature of the maximum of C_M is taken as the thermodynamical definition of T_C, it corresponds to the maximum slope of M_s. Many authors define T_C (T_C^* for us) as the temperature where the maximum slope of M_s extrapolates to $M_s = 0$. Although this is not a thermodynamical definition, the difference: $\Delta T_C = T_C^* - T_C$ is a measure of the fluctuations above T_C as was mentioned in sect. 2.

The magnetic susceptibility of a free-electron gas (or the Pauli susceptibility) χ_0, is related with the specific heat through the density of states, having a ratio of $\chi_0/\gamma = 0.014$ emu $K^2 J^{-1}$. In a Fermi liquid, where the effective mass of the electrons increases significantly, such a ratio retains its validity after renormalizing by the effective magnetic moment and the ground-state degeneracy: $\chi_0/\gamma = (\mu_{eff}/\pi k)^2 (1 + \frac{1}{2}J)$, where $\mu_{eff}^2 = g_J^2\mu^2 J(J+1)$, g_J being the g factor and J the total angular momentum (Wilson 1975, Newns et al. 1982). Experimentally μ_{eff}^2 can be evaluated from the Curie constant, then $(\mu_{eff}/\pi k)^2 = 3C/\pi^2 R$. We can therefore evaluate the χ_0/γ ratio for cubic Ce compounds as: 0.035 and 0.014 emu $K^2 J^{-1}$, for IV ($J = \frac{5}{2}$) and HF($J = \frac{1}{2}$) compounds, respectively. For lower symmetries the C value corresponding to the doublet ground state has to be evaluated for each case. An extensive

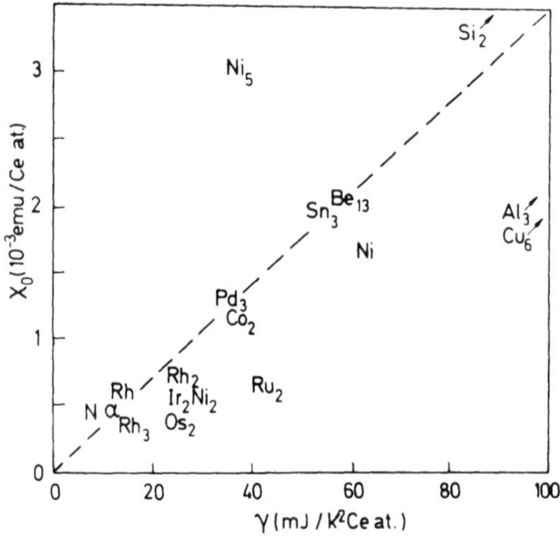

Fig. 28. Correlation between χ_0 and γ for Ce_jX_k compounds from table 5, labelled as $_jX_k$. The dashed line is the expected value for a Γ_7–CF ground state, see the text.

application of this ratio to cubic IV Ce compounds, made by Besnus et al. (1985), confirms the predicted value. The experimental values for the χ_0/γ ratio are shown in table 5. In fig. 28, we show some new data for small γ- and χ_0-values. The deviation from the expected values is large because of the reduced contribution of the f states to γ and χ_0, which becomes comparable to that of the conduction band (particularly in $CeNi_5$ and $CeRu_2$). In addition, $CeCu_6$ and $CeAl_3$ are shown in the figure, whose χ_0/γ ratio is closer to that expected for a $J = \frac{1}{2}$ system. A phenomenological correlation for χ and γ, including a large number of Ce and actinide compounds has been presented by De Long (1986).

10.2. *Electrical resistivity*

It was found that for a variety of systems the resistivity behaves as $\rho(T) \simeq \rho_0 + AT^2$ in the limit of $T \to 0$, where ρ_0 is the residual resistivity. The T^2 term is usually attributed to the Umklapp process of the electron–electron collision (Abrikosov 1972). In the case of the so-called concentrated Kondo materials, the A factor has much larger values than those observed in pure TM, as happens with the γ term. At high temperatures, the resistivity of these materials increases as $-\ln T$ with decreasing temperature. After a maximum, it decreases rapidly showing what is known as a coherent behaviour, which is connected with the regularity of the magnetic lattice. On the other hand, the impurities show the characteristics of incoherent Kondo scattering. It is surprising that materials which have different ground states with different low-temperature scattering mechanisms follow the same T^2 relationship in the resistivity (Kadowaki and Woods 1986).

Regardless of the large values of A and γ in the HF compounds, the ratio A/γ^2 is constant and equals $1 \times 10^{-5}\,\Omega\,\mathrm{cm}\ (\mathrm{J\,K^{-1}/Ce\ atom})^{-2}$. In fig. 29 we reproduce this universal relationship shown by Kadowaki and Woods (1986) for Ce and some U compounds, adding some other IV systems, like $CeBe_{13}$, $CeNi_5$ and CeRh. As the IV regime is reached, the A/γ^2 ratio tends to the values shown by pure TM, such as Pd, Pt and Ni [$A/\gamma^2 \simeq 0.9 \times 10^{-6}\,\mu\Omega\,\mathrm{cm}\,(\mathrm{mJ\,K^{-1}/mol})^{-2}$] where a Baber's type of scattering is expected (Rice 1968). We have to be reminded that in the case of CeRh, where γ has very low values, the conduction band and f contributions to the density of states become comparable. Note, however, that Ce compounds with a spin fluctuation character ($CeSn_3$ and $CeBe_{13}$) behave closer to the TM elements.

10.3. *Neutron scattering*

Although the magnetic internal energy, $\mathscr{U}_M$, is not a parameter that is much used, it becomes an important tool for recognizing systems with low dimensionality. For example, $\mathscr{U}_M(T_N)/\mathscr{U}(\infty) = 0.71$ in the two-dimensional Ising model and 0.27 in the three-dimensional one (de Jongh and Miedema 1974). Specific heat and neutron scattering measurements can be related through $\mathscr{U}_M$ by the eqs. (4) and (5) (cf. sect. 2). Experimental results on $CeAl_2$ and $CeIn_3$ were compared by Peyrard (1980), as shown in fig. 30.

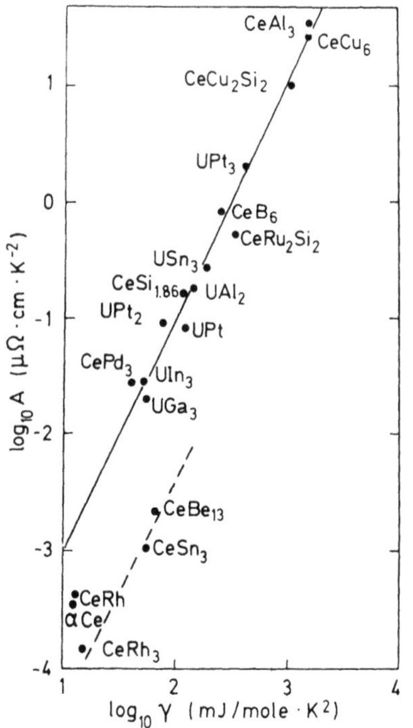

Fig. 29. Correlation between the T^2 coefficient of the resistivity, A, and γ for HF and IV compounds. Full and dashed lines represent $A/\gamma^2 = 10$ and 0.9×10^{-6} [$\mu\Omega\,\mathrm{cm}\,(\mathrm{mJ\,K^{-1}/mol})^{-2}$] respectively, see the text. Part of the data are taken from Kadowaki and Woods (1986).

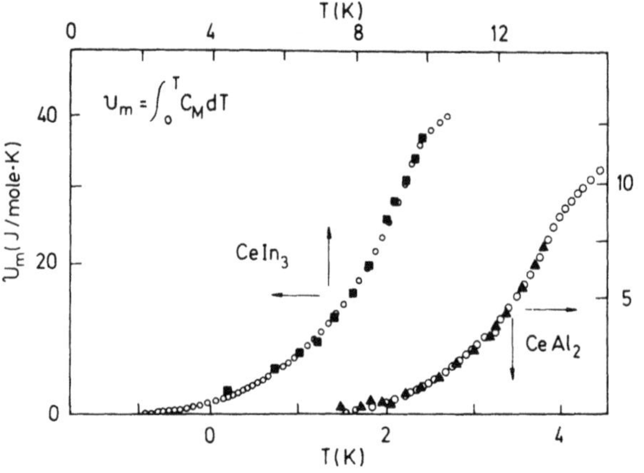

Fig. 30. Internal magnetic energy of $CeIn_3$ and $CeAl_2$ as obtained from specific heat (○) and neutron scattering (■ and ▲), after Peyard (1980).

11. Concluding remarks

Different Ce compounds have received in the past different degrees of attention. During the last few years, the race to discover new HF compounds has focused the attention of many researchers on that particular phenomenon to the detriment of systematic studies of the Ce behaviour as a function of its environment. In that sense, it became much more important to study, e.g., how the effective mass of the electrons increases rather than to understand how they behave in a highly hybridized f orbital, whose density of states approaches that of the conduction band. However, as we have seen in this work, the amount of information accumulated at present allows us to extract phenomenological correlations, because each particular behaviour found in a compound has a good probability of also being observed in other compounds. If we take into account that we have correlated the magnetic properties of about eighty binary compounds, while *Pearson's Handbook* doubles that number in known crystal structures, one can conclude that the present "mass of information" should grow rapidly in the future.

Although the choice of the characteristic temperatures, like T_C, T_N and T_m, versus the minimum Ce–Ce spacing has been shown to be appropriate for our proposed general systematic study, one can imagine other sets of parameters for more specific analysis. One can predict, e.g., that the CF parameters of Ce in the bcc compounds should be correlated with their low-temperature magnetic behaviour. It was shown that the Ce–partner spacing also plays an important role in the Ce behaviour. There, e.g., it was found that in the Laves compounds CeM_2, Ce behaves as IV when the Ce–M contacts occur, whereas Ce orders magnetically when there are no Ce–M contacts (Sereni 1982a). Following this evidence, a further generalization of the interatomic spacing parameter can be made by taking the Ce–nearest-neighbour spacing (whether Ce or not) in each sublattice together with the Ce coordination. Another example is given by the correlation between the volume contraction $\Delta V/V$ arising from the formation of the compound and the electrostatic energy $\mathscr{E}_4$ of the valence electrons immersed in the screened Coulomb potential of the total charge distribution. In this case, it was found that Ce has to change its valence to fit into the actual $\Delta V/V$ value (Sereni 1982b).

The most significant conclusions that we extract from our systematic study are:

(i) Ce–Ce compression, $D < 2r(\mathrm{Ce}^{3+})$, does not necessarily induce valence instability. Cubic local symmetry with a high coordination number is required for a strong f electron hybridization. Specific heat results show, from the large γ_{LT} and γ_{HT} values and the entropy deficit, that the f conduction band hybridization is important anyway.

(ii) The ferromagnetic behaviour of Ce is related to its local symmetry (it only occurs in noncubic systems) and with the Ce–Ce spacing (there is a reduced range for D between about 4.1 Å and 3.65 Å).

(iii) The cubic compounds where Ce has a quartet as CF ground state, undergo quadrupolar transitions at or about T_{N}.

(iv) The characteristic temperature T_{m} of the nonmagnetically ordered compounds has two different ranges of values: $T_{\mathrm{m}} < 10$ K (HF) and $T_{\mathrm{m}} > 100$ K (IV).

(v) Specific heat measurements show that spin fluctuations are present in IV compounds with large γ values ($\geqslant 50$ mJ K^{-2}/Ce atom) and small volume contractions ($\Delta V/V \leqslant 1.5\%$).

As we have stated, the amount of experimental data on Ce compounds had already reached the "critical mass" to allow us, at present, to correlate some apparently independent systems. However, as the experimental information grows, the need for a more thorough condensation of the information becomes apparent. In that sense we see that the literature is richer in theoretical reviews (see, e.g., Newns and Read 1987, Fulde 1988, Schlottmann 1989) than in experimental ones. It is therefore necessary to have more experimental reviews concerning, e.g., the magnetic and transport properties, in order to increase the applications of the theoretical models to real systems.

Acknowledgements

The author is indebted to Drs J.P. Kappler and G.L. Nieva, who have been closely involved with many experimental aspects of these studies. He gratefully acknowledges Ms N. Badino for reading this review, Ms R. Cohen and A. Cohen for typing the manuscript, and Ms M. Rangone for drawing the illustrations.

References

Abrikosov, A.A., 1972, Introduction to the Theory of Normal Metals (Academic Press, New York).

Aeby, A., F. Hulliger and B. Natterer, 1973, Solid State Commun. **13**, 1365.

Allen, A.C., and D.T. Cromer, 1961, Acta Crystallogr. **14**, 545.

Allen, J.W., S.-J. Oh, M.B. Maple and M.S. Torikachvili, 1983, Phys. Rev. B **28**, 5347.

Amato, A., D. Jaccard, J. Flouquet, F. Lapierre, J.L. Tholence, R.A. Fisher, S.E. Lacy, J.A. Olsen and N.E. Phillips, 1987, J. Low Temp. Phys. **68**, 371.

Andraka, B., G. Fraunberger, J.S. Kim, C. Quitmann and G.R. Stewart, 1989, Phys. Rev. B **39**, 6420.

Andres, K., J.E. Graebner and H.R. Ott, 1975, Phys. Rev. Lett. **35**, 1779.

Armbrüster, H., and F. Steglich, 1987, Solid State Commun. **27**, 873.

Asada, V., 1976, J. Phys. Soc. Jpn. **41**, 26.

Atoji, M., 1967, J. Chem. Phys. **46**, 4148.

Atoji, M., 1971, J. Chem. Phys. **54**, 3226.

Atoji, M., and D.E. Williams, 1961, J. Chem. Phys. **35**, 1960.

Bauer, E., E. Gratz and C. Schmitzer, 1987, J. Magn. & Magn. Mater. **63–64**, 37.

Bauer, E., E. Gratz, N. Pillmayr, D. Gignoux and D. Schmitt, 1988, J. Magn. & Magn. Mater. **76–77**, 131.

Baurepaire, E., P. Panissod and J.P. Kappler, 1983, Proc. 6ième Journées de la R.C.P. 520 sur les Composés de Terres Rares et Actinides à Valence Anomale, Strasbourg, France, 6–7 Oct., p. 44.

Bècle, C., and R. Lemaire, 1967, Acta Crystallogr. **23**, 840.

Belov, K.P., 1959, in: Magnetic Transitions (State Publishing House for Physicomathematical Literature, Moscow, USSR) ch. 1.

Benoit, A., J.X. Boucherle, J. Buevoz, J. Flouquet, B. Lambert, J. Palleau and J. Schweizer, 1979, J. Magn. & Magn. Mater. **14**, 286.

Berton, A., J. Chaussy, G. Chouteau, B. Cornut, J. Peyrard and R. Tournier, 1977, in: Valence Instabilities and Related Narrow Band Phenomena, ed. R. Parks (Plenum, New York) p. 471.

Berton, A., J. Chaussy, B. Cornut, J.L. Lasjaunias, J. Odin and J. Peyrard, 1980, J. Magn. & Magn. Mater. **15–18**, 379.

Besnus, M.J., J.P. Kappler, G. Krill, M.F. Ravet, N. Hamdani and A. Meyer, 1982, in: Valence Instabilities, eds. P. Wachter and H. Boppart (North-Holland, Amsterdam) p. 165.

Besnus, M.J., J.P. Kappler and A. Meyer, 1983a, Solid State Commun. **48**, 835.

Besnus, M.J., J.P. Kappler and A. Meyer, 1983b, J. Phys. F **13**, 597.

Besnus, M.J., J.P. Kappler and A. Meyer, 1985, Physica B **130**, 127.

Birgenau, R.J., E. Bucher, J.P. Maita, L. Passel and K.C. Turberfield, 1973, Phys. Rev. B **8**, 5345.

Bonnet, M., J.X. Boucherle, F. Givord, P. Lejay, J. Schweizer and A. Stunault, 1988, J. Phys. (Paris) **49**, C8-785.

Boucherle, J.X., G. Fillion, J. Flouquet, F. Givord, P. Lejay and J. Schweizer, 1986, J. Magn. & Magn. Mater. **62**, 277.

Boucherle, J.X., F. Givord, P. Lejay, J. Schweizer and A. Stunault, 1987a, J. Magn. & Magn. Mater. **63–64**, 575.

Boucherle, J.X., F. Givord, F. Lapierre, P. Lejay, J. Peyrard, J. Odin, J. Schweizer and A. Stunault, 1987b, in: Theoretical and Experimental Aspects of Valence Fluctuations and Heavy Fermions, eds L.C. Gupta and S.K. Malik (Plenum, New York) p. 485.

Bredl, C.D., 1985, PhD. Thesis (Technische Hochschule Darmstadt) unpublished.

Bredl, C.D., 1987, J. Magn. & Magn. Mater. **63–64**, 355.

Bredl, C.D., and F. Steglich, 1978, J. Magn. & Magn. Mater. **7**, 286.

Bredl, C.D., F. Steglich and K.D. Schotte, 1978, Z. Phys. B **29**, 327.

Bredl, C.D., S. Horn, F. Steglich, B. Luethi and R.M. Martin, 1984, Phys. Rev. Lett. **52**, 1982.

Brochado Oliveira, J.M., and I.R. Harris, 1983, J. Mater. Sci. **118**, 3649.

Brodale, G.E., R.A. Fisher, N.E. Phillips and J. Flouquet, 1986a, Phys. Rev. Lett. **56**, 390.

Brodale, G.E., R.A. Fisher, N.E. Phillips, J. Flouquet and C. Marcenat, 1986b, J. Magn. & Magn. Mater. **54–57**, 419.

Bucher, E., J.P. Maita, G.W. Hull, R.C. Fulton and A.S. Cooper, 1975, Phys. Rev. B **11**, 440.

Bucher, E., K. Andres, J.P. Maita and G.W. Hull, 1986, Helv. Phys. Acta **41**, 723.

Burlet, P., M.A. Frémy, D. Gignoux, G. Lapertot, S. Quezel, L.P. Reynault, J. Rossat-Mignod and E. Roudaut, 1987, J. Magn. & Magn. Mater. **63–64**, 34.

Busch, G., W. Stutius and D. Vogt, 1971, J. Appl. Phys. **42**, 1493.

Buschow, K.H.J., 1977, Rep. Prog. Phys. **40**, 1179.

Buschow, K.H.J., 1979, Rep. Prog. Phys. **42**, 1373.

Buschow, K.H.J., and J.F. Fast, 1966, Phys. Status Solidi **16**, 467.

Buschow, K.H.J., H.W. de Wijn and A.M. van Diepen, 1969, J. Chem. Phys. **50**, 137.

Buschow, K.H.J., R.C. Sherwood and S.L. Hsu, 1978, J. Appl. Phys. **49**, 1510.

Canepa, F., G.L. Olcese, R. Eggenhoffner and M.G. Fossatti, 1989, Physica B **154**, 390.

Chen, Y.Y., J.M. Lawrence, J.D. Thompson and J.O. Willis, 1989, Phys. Rev. B **40**, 10766.

Chouteau, G., and J. Palleau, 1980, J. Magn. & Magn. Mater. **15–18**, 311.

Cooper, J.R., C. Rizzuto and G.L. Olcese, 1971, J. Phys. (Paris) **32**, C1–1136.

Costa, G.A., F. Canepa and G.L. Olcese, 1982, Solid State Commun. **44**, 67.

Croft, M.C., R.P. Guertin, L.C. Kupferberg and R.D. Parks, 1979, Phys. Rev. B **20**, 2073.

Cromer, D.T., A.C. Larson and R.B. Roof Jr, 1973, Acta Crystallogr. B **29**, 564.

Danan, J., C. de Horin and R. Lallement, 1969, Solid State Commun. **7**, 1103.

de Boer, F.R., J. Klaasse, J. Aarts, C.D. Bredl, W. Lieke, U. Rauchschwalbe, F. Steglich, R. Felten, U. Umhofer and G. Weber, 1985, J. Magn. & Magn. Mater. **47–48**, 60.

de Jongh, L.J., and A.R. Miedema, 1974, Adv. Phys. **23**, 1.

De Long, L.E., 1986, Phys. Rev. B **33**, 3556.

De Long, L.E., M.B. Maple and M. Tovar, 1978, Solid State Commun. **26**, 469.

De Long, L.E., J.G. Huber and K. Bedell, 1990a, J. Appl. Phys. **67**, 5222.

De Long, L.E., J.G. Huber and K. Bedell, 1990b, Proc. 5th Int. Conf. on Physics of Magnetic Materials, Madralin, Poland, October 9–12, 1990.

Debray, D., M. Sougi and P. Meriel, 1972, J. Chem. Phys. **56**, 4325.

Desgranges, H.-U., and J.W. Rasul, 1985, Phys. Rev. B **32**, 6100.

Desgranges, H.-U., and K.D. Schotte, 1982, Phys. Lett. A **91**, 240.

Dhar, S.K., and K.A. Gschneidner Jr, 1989, J. Magn. & Magn. Mater. **79**, 151.

Dhar, S.K., S.K. Malik and R. Vijayaraghavan, 1981a, J. Phys. C **14**, L321.

Dhar, S.K., S.K. Malik and R. Vijayaraghavan, 1981b, Phys. Rev. B **24**, 6182.

Dhar, S.K., K.A. Gschneidner Jr and O.D. McMasters, 1987a, Phys. Rev. B **35**, 3291.

Dhar, S.K., K.A. Gschneidner Jr, W.H. Lee, P. Klavins and R.N. Shelton, 1987b, Phys. Rev. B **36**, 341.

Dhar, S.K., K.A. Gschneidner Jr, C.D. Bredl and F. Steglich, 1989, Phys. Rev. B **39**, 2439.

Dijkman, W.H., 1982, PhD. Thesis (University of Amsterdam, The Netherlands) unpublished.

Doniach, S., 1977, Physica B **91**, 231.

Dwight, A.E., R.A. Conner Jr and J.W. Downey, 1965, Acta. Crystallogr. **18**, 837.

Effantin, J.M., J. Rossat-Mignod, P. Burlet, H. Bartholin, S. Kunii and T. Kasuya, 1985, J. Magn. & Magn. Mater. **47–48**, 145.

Elenbaas, R.A., C.J. Schinkel, S. Strom van Leeuwen and C.J. Deudekom, 1980, J. Magn. & Magn. Mater. **15–18**, 1218.

Felten, R., 1987, PhD. Thesis (Technische Hochschule Darmstadt, FRG) unpublished.

Ferreira da Silva, J., M.A. Frey Ramos, D. Gignoux and J.C. Gomez-Sal, 1989, J. Magn. & Magn. Mater. **80**, 149.

Fillion, G., M.A. Frémy, D. Gignoux, J.C. Gomez-Sal and B. Gorges, 1987, J. Magn. & Magn. Mater. **63–64**, 117.

Flouquet, J., J.L. Lasjaunias, J. Peyrard and M. Ribault, 1982, J. Appl. Phys. **53**, 2127.

Fortune, N.A., G.M. Schmiedeshoff, J.S. Brooks, C.L. Lin, J.E. Crow, T.W. Mihalisin and G.R. Stewart, 1987, Jpn. J. Appl. Phys. **26**, S3–541.

Fujii, H., M. Akayama, Y. Uwakoto, V. Hashimoto and T. Kitai, 1988, J. Phys. (Paris) **49**, C8–417.

Fujita, T., M. Suzuki, T. Komatsubara, S. Kunii, T. Kasuya and T. Ohtsuka, 1980, Solid State Commun. **35**, 569.

Fulde, P., 1988, J. Phys. F **18**, 601.

Furrer, A., W. Buhrer, H. Herr, W. Hälg, J. Benes and O. Vogt, 1972, in: Neutron Elastic Scattering 1972 (IAEA, Vienna) p. 563.

Furrer, A., W. Hälg, H. Herr and O. Vogt, 1979, J. Appl. Phys. **50**, 2040.

Galera, R.M., A.P. Murani and J. Pierre, 1981, J. Magn. & Magn. Mater. **23**, 317.

Galera, R.M., J. Pierre, J. Voiron and G. Dampne, 1983, Solid State Commun. **46**, 45.

Gamari-Saele, H., 1980, J. Less-Common Met. **75**, P43.

Germann, A., A.K. Nigam, J. Dutzi, A. Schroeder and H. v. Löhneysen, 1988, J. Phys. (Paris) **49**, C8–755.

Gignoux, D., and J.C. Gomez-Sal, 1984, Phys. Rev. B **30**, 3967.

Gignoux, D., and J. Voiron, 1985, Phys. Rev. B **32**, 4822.

Gignoux, D., F. Givord and R. Lemaire, 1982, J. Phys. (Paris) **43**, 173.

Gignoux, D., F. Givord, R. Lemaire and F. Tasset, 1983, J. Less-Common Met. **94**, 165.

Gignoux, D., D. Schmitt and M. Zerguine, 1987, J. Magn. & Magn. Mater. **66**, 373.

Gomes de Mesquita, A.H., and K.H.J. Buschow, 1967, Acta Crystallogr. **22**, 497.

Gopal, E.S.R., 1966, Specific Heats at Low Temperatures (Plenum, New York).

Gratz, E., E. Bauer, B. Barbara, S. Zemirli, F. Steglich, C.D. Bredl and W. Lieke, 1985, J. Phys. F **15**, 1975.

Gschneidner Jr, K.A., and K. Ikeda, 1983, J. Magn. & Magn. Mater. **31–34**, 265.

Gschneidner Jr, K.A., S.K. Dhar, R.J. Sherman, T.-W.E. Tsang and O.D. McMasters, 1985, J. Magn. & Magn. Mater. **47–48**, 51.

Gschneidner Jr, K.A., J. Tang, S.K. Dhar and A. Goldman, 1990, Physica B **163**, 507.

Herr, H., A. Furrer, W. Hälg and O. Vogt, 1979, J. Phys. C **12**, 5207.

Hill, H.H., 1970, in: Plutonium 1970 and other Actinides, ed. W. Miner (The Metallurgical Society of the AIME, New York).

Hill, R.W., J. Cosier and D.A. Hukin, 1983, J. Phys. C **16**, 2871.

Hirschfeld, P., D. Wollhardt and P. Wolfe, 1986, Solid State Commun. **59**, 111.

Ho, J.C., S.M. Taher, J.B. Gruber and K.A. Gschneidner Jr, 1982, Phys. Rev. B **26**, 1369.

Hohnke, D., and E. Parthé, 1966, Acta Crystallogr. **20**, 572.

Holt, B.J., J.D. Ramsden, H.H. Sample and J.G. Huber, 1981, Physica B **107**, 255.

Huber, J.G., 1989, Bull. Am. Phys. Soc. **34**, 566.

Huber, J.G., and C.A. Luengo, 1978, J. Phys. (Paris) **39**, C6–781.

Hulliger, F., and H.R. Ott, 1978, Z. Phys. B **29**, 47.

Hulliger, F., M. Landolt, H.R. Ott and R. Schmelczer, 1975, J. Low Temp. Phys. **20**, 269.

Hulliger, F., B. Natterer and H.R. Ott, 1978, J. Magn. & Magn. Mater. **8**, 87.

Hungsberg, R.E., and K.A. Gschneidner Jr, 1972, J. Phys. & Chem. Solids **33**, 401.

Iandelli, A., and E. Botti, 1937, Rend. Accad. Naz. Lincei **25**, 129.

Iandelli, A., and A. Palenzola, 1979, in: Handbook on the Physics and Chemistry of Rare Earth, Vol. 2, eds K.A. Gschneidner Jr and L. Eyring (North-Holland, Amsterdam) ch. 13, p. 1.

Ikeda, K., and K.A. Gschneidner Jr, 1980, Phys. Rev. Lett. **45**, 1341.

Ikeda, K., and K.A. Gschneidner Jr, 1982, Phys. Rev. B **25**, 4623.

Ikeda, K., K.A. Gschneidner Jr, B.J. Beaudry and U. Atzmany, 1982, Phys. Rev. B **25**, 4604.

Isobe, A., A. Ochiai, S. Onodera, K. Yamada, M. Kohgi, Y. Endoh, T. Suzuki and T. Kasuya, 1987, J. Magn. & Magn. Mater. **70**, 391.

Johansson, B., 1978, J. Phys. & Chem. Solids **39**, 467.

Joseph, R.R., K.A. Gschneidner Jr and R.E. Hungsberg, 1972a, Phys. Rev. B **5**, 1878.

Joseph, R.R., K.A. Gschneidner Jr and D.C. Koskimaki, 1972b, Phys. Rev. B **6**, 3286.

Kadomatsu, H., H. Tanaka, M. Kurisu and G. Fujiwara, 1986, Phys. Rev. B **33**, 4799.

Kadowaki, K., and S.S. Woods, 1986, Solid State Commun. **58**, 507.

Kappler, J.P., 1990, private communication.

Kappler, J.P., and A. Meyer, 1979, J. Phys. F **9**, 143.

Kappler, J.P., and J.G. Sereni, 1990a, unpublished.

Kappler, J.P., M.J. Besnus, A. Meyer, J.G. Sereni and G.L. Nieva, 1983, Proc. 6ième Journées de la R.C.P. 520 sur les Composés de Terres Rares et Actinides à Valence Anomale, Strasbourg, France, 6–7 Oct., p. 23.

Kappler, J.P., M.J. Besnus, P. Lehman, G. Krill and A. Meyer, 1984, Proc. 7ième Journées de la R.C.P. 520 sur les Composés de Terres Rares et Actinides à Valence Anomale, Grenoble, France, 7–8 Nov., p. 73.

Kappler, J.P., M.J. Besnus, P. Lehman, A. Meyer and J.G. Sereni, 1985, J. Less-Common Met. **111**, 261.

Kappler, J.P., P. Lehman, G. Schmerber, G.L. Nieva and J.G. Sereni, 1988a, J. Phys. (Paris) **49**, C8–721.

Kappler, J.P., E. Baurepaire, G. Krill, C. Godart, G.L. Nieva and J.G. Sereni, 1988b, J. Phys. (Paris) **49**, C8–723.

Kappler, J.P., G. Schnerber and J.G. Sereni, 1988c, J. Magn. & Magn. Mater. **76–77**, 185.

Kappler, J.P., G.L. Nieva, J.G. Sereni and J. Souletie, 1988d, J. Phys. (Paris) **49**, C8–725.

Kappler, J.P., M.J. Besnus, A. Herr, A. Meyer and J.G. Sereni, 1990b, Proc. VI Internat. Conf. on Valence Fluctuations, Rio de Janeiro, Brasil, July 9–13, to be published in Phys. B (1991).

Kissel, F., and W.E. Wallace, 1966, J. Less-Common Met. **11**, 417.

Kitazawa, H., V.S. Kwon, A. Oyamada, N. Takeda, H. Suzuki, S. Sakatsume, T. Stoh, T. Suzuki and T. Kasuya, 1988, J. Magn. & Magn. Mater. **76–77**, 40.

Kohlmann, J., E. Bauer and K. Winzer, 1989, J. Magn. & Magn. Mater. **82**, 169.

Kohlmann, J., E. Bauer and K. Winzer, 1990, Phys. B **163**, 188.

Komatsubara, T., T. Suzuki, M. Kawakami, S. Kunii, T. Fujita, Y. Isikawa, A. Takase, K. Kojima, M. Suzuki, Y. Aoki, K. Takegahara and T. Kasuya, 1980, J. Magn. & Magn. Mater. **15–18**, 963.

Kontani, M., M. Senda, M. Nakano, J.M. Law-

rence and K. Adachi, 1987, J. Magn. & Magn. Mater. **70**, 378.
Koskenmaki, D.C., and K.A. Gschneidner Jr, 1978, in: Handbook on the Physics and Chemistry of Rare Earths, Vol. 1, eds K.A. Gschneidner Jr and L. Eyring (North-Holland, Amsterdam) ch. 4.
Kurisu, M., H. Tanaka, H. Kadomatsu, K. Sekizawa and H. Fujiwara, 1985, J. Phys. Soc. Jpn. **54**, 3548.
Lahiouel, R., R.M. Galera, J. Pierre and E. Siaud, 1986, Solid State Commun. **58**, 815.
Larson, A.C., and D.T. Cromer, 1961, Acta Crystallogr. **14**, 73.
Larson, A.C., and D.T. Cromer, 1962, Acta Crystallogr. **15**, 1224.
Lawrence, J.M., and S.M. Shapiro, 1980, Phys. Rev. B **22**, 4379.
Lee, K.N., and B. Bell, 1972, Phys. Rev. B **6**, 1032.
Lethuiller, P., 1976, J. Phys. (Paris) **37**, 123.
Lin, C.L., J. Teter, J.E. Crow, T. Mihalisin, J. Brooks, A.I. Abou-Aly and G.R. Stewart, 1985, Phys. Rev. Lett. **54**, 2541.
Lin, C.L., J.E. Crow, T. Mihalisin, J. Brooks, G.R. Stewart and A.I. Abou-Aly, 1986, J. Magn. & Magn. Mater. **54–57**, 379.
Lucken, H., M. Meier, G. Klesses, W. Bronger and J. Fleischhauer, 1979, J. Less-Common Met. **63**, 35.
Luengo, C.A., J.G. Huber, M.B. Maple and M. Roth, 1975, J. Low Temp. Phys. **21**, 129.
Luo, J.L., M.B. Maple, I.R. Harris and T.F. Smith, 1968, Phys. Lett. A **27**, 519.
Machado da Silva, J.M., and R.W. Hill, 1972, J. Phys. C **5**, 1584.
Mader, K.H., and W.M. Swift, 1968, J. Phys. & Chem. Solids **29**, 1759.
Mahony, J.V., V.U.S. Rao, W.E. Wallace and R.S. Craig, 1974, Phys. Rev. B **9**, 154.
Maple, M.B., and D. Wohlleben, 1973, AIP Conf. Proc. **18**, 447.
Maury, A., R. Freitag, J.E. Crow, T. Mihalisin and A.I. Abou-Aly, 1982, Phys. Lett. A **92**, 411.
Mihalisin, T., P. Scorboria and J.A. Ward, 1981, Phys. Rev. Lett. **46**, 862.
Moreau, J.M., D. Paccard and D. Gignoux, 1974, Acta Crystallogr. B **30**, 2122.
Mori, H., H. Yashima and N. Sato, 1985, J. Low Temp. Phys. **58**, 513.
Morin, P., 1988, J. Magn. & Magn. Mater. **71**, 151.
Morin, P., J. Rouchy, Y. Miyako and T. Nishioka, 1988, J. Magn. & Magn. Mater. **76–77**, 319.
Mueller, T., W. Doss, J.M. van Ruitenbeek, U. Welp and P. Wyder, 1988, J. Magn. & Magn. Mater. **76–77**, 35.
Muraoka, V., S.M. Shiga and Y. Nakamura, 1976, J. Phys. Soc. Jpn. Lett. **40**, 905.
Nakazato, M., N. Wakabayashi and T. Kitai, 1988, J. Phys. Soc. Jpn. **57**, 953.
Narasimham, K.S., V.D. Rao and R.A. Butera, 1974, AIP Conf. Proc. **10**, 1081.
Nasu, S., H.H. Neumann, N. Marzauk, R.S. Craig and W.E. Wallace, 1971, J. Phys. & Chem. Solids **32**, 2779.
Newns, D.M., and N. Read, 1987, Adv. Phys. **36**, 799.
Newns, D.M., A.C. Hewson, J.W. Rasul and N. Read, 1982, J. Appl. Phys. **53**, 7877.
Nieva, G.L., 1988, PhD. Thesis (Universidad de Cuyo, Argentina) unpublished.
Nieva, G.L., J.G. Sereni and J.P. Kappler, 1987, Phys. Scr. **35**, 201.
Nieva, G.L., J.G. Sereni, M. Afyoumi, G. Schmerber and J.P. Kappler, 1988, Z. Phys. B **70**, 181.
Olcese, G.L., 1966a, Rend. Accad. Naz. Lincei **40**, 1052.
Olcese, G.L., 1966b, Boll. Sci. Fac. Chim. Ind. Bologna **24**, 165.
Olcese, G.L., 1973, J. Less-Common Met. **33**, 71.
Olcese, G.L., 1979, J. Phys. F **9**, 569.
Olcese, G.L., 1980, Solid State Commun. **35**, 87.
Ōnuki, Y., Y. Shimizu, T. Komatsubara, A. Sumiyama, Y. Oda, H. Nagano, T. Fujita, Y. Maeno, K. Satoh and T. Ohtsuka, 1985a, J. Magn. & Magn. Mater. **52**, 344.
Ōnuki, Y., Y. Machii, Y. Shimizu, T. Komatsubara and T. Fujita, 1985b, J. Phys. Soc. Jpn. **54**, 3562.
Ott, H.R., F. Hulliger and F. Stucki, 1978, Inst. Phys. Conf. Ser., Vol. 37, ch. 3, p. 72.
Ott, H.R., J.K. Kjems and F. Hulliger, 1979, Phys. Rev. Lett. **42**, 1378.
Ott, H.R., H. Rudigier, Z. Fisk, J.O. Willis and G.R. Stewart, 1985, Solid State Commun. **53**, 235.
Parthé, E., and B. Chabot, 1984, in: Handbook on the Physics and Chemistry of Rare Earths, Vol. 6, eds K.A. Gschneidner Jr and L. Eyring (North-Holland, Amsterdam) ch. 48.
Pearsons Handbook of Crystallographic Data for Intermetallic Phases, 1985, eds P. Villars and L.D. Calvet (American Society for Metals, Metals Park, OH).
Penney, T., F.P. Milliben, F. Holzberg and Z. Fisk, 1987, in: Valence Fluctuations and Heavy

Fermions, eds L.C. Gupta and S.K. Malik (Plenum, New York) p. 77.
Peyrard, J., 1980, PhD. Thesis (University of Grenoble, France) unpublished.
Peysson, Y., C. Ayache, J. Rossat-Mignod, S. Kunii and T. Ohtsuka, 1986, J. Phys. (Paris) **47**, 113.
Phillips, N.E., R.A. Fisher, S.E. Lacy, C. Marcenat, J.A. Olsen, J. Flouquet, A. Amato, D. Jaccard, Z. Fisk, A.L. Giorgi, J.L. Smith and G.R. Stewart, 1987, in: Valence Fluctuations and Heavy Fermions, eds L.C. Gupta and S.K. Malik (Plenum, New York) p. 141.
Pierre, J., and A.P. Murani, 1980, in: Crystalline Electric Field and Structural Effects in f-Electron Systems, eds J.E. Crow, R.P. Guertin and T. Mihalisin (Plenum, New York) p. 607.
Pierre, J., A.P. Murani and R.M. Galera, 1981, J. Phys. F **11**, 679.
Pierre, J., R.M. Galera and E. Siaud, 1985, J. Phys. (Paris) **46**, 621.
Pountney, J.M., J.M. Winterbottom and I.R. Harris, 1978, Inst. Phys. Conf. Ser., Vol. 37, ch. 3, p. 85.
Rahman, S., J.E. Crow, T. Mihalisin and P. Schlottmann, 1989, Solid State Commun. **71**, 379.
Rajan, V.T., J.H. Löwenstein and N. Andrei, 1982, Phys. Rev. Lett. **49**, 497.
Rauchschwalbe, U., U. Gottwick, U. Ahlheim, H.M. Mayer and F. Steglich, 1985, J. Less-Common Met. **111**, 265.
Ray, D.K., and J. Sivardiere, 1976, Solid State Commun. **19**, 1053.
Ribault, M., A. Benoit, J. Flouquet and J. Palleau, 1979, J. Phys. (Paris) Lett. **40**, L-413.
Rice, M.J., 1968, Phys. Rev. Lett. **20**, 1439.
Röhler, J., 1987, in: Handbook on the Physics and Chemistry of Rare Earths, Vol. 10, eds K. Gschneidner Jr and L. Eyring (North-Holland, Amsterdam) ch. 71.
Rossat-Mignod, J., P. Burlet, H. Bartholin, D. Vogt and R. Lagnier, 1980, J. Phys. C **13**, 6381.
Sakai, T., G. Adachi, T. Yoshida and J. Shiokawa, 1981, J. Chem. Phys. **75**, 3027.
Sakurai, J., T. Matsuura and Y. Komura, 1988, J. Phys. (Paris) **49**, C8-783.
Sales, B.C., 1974, PhD. Thesis (University of California, USA) unpublished.
Sales, B.C., 1977, J. Low. Temp. Phys. **28**, 107.
Sato, N., M. Kohgi, T. Satoh, Y. Ishikawa, H. Hiroyoshi and H. Takei, 1985, J. Magn. & Magn. Mater. **52**, 360.
Sato, N., M. Kontani, H. Abe and K. Adachi, 1987, J. Magn. & Magn. Mater. **70**, 372 .
Sato, N., M. Kontani, H. Abe and K. Adachi, 1988a, J. Phys. Soc. Jpn. **57**, 1069.
Sato, N., H. Mori, T. Satoh, T. Miura and H. Takei, 1988b, J. Phys. Soc. Jpn. **57**, 1384.
Satoh, K., T. Fujita and Y. Maeno, 1985, Solid State Commun. **56**, 237.
Satoh, K., M. Kato, Y. Maeno, T. Fujita, Y. Ōnuki and T. Komatsubara, 1987, Jpn. J. Appl. Phys. **26**, Supp. 26-3, 521.
Satoh, K., T. Fujita, Y. Maeno, Y. Ōnuki and T. Komatsubara, 1988, J. Magn. & Magn. Mater. **76–77**, 128.
Schlottmann, P., 1989, Phys. Rep. **181**, 1.
Schmitt, D., P. Morin and J. Pierre, 1978, J. Magn. & Magn. Mater. **8**, 249.
Schroeder, A., R. Van den Bag, H. v. Löhneysen, W. Paul and H. Lucken, 1988, Solid State Commun. **65**, 99.
Sera, M., T. Satoh and T. Kasuya, 1987, J. Magn. & Magn. Mater. **63–64**, 82.
Sereni, J.G., 1980a, J. Phys. F **10**, 2831.
Sereni, J.G., 1980b, in: Journées de Calorimetrie et d'Analyse Thermique, Barcelona, Spain, June 4–6, unpublished.
Sereni, J.G., 1981, unpublished.
Sereni, J.G., 1982a, J. Less-Common Met. **84**, 1.
Sereni, J.G., 1982b, J. Less-Common Met. **86**, 1.
Sereni, J.G., 1985, J. Magn. & Magn. Mater. **47–48**, 228.
Sereni, J.G., and J.P. Kappler, 1989, unpublished.
Sereni, J.G., and J.P. Kappler, 1990, unpublished.
Sereni, J.G., and G.L. Olcese, 1979, unpublished.
Sereni, J.G., T.E. Huber and C.A. Luengo, 1979, Solid State Commun. **29**, 671.
Sereni, J.G., G.L. Nieva, J.P. Kappler, M.J. Besnus and A. Meyer, 1986, J. Phys. F **16**, 435.
Sereni, J.G., G.L. Nieva, G. Schmerber and J.P. Kappler, 1989a, Mod. Phys. Lett. B **3**, 1225.
Sereni, J.G., G.L. Olcese, G. Krill and J.P. Kappler, 1989b, in: Conf. on Intermediate Valence Compounds, Brasilia, Brasil, April 3–7, unpublished.
Sereni, J.G., G.L. Nieva, J.P. Kappler and P. Haen, 1990a, submitted to J. de Phys. (France).
Sereni, J.G., G.L. Nieva, G.L. Olcese, A. Herr and J.P. Kappler, 1990b, Proc. VI Internat. Conf. on Valence Fluctuations, Rio de Janeiro, Brasil, July 9–13, to be published in Phys. B (1991).
Sereni, J.G., O. Trovarelli and J.P. Kappler, 1990c, unpublished data.

Sereni, J.G., O. Trovarelli, G. Schmerber and J.P. Kappler, 1991, to be published.

Shaheen, S.A., J.S. Schilling, S.H. Liu and O.D. McMasters, 1983, Phys. Rev. **27**, 4325.

Souletie, J., 1988, J. Phys. (Paris) **49**, 1211.

Starovoitov, A.T., V.I. Ozhogin, G.M. Loginov and V.M. Sergeeva, 1970, Sov. Phys.-JETP **30**, 433.

Stassen, W.N., M. Sato and L.D. Calvert, 1970, Acta Crystallogr. B **26**, 1534.

Steglich, F., 1985, in: Theory of Heavy Fermions and Valence Fluctuations, Springer Series in Solid State Sciences, Vol. 62, eds T. Kasuya and T. Sato (Springer, Berlin).

Stewart, G.R., Z. Fisk and M.S. Wire, 1984, Phys. Rev. B **30**, 482.

Stunault, A., 1988, Ph.D. Thesis (University of Grenoble I, France) unpublished.

Sundström, L.J., 1978, in: Handbook of the Physics and Chemistry of the Rare Earths, Vol. 1, eds K.A. Gschneidner Jr and L. Eyring (North-Holland, Amsterdam) ch. 5.

Suzuki, T., Y. Nakabayashi, A. Ochiai, T. Kasuya, K. Suziyama and M. Date, 1987, J. Magn. & Magn. Mater. **63–64**, 58.

Suzuki, T., O. Nakamura, N. Tomonaga, S. Ozeki, Y.S. Kwon, A. Ochiai, T. Takeda and T. Kasuya, 1990, Physica B **163**, 131.

Swift, W.M., and W.E. Wallace, 1968, J. Phys. & Chem. Solids **29**, 2053.

Taher, S.M., J.C. Ho, J.B. Gruber, B.J. Beaudry and K.A. Gschneidner Jr, 1980, in: Rare Earths in Modern Physics, Vol. 2, eds G.J. McCarthy, H.B. Silber and J.J. Rhyne (Plenum, New York) p. 423.

Takahashi, M., H. Tanaka, T. Satoh, M. Kohgi, Y. Ishikawa, T. Miura and H. Takei, 1988, J. Phys. Soc. Jpn. **57**, 1377.

Takke, R., N. Dolezal, W. Assmun and B. Luethi, 1981, J. Magn. & Magn. Mater. **23**, 247.

Tang, J., and K.A. Gschneidner Jr, 1988, J. Magn. & Magn. Mater. **75**, 355.

Tang, J., and K.A. Gschneidner Jr, 1989, J. Less-Common Met. **149**, 341.

Tang, J., K.A. Gschneidner Jr, R. Caspary and F. Steglich, 1990, Physica B **163**, 201.

Thomson, J.R., 1963, Acta Crystallogr. **16**, 320.

Torikachvili, M.S., and M.B. Maple, 1984, Proc. LT-17, Karlsruhe, FRG, eds U. Eckern, A. Schmid, W. Weber and H. Wühl (North-Holland, Amsterdam) p. 329.

Tsang, T.-W.E., K.A. Gschneidner Jr, O.D. McMasters, R.J. Stierman and S.K. Dhar, 1984, Phys. Rev. B **29**, 4185.

Tsuchida, T., and W.E. Wallace, 1965a, J. Chem. Phys. **43**, 2885.

Tsuchida, T., and W.E. Wallace, 1965b, J. Chem. Phys. **43**, 3811.

Tsuchida, T., S. Onga and T. Kawai, 1973, J. Phys. Soc. Jpn. **35**, 1788.

Van Daal, H.J., and K.H.J. Buschow, 1970, Phys. Lett. A **31**, 103.

Veyssie, J.J., J. Chaussy and A. Berton, 1964, Phys. Lett. **13**, 29.

Vijayaraghavan, R., V.U.S. Rao, S.K. Malik and V. Marathe, 1968, J. Appl. Phys. **39**, 1086.

Walline, R.E., and W.E. Wallace, 1965, J. Chem. Phys. **42**, 604.

Wegener, W., A. Furrer, W. Bührer and S. Hautecler, 1981, J. Phys. C **14**, 2387.

Weidner, P.C., 1981, Diplomarbeid (University of Cologne, FRG) unpublished.

Weidner, P.C., 1984, PhD. Thesis (University of Cologne, FRG) unpublished.

Weidner, P.C., K. Krengler, R. Löhe, B. Roden, J. Röhler, B. Wittershagen and D. Wohlleben, 1985, J. Magn. & Magn. Mater. **47–48**, 75.

Westrum, E.F., and B.H. Justice, 1968, Proc. 7th Rare Earth Research, Coronado, CA, Oct. 28, p. 821.

Wilhelm, M., and B. Hillebrand, 1970, J. Phys. & Chem. Solids **31**, 559.

Willis, J.O., R.H. Aiken, Z. Fisk, E. Zirngibel, J.D. Thompson, H.R. Ott and B. Battlog, 1987, in: Valence Fluctuations and Heavy Fermions, eds L.C. Gupta and S.K. Malik (Plenum, New York) p. 57.

Wilson, K.G., 1975, Rev. Mod. Phys. **47**, 773.

Wohlleben, D., 1976, J. Phys. (Paris) **10**, C4-231.

Wohlleben, D., and B. Wittershagen, 1985, J. Magn. & Magn. Mater. **52**, 32.

Yashima, H., and T. Satoh, 1982, Solid State Commun. **41**, 723.

Yashima, H., N. Sato, H. Mori and T. Satoh, 1982, Solid State Commun. **43**, 595.

Yashima, H., C. Feng-Lin, T. Satoh, H. Hiroyoshi and K. Kohn, 1986, Solid State Commun. **57**, 793.

Zalkin, A., and D.H. Templeton, 1953, Acta Crystallogr. **6**, 269.

Zemirli, S., and B. Barbara, 1985, Solid State Commun. **56**, 385.

Handbook on the Physics and Chemistry of Rare Earths, Vol. 15
edited by K.A. Gschneidner, Jr. and L. Eyring

Chapter 99

RARE EARTH CARBIDES

GIN-YA ADACHI, NOBUHITO IMANAKA and ZHANG FUZHONG
Department of Applied Chemistry, Osaka University, Suita, Osaka 565, Japan

---

Contents

1. Introduction

Binary rare earth carbides have long been known to exist. Pettersson (1895) first prepared the dicarbides of lanthanum and yttrium in an electric arc furnace from the oxide by reducing with carbon. Moissan's (1896a–c, 1900a, b) pioneering observations in this field have stimulated many examinations of the preparation of the carbides of the rare earth elements. By 1900, Moissan (1896a–c, 1900a, b) had prepared the dicarbides of yttrium, lanthanum, cerium, praseodymium, neodymium and samarium by reducing their oxides with carbon, sugar charcoal and graphite in an electric arc furnace. Most subsequent investigators have used this procedure or modifications of it, although the product is invariably contaminated with oxygen, nitrogen or carbon. Spedding and his co-workers prepared the carbide by direct reaction of metallic lanthanum and carbon in an electric arc furnace under helium or argon atmospheres, showing that two rare earth carbide phases exist (Spedding et al. 1958), one of which was identified as the lanthanum sesquicarbide (Atoji et al. 1958), being isostructural with the cerium sesquicarbide as reported by Brewer and Krikorian (1956). Up to date, the existence of several binary rare earth carbides has been confirmed. Their crystal structures have been determined by X-ray and neutron diffraction analyses (von Stackelberg 1930, Atoji et al. 1958). According to their crystal structures and compositions, these compounds could be classified as follows:

α-R_2C (R = Sc, Y, Gd, Tb, Dy, Ho, Yb)

R_3C (R = Y, Sm through Lu)

Sc_4C_3 and $Sc_{13}C_{10}$

$R_{15}C_{19}$ (R = Sc, Y, Er through Lu)

R_2C_3 (R = Y, La through Lu)

α-RC_2 (R = Y, La through Yb)

β-RC_2 (R = Y, La through Lu, except for Yb)

α'-RC_2 (R = Y, Tb through Tm, Lu)

RC_6 (R = Eu, Yb)

Yb_4C_5 etc . . . .

In addition, the rare earths have been known to form a series of stable gaseous carbides through the pioneering works by Chupka et al. (1958), DeMaria et al . (1967), DeMaria and Balducci (1972), and Stearns and Kohl (1971). These stable gaseous carbides include the dicarbides, tetracarbides and the first known gaseous metal tricarbides as well as the recently found gaseous carbides with more than four carbon atoms, namely up to six for cerium (Gingerich et al. 1976a) and yttrium (Gingerich and Haque 1980), and eight for lanthanum (Gingerich et al. 1982). The detailed thermodynamic studies have been carried out by combining the use of a mass spectrometer and a high-temperature Knudsen effusion cell and are expected to yield atomization energies, vaporization energies, dissociation energies, enthalpies of formation and, possibly, structural information, on which empirical correlations may be based. This will permit the prediction of the stability and possibly the molecular structural properties of a large number of yet unknown gaseous metal carbides.

Ternary rare earth carbon compounds are also receiving increased attention because of their unique magnetic and electrical properties as well as their potential importance in nuclear technology and in the construction of permanent magnets. A large number of investigators focus their attention on the superconductivity of the rare-earth–(thorium or uranium)–carbon systems, the magnetic properties of the rare-earth–transition-metal–carbon systems and the electrical properties of the rare-earth–boron–carbon systems. However, almost all the results of investigations have not yet been adopted in a practical application.

In this work, we shall give a review of the fundamental research on the rare earth carbides by focussing our attention on the phase diagram and the thermodynamics concerned with the formation of carbides, and on the crystal structures and the chemical and physical properties. In addition, we also pay attention to those topics that need further investigation.

2. Binary rare-earth–carbon phase diagrams

2.1. *Two prototypes of binary rare-earth–carbon phase diagrams*

Only two nearly complete and one partial binary rare-earth–carbon phase diagrams have been reported. Almost complete phase diagrams were proposed by Spedding et al. (1959) for the lanthanum–carbon system, and by Calson and Paulson (1968), refined by Storms (1971), for the yttrium–carbon system. The partial cerium–carbon phase diagram reported by Stecher et al. (1964) included only the data on the existence and crystal structure of the phases from thermal analysis and metallographic observation. The entire liquidus and the peritectic and eutectic reactions were not determined and are thus shown as dashed lines in their diagram. As is well known, the lanthanides can be divided into two groups, the light rare earths and the heavy rare earths. Lanthanum behaves much like cerium, praseodymium and neodymium, and yttrium behaves like dysprosium and holmium. Both lanthanum and yttrium could be regarded as the representative elements of the light rare earths and the heavy rare earths, respectively. Therefore, both the phase diagrams of the lanthanum–carbon and yttrium–carbon systems may be regarded as the prototypes of the phase diagrams of the light-rare-earth–carbon and heavy-rare-earth–carbon systems, respectively. As is described later, the lanthanum–carbon phase diagram is available for all the light lanthanide systems with carbon, but the yttrium–carbon phase diagram may be available for only some of the heavy-rare-earth–carbon systems, because the types, compositions and structures of the rare earth carbides in each of the heavy-rare-earth–carbon systems change as the atomic number of the rare earth increases. In particular, at about holmium the rare earth sesquicarbides with the Pu_2C_3-type body-centered cubic structure disappear in the systems and are substituted by the tetragonal $Sc_{15}C_{19}$-type carbides in the systems of holmium, erbium, thulium and lutetium with carbon (Bauer and Nowotny 1971). In addition, it was shown (Krikorian et al. 1967) that the lutetium dicarbide has a different stable room temperature modification from the other heavy lanthanide dicarbides.

2.1.1. *Phase diagram of the lanthanum–carbon system*

A phase diagram of the La–C system has been proposed on the basis of thermal metallographic, X-ray, dilatometric and electrical resistance analyses (Spedding et al. 1959), as is shown in fig. 1. Using La metal, which was prepared by the metallothermic reduction of the fluoride with Ca metal, and high-purity graphite, alloys were prepared by arc melting under an atmosphere of purified helium or argon. Photomicrographs of quenched low-carbon alloys indicated that the solid solubility of C in La lies between 1.6 and 3.4 at.% at 775°C, and slightly below 1.6 at.% at 695°C. The eutectic between La and La_2C_3 occurs at 806 $\pm$ 2°C at a composition of 20.6 at.% C. The solubility limit of C in β-La at the eutectic temperature is 3.4 $\pm$ 0.6 at.% C. The solid solution range of La_2C_3 at room temperature was found to extend from 56.2 to 60.2 at.% C. The C content at the La–La_2C_3 phase boundary was found to vary from 54 $\pm$ 0.7 at.% at 800°C to 57.1 $\pm$ 1.2 at.% at 1000°C. La_2C_3 forms peritectically at 1415 $\pm$ 3°C on cooling from the melt and LaC_2. The crystal structure of this compound was found to be body-centered cubic of the Pu_2C_3 type with the $I\bar{4}3d$ space group (Atoji et al. 1958). The melting point of LaC_2 is 2356 $\pm$ 25°C. This compound is dimorphic, with tetragonal CaC_2-type structure at room temperature and cubic CaF_2-type structure at high temperatures. The polymorphic transformation occurs at 1800 $\pm$ 100°C, which is in good agreement with the value of 1750°C reported by Bredig (1960). However, more recent investigators have reported considerably lower temperatures for this transition, Krikorian et al. (1967) gave a value of 1060°C. They speculated that the definite change in slope of the electrical resistance curves between 1750 and 1800°C observed by Spedding et al. (1959) may have been related to some change in carbon content at that temperature. In fig. 1 this temperature is shown as 1060°C.

The temperature of the eutectic reaction between LaC_2 and C is also shown in fig. 1 at 2250°C, which is the average of the two reported values by Spedding et al. and by Krikorian et al.

α-LaC_2 contains less than the stoichiometric amount of C at room temperature, and the lattice parameter of the alloy around the LaC_2 composition varies slightly with C

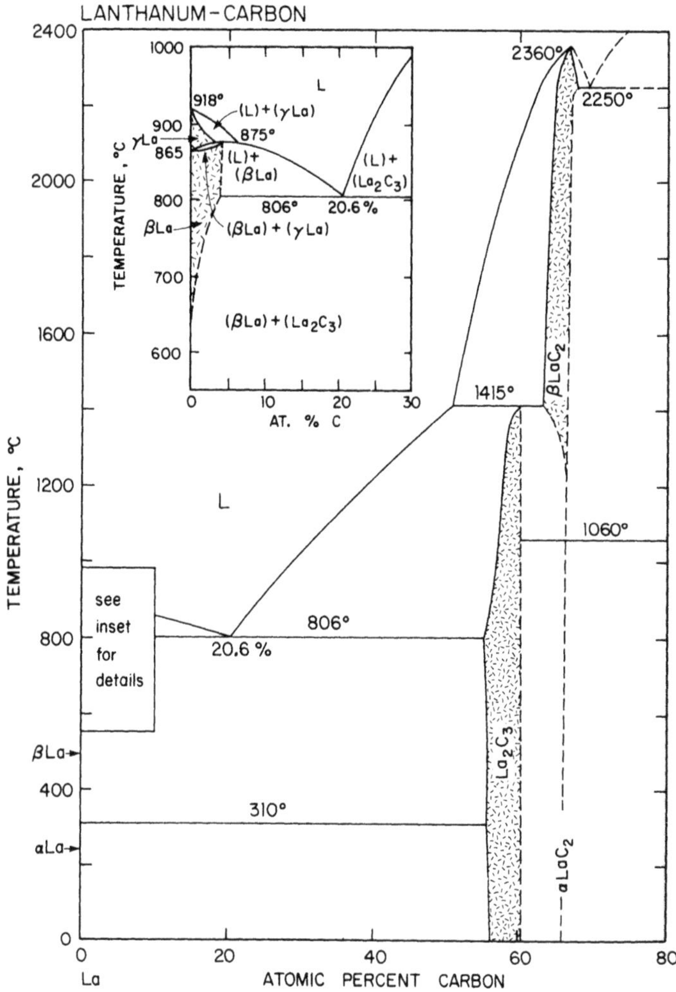

Fig. 1. Assessed La–C phase diagram (Gschneidner and Calderwood 1986).

content, indicating a solid solubility range from about 2.3 to 3.4 at.%. It reaches the stoichiometric dicarbide composition at higher temperatures.

The microscopic examinations showed that a 64.4 at.% C specimen revealed a single-phase structure at 1385 ± 25°C, from which La_2C_3 precipitated during furnace cooling to room temperature, and a 62.5 at.% C sample consisting of two phases on quenching from 1400 ± 25°C, but on slow cooling the amount of the La_2C_3 phase increased further at the expense of the LaC_2 phase.

The study of the La–C phase diagram showed that only two carbides, LaC_2 and La_2C_3, exist, which are the compounds with a solid solubility range instead of a line compound. La_2C_3 forms by a peritectic reaction while LaC_2 melts congruently.

2.1.2. *Phase diagram of the yttrium–carbon system*

The phase diagram of the yttrium–carbon system was proposed by Carlson and Paulson (1968) and refined by Storms (1971), is shown in fig. 2.

Carlson and Paulson used Y of 99.9 wt.% purity prepared by Ca reduction of YCl_3. Samples were prepared by arc melting a Y sponge with pieces of high-purity graphite in a purified argon atmosphere. The solidus was determined by observations with an optical pyrometer focussed on a small hole in the specimen. Thermal analysis was employed in the Y-rich end of the system. Specimens were also examined under the microscope after quenching in an oil bath.

Storms (1971) studied the phase relationship at temperatures between 1027 and 1827°C over a wide composition range in the Y–C system using a combination of mass spectrometric and thermal analysis techniques. Samples were prepared by arc melting purified crystal bar yttrium metal and a spectroscopically pure graphite rod.

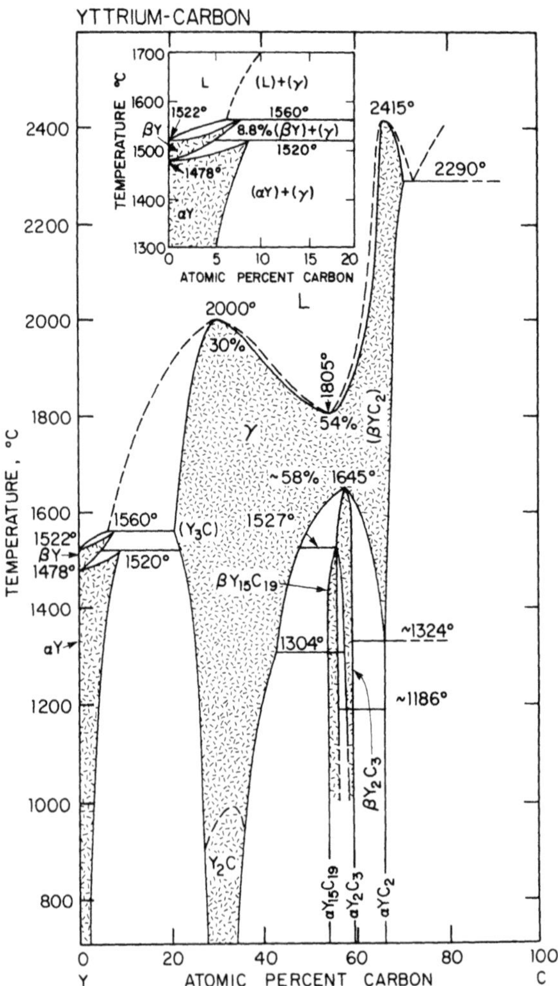

Fig. 2. Assessed Y–C phase diagram (Gschneidner and Calderwood 1986).

The as-supplied yttrium metal was purified by heating the metal in vacuum just below the melting point to reduce the YF^+/Y^+ ratio from 10 to 10^{-4}. The phase boundaries were obtained from a plot of the logarithm of Y activity at various temperatures against the C to Y ratio by observing the intersections of the curve through single-phase regions with the horizontal line through two-phase regions.

According to Carlson and Paulson, the Y–C system consists of three compounds; Y_2C, Y_2C_3, and YC_2. At high temperatures, Y_2C and YC_2 are face-centered cubic and join in a single-phase region above 1645°C. On the other hand, β-Y_2C_3 decomposes congruently into this phase. The phase diagram proposed by Storms is in good agreement with that of Carlson and Paulson but with two major exceptions. Storms found a higher C concentration at the low-C phase boundary of the hypocarbide phase and the Y_5C_6 phase, which exists in at least two crystal forms and decomposes at about 1527°C into β-Y_2C_3 and Y_2C. On the basis of the review of Gschneidner and Calderwood (1986), this new phase was regarded as $Y_{15}C_{19}$ with the $Sc_{15}C_{19}$-type tetragonal structure (Bauer and Nowotny 1971).

Figure 2 shows the finally refined Y–C phase diagram. The solvus data for the Y-rich end are taken from Carlson et al. (1974). The low-C phase boundary for the γ-(Y_3C) phase is midway between the boundary presented by Carlson and Paulson, and Storms. Above 900°C, the Y-rich γ phase has a cubic structure of the Fe_4N type. Below about 900°C, the ordering of C atoms creates the $CdCl_2$-type trigonal structure for Y_2C. The solidus line of the γ phase exhibits a maximum at 2000 ± 15° C at about 30 at.% C, and a minimum at 1805 ± 5°C near 54 at.% C. The $Y_{15}C_{19}$ phase decomposes into the cubic Fe_4N-type Y_3C phase and β-Y_2C_3 phase at 1527°C, and undergoes a polymorphic transformation at 1304°C. The α phase, for which no structural data are available, has a tetragonal $Sc_{15}C_{19}$-type structure. The Y_2C_3 phase forms from the γ phase solid solution at 1645°C and the $\alpha \rightleftarrows \beta$ transformation occurs at 1186°C. In addition, a metastable yttrium sesquicarbide with the bcc Pu_2C_3-type structure formed at a high temperature under a high pressure (Krupka et al. 1969a). The YC_2 phase melts at 2415°C, and the eutectic between YC_2 and C is placed at 2290°C, the mean of the acceptable data (Carlson and Paulson 1968, Storms 1971, Krikorian et al. 1967, Kosolapova and Makarenko 1964). The α-YC_2 phase has been observed to have tetragonal CaC_2-type structure. The high-temperature form was stated by Storms (1971) to be cubic, assuming that β-YC_2 has cubic CaF_2-type structure similar to the β-RC_2 phase for R = La, Ce, Eu, Tb, Lu. The $\alpha \rightleftarrows \beta$ transformation occurs at 1324°C, the average of the acceptable values (Krikorian et al. 1967, Adachi et al. 1976, Carlson and Paulson 1968, Storms 1971).

2.2. *General characteristics of phase diagrams of the light-lanthanide–carbon systems*

Except for the phase diagram of the lanthanum–carbon system, no complete phase diagram is available for the light-lanthanide–carbon systems. Only some data on the formation of the carbides have been reported. However, on the basis of these data, the general characteristics of the binary light-lanthanide–carbon phase diagram can be deduced.

In these systems, two types of rare earth carbides have been confirmed to exist; the rare earth dicarbides and the rare earth sesquicarbides. The existence of the monocarbides of cerium and praseodymium (Warf 1955, Brewer and Krikorian 1956, Dancy et al. 1962) has been discredited. It was found on the basis of an X-ray study (Spedding et al. 1958) that the earlier reported cerium monocarbide was most probably a solid solution of carbon in cerium; the lattice parameter reported by Brewer and Krikorian (1956) for the cerium monocarbide was identical with that of cerium metal saturated with carbon.

The sesquicarbides of cerium, neodymium and praseodymium were shown as forming peritectically from the melt, the corresponding rare earth dicarbide just like the lanthanum sesquicarbide (see fig. 1) at 1505°C [the average of the values reported by Stecher et al. (1964), Anderson et al. (1969), Paderno et al. (1969) and Kosolapova et al. (1971), discarding 1800°C of Stecher et al. (1964)], 1560°C of Kosolapova et al. (1971) and 1620°C of Paderno et al. (1969a, b). These compounds, except for the cerium sesquicarbide, exhibit a solid solubility range (Spedding et al. 1958) with different lattice parameters for the metal-rich and the carbon-rich phases. All of the lighter lanthanide sesquicarbides have the cubic Pu_3C_3-type structure with the $I\bar{4}3d$ space group (for Ce, Spedding et al. 1958, Atoji and Williams 1961, Anderson et al. 1968, Baker et al. 1971; for Pr and Nd, Spedding et al. 1958).

The dicarbides of cerium, neodymium and praseodymium as well as lanthanum crystallize congruently from the melt and form a eutectic mixture with carbon. The melting points of the dicarbides were given by Kosolapova and Makarenko (1964) as 2540 ± 100°C for the cerium dicarbide, and 2535 ± 100°C for the praseodymium dicarbide. For the neodymium dicarbide, the reported melting point temperatures range from 2207 to 2280 ± 15°C, whereas Krikorian et al. (1967) found the eutectic temperature to be 2275 ± 20°C. Because of this discrepancy and the lack of the available data, the melting temperature of the neodymium dicarbide would be certainly estimated to be greater than 2260°C. The analogous situation also occurs for the praseodymium dicarbide. The reported melting temperatures by Paderno et al. (1966) and Makarenko et al. (1965) were in the range from 2147 to 2160°C, which is about 100°C below the PrC_2–C eutectic temperature found by Krikorian et al. (1967). Thus, these data were discarded in the case of determining a reasonable value. The compositions for the eutectic of the light lanthanide dicarbide with carbon were not found in the literature. The temperatures of the LaC_2, CeC_2, PrC_2 and NdC_2–C eutectic reactions are at 2250, 2260 [the average of 2270 ± 20°C (Winchell and Baldwin 1967) and 2245 ± 20°C (Krikorian et al. 1967)], 2255 and 2075 ± 20°C (Krikorian et al. 1967), respectively.

Like the lanthanum dicarbide, the cerium dicarbide exists in two types of modification, the high-temperature β-CeC_2 having a cubic CaF_2-type structure with the space group Fm3m, and the room-temperature α-CeC_2 having a tetragonal structure of CeC_2 type with the space group I4/mmm. The $\alpha \rightleftarrows \beta$ transformation occurs at 1100 $\pm$ 20°C (Winchell and Baldwin 1967). Other reported values for this transformation temperature range from 1090 to 1107°C (Krikorian et al. 1967, McColm et al. 1973, Loe et al. 1976, Adachi et al. 1978). The transformation temperatures for the praseodymium dicarbide and the neodymium dicarbide have also been reported to be at 1130 ± 30°C and 1150 ± 20°C (the latter being the average of 1120 ± 10°C to 1164 ± 6°C), respectively, and their room-temperature modifications have been found to be of the tetragonal CeC_2-type structure. However, no information is available on the structure of their high-temperature modifications. Because the known high-temperature modifications of the dicarbides of lanthanum and cerium have the cubic CaF_2-type structure, with the Fm3m space group, it seems reasonable to expect β-PrC_2 and β-NdC_2 to follow this pattern. In addition, El-Makrini et al. (1980) prepared the compounds RC_6 (R = Ce, Pr, Nd, Sm, Eu, Yb) by compression of powder. However, the structures were not determined at that time.

According to the general characteristics of the light-lanthanide (La, Ce, Pr, Nd)–carbon phase diagrams, the hypothetical ideal phase diagram for misch metal–carbon has been drawn (Gschneidner and Calderwood 1986, fig. 3). They assumed that the ideal misch metal contains 27 at.% La, 48 at.% Ce, 5 at. % Pr, 16 at.% Nd, and 4 at.% other rare earths that behave as 2 at.% Gd and 2 at.% Y. The sesquicarbide of each of these lanthanides decomposes peritectically at temperatures that increase systematically with atomic number. By applying the method of Palmer et al. (1982) to calculate the misch metal properties from the reported data for the light-lanthanide–carbon systems involved, a peritectic reaction is calculated to occur at 1510°C, where the misch metal sesquicarbide decomposes to the melt and β-misch metal dicarbide, which melts at about 2425°C and undergoes an $\alpha \rightleftarrows \beta$ transformation at 1100°C. Below this temperature the misch metal dicarbide exists in the tetragonal CaC_2-type structure and above this temperature it transforms to the cubic CaF_2-type structure. The misch metal–carbon eutectic composition is not known, because no composition was given for any of the individual rare-earth–carbon systems. A eutectic temperature of about 2260°C was also estimated by Gschneidner and Calderwood (1986). As for the solid solubility region for the dicarbide and sesquicarbide of misch metal, the sesquicarbides of lanthanum, praseodymium and neodymium exhibit solid solubility while the cerium sesquicarbide does not, and the cerium is 48 at.% of the misch metal. The solid solubility region for the misch metal sesquicarbide was narrowed from that of the lanthanum–carbon phase diagram. The solid solubility for the misch metal dicarbide was suggested to be lower on the basis of only the data of the lanthanum dicarbide because no information is available about the Ce–C, Pr–C, and Nd–C systems.

2.3. *General characteristics of phase diagrams of the heavy-lanthanide–carbon systems*

Here, the so-called heavy lanthanides include the elements from samarium to lutetium, except for ytterbium and europium which behave like bivalent metals and have unique properties. For these heavy-lanthanide–carbon systems, no complete phase diagram was found, only some information about the formation and the crystal structure of the carbides is available. On the basis of these data the general characteristics of the phase diagrams of the heavy-rare-earth–carbon systems can be summarized. In this case the yttrium–carbon phase diagram may be regarded as the best prototype available for compounds of the heavy lanthanide systems with carbon.

The existence of the hypocarbides of the heavy lanthanides has been established (Spedding et al. 1958, Huber et al. 1973, Atoji 1981, Aoki and Williams 1979, Bacchella et al. 1966). Like the yttrium hypocarbides, the cubic tri-rare-earth carbides β-R_3C with the Fe_4N-type structure are the high-temperature forms of the heavy lanthanide hypocarbides, which can exist at room temperature in a metastable state and have a solid solubility range as reported for R = Gd (Spedding et al. 1958), Dy (Aoki and Williams, 1979) and Er (Atoji 1981). This form was obtained immediately from the melt. The stable room-temperature form has the approximate composition R_2C and the rhombohedral $CdCl_2$-type structure for R = Gd, Tb, Dy, Ho. For the erbium hypocarbide Er_xC, if x is close to 2, the cubic form transforms at lower temperatures to a trigonal $CdCl_2$-type structure, and if x is much different from 2, the cubic structure is retained at all temperatures (Atoji 1981). For the hypocarbide of thulium and lutetium, no report of a trigonal compound with the $CdCl_2$-type structure was found. There is some difference of opinion concerning the existence of the tri-samarium carbide between Haschke and Deline (1980) and earlier investigators. The latter reported the existence of Sm_3C (Spedding et al. 1958, Börner et al. 1972, Hackstein et al. 1971), while Haschke and Deline insisted that binary Sm_3C was not formed on the basis of their reinvestigation of phase equilibria in the Sm–C system, and suggested that the phases described by Hackstein et al. (1971) and Börner et al. (1972) actually were oxycarbides, because the colors, the C-to-Sm ratio, the structure, and the lattice parameters reported for the Sm_3C phase were nearly identical with the corresponding properties observed by them for $SmO_{0.5}C_{0.4}$. However, they could not explain the reason why the lattice parameter reported by Spedding et al. (1958) for Sm_3C is about 0.01 Å larger than that observed by Haschke and Deline for the oxycarbide.

In addition to the hypocarbides, the dicarbides of the heavy lanthanides were reported to exist in all the heavy-lanthanide–carbon systems. Until now, three forms of the dicarbide have been reported, the tetragonal CaC_2-type room-temperature structure, the cubic CaF_2-type high-temperature structure and the orthorhombic LuC_2-type structure. These structures formed under high pressure or during long annealing at high temperatures from α-RC_2 with the CaC_2-type structure for the systems of R = Tb, Dy, Y or R = Ho, Er, Tm (Krupka et al. 1968). From the viewpoint of phase equilibrium, these compounds of holmium, erbium and thulium with such an orthorhombic LuC_2-type structure should be the more stable phase with respect to the β-RC_2 phase at least at about 1150, 1305 and 900 to 1250°C, respectively.

The room-temperature forms of the heavy-lanthanide dicarbides possess the body-centered tetragonal CaC_2-type structure (Spedding et al. 1958) with the exception of the lutetium dicarbide, which has the low-symmetry orthorhombic structure with a large cell volume (Krupka et al. 1968) but was also designated as the α-RC_2 compound. These α-RC_2 compounds transform to the high-temperature cubic structure. These temperatures have been assessed by Gschneidner and Calderwood (1986) and a plot of the average transition temperature against the atomic number of the lanthanides has been given in combination with the data for the light rare earth dicarbides. As shown in fig. 3, it can be seen that the transition temperatures of the

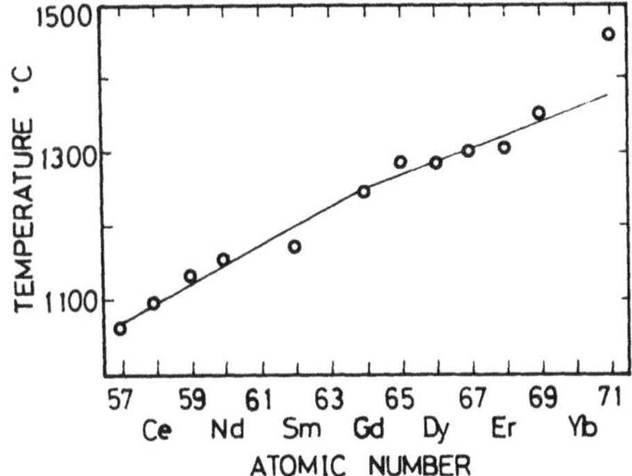

Fig. 3. The transformation temperatures of the rare earth dicarbides (Gschneidner and Calderwood 1986).

RC_2 compounds increase with increasing lanthanide atomic number, roughly following two straight lines which intersect at gadolinium, with the exception of the dicarbides of europium, ytterbium (with a mixed-valence state) and lutetium (α-LuC_2 has a different stable crystal structure from the other α-RC_2 compounds). The distinction between the light rare earths and the heavy rare earths may be noted from the YC_2 transformation temperature, which may be placed between ErC_2 and HoC_2 in fig. 3.

Krikorian et al. (1967) systematically investigated the phase relationship of the high-carbon portion of the lanthanide–carbon systems, including the $\alpha \rightleftarrows \beta$ transformation of the rare earth dicarbides, and provided structural data on the β form, which has the cubic KCN-type structure with Fm3m space group (Bowman et al. 1966). The lattice parameters of the β-dicarbides, LaC_2, TbC_2, LuC_2 and YC_2 (Bowman et al. 1967) have been determined. However, the existence of the β structured form for the β-dicarbides of Sm, Gd, Dy, Ho, Er, and Tm were established mainly by thermal analysis data on the $\alpha \rightleftarrows \beta$ transformation and their structures were reasonably assumed to be the same as the KCN-type cubic structure.

Krikorian et al. also reported the temperatures of the β-RC_2–C eutectic reactions for all the rare earths except for Eu but no eutectic composition was given (table 1). In addition, there are few other data to confirm or contradict their results.

Gschneidner and Calderwood (1986), in a systematic investigation with respect to the melting points of the rare earth dicarbides and the RC_2–C eutectic temperatures, treated the values reported by Russian investigators (Yupko et al. 1974, Kosolapova et al. 1971) as the melting points of the dicarbides of gadolinium, dysprosium and erbium, or as the eutectic temperatures of the corresponding RC_2–C eutectic reaction. Including these values in their calculation of the average values, the average eutectic temperatures of the RC_2–C eutectic (with R = Gd, Dy, Er) were given as is listed in table 1 together with the values reported by Krikorian et al. (1967). Compared with the data of Krikorian et al., the average values of Gschneidner and Calderwood vary in a rather regular manner; as a function of atomic number, however, the data treatment they made was based on some not very strict assumptions.

According to the data in table 1, the eutectic temperature increases with increasing atomic number of the light rare earth in the dicarbide, and reaches a maximum at

TABLE 1
Temperatures (°C) of the RC_2–C eutectic reaction for the rare earth dicarbides (Krikorian et al. 1967).

Dicarbide	RC_2–C Eutectic (°C)	
	Value of Krikorian et al. (1967)	Obtained from systematic studies (Gschneidner and Calderwood 1986)
LaC_2	2230 ± 20	2250
CeC_2	2245 ± 20	2260
PrC_2	2255 ± 20	2255
NdC_2	2275 ± 20	2260
SmC_2	2240 ± 30	2240
EuC_2	–	–
GdC_2	2315 ± 25	2280
TbC_2	2275 ± 20	2275
DyC_2	2290 ± 25	2260
HoC_2	2270 ± 20	2270
ErC_2	2255 ± 25	2260
TmC_2	2245 ± 35	2245
YbC_2	2215 ± 40	2215
LuC_2	2230 ± 20	2230
YC_2	2275 ± 25	2290

GdC_2, then decreases as the atomic number of the heavy rare earth in the dicarbide increases. Gschneidner and Calderwood (1986) suggested that these data follow two straight lines that connect the eutectic temperatures of LaC_2 and GdC_2, and GdC_2 and LuC_2, respectively. The values for SmC_2 and YbC_2 deviated from the straight lines of the eutectic temperature against the atomic number and could be attributed to the variable valence of samarium and ytterbium.

Except for YC_2, whose melting point has been determined in extensive studies (Carlson and Paulson 1968), there are not enough reliable data for the melting temperatures of the heavy rare earth dicarbide systems to discern any trends.

In addition to the hypocarbides and the dicarbides, there exist in the heavy-rare-earth–carbon systems the sesquicarbides and other carbides near 60%, as reported for the yttrium–carbon system, which has the same pseudocubic tetragonal cell as $Sc_{15}C_{19}$ (Jedlicka et al. 1971).

The sesquicarbides of the lanthanides and yttrium have the body-centered cubic Pu_2C_3-type structure, with the space group $I\bar{4}3d$. For the heavy-rare-earth–carbon system the sesquicarbides can be prepared in the b.c.c. form either directly by arc melting (Spedding et al. 1958) for R = Sm, Gd, Tb, Dy, Ho, or by using a high-pressure, high-temperature technique for R = Er through Lu (Krupka and Krikorian 1970, Novokshonov 1980) and yttrium (Krupka et al. 1969a). Correspondingly, the low-symmetry compounds with the $R_{15}C_{19}$ composition and the $Sc_{15}C_{19}$-type structure exist in the latter systems as the phase stable at room temperature. They can also be obtained directly by arc melting (Spedding et al. 1958). In the case of Ho_2C_3, it was found (Spedding et al. 1958) that this phase is dimorphic and occurs in both a cubic structure of the Pu_2C_3 type and a lower-symmetry form that has not been indexed on the basis of a $Sc_{15}C_{19}$-type tetragonal cell. It is now obvious that this lower-symmetry form observed by Spedding et al. (1958) was the $Ho_{15}C_{19}$ phase. As a result, it can be seen that with increasing lanthanide atomic number the stability of the sesquicarbides with the cubic Pu_2C_3-type structure gradually decreases from Sm through Lu. Another form begins to appear with the low-symmetry $Sc_{15}C_{19}$-type structure when the lanthanide is Ho. At the same time, the cubic form becomes metastable at room temperature and can be obtained only at high temperatures under a high pressure.

In general, the sesquicarbides of Sm, Gd, Tb, Dy and Ho have the cubic Pu_2C_3-type structure and exhibit a range of solid solubility (Spedding et al. 1958), like the yttrium sesquicarbide. For example, the lattice parameter of the C-rich Gd sesquicarbide is 0.22% larger than that of the Gd-rich compound (Huber et al. 1973). No melting information was found for these heavy lanthanide sesquicarbides. The metastable cubic sesquicarbides of Er through Lu can be formed only by a high-temperature, high-pressure technique. These values are 30–90 kbar and 1200–1400°C for the erbium sesquicarbide (Krupka and Krikorian 1970, Novokshonov 1980), and 40–90 kbar and 1100–1500°C for the lutetium sesquicarbide (Vereshchagin et al. 1978, Novokshonov 1980).

Several investigators showed that the "sesquicarbides" of Ho, Er, Tm and Lu as reported by Spedding et al. (1958) are an intermediate $R_{15}C_{19}$ phase with a carbon content ranging from about 55 to about 58 at.% instead of a true sesquicarbide (Bauer and Nowotny 1971, Bauer 1974, Bauer and Bienvenu 1980). They determined the lattice parameters for $Er_{15}C_{19}$ (Bauer 1974) and $Lu_{15}C_{19}$ (Bauer and Bienvenu 1980) as well as for $Sc_{15}C_{19}$ and $Yb_{15}C_{19}$ (Hájek et al. 1984a–d). At present, the crystal structures of $Ho_{15}C_{19}$ and $Tm_{15}C_{19}$ have also been indexed on the basis of a $Sc_{15}C_{19}$-type tetragonal pattern (Bauer and Ansel 1985).

Summarizing the phase relationship in the heavy-lanthanide–carbon systems and the general characteristics with respect to formation of the carbides, it can be concluded that the phase diagram of the yttrium–carbon system is a good prototype of the heavy-lanthanide–carbon systems, which have not yet been studied in detail. In addition, on the basis of the yttrium–carbon phase diagram the unknown information about the R–C phase diagrams could be deduced.

2.4. *Phase relationship and formation of carbides in the Eu–C and Yb–C systems*

Many investigators have studied the phase relationship and the formation of the carbides in both the Eu–C and the Yb–C systems, but no complete phase diagram was reported although some data are available on carbides that are formed in the systems. Although europium and ytterbium exhibit variable-valence tendencies, the properties and lattice parameters of their carbides do not follow the systematic variation encountered between the individual lanthanide–carbon system. In addition, an additional new phase, RC_6 (R = Eu and Yb) forms in the two systems, which is reported to be hexagonal with a $P6_3/mmc$ space group (Guerard and Hérold 1975, El-Makrini et al. 1980).

The carbides of europium have been prepared by either reducing Eu_2O_3 (high purity) with spectrographic grade graphite or synthesizing the elements, europium and graphite, with a high purity under vacuum or an inert atmosphere (Gebelt and Eick 1964, Sakai et al. 1982a, b). Despite the various methods, the structural data from the samples prepared by these methods were found to be identical.

The structure of the hypocarbide EuC_{1-x} has been identified to be cubic NaCl-type, over a range of solid solubility (Laplace and Lorenzelli 1970). The lattice parameters of both the Eu-rich phase and the C-rich phase are in good agreement with those of the cubic Fe_4N-type R_3C structure.

The existence of the europium sesquicarbide has been established by X-ray diffraction (Colquhoun et al. 1972), and a cubic Pu_2C_3-type structure was also determined with a lattice parameter, following the systematic variation of that of the neighboring sesquicarbides.

The europium dicarbide was reported to have an anomalous X-ray pattern, more complex than that of LaC_2, CeC_2, or NdC_2 (Faircloth et al. 1968), which has been tentatively indexed as a body-centered tetragonal large unit cell with $a = 12.15$ Å and $c = 7.290$ Å. Gebelt and Eick (1964) and Sakai et al. (1982) examined the crystal structure of the EuC_2 samples prepared by two different methods and found that the samples consisted of a body-centered tetragonal phase and a second phase. A larger amount of the phase could be obtained in a sample containing a lower C-to-Eu ratio and heated to a somewhat lower temperature (Gebelt and Eick 1964) and could not be removed by annealing (Sakai et al. 1982a, b). This body-centered tetragonal phase has been identified to be the CaC_2-type with larger lattice parameters than that of SmC_2 and GdC_2, exhibiting the unique behavior of a variable valence. The second phase was determined to be orthorhombic with the lattice parameters $a = 8.76$ Å, $b = 11.23$ Å and $c = 7.19$ Å, but its composition was not reported (Gebelt and Eick 1966a) and no other structural data were provided.

Sakai et al. (1982a, b) found that the bct α-EuC_2 form transforms into the fcc CaF_2-type β-EuC_2 at 350°C, where Krikorian et al. (1967) had observed a change at 335°C. Owing to the high vapor pressure of europium over EuC_2, the EuC_2–C eutectic was not determined (Krikorian et al. 1967).

Formation of the compounds in the ytterbium–carbon system over a composition range of 25 to 70 at.% C has been studied by an X-ray diffraction and chemical analysis (Haschke and Eick 1970a, b). The carbides were prepared by synthesizing 99.9% pure ytterbium and the CP grade graphite which was degassed under vacuum at 1800°C, and annealed at 800–1100°C and under 10^{-6}–10^{-7} Torr for 1 h before quenching. Four phases have been identified: tetragonal $YbC_{2.0}$ ($a = 3.639 \pm 0.003$ and $c = 6.110 \pm 0.008$ Å), monoclinic $YbC_{1.25+y}$ with $0 \leqslant y \leqslant 0.16$ ($a = 7.070 \pm 0.005$, $b = 7.850 \pm 0.004$, $c = 5.623 \pm 0.005$ Å and $\beta = 90.99 \pm 0.09°$, for samples quenched from 1450 K), $YbC_{0.95}$ and dimorphic $YbC_{0.50+z}$ (fcc, $a = 5.001 + 0.003$ Å and rhombohedral, $a = 6.167$ Å and $\alpha = 33.33°$). A linear variation of both the carbon-rich $YbC_{1.25+y}$ phase boundary and the monoclinic b parameter with temperature has been observed. The carbide, called $YbC_{0.95}$, has a complex structure that could not be indexed on either the orthogonal or the trigonal symmetry.

Tetragonal YbC_2 has a CaC_2-type structure similar to the room-temperature form of the other RC_2 phases. The YbC_2–C eutectic occurs at 2215 ± 40°C but the tetragonal⇄cubic phase transformation for the YbC_2 carbide was not observed by thermal analysis (Krikorian et al. 1967). However, a low-symmetry ytterbium dicarbide had been prepared by annealing the tetragonal form at 1000°C for 100 h and is isostructural with the arc-melted LuC_2. However, the yield was too low to determine the lattice parameter (Krupka et al. 1968). As a result, this low-symmetry form might be the high-temperature equilibrium form of the ytterbium dicarbide, at least at 1000°C.

The existence of the ytterbium sesquicarbide with the cubic Pu_2C_3-type structure was not established (Spedding et al. 1958) under common conditions. However, this type of ytterbium carbide can be prepared at high pressures, like the other high-pressure heavy-lanthanide sesquicarbides (Krupka and Krikorian 1970).

The carbide, $YbC_{1.25+y}$, was reported by Haschke and Eick (1970a, b) to be of the monoclinic structure, and its cell volume decreases with the enhancement of the preparation temperature. However, Bauer and Bienvenu (1980) investigated alloys in this composition range of the Yb–C system and concluded that the phase that exists in this range has the $R_{15}C_{19}$ stoichiometry and the tetragonal $Sc_{15}C_{19}$-type structure. Their samples were prepared by synthesizing 99.99% pure carbon and ytterbium metal with 99.9% purity at 2000–2100°C under an inert atmosphere and annealing at 1650°C for an hour and subsequently quenching. These authors suggested that the differences between the two groups can probably be explained by the different annealing and cooling procedures. It is not clear whether the two carbides described by the two groups are two different equilibrium phases, or two modifications of a compound, or whether one of these carbides is metastable.

The $YbC_{0.95}$ carbide reported by Haschke and Eick (1970a) was not verified to exist in the ytterbium–carbon system by other workers. Its crystal structure was not determined by these authors. In contrast, the two forms of the $YC_{0.5\pm z}$ compound, the face-centered cubic Fe_4N-type form and the rhombohedral CdCl-type form, have been well determined. Their lattice parameters are in good agreement with the systematic variation between those of the other heavy lanthanide hypocarbides, although the composition range for this diphase mixture was not determined. The only information about this material was provided by Spedding et al. (1958). They reported that a carbon-rich Yb_3C compound exists.

In addition to these carbides mentioned above, the intercalation compounds of europium or ytterbium with graphite have been found. El-Makrini et al. (1980) prepared these intercalation compounds by directly reacting the Eu or Yb metal vapor on graphite under vacuum. The yields of intercalation with Eu and Yb are 25% at 500°C after 20 days and 18% at 450°C after 20 days, respectively; the cores of the samples remain pure graphite. The structure of the first-stage compounds EuC_6 and YbC_6 was determined with oriented samples by a separated study of the different families of reflection. The hexagonal unit cell of these phases belongs to the $P6_3/mmc$ space group. The parameters are $a = 4.314 \pm 0.003$ Å and $c = 9.745 \pm 0.008$ Å (EuC_6) and $a = 4.320 \pm 0.004$ Å and $c = 9.147 \pm 0.004$ Å (YbC_6).

In comparison with the carbide-forming characteristics of the other heavy-lanthanide–carbon systems, the europium–carbon and ytterbium–carbon systems behave like their neighboring systems. In particular, the types and structures of these carbides are identical. The main difference is the larger lattice parameter of the europium and ytterbium dicarbides compared to their neighboring rare earth dicarbides. Probably, also with respect to phase transformations and phase relationships there exist many differences, but little information has been reported.

2.5. *Formation of the scandium carbides*

No phase diagram of the scandium–carbon system is available. Many investigators have prepared and studied the scandium carbides in this system showing that four types of the carbides exist and that two of them have novel stoichiometric compositions and structures, which have not been reported previously for other rare-earth–carbon systems. The two scandium carbides are the bcc anti-Th_3P_4-type Sc_4C_3 carbide with a $I\bar{4}3d$ space group (Krikorian et al. 1969a) and the Ge stabilized cubic $Sc_{13}C_{10}$ carbide (Krikorian et al. 1969b).

Rassaerts et al. (1967) reported that samples with 85 to 90 at.% Sc were heterogeneous and consisted of Sc and a face-centered cubic phase. Samples containing 60 to 80 at.% Sc had a face-centered cubic structure and were practically single-phase. Atoji and Kikuchi (1968) examined their data on the basis of the calculation of intensities in the X-ray pattern, and concluded that the data for $Sc_{2\sim 3}C$ were more consistent with a trigonal Y_2C structure than with the face-centered cubic structure. Krikorian et al. (1969a) confirmed experimentally the existence of $Sc_{2\sim 3}C$, as reported by Rassaerts et al. (1967), and the trigonal Sc_2C structure, as calculated by Atoji and Kikuchi (1968).

Rassaerts et al. (1967) first reported the existence of a carbon-rich phase, but did not succeed in indexing its complex structure. Krikorian et al. (1969a) also found the phase in the composition range of $ScC_{1.2}$ to $ScC_{1.3}$. Jedlicka et al. (1971) determined the structure of the $ScC_{1.2\sim 1.3}$ single crystal on the basis of X-ray structural analysis and found a new tetragonal structure with two $Sc_{15}C_{19}$ formula units per cell. Hájek et al. (1984a–d) gave its lattice parameters and reported a decreasing cell volume with increasing preparation temperature for the compound $Sc_{15}C_{19}$. Recently, it has become well known that the $Sc_{15}C_{19}$-type rare earth carbides exist widely in the heavy-rare-earth–carbon systems, where the heavy rare earth is yttrium, holmium, erbium, thulium, ytterbium or lutetium.

The Sc_4C_3 carbide was prepared by the arc melting method with a high-pressure, high-temperature belt-type press (Krikorian et al. 1969a). This phase has a body-centered cubic structure with the $I\bar{4}3d$ space group, and possesses a narrow range of homogeneity. This phase remained stable up to 1500°C upon heating and changed into another unknown phase at 1600°C.

The body-centered cubic scandium carbide with the composition approximated to $Sc_{13}C_{10}$ (≈ 43.5 at.% C) has been prepared by annealing the arc-melted Sc_4C_3 compound with a minor addition of germanium (Krikorian et al. 1969b). The lattice parameters have been reported. The authors believed that $Sc_{13}C_{10}$ is a binary carbide stabilized by a minor addition of germanium that does not enter into the crystal lattice of the $Sc_{13}C_{10}$ compound.

Rassaerts et al. (1967) also reported a nearly pure Pu_2C_3-type structural phase with 42 at.% C. However, no information about the lattice parameters was provided, and the existence of this compound remains to be confirmed.

In addition, Rassaerts et al. (on the basis of their observations) concluded that a series of peritectic reactions take place in the scandium–carbon system, but they did not give the data about peritectic reactions. Only information about phase relationships in the scandium system was provided by Krikorian et al. (1969a) on the basis of thermal analysis cooling curves, showing that the eutectic reaction of the $Sc_{15}C_{19}$ phase and free carbon occurs at 1780°C. Gschneidner and Calderwood (1986) suggested that in the region between 55 and 60 at.% C, the phase relationship in the Sc–C phase diagram may be quite similar to those observed in the Y–C system.

2.6. *Summary of the formation of the binary rare earth carbides*

According to the reported information on the binary rare-earth–carbon phase diagrams, a survey on formation of the binary rare earth carbides has been made and is given in table 2.

2.6.1. *Decomposition of the rare earth sesquicarbides*

It has been reported that the yttrium sesquicarbide decomposes congruently at 1645°C (Carlson and Paulson 1968) and 1652°C (Storms 1971) for the alloys with 58 at.% C and the stoichiometric composition, respectively. These values are more reliable than the value of 1800°C reported by Paderno et al. (1969b), Kosolapova et al. (1971) and Yupko et al. (1974) because the former are based on extensive phase diagram studies. In contrast to the decomposition of Y_2C_3, the decomposition of La_2C_3 (Spedding et al. 1959) and Ce_2C_3 (Anderson et al. 1969, Stecher et al. 1964) is peritectic. Russian investigators (Paderno et al. 1969b, Kosolapova et al. 1971, Yupko et al. 1974) also reported the “melting points” of La_2C_3, Ce_2C_3, Pr_2C_3 and Nd_2C_3, in good agreement with the decomposition temperatures obtained by Spedding et al. (1959) and Anderson et al. (1969). However, they did not point out whether it is congruent or incongruent melting. In addition, Stecher et al. (1964) also reported the temperature range of 1800 to 1900°C as the temperatures of peritectic decomposition for the cerium sesquicarbide, which is too high to be accepted for calculation of the average value. On the basis of the acceptable data, the average decomposition temperatures were calculated and listed in table 3 together with these acceptable original values, and a plot of the average decomposition temperature against the atomic number of the light lanthanides was given (fig. 4). It can be seen that there is a good linear relationship between the decomposition temperature of the sesquicarbide and the atomic number of the light lanthanides.

2.6.2. *Melting points and the tetragonal–cubic transformations of the rare earth dicarbides*

2.6.2.1. *Melting point of rare earth dicarbides.* As shown in figs. 1 and 2, the dicarbides of lanthanum and yttrium have a homogeneous phase region at high temperatures. A solid solution was obtained in the case of yttrium. Therefore, it is obvious that their melting temperatures vary with the composition of the solid solution, and only at a certain composition could congruent melting occur. Any deviation in composition of the carbide will give rise to a dramatic decrease in the

TABLE 2
Binary rare earth carbides and their lattice parameters (Å)[a].

	Type of carbide									
	R_3C	α-R_2C	R_4C_3	$R_{13}C_{10}$	$R_{15}C_{19}$	R_2C_3	α-RC_2	β-RC_2	α-RC_2	RC_6
Crystal structure	Cubic Fe_4N	Rhombohedral $CdCl_2$	bcc anti-Th_3P_4	Cubic	Tetragonal $Sc_{15}C_{19}$	bcc Pu_2C_3	Tetragonal CaC_2	Cubic KCN[c]	Orthorhombic LuC_2	Hexagonal EuC_6
Space group	Fm3m	R3m	I43d		P42$_1$d	I43d	I4/mmm	Fm3m		P6$_3$/mmc
Scandium		$a =$ 5.5753 $\alpha = 33.53 \pm 3°$	$a = 7.207$	$a = 8.526$	$a = 7.503$ $c = 15.00$					
Rare Earth										
Lanthanum						$a = 8.805$–8.815	$a = 3.935$ $c = 6.571$	$a = 6.011$		
Cerium						$a = 8.445$–8.447	$a = 3.877$ $c = 6.485$	$a = 5.939$		Possible
Praseodymium						$a = 8.573$–8.607–	$a = 3.847$ $c = 6.430$	$(\alpha \rightleftarrows \beta)$		Possible
Neodymium						$a = 8.521$–8.548	$a = 3.817$ $c = 6.391$	$(\alpha \rightleftarrows \beta)$		Possible
Samarium	$a = 5.172$					$a = 8.399$–8.426	$a = 3.767$ $c = 6.324$	$(\alpha \rightleftarrows \beta)$		Possible
Europium	$a = 5.145$–5.141					$a = 8.368$	$a = 4.082$[b] $c = 6.701$	$a = 5.961$		$a = 4.314$ $c = 9.745$
Gadolinium	$a = 5.126$	$a =$ 6.315 $\alpha = 33.57°$				$a = 8.322$–8.343	$a = 3.718$ $c = 6.275$	$(\alpha \rightleftarrows \beta)$		
Terbium	$a = 5.107$	$a =$ 6.359 $\alpha = 32.73°$				$a = 8.243$–8.262	$a = 3.690$ $c = 6.216$	$a = 5.691$	$a = 13.45$ $b = 27.65$ $c =$ 7.130 (HP)	

Dysprosium	$a = 5.001$– 5.079	$a = 6.312$ $\alpha = 32.99°$		$a = 8.198$– 8.215	$a = 3.669$ $c = 6.171$	Existence	$a = 13.55$ $b = 26.97$ $c = 7.100$ (HP)	
Yttrium	$a = 5.102$– 5.127	$a = 6.339$ $\alpha = 33.15 \pm 7°$	$a = 7.940$ $c = 15.88$	$a = 8.191$– 18.248 (HT, HP)[f]	$a = 3.664$ $c = 6.169$	$(\alpha \rightleftarrows \beta)$	$a = 13.42$ $b = 27.65$ $c = 7.310$ (HP)	
Holmium	$a = 5.061$	$a = 6.248$ $\alpha = 33.07°$	Existence	$a = 8.172$– 8.175	$a = 3.648$ $c = 6.144$	$(\alpha \rightleftarrows \beta)$	$a = 13.07$ $b = 27.02$ $c = 7.530$ (1155°, 100–270 h)	
Erbium	$a = 5.034$	Existence	$a = 7.989$ $c = 15.79$	$a = 8.137$ (HT, HP)	$a = 3.619$ $c = 6.097$	$(\alpha \rightleftarrows \beta)$	$a = 13.28$ $b = 27.24$ $c = 7.020$ (1150°, long time)	
Thulium	$a = 5.016$		Existence	$a = 8.083$– 8.097 (HT, HP)	$a = 3.602$ $c = 6.040$	$(\alpha \rightleftarrows \beta)$	$a = 13.25$ $b = 26.55$ $c = 7.360$ (900–1250°)	Possible
Ytterbium[c]	$a = 5.001$– 4.993	$a = 6.167$ $\alpha = 33.33°$	$a = 7.906$ $c = 15.54$	$a = 8.050$– 8.073 (HT, HP)	$a = 3.638$ $c = 6.112$		Existence	$a = 4.320$ $c = 9.147$
Lutetium	$a = 4.965$		$a = 7.873$ $c = 15.52$	$a = 8.020$– 8.036 (HT, HP)	$a = 3.563$[d] $c = 5.964$	$a = 5.505$	$a = 12.40$ $b = 27.30$ $c = 6.870$	

[a] The lattice parameters given by Gschneidner and Calderwood (1986).
[b] Other structural parameters have been reported: tetragonal structure, $a = 12.17$ Å, $c = 7.29$ Å (Faircloth et al. 1968).
[c] A possible phase was reported, $a = 7.070$ Å, $\beta = 90.99 \pm 9°$, $b = 7.850$ Å, $c = 5.623$ Å (Haschke and Eick 1970a).
[d] Determined by neutron diffraction (Atoji 1961).
[e] Bowman et al. (1968).
[f] HP, high pressure; HT, high temperature.

TABLE 3
Decomposition temperature of the rare earth sesquicarbides.

Compound	Decomposition temperature (°C)		
	Reported values		Mean value
La_2C_3	1415 ± 3	(Spedding et al. 1959); 1430 ± 20 (Paderno et al. 1969b)	1423
Ce_2C_3	1450	(Anderson et al. 1969); 1530 ± 30 (Paderno et al. 1969b)	1490
Pr_2C_3	1560	(Kosolapova et al. 1971)	1560
Nd_2C_3	1620 ± 30	(Paderno et al. 1969b)	1620
Y_2C_3	1645	(Carlson and Paulson 1968); 1652 (Storms 1971)	1648

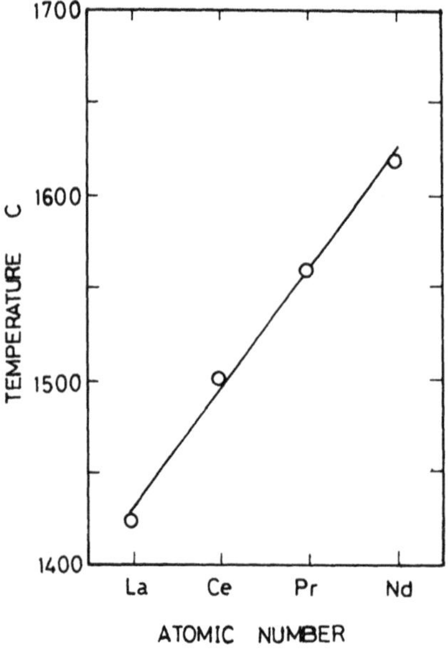

Fig. 4. Decomposition temperature of the light lanthanide sesquicarbides (Gschneidner and Calderwood 1986).

melting temperature because the liquidus falls off sharply on the metal-rich side of the dicarbide. For other rare earth dicarbides, there is probably identical behavior. As a result, it is not surprising that certain reported melting temperatures of the rare earth dicarbides were lower than the eutectic temperature reported by Krikorian et al. (1967) for the corresponding RC_2–C eutectic.

On the basis of a comprehensive study for the lanthanum–carbon system and the yttrium–carbon system, Spedding et al. (1959) and Carlson and Paulson (1968) determined the congruent melting temperature of LaC_2 and YC_2 as 2356 ± 25 and 2416 ± 25°C, respectively. Several Russian publications (Makarenko et al. 1965, Paderno et al. 1966, Yupko et al. 1974, Kosolapova et al. 1971) gave melting temperatures for PrC_2 ranging from 2147 to 2160°C, about 100°C below the eutectic temperature 2255 ± 60°C reported by Krikorian et al. (1967) for the PrC_2–C eutectic.

The earlier work of Kosolapova and Makarenko (1964) listed 2438, 2540, and 2535°C (each with error limits of ±100°C) as the melting temperatures of LaC_2, CeC_2, and PrC_2, respectively, but exhibited contradictory data for PrC_2. However, as is described above, the inconsistency in the melting temperature of PrC_2 might be explained as the difference in composition of the PrC_2 samples used in various measurements, and the reported higher temperatures could be regarded as the congruent melting point of PrC_2.

Gschneidner and Calderwood (1986) discarded some of the lower temperatures reported for the melting points of PrC_2 and regarded other reported melting temperatures for several other RC_2 compounds as eutectic temperature data. They included in their calculation of the average values for these eutectic temperatures only four data, 2360 ± 25, 2420 ± 180, 2540 ± 100 and 2415 ± 25°C, as the mean values of the melting points of LaC_2, CeC_2, PrC_2 and YC_2, respectively. Obviously, such a data evaluation is open to question.

These data, discarded or regarded as the eutectic temperature by Gschneidner and Calderwood (1986), include the melting temperatures for TbC_2 (2100 ± 60°C) (Yupko et al. 1974), for TmC_2 (2180°C) (Kosolapova et al. 1971) and for PrC_2 (from 2147 to 2160°C) (Makarenko et al. 1965, Paderno et al. 1966, Yupko et al. 1974, Kosolapova et al. 1971), as well as for CeC_2 (2250 ± 20°C) and NdC_2 (2290 ± 10°C) from Faircloth et al. (1968), for NdC_2 (2207°C) from Makarenko et al. (1965), for NdC_2 (2260 ± 60°C), for GdC_2 (2260 ± 60°C), for DyC_2 (2250 ± 60°C) and for ErC_2 (2280 ± 60°C) from Yupko et al. (1974), for NdC_2 (2207°C), for GdC_2 (2265°C), for DyC_2 (2245°C) and for ErC_2 (2230°C) from Kosolapova et al. (1971); and for YC_2 (2300 ± 50°C) from Kosolapova and Makarenko (1964). From the point of solid solution melting, the composition of the alloys could be estimated from these data.

2.6.2.2. *Tetragonal–cubic transformation of rare earth dicarbides.* Several investigators systematically studied the tetragonal–cubic transformation of the rare earth dicarbides and gave the transformation temperatures. No data were found for the transition temperature of ScC_2 and YbC_2. In the scandium–carbon system there is no dicarbide (Rassaerts et al. 1967, Jedlicka et al. 1971) and in the ytterbium–carbon system the tetragonal–cubic transformation of the dicarbide was not detected by thermal analysis although the tetragonal lattice parameters have been provided for YbC_2.

Krikorian et al. (1967) and Loe et al. (1976) investigated this transformation by thermal analysis measurements for twelve and nine rare earth dicarbides, respectively. Their data have been listed in table 4, together with other acceptable results (McColm et al. 1973, Bowman et al. 1967, Adachi et al. 1974, 1976, 1978, Winchell and Baldwin 1967, Sakai et al. 1982a, b, Carlson and Paulson 1968, Storms 1971). Paderno et al. (1969a) measured the electrical resistivity, the thermal e.m.f. and the thermal expansion as a function of temperature for the dicarbides of eleven rare earths. They determined the onset temperature of the tetragonal–cubic transformation for LaC_2 (950 ± 10°C), CeC_2 (975 ± 25°C), PrC_2 (980 ± 45°C), NdC_2 (1005 ± 70°C), SmC_2 (1070 ± 85°C), GdC_2 (1100 ± 85°C), TbC_2, DyC_2, and ErC_2 (each 1120 ± 80°C), TmC_2 (1172 ± 65°C), and YC_2 (1090 ± 80°C). Their values are considerably below those given in table 4. Another value reported by Spedding et al. (1959) on the basis of electrical resistivity measurements for the $\alpha \rightleftarrows \beta$ transformation in LaC_2, 1800°C, is too high to be regarded as the temperature in question, and should be responsible for some other phenomena at high temperatures (Krikorian et al. 1967).

Gschneidner and Calderwood (1986) calculated the average value of the $\alpha \rightleftarrows \beta$ transition temperature for each individual rare earth dicarbide on the basis of the data listed in table 4. A plot of these averaged values against the atomic number of the lanthanides has been given in fig. 3. As shown in fig. 3, the transformation temperatures for the lanthanide dicarbides, except for those of Pm, Eu, Yb and Lu, lie close to two straight lines, which connect the transformation temperatures of LaC_2 with GdC_2 as well as of GdC_2 and HoC_2. The lanthanides could be reasonably divided into two groups, the lighter and the heavier, in good agreement with the dependence of the RC_2–C eutectic temperature with the atomic number of the lanthanides.

2.6.3. *Metastable phases and the effect of pressure*

As indicated above, the rare earth sesquicarbides of lanthanum through dysprosium have a Pu_2C_3-type cubic structure after arc melting, while for the heavier lanthanides, erbium through lutetium and yttrium, a compound with lower than cubic symmetry forms after synthesizing, which was identified to be isotropic with a tetragonal $Sc_{15}C_{19}$-type structure (Bauer and Nowotny 1971). In the correspondingly prepared samples for holmium, the two forms of compounds mentioned above coexist. This fact showed that the rare earth sesquicarbides could be stable only for rare earths lighter than holmium. For rare earths heavier than holmium, there are no stable sesquicarbides at ambient temperature and pressure.

Krupka et al. (1969a, b) first prepared the body-centered cubic yttrium sesquicarbide with the Pu_2C_3-type structure by using a high-pressure, high-temperature technique. They reacted the alloys with about 38 to 66 at.% C at 15 to 25 kbar at 1200 to 1500°C for 2 to 6 min, and obtained bcc yttrium sesquicarbide in all the alloys except those closest to the Y_2C and YC_2 compositions. The highest yields were obtained from samples in the composition range from 55.9 to 59.7 at.% C. Subsequently, Krupka and Krikorian (1970) applied this technique to other heavy lanthanide carbides and prepared bcc Er_2C_3, Tm_2C_3 and Yb_2C_3 compounds successfully. Other investigators also reported the synthesis of cubic Pu_2C_3-type Y_2C_3 (Vereshchagin et al. 1979) and Lu_2C_3 (Vereshchagin et al. 1978, Novokshonov 1980) compounds under 30 to 90 kbar at 1100 to 1700°C and under 40 to 90 kbar at 1100 to 1500°C. These synthesized R_2C_3 compounds are unstable and the cubic Pu_2C_3-type structure will be destroyed during high-temperature annealing at ambient pressure (Krupka et al. 1969).

In the heavier lanthanides, such as terbium, dysprosium and yttrium, there also exist the high-pressure dicarbides with orthorhombic LuC_2-type structure. These were prepared from the tetragonal dicarbides with CaC_2-type structure by the high-pressure technique (Krupka et al. 1968). For the dicarbides of holmium through ytterbium, the orthorhombic LuC_2-type compounds can be formed only by direct annealing of the corresponding tetragonal dicarbides at temperatures ranging from

TABLE 4
Tetragonal to cubic transformation temperature of rare earth dicarbides (°C).

Compound	Krikorian et al. (1967)	Loe et al. (1976)	McColm et al. (1973)	Bowman et al. (1968)	Adachi et al.			Other authors
					(1974)	(1976)	(1978)	
LaC_2	1060	1075	995	1060	1081 ± 2			
CeC_2	1090	1095	1090	1090			1107 ± 2	1100 ± 10 (Winchell and Baldwin 1967)
PrC_2	1135	1140	1100	1130				
NdC_2	1150	1165	1150	1150	1159 ± 2			
SmC_2	1170							
EuC_2								350 (Sakai et al. 1982b)
GdC_2	1270	1270	1218	1365	1247 ± 2			
TbC_2	1285				1291 ± 2			
DyC_2	1295	1300	1250	1290	1318 ± 2			
HoC_2	1305	1320	1280	1300				
ErC_2	1325	1330	1275	1320	1327 ± 2			
TmC_2	1355							
LuC_2	1500	1450	1390	1500				
YC_2	1320					1329 ± 2		1327 (Storms 1971) 1320 (Carlson and Paulson 1968)

900 to 1250°C for a period of 100 to 270 h. For example, the transformation from the tetragonal to the orthorhombic form for HoC_2 and ErC_2 occur at 1155 ± 10 and 1305 ± 10°C, respectively. Their observations showed that when progressing from the heavier to the lighter lanthanide dicarbides, higher pressures, longer reaction-time periods, and lower temperatures are required. The orthorhombic LuC_2-type dicarbides are stable at ambient pressure and temperature but need to be stored in an inert atmosphere.

2.6.4. *On the existence of the rare earth carbides R_5C_6 and RC*

2.6.4.1. *R_5C_6 Compounds.* In the yttrium–carbon system, the Y_5C_6 compound has been reported by Storms (1971), indicating that this phase exists in at least two crystal forms and decomposes at about 1527°C into β-Y_2C_3 and Y_2C. The polymorphic transition temperature of the Y_5C_6 compound has been determined to be about 1304 ± 20°C on the basis of a break in the slope of the activity curve. Y_5C_6 formed readily at about 1527°C on cooling from the corresponding compositions of the hypocarbides. X-ray patterns in the 54.5 to 57.5 at.% C region showed no lines attributable to known phases. No crystal structural data were presented for either form of this compound. Although Bauer and Nowotny (1971) found a compound with 55.9 at.% C to be isotypic with tetragonal $Sc_{15}C_{19}$, it is not possible to be certain that this $Y_{15}C_{19}$ compound is the low-temperature form of Y_5C_6.

At present, Bauer and Ansel (1985), in investigating the preparation of the $Ho_{15}C_{19}$ compound, discovered the Ho_5C_6 compound, which is isostructural with Y_5C_6 as shown by the nearly identical X-ray patterns. They claimed that this pure phase could be synthesized and they have succeeded in indexing it on the basis of a tetragonal cell of dimensions $a = 8.080$ Å and $c = 19.598$ Å. In comparison with the $Ho_{15}C_{19}$ structure, the c parameter of Ho_5C_6 corresponds to four subsequent metal octahedra along this axis.

2.6.4.2. *Monocarbides.* Brewer and Krikorian (1956) and Dancy et al. (1962) had reported the monocarbide of cerium, but its existence has been discredited. Spedding et al. (1958) found that the lattice parameter reported by Brewer and Krikorian (1956) for CeC was identical with that of Ce metal saturated with C, and the reported CeC was most likely a solid solution of C in Ce. Anderson et al. (1969) re-examined the binary cerium–carbon system, and gave no evidence for the existence of any phase structurally based on monoatomic (methanide) anions.

Russian investigators (Samsonov et al. 1961) prepared yttrium monocarbide by heating a mixture of the stoichiometric composition at 1900°C and measured the physical properties of this yttrium carbide (melting point at 1950 ± 20°C). However, no structural data were given.

All reports concerning the existence of the rare earth monocarbides were published before the middle 1960s, e.g. Auer-Welsbach and Nowotny (1961) reported the lattice parameter of ScC to be $a = 4.45$ Å. However, after that no data were obtained to prove the existence of the rare earth monocarbides.

Schwetz et al. (1979) regarded the compound with a composition range of $RC_{0.35}$ to $RC_{0.65}$ as the monocarbide. They pointed out that this compound exists for the elements from samarium to the end of the rare earth period and for scandium as well as yttrium, and the lattice is face-centered cubic. The carbon atoms are isolated and appear no longer as pairs. Since a complete filling of the octahedron gaps with carbon atoms for rare earth metals is never achieved, the structure always remains strongly nonstoichiometric. Therefore, this type of compound is not a "true" monocarbide.

As will be described later, the monocarbides of the rare earths could be formed only after adding a third element such as N, O, etc. In these cases, the compound is a ternary one, instead of a binary one.

3. Crystal structure of the binary rare earth carbides

The formation of the binary rare earth carbides has been summarized in table 2. The crystal structure and lattice parameter data listed in this table were quoted from the review "Critical evaluation of binary rare earth phase diagrams" (Gschneidner and Calderwood 1986). The listed lattice parameters were assessed by them and are the mean values when more than one acceptable set of data were presented for an individual compound. In this section, the crystal structures of each binary rare earth carbide will be evaluated in detail.

3.1. *Crystal structure of the RC_2 compounds*

Rare earth dicarbides are commonly formed in the rare-earth (except for europium)–carbon systems. Many investigators, in particular Atoji, have made a great contribution to the determination of the structure of RC_2. The neutron diffraction investigations on the structure of the rare earth dicarbide, first by Atoji and Medrud (1959) showed that for the lanthanum dicarbide the atomic coordinates in the unit cell are: $(000, \frac{1}{2}\frac{1}{2}\frac{1}{2}) \pm (00z)$ for carbon atoms and $z = 0$ for lanthanum atoms, and the carbon positional parameter $z(\text{Å})$ is 0.403 ± 0.002, corresponding to a well-defined minimum at a C–C distance of 1.28 ± 0.03 Å. LaC_2 can probably be described approximately in terms of $La^{3+}C_2^{2-}$ ions, with the extra electron in a conduction band and thus with a metallic conductivity. In LaC_2, the C_2 group appears to be acetylene-like with one nearest lanthanum at 2.65 Å from the carbon on a line collinear with the C–C bond. Four other lanthanum atom sites are located around the C_2 group as though π-bonded to it. Bond numbers (n), distances and coordination numbers are as follows:

C–C	1.28 Å,	$n = 2.73$,	La–2C	2.65 Å,	$n = 0.48$,
C–La	2.65 Å,	$n = 0.48$,	La–8C	2.85 Å,	$n = 0.22$,
C–4La	2.85 Å,	$n = 0.22$,	La–4La	3.93 Å,	$n = 0.12$.

Bowman et al. (1968) determined the high-temperature structures of the tetragonal LaC_2 phase at 900 and 1150°C by means of high-temperature neutron diffraction.

The tetragonal-to-cubic transformation was observed at 1060°C with the lattice parameters $a_0 = 4.00$, $c_0 = 6.58$ Å at 900°C and $a_0 = 6.02$ Å at 1150°C. The structure of the tetragonal phase was found to be Clla type, in agreement with the previous room-temperature results, with a C–C distance of 1.26 ± 0.03 Å. The cubic phase was found to be isomorphous with the cubic KCN-type, instead of the CaF_2-type, uranium dicarbide (Bowman et al. 1966), space group Fm3m. In this structure, the lanthanum atomic coordinates are ($\frac{1}{2}$, $\frac{1}{2}$, $\frac{1}{2}$) and C_2 group with centers at (0, 0, 0) randomly oriented along the [111] directions. The observed C–C distances are in reasonable agreement with the room-temperature value of 1.30 Å (Atoji 1961), the actual values may be somewhat larger owing to the effect of thermal motion of the carbon atoms. Both the tetragonal and high-temperature cubic structures of the compounds RC_2 are shown in fig. 5 and the difference between both structures is obvious.

In addition to LaC_2, the crystal structures of CeC_2 and TbC_2 (Atoji 1961, 1962, 1967a), PrC_2 and NdC_2 (Atoji 1967a), DyC_2 (Atoji 1968), HoC_2 (Atoji 1967a), ErC_2 (Atoji 1972), TmC_2 (Atoji 1970), YbC_2 (Atoji 1961, 1962, Atoji and Flowers 1970), LuC_2 (Atoji 1961, 1962) and YC_2 (Atoji 1961, 1962, Bowman et al. 1967) have also been established by neutron diffraction. The coherent nuclear reflection analysis technique (Atoji 1967a) verified the previous X-ray diffraction results (Spedding et al. 1958) that the chemical unit cell contains two molecules and possesses the body-centered tetragonal symmetry, D_{4h}^{17}–I4/mmm, as reported above for LaC_2. Table 5 lists the lattice parameters of the compounds RC_2 obtained by the neutron diffraction method at 300 and 5 K. Only one set of data was adopted when more than one set of data was present in different works by the same author. Corresponding to the lattice parameters, the average linear thermal expansion coefficients in the temperature range of 5–300 K are also listed. In the light rare earth dicarbides the expansion along the *a* axis is about a half of that along the *c* axis, while the relation is reversed in the heavy rare earth dicarbides. However, the expansion along the *c* axis is several times larger than that along the *a* axis in the case of DyC_2 and YbC_2.

Table 5 also summarized other pertinent crystallographic data: the Debye–Waller temperature factor coefficients, B in $\exp[-2B(\sin\theta/\lambda)^2]$ at 300 K and 5 K; the carbon positional parameters, z in (000, $\frac{1}{2}\frac{1}{2}\frac{1}{2}$) ± (00$z$), ($z = 0$ for the R atoms); the intramolecular C–C distances and the nearest C–R distances.

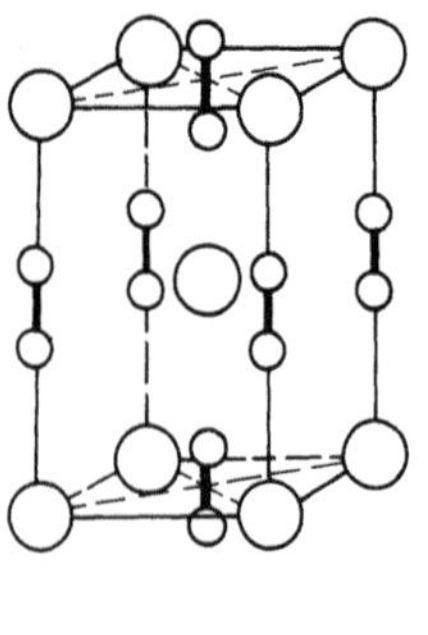

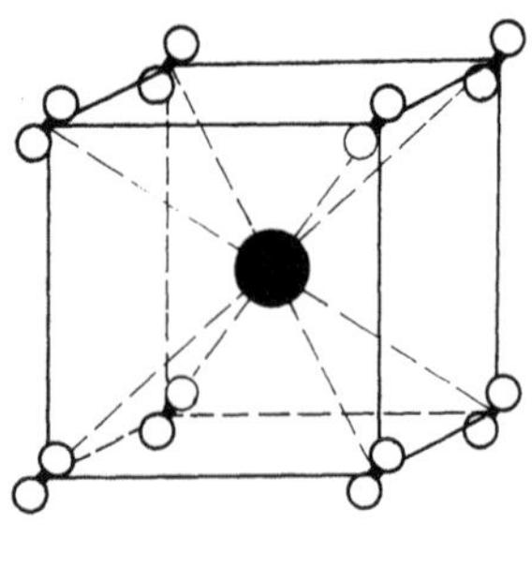

Fig. 5. RC_2 with the tetragonal CaC_2-type structure at room temperature (a) and with cubic KCN-type structure at high temperature (b).

The paramagnetic scattering, the interatomic distances, and the C–C energy level indicate that these compounds are composed of R^{n+}, C_2^{2-}, and an $(n-2)$ conduction electron with $n = 3$ (except for $n = 2.8$ in YbC_2) and the R ions are in their tripositive Hund ground states (Atoji 1961). Despite the presence of the negative ion C_2^{2-}, the RC_2 is typically metallic. This aspect has been reasonably interpreted in terms of the 5d-π_g2p conduction band that correlates self-consistently with the valency wave-functions (Atoji 1961, 1962, 1967a, b).

The C–C distances in the C_2 groups for the rare earth dicarbides are insignificantly different from their weighted average value of 1.286 Å. In particular, the C–C distance in YbC_2 revealed no substantial effect due to possible anomalous valence of Yb in this compound. The average value in RC_2 is considerably larger than the 1.20 Å in alkaline earth dicarbides (Atoji 1961, 1962, Atoji and Medrud 1959) which usually exhibit nonmetallic characteristics.

Other interatomic distances in the rare earth dicarbides show unusual features. Pauling's bond number (Pauling 1960) for the C–R and R–R bonds in these dicarbides increases roughly with the atomic number of the rare earth atom: 0.5 (C–La) to 0.9 (C–Lu) (Atoji 1961) for the nearest C–R distance; 0.2 to 0.35 for the next nearest C–4R distances; 0.1 to 0.2 for the nearest R–R distances which are equal to the lattice parameters; 3 to 5 and 3.5 to 4.5, respectively, for the total bond numbers of the carbon and rare earth atoms. The bond numbers for YC_2 fall between Ho and Lu, indicating that Y in YC_2 behaves as a heavy lanthanide (Atoji 1961, 1962).

3.2. *Crystal structure of the R_2C_3 compounds*

The lanthanide sesquicarbide is body-centered cubic with Pu_2C_3($D5_c$)-type structure and $I\bar{4}3d$ space group with eight formula units per unit cell. This structure has been confirmed by neutron diffraction studies carried out at 296 K for La_2C_3 and Pr_2C_3 (Atoji and Williams 1961), at 296 to 5 K for Ce_2C_3 (Atoji 1967b), Tb_2C_3 (Atoji 1971), Ho_2C_3 (Atoji and Tsunoda 1971), and Pr_2C_3, Nd_2C_3 and Dy_2C_3 (Atoji 1978). The metal and carbon atoms occupy the sites 16(c) (uuu) and 24(d) (v o $\frac{1}{4}$), respectively. The least-squares refinement to the neutron diffraction data yielded the position parameters, temperature factor coefficients (B) and interatomic distances. The configurations of both lanthanide atoms about C_2 groups and the C_2 groups about lanthanide atoms (point symmetry C_3–3) are shown in fig. 6. The point symmetry of the midpoint of the C_2 group is S_4–$\bar{4}$ and approximates to D_{2d}–$\bar{4}$m2. Referring to fig. 6a, $\angle R_1C_1R_2$ is 79.1°, while $\angle R_3C_1R_4$ is 180°. The angle between R_1R_2 and R_3R_4 is 84.9° (Atoji et al. 1958) and each R atom also has three nearest R neighbors at only 3.630 Å for R = La (Atoji 1961), compared with 3.75 Å for nearest neighbors in β-La.

All Pu_2C_3-type rare earth sesquicarbides, except Ce_2C_3, have a C–C distance of 1.246 Å (the average for La_2C_3, Pr_2C_3, Tb_2C_3 and Ho_2C_3). The distance is shorter than the average C–C distance in the rare earth dicarbides, 1.286 Å. In Ce_2C_3 the C–C distance, 1.276 ± 0.005 Å, is significantly different from that in other rare earth sesquicarbides and U_2C_3 has a much longer value, C–C = 1.42 Å (Novion et al. 1966). Therefore, these C–C distances seem to indicate that the intramolecular C–C distance in the metal acetylide should become longer as the ionic charge of the metal atom

TABLE 5
Crystallographic data of RC_2 obtained by neutron diffraction at 5 and 297 ± 3 K.

	Lattice parameters (Å)				Linear thermal expansion coefficients (10^{-6}/K)		Temperature factor coefficients B(Å)		Carbon positional parameter (Å)	C–C distance (Å)	R(0, 0, 0)–C distance (Å)	$R(\frac{1}{2},\frac{1}{2},\frac{1}{2})$–C distance (Å)
	At 5 K		At 297 ± 3 K									
RC_2	a	c	a	c	∥ a	∥ c	300 K	5 K				
LaC_2[a]			3.934 ± 0.003	6.572 ± 0.0015			0.18 ± 0.16		0.4009 ± 0.0009	1.303 ± 0.012	2.635 ± 0.006	2.857 ± 0.007
CeC_2[b]	3.875 ± 0.001	6.477 ± 0.003			3 ± 1	6 ± 2	0.60 ± 0.04	0.40 ± 0.05	0.4011* ± 0.0005	1.281* ± 0.007	2.598* ± 0.003	2.814* ± 0.001
PrC_2[b]	3.852 ± 0.001	6.425 ± 0.002			3 ± 2	5 ± 2	0.98 ± 0.14	0.64 ± 0.11	0.3993* ± 0.0005	1.294* ± 0.006	2.566* ± 0.003	2.800* ± 0.001
NdC_2[b]	3.820 ± 0.001	6.390 ± 0.002			3 ± 1	8 ± 2	0.52 ± 0.10	0.34 ± 0.08	0.3990* ± 0.0007	1.291* ± 0.009	2.550* ± 0.004	2.777* ± 0.002
SmC_2												
EuC_2												
GdC_2												
TbC_2[b]	3.678 ± 0.001	6.206 ± 0.001			11 ± 2	6 ± 3	0.67 ± 0.06	0.30 ± 0.05	0.3960* ± 0.0007	1.291* ± 0.009	2.458* ± 0.004	2.680* ± 0.002

DyC_2[c]	3.656	6.171	3.668	6.176	11 ± 1	2.7 ± 0.8	0.73	0.25	0.3963*	1.287*	2.400*	2.634*
	± 0.001	± 0.001	± 0.001	± 0.001			± 0.16	± 0.10	± 0.0005	± 0.006	± 0.003	± 0.001
HoC_2[b]	3.633	6.132			9 ± 2	4 ± 2	0.88	0.35	0.3957*	1.279*	2.426*	2.647*
	± 0.002	± 0.002					± 0.02	± 0.10	± 0.0007	± 0.009	± 0.004	± 0.003
ErC_2[d]	3.612	6.087	3.620	6.095	7.5 ± 1.3	4.5 ± 1.2	0.72	0.32	0.3943	1.288	2.403	2.640
	± 0.001	± 0.001	± 0.001	± 0.001			± 0.18	± 0.15	± 0.0005	± 0.006	± 0.003	± 0.001
TmC_2[e]	3.589	6.054	3.601	6.062	11 ± 2	4 ± 2	0.96	0.42	0.3945	1.279	2.392	2.625
										(1.277*)	(2.388*)	(2.617)*
	± 0.002	± 0.004	± 0.001	± 0.002			± 0.05	± 0.09	± 0.005	± 0.006	± 0.003	± 0.001
										(0.006)	(0.003)	(0.002)
YbC_2[f]	3.625	6.009	3.630	6.101	5 ± 1	1 ± 1	0.7	0.3	0.3942	1.29	2.405	2.647
										(1.291*)	(2.404*)	(2.643)*
	± 0.001	± 0.001	± 0.001	± 0.001			± 0.10	± 0.10	± 0.005	± 0.006	± 0.003	± 0.001
LuC_2[a]			3.563	5.964			0.75		0.393	1.276	2.344	2.599
			± 0.0015	± 0.009			± 0.10		± 0.001	± 0.012	± 0.007	± 0.002
YC_2[a]			3.664	6.169			0.36		0.3967	1.275	2.447	2.668
			± 0.0015	± 0.006			± 0.02		± 0.0002	± 0.002	± 0.003	± 0.001

[a] Atoji (1961),
[b] Atoji (1967a),
[c] Atoji (1968),
[d] Atoji (1972),
[e] Atoji (1970),
[f] Atoji and Flowers (1970),
* At 5 K.

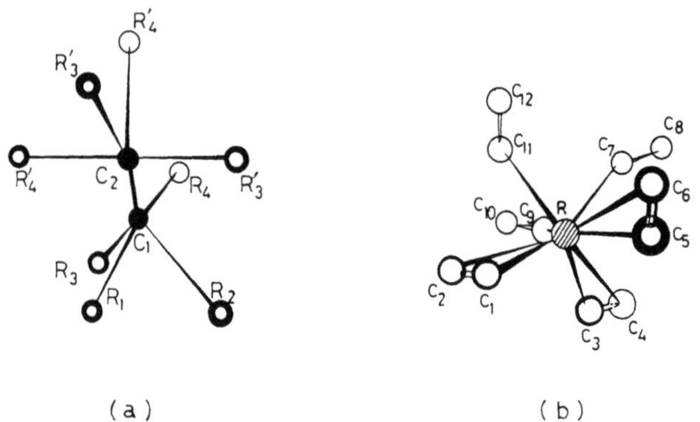

Fig. 6. (a) Configuration of the R atoms about the C_2 group in R_2C_3; (b) Configuration of the C_2 groups about an R atom in R_2C_3 (Atoji et al. 1958). (Reprinted by permission of the publisher, The American Chemical Society, Inc.)

increases, as in the metal dicarbides, MC_2. In the metal sesquicarbides, M_2C_3, the C–C bond distances are 1.246, 1.276 and 1.42 Å when M is a tripositive rare earth, $Ce^{3.4+}$ (Atoji 1967b) and U^{4+}, respectively.

The shortest R–R distances in the rare earth sesquicarbides, R_0–$3R_1$, are nearly 10% shorter than the shortest R–R distances in the dicarbides, which are equal to their lattice parameters. Other R_0–$2R_2$ and R_0–$6R_3$ distances are also shorter than this value, indicating that there are stronger R–R interactions in the sesquicarbides than in the dicarbides. The interatomic distances in Ce_2C_3, particularly the Ce–Ce distances are markedly smaller than the expected values for Ce_2C_3 with the pure trivalent Ce atoms. This is in accordance with the result obtained from the paramagnetic scattering analysis (Atoji and Williams 1967).

In the sesquicarbides with the trivalent rare earth metal ions, the number of delocalized electrons is $\frac{4}{3}$ per C_2 group, while the dicarbides with the trivalent lanthanide ions have only one delocalized electron per C_2 group. If the C_2 antibonding orbital ($\pi_g 2p$) (Atoji and Medrud 1959) participates equally in forming the lowest conduction bands of the sesquicarbides and the dicarbides, the C–C bond distances in the sesquicarbides should be longer than those in the dicarbides, in contradiction to the experimental results. Thus, the $\pi_g 2p$ contribution of the C_2 group to the conduction bands must be smaller in the sesquicarbides than in the dicarbides, i.e. the conduction electrons in the sesquicarbides are associated with the metal orbitals with a higher order than those in the dicarbides. In Ce_2C_3, more electrons occupy the C_2 antibonding orbitals than in other sesquicarbides, and thus a markedly shorter C–C distance was observed.

The neutron diffraction study on the crystal structure of Ce_2C_3 at temperatures between 300 and 4 K (Atoji 1967b) showed an anomalous temperature dependence of the Ce_2C_3 lattice parameter, suggesting a gradual "lattice collapsing" transformation, probably due to a 4f–5d transition, ending at about 90 ± 5 K and that Ce in Ce_2C_3 below 90 K is essentially tetravalent.

3.3. *Crystal structure of the $R_{15}C_{19}$ compounds*

Since 1958, seven representatives of the intermediate carbides (Spedding et al. 1958), $Sc_{15}C_{19}$ (Jedlicka et al. 1971), $Y_{15}C_{19}$ (Bauer and Nowotny 1971), $Er_{15}C_{19}$ (Bauer 1974), $Yb_{15}C_{19}$ (Haschke and Eick 1970a, Bauer and Bienvenu 1980, Hájek et al. 1984), $Lu_{15}C_{19}$ (Bauer and Bienvenu 1980), $Ho_{15}C_{19}$ and $Tm_{15}C_{19}$ (Bauer and Ansel 1985) have been reported. The existence of $Gd_{15}C_{19}$ has also been noted in the preparation of $Gd_5Si_3C_{0.95}$ (Al-Shahery et al. 1983). The crystal structure of $Sc_{15}C_{19}$ has been determined using a single-crystal X-ray method and a powder method (Jedlicka et al. 1971) to be a pseudocubic tetragonal, $a = 7.50$ Å and $c = 15.00$ Å, with space group $P\bar{4}2$, c–D_{2d}. In $Sc_{15}C_{19}$ the scandium atoms are arranged in an orderly way in six almost equidistant layers perpendicular to the c-axis and they form octahedral structural elements [Sc_6C] with carbon atoms. In the unit cell with two formulae units of $Sc_{15}C_{19}$ the atomic position parameters have also been determined.

From fig. 7, it can be seen that the structure in $Sc_{15}C_{19}$ can be accordingly considered as [Sc_6C] and [Sc_6C_2] groups in the c-axis direction:

in $z \sim 0$ [Sc_6C];
in $z \sim \frac{1}{6}$ [Sc_6C] and [Sc_6C_2];
in $z \sim \frac{1}{3}$ [Sc_6C] and [Sc_6C_2];
in $z \sim \frac{1}{2}$ [Sc_6C];
in $z \sim \frac{2}{3}$ [Sc_6C] and [Sc_6C_2];
in $z \sim \frac{5}{6}$ [Sc_6C] and [Sc_6C_2].

The number of carbon atoms in the structure depends on the small metal octahedra surrounding a carbon atom. The size of a C_2 pair and the size of the octahedral gap, and the distortions in the metal octahedron are obvious because in the planes perpendicular to the [001] direction the edges deviate from the ideal positions. In this

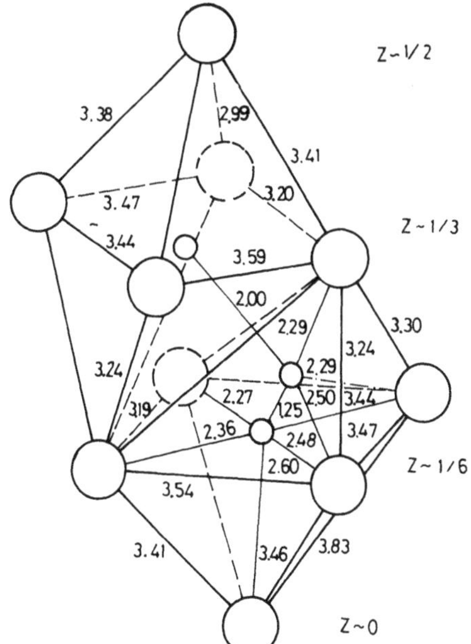

Fig. 7. Characteristic [Sc_2C] and [Sc_6C_2] groups in $Sc_{15}C_{19}$ (Jedlicka et al. 1971). (Reprinted by permission of the publisher, Institüt für Anorganische Chemie, Inc.)

way, a coordination number of seven arises at the center of the prismatic face in accordance with the large deformation of the trigonal prism.

The radius ratio r_C/r_{Sc} of 0.48 is the limiting value for the C_2 pair to occupy the octahedron gap. As described above, the occupation of C_2 pairs has led to such a large distortion of the structure element that a trigonal prismatic environment occurs.

For a compound with higher carbon content than this compound, no indication that the additional occupation of the C_2 pairs in the metal octahedra occurs was found although the octahedron at $z \approx 0$ exhibits a relatively large Sc–Sc distance. The shortest interatomic distances, 2.99, 3.11 and 3.18 Å for Sc–Sc, as well as 2.24 and 2.26 Å for Sc–C, are related to the noticeable ionic portion of the bond. The C–C distance in the C_2 pair, 1.25 Å, is in the range of that for the dicarbides of Ca, Y and the rare earths. All the interatomic distances have been reported by Jedlicka et al. (1971), showing that the $Sc_{15}C_{19}$ compound has relatively large values, 3.30 and 2.55 Å, for the Sc–Sc and Sc–C distances.

Figure 8 shows the plot of the lattice parameters of the seven $R_{15}C_{19}$ compounds against the metallic–covalent radii of the lanthanides, which were computed from the lattice parameters of the RC_x phases at high carbon concentrations. The values of the radii are intermediate between the covalent and metallic radii of the elements, taking into account the presumed metallo–covalent bond in these carbides. There is a linear relationship between the metal size and the parameters with the exception of yttrium (Bauer and Ansel 1985).

Hájek et al. (1984a) related the products of the hydrolysis of $Sc_{15}C_{19}$ to a carbide structure containing isolated carbon atoms and three-atom groups of the type C_2–C_1. The substructure of carbon atoms of the elementary cell of $Sc_{15}C_{19}$ has been calculated from the atom position of the asymmetrical unit (Jedlicka et al. 1971). Of the 38 carbon atoms in the basic carbon substructure of the carbide, 14 are totally isolated with a minimum distance from any carbon neighbor of 3.30 Å. The remaining 24 carbon atoms are arranged in groups of three, with internuclear separations of 1.25 Å, 2.00 Å and 2.93 Å. The following distribution of carbon atom groups along the z axis can be assigned to the elementary cell: $z \approx 0$, $5C_1$; $z \approx \frac{1}{6}$, C_1 and $2(C_2$–$C_1)$; $z \approx \frac{1}{3}$, C_1 and $2(C_2$–$C_1)$; $z \approx \frac{1}{2}$, $5C_1$; $z \approx \frac{2}{3}$, C_1 and $2(C_2$–$C_1)$; $z \approx \frac{5}{6}$, C_1 and $2(C_2$–$C_1)$.

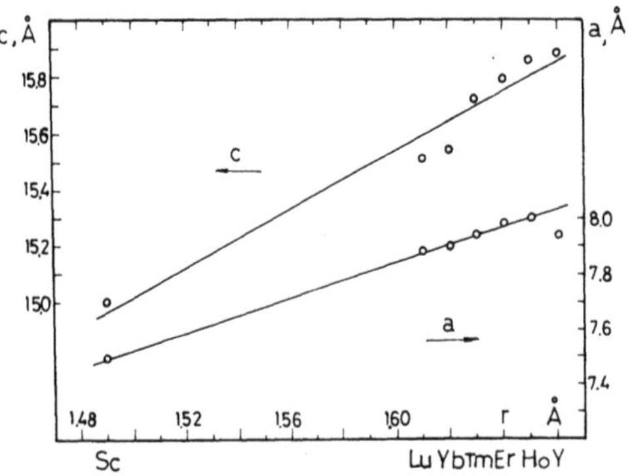

Fig. 8. Lattice parameters of the $R_{15}C_{19}$ phases plotted against metallic–covalent radii of the rare earth elements (Bauer and Ansel 1985). (Reprinted by permission of the publisher, Elsevier Sequoia S.A., Inc.)

3.4. *Crystal structures of the R_3C and R_2C compounds*

The rare earth and carbon systems contain the hypocarbides, R_3C (RC_x, $0.25 < x < 0.65$) for R = Sm through Lu, and Y and R_2C for R = Sm through Lu, Sc and Y. The crystal structures of the trigonal R_2C and the cubic RC_x compounds have been determined. The transformation mechanism between the trigonal and cubic structures was revealed by using a single crystal in which the transient state at the phase transition was arrested (Atoji and Kikuchi 1969).

3.4.1. *Crystal structure of the R_3C compounds*

All of the carbides R_3C have the cubic C-deficient NaCl-type structure, commonly known as the Fe_4N-type. This structure is the high-temperature one and can also exist at room temperature in a metastable state. In this structure, rare earth atoms are arranged in a cubic close-packed structure with C atoms distributed randomly over the octahedral sites. This phase exists over a wide solid solubility range and the reported lattice parameters have been listed in table 2.

The neutron diffraction study of Atoji (1981b) on the crystal and magnetic structures of the cubic $ErC_{0.6}$ compound in the temperature range 1.6–296 K confirmed the previously reported results (Spedding et al. 1958, Lallement 1966). Its cubic structure is illustrated in fig. 9, in trigonal coordinates in order to compare it with the trigonal structure of R_2C (Atoji and Kikuchi 1969).

Aoki and Williams (1979) pointed out that the NaCl defect structure is stable in the approximate composition range $0.33 \leqslant x \leqslant 0.45$. They inferred that in the cubic close-packed lattice built up of atoms of radius r, the octahedral holes will just contain

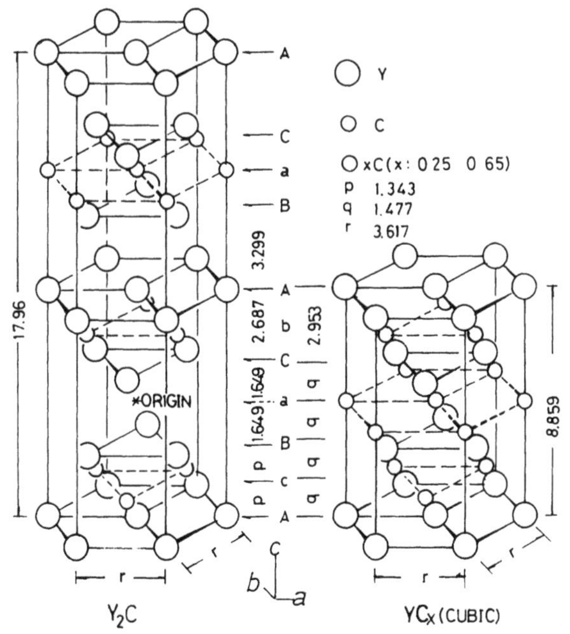

Fig. 9. Schematic representations of the cubic and trigonal structures of yttrium hypocarbide (Atoji and Kikuchi 1969). (Reprinted by permission of the publisher, The American Institute of Physics.)

spheres of radius $0.41r$, i.e., the smaller atoms will exactly fill the octahedral sites. Carbides with the interstitial cubic close-packed lattice are actually formed when the radius ratio is between 0.41 and 0.59. By using $r(C) = 0.772$ Å, which corresponds to the nearest-neighbor atomic distance of diamond, values of $r(C)/r(R)$ of 0.433–0.443 are obtained for the lanthanides, Gd–Tm, respectively; these are within the radius ratio range.

3.4.2. *Crystal structure of the R_2C compounds*

Atoji and his co-workers, as well as Lallement and Bacchella and co-workers have studied Y_2C (Atoji and Kikuchi 1969), $Tb_{2.1}C$ (Atoji 1969), Ho_2C (Atoji 1981a, Lallement 1966, Bacchella et al. 1966), Tb_2C and Dy_2C (Atoji 1981c) by neutron powder diffraction. The anti-$CdCl_2$-type trigonal R_2C structure can be described by space group $D^5_{3d} - R\bar{3}m$ and the atomic coordinates $(0, 0, 0; \frac{2}{3}, \frac{1}{3}, \frac{1}{3}; \frac{1}{3}, \frac{2}{3}, \frac{2}{3}) \pm (0, 0, z)$ for rare earth atoms and $\pm(0, 0, 0)$ for carbon atoms and three R_2C formula units, as shown in fig. 9. For Y_2C (Atoji and Kikuchi 1969), the intensity analysis gave $z = 0.2593(6)$ for Tb_2C (Atoji 1981) and $Tb_{2.1}C$ (Atoji 1969); 0.2575(8) for Dy_2C (Atoji 1981); 0.256(1) (Lallement 1966, Bacchella et al. 1966) and 0.2564(8) (Atoji 1981a) for Ho_2C; 0.2587(0.2585) for Y_2C (Atoji and Kikuchi 1969). Other crystal data have also been provided by neutron reflections. In R_2C, the number of C–C pairs remains negligibly small, while those for the R–C pairs are larger than those in the corresponding cubic structure. It suggests that the bond number may be neglected for C–C pairs in both the cubic and in the trigonal structures.

3.4.3. *The orientation relationship of the cubic–trigonal structural transformation*

Atoji and Kikuchi (1969) have given the orientation relationship of the cubic–trigonal transformation for the Y_3C and Y_2C compounds by using a single-crystal X-ray analysis. In the cubic structure, the layer sequence along the [111] axis is the repetition of [Ac$^{\square}$ Ba$^{\square}$ Cb$^{\square}$], where A, B, and C signify the Y layers and a$^{\square}$, b$^{\square}$, and c$^{\square}$ represent the partly occupied carbon layers. The layer repetition unit of the trigonal structure is given by [AcB$^{\square}$ Cb$^{\square}$ BaC$^{\square}$], where a, b and c designate the fully occupied carbon layers and □ denotes vacant layers. Therefore, the cubic-to-trigonal transformation could be accomplished by a short-range transport of the carbon atoms. Upon the ordering of the carbon atoms, the statistical high symmetry in Y_3C breaks down to considerably lower symmetries while no change takes place in the symmetry and the interatomic dimensions within the layers. In the cubic structure, the statistical point symmetry at the Y and C sites is the octahedral m3m group, while in Y_2C, that of C becomes the centric 3 m group, representing a distorted octahedron (CY_6) with C–6Y $= 2.483 \pm 0.003$ Å, $\angle Y_I C_I Y_{II} = 93°30' \pm 10'$ and $\angle Y_{II} C_I Y_{III} = 86°30 \pm 10'$; and that of Y is the highly asymmetric 3m group and the Y–C bonding configuration is a trigonal pyramid with Y at its apex. The orientation relationship between the two phases is shown clearly in fig. 9, $[111]_{cubic} \parallel [001]_{trigonal}$, and $(111)_{cubic} \parallel (001)_{trigonal}$. When choosing the (111) plane of Y_3C(cubic) and the (001) plane of Y_2C(trigonal) as coherent interfaces, the minimum interface energy could be attained (Atoji and Kikuchi 1969).

The cubic-to-trigonal or disorder-to-order transformation accompanies a small unit cell volume expansion of 1.32%, apparently contrary to a commonly conceived concept. Despite this, the total bond energy in Y_2C appears to be considerably larger than that in Y_3C(cubic). Here, the relative bond strength was estimated using Pauling's formula (Pauling 1960), which gave the total bond numbers of C and Y, 3.2 and 4.2 in Y_3C(cubic), and 4.2 and 5.3 in Y_2C, respectively (Atoji and Kikuchi 1969).

3.4.4. *Bonding character of the hypocarbides*

As has been noted, the trigonal R_2C structure can be composed of the stacking of the hexagonal layers along the *c* axis in the sequence AcB□ CbA□ BaC□ .... Thus, this structure may be viewed as the cubic closest packing of the rare earth atoms interleaved alternately by the filled and vacant carbon layers. The perpendicular distance between the adjacent rare earth layers across the carbon layer is much shorter than that across the vacant layer as indicated by the values of 2.68(2) and 3.35(2) Å in Tb_2C at 4 K, 2.71(3) and 3.25(3) Å in Dy_2C at 4 K, 2.73 and 3.16 Å in Ho_2C at 4 K, and 2.687 and 3.299 Å in Y_2C at room temperature, respectively, in contrast to the transition-metal carbide, e.g. in α-Ta_2C the corresponding distances are 2.505 and 2.432 Å (Bowman et al. 1965).

The metal–metal, metal–carbon and carbon–carbon bond distances (in angstroms) in R_2C are given in table 6. The R(I)–R(III) distance across the carbon layer is considerably shorter than the bonding distance R(I)–R(IV) across the vacant layer. For comparison, the nearest-neighbor distances for Tb, Dy, Ho and Y metal are 3.56, 3.55 (Atoji 1968), 3.486 (Atoji 1981a) and 3.556 Å (Gschneidner 1961), respectively. These distances indicate that the carbon atom strengthens the metallic bond in R_2C, also in contrast to the transition-metal carbide (Bowman et al. 1965).

The carbon atom is bonded to the rare earth atoms at a comparable distance to that in the RC_2 compounds, however, the C–C bonding in the R_2C compound remains negligibly small. This suggests that the carbon atoms exist in the form of isolated atoms instead of C–C pairs and that they provide a considerable number of bonding electrons to the s–d bond orbitals among the rare earth atoms so that the lattice parameter of the hypocarbides, either the cubic R_3C compound or the trigonal R_2C compound, decreases as the carbon content increases (Atoji 1981c).

Aoki and Williams (1979) have estimated the bonding characters of the cubic R_3C and the trigonal R_2C compounds from Pauling's equation (Pauling 1960). The bond

TABLE 6

Bond distance	Tb_2C	Dy_2C	Ho_2C	Y_2C
R(I)–3R(III) across the C layer	3.381(9)	3.41(1)	3.416(15)	3.094(8)
R(I)–GR(II) within the layer	3.570(3)	3.578(8)	3.550(8)	3.617(2)
R(I)–3R(IV) across the vacant layer	3.94(2)	3.85(3)	3.76(3)	3.402(7)
C–6R	2.458(6)	2.470(9)	2.46(1)	2.483(3)
C–6C	[a]	[a]	3.550(8)	3.617

[a] Not determined.

numbers *n* have been given for $DyC_{0.33}$ and compared with those in $YC_{0.48}$ (Atoji and Kikuchi 1969), indicating that the *n* value comes from bonding between rare earth and carbon atoms rather than between carbon atoms.

3.5. *Structure of the first-stage intercalation compounds EuC_6 and YbC_6*

It has been determined from the intensities of *hk*0, *hkl* reflections, and the presence of systematic extinctions, that the hexagonal unit cell of the first-stage phases belongs to the space group $P6_3/mmc$ with two metal atoms at position b and 12 carbon atoms at i. The parameters are $a = 4.314 \pm 0.003$ Å, $c = 9.745 \pm 0.008$ Å (EuC_6) and $a = 4.320 \pm 0.004$ Å, $c = 9.147 \pm 0.004$ Å (YbC_6) (El-Makrini et al. 1980). Figure 10 shows this structure.

Suematsu et al. (1981) also determined the *c* lattice parameter of C_6Eu as 9.73 Å from the (001) reflections. They suggested that the *c*-axis alignment of crystallites in C_6Eu is remarkably reduced during intercalation. In this structure, europium atoms are located on centers of graphite hexagons and form a triangular lattice with an atomic distance three times the unit vector of graphite. The europium layers are separated by one graphite layer with the interlayer distance being 4.87 Å. Thus a europium atom is surrounded by six nearest-neighbor europium atoms in the same layer and by six next-nearest neighbors separated by the adjacent graphite layers. The nearest intraplane distance of the Eu–Eu atoms is 4.31 Å while the next-nearest interlayer distance is 5.47 Å (Suematsu et al. 1981).

3.6. *Structure of the Sc_4C_3 compound*

The structure of Sc_4C_3 was determined (Krikorian et al. 1969) by X-ray and neutron diffraction to be cubic, with space group $I\bar{4}3d$, with 16 scandium atoms in 16(c)(x, x, x), $x = 0.0496 \pm 0.0006$ Å and 12 carbon atoms in 12(a)($\frac{3}{8}, 0, \frac{1}{4}$). The lattice parameter on the metal-rich side is $a_0 = 7.2081 \pm 0.0004$ Å in equilibrium with Sc_2C (trigonal) and $a_0 = 7.2067 \pm 0.0002$ on the carbon-rich side in equilibrium with the higher carbides phase ($ScC_{\sim 1.2}$, which was later identified as $Sc_{15}C_{19}$). Based on the intensity analysis of the neutron diffraction pattern, the structure may be considered to be anti-Th_3P_4 type, or perhaps even better, Pu_2C_3-type, with the C_2 groups

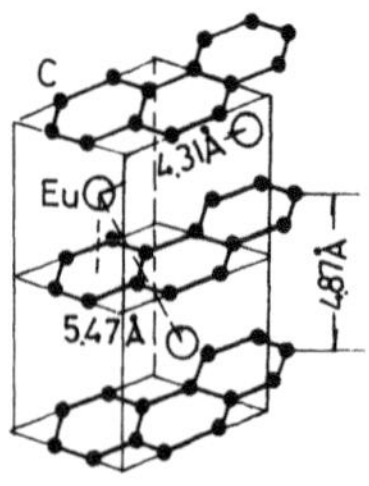

Fig. 10. The crystal structure of the first-stage compound C_6Eu (Suematsu et al. 1981). (Reprinted by permission of the publisher, Pergamon Press, Inc.)

replaced by single carbon atoms. The interatomic distances are listed as follows:

Sc–C (3) 2.255(8) Å	Sc–Sc (3) 2.976(18) Å	C–C(8) 3.371(2) Å
(3) 2.227(8) Å	(2) 3.121(2) Å	
	(6) 3.280(16) Å	

3.7. *Summary of the crystal structure of the binary rare earth carbides*

3.7.1. *Lattice parameter as a function of the R^{3+} ionic radius*

In general, the decrease in lattice parameters of the rare earth carbides from La to Lu obeys the lanthanide contraction law, e.g. the plots of the lattice parameter against the rare earth ion radius for the dicarbides (tetragonal) and the hypocarbides (cubic) show a linear relationship, as shown in figs. 11 and 12.

From these curves the valence state of the rare earth elements, in particular, europium and ytterbium in the carbides can be deduced. It is obvious that europium exhibits the divalent and trivalent states in the tetragonal dicarbide and the cubic hypocarbide, respectively, while ytterbium is only partially divalent (17 at.%

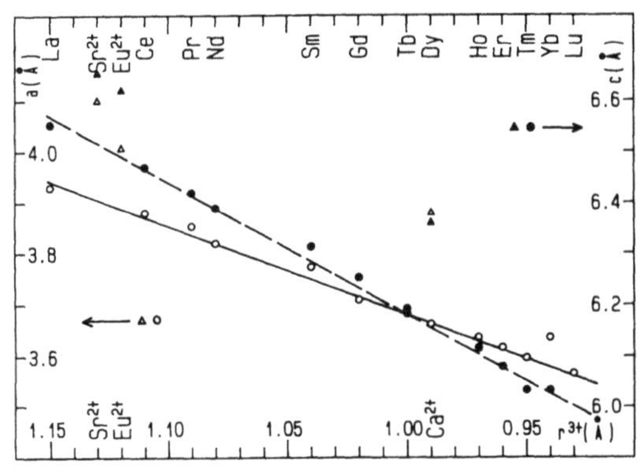

Fig. 11. The *a* and *c* lattice parameters of the dicarbides as a function of the ionic radius (Schwetz et al. 1979).

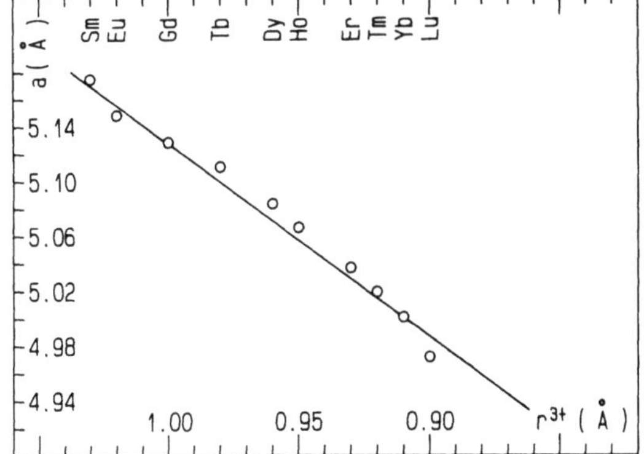

Fig. 12. The lattice parameter of the hypocarbide (cubic) as a function of the ionic radius (Schwetz et al. 1979).

Yb^{2+}–83 at.% Yb^{3+}) when present in the tetragonal dicarbide, which was confirmed by magnetic susceptibility measurements (Flahaut 1969), and trivalent in the cubic hypocarbide. In the sesquicarbide, the behavior of europium and ytterbium is anomalous. Ytterbium forms a monoclinic Yb_2C_3 phase (Haschke and Eick 1970a) while europium forms a tetragonal phase with a Ca_2C_3-type structure (Pasto and Morgan 1976, Schwetz et al. 1979).

3.7.2. *Chemical bonds of the binary rare earth carbides*

As described above, in the dicarbides and the sesquicarbides of the rare earth elements there are pairs of carbon atoms, while in the hypocarbides, either the cubic structure or the trigonal one, the carbon atoms are isolated and no longer appear as pairs. In the intermediate carbides, $R_{15}C_{19}$, the tetragonal structure is made up of distorted metal octahedra, carbon pairs and also single carbon atoms.

All dicarbides of the rare earth elements have a significantly longer C–C bond distance than the alkaline earth dicarbides, e.g., 1.303 to 1.276 Å for LaC_2 to LuC_2 and 1.191 Å for CaC (Atoji 1961). Presumably an electron from the rare earth atom is donated to the antibonding orbitals of $(C{\equiv}C)^{2-}$ ion, thus lengthening the bond. On hydrolysis, acetylene is no longer the major product; hydrogen and a variety of other organic compounds are produced. For the hydrolysis of YbC_2, no hydrogen is liberated (Halliday et al. 1973). This fact, together with the anomalous lattice parameters, indicates that it is probably more closely related to the alkaline earth dicarbides and may contain Yb(II).

In carbides of type R_2C_3, there is some lengthening of the $(C{\equiv}C)^{2-}$ bond (Atoji and Williams 1961), although the effect is not as marked as for RC_2 carbides. For the cubic R_3C compounds, hydrolysis gave methane and hydrogen, indicating some resemblance to the methanide corresponding to the isolated carbon atoms present in these compounds.

The dimeric C_2 unit has been found in many binary metal–carbide systems. The formation of C_2 units is an important factor affecting the structural stability as well as the other physical properties. Li and Hoffmann (1989) have calculated the densities of states for UC_2 and CaC_2, which contain and do not contain a short C–C bond, respectively. They showed that in the UC_2 structure, both uranium–carbon bonding and carbon–carbon bonding are enhanced upon formation of such dimeric C_2 units and the system is greatly stabilized. Their calculations also indicated that UC_2 is metallic, whereas CaC_2, a structure belonging to the same crystal family, has a substantial gap between the valence and conduction bands. Their analysis on the structural stabilities of UC_2 and CaC_2 is also applicable for RC_2.

On the electronic structure of the carbides of the group III elements (scandium, yttrium and the lanthanides), only a few theoretical investigations are available. The bond structure of ScC was calculated by Schwarz et al. (1969) using the augmented-plane-wave method. Ivashchenko et al. (1984) applied the same method to YC and presented the densities of states for YC_x, $x = 0.8$, 0.7 and 0.6, which were calculated by using the coherent potential approximation. The results have been used to analyze the stability of the compounds. However, both calculations (Schwarz et al. 1969, Ivashchenko et al. 1984) were not performed self-consistently. Zhukov et al. (1987)

calculated the energy band structures of ScC_x and YC_x self-consistently for $x = 1.0$, 0.75, and 0.5 by the linear-muffin-tin orbital method in the atomic sphere approximation. From the total energies, the theoretical values of the lattice constants and the energies of the vacancy formation were determined. It is shown that in YC_x for $x > 0.7$ the metalloid vacancies are stabilized energetically. It is suggested that vacancies in YC_x for $x < 0.7$ and in ScC_x for all x are formed when the compounds are synthesized at high temperature and these vacancies persist at low temperatures as a result of quenching. These conclusions obtained from the theoretical calculations are in agreement with the experiments showing that the samples of YC_x always contain a large number of vacancies (Atoji and Kikuchi 1969). In the case of scandium carbide, ScC_x was reported to be grown as almost perfect crystals (Nowotny and Auer-Welsbach 1961).

Despite these theoretical investigations, a systematic study of the relationship of the chemical bonds with the formation and the structural stability of the binary rare earth carbides remains to be carried out.

4. Pseudobinary system: mixed rare earth carbides and solid solutions of the rare earth carbides and the actinide carbides

4.1. *Mixed rare earth dicarbides* $(R_{1-x}R'_x)C_2$ *with different structures*

4.1.1. *Formation of* $R_{1-x}R'_xC_2$

In the mixed rare earth dicarbide and the rare-earth–actinide carbide systems, the ternary compounds were found to be present in the form of a solid solution with the same structure as the component carbide. Thus, these carbides have also been called pseudobinary carbides.

Adachi et al. (1970) studied the mixed rare earth dicarbides containing two different rare earth ions with either similar or very different radii, and pointed out (i) that in the former case the solid solutions also have a body-centered tetragonal structure like the pure components and (ii) that the relation between the lattice constants and the mole fraction of the rare earths obeys Vegard's law over the whole region. Such solid solutions include LaC_2–(Ce, Tb)C_2, CeC_2–(Pr, Ho, Y)C_2 and PrC_2–(Nd, Tm)C_2 (Adachi et al. 1973). In the latter case, e.g. in the systems LaC_2–(Dy, Tm, Y)C_2, CeC_2–(Er, Lu)C_2 and PrC_2–LuC_2, the face-centered cubic phases were observed in a 1:1 mole ratio of two-component carbides (Adachi et al. 1970, 1973). The cubic-to-tetragonal transformation observed for the one-component dicarbides seems not to occur in these systems. The limit of the ionic radius difference for the formation of a fcc phase is about 14% in the La–R–C system or 15% in the Ce–R–C system. A mixed dicarbide of lanthanum with lutetium (La:Lu = 1:1) appeared to be hexagonal with lattice parameters $a = 5.78$ Å, $c = 8.32$ Å (Adachi et al. 1973).

4.1.2. *Structure of* $R_{1-x}R'_xC_2$

As described above, the mixed rare earth dicarbides were found to form in three different structures, bct, fcc and hexagonal (Adachi et al. 1973). With respect to the tetragonal structure of some mixed rare earth dicarbides, crystal structure line profile refinements have been made for neutron powder diffraction data collected at room temperature from $Y_xHo_{1-x}C_2$ and $Ce_xNd_{1-x}C_2$ ($x = 0.25, 0.50, 0.75$) (Jones et al. 1984, 1986). The results show that in both $Y_xHo_{1-x}C_2$ and $Ce_xNd_{1-x}C_2$, the unit cell dimensions a and c progressively decrease with decreasing x. The C–C bond lengths are essentially the same, 1.28 Å for the entire composition range in $Y_xHo_{1-x}C_2$ (although the actual values are slightly higher, 1.281–1.282 Å, in the range $x = 0.25$–0.50), while in $Ce_xNd_{1-x}C_2$ they pass through a minimum, 1.275(5) Å, in the middle of the composition range, presumably indicative of rather stronger bonding at this composition.

4.1.3. *Cubic-to-tetragonal transformation in rare earth dicarbide solid solutions*

The cubic-to-tetragonal phase transformation temperature of various binary dicarbide solid solutions have been measured (McColm et al. 1973, Adachi et al. 1974, 1976, 1978, Loe et al. 1976). A typical example is shown in fig. 13.

The results showed that the transformation temperatures for pure dicarbides exceed 1000°C but those for the solid solutions fall rapidly as the concentration of the NdC_2, GdC_2, or HoC_2 increases with respect to that of the LaC_2 or PrC_2 up to nearly 50 mol.% NdC_2, GdC_2 or HoC_2, respectively.

This type of lowering in the transformation temperature has been explained by McColm et al. (1973) using a strain model. According to this model, a relationship was established between the depression of T_t and the difference in the unit cell volumes ΔV of the dicarbides, $\Delta T_t = K(\Delta V)^2$, for 50 mol.% solid solutions. These authors emphasized that the solvent should be defined as that dicarbide possessing the smaller unit cell volume, regardless of the relative concentrations. Loe et al. (1976) found that the same relationship holds for other compositions and proposed that a local strain on the C_2^{2-} ions, caused by local variations in ion size and represented by an X-ray unit cell volume difference, is responsible for lowering the nucleation temperature of a

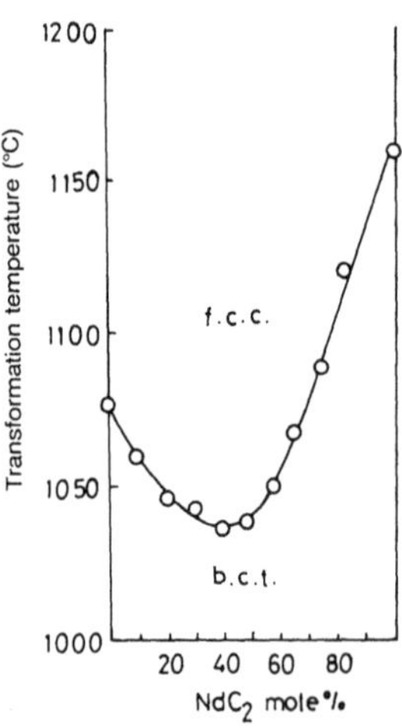

Fig. 13. Transformation temperature of the LaC_2–NdC_2 solid solution (Adachi et al. 1976).

tetragonal nucleus in a cubic matrix. They derived the strain energy, E_s, from DTA results and found it to be as large as 13.8 kJ mol^{-1} in the HoC_2–NdC_2 and GdC_2–LaC_2 systems. They estimated that when the strain exceeds a value of 16 kJ mol^{-1}, the cubic phases are stabilized down to ambient temperature.

In McColm's model, the entropy of transformation for the dicarbides ΔS_{trans} was presumed to be constant for all CaC_2-type carbides. However, as was evident from the thermal data of some rare earth dicarbides (Adachi et al. 1974, 1976, 1978), the observed value of ΔS_{trans} for the solid solution changes with its composition.

Adachi et al. (1974, 1976, 1978) measured the values of the heat of transformation in the systems LaC_2–NdC_2, LaC_2–GdC_2, LaC_2–CeC_2, LaC_2–TbC_2 and LaC_2–DyC_2, as well as those of the pure dicarbides, showing that the values for the solid solutions are smaller than those of the pure dicarbides (about 16.72 kJ mol^{-1}). In the case of $La_{0.47}Gd_{0.53}C_2$ it becomes as small as about 3.34 kJ mol^{-1}, as well as in the case of the LaC_2–DyC_2 solid solution, where the apparent heat of transformation becomes almost zero at both 26 mol% DyC_2 and 69 mol% DyC_2. They suggested that the volume difference between a pure dicarbide and a solid solution, ΔV, would lead to the strain energy "E_s" of the solid solution and if the amount of the strain energy evolved is equal to the heat of transformation, no thermal effect will be observed at all. Within the range of $26 < x < 69$ mol% DyC_2, a face-centered cubic phase appeared even at room temperature since the strain energy in the solid solution was probably much greater than the heat of the transformation.

The strain energy "E_s" in the systems LaC_2–CeC_2, LaC_2–TbC_2 (Adachi et al. 1978), LaC_2–NdC_2, LaC_2–GdC_2 and LaC_2–DyC_2 (Adachi et al. 1976, 1974), as well as NdC_2–HoC_2 (Loe et al. 1976) has been determined by using the following equation:

$$E_s = \Delta H_{hypo} - \Delta H_{obs}, \tag{1}$$

where ΔH_{hypo} is the heat of transformation in a hypothetical strain-free solid solution and H_{obs} is the observed value. H_{hypo} can be calculated from eq. (2) by assuming that the solid solution is ideal:

$$\Delta H_{hypo} = x\Delta H_{RC_2} + (1-x)\Delta H_{R'C_2}, \tag{2}$$

where x is the mole fraction of RC_2 in the RC_2–$R'C_2$ solid solution, and ΔH_{RC_2} and $\Delta H_{R'C_2}$ are the heats of the transformation of RC_2 and $R'C_2$, respectively.

In order to find the relationship between E_s and volume difference, Adachi et al. (1978) introduced a new term (a ratio of ΔV to the volume of a major component dicarbide of the solid solution V, namely $\Delta V/V$), where ΔV = [unit cell volume of a given RC_2–$R'C_2$ solid solution] – [unit cell volume of pure RC_2 (>50 mol.% RC_2) or $R'C_2$ (>50 mol.% $R'C_2$)]. Finally, they discovered that in the light–light lanthanide dicarbide solid solution systems, such as LaC_2–CeC_2, LaC_2–NdC_2, $\Delta V/V$ is proportional to E_s, and in the heavy–light lanthanide systems, LaC_2–GdC_2, LaC_2–TbC_2 and LaC_2–DyC_2, straight lines are given only in the high-E_s region and curves in the low-E_s region. The intercepts obtained by an extrapolation of the straight-line portions are different. This difference in the intercepts may reflect the difference in bonding energy between La^{3+}–C_2^{2-} and the other R^{3+}–C_2^{2-} compounds (R′ = Ce, Nd, Gd, Tb, Dy).

These facts suggest that in light–light lanthanide systems such as LaC_2–CeC_2 and LaC_2–NdC_2 the effect of the volume difference ratio $\Delta V/V$ on the strain energy is predominant, while in light–heavy rare earth systems the bonding energy difference between the two rare earth ions to C_2^{2-}, as well as the volume difference, is significant (Adachi et al. 1978).

4.2. *Mixed rare earth carbide* $(R_{1-x}R_x)_{15}C_{19}$

Hájek et al. (1984b) remelted the carbon-reduced product containing Sc and Dy in a molar ratio of $R_{15}C_{19}$ at nearly 1800°C and obtained a $Sc_{15}C_{19}$-type mixed scandium–dysprosium phase, $(Sc_{0.94}Dy_{0.06})_{15}C_{19}$, in amounts detectable by X-ray diffraction ($P\bar{4}2_1c$, $a = 7.53 \pm 0.01$ Å, $c = 1.506 \pm 0.03$ Å). The presence of C_3 hydrocarbons in the products of hydrolysis of the mixed phases, in amounts of about 1–2 vol.% provides evidence of the occurrence of 6–12 wt.% $R_{15}C_{19}$ phases. No report with respect to the formation of other $Sc_{15}C_{19}$-type mixed crystals was found.

4.3. *Solid solutions of the rare earth carbides and the uranium carbides*

The pseudobinary systems UC_2–CeC_2 and UC_2–LaC_2 have been investigated (McColm et al. 1972), as well as the high-carbon portion of the uranium–gadolinium–carbon system (Wallace et al. 1964) and the solid solubilities of cerium, lanthanum and neodymium in UC and U_2C_3 (Lorenzelli and Marcon 1972, Stecher et al. 1964, Haines and Potter 1970). The phase diagrams from these investigations have been presented by Wallace et al. (1964), and McColm et al. (1972).

4.3.1. *Formation and the cubic-to-tetragonal transformation of* $(U_{1-x}R_x)C_2$

As in the mixed rare earth dicarbide systems, the rare earth dicarbides also form a solid solution with uranium dicarbide and there exists a composition dependence of the cubic-to-tetragonal phase transformation for the UC_2–LaC_2 or the UC_2–CeC_2 solid solutions (McColm et al. 1972), as well as for the UC_2–GdC_2 solid solution (Wallace et al. 1964).

For the U–Ce–C and the U–La–C systems, the existence of cubic phases down to room temperature has been established for ternary carbides within the composition limits $La_{0.10}U_{0.90}C_2$–$La_{0.85}U_{0.15}C_2$ and $Ce_{0.16}U_{0.84}C_2$–$Ce_{0.81}U_{0.19}C_2$ (McColm et al. 1972). In the U–Gd–C system the situation is more complex. UC_2 and GdC_2 form a continuous series of solid solutions above 1785°C, with the solid solution solidus and carbon eutectic temperatures decreasing in a regular manner from the UC_2 boundary to the GdC_2 boundary. In addition, there is a narrow two-phase region [tetragonal-(U, Gd)C_2 + cubic-(U, Gd)C_2] lying along a line connecting UC_2 at 1785°C and approximately $(U_{0.8}Gd_{0.2})C_2$ at 1510°C (fig. 14). At 1300°C the solid solution is still stable over the region from $(U_{0.12}Gd_{0.88})C_2$ to GdC_2. The lattice parameter of the quenched tetragonal (U, Gd)C_2 solid solution phase varies continuously with composition from $a_0 = 3.522$ Å and $c_0 = 5.982$ Å at UC_2 to $a_0 = 3.717$ Å and $c_0 = 6.264$ Å at GdC_2. The tetragonal-to-cubic transformation temperature of 1785 ± 20°C for UC_2 was found to decrease with the addition of GdC_2; likewise, this

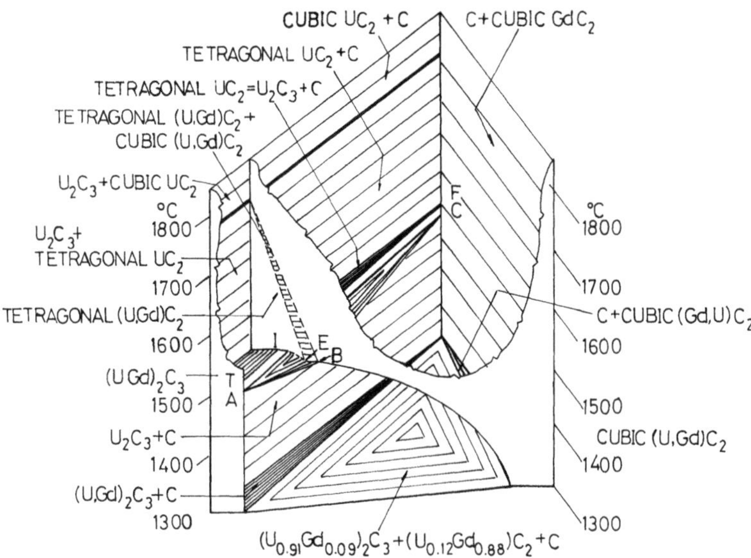

Fig. 14. Perspective drawing of the high-carbon portion of the Gd–U–C ternary system (Wallace et al. 1964). (Reprinted by permission of the publisher, The Electrochemical Society, Inc.)

transformation temperature of 1275 ± 20°C for GdC_2 was also found to decrease with the addition of UC_2, and thus a minimum occurs at 1155 ± 20°C and a composition of $(U_{0.34}Gd_{0.66})C_2$. However, such a minimum does not occur in the U–Ce–C and the U–La–C systems. McColm et al. found that there are two face-centered cubic phases present in the same melt over a composition range. One based on fcc UC_2 with a parameter increasing with increasing lanthanum or cerium content of the solid solution, and the other based on fcc LaC_2 or CeC_2 with a parameter falling with increasing uranium content. Thus, McColm et al. indicate a miscibility gap in the phase diagram, although the shape and the temperature to which it extends have not been fixed to date.

4.3.2. *Decomposition of* $(U_{1-x}R_x)C_2$

McColm et al. found that when some samples of the cubic phase $(La, U)C_2$ were annealed at 1720°C for 6 h in vacuum and cooled to below 800°C within about 3 min, the precipitation of UC occurred, showing that precipitation of UC begins from the cubic β phase instead of the tetragonal α phase. In addition, the experiments also showed that the ternary fcc solid solutions containing C_2 units do not occur when the overall carbon/total metal ratio falls below 2:1. In such preparations the phases in the melt are only UC and the lanthanide carbide expected from the melt stoichiometry (McColm et al. 1972).

In contrast to the decomposition products of the U–Ce (or La) dicarbide, $(U_{0.95}Gd_{0.05})C_2$ decomposed between 1515 and 1550°C, and $(U_{0.89}Gd_{0.11})C_2$ between 1515 and 1535°C, to give $(U, Gd)_2C_3$ and C, while the samples with a higher GdC_2 content decomposed into three phases: $(U, Gd)_2C_3$, $(U, Gd)C_2$ and C (Wallace et al. 1964). These results are in good agreement with the eutectoid decomposition of the pure uranium dicarbide to $U_2C_3 + C$ at 1500 ± 25°C (Langer 1963). The gadolinium-carbide-rich phase boundary of the $(U, Gd)_2C_3$ phase was determined to be $(U_{0.91}Gd_{0.09})_2C_3$, and is essentially constant in the temperature range 1300–1500°C, and the decomposition curve progresses from the UC_2 boundary (1525°C) to a composition of $(U_{0.12}Gd_{0.88})C_2$ at 1300°C.

4.3.3. *Solid solubility of R in uranium carbides*

The limiting solubilities of the rare earth elements in the uranium carbides and of uranium in the rare earth carbides have been determined, those of Ce in UC and of U in Ce_2C_3 are 4.5 and 5.5 at.% total metal, respectively (McColm et al. 1972). However, these results obtained from McColm et al. are significantly smaller than those reported by Stecher et al. (1964) and Haines and Potter (1970) (30 at.% Ce at 1600°C and 9.5 at.% Ce at 1450°C in UC, respectively), but in agreement with those obtained from an accurate determination of solubility limits in various phase fields of the U–Ce–C, U–Nd–C and U–La–C systems (Lorenzelli and Marcon 1972). The isothermal sections of the phase diagram at 1250°C and 1600°C have been carefully investigated for the system described above. At 1600°C the limited solubilities of Ce in the solid solution $(Ce_xU_{1-x})C$ in equilibrium with $U_2C_3 + CeC_2$, $Ce_2C_3 + CeC_2$, and $Ce_2C_3 + Ce$ are 1%, 7%, and 14%, respectively; those of Ce in U_2C_3 and U in Ce_2C_3 are 4.2% and 2.2%, respectively. However, the value of U in CeC_2 is very low. Similar values were obtained for the U–La–C and the U–Nd–C system, e.g. the values of R (La and Nd) in $(R_xU_{1-x})C$ in the presence of $U_2C_3 + RC_2$ and $RC_2 + R_2C_3$ are 0.5 to 1% and 2 to 8%, and those of La and Nd in U_2C_3 are 0.4 to 2.5%.

4.3.4. *Mechanism of decrease in transition temperature*

The mechanism of the decrease in the tetragonal-to-cubic transformation temperature by adding the rare earth atom to UC_2 has been discussed in terms of a strain energy model, which was also applied to the mixed rare earth dicarbide system, see sect. 4.1.3 (McColm et al. 1973, Adachi et al. 1974, 1976, 1978). The cubic-to-tetragonal transformation was described by Chang (1961) as a diffusionless transformation. McColm et al. (1972) suggested that this is an isothermal martensite-type transformation for $(U_{1-x}R_x)C_2$, and in such a case nucleation rather than growth is the rate-determining step. However, nucleation of the tetragonal phase within the cubic dicarbide must involve the coherent alignment of a considerable number of C_2 units over a number of unit cells to provide a viable domain of the tetragonal carbide. The random ⟨111⟩ orientation must be replaced by an ordered orientation along the tetragonal ⟨001⟩ direction. However, in a dicarbide containing solute ions with a different size from the host, this preferred orientation will be influenced by the presence of a nearest-neighbor metal ion larger or smaller than the others. C_2 units will tend to orient "side on" to a larger ion and "end on" to a smaller ion because this minimizes the strain in the lattice. Thus, each different-sized ion gives a preferred orientation. However, as the solute ions are distributed randomly, the preferred orientations would be incoherent and make more difficult any concerted action to align C_2 units parallel to each other and thus form a tetragonal nucleus. Only by further lowering the temperature to allow the M–C–C–M interactions to become important compared to the thermal energy, would the C_2 units be able to align in a given orientation.

This model gives a satisfactory explanation for the effect of size difference on the cubic-to-tetragonal transformation (McColm et al. 1972), and has been verified by the neutron diffraction of $Ce_{0.33}U_{0.67}C_2$ and $Ce_{0.67}U_{0.33}C_2$ over 296 to 4 K (Atoji 1980). In the two ternary CaC_2-type compounds the cubic NaCl-type structure can be arrested down to 4 K because of the large strain energy of the disordered lattice. A random distribution of Ce and U atoms as well as the orientational disordering of C_2 units are also retained down to 4 K. The lattice parameters indicate no valency change over the range 296–4 K, such as $Ce^{3+} \rightarrow Ce^{4+}$.

4.4. *Solid solutions (Ca, Y)C_2 and (Th, Y)$_2C_3$*

The solid solutions in the system CaC_2–YC_2 can be formed over the entire composition range; however, no Vegard's law behavior with respect to the composition dependence of the lattice parameters was found (Hájek et al. 1971), e.g., the solid solution $Ca_{0.33}Y_{0.66}C_2$ has essentially the same lattice parameters ($a = 3.62$ Å, $c = 6.06$ Å) as the yttrium dicarbide ($a = 3.62$ Å, $c = 6.05$ Å) (Brozek et al. 1970). The lattice parameters of the solid solution with 33 mol% YC_2 ($a = 3.64$ Å, $c = 6.16$ Å) are only slightly larger than those of YC_2 but markedly lower than those of CaC_2 ($a = 3.87$ Å, $c = 6.38$ Å), indicating that the anomalous valence state of the yttrium ion in these solid solutions leads to an irregular change in lattice parameters.

At high temperatures and under high pressures, Th formed a single-phase solid solution series with Y_2C_3 over a wide ternary composition field (Krupka et al. 1969). The new phase, crystallizing into the body-centered cubic Pu_2C_3-type structure, was superconducting over its entire range of homogeneity with a variable transition temperature $< 4 \sim 17$ K. Annealing at high temperature and ambient pressure destroyed both the bcc structure and the superconductivity.

5. Thermodynamic properties of binary rare earth carbides

5.1. *Thermodynamic stability of gaseous rare earth carbides*

Since the first observation of stable gaseous rare earth metal carbides by Chupka et al. (1958), the vaporization behavior and the composition of the vapor in metal–graphite systems have been extensively studied (Gingerich 1980, De Maria and Balducci 1972). In their initial work, Chupka et al. pointed out the pseudo-oxygen character of the C_2 radical, and showed the similarity between the bonding enthalpies of R–O and R–C_2. This pseudo-oxygen concept has been extended to the gaseous rare earth tetracarbides by Balducci et al. (1965). More recently, the molecules YC_n ($n = 2$–8) (Pelino et al. 1988a, b, Gingerich and Haque 1980) and Y_2C_n ($n = 1$–7) (Pelino et al. 1988b), CeC_n ($n = 1$–6) and Ce_2C_n ($n = 1$–6) (Gingerich et al. 1976, Kingcade et al. 1983), LaC_n ($n = 1$–8) (Gingerich et al. 1981b, 1982, Pelino and

Gingerich 1989) and La_2C_n ($n = 1$–6, 8) (Pelino et al. 1984, Pelino and Gingerich 1989), as well as ScC_n ($n = 2$–6) (Haque and Gingerich 1981), have been observed in the equilibrium vapor above the metal–graphite systems at high temperatures. Their thermodynamic characterizations, such as atomization energies etc., have been determined by the Knudsen effusion technique combined with mass spectrometry. The structure of the molecules have been inferred from the assumed models and consistency of the thermodynamic results based on various methods of evaluation.

The experimental work on ternary gaseous rare earth carbides has been pioneered by Guido and Gigli (1973) for a molecule containing one metal atom, CeSiC, and by Gingerich (1974) for molecules containing two different metals, $RhCeC_2$ and $PtCeC_2$. Since 1976, these mixed-metal carbide studies have been extended to a number of rare earth metals, including molecules of the type $M(R)C_n$ ($n = 1$–4), (Gingerich and Cocke 1979, Haque and Gingerich 1979, Gingerich et al. 1981a, Pelino et al. 1986).

5.1.1. *The species RC_n*

The high-temperature Knudsen effusion mass spectrometric technique has been used to study equilibria involving gaseous rare-earth-containing molecules. The existence of a large number of stable gaseous rare earth carbides has been established. The enthalpies ΔH°_0 of the reactions

$$R(g) + nC(\text{graphite}) \rightarrow RC_n(g) \quad (n = 1, 2, 3, \ldots, \text{etc.}) \tag{1}$$

were evaluated by the second-law method and the third-law method, as well as the atomization enthalpies, ΔH°_a and the standard enthalpies of formation, ΔH°_f. Table 7 shows the selected thermodynamic data of RC_n (gas) species obtained from the averages, with equal weight, of the second- and third-law values.

Some investigators also reported enthalpy changes for the reactions

$$RC_2(g) + R'(g) \rightarrow R'C_2(g) + R(g) \tag{2}$$

and

$$RC_2(g) + 2C(s) \rightarrow RC_4(g) \tag{3}$$

and further calculated the dissociation energies of the gaseous dicarbides of all lanthanides at 0 K (Filby and Ames 1971a, b, 1972) from the reaction (2) and the atomization energies of $RC_4(g)$ (R = Ce, Nd, Dy, Ho, Lu) (table 7) from the reaction (3) (Balducci et al. 1968, 1969a, b, Guido et al. 1972).

According to De Maria and co-workers (1972), the dicarbide is always the most abundant gaseous molecular species in the equilibrium vapor over the corresponding rare-earth-carbide–graphite system, followed by the tetracarbide. Table 8 shows examples of equilibrium partial-pressure ratios, $P(RC_2)/P(R)$ and $P(RC_4)/P(R)$, at the specified temperatures.

As to the partial pressures of the monolanthanum and monocerium carbides, LaC_n and CeC_n, in the equilibrium vapor above the rare-earth-metal–platinum-metal–graphite systems, the relative ion currents, which represent more than 90% ionization from the parent neutral molecules, and thus, the relative partial pressures of the latter, have been measured by means of the mass spectrometric method –

at 2835 K for LaC_n, LaC^+, 2.26×10^{-2}; $LaC_2{}^+$, 1.36; $LaC_3{}^+$, 1×10^{-2}; $LaC_4{}^+$, $7.15 + 10^{-2}$; $LaC_5{}^+$, 9.0×10^{-4}; $LaC_6{}^+$, 4.8×10^{-4}; $LaC_7{}^+$, 5.7×10^{-6}; $LaC_8{}^+$, 4.6×10^{-6} (Gingerich et al. 1981a, b, 1982, Gingerich 1985) and at 2733 K for CeC_n, CeC^+, 1; $CeC_2{}^+$, 5.3×10^{-1}; $CeC_3{}^+$, 9.09; $CeC_4{}^+$, 1.6×10^{-2}; $CeC_5{}^+$, 3.2×10^{-2}; $CeC_6{}^+$, 1.6×10^{-4}; $CeC_7{}^+$ 8.1×10^{-5} (Gingerich et al. 1976a, b). It can be seen from these data that the molecules LaC_2 (Chupka et al. 1958, Stearns and Kohl 1971), LaC_3 and LaC_4 (Stearns and Kohl 1971) are the most abundant molecular lanthanum carbides. The additional equilibrium species have a concentration of the order of 0.1% or less of the vapor. In addition, it is also apparent that up to RC_5 the molecules with an odd number of atoms, in fact, those containing the C_2 radical, have a larger partial pressure than the molecules with the preceding even number of atoms.

The first gaseous rare earth monocarbide, CeC, was observed in the study on the chemical equilibria of the reaction

$$CeC_2(g) + Ce(g) \rightarrow 2CeC(g),$$

by the mass spectrometer and Knudsen effusion cell method. Its dissociation energy D_0° is 301.2 ± 19.6 kJ mol.$^{-1}$ (Balducci et al. 1967). Subsequently, Kingcade et al. (1983) studied the reaction

$$CeC(g) \rightarrow Ce(g) + C(s)$$

and obtained the values $\Delta H^\circ_{f,298.15} = 694 \pm 12$ and $\Delta H^\circ_{a,0} = 441 \pm 12$ kJ mol^{-1}, as listed in table 7. More recently, the monocarbide of lanthanum, LaC, was also found by Pelino and Gingerich (1989); its standard enthalpy of formation $\Delta H^\circ_{f,298.15} = 685 \pm 20$ kJ mol^{-1} and atomization enthalpy, $\Delta H^\circ_{a,0} = 458 \pm 20$ kJ mol^{-1} are close to the corresponding values of the cerium monocarbide. However, these monocarbides of rare earths are stable only in the gaseous phase but fail to be found in the solid state as noted earlier, see sect. 2.6.4.

The high-temperature Knudsen effusion mass spectrometric studies of the vaporization of the rare-earth-carbide–carbon systems also lead to a good knowledge of the bond dissociation energies D_0° (R–C_2), D_0° (C_2–R–C_2), D_0° (C–R) and D_0° (R–C_3), etc. (table 9). These values are in good agreement with the dissociation energies of RC_2(g) as determined from the exchange reaction (Filby and Ames 1971a, b), although different thermodynamic functions were used and such a comparison may lead to a misleading conclusion.

The R–C_2 bond energy for the dicarbides and the tetracarbides are almost equal and the R–C_2 bond strength decreases on going from La to Y to Sc, analogous to the trend observed for the oxides LaO, YO, and ScO (Drowart and Goldfinger 1967). These findings tempt many investigators to make predictions about the existence and physical stability of the RO_2(g) molecule although no isoelectronic RO_2 molecule has been observed. In addition, from a comparison of the bond energies between R–C, R–C_2, R–C_3, R–C_4 and R–C_5, the most preferable structure of the polyatomic carbides could be found; for instance, the linear structures C_2–R–C_3 and C_3–R–C_3 for RC_5 and RC_6 (R = Ce, Sc), respectively (Haque and Gingerich 1981).

Of course, it is best to determine the molecular and electronic structures of these carbides directly by optical spectroscopy, however, no information is available.

TABLE 7

Selected reaction enthalpies, ΔH°_0 and ΔH°_{298} for the reaction R(g) + nC (graphite) = RC_n(g); atomization enthalpies. $\Delta H^\circ_{a,0}$ and $\Delta^\circ_{a,298}$, standard enthalpies of formation, $\Delta H^\circ_{f,298}$ of RC_n (g) species (kJ mol^{-1}).

Species	ΔH°_0	ΔH°_{298}	$\Delta H^\circ_{a,0}$	$\Delta H^\circ_{a,298}$	$\Delta H^\circ_{f,298}$	Temperature range (K)	References
CeC(g)	272 ± 3.6		441 ± 12		694 ± 12	2200–2800	Kingcade et al. (1983)
LaC(g)			458 ± 20		685 ± 20	2286–2609	Pelino and Gingerich (1989)
LaC_2(g)	159.3 ± 40	161.2 ± 4.0	1263 ± 5	1272.2 ± 5.0	592.2 ± 6.0	2286–2609	Gingerich et al. (1981)
LaC_2(g)		162.4 ± 15				2240–2564	Chupka et al. (1958)
LaC_2(g)			1257 ± 8			2267–2600	Stearns and Kohl (1971)
CeC_2(g)	145.2 ± 4.5 (138 ± 29)[b]		1274 ± 6			1990–2300	Balducci et al. (1969a)
CeC_2(g)	154.2 ± 1.8		1268 ± 5		587 ± 2.7	2200–2800	Kingcade et al. (1983)
PrC_2(g)	184.2(165 ± 17)[b]		1234.2				Balducci et al. (1969c)
NdC_2(g)	205		1213.5				Balducci et al. (1969c)
NdC_2(g)	207.6 ± 11.7		1218 ± 16.8			1800–2300	de Maria et al. (1967)
EuC_2(g)	284.4 ± 12.6		1134 ± 17		539.8 ± 20	1800–2300	Balducci et al. (1972)
GdC_2(g)	(ΔH_{2150}158.7 ± 2)						Jackson et al. (1963)
GdC_2(g)	184.2		1234				Balducci et al. (1969c)
DyC_2(g)	266.4		1153				Balducci et al. (1969c)
DyC_2(g)	268 ± 8.1		1151 ± 9.3			2275	Balducci et al. (1969b)
HoC_2(g)	270 ± 4 (242 ± 33)[b]		1149.3 ± 6			2340	Balducci et al. (1969b)
ErC_2(g)	261.6 ± 11		1157.4			2185–2490	Balducci et al. (1969d)
LuC_2(g)	213.3 ± 12.5		1205 ± 17			2250–2543	Guido et al. (1972)
ScC_2(g)	264.3 ± 0.9		1155 ± 21			2165–2328	Kohl and Stearns (1971b)
ScC_2(g)	237.2 ± 8.4		1185 ± 14	1195 ± 1	617 ± 12	1960–2530	Haque and Gingerich (1981)
YC_2(g)	197 ± 5	199 ± 5	1225 ± 8	1235 ± 8	623 ± 8	2068–2739	Pelino et al. (1988)
YC_2(g)	165.3 ± 4		1257 ± 15			1824–2530	Gingerich and Haque (1980)
YC_2(g)	190 ± 17		1229 ± 17			2274–2552	Kohl and Stearns (1970)
YC_2(g)	174		1247 ± 17			2075–2340	de Maria et al. (1965)
LaC_3(g)	396 ± 31		1759 ± 31			2397–2600	Stearns and Kohl (1971)

$LaC_3(g)$	314.5 ± 5	317 ± 5	1819 ± 7	1833 ± 7	748 ± 7	2331–2609	Gingerich et al. (1981a,b)
$CeC_3(g)$	300.8 ± 5		1833 ± 11		740 ± 11	2200–2800	Kingcade et al. (1983)
$ScC_3(g)$	355.0 ± 16		1780 ± 23	1793.9 ± 23	735 ± 20	2318–2530	Haque and Gingerich (1981)
$YC_3(g)$	377 ± 10	379 ± 10	1757 ± 12	1711 ± 12	804 ± 12	2403–2792	Pelino et al. (1988)
$YC_3(g)$	328.8 ± 8.8		1805 ± 30			2392–2530	Gingerich et al. (1981a,b)
$LaC_4(g)$	317.5 ± 5	322.6 ± 5	2527 ± 7	2544 ± 7	754 ± 8	2286–2609	Gingerich and Haque (1980)
$LaC_4(g)$	328.4 ± 18.6		2500 ± 16			2417–2600	Stearns and Kohl (1971)
$CeC_4(g)$	333.5 ± 3		2512 ± 9		759 ± 9	2200–2800	Kingcade et al. (1983)
$CeC_4(g)$	(295.8 ± 29)[b]		2548 ± 21[a]			1800–2500	Balducci et al. (1969a)
$NdC_4(g)$	344 ± 0.8[a]		2493.5 ± 21[a]			2210–2350	Balducci et al. (1968)
$DyC_4(g)$			2409.5 ± 22[a]			2000–3000	Balducci et al. 1969b)
$HoC_4(g)$	(389 ± 50)[b]		2412.8 ± 9.6[a]			2000–3000	Balducci et al. (1969b)
$LuC_4(g)$			2452 ± 38[a]			2429–2517	Guido et al. (1972)
$ScC_4(g)$	441 ± 21		2397 ± 21			2267–2328	Kohl and Stearns (1971b)
$ScC_4(g)$	390.2 ± 14		2455 ± 22	2473 ± 22	772 ± 18	2297–2530	Haque and Gingerich (1981)
$YC_4(g)$	361 ± 8	365 ± 8	2484 ± 10	2502 ± 10	790 ± 10	2360 ± 2792	Pelino et al. (1988)
$YC_4(g)$	314.5 ± 4		2530 ± 24			2228–2530	Gingerich and Haque (1980)
$YC_4(g)$	377.3 ± 17.6		2461 ± 18			2739–2552	Kohl and Stearns (1970)
$LaC_5(g)$	435 ± 42	440 ± 42	3121 ± 45	3144 ± 45	870 ± 42	2000–3200	Gingerich et al. (1981a,b)
$CeC_5(g)$	444.6 ± 2.4		3111 ± 15		872 ± 15	2200–2800	Kingcade et al. (1983)
$ScC_5(g)$	416.6 ± 20		3140 ± 29	3163 ± 29	798 ± 24	2318–2530	Haque and Gingerich (1981)
$YC_5(g)$	499 ± 10	500 ± 10	3057 ± 15	3084 ± 15	924 ± 15	2532–2792	Pelino et al. (1988)
$YC_5(g)$	475 ± 8		3081 ± 35			2392–2498	Gingerich and Haque (1980)
$LaC_6(g)$	499.8 ± 42		3795 ± 45		936 ± 42	2000–3200	Gingerich et al. (1981a,b)
$CeC_6(g)$	512.4		3748 ± 14		945.7 ± 14	2200–2800	Kingcade et al. (1983)
$ScC_6(g)$	490.6 ± 24		3776.6 ± 34	3803.9 ± 34	874.2 ± 28	2318–2530	Haque and Gingerich (1981)
$YC_6(g)$	536 ± 15	541 ± 15	3731 ± 20	3759 ± 20	966 ± 20	2532–2792	Pelino et al. (1988)
$YC_6(g)$	483.7		3780 ± 60			2445	Gingerich and Haque (1980)
$LaC_7(g)$	656.5 ± 7.8	664.1 ± 7.8	4320 ± 60	4351 ± 60	1095 ± 50	2668–2840	Gingerich et al. (1982)
$YC_7(g)$	684 ± 35	690 ± 35	4294 ± 35	4327 ± 35	1105 ± 35	2739–2792	Pelino et al. (1988)
$LaC_8(g)$	700.4	708.9	4989 ± 70	5024 ± 70	1140 ± 60	2668–2840	Gingerich et al. (1982)
$YC_8(g)$	693 ± 35	699 ± 35	4727 ± 35	5034 ± 35	1124 ± 35	2739–2792	Pelino et al. (1988)

[a] These data were calculated from the reaction ΔH_0° for the reaction $RC_2(g) + 2C(s) \rightarrow RC_4(g)$.

[b] Taken from Balducci et al. (1965).

TABLE 8

Partial-pressure ratios $P(RC_2)/P(R)$ and $P(RC_4)/P(R)$ of rare-earth-metal–carbide–graphite systems.

Rare earth metal	T(K)	$P(RC_2)/P(R)$	$P(RC_4)/P(R)$	Reference
Sc	2319	2.6×10^{-3}	6.6×10^{-6}	Kohl and Stearns (1971b)
Y	2504	1.2×10^{-1}	1.3×10^{-3}	Kohl and Stearns (1970)
La	2561	1.2	2.4×10^{-2}	Stearns and Kohl (1971)
Ce	2500	1.0	4×10^{-2}	Balducci et al. (1969a)
Pr	2500	3.7×10^{-1}	1.3×10^{-4}	Balducci et al. (1970)
Nd	2500	1.5×10^{-1}	7.5×10^{-3}	Balducci et al. (1970)
Gd	2500	3.1×10^{-1}		Balducci et al. (1970)
Dy	2500	6.2×10^{-3}	1×10^{-4}	Balducci et al. (1970)
Ho	2500	5.8×10^{-3}	1×10^{-4}	Balducci et al. (1970)
Er	2500	1.1×10^{-2}		Balducci et al. (1970)
Eu	2320	1.3×10^{-3}		Balducci et al. (1972)

TABLE 9

Bond dissociation energies[a] of R–C_2, C_2–R–C_2 and R–C D_0°(kJ mol^{-1}).

R–C_2	D_0°(R–C_2)	Reference	R–C_2	D_0°(R–C_2)	Reference
Sc–C_2	565 ± 21	Verhagen et al. (1965)	Nd–C_2	619 ± 21	de Maria et al. (1965)
	566 ± 21	Kohl and Stearns (1971b)	Eu–C_2	539.7 ± 21	Balducci et al. (1972)
Y–C_2	634 ± 19	Kohl and Stearns (1970)	Dy–C_2	556.5 ± 12	Balducci et al. (1969b)
	653 ± 21	de Maria et al. (1965)	Ho–C_2	556.5 ± 9	Balducci et al. (1969b)
La–C_2	664	Gingerich et al. (1981b)	Er–C_2	565 ± 13	Balducci et al. (1969d)
	668 ± 8	Stearns and Kohl (1971)	Lu–C_2	610 ± 21	Guido et al. (1972)
Ce–C_2	678 ± 8	Balducci et al. (1969a)			
Pr–C_2	661 ± 25	Balducci et al. (1965)			

C_2–R–C_2	D_0°(C_2–R–C_2)	Reference	C_2–R–C_2	D_0°(C_2–R–C_2)	Reference
C_2–Sc–C_2	1217 ± 22	Kohl and Stearns (1971b)	C_2–Ce–C_2	1427 ± 46	Balducci et al. (1965)
C_2–Y–C_2	1271 ± 21	Kohl and Stearns (1970)	C_2–Ho–C_2	1259 ± 80	Balducci et al. (1965)
C_2–La–C_2	1329 ± 16	Gingerich et al. (1981b)			

R–C	D_0°(Ce–C) $= 452\pm29$ kJ mol^{-1} Gingerich (1969)
	D_0°(Sc–C) $= 439.8\pm21$ kJ mol^{-1} (Haque and Gingerich 1981)
R–C_3, C_4, C_5	D_0°(Ce–C_3) $= 530$ kJ mol^{-1}; D_0°(Ce–C_5) $= 533$ kJ mol^{-1} (Gingerich et al. 1976a)
	D_0°(La–C_3) $= 505$ kJ mol^{-1}; D_0°(La–C_4) $= 745$ kJ mol^{-1} (Gingerich et al. 1981)

[a] These values are in good agreement with the dissociation energies of RC_2(g) as determined from the exchange reaction (Filby and Ames 1971b), despite the different thermodynamic functions used, but such a comparison may lead to misleading conclusions.

Therefore, assumptions have to be made concerning the low-lying electronic states and their multiplicities, the molecular geometry model and the force constants in order to calculate the thermal functions needed in the evaluation of the bond energies from the mass spectrometric data. In general, indirect information as to the probable favorable geometry of a new molecule can be obtained if reliable second- and third-law evaluations can be performed. For example, for the molecules LaC_3 and LaC_5, the linear chain structure with the lanthanum atom at the end of the carbon chain appears to be favored on the basis of the best agreement between the second and third laws, and also in terms of the smallest standard deviation in the third-law reaction enthalpy.

Concerning the possible structure of the higher complex carbides (RC_n), the best agreement between the two thermodynamic treatments of the data and the smallest associated standard deviation were achieved with the linear symmetric structure for RC_2 and for RC_4 (Pelino et al. 1988a, b, Gingerich et al. 1981b, Kohl and Stearns 1971a, b), confirming the assumptions made by De Maria et al. (1965) and Stearns and Kohl (1971) about the structure of these molecules. In the same way with the linear asymmetric structure ($R–C_n$) for RC_3, RC_5 and RC_6 (R = La, Y), one finds that the $R–C_n$ structure is preferred to the linear structure $C_n–R–C_m$ with the R atom in the center of the carbon chain (Pelino et al. 1984, 1988a, b, Gingerich et al. 1981b). However, for RC_3, RC_5 and RC_6 (R = Ce, Sc), the linear symmetric structures $C_1–R–C_2$, $C_2–R–C_3$ and $C_3–R–C_3$, respectively, were found to be the ones that are most preferred (Kingcade et al. 1983, Haque and Gingerich 1981).

Finally, it should be pointed out that for the molecular series LaC_n (Gingerich et al. 1981b), CeC_n (Kingcade et al. 1983) and YC_n (Pelino et al. 1988a, b), the experimental atomization energies show an alternation of bond energies for each added carbon, with the molecules containing an even number of carbon atoms being more stable and, correspondingly, the relative abundances of these gaseous carbides that contain an even number of carbon atoms are higher than those of the corresponding species with an odd number of carbon atoms in the molecule. These facts further confirm the assumption made by Chupka et al. (1958) on the pseudo-oxygen character of the C_2 radical in the gaseous rare earth carbide molecules.

5.1.2. *The species R_2C_n*

The high-temperature Knudsen effusion mass spectrometric technique has also been used to identify and characterize the La_2C_n (n = 2–6, 8) (Pelino et al. 1984), Ce_2C_n (n = 1–6) (Kingcade et al. 1984) and Y_2C_n (n = 2–8) (Pelino et al. 1988a, b) molecules in the equilibrium vapor above the La–Ir–graphite system at 2220–2831 K, the cerium–graphite system at 2100–2800 K and the Y–Ir–Au–graphite system at 2532–2792 K, respectively. Recently, the stable gaseous dilanthanum monocarbide La_2C has also been observed (Pelino and Gingerich 1989). The enthalpies ΔH°_0 of the reaction $2R(g) + nC(graphite) \rightarrow R_2C_n(g)$ were evaluated using the third law, and only for two dilanthanum carbides, $La_2C_3(g)$ and $La_2C_4(g)$, a second-law evaluation was also performed, indicating the linear symmetric structure, $R–C_n–R$, as being the most probable.

The selected values for the reaction enthalpies were combined with ancillary literature data to yield the atomization energies $\Delta H^\circ_{a,0}$ and the standard heats of formation $\Delta H^\circ_{f,298.15}$ of the gaseous di-rare-earth carbides (table 10).

In the equilibrium vapor above the rare-earth-metal–graphite systems described above at high temperature, the concentration of the molecules R_2C_4 is greater than

TABLE 10
Selected reaction enthalpies ΔH°_0, ΔH°_{298}, for the reaction $2R(g) + nC(graphite) \rightarrow R_2C_n(g)$ and the atomization energies, $\Delta H^\circ_{a,0}$ and $\Delta H^\circ_{a,298}$, and the standard heats of formation, $\Delta H^\circ_{f,218}$ of gaseous R_2C_n ($kJ\,mol^{-1}$).

Species	ΔH°_0	ΔH°_{298}	$\Delta H^\circ_{a,0}$	$\Delta H_{a,298}$	$\Delta H^\circ_{f,298}$	Reference
La_2C			945±30		625±30	Pelino and Gingerich (1989)
La_2C_2	−247.2±40	−247.9±40	1670±40	1682±40	621±40	Pelino et al. (1984)
La_2C_3	−205.5±27	−203.9±27	2339±27	2355±27	658±27	Pelino et al. (1984)
La_2C_4	−214.5±21	−213.4±21	3060±21	3081±21	648±21	Pelino et al. (1984)
La_2C_5	−146.8±50	−144.9±50	3703±50	3728±50	717±50	Pelino et al. (1984)
La_2C_6	−127.7±50	−125.1±50	4395±50	4425±50	737±50	Pelino et al. (1984)
La_2C_8	31.0±60	35.8±60	5658±60	5697±60	898±60	Pelino et al. (1984)
Ce_2C	336±4	339	1047±26	1056	506±26	Kingcade et al. (1984)
Ce_2C_2	268±7	269	1690±25	1702	576±25	Kingcade et al. (1984)
Ce_2C_3	198±1	198	2332±28	2348	647±28	Kingcade et al. (1984)
Ce_2C_4	231±8.6	230	3076±25	3097	615±25	Kingcade et al. (1984)
Ce_2C_5	98±9	95	3654±32	3679	750±32	Kingcade et al. (1984)
Ce_2C_6	79±4	75	4346±36	4375	4770±36	Kingcade et al. (1984)
Y_2C_2						
Y_2C_3						
Y_2C_4						
Y_2C_5						
Y_2C_6						
Y_2C_7						
Y_2C_8						

that of the R_2C_3 species and is approximately equal to that of the pentacarbides and hexacarbides. The remaining dimetal carbides have a concentration approximately an order of magnitude less than that of the dimetal tetracarbides. For example, the relative ion currents over the La–Ir–graphite system measured with 20 V electrons at 2829 K are: 1.5×10^{-5} (La_2C^+), 4.6×10^{-5} ($La_2C_2{}^+$), 3.9×10^{-5} ($La_2C_3{}^+$), 3.4×10^{-4} (La_2C_4), 6.5×10^{-5} ($La_2C_5{}^+$), 1.3×10^{-4} ($La_2C_6{}^+$), 3.0×10^{-6} (La_2C_7) and 4.6×10^{-6} (La_2C_8) (Gingerich 1985). Therefore, for these dimetal carbides it is interesting to note that the ones with an even number of carbon atoms appear to be favored; in particular, the R_2C_4 molecules, which may be considered as the dimer of the very stable (abundant) RC_2 molecules, are most abundant. In conclusion, it can be said that the pseudo-oxygen character of the C_2 group bonded with an electron donor atom forms a stronger bond for an even number of carbon atoms while the covalent bonds in a carbon chain are more stable for an odd number of atoms (Pelino et al. 1988a, b).

5.1.3. *Ternary gaseous rare-earth-containing carbides*

The ternary rare-earth-metal–platinum-metal carbides have been observed in the Knudsen effusion mass spectrometric investigations of the gaseous species over the (La, Ce, Y, Sc)–(Pt, Rh, Ir, Ru, etc.)–graphite systems. The existence of RhCeC, $RuCeC_2$, RuCeC, $RuCeC_2$ and $PtCeC_2$ (Gingerich et al. 1976b, Gingerich 1974),

TABLE 11
Atomization energies, $\Delta H^{\circ}_{a,0}$, of ternary gaseous rare-earth-containing carbides ($kJ\,mol^{-1}$).

Molecule	$\Delta H_{a,0}$	Reference
CeSiC	1046 ± 42	Guido and Gigli (1973)
LaIrC	1027 ± 30	Pelino et al. (1986)
YIrC	996 ± 33	Gingerich et al. (1981a)
CeRhC	1031 ± 40	Gingerich and Cocke (1979)
ScRhC	1010 ± 40	Haque and Gingerich (1979)
CeRuC	1109 ± 40	Gingerich and Cocke (1979)
$ScRhC_2$	1629 ± 50	Haque and Gingerich (1979)
$YRhC_2$	1672 ± 50	Haque and Gingerich (1979)
$CePtC_2$	1695 ± 50	Gingerich (1974)
$YIrC_2$	1678 ± 33	Gingerich et al. (1981a)
$LaIrC_2$	1779 ± 25	Pelino et al. (1986)
$LaIrC_3$	2344 ± 35	Pelino et al. (1986)
$LaIrC_4$	2966 ± 35	Pelino et al. (1986)

RhScC, $RhScC_2$ and $RhYC_2$ (Haque and Gingerich 1979), etc., have been established. The relative concentrations (relative ion currents) of the molecules, $LaIrC^+$, $LaIrC_3{}^+$ and $LaIrC_4{}^+$ over the La–Ir–graphite system measured with 20 V electrons at 2829 K have been reported to be 2.2×10^{-4}, 7.9×10^{-6} and 3.9×10^{-6}, respectively (Gingerich 1985). Owing to the presence of a C–C bond (≈ 600 kJ mol.$^{-1}$) and a R–C_2 bond (≈ 650 kJ mol.$^{-1}$) and multiple R–M bonds (400–500 kJ mol.$^{-1}$), these compounds are highly stable. Their atomization energies, $\Delta H^{\circ}_{a,0}$, have been determined (table 11).

5.2. *Thermodynamic properties of the solid state rare earth carbides*

The vaporization studies of the solid rare earth carbides by using the mass spectrometry Knudsen effusion method have succeeded in determining the standard enthalpy of formation, $\Delta H^{\circ}_{f,298}$, for some of the solid rare earth dicarbides through the reaction

$$RC_2(s) \rightarrow R(g) + 2C(s),$$

but it is necessary to use estimated thermal functions when reducing the equilibrium data taken at high temperatures upto 298.15 K.

The entropy of formation at 298 K for several RC_2 phases and a few R_2C_3 phases were calculated by Gschneidner and Calderwood (1986) in their progress report using, respectively, the known values of 70.29 J mol.$^{-1}$ K^{-1} for CaC_2 and of 47.46 J mol.$^{-1}$ K^{-1} for Ho_2C_3 (Wakefield et al. 1965), as well as assuming the validity of the following relationships:

$$S_{RC_2} = S_{CaC_2} + S_R - S_{Ca}$$

and

$$S_{R_2C_3} = S_{Ho_2C_3} + 2S_R - 2S_{Ho}.$$

Values of S_R, S_{Ca} and S_{Ho} were taken from Hultgren et al. (1973).

On the basis of the experimental high-temperature mass spectrometric thermodynamic data and the estimated free-energy functions and heat content data for the dicarbides of the rare earth metals, the enthalpies of formation of solid RC_2 have been reported by a number of scientists (table 12). It should be noted that the data reported by Anderson and Bagshaw (1970, 1972) were obtained from a study by using a solid state galvanic cell with a calcium fluoride electrolyte. The reduction of their data by the second law gave erroneous results but that by the third law gave values for CeC_2, LaC_2 and YC_2 that compared favorably with those obtained by mass spectrometry. In addition, the data from Baker et al. (1971) were obtained by using bomb calorimetry.

Anderson and Bagshaw (1970) reported high-temperature e.m.f. data for La_2C_3, Ce_2C_3, Pr_2C_3 and Nd_2C_3. Gschneidner and Calderwood (1986), utilizing free-energy function estimates and the $\Delta G^{\circ}_{f,T}$ equations presented by Anderson and Bagshaw, calculated $\Delta G^{\circ}_{f,298}$ values, and thus the $\Delta H^{\circ}_{f,298}$ values ($\Delta H^{\circ}_{f,T} = \Delta G^{\circ}_{f,T} - \Delta S^{\circ}_{f,T}T$).

TABLE 12
Enthalpy of formation of solid RC_2 (kJ mol^{-1}).

Species	Anderson and Bagshaw (1972)	Faircloth et al. (1968)	Other references
LaC_2	−87.4	−108 (second law) −75 (third law)	−89±24 (Stearns and Kohl 1971)
CeC_2	−88.7	−104.6	−97±5.4 (Baker et al. 1971), −81.6 (Winchell and Baldwin 1967), −63±20 (Balducci et al. 1969a, b, c, d)
PrC_2	−84.5		
NdC_2	−88.7	−52.3±10.5	
SmC_2		−65.2±6.7	−97.9 (Stout et al. 1969)
EuC_2		−66.9±5.4	−38.3 (Gebelt and Eick 1966a, b), −67.5 (Cuthbert et al. 1967)
GdC_2	−101.2		−82 (Jackson et al. 1963), −125.5 (Hoenig et al. 1967) −88 (Balducci et al. 1969b), −110±15 (Huber et al. 1973)
DyC_2	−94.1		−46.2 (Balducci et al. 1969b)
HoC_2			−88±4 (Balducci et al. 1969b), −82 (Wakefield et al. 1965)
ErC_2	−110.8		−86.4 (Balducci et al. 1969d)
TmC_2			−98.7 (Seiver and Eick 1971)
YbC_2			−75.3±4 (Haschke and Eick 1968)
LuC_2			−117±21 (Guido et al. 1972)
YC_2	−105.0		−91±17 (Kohl and Stearns 1970), −119 (Storms 1971), −98 (de Maria et al. 1965)

TABLE 13
Entropy, free energy and enthalpy of formation of solid R_2C_3 at 298 K.

Species	$\Delta S_{f,298}$ ($J\,mol^{-1}$)	$-\Delta G_{f,298}$ ($kJ\,mol^{-1}$)	$-\Delta H_{f,298}(T\Delta S - \Delta G)$ ($kJ\,mol^{-1}$)	Reference
La_2C_3	−24.18[a]	108.69[a]	115.89	Anderson and Bagshaw (1970)
Ce_2C_3	40.77[a]	111.72[a]	105.99	Anderson and Bagshaw (1970), Baker et al. (1971)
Pr_2C_3	46.14[a]	97.65[a]	83.91	Anderson and Bagshaw (1970)
Nd_2C_3	42.72[a]	114.99[a]	102.27	Anderson and Bagshaw (1970)
Sm_2C_3	40.83[a]	123.18	111	Haschke and Deline (1982)
Ho_2C_3	47.40	48.93	34.8	Wakefield et al. (1965)

[a] These data are obtained from Gschneidner and Calderwood (1986).

The calculated values are listed in table 13, together with a few experimental values for $SmC_{1.43}$ (Haschke and Deline 1982) and Ho_2C_3 (Wakefield et al. 1965).

For other binary rare earth carbides, a few thermodynamic data have also been determined. For example, Storm (1971) reported the free energy of formation at 1700 K, $-64.3\,kJ\,mol^{-1}$ for Y_2C; $-112.5\,kJ\,mol^{-1}$ for Y_5C_6; $-125.44\,kJ\,mol^{-1}$ for Y_2C_3 and $-143.22\,kJ\,mol^{-1}$ for YC_2, but these values have not been reduced to 298.15 K. In addition, several investigators reported only high-temperature ΔH_v values for the RC_2 phases. These values are: $\Delta H_{v,2200} = 543.0\,kJ\,mol^{-1}$ for LaC_2 (Jackson et al. 1963), $\Delta H_{v,2145} = 149.1 \pm 13.8\,kJ\,mol^{-1}$ for NdC_2 (de Maria et al. 1967), and $\Delta H_{v,1690} = 272.7 \pm 16.8\,kJ\,mol^{-1}$ for SmC_2 (Avery et al. 1967) for the reaction $RC_2(s) \rightarrow R(g) + 2C(s)$; $\Delta H_{v,2150} = 596.7 \pm 2.4\,kJ\,mol^{-1}$ for GdC_2 (Jackson et al. 1963) for the reaction $GdC_2(s) \rightarrow GdC_2(g)$; $\Delta H_{v,1607} = 509.0 \pm 33.5\,kJ\,mol^{-1}$ for Sm_2C_3 (Avery et al. 1967) for the reaction $Sm_2C_3(s) \rightarrow 2Sm(g) + 3C(s)$.

In addition to the experimental thermodynamic values, Niessen and de Boer (1981) calculated the enthalpies of formation at 0 K according to a semiempirical model proposed by Miedema et al. (1980) for the carbides R_2C, RC and RC_2, where R = La, Sc, Y, $\Delta H_{f,\text{room temperature}} = 99$, 132 and 111 for La_2C, Sc_2C and Y_2C; 100, 132 and 112 for LaC, ScC and YC; 180, 210 and 192($kJ\,mol^{-1}$) for LaC_2, ScC_2 and YC_2, respectively. Comparison of the predicted heats of formation for LaC_2 and YC_2 with the experimental values shows these predicted values to be inaccurate by about 100%.

6. Ternary rare-earth–X–carbon phase diagrams and ternary carbides

6.1. *Phase diagrams and formation of ternary carbides in R–B–C systems*

6.1.1. *Ternary R–B–C phase diagrams*

Although only a few phase diagrams of the R–boron–carbon systems [R = Y (Bauer and Nowotny 1971), Eu (Schwetz et al. 1979), Gd (Smith and Gilles 1967), and Ho (Bauer et al. 1985)] have been published, several studies on the formation of the rare earth borocarbides have been carried out by researchers. In ternary R–B–C systems, there are nine compounds, i.e. the RB_2C_2, RB_2C, RBC_3, RBC, RB_2C_4, $R_5B_2C_5$, $R_5B_2C_6$, $R_{15}B_2C_{17}$ and R_2BC_2 compounds. Their existences and crystal structures have been reported.

6.1.1.1. *The yttrium–boron–carbon phase diagram.*
Bauer and Nowotny (1971) investigated the ternary system Y–B–C. The arc-melted alloys were quenched without additional annealing. Four ternary phases YB_2C_2, YB_2C, YBC and $YB_{0.5}C$ were found, all of which melt congruently. The phase diagram of the ternary system Y–B–C in the as-quenched condition is shown in fig. 15.

According to their X-ray diffraction data, three of the four ternary compounds have been identified (table 14), the crystal structures have been determined and one of these structures is shown in fig. 16.

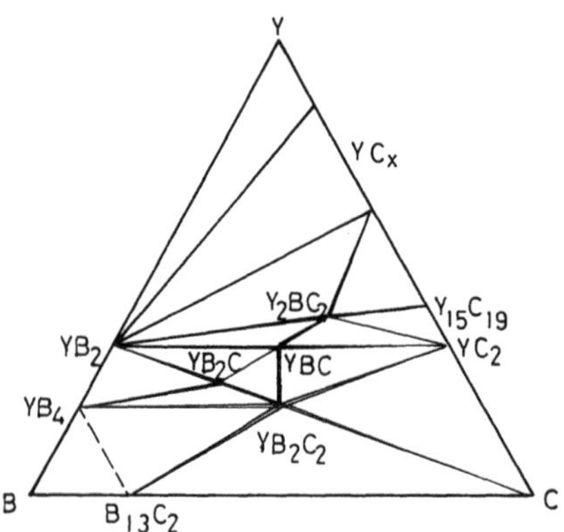

Fig. 15. Phase diagram of the Y–B–C system in the as-quenched condition (Bauer and Nowotny 1971). (Reprinted by permission of the publisher, Institut für Anorganishe Chemie, Inc.)

TABLE 14
Crystal structures of ternary yttrium borocarbides.

Compound	Crystal structure	Lattice parameters (Å)	Interatomic distance (Å)					
			Y–Y	Y–B	Y–C	B–C	C–C	B–C
YB_2C_2	Tetragonal	$a = 3.79(6)$ $c = 7.12(4)$ $c/a = 1.876$	3.68[a]	2.75	2.68	1.76	1.28	1.62
YB_2C	Tetragonal	$a = 6.76(9)$ $c = 7.430$ $c/a = 1.096$	3.60[a]	2.73	2.55	1.75	–	1.64
YBC	Orthorhombic	$a = 3.38(8)$ $b = 13.69(3)$ $c = 3.62(7)$	3.71[a]	2.70[a]	2.55	1.98	–	1.65

[a] The average value.

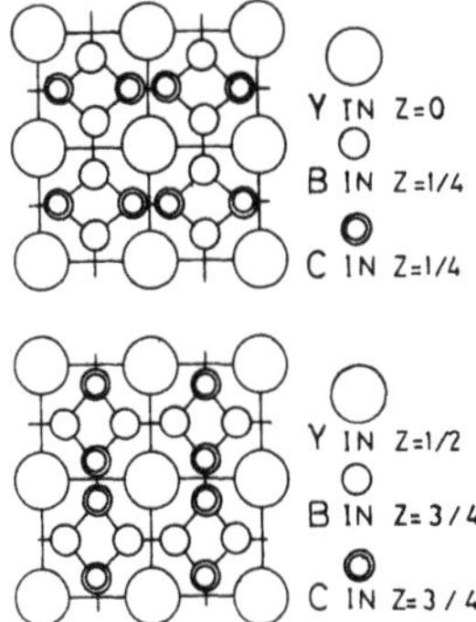

Fig. 16. Crystal structure of the YB_2C_2 phase with a $D_{2d}(P42c)$ space group (Bauer and Nowotny 1971). (Reprinted by permission of the publisher, Institut für Anorganische Chemie, Inc.)

As for the $YB_{0.5}C$ compound, they found that this compound was present in the mixture with a composition of 40:20:40 at.% (Y:B:C) as a homogeneous form, and its X-ray diffraction pattern was similar to that of the $Y_{15}C_{19}$ carbide. More recently, Hájek et al. (1984a–d) have determined the composition to be $Y_{15}B_2C_{17}$.

6.1.1.2. *The europium–boron–carbon phase diagram.* The ternary system Eu–B–C has been studied, with special emphasis on the technologically important sections EuB_6–C and EuB_6–B_4C (Schwetz et al. 1979). Samples were synthesized from the pure elements at 1500°C in sealed molybdenum capsules and annealed for 20 h. The X-ray diffraction pattern of EuB_2C_2 has been indexed on the basis of a tetragonal unit cell with $a = 3.77$ Å and $c = 4.03$ Å.

The pseudobinary sections of EuB_6–C and EuB_6–B_4C have been studied. The solubility of carbon in europium hexaboride was measured by using equilibrate samples annealed at various temperatures. Schwetz et al. found that the lattice parameters of the $EuB_{6-x}C_x$ solid solutions decrease linearly with increasing carbon substitution x, in excellent agreement with the results of Kasaya et al. (1978). The solid solubility limits are 0.82 wt.% C ($x = 0.15$) at 1400°C, 0.87 ($x = 0.16$) 1600°C, 1.04 ($x = 0.19$) 1800°C and 1.08 ($x = 0.20$) 2000°C, respectively, and the maximum solubility 1.4 wt.% C ($x = 0.25$) at the eutectic temperature was obtained from the sample containing more than 50 at.% C and quenching from the melt. In carbon-rich (>3 wt.% C) samples equilibrated at temperatures higher than 2000°C, a new phase coexists with $Eu(B, C)_6$.

Along the EuB_6–B_4C section again the formation of the solid solution $EuB_{6-x}C_x$ occurs, but no ternary compound was found. The isothermal section of the tentative phase diagram of the europium–boron–carbon system at 1500°C has been deduced (see fig. 17). To give a better overview, the $Eu(B, C)_6$ and $B_{12+x}C_{2-x}$ solid solution has not been included within the diagram.

6.1.1.3. *The gadolinium–boron–carbon phase diagram.* The Gd–B–C system in the range 2000–3000°C was determined (Smith and Gilles 1967, fig. 18). The upper portion of this diagram is less well-established than the lower part. This diagram shows only the approximate composition of the phases and does not necessarily show

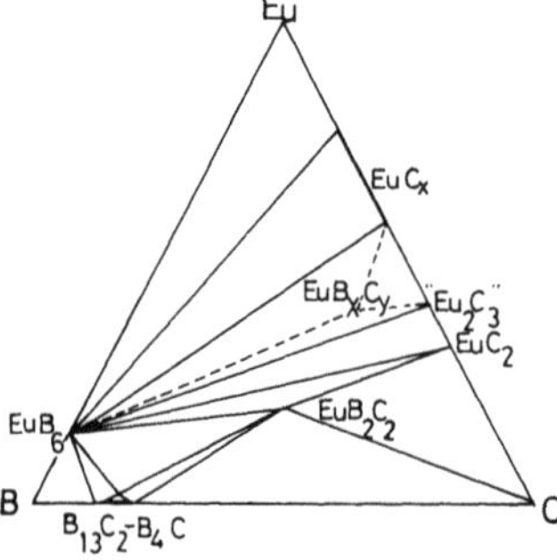

Fig. 17. Isothermal section of the Eu–B–C system at 1500°C (Schwetz et al. 1979).

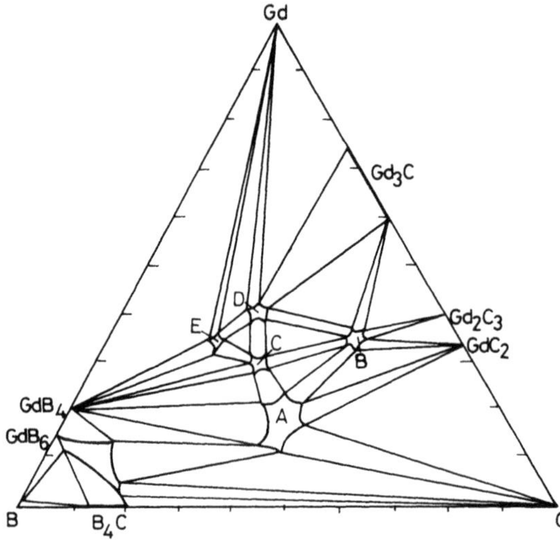

Fig. 18. Phase diagram of the Gd–B–C system in the range of 2000–3000°C (Smith and Gilles 1967). (Reprinted by permission of the publisher, Pergamon Press, Inc.)

the isothermal relationships among these phases, owing to the very rapid quenching of the samples in the arc-melting apparatus. As shown in fig. 18, the solid solubility of carbon in GdB_6 is surprising and it is not clear how this phase can have such a great carbon solubility.

The phase GdB_2C_2 was prepared by the reaction between graphite and the tetraboride or hexaboride of gadolinium and does not melt in graphite crucibles in vacuum up to 2400 K. The lattice parameters of the tetragonal GdB_2C_2 compound are $a = 3.792 \pm 0.001$ Å and $c = 3.640 \pm 0.001$ Å, in agreement with those of GdB_x ($a = 3.79$ Å and $c = 3.63$ Å) (Post et al. 1956). In addition to the GdB_2C_2 phase, there exist four other ternary phases, which have been assigned within 10% of each atomic fraction; these are the compositions $Gd_{0.35}B_{0.19}C_{0.46}$ ("Gd_2BC_2"), $Gd_{0.35}B_{0.45}C_{0.20}$, $Gd_{0.40}B_{0.35}C_{0.25}$ and $Gd_{0.30}B_{0.40}C_{0.30}$. They are stable in air at room temperature except for the first, which oxidized at room temperature at a rate less than that of GdC_2. The X-ray pattern of $Gd_{0.35}B_{0.19}C_{0.46}$ was partially indexed as orthorhombic with $a = 6.09$ Å, $c = 8.56$ Å and b greater than 8 Å. A wide variation in the spacing of the diffraction lines in the two patterns obtained for this phase indicate a wide range of solid solution. The diffraction patterns of the other phases are very complex and have not been indexed.

6.1.1.4. *Isothermal section of the ternary system holmium–boron–carbon at 1500°C.* The isothermal section at 1500°C of the ternary system holmium–boron–carbon has been investigated by Bauer et al. (1985). Samples were prepared by arc melting and subsequent annealing for 100 h under an argon atmosphere at 1500°C. Figure 19 presents the phase equilibria as observed at 1500°C. The carbon solubility was found to be largest in HoB_2 (about 6 at.% at 1500°C) and diminished with increasing boron content in the other phases.

Three carboborides are observed in the system: two "sandwich" compounds with two-dimensional boron–carbon networks, HoB_2C and HoB_2C_2. Both exhibit tetragonal symmetry – HoB_2C: $P4_2/mbc$, $a = 6.773(3)$ Å, $c = 7.399(3)$ Å; HoB_2C_2: $P\bar{4}2c$, $a = 3.780$ Å, $c = 7.074$ Å.

Ho–B–C shows two modifications. In arc-melted samples, the Y–B–C-type (Bauer and Nowotny 1971) structure is present with $a = 3.384$ Å, $b = 13.09$ Å and $c = 3.594$ Å, space group Cmmm. After annealing, a more complex structure appears. The powder diffraction pattern could be indexed with a tetragonal cell of $a = 3.546$ Å and $c = 46.40$ Å. This very long c axis is approximately double that of Th$\bar{\text{B}}$–C (Rogl 1978) or four times as large as that of U–B–C (Toth et al. 1961). The crystal structure is not known, but the basic structural element, i.e. infinitesimal zigzag boron chains with a carbon branch is again present in the structure.

Bauer et al. (1985) suggested that from a viewpoint of their crystalline structure and chemistry, the ternary R–B–C compounds can be divided into two parts, carboborides and borocarbides. The lattices of carboborides are formed by three- and two-dimensional covalently bonded frameworks of the boron and carbon atoms and the

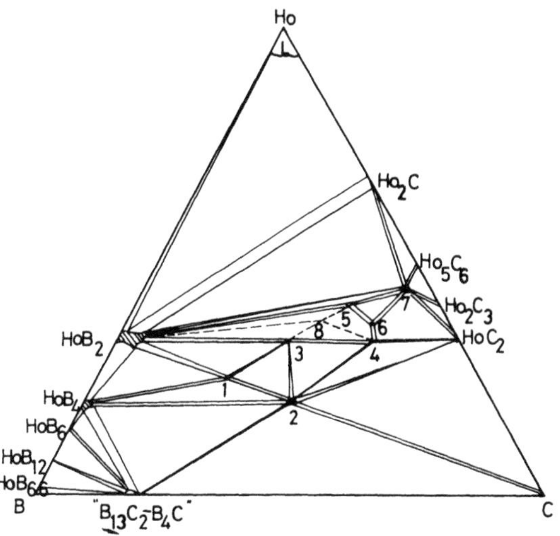

Fig. 19. Isothermal section at 1500°C of the Ho–B–C system (Bauer et al. 1985). (Reprinted by permission of the publisher, Elsevier Sequoia S.A., Inc.)

metal atoms are located in the interstitial positions, whereas those of borocarbides are formed by the metal atom lattice with the boron and carbon atoms occupying the interstices. Of the latter, a new phase was observed at the composition Ho_2BC_3 (topochemically: HoBC + HoC_2). The powder pattern of the arc-melted sample showed I-centered tetragonal symmetry with $a = 3.561$ Å and $c = 12.455$ Å. The c parameter is twice as large as that of HoC_2, and the a parameters are nearly equal, showing a close structural relationship. The low-temperature modification, stable at 1500°C, has a primitive tetragonal lattice and a doubled c parameter with $a = 3.567(1)$ Å and $c = 24.514(8)$ Å.

$Ho_5B_2C_6$ is isostructural with the lanthanum borocarbide $La_5B_2C_6$ (Bauer and Bars 1983, Bauer and Politis 1982). The parameters at the nominal composition are $a = 3.981(2)$ Å and $c = 11.561(3)$ Å, space group P4. The c parameter increases smoothly with increasing carbon concentration up to $c = 12.145(6)$ Å at $Ho_5B_2C_7$, and with decreasing carbon content the c parameter sharply drops to 10.812 Å with a larger a parameter of 8.033(6) Å and small intensity variations, indicating the formation of $Ho_5B_2C_5$, which can be explained by the filling of the octahedral voids on the c axis with single carbon atoms instead of carbon pairs in the $Ho_5B_2C_6$ structure.

The diffraction pattern in the neighborhood of $Ho_3B_2C_3$ – which was in analogy with $Th_3B_2C_3$ (Rogl 1979) – was very complex and could not be indexed. Thus, the hypothetical equilibria in this region are drawn with dotted lines.

Small amounts of boron (3–6 at.%) in substitution for carbon stabilize the $Ho_{15}C_{19}$ structure at high temperatures. A single-phase compound can be obtained directly from the melt. An analogous behavior has been observed in the Y–B–C system (Bauer and Nowotny 1971). This can be explained by the formation of a ternary phase $Ho_{15}B_2C_{17}$, where the central carbon atom in the C–C–C chains (Hájek et al. 1984a, b) is substituted by a boron atom, in this way forming more stable C–B–C chains. Lattice parameters with respect to $Ho_{15}C_{19}$ are insignificantly changed and were found to be $a = 8.004(2)$ Å and $c = 15.984(4)$ Å. Small diffraction intensity variations indicate slight rearrangement of the atoms; a superstructure with a doubled a-axis is possible.

6.1.2. *Ternary rare earth borocarbides*

6.1.2.1. *The RB_2C_2 phase.* Smith and Gilles (1967) prepared this phase for Nd, Gd, Tb, Dy, Ho, Er and Yb by the reaction between graphite and the corresponding rare earth tetraboride, identified the structure of RB_2C_2 phases on the basis of a primitive tetragonal cell, and established that previously reported RB_x phases (R = La, Pr, Gd, Er, Yb, Y) (Post et al. 1956, Binder 1956, 1960) are also members of the RB_2C_2 series. Later work by Nordine et al. (1964) for LuB_2C_2, Fishel and Eick (1969) for RB_2C_2 (R = La, Ce, Sm, Eu, Gd, Tm, Yb, Lu, Y), Bauer and Nowotny (1971) for YB_2C_2, Bauer and Debuigne (1972) for DyB_2C_2, Bezruk and Markovskii (1972) for RB_2C_2 (R = Tb, Tm, Lu, Yb) and Schwetz et al. (1979) for EuB_2C_2 have confirmed that all the rare earths, including yttrium, form this phase. The crystal structure of the RB_2C_2 (R = Y and La through Lu) phases had been studied in detail by Smith et al. (1964), but was finally established by Bauer and Nowotny (1971). According to their results, this phase is tetragonal with the P$\bar{4}$2c space group and the structure consists of two planar infinite layers which alternate along the *c* axis. Within the ordered B–C network, each atom is bonded to three other atoms so as to form aromatic-like, fused four- and eight-membered rings. Each four-membered ring contains two B and two C atoms in opposite positions, each eight-membered ring contains four B and four C atoms. The R atoms are located in the interstices between the eight-membered rings. Bauer and Bars (1980), on the basis of the intensity analysis of X-ray powder photographs for LaB_2C_2, calculated the bond distances: La–B, 2.885(5) Å; La–C, 2.831(4) Å; B–C, 1.628(19) Å; C–C. 1.321(13) Å; B–B, 1.725(13) Å; and bond angles: C–B–B, 130.0(5)°; B–C–C, 140.0(5)° for this phase. They also pointed out that for the ordered arrangement of the nonmetallic atoms in the B–C networks, the *c* axis of the small tetragonal cell has to be doubled. According to this analysis, they corrected the previously reported unit cell parameters of these phases. If the lattice parameters of the MB_2C_2 (M = R, Ca) phases are plotted against the M ionic radii, it can be seen that the relationship between the *a*-axis parameters and the ionic radii is strongly linear, whereas for YbB_2C_2, EuB_2C_2 and CaB_2C_2 anomalies occur in the graph of the *c*-axis parameters. This certainly indicates that ytterbium and europium are partially divalent in their diboride dicarbide phases.

The ScB_2C_2 phase has also been prepared by arc melting a mixture of the elements and annealing at 2100°C under a vacuum of 10^{-15} Torr (Smith et al. 1965). However, the crystals are orthorhombic with a Pbam space group instead of tetragonal with a P$\bar{4}$2c space group as reported for the other RB_2C_2 phase. The lattice parameters obtained from single-crystal measurements are: $a = 5.175 \pm 0.005$ Å, $b = 10.075 \pm 0.007$ Å and $c = 3.440 \pm 0.005$ Å. In the crystal structure, the B–C networks are composed of fused five- and seven-membered rings. Scandium atoms are situated above and below the seven-membered rings. In addition to these 14 light atoms, each Sc is surrounded by five other Sc atoms at distances that are nearly the same as in metallic scandium. The authors also gave the bond data for the ScB_2C_2 phase on the basis of the intensity analysis of 47 independent X-ray reflections.

As mentioned above, the ScB_2C_2 compound is not member of the tetragonal RB_2C_2 series, thus it seems likely that there is a small size limit for the rare earth ion in the tetragonal diboride dicarbide structure.

These RB_2C_2 compounds were synthesized at high temperatures, e.g. 2000°C for R = La, Ce (Markovskii et al. 1972), and melt congruently at 205° ± 100°C for R = Lu, Tm, Tb, Yb (Bezruk and Markovskii 1972).

6.1.2.2. *The R–B–C phase.*

The structure of the Y–B–C phase (Bauer and Nowotny 1971) consists of the B–B zigzag chains with C-branches and trigonal metal prisms, which is similar to that of the phases U–B–C (Toth et al. 1961) and Mo_2BC (Jeitschko et al. 1963). Later work by Bauer and Debuigne (1972, 1975) also found phases with this type of structure in the R–B–C (R = Dy, Gd, Tb) systems. The Dy–B–C phase has lattice parameters $a = 3.384$ Å, $b = 13.727$ Å and $c = 3.647$ Å. The positions of atoms in the D_{2h}^{19}–Cmmm space groups are: four dysprosium atoms at 4i sites, $y = 0.135$; four boron atoms at 4j sites, $y = 0.711$; and four carbon atoms at 4j sites, $y = 0.591$.

In the study of the Gd–B–C system, Smith and Gilles (1967) found a phase with the composition of $Gd_{0.30}B_{0.40}C_{0.30}$ near to that of the phases Y–B–C and Dy–B–C. Later, Bauer and Debuigne (1972) prepared this phase and, as expected, this compound is isotypic with Y–B–C and Dy–B–C and with lattice parameters $a = 3.456$ Å, $b = 13.796$ Å and $c = 3.697$ Å.

Bauer et al. (1985) and Bauer and Debuigne (1975) reported the existence of Ho–B–C ($a = 3.384$ Å, $b = 13.697$ Å and $c = 3.594$ Å) and Tb–B–C, but for the latter no structural data were given.

No reports were found about the existence of other R–B–C phases. However, Johnson and Daane (1961) reported a ternary phase in the La–B system with a composition estimated to be LaBC, but neither the crystal structure nor the exact composition were given. Its diffraction pattern was too complex to index.

6.1.2.3. *The compounds LaB_2C_4 and CeB_2C_4.* Markovskii et al. (1965) were the first to find that lanthanum and cerium form compounds with the composition of MB_2C_4. After the RC_2B_2 phases were discovered (Smith and Gilles 1967), they again investigated the formation of rare earth borocarbides and attempted to establish which of the rare earths form the RC_2B_2 borocarbide phase, and which of them yield the previously described RB_2C_4 phase. They synthesized these rare earth borocarbides at 2000°C in graphite crucibles by using starting materials of 99.5% purity. The pure phase of the composition LaB_2C_4 was still obtained. Cerium also forms the phase, but only in mixtures with the more stable CeC_2B_2 compound. Of the other rare earths, including praseodymium, neodymium and samarium, they were found to form only the RC_2B_2 phase instead of the RB_2C_4 phase. The structure of the LaB_2C_4 and CeB_2C_4 phases has not yet been determined, only the powder diffraction patterns were given, which indicated a hexagonal structure.

6.1.2.4. The *$R_5B_2C_6$ phase.* Alloys of the 2:1:2 (R:B:C) composition, namely Y_2BC_2 (Bauer and Nowotny 1971), "Gd_2BC_2" ($Gd_{0.35}B_{0.19}C_{0.46}$) (Smith and Gilles 1967), Dy_2BC_2 (Bauer and Debuigne 1972), La_2BC_2, Ho_2BC_2 and Er_2BC_2 (Bauer and Debuigne 1975), have been shown to be isostructural with the $R_5B_2C_6$ compounds (Bauer and Bars 1982). The $R_5B_2C_6$ compound seems to exist for all the rare earth elements. The structure of $R_5B_2C_6$ for R = Ce (Bauer and Bars 1982) and La (Bauer and Bars 1983) is tetragonal with space group P4 and four formulae per unit cell. The composition dependence of lattice parameters for $Ho_5B_2C_6$ (Bauer et al. 1985) shows that $R_5B_2C_6$ has wide homogeneity region with a composition range of $R_5B_2C_5$ to $R_5B_2C_7$. However, $Ho_5B_2C_5$ is formed by the filling of the octahedral void on the *c* axis with single carbon atoms instead of carbon pairs in $Ho_5B_2C_6$. Thus the assignment of the correct stoichiometric formulae to these phases is rather difficult.

The crystal structure of $R_5B_2C_6$ is built up from chains of metal octahedra running parallel to the *c* axis and containing double-bonded carbon pairs. These carbon pairs are randomly distributed like the carbon pairs in the high-temperature cubic phase of RC_2 (Bowman et al. 1967) or in the cubic modification of $(R, R')C_2$ (Adachi et al. 1973). The boron atoms and their neighboring rare earth atoms form an octahedral

TABLE 15
The lattice parameters of known phases of $R_5B_2C_6$.

Phase	Parameter (Å)	
$La_5B_2C_6$	$a = 8.585(3)$	$c = 12.313(2)$
$Ce_5B_2C_6$	$a = 8.418(3)$	$c = 12.077(4)$
$Ho_5B_2C_6$	$a = 7.981(2)$	$c = 11.561(3)$
$Ho_5B_2C_7$	$a = 7.981$	$c = 12.145(6)$
$Ho_5B_2C_5$	$a = 8.0336$	$c = 10.81(2)$

environment sufficiently small to accommodate a single carbon atom. These carbon atoms are covalently bonded to the boron atoms to form C–B–C chains similar to those in boron carbide. This structure is a typical transition carbide structure with the slightly distorted R_6 octahedra containing carbon pairs and the more distorted R_5B octahedra containing a single carbon atom. The crystal structure of $R_5B_2C_6$ is closely related to the structure of $R_{15}C_{19}$ carbides. Both structures are composed of chains of octahedra running parallel to the c axis with the difference being that in the $R_{15}C_{19}$ structure subsequent octahedra are rotated by 30° and there are three octahedra instead of two in $R_5B_2C_6$ to give the length of the c axis.

The lattice parameters of the known phases $R_5B_2C_6$ are listed in table 15.

6.1.2.5. *The* RB_2C *phase.* The compound RB_2C was first found in the Y–B–C system and then in the Dy–B–C system (Bauer and Nowotny 1971, Bauer and Debuigne 1972). However, Smith and Gilles (1967) did not find this compound with a composition near to RB_2C in the Gd–B–C system.

A systematic study was carried out by Bauer and Debuigne (1975). Samples with the stoichiometry of RB_2C were prepared by using the sintering mixtures of B (99.999% purity), C (99.999% purity), and R (from 99.5 ~ 99.9% purity) powders in a high-frequency induction furnace under a purified argon atmosphere. The melting temperature was estimated at 1900 ± 100°C. Although Y, La, Pr, Nd, Sm, Gd, Tb, Dy, Ho, Er, Tm and Yb were used to prepare the samples, only for Tb through Yb and Y were the X-ray diffraction diagrams isotypic with RB_2C, and the pure phases could be obtained. But for the synthesized products with La, Pr, Nd, Sm and Gd the diffraction diagrams were too complex to index. Metallographic examinations showed that polyphase structures were present in these samples. The lattice parameters of the RB_2C phases for R = Tb through Yb have been measured and the variation of the lattice parameters a and c of the RB_2C phases as a function of the radius of the trivalent rare earth ions follows a straight line. This shows that the ytterbium compound has the normal lattice parameters and thus suggests that in this compound ytterbium is trivalent.

In fact, the structure of the RB_2C phases is very similar to that of the RB_4 phases, as shown by their analogous X-ray powder diffraction diagrams and the same indexes being identified. The difference between the two phases represents only the changes of the lattice parameters, the parameters a and c diminish by about 4.5% and 7 to 9%, respectively, from the RB_4 phases to the RB_2C phases for a given rare earth metal. These changes could be attributed to the rearrangement of the networks of non-metallic atoms, suppressing the boron atoms at the ± 002 positions from the tops of the octahedra and substituting two boron atoms by two carbon atoms. However, the space group of RB_4 (D_{4h}^5–P4/mbm) is actually not suitable as an ordering arrangement for the carbon and boron atoms. In terms of only a possible arrangement in this space group, the boron atoms at the 8j position and the carbon atoms at the 4h position will lead to an unacceptable interatomic distance. Therefore, it is necessary to establish the space group again and the corresponding atomic arrangement in this space group.

On the basis of geometric considerations and remaining in the tetragonal symmetry, a choice of space D_{4h}^{13}–$P4_2$/mbc was made by introducing a double c axis for the RB_2C phase. In this space group, the atoms are at the following positions:

8R at 8g, $x = 0.313$;

$8B_I$ at 8h, $x = 0.095$, $y = 0.595$;

$8B_{II}$ at 8h, $x = 0.140$, $y = 0.035$;

$8B_C$ at 8h, $x = 0.456$, $y = 0.322$.

From this atom arrangement, the stack of the boron–carbon network and the rare earth metal layer was also derived. The boron–carbon network consists of a heterocyclic lozenge and a heptagonal ring, and the other plane can successfully be deduced by a rotation of 90°.

6.1.2.6. *Summary of the formation of the rare-earth–boron–carbon compounds.* As described above, the rare-earth–boron–carbon compounds can be divided into two groups, the carboborides and the borocarbides (Bauer et al. 1985). The carboborides include the RB_2C_2, RBC and RB_2C compounds, and the compounds R_2BC_3, $R_5B_2C_6$, $R_5B_2C_5$, $R_{15}B_2C_{17}$, and possibly $R_3B_2C_3$, belong to the borocarbides. For the compounds RB_2C_4, no structural data were given to indicate which of the groups this compound belongs to. These compounds have various individual ranges of existence among the lanthanides. For example, the lower limit of existence for the RB_2C_2 phase is at scandium, which forms a compound of the same composition but with another type of structure, and the upper limit corresponds to the divalent europium compound (Fishel and Eick 1969). For the RB_2C compound, the analogous limits also occur but with an exception for R = Gd. On the basis of the general regularity that the rare earths that form both RB_2 and RB_2C_2 phases also form the RB_2C phases (Bauer and Debuigne 1975), the GdB_2C phase probably exists at high temperatures, nevertheless, this phase has not been successfully synthesized and stabilized by annealing. Two phases similar to GdB_2C reported by Smith and Gilles (1967) were probably deduced erroneously from the Gd_2B_5 phase (Bauer and Debuigne 1975). Thus, for the formation of this type of compound, the size of the rare earth atoms (or ion) is also an important factor.

All the known rare earth carboborides, RB_2C_2, RBC and RB_2C, have a laminar structure. Their lattices are formed by two-dimensional covalently bonded boron–carbon atom frameworks and the rare earth atoms are located at the interstitial positions. For the RB_2C_2 compound, the variation in the c parameter over the lanthanide series is about an order of magnitude larger than that in the a parameter. This type of variation in the lattice parameters a and c also occurs in other compounds e.g. the diboride, in which sheets (layers) of tightly bonded non-metallic atoms are separated by a metal atom layer. As one rare earth atom (or ion) is replaced by another, the separation of the layers and the c value are more obviously changed. However, within the layers themselves the a parameter hardly varies. From a comparison between the two phases it is apparent that the RB_2C_2 compounds also have a laminar structure consisting of alternating rare earth atom layers and covalently-bonded C–B networks. The latter accounts for the fixed distance within the rare earth atom layer and the B–C network. For the rare earth diboride carbides RB_2C, the change in the c parameter from terbium through lutetium is about four times larger than that in the a parameter. In the case of the RBC compound, not enough data are available to reveal the above-mentioned parameter variation relationship, however, it is true that this compound also has a laminar structure.

The main structural difference between the three above-described types of the rare earth carboborides are in the different constructions of the B–C network and the correspondingly changed arrangements of the rare earth atoms within the metal layers. RB_2C_2, except for ScB_2C_2, the ordered B–C network, consists of the fused four- and eight-membered rings, and the rare earth atoms located at the interstices between the eight-membered rings. ScB_2C_2 has the fused five- and seven-membered B–C ring and the scandium atom is located at the interstice between the seven-membered rings. For RBC, the B–C network is composed of B–B zigzag chains with carbon branches and the rare earth atoms form a trigonal prism. For RB_2C the boron–carbon network is composed of heterocyclic lozenges and heptagonal rings, and the rare earth atoms are located at the interstices between the heptagonal rings.

The rare earth borocarbides, R_2BC_3, $R_5B_2C_6$, $R_5B_2C_5$, $R_{15}B_2C_{17}$ and possibly $R_3B_2C_3$, have a metal atom lattice with the carbon and boron atoms occupying the interstices. In fact, the structures of these phases are closely related to the various types of rare earth carbides, e.g., the c parameter of Ho_2BC_3 is twice as large as that of HoC_2 and the a parameters are nearly equal; for $R_{15}B_2C_{17}$, small amounts of boron are substituted for carbon to stabilize the $R_{15}C_{19}$ structure at high temperatures, where the central carbon atom in the C–C–C chains is substititued by a boron atom forming more stable C–B–C chains; for $R_5B_2C_6$ and $R_5B_2C_5$, as described above, a typical transition carbide structure is the common basis of their crystal structures, which are closely related to the $R_{15}C_{19}$ structure. Summing up what has been mentioned above, all rare-earth–boron–carbon compounds are derived from either the rare earth borides or the rare earth carbides and their respective crystal structures are closely related.

6.1.3. *Solid solutions of the $EuB_{6-x}C_x$ system*

Schwetz et al. (1979) had determined that the limits of the solid solubility of carbon in EuB_6 increases with increasing equilibrium temperatures and at the eutectic temperature the maximum solubility is 1.4 wt.% C ($EuB_{5.35}C_{0.25}$). Kasaya et al.

(1978) also determined the composition of the single-phase $EuB_{6-x}C_x$ by X-ray diffraction on the samples prepared at 1700°C, revealing a homogeneity range of $0 < x < 0.21$. The lattice parameter of the unit cell decreases with increasing x from $a = 4.1855$ Å ($x = 0$) to $a = 4.1703$ Å ($x = 0.21$).

The cause of the changes in the lattice parameter of $EuB_{6-x}C_x$ and $YbB_{6-x}C_x$ with carbon content has been discussed (Kasaya et al. 1978). In contrast with $SmB_{5.945}C_{0.055}$, in which no change of the lattice parameters was found, the reduction of the lattice parameters for $EuB_{6-x}C_x$ and $YbB_{6-x}C_x$ seems to suggest that the effect of the smaller size of the carbon atom is not decisive, and this reduction may be the result of an increase in metallic bonding. This deduction is in good agreement with their resistivity measurements (see sect. 8.3.3).

6.2. *The R–(Al, Ga, In, Tl)–C systems*

Nasonov and co-workers have investigated the phase equilibria of the aluminium–rare-earth-metal–carbon systems and given partial phase diagrams for the aluminium–(scandium, yttrium, lanthanum, gadolinium and erbium)–carbon systems (Nasonov 1981, Nasonov et al. 1988) in the Al-rich region at 572°C. However, we were unable to obtain their reports because of lack of international academic exchanges.

The compounds R_3MC and RM_3C_3 are the only known compositions that exist in the rare-earth-metal–(aluminium, gallium, indium and thallium)–carbon systems at the present time. A number of reports on the preparation of cubic perovskite-type carbides containing rare earth elements with the general formula $R_3MC_{1.00}$ (Jeitschko et al. 1964, Rosen and Sprang 1965, Haschke et al. 1966a, b, Nowotny 1968) contain little information about their properties other than their lattice parameters.

Jeitschko et al. (1964) were the first to prepare the R_3M phase (R = cerium and yttrium; M = indium, thallium and the group IVB elements tin and lead). The compound Y_3TlC_x is isotypic with the phase Y_3Tl, which has the cubic Cu_3Au-type structure, and in the same way, the compounds Ce_3MC_x (M = indium and thallium as well as tin and lead) have the filled Cu_3Au-type structure, like their corresponding phases Ce_3M. This suggests that when forming the compounds R_3MC_x from the phases R_3M, the carbon atoms enter the R octahedron of the cubic Cu_3Au-type R_3M phase. Only small changes in the lattice parameters of these perovskites with changes in the carbon content were noted.

Haschke et al. (1966a, b) comprehensively prepared the R_3MC compounds and determined their lattice parameters. All the samples were prepared by annealing in the temperature range 800–1000°C for times ranging from 30–180 h. The single phases exist in a few but not in all of these preparations. In the Pr–Tl–C system a typical homogeneous phase Pr_3TlC exists, while in the Pr–Ga–C system no single phase was obtained with a composition of Pr_3GaC.

For all preparations of the dysprosium and ytterbium compounds as well as the compounds of M = Ga, Pb, second phases appear in the samples, showing that the compounds R_3MC_x have an x value less than unity. From observations of the changes in lattice parameters with composition in the compound Nd_3InC_x, the limiting composition is found to be $Nd_3InC_{0.68}$, beyond which the second phase Nd_3C occurs. In addition, the compound Nd_3InC_x also exhibits a wide homogeneity range of the Nd:In atomic ratio, in which neodymium atoms can be substituted partially by indium atoms and no second phase occurs.

The above-mentioned facts seem to indicate that the compound R_3MC_x is a solid solution of carbon in the phase R_3M, however, this is questioned by the observation of a limiting carbon content less than unity for Nd_3InC_x. McColm et al. (1971) have provided a covalent model to explain why the limiting carbon content is less than unity in some of the R_3MC_x systems. According to this model, M atoms could be divided into two groups, the carbide-forming elements such as aluminium, and the noncarbide-forming elements such as indium and group IV elements tin and lead. In the R_3MC_x compound of the latter, the R–C bond instead of the M–C bond is a governing factor in the formation of these methanide carbides, and the rare earth atoms and carbon atoms are randomly distributed throughout the sites of the lattice. This would lead to a fractional limiting carbon content (see sect. 6.3.1), while for the R_3AlC compounds, the carbon atoms can gradually fill in the octahedral sites at the center of the unit cell and the structure appears to be ordering. Thus the x values approach close to 1.0. Owing to the lack of enough reliable information about the x values in various R_3MC_x compounds, it is difficult to distinguish exactly the two groups of the compounds R_3MC_x.

The lattice parameter of the compounds R_3MC_x in the lanthanide series for a given element M changes linearly with the radius of the trivalent rare earth ions, following the lanthanide contraction regularity, except for europium and ytterbium which appear to have a variable-valence state. In the ytterbium–M (Al, Ga, In, Tl)–carbon system, the unit cell volume of the compound Yb_3MC_x is larger than the value expected from the relationship mentioned above. In the europium–M(Al, Ga, In, Tl)–carbon system no compound was found. For the compounds of yttrium, as expected, the lattice parameter falls between those of the respective terbium and dysprosium compounds.

Recently, another series of hexagonal compounds RAl_3C_3 with space group $P6_3/mmc$ have been found for R = Sc, Y, Gd through Lu except for Yb, and the lattice parameters for $ScAl_3C_3$ have been determined as $a = 3.355(1)$ Å, $c = 16.776(3)$ Å, $Z = 1$ (Tsokol et al. 1986).

6.3. *The R–(Si, Ge, Sn, Pb)–C systems*

6.3.1. *The R_3MC_x compounds*

A series of compounds R_3MC_x for M = Sn, Pb have been discussed by Haschke et al. (1966a, b). McColm et al. (1971) extensively investigated the binary system R_3Sn (R = La, Ce, Pr, Sm, Ho, Yb) and the reaction of the cubic binary phases or eutectic alloys of R_3Sn composition with carbon between 950°C and greater than 2500 °C; with greater emphasis on the lighter lanthanides because of the need to investigate the fact that carbides of composition R_3C have only been reported for lanthanides heavier than samarium, while the small atomic number elements in this series have failed to form such a phase.

Face-centered cubic phases were prepared in sealed silica tubes at 950°C for two to three weeks. Only cerium and praseodymium gave satisfactory homogeneous phases; for lanthanum, holmium, samarium and ytterbium compounds varying amounts of the fcc phase were obtained. According to composition analysis, the compositions of the samples correspond to Ce_3Sn, $La_3SnC_{0.56}$ and $Yb_3SnC_{0.47}$. As estimated from X-ray and metallographic examinations, this major phase is face-centred cubic and represents 60–70% of the preparation. The lattice parameters of the R_3Sn and R_3SnC_x compounds and the amounts of the fcc phases have been measured. The results suggest that the fcc phase becomes more difficult to prepare as the atomic number of the lanthanide increases, in contrast to the observations about the preparation and the existence of R_3C. Mössbauer spectra of the R_3Sn and R_3SnC_x phases showed that the R_3SnC_x compounds have the same isomer shift as the R_3Sn alloys and it appears to be unaffected by large changes in carbon content. It is about 0.5 $mm\,s^{-1}$ less than that found for metallic tin, showing that the net s-electron density at the nucleus in R_3Sn and R_3SnC_x compounds is less than that in metallic tin, and when forming the alloy containing lanthanides tin is not completely ionized.

These results also suggest that no tin–carbon bonding exists in these carbides, and it can be deduced that only the carbon bonds to the lanthanide are present. The proportion of lanthanide in an alloy would therefore appear to be a governing factor for the formation of these carbides.

The R_3Sn and R_3SnC_x compounds are not ordered, and tin behaves like indium instead of alumium which itself forms carbide. In the case of aluminium, the x-value of the R_3AlC_x compound approaches very close to 1.0 and the X-ray superstructure lines denoting an ordering in the Al compound have been observed (Nowotny 1968).

As showed by the hydrolysis studies the carbides R_3SnC_x belong to methanides, which are characterized by C_1 units in the structure when $x < 0.6$. As more carbon is added a greater proportion of C_2 hydrocarbons is found, consistent with the conversion of the perovskite carbides to RC_2 or R_2C_3 and R_xSn. These results are consistent with those obtained by X-ray and metallographic data. The limiting carbon content of the fcc phase ($x = 0.66$) has also been determined from a plot of the percentage of methane evolved against the carbon content of the samples, confirming the difficulty in preparing single-phase cubic carbides for the heavier lanthanides.

The lower-temperature magnetic susceptibility data for the compounds Ce_3Sn, and Pr_3Sn as well as their carbides show that the proportion of Ce or Pr in the valency state IV increases as the carbon content increases. Thus the ionization tendency of cerium is enhanced as the Ce–C bond is progressively polarized, with carbon attaining some negative charge and so promoting a greater degree of ionization of cerium. Thus, as the carbon content in these carbides increases, the ionic or substantially ionic compounds such as RC_2 and R_2C_3 would tend to replace the R_3SnC_x carbides formed in the preparation.

A covalent model has been provided to explain the limiting range of carbon composition and the possible structure of the R_3SnC_x compounds (McColm et al. 1971). The perfect perovskite carbide structure would have a carbon atom in the octahedral site at the center of the unit cell. However, a random distribution of lanthanide and tin atoms occurs over all the fcc sites. In this situation, a probability distribution of atoms around the octahedral sites can be derived. Since it is well known that the R_3C phase can form for the heavier lanthanides and not for the lighter ones, $CeSn_3$ has a carbon content much below $CeSn_3C_{0.1}$ if it forms a carbide at all. Thus, the interstitial environments consisting of six R and six Sn are not favorable to three R and three Sn for carbon occupation with the resultant stable R–C bond formation in the lighter and the heavier lanthanide compounds, respectively. This suggests that for R_3Sn phases formed from the heavier lanthanides 86.9% of the interstitial sites would be favorable for bonding, while the percentage for Ce, Pr or Nd alloys would be 69.2%. Considering each carbon site must be surrounded by empty octahedral sites (otherwise, two carbon atoms at adjacent interstitial sites are capable of forming a C_2 unit), these compositions would become $R_3SnC_{0.869}$ and $R_3SnC_{0.691}$, respectively. This is in agreement with the value of 0.66 for Ce_3SnC_x and in excellent agreement with Nowotny's value of $x = 0.68$ for Nd_3InC_x, where indium is considered to be analogous to tin, and $x = 0.2$–0.6 for Nd_3PbC_x (Haschke et al. 1966a, b). For the heavier lanthanides the x value has not been sufficiently tested.

The sp^3 hybridization of carbon atoms bonding to either five or four lanthanide atoms would provide strong covalent bonds. The bond strength of each R–C bond would be $\frac{4}{5}$ and 1 electron per bond, respectively, and is sufficient to overcome the catenation tendency of carbon with the formation of C_2 units. When carbon is present in amounts greater than the limiting value, the occupation of adjacent octahedral interstices with poor carbon–lanthanide bond properties results in the formation of R_2C_3 and R_2C_2 phases (McColm et al. 1971).

In the R_3SnC_x and R_3PbC_x compounds tin and lead atoms play a role of a structural modifier, leading to geometrical situations where covalent bonding between lanthanide atoms and carbon atoms can occur; significant differences are expected where the group IIIA atoms, e.g. aluminium, itself can form a carbide.

6.3.2. *The $R_5M_3C_x$ compounds (M = Si, Ge)*

The existence of the R_5M_3-type rare earth and scandium as well as yttrium silicides and germanides has been reported in a number of works. The R_5M_3-type rare earth silicides crystallize in Cr_5B_3-type tetragonal structures for the rare earths lanthanum through neodymium, and in $D8_8\,Mn_5Si_3$-type hexagonal structures for samarium through lutetium, yttrium and scandium, except for europium. The corresponding germanides are all (except europium and ytterbium) of the Mn_5Si_3 type (Mayer and Shidlovsky 1969, Arbuckle and Parthé 1962, Parthé 1960). There are many compounds formed when small nonmetal atoms such as carbon, nitrogen and boron are dissolved in the structure with little apparent structural effect.

Mayer and Shidlovsky (1969) first studied the effect of the addition of carbon on these compounds and prepared ternary phases of the R_5M_3-type compounds with varying amounts of carbon. In the case of Cr_5B_3-type silicides, the addition of carbon caused the disappearance of the tetragonal structure and a complex pattern was obtained. An exception was Nd_5Si_3, in which, after the addition of carbon, the $D8_8\,Mn_5Si_3$-type hexagonal structure was formed. When carbon was added to the Mn_5Si_3-type silicides and germanides, the hexagonal structure remained until the composition of $R_5M_3C_{1.0}$. At higher carbon content, the hexagonal phase begins to disappear, and at $R_5M_3C_{2.0}$ a completely different but undetermined structure was obtained. It should be pointed out that in the cases of the germanides of lanthanum, terbium, lutetium and yttrium, additional phases were obtained even at the $R_5Ge_3C_{0.5}$ composition. In the case of $La_5Ge_3C_{1.5}$, a well-defined single phase was formed with a body-centered cubic lattice (Mayer and Shidlovsky 1968).

The addition of carbon to Gd_5Si_3, Ho_5Si_3 (Al-Shahery et al. 1983), Er_5Si_3 (Al-Shahery et al. 1982a) and Y_5Si_3 (Button and McColm 1984) led to unexpectedly complex compounds. A carbon solid solution evidently occurs with an accommodation in a large unit cell that is a superstructure variant of the $D8_8$ structure and is fully ordered at $x = 0.5$ and 0.95 for $Gd_5Si_3C_x$ and $Ho_5Si_3C_x$, respectively and for $x = 0.5$ and 1.0 for $Er_5Si_3C_x$ and $Y_5Si_3C_x$, respectively. It cannot be described simply as an interstitial solid solution of carbon at octahedral sites. The ordering giving rise to the superlattice probably involves partial occupancy of octahedral sites by carbon, and this is accompanied by defects in the E_5 octahedra and clustering of silicon nearer to the vacant sites. The limit of carbon solubility is $x = 1.0$ in all cases, and then above this concentration new ternary carbides occur, such as the orthorhombic $R_5Si_3C_{1.8}$ (R = Y, Er) and $Er_5Si_3C_{2.0}$ compounds. At $x = 1.8$ and $x = 2.0$, the carbon atoms are close together so that the hydrocarbons evolved on hydrolysis change markedly (Al-Shahery et al. 1982a, b). Throughout the composition range the compounds become less chemically stable as the carbon content increases, which suggests that the charge localized at the carbon atoms is increasing.

Al-Shahery et al. (1983) pointed out that between $x = 0.2$ and $x = 0.8$ carbon is accommodated via a nonstoichiometric phase at the order–disorder boundary and that the excess metal arising from vacancy production combines with carbon during cooling to form the carbides. The experimental facts have proved this hypothesis: for the compound systems of the more volatile Er and Ho the $R_{15}C_{19}$ phase is precipitated when $x > 0.4$, and for the compounds of the less volatile Gd and Y this phase occurs at $x = 0.5$ and 0.7, respectively. In particular, $Gd_{15}C_{19}$ has not been reported before but the X-ray pattern obtained is very similar to that from $Er_{15}C_{19}$ (Bauer 1974).

The first superstructure at composition $Er_5Si_{2.86}C_{0.71}$ has a hexagonal unit cell in which the a axis is a factor of $\sqrt{3}$ larger than the parent Er_5Si_3 axis but there is only a slightly expanded c axis. The second superstructure at composition $Er_5Si_3C_{1.0}$ has the parameter $\sqrt{3}a$ together with a triple Er_5Si_3 c axis (Al-Shahery et al. 1983). In the type-1 superstructure the carbon is in a complete erbium octahedron at 0, 0, z and fully occupies these sites whereas the $\frac{1}{3}, \frac{2}{3}, z$ octahedra are only partially occupied. It is postulated that occupation of the $\frac{1}{3}, \frac{2}{3}, z$ sites by carbon causes a metal atom vacancy in the octahedron that is partialy filled by relaxation of two neighboring silicon atoms towards it. Ordering of these defect complexes is responsible for the appearance of the type-1 superstructure.

According to Al-Shahery et al. (1983), the partial occupancy of the $\frac{1}{3}, \frac{2}{3}, z$ octahedra must be related to the nonstoichiometry of the phase, and this can be done through the formula $R_{5-x/n}Si_{3-x/n}C_x$ where n is half the number of partially occupied octahedra. This general formula predicts a maximum carbon solubility of $x = 0.78$, in good agreement with $x = 0.8$ beyond which a structural transformation occurs. The first structural change after $x = 0.8$ is a stable transition structure. In this structure the c axis becomes indeterminate as ordering of the carbon-occupied octahedra changes from type 1 to type 2. The type-2 superstructure is found when carbon is close to $x = 1.0$. Not all the carbon is present in octahedral sites, and the unfilled octahedra in this structure are systematically ordered.

$Sc_5Si_3C_x$ carbides ($0.01 < x < 1.0$) have been prepared by arc melting Sc_5Si_3 with carbon. X-ray powder diffractions have failed to reveal any $D8_8$ superstructure phases (Kotroczo et al. 1987). A small effect of carbon content on the an-axial length is notable, together with the rapid increase in the c axis after $Sc_5Si_3C_{0.5}$.

In the work of Al-Shahery et al. (1982a, b, 1983) and Button and McColm (1984) the formation of the heavy rare earth metal silicides $R_5Si_3C_x$ has been studied. A pseudobinary diagram of the Er_5Si_3–C system has been constructed. The possible congruent melting alloys are $Er_5Si_3C_{1.3}$ and $Er_5Si_3C_2$ with peritectics around the compositions $Er_5Si_3C_{0.6}$ and $Er_5Si_3C_{0.85}$. Alloys of the $Er_5Si_3C_{0.5}$ compositions were always microscopically pure single-phase material.

Metallographically, the composition at $x = 1.0$ was also always single-phase and at $x = 0.8$ a carbon-rich phase precipitates in the grain boundaries because of the volatility of the erbium. In the two phase regions (from $x = 0$ to $x = 0.4$ and from 0.6 to 0.7) the $Er_5Si_3C_x$ phase coexists with Er_5Si_4.

In the yttrium system the melting point of the silicide–carbides decreased as the carbon content increased from 1710°C for Y_5Si_3 to 1539°C for $Y_5Si_3C_{2.0}$. In specimens of composition $Y_5Si_3C_{0.5}$, a 24 h annealing in a vacuum at 1200°C led to a single-phase structure (Button and McColm 1984). In the Gd_5Si_3–C system the solidification temperatures showed a steady decrease from 1545°C for Gd_5Si_3 through to 1445°C for $Gd_5Si_3C_{2.0}$. Cooling curve analyses gave evidence for the appearance of a second phase only above $x = 0.8$. A subsequent X-ray examination showed that this second phase was probably the $Gd_{15}C_{19}$ carbide. In the Ho_5Si_3–C system the melting point generally decreased as carbon dissolved in Ho_5Si_3. The special position of the composition $Ho_5Si_3C_{0.5}$ was indicated by the maximum melting point and a relatively pure phase was obtained. As a result of heavy evaporation of holmium, the Ho_5Si_4 phase also occurred in some specimens.

Because the $D8_8$ structure is capable of absorbing nonmetal atoms to produce a series of superstructure phases, the Y_5Si_3 phase has been used to study hydrogen uptake and desorption. There was some indication that the highest hydride is $x = 6$ with an orthorhombic structure like $Y_5Si_3C_{1.8}$. However, for the $Y_5Si_3C_x$ phases with x in the range 0–1.0 the hydrogen content decreased as the carbon content increased and rapid reversible behavior was lost when $x = 0.15$ (McColm et al. 1986).

6.4. *The rare-earth–(iron, cobalt, nickel)–carbon systems*

6.4.1. *The ternary R–Fe (Co, Ni)–C phase diagrams*

Phase diagrams of the Ce–Fe–C, Gd–Fe–C, Sm–Co–C and Y–Ni–C systems have been reported. The Ce–Fe–C and Gd–Fe–C systems are, respectively, representative of the light and heavy lanthanide systems (Park et al. 1982, Stadelmaier and Park 1981), forming two and six ternary compounds, respectively.

6.4.1.1. *The Ce–Fe–C systems.* In the ternary Ce–Fe–C system, the liquidus projection and the isothermal sections at 800 and 950°C were determined (fig. 20a, b). The liquidus projection was determined from the as-cast specimens.

X-ray diffraction analysis shows that the ternary phase does not crystallize from the melt and the primary phase fields consist of only the elements and binary compounds, in particular, the CeC_2 phase persists far into the iron corner of the diagram. The four-phase reactions were listed as follows:

1. (a) Metastable $L \rightarrow \gamma$-Fe + CeC_2 + Fe_3C,
   (b) Stable $L \rightarrow \gamma$-Fe + CeC_2 + C (not observed),
2. L + C $\rightarrow Fe_3C + CeC_2$ (existence not certain),
3. L + $CeC_2 \rightarrow \gamma$-Fe + Ce_2C_3,
4. L + γ-Fe $\rightarrow Fe_{17}Ce_2 + Ce_2C_3$,
5. L + $Fe_{17}Ce \rightarrow Fe_2Ce + Ce_2C_3$,
6. L $\rightarrow Fe_2Ce + Ce_2C_3$ + Ce.

The section at 800°C contains two ternary compounds $Fe_2Ce_2C_3$ and $Fe_4Ce_4C_7$ which are in equilibrium with α-Fe and γ-Fe, respectively. The powder diffraction data are: $Ce_2Fe_2C_3$, hexagonal, $a = 8.647$ Å, $c = 10.53$ Å; $Ce_4Fe_4C_7$, tetragonal, $a = 7.22$ Å, $c = 5.82$ Å. $Ce_2Fe_2C_3$ tends to decompose in water, $Fe_4Ce_4C_7$ does not. The 950°C isothermal section differs from that at 800°C mainly by the intrusion of liquid Fe–Ce and the absence of both Fe_2Ce and $Fe_2Ce_2C_3$. To account for $Ce_2Fe_2C_3$ at 800°C, one must assume the reaction $2Fe + Ce_2C_3 \rightarrow Fe_2Ce_2C_3$ at a temperature between 800 and 950°C. On the basis of the line shift in the X-ray

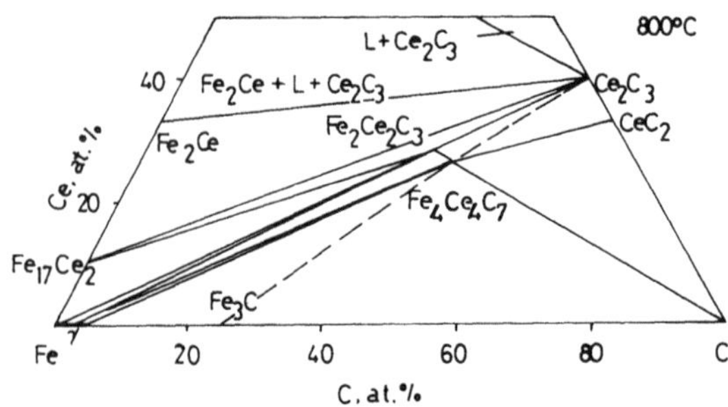

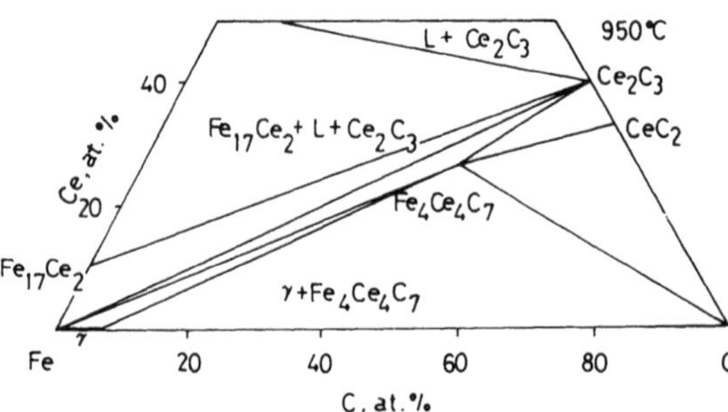

Fig. 20. The isothermal sections of the ternary Ce–Fe–C system at 800°C (a) and 950°C (b) (Park et al. 1982).

diffraction pattern, it was concluded that Ce_2Fe_{17} dissolves carbon, but the amount was not known.

6.4.1.2. *The Gd–Fe–C system.* In the iron–gadolinium–carbon system, the liquidus projection and the 900°C isothermal section (fig. 21) were determined (Stadelmaier and Park 1981). Six ternary carbides were found in this system. Unlike the binary gadolinium carbides, they do not decompose readily in water. Two ternary carbides are metal-rich and the structures are derivatives of binary Fe–Gd structures. The crystal structures of the six ternary compounds are listed in table 16.

From the liquidus projction it is known that three ternary phases, GdFeC, $Gd_2Fe_2C_3$ and $Gd_4Fe_4C_7$, crystallized from the melt. Only GdFeC melts congruently, and $Gd_2Fe_2C_3$ and $Gd_4Fe_4C_7$ have narrow primary phase fields which do not include the composition of the phase itself. Eleven four-phase reactions are recognized in the microstructures:

1. (a) Metastable $L + GdC_2 \rightarrow Fe_3C + Fe_4Gd_4C_7$,
   (b) Stable $L + GdC_2 \rightarrow Fe_4Gd_4C_7 + C$,
2. (a) Metastable $L \rightarrow \gamma\text{-Fe} + Fe_4Gd_4C_7 + Fe_3C$,
   (b) Stable $L \rightarrow \gamma\text{-Fe} + Fe_4Gd_4C_7 + C$,
3. $L + Gd_4Fe_4C_7 \rightarrow \gamma\text{-Fe} + Fe_2Gd_2C_3$,
4. $L + Gd_2Fe_2C_3 \rightarrow \gamma\text{-Fe} + Gd_2C_3$,
5. $L \rightarrow \gamma\text{-Fe} + Gd_2C_3 + FeGdC$,
6. $L + \gamma\text{-Fe} \rightarrow Fe_{17}Gd_2 + FeGdC$,
7. $L + Fe_{17}Gd_2 \rightarrow Fe_3Gd + FeGdC$,
8. $L \rightarrow Fe_3Gd + GdFeC + Gd_3C$,
9. $L + C \rightarrow Fe_3C + GdC_2$ (existence not certain),
10. $L + GdC_2 \rightarrow Fe_4Gd_4C_7 + Gd_2C_3$,
11. $L + Fe_4Gd_4C_7 \rightarrow Fe_2Gd_2C_3 + Gd_2C_3$.

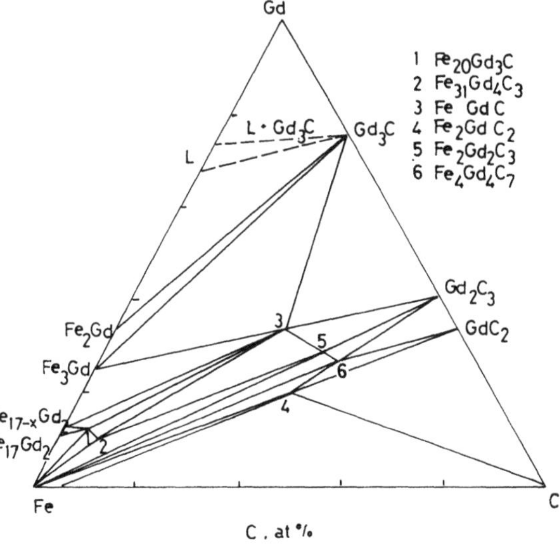

Fig. 21. The isothermal section of the ternary Gd–Fe–C system at 900°C (Stadelmaier and Park 1981).

TABLE 16
Crystal structures of six Gd–Fe–C ternary compounds.

Compound	Structure	Lattice parameters (Å)		
		a	*b*	*c*
GdFeC	p hexagonal	9.12	–	5.97
$Gd_4Fe_4C_7$	p tetragonal	7.045	–	10.23
$Gd_2Fe_2C_3$	p hexagonal	8.388	–	10.260
$GdFe_2C_2$	p orthorhombic	3.678	2.995	14.5
$Ge_3Fe_{20}C$	p tetragonal	8.76	–	11.81
$Ge_4Fe_{31}C_3$	rhombohedral	8.647	–	12.461

In the isothermal section at 900°C. with the exception of L + Gd_3C and γ-Fe + Fe_2GdC_2, the two phase fields are shown to degenerate to a single line based on the assumption that the ternary phases are stoichiometric. Actually, GdFeC must have a small homogeneity range because the precipitation of Fe_3Gd was observed and the lines of the diffraction patterns were consistently broadened after 300 h at 900°C. The iron-rich ternary compounds, $Gd_3Fe_{20}C$, $Gd_4Fe_{31}C_3$ and $GdFe_2C_2$, were not found in the as-cast alloys, so it is concluded that they formed in the solid. With the appearance of Fe_2GdC_2 in the annealed microstructure, the moisture-sensitive binary carbide GdC_2 completely disappeared in alloys located between the iron corner and the composition of $GdFe_2C_2$. Thus, it is apparent that the reaction $2Fe + GdC_2 \rightarrow GdFe_2C_2$ leads to the formation of the ternary $GdFe_2C_2$ carbide, which does not decompose in water.

The space group and structural type of the six ternary carbides were not determined. Some postulated structural types have been proposed but not enough data were given to confirm these assumptions. Some of these compounds were found to have another stoichiometry. For example, the compounds $GdFe_2C_2$ and $Gd_3Fe_{20}C$ have been redetermined to have the $GdFeC_2$ (Jeitschko and Gerss 1986) and $Gd_2Fe_{14}C$ (Stadelmaier and El-Masry 1985) compositions, respectively.

6.4.1.3. *The Sm–Co–C system.* The liquidus projection and the isothermal section at 900°C of the Sm–Co–C ternary phase diagram have been determined (Stadelmaier and Liu 1985). Nine four-phase reactions occur during crystallization:

1. $L \rightarrow Co + C + CoSmC_2$,
2. $L + Co_{17}Sm_2 \rightarrow Co + CoSmC_2$,
3. $L + Co_5Sm \rightarrow Co_{17}Sm_2 + CoSmC_2$,
4. $L \rightarrow [Co_5Sm] + Co_7Sm_2 + CoSmC_2$,
5. $L \rightarrow Co_7Sm_2 + Co_3Sm + CoSmC_2$,
6. $L \rightarrow Co_3Sm + Co_{11}Sm_5C_2 + CoSmC_2$,
7. $L \rightarrow Co_2Sm + Co_{11}Sm_5C_2 + CoSmC_2$,
8. $L \rightarrow Co_3Sm + Co_{11}Sm_5C_2$,
9. $L \rightarrow Co_2Sm + Sm_3C + CoSmC_2$.

The isothermal section of the Co–Sm–C system at 900°C is shown in fig. 22. From a comparison of the liquidus projection and the isothermal section at 900°C, $Sm_5Co_{11}C_2$ occurs on the liquidus projection and decomposes eutectoidally above 900°C, and $SmCoC_2$ melts congruently. The compound $Sm_5Co_{11}C_2$ is hexagonal

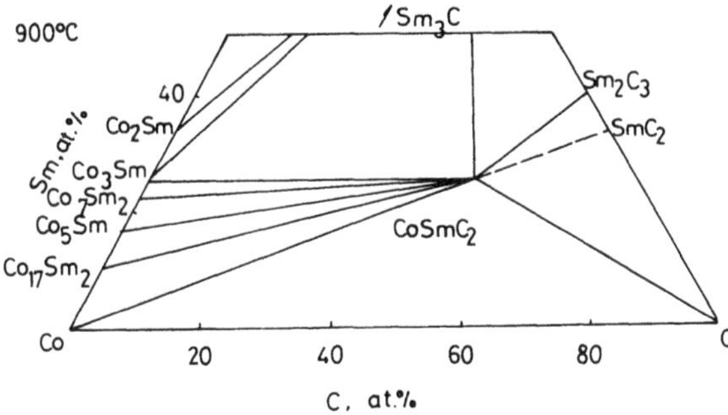

Fig. 22. The isothermal section of the ternary Co–Sm–C system at 900°C (Stadelmaier and Liu 1985).

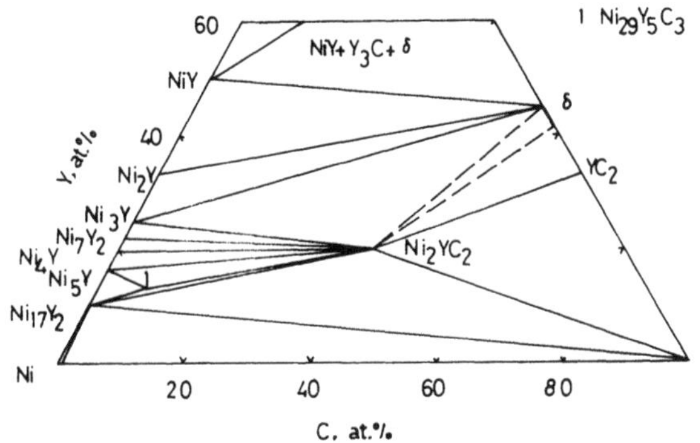

Fig. 23. The isothermal section of the Y–Ni–C system at 800°C (Stadelmaier and Kim 1984).

with lattice parameters $a = 5.148$ Å and $c = 10.744$ Å and is a member of the homologous series $R_{2n+1}T_{4n+3}X_2$, where R, T and X are lanthanide, transition metal and metalloid, respectively, and obtained by stacking n units of R_2T_4 on one unit of RT_3X_2 (El-Masry et al. 1983). The compound $CoSmC_2$ has a monoclinic structure of $CoCeC_2$ type with $a = 5.138$ Å, $b = 7.389$ Å, $c = 5.312$ Å. $\gamma = 101.4°$, and it has the space group Bb (Bodak et al. 1980). This compound exhibits a phase transformation at high temperatures, and the X-ray diffraction pattern of as-cast $SmCoC_2$ is similar but not identical to that of the annealed compound. The annealed structure was still found in a specimen quenched from 1400°C so that the transformation temperature appears to be above 1400°C.

6.4.1.4. *The Y–Ni–C system.* The liquidus projection and 800°C isothermal section of the yttrium–nickel–carbon ternary have been determined by means of microprobe, X-ray diffraction and metallographic techniques (Stadelmaier and Kim 1984). 75 specimens were annealed for 300 h at 800 ± 10°C to construct the 800°C isothermal section which is shown in fig. 23.

From liquidus projection, it is known that two ternary phases with composition of YNi_2C_2, probably, $Y_5Ni_{29}C_3$ are present in this system. YNi_2C_2 melts congruently and has a large field of primary crystallization from the melt, and the phase with a probable composition $Y_5Ni_{29}C_3$ is located between the fields of primary Y_2Ni_{17} and

YNi_5. The four-phase reactions found below 50 at.% Y are as follows:

1. $L(+C) \to Ni + YNi_2C_2$,
2. $L \to Ni + Y_2Ni_{17} + YNi_2C_2$,
3. $L + Y_5Ni_{29}C_3 \to Y_2Ni_{17} + YNi_2C_2$,
4. $L + Y_2Ni_{17} \to Y_5Ni_{29}C_3 + YNi_5$,
5. $L \to YNi_5 + Y_5Ni_{29}C_3 + YNi_2C_2$,
6. $L \to YNi_4 + YNi_5 + YNi_2C_2$,
7. $L \to Y_2Ni_{17} + YNi_4 + YNi_2C_2$,
8. $L \to YNi_3 + Y_2Ni_{17} + YNi_2C_2$,
9. $L \to YNi_2 + YNi_3 + YNi_2C_2$,
10. $L \to YNi + YNi_2 + \gamma$ (Y–C),
11. $L \to YNi_2 + \gamma + YNi_2C_2$,
12. $L + \gamma \to YNi_2C_2 + YC_2$,
13. $L + YC_2 \to YNi_2C_2 + C$.

The isothermal section shows that YNi_2C_2 is in equilibrium with the binary Y–Ni phases from Ni to Ni_3Y but not with YNi_2 and that Y_2Ni_{17} does not coexist with graphite upon solidification but does coexist at 800°C. This requires two reactions

$$YNi_2 + YNi_2C_2 \to YNi_3 + \delta(\text{Y–C}) \quad \text{and} \quad Ni + YNi_2C_2 \to Y_2Ni_{17} + C$$

to take place above 800°C. In addition, it can be shown that during the annealing process at 800°C, YNi_2C_2 is dissolved while the amount of $Y_5Ni_{29}C_3$ increases, leaving a two-phase mixture of YNi_5 and $Y_5Ni_{29}C_3$.

The phase with a probable composition of $Y_5Ni_{29}C_3$ has a hexagonal structure, $a = 8.510$ Å, $c = 16.62$ Å. The lattice parameters suggest that this compound is a derivative of a binary type R_5T_{32} (where R and T are the rare earth and transition metals respectively). YNi_2C_2 has the same powder diffraction pattern as reported for $GdFe_2C_2$ (Stadelmaier and Park 1981). According to the work of Jeitschko and Gerss (1986), these RFe_2C_2 phases actually have the stoichiometric composition of RTC_2.

6.4.2. *Ternary carbides with the $CeCoC_2$- and $CeNiC_2$-type structures*

Semonenko et al. (1983) prepared the compounds $RNiC_2$ (R = yttrium and all the lanthanides with the exceptions of promethium and europium) at high pressures and temperatures. Stadelmaier and Kim (1984) reported the compound YNi_2C_2 to be isotypic with $GdFe_2C_2$. The powder pattern of the latter compound (Stadelmaier and Park 1981) corresponds to that found for $GdFeC_2$ in the work of Jeitschko and Gerss (1986). Thus, the existence of the compounds $YNiC_2$ and $GdFeC_2$ has been confirmed.

Jeitschko and Gerss (1986) prepared 32 RTC_2 carbides (R = rare earth metal, T = iron group elements of iron, cobalt and nickel), 18 for the first time, by arc melting of the elemental components and subsequent annealing in an evacuated silica tube, and then determined the crystal structures of these compounds. In the course of preparation, the starting composition of the cobalt and nickel compound corresponded to the ideal atomic ratios of the rare earth metals:(cobalt or nickel):C = 1:1:2. For the iron compounds the nonideal ratios 3:8:9 were used because the X-ray powder patterns of such samples contain only two impurity lines of iron, whereas samples prepared with the ideal starting ratios showed many reflections because of second- and third-phase ternary compounds with similar compositions, which did not disappear after annealing at 900°C for 10 to 12 days. Thus, the cobalt and nickel compounds were obtained in a well-crystallized form directly after the arc melting, indicating a congruent melt of these compounds, whereas the iron compounds were observed only after the annealing procedure, and obviously, were in equilibrium with the iron element.

Indices of a Guinier powder pattern of the samples were assigned on the basis of the orthorhombic $CeNiC_2$-type (Bodak and Marusin 1979) and the monoclinic $CeCoC_2$-type (Bodak et al. 1980) unit cells, respectively, in comparison with the calculated patterns provided by Yvon et al. (1977). The crystal parameters of these compounds can be obtained from the reports of Bodak et al. (1980), Stadelmaier and Liu (1985), Semonenko et al. (1983), Bodak and Marusin (1979). The compounds $RCoC_2$ (R = lanthanum through neodymium) were identified to be isotypic with the monoclinic $CeCoC_2$-type structure, whereas all the other compounds $RFeC_2$, $RCoC_2$ and $RNiC_2$ have the orthorhombic $CeNiC_2$-type structure. The composition of $SmCoC_2$ is remarkable; it crystallized in the orthorhombic form in the as-cast alloys, while the monoclinic form was observed in the annealed samples. Thus the compound $SmCoC_2$ has two modifications, with either the $CeCoC_2$-type structure at low temperatures or the $CeNiC_2$-type structure at high temperatures. For the iron compounds the $RFeC_2$ phase was found in only those lanthanides which are heavier than samarium, in agreement with the results of the investigation of the Ce–Fe–C phase diagram (Park et al. 1982).

The crystal structures of $DyCoC_2$ and $DyNiC_2$ were refined from single-crystal X-ray data to residuals of R = 0.012 (581 structure factors) and R = 0.023 (971 structure factors) for 17 variable parameters. The space group Amm2 was found to be correct during the structure refinements.

The crystal structures of the ternary carbides with the orthorhombic $CeNiC_2$-type and the monoclinic $CeCoC_2$-type structures were first determined by Bodak and Marusin (1979) and Bodak et al. (1980). In the $CeNiC_2$-type structure the large rare earth atoms form a somewhat distorted hexagonal primitive lattice, while the nickel atoms and the carbon pairs fill the trigonal prismatic voids of this lattice. This structure may thus be derived from the hexagonal AlB_2 structure. The monoclinic $CeCoC_2$-type structure is closely related to the $CeNiC_2$ structure and may be regarded as distorted version of the latter. The greatest difference between the two arises through the tilting of the carbon pairs in $CeCoC_2$ that are aligned parallel to the y axis in $CeNiC_2$.

The phase transformation of $SmCoC_2$ from the high-temperature orthorhombic form to the low-temperature monoclinic form can be better regarded as a reconstructive transformation as opposed to a displacive one because this orthorhombic modification can be obtained by quenching to room temperature, however, the mechanism of this phase transformation remains to be studied in detail.

The formation of the ternary rare earth cobalt and nickel carbides is surprising because cobalt and nickel do not form any thermodynamically stable binary carbides. However, it can be made clear by considering that the rare earth and iron group metals together contribute as many valence electrons as a transition metal with an intermediate-valence electron count; e.g. in $LaCoC_2$, lanthanum with three and cobalt with nine valence electrons together contribute six valence electrons per metal atom and thus the valence electron density of $LaCoC_2$ is similar to that of MoC. Therefore, the carbon atoms are able to locate themselves in voids formed by the rare earth and iron group metals.

The plots of the cell volume of the ternary carbides against the atomic number of the lanthanides showed that only small deviations from the smooth curves occur for some of the cerium or ytterbium compounds, indicating that these elements are essentially trivalent in the compounds. No europium compound was found with the composition RTC_2 and the above-mentioned structures are, obviously, the result of its divalent state. The cell volume of the ternary yttrium carbides are close to those of the dysprosium compounds, showing that yttrium is characteristic of the heavy lanthanides.

It has been shown that the cobalt and nickel compounds with the magnetic rare earth ions (not the yttrium, lanthanum and lutetium compounds) are strongly attracted by a magnet at room temperature (Jeitschko and Gerss 1986). Kotsanides et al. (1989) have investigated the magnetic properties of the $RNiC_2$ compounds (R = Pr, Nd, Gd, Tb, Dy, Ho, Er, Tm, Y) and obtained some interesting results (see sect. 8.3.4).

Marusin et al. (1985a, b) reported the existence of the scandium compounds $ScFeC_2$, $ScCoC_2$ and $ScNiC_2$, which belong to PbFCl-type or $UCoC_2$-type (Li and Hoffmann 1989). The lattice parameters were determined to be $a = 3.3344(7)$ Å, $c = 7.292(2)$ Å and $Z = 2$ for $ScCoC_2$.

6.4.3. *Ternary $R_2Fe_{14}C$ carbides with $Nd_2Fe_{14}B$-type structure*

Rare earth compounds of the type $R_2Fe_{14}B$ have received widespread attention owing to their interesting magnetic properties and their potential as starting materials for permanent magnets. There are indications that the $R_2Fe_{14}C$ compounds will be equally interesting (Denissen et al. 1988a, b, de Boer et al. 1988a, b). Bolzoni et al. (1985), Abache and Oesterreicher (1985), Pedziwiatr et al. (1986) have prepared $R_2Fe_{14}C$ for Gd, Dy, Er. Liu et al. (1987), Gueramian et al. (1987) and Sagawa et al. (1987) have synthesized a series of ternary carbides $R_2Fe_{14}C$ (R = Pr, Sm, Gd, Tb, Dy, Ho, Er, Tm, Lu) with the $Nd_2Fe_{14}B$-type structure, and have studied the magnetic properties of these compounds. Luo Sheng et al. (1987) have reported the structure of the closely related compound $Nd_3Fe_{20}C_{1.3}$ ($\approx Nd_2Fe_{14}C$) and Buschow et al. (1988a) prepared $Nd_2Fe_{14}C$ by annealing in a fairly narrow temperature range after casting the alloy. Helmholdt and Buschow (1988) employed neutron diffraction to establish firmly the occurrence of a tetragonal compound $Nd_2Fe_{14}C$ isotypic with $Nd_2Fe_{14}B$, and showed that the magnetic structure of $Nd_2Fe_{14}C$, both at room temperature and at 4.2 K, is similar to that of the corresponding boride. Several investigations have shown that the heavy rare earth compounds form comparatively easily (de Mooij and Buschow 1988, Verhoef et al. 1989) and the formation of the light rare earth compounds, in particular the cerium compound, is difficult as a consequence of a solid state transformation. However, Jacobs et al. (1989) have also succeeded in preparing $Ce_2Fe_{14}C$ by substituting 2% manganese for iron. These results, in combination with those of previous works for lanthanum (Marusin et al. 1985b), gadolinium (Abache and Oesterreicher 1985), erbium (Jeitschko and Gerss 1986, Panek and Hoppe 1972, Pedziwiatr et al. 1986) and dysprosium (Liu and Stadelmaier 1986, Pedziwiatr et al. 1986), showed that all the lanthanides could form this type of compound except for europium and ytterbium. Up to date, investigations with respect to the magnetic properties and magnetic structures of these compounds have been carried out comprehensively, and important information about the spin reorientation in $Nd_2Fe_{14}C$ (Denissen et al. 1988b), the magnetovolume effect in $R_2Fe_{14}C$ (Buschow 1988) and the hyperfine fields in $Gd_2Fe_{14}C$ (Erdmann et al. 1989) have been published.

For the Ce–Fe–C system the extrapolated transformation temperature from $Ce_2Fe_{17}C_x$ to $Ce_2Fe_{14}C$ is found to be about 750°C. The formation of $Ce_2Fe_{14}C$ can be expected only when the annealing of the sample is performed below this temperature. However, annealing below 750°C did not produce the $Ce_2Fe_{14}C$ phase, probably because the reaction rates were too low. The existence of this compound was confirmed by extrapolating the compounds $Ce_2Fe_{14-x}Mn_xC$ with $x = 0.8, 0.3$ to $x = 0$. The samples with $x = 0.3$ were vacuum-annealed at 800°C for four weeks after arc melting, and their ^{57}Fe Mössbauer spectrum at 300 K and 20 K is of the same type as observed for the other $R_2Fe_{14}C$ compounds but the Fe moments are substantially smaller than in the latter. This suggests that this cerium compound has the $Nd_2Fe_{14}B$-type structure; however, cerium in this compound is close to the tetravalent state instead of the trivalent state (Jacobs et al. 1989).

For the Nd–Fe–C system the preparation of the $Nd_2Fe_{14}C$ compound is more difficult. Only when samples were annealed at temperatures below about 890°C (e.g. at 870°C for 500 h) after arc melting could the single-phase compound $Nd_2Fe_{14}C$ be formed. At higher temperatures the compound $Nd_2Fe_{14}C$ decomposes by means of a solid state transformation into the compound $R_2(Fe, C)_{17}$ (or $R_2Fe_{17}C_x$), possibly accompanied by small quantities of some other ternary R–Fe–C compound not detected in X-ray diagrams. When a sample (annealed at 870°C) consisting of $Nd_2Fe_{14}C$ was heated to 1000°C for 100 h and quenched to room temperature, the tetragonal $Nd_2Fe_{14}C$ phase disappeared. Instead, the rhombohedral Th_2Zn_{17}-type structure and a phase resembling Nd_2C_3 were observed. When the samples were annealed at comparatively low temperatures, a mixture of $Nd_2Fe_{14}C$ and $Nd_2(Fe, C)_{17}$ was found (Buschow et al. 1988a). Thus, the formation of the compound $Nd_2Fe_{14}C$ requires an annealing treatment of the as-cast alloy in the appoximate temperature range 830–880°C. It has been shown that this compound $Nd_2Fe_{14}C$ is ferromagnetic with a Curie temperature $T_C = 535$ K, and the moments of the iron atoms in it are roughly 5% lower than in $Nd_2Fe_{14}B$. The neutron diffraction study of the crystallographic and magnetic structure of $Nd_2Fe_{14}C$ by Helmholdt and Buschow (1988) showed that the atomic positional parameters obtained after refinement of the neutron diagrams of $Nd_2Fe_{14}C$ are close to those reported for $Nd_2Fe_{14}B$ by Herbst et al. (1984). This means that the arrangement of the various atoms in the tetragonal unit cell, space group $P4_2/mnm$, is almost identical for the carbides and borides. At room temperature, $Nd_2Fe_{14}C$ has an easy magnetization direction along [001] and at 4.2 K the easy magnetization direction no longer coincides with the [001] direction but deviates from it. However, Luo Sheng et al. (1987) previously determined the structure of $Nd_3Fe_{20}C_{1.3}$ ($\simeq Nd_2Fe_{14}C$) by neutron diffraction to be rhombohedral, which is related to that of Nd_2Fe_{17}. These conflicting results arise from the occurrence of a solid state transformation (Buschow et al. 1988a) that leads to a decomposition of the tetragonal compound $Nd_2Fe_{14}C$ when heated at 940°C, as done by Luo Sheng et al. (1987).

The above-mentioned solid state transformation in respect to $Nd_2Fe_{14}C$ was also found for other rare earth systems (de Mooij and Buschow 1988). The transformation temperatures T_t determined from the annealing experiments on the various $R_2Fe_{14}C$ samples are shown as a function of the rare earth component in fig. 24. As shown in this figure, only when the annealing was performed at temperatures that lie between the full curve and 830°C, could single-phase samples of $R_2F_{14}C$ be obtained. When single-phase samples obtained in this way were subsequently subjected to annealing at temperatures higher than those corresponding to the full curve, the 2:17 phase formed at the expense of the tetragonal phase, but subsequently, annealing below T_t could not lead to the opposite transformation of the 2:17 phase and only negligibly small amounts of the tetragonal phase were formed. According to investigations by Liu et al. (1987), two different schemes could account for the formation of $R_2Fe_{14}C$:

$$R_2Fe_{17} + RFeC \rightleftharpoons R_2Fe_{14}C + \text{R-rich phase},$$

$$R_2Fe_{17}C + RFeC \rightleftharpoons R_2Fe_{14}C + RFe_2C_2,$$

and if one assumes that the solubility limit of carbon in $Nd_2Fe_{17}C_x$ can increase to a value slightly higher than $x = 0.6$, the decomposition reaction of $R_2Fe_{14}C$ may involve only three phases (de Mooij and Buschow 1988),

$$17R_2Fe_{14}C \rightleftharpoons 14R_2Fe_{17}C_{0.64} + 3R_2C_3.$$

De Mooij and Buschow (1988) pointed out that there is a strong decrease in T_t towards the beginning of the lanthanide series. For the light lanthanides, the value of T_t is expected to fall into the temperature range where the diffusion-limited reaction rate is too low to allow formation of the compound $R_2Fe_{14}C$. In general, therefore, their preparations are more difficult to perform. However, the reaction rate below T_t can be enhanced by other means, and the formation of the $R_2Fe_{14}C$ phase in these systems still remains possible, e.g. $La_2Fe_{14}C$ (Marusin et al. 1985), $Ce_2Fe_{14}C$ (Jacobs et al. 1989) and $Pr_2Fe_{14}C$ (Gueramian et al. 1987) have been prepared.

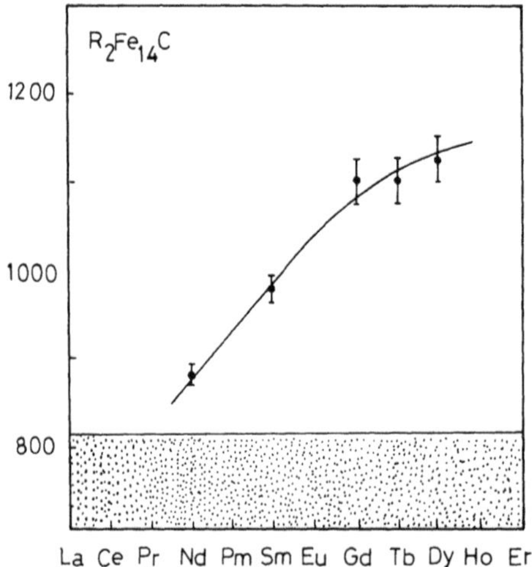

Fig. 24. Transformation temperatures T_t for various R–Fe–C systems showing the decomposition of the $R_2Fe_{14}C$ phase (full line). The shaded region indicates the temperature range in which the reaction rate is too slow for the formation of the $R_2Fe_{14}C$ phase from the nonequilibrium phases of the as-cast melt (de Mooij and Buschow 1988).

The annealed samples consisted of a majority phase having the tetragonal $Nd_2Fe_{14}B$-type structure for R = Sm, Gd, Dy, Tm, and those based on Pr and Ho contained significant amounts of impurity phases (mainly Fe and/or $R_2Fe_{17}C_{3-x}$). In the samples containing yttrium, no trace of this phase was found in the X-ray patterns (Gueramian et al. 1987). The absence of a member with yttrium is surprising in view of the existence of the corresponding borides (Sinnema et al. 1984) and the similarity between behaviors of yttrium and dysprosium with respect to the formation of the ternary compounds. The lattice parameters of this type of compounds are listed in table 17; the results for lanthanum and cerium were taken from the works of Marusin et al. (1985a, b) and Jacobs et al. (1989), respectively.

The crystallographic parameters of the compound $Nd_2Fe_{14}C$ derived from fitting the neutron diffractograms obtained at 4.2 K and 300 K are provided by Helmholdt and Buschow (1988).

6.4.4. *The $R_2Fe_{17}C_x$ compounds*

As has been pointed out in the previous section, $R_2Fe_{17}C_x$ compounds are present as the main phase when samples of the type $R_2Fe_{14}C$ are heated above their decomposition temperatures (de Mooij and Buschow 1988). These compounds can also be synthesized directly by arc melting (Gueramian et al. 1987), having either the rhombohedral Th_2Zn_{17}-type structure (R = Y, Ce, Pr, Sm, Gd, Tb, Dy) or the hexagonal Th_2Ni_{17}-type structure (R = Dy, Ho, Er, Tm, Lu). In the Dy compound sample, phases with both types of structures are found. The lattice parameters are summarized in table 18.

It is seen that the unit cell volumes of the carbon-containing compounds are expanded with respect to those of the corresponding binary R_2Fe_{17} compounds

TABLE 17

Lattice parameters of the tetragonal $R_2Fe_{14}C$ series (Å), space group $P4_2/mnm$ [Gueramian et al. (1987), except for La, Ce and Nd].

	Y	La	Ce	Pr	Nd	Sm	Gd
a	–	8.819(2)[c]	8.74[d]	8.816(1)	8.8280(5)[g]	8.802(1)	8.795(1)
c	–	12.142(6)[c]	1.184[d]	12.044(1)	12.0343(5)[g]	11.952(2)	11.902(1)
a				8.82[a]			
c				12.05[a]			

	Tb	Dy	Ho	Er	Tm	Lu
a	8.771(1)	8.763(1)	8.752(1)	8.742(1)	8.732(1)	8.719(1)
c	11.864(1)	11.836(1)	11.813(1)	11.791(1)	11.766(1)	11.726(1)
a		8.78[e]	8.756[f]	8.723[f]		8.709[b]
c		11.83[e]	11.801[f]	11.786[f]		11.713[b]

[a] Verhoef et al. (1988).
[b] Denissen et al. (1988a).
[c] Marusin et al. (1985b).
[d] Jacobs et al. (1989).
[e] Liu and Stadelmaier (1986).
[f] Pedziwiatr et al. (1986).
[g] Helmholdt and Buschow (1988).

TABLE 18
Lattice parameters of various $R_2Fe_{17}C_x$ compounds.

	$Y_2Fe_{17}C_x$	$Ce_2Fe_{17}C_x$	$Pr_2Fe_{17}C_x$	$Nd_2Fe_{17}C_x$	$Sm_2Fe_{17}C_x$	$Gd_2Fe_{17}C_x$	$Tb_2Fe_{17}C_x$	$Dy_2Fe_{17}C_x$
a (rhombohedral) (Å)	8.567(1)	8.534(1)	8.615(1)	8.633(1)	8.562(1)	8.562(1)	8.618(1)	8.507(1)
c (Å)	12.489(2)	12.436(2)	12.478(2)	12.479(1)	12.450(2)	12.501(2)	12.469(1)	12.441(1)
	$Dy_2Fe_{17}C_x$	$Ho_2Fe_{17}C_x$	$Er_2Fe_{17}C_x$	$Tm_2Fe_{17}C_x$	$Lu_2Fe_{17}C_x$			
a (hexagonal (Å)	8.115(2)	8.507(1)	8.493(1)	8.476(1)	8.461(1)			
c (Å)	8.330(2)	8.328(2)	8.319(1)	8.307(1)	8.311(1)			

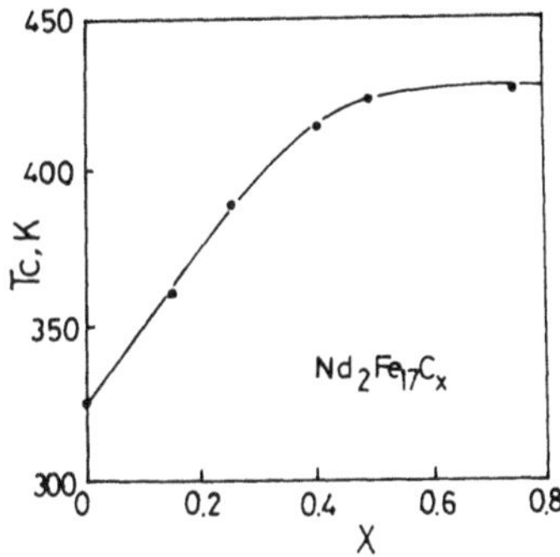

Fig. 25. Composition dependence of the Curie temperature in $Nd_2Fe_{17}C_x$ (de Mooij and Buschow 1988).

(*Pearson's Handbook of Crystallographic Data for Intermetallic Phases*, Vol. 2 (1985) eds. P. Villars and L. D. Calvert, American Society for Metals), indicating that carbon is at least partially dissolved in the structure. Thus, these carbides can be considered as pseudobinary compounds. The investigation of de Mooij and Buschow (1988) also found that in $Nd_2Fe_{17}C_x$ the carbon atom does not replace iron in Nd_2Fe_{17}, but occupies a separate lattice position, 9e, in the Th_2Zn_{17} structure type ($R\bar{3}m$), and the lattice parameters of $Nd_2Fe_{17}C_{0.4}$ are $a = 8.625$ Å and $c = 12.475$ Å in accordance with those reported by Gueramian et al. (1987). They further determined from the concentration dependences of the lattice parameters and the Curie temperature (fig. 25) that the solubility limit of carbon in $Nd_2Fe_{17}C_x$ at 900°C is about $x = 0.6$.

Liu et al. (1987) investigated the phase relationship between $R_2Fe_{17}X$, RFeX, $R_2Fe_{14}X$ and RFe_2X_2 ($X = B_xC_{1-x}$) and proposed a reaction equation

$$R_2Fe_{17}X + RFeX \rightleftharpoons R_2Fe_{14}X + RFe_2X_2,$$

which can be used to describe the transformation between $R_2Fe_{17}X$ and $R_2Fe_{14}X$ (de Mooij and Buschow 1988). If we assume that the solubility limit of carbon in $Nd_2Fe_{17}C_x$ can increase to a value slightly higher than $x = 0.6$, the decomposition reaction may involve only three phases and proceed via the reaction

$$17R_2Fe_{17}C \rightleftharpoons 14R_2Fe_{17}C_{0.64} + 3R_2C_3.$$

6.4.5. *The R_2FeC_4 compounds*

The R_2FeC_4 compounds were formed by a peritectic reaction of the RC_2 carbides with iron at 900°C for ten days (Gerss et al. 1987). The carbides R_2FeC_4 are gray with a metallic luster and show good electrical conductivity. The carbide Y_2FeC_4 exhibits a transition to a superconducting state at $T_c = 3.6$ K. Magnetic and ^{57}Fe Mössbauer measurements at 295 and 4 K showed that Y_2FeC_4 is essentially nonmagnetic while Er_2FeC_4 is strongly paramagnetic, but neither a magnetic hyperfine structure nor magnetic ordering were found above 4 K.

Their crystal structures were determined from X-ray powder diffraction data for Tm_2FeC_4 and from neutron diffraction data for Er_2FeC_4 to be body-centered orthorhombic, space group Ibam with $Z = 4$ formula units per cell. The lattice parameters of R_2FeC_4 (R = Y, Tb, Dy, Ho, Er, Tm, Yb, Lu) can be found in the work by Gerss et al. (1987). The values for the ytterbium compound fit smoothly between those of the thulium and the lutetium compounds indicating the trivalent nature of ytterbium in this structure. The values of the yttrium compound are close to those of the dysprosium compound, as for other ternary rare earth metal carbides (Jeitschko and Gerss 1986, Gerss and Jeitschko 1986, Jeitschko and Behrens 1986). The structure of Er_2FeC_4 was determined by a Rietveld peak shape refinement of the neutron diffraction data to be new, with carbon pairs (bond distance 1.33 Å) in an environment of five rare earth atoms and two adjacent iron atoms. A projection of the structure and the coordination polyhedra, as well as the positional parameter can be found in Gerss et al. (1987). The most remarkable feature of this structure is the empty channels along the z axis adjacent to the iron chains. However, these channels are too small to accommodate additional carbon atoms and the empty position within a channel farthest away from the metal atoms is at $\frac{1}{2}, \frac{7}{10}, \frac{1}{4}$. Its distances from the metal atoms (void–Fe is 1.90 Å; on the opposite site, void–2Er is 2.16 Å) are all shorter than the corresponding bond distances of the structure. Gerss et al. have given the interatomic distances in Er_2FeC_4 and pointed out that the carbon atoms form pairs with a bond distance of 1.33 Å, corresponding to the C=C bond distance of 1.34 Å found in olefins, and strong Fe–Fe bonds are formed within the iron chain.

6.4.6. *The R–Fe–Si–C compounds*

In order to synthesize new compounds for high-performance permanent magnets in the Dy–Fe–Si–C system, Paccard et al. (1987) and Paccard and Paccard (1988) as well as Le Roy et al. (1987) prepared the compounds $DyFe_2SiC$ (Re_3B-type) and $Dy_2Fe_2Si_2C$ (Ge_2Os-type) as well as $RFe_{10}SiC_{0.5}$ ($LaMn_{11}C_{2-x}$-type), respectively. The formations of the three compounds are commonly associated with the structure-stabilizing effect of carbon addition.

6.4.6.1. *$DyFe_2SiC$*. X-ray diffraction studies of single-crystal $DyFe_2SiC$ showed that the compound $DyFe_2SiC$ has the $YNiAl_2$-type (Re_3B-type derivative) structure with space group Cmcm and lattice parameters $a = 3.712(0)$ Å, $b = 10.531(3)$ Å and $c = 6.863(1)$ Å. In this crystal structure, the columns of Dy_2Fe_4 prisms are centered by silicon atoms with formula R_2T_4M but these series of columns are separated by a chain of carbon atoms at $(0, \frac{1}{2}, 0)$. This is a new compound. Its positional and thermal parameters as well as the interatomic distances have been given by Paccard et al. (1987).

6.4.6.2. *$R_2Fe_2Si_2C$*. Single-crystal X-ray diffraction studies determined $Dy_2Fe_2Si_2C$ to have a monoclinic cell with space group C2/m. The crystal structure is derived from the Ge_2Os-type with an ordered replacement of germanium atoms by dysprosium and iron atoms, and of osmium atoms by silicon atoms, and with an addition of carbon atoms between the isolated infinite columns of silicon-centered double prisms. One carbon atom is at the center of an octahedron, Dy_4Fe_2, where the six vertices are occupied by four dysprosium and two iron atoms, as in $DyFe_2SiC$ (fig. 26). The atomic positional and thermal parameters for $Dy_2Fe_2Si_2C$ have been given by Paccard and Paccard (1988), as well as the interatomic distances in $Dy_2Fe_2Si_2C$.

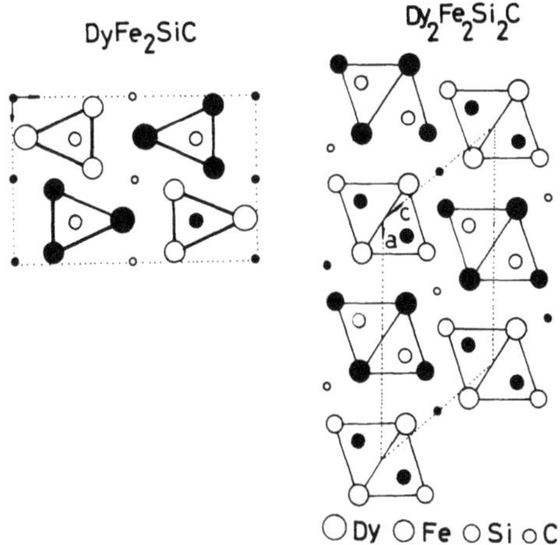

Fig. 26. Structures of $DyFe_2SiC$ (a) and $Dy_2Fe_2Si_2C$ (b) (Paccard and Paccard 1988). (Reprinted by permission of the publisher, Elsevier Sequoia S.A., Inc.)

In addition to $Dy_2Fe_2Si_2C$, $R_2Fe_2Si_2C$ (R = Pr, Nd, Y, Gd, Tb, Ho, Er) were also prepared and were recognized to be isotypic by examination of X-ray photographs of powdered samples. The variation of the cell parameters is a consequence of the normal lanthanide contraction.

Summarizing the above, in the structure of both $Dy_2Fe_2Si_2C$ and $DyFe_2SiC$, the carbon atoms completely fill the octahedral voids formed by four iron atoms and two rare earth atoms and thus stabilize their structures, since no ternary compound R–T–M (R = rare earth, T = transition metal, M = Si, B, C) with either the structures of Ge_2Os or Re_3B was found.

6.4.6.3. *$RFe_{10}SiC_{0.5}$*. Another compound $RFe_{10}SiC_{0.5}$was also obtained by the addition of carbon (Le Roy et al. 1987), however, the carbon atoms fill the octahedral voids formed by four iron atoms and two rare earth atoms only with an occupancy of 25% instead of a complete occupancy, as found in $Dy_2Fe_2Si_2C$ and $DyFe_2SiC$.

X-ray diffraction studies on single crystals of $NdFe_{10}SiC_{0.5}$ have established its structure with the space group $I4_1/amd$ and $Z = 4$ formula units per unit cell. The positions of the neodymium, iron and silicon atoms correspond to the $BaCd_{11}$ structure. The positional and thermal parameters for $NdFe_{10}SiC_{0.5}$ of $BaCd_{11}$ type structure have been given by Le Roy et al. (1987), as well as the interatomic distances in this compound. This structure is a disordered version of $LaMn_{11}C_{1.52}$ (Ross et al. 1981). Each neodymium atom is surrounded by a polyhedron of 26 atoms, i.e. 18 iron atoms, four carbon atoms and four atoms which may be iron or silicon. Fe(1) is surrounded by 14 atoms, Fe(2) by 13 atoms, and the (Fe–Si) site by 12 atoms. As has been pointed out, the carbon atoms fill octahedral voids formed by four iron atoms and two neodymium atoms, with an occupancy of 25%.

Several compounds of the type $NdFe_9(Fe_{1-y}Si_y)_2C_{2-x}$ were prepared with $0.5 < y < 1$ and $0 < x < 2$, and the lattice parameters, given in table 19, showed small deviations from the expected values because of the lanthanide contraction. These deviations are probably caused by variations of the iron-to-silicon ratio and the

TABLE 19
The lattice parameters of $RFe_{10}\ SiC_{0.5}$.

Parameter	$CeFe_{10}SiC_{0.5}$	$PrFe_{10}SiC_{0.5}$	$NdFe_{10}SiC_{0.5}$	$SmFe_{10}SiC_{0.5}$
a (Å)	10.049(4)	10.107(3)	10.083(3)	10.092(7)
c (Å)	6.528(3)	6.534(2)	6.529(2)	6.538(4)

smaller values for the cerium compound indicate that the cerium ion is in a valence state higher than three.

Concerning the formation of compounds $RFe_{10}SiC_{0.5}$, it should be emphasized that the addition of carbon is so important that the preparation of samples without carbon will lead to only the formation of the $Nd_2(Fe\text{–}Si)_{17}$ phase with a different structure from $RFe_{10}SiC_{0.5}$. Only when synthesizing the samples from the binary compound of carbon can the compounds $RFe_{10}SiC_{0.5}$ be formed.

6.4.7. *Other ternary R–Ni–C compounds*

Except for the above-described ternary nickel compounds $RNiC_2$ (Semonenko et al. 1983), YNi_2C_2 and $Y_5Ni_{29}C_3$ (probable composition) (Stadelmaier and Kim 1984), several types of the ternary nickel compounds have been reported by Russian investigators. In particular, it is well known that nickel itself does not form stable carbides, so the possible existence of stable rare-earth–nickel–carbon compounds is of interest in respect to either fundamental science or practical applications.

The $R_2Ni_5C_3$ (R = Ce, La) compounds are tetragonal with space group P4/mbm and two formula units per cell, $a = 8.331(2)$ and 8.307(5) Å, $c = 4.0283(9)$ and 3.979(3) Å, respectively, for R = La and Ce (Tsokol et al. 1986a).

The RNi_8C_2 (R = the heavy rare earth metals) compounds are orthorhombic and their lattice parameters have been determined for R = Gd–Lu (Putyatin and Sevastyanova 1987).

The $R_{11}Ni_{60}C_6$ compounds (R = Y, Er, Tm, Yb, Lu) have a cubic structure with space group Im3m. Their structural packing has been reported (Putyatin 1987) and the lattice parameters are 12.499(1), 12.472(1), 12.466(2), 12.425(2) and 12.422(2) Å for R = Y, Er, Tm, Yb, Lu. In comparison with the hexagonal compound $Y_5Ni_{29}C_3$ (Stadelmaier and Kim 1984), the cubic compounds $R_{11}Ni_{60}C_6$ appear to have an approximate stoichiometric composition. Because Stadelmaier and Kim have taken notice of the uncertainty of composition for the compound $Y_5Ni_{29}C_3$, it is thus likely that the compounds $R_{11}Ni_{60}C_6$ and $Y_5Ni_{29}C_3$ belong to the same type of compound.

$R_2Ni_{22}C_3$ compound has an orthorhombic structure with space group Cmca and eight formula units per unit cell. The structure was solved by Patterson and Fourier methods and refined by a least-squares calculation to $R = 0.082$. The atomic position parameters have been determined (Bodak et al. 1982). The lattice parameters of the compounds $La_2Ni_{22}C_3$, $Ce_2Ni_{22}C_3$ and $Pr_2Ni_{22}C_3$ are $a = 11.45(5)$, 11.384(3) and 11.38(4) Å; $b = 15.12(4)$, 15.024(7) and 15.05(3) Å; $c = 14.63(4)$, 14.671(5) and

14.59(3) Å, respectively. Sc_3NiC_4 has an orthorhombic structure which is isotypic with Sc_3CoC_4 (Tsokol et al. 1986b), space group I2,2,2.

Summarizing the compounds that have been described in the rare-earth-metal–nickel–carbon systems, we have the following: $RNiC_2$, $R_2Ni_5C_3$ (R = La, Ce), RNi_8C_2 (R = the heavy rare earth metals), $R_2Ni_{22}C_3$ (R = La, Ce, Pr), $R_{11}Ni_{60}C_6$ and Sc_3NiC_4. However, no information on the phase relationships related to their formation was reported.

6.4.8. *The R–T–C compounds and structural relation between the RTC and RTC_2 compounds*

The RTC compounds were found in the Gd–Fe–C system (Stadelmaier and Park 1981), and the GdFeC compound has a hexagonal structure, $a = 9.12$ Å and $c = 5.97$ Å.

The compound YCoC has also been reported (Gerss and Jeitschko 1986) to have a layered structure with alternating CoC and Y sublayers. The CoC sheet is actually composed of parallel --Co–C–Co–C–Co-- chains, and the separations between these chains are large. The chain axes of the neighboring CoC layers are perpendicular to each other. Figure 27 shows a unit cell of YCoC formed by four yttrium atoms in equatorial positions and two cobalt atoms in axial sites. Only one half of these holes are occupied by carbon atoms; the other half are vacant. This compound does not possess any short carbon–carbon bonds (the C–C distance is 3.65 Å) (Li and Hoffmann 1989). Its electronic structure has been studied by Hoffmann (1987).

In contrast to the compound YCoC, $YCoC_2$ with a $CeNiC_2$-type structure has quite a short C–C distance and the structure is base-centered orthorhombic with

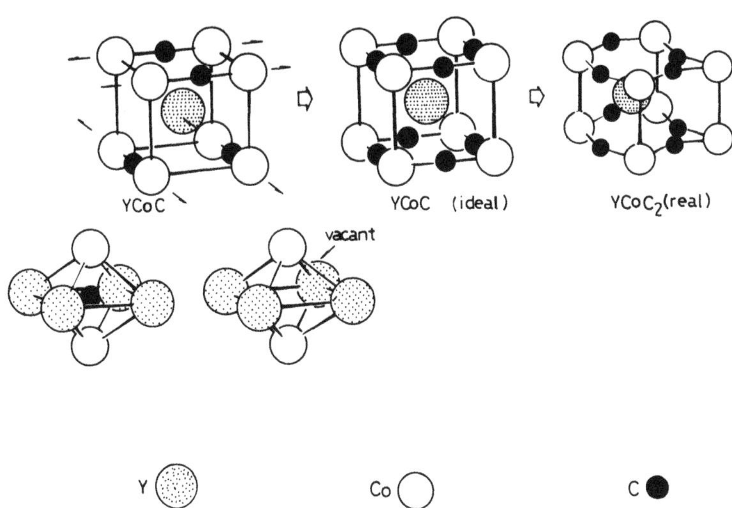

Fig. 27. Change of the structure from YCoC to $YCoC_2$ (Li and Hoffman 1989). (Reprinted by permission of the publisher, The American Chemical Society, Inc.)

space group Amm2. In fact, the $CeNiC_2$-type structure is closely related to the YCoC-type (Gerss and Jeitschko 1986) and can also be considered as a layered one with alternating CoC and Y sublayers, but the carbon filling up the other half of the holes in the YCoC-type structure are accompanied by the formation of the C–C pairs and the distortion of the lattice (fig. 27).

Therefore, the two compounds YCoC and $YCoC_2$ (the $CeNiC_2$-type) can be distinguished in terms of the existence of either the bonded carbon pairs or the isolated C atoms. The former has no carbon pairs in its structure, while the latter possesses the closely bonded carbon pairs. A close look at the $CeNiC_2$ structure revealed (Li and Hoffmann 1989) that the system is selective in electron count, only 19–21 electron species in the carbon–carbon bonding and antibonding states have been found so far, and the transition metals are limited to group VIII Fe, Co, Ni atoms.

6.4.9. *Summary: two categories of the rare-earth-metal–transition-metal carbides*

Like the binary systems, the ternary carbides can be classified into two categories: (1) structures containing bonded carbon pairs and (2) those with isolated C atoms, as found for the YCoC and $YCoC_2$ compounds. The tetragonal $UCoC_2$-type compounds $ScCoC_2$, $ScFeC_2$ and $ScNiC_2$ (Marusin et al. 1985a), the base-centered orthorhombic $CeNiC_2$-type compounds, $RFeC_2$, $RNiC_2$ and $YCoC_2$, $RCoC_2$ (R = Gd–Lu), the high-temperature $SmCoC_2$ phase (Jeitschko and Gerss 1986), the orthorhombic R_2FeC_4 (R = Y, Tb–Lu) phases (Gerss et al. 1987), the tetragonal $R_2Ni_5C_3$ (R = La, Ce) phases (Tsokol et al. 1986a), and the monoclinic $RCoC_2$ (R = La–Nd, Sm) compounds (Bodak et al. 1980) belong to group (1) materials. In each structure the carbon atoms form dimers and the C–C distances vary over a wide range, e.g. from 1.26 Å in $ScCoC_2$ (Marusin et al. 1985a) to 1.42 Å in $La_2Ni_5C_3$ (Tsokol et al. 1986a). Other selected values are 1.37 Å in $DyCoC_2$ and $DyNiC_2$ (Jeitschko and Gerss 1986), 1.36 Å in $CeCoC_2$ (Bodak et al. 1980) and 1.33 Å in Er_2FeC_4 (Gerss et al. 1987). In addition, there are short transition metal–metal bonds in these structures, i.e. Fe–Fe = 2.50 Å in Er_2FeC_4.

A large number of compounds of the second group are metal-rich, such as $R_2Ni_{22}C_3$ (R = Ce, La, Pr), $La_2Fe_{14}C$ (Marusin et al. 1985b), however, YCoC also belongs to this group. The C–C distance is 3.65 Å and no short carbon–carbon bonds are formed in YCoC.

Li and Hoffmann (1989), on the basis of calculations in respect to the density of state and band structure of the carbides, determined that the RTC_2 compounds containing short C–C bonds exist in two stable forms, the $CeNiC_2$-type and the $UCoC_2$-type structures. The former should be favorable for electron counts of 19–21 with late transition metal elements, and the latter is favorable for a valence electron count of 23 or more.

6.5. *Ternary rare-earth-metal–platinum-group-metal–carbon systems*

6.5.1. *The Ce–Ru–C phase diagram and information on other phase diagrams.*

Solid state relationships in the Ce–Ru–C system have been examined by annealing arc-cast samples in the temperature range 600–1250°C and using X-ray diffraction, electron probe microanalysis and reflected light microscopy (Parnell et al. 1985). The phase equilibria in the ruthenium corner are in agreement with those reported by Holleck (1977). The existence of the perovskite-type compound $CeRu_3C_{1-x}$ reported by Holleck (1977) was also confirmed. This compound is face-centered cubic and the lattice parameter ($a = 4.1515 \pm 0.0005$ Å) is slightly larger than that reported by Holleck (1972) ($a = 4.137 \pm 0.002$ Å). In addition, a new ternary compound $Ce_3Ru_2C_5$ was reported to be hexagonal with $a = 8.7920$ Å and $c = 5.2983$ Å. The two ternary carbides form peritectically and a long annealing period (i.e., 20 days at 900°C for $CeRu_3C$ and 15 days at 950°C for $Ce_3Ru_2C_5$) was necessary to allow their formation.

Krikorian (1971) investigated the reaction of the dicarbides of yttrium, cerium and erbium with platinum and iridium and gave the ternary phase diagram of the Y–Pt–C system at 1000–1200°C, the Ce–Pt–C and Er–Pt–C systems at 1100–1400°C as well as the Y–Ir–C system at 1100–1500°C, and the Ce–Ir–C and Er–Ir–C systems at 1100–1400°C. No ternary carbide was found.

In the Ce–Ir–C system the existence of compatibility ties from carbon to the binary metal edge was also confirmed (cited by Parnell et al. 1985). However, in the Ce–Ru–C system carbon atoms do not tie themselves to Ru–Ce binary edges, rather CeC_2 ties to ruthenium metal in contrast to the result reported by Holleck (1977) for this system.

6.5.2. *Crystal and electronic structures of the $R_8Rh_5C_{12}$ compounds*

Recently, Hoffmann et al. (1989) reported a series of the $Er_8Rh_5C_{12}$-type compounds present in the R–Rh–C system. These compounds $R_8Rh_5C_{12}$ (R = Y, Gd–Tm) are thermodynamically stable at 900°C and have a monoclinic structure, space group C2/m, with $Z = 2$ formula units per unit cell. The lattice parameters of these compounds have been measured by single-crystal X-ray diffraction. The structure contains a finite chain-like centrosymmetric polyanion $[Rh_5C_{12}]^{24-}$ with two Rh–Rh bonds (2.708 Å) and six pairs of carbon atoms. The shortest distances between adjacent Rh_5C_{12} clusters are the Rh–C distances of 2.94 Å and the Rh–Rh distances of 3.27 Å, and thus the Rh_5C_{12} units may be treated as isolated from each other. This structure is characterized by three different kinds of C_2 pairs. The C–C bond distances of 1.27, 1.32 and 1.33 Å are between those of a triple bond (1.20 Å) and a double bond (1.34 Å) in hydrocarbons, and are the shortest found so far in ternary carbides of the rare earth metals with transition metals.

The studies of Hoffmann et al. (1989) showed that $Y_8Rh_5C_{12}$ is a metallic conductor but does not become superconducting, at least above 1.9 K. Magnetic susceptibility measurements of $Y_8Rh_5C_{12}$ show temperature-independent weak paramagnetism, while the other $Er_8Rh_5C_{12}$-type compounds are Curie–Weiss paramagnets with magnetic moments corresponding to those of the rare earth metal ions.

The special crystal structure of the $Er_8Rh_5C_{12}$-type compounds has led to a deep study of the electronic and band structures by using standard one-electron (extended Hückel) calculations (Lee et al. 1989). Their calculations have rationalized the varying Rh–C distances within the $[Rh_5C_{12}]^{24-}$ polyanion as well as the various Rh–C–Rh and Rh–C–C bond angles. They pointed out that the interactions that bind the Rh and C atoms to one another remain of a two-fold nature. Firstly, the C_2 bonding and nonbonding orbitals interact with the s and p orbitals of Rh and secondly the $C_2\pi g$ orbitals interact with the d orbitals of Rh.

For comparison, the electronic structures of several hypothetical and real $RhC_2{}^{3-}$ systems have also been studied (Lee et al. 1989), including the polyanion of $SmRhC_2$ (Hoffmann and Jeitschko 1988) with the $CeNiC_2$-type structure.

6.5.3. *The RT_3C-type compounds and other ternary carbides*

In the rare-earth–metal–platinum-group-metal–carbon systems a large number of ternary compounds have been found, $RRhC_2$ (R = La, Ce, Sm) (Hoffmann and Jeitschko 1988), $Er_{10}Ru_{10}C_{19}$ (Hoffmann and Jeitschko 1987), Ce_3RhC_5 (Parnell et al. 1985) and $Er_8Rh_5C_{12}$ (Hoffmann et al. 1989), as well as ScT_3C (T = Ru, Rh, Ir), CeT_3C (T = Ru, Rh) and RRh_3C (R = La, Er) (Holleck 1977). The latter type of compounds is characterized by the solution of an isolated carbon atom in the binary compound RT_3 and was determined to be the carbon-stabilized ordered compound with a structure of the filled Cu_3Au type. Based on the variation of the lattice parameters with the measured carbon content for the phase $ScRh_3C_{1-x}$, the carbon content tends closely to the formula ScT_3C.

An empirical rule has been proposed to relate the valence states of the transition metals to the formation of these compounds (Holleck 1975). In different RT_3C compounds scandium appears to have different valence states (Holleck 1977), e.g. in the RRu_3C compound Sc is tetravalent, like Ce; while in RRh_3C and RM_3C (M = In, Tl, Sn, Pb) it is trivalent like La and Er in the corresponding compounds. The atomic radius of the tetravalent scandium was deduced from the lattice parameters of phases with the composition of RRu_3C and RRh_3 (R = Sc, Ce) to be about 1.52 Å, which is less than the value of trivalent scandium, 1.64 Å, deduced from the compounds RT_3C (R = La, Er, Sc; T = In, Tl, Sn, Pb, Rh). The lattice parameters of the compounds $ScRu_3C_{1-x}$, $ScRh_3C_{1-x}$ and $ScIr_3C_{1-x}$ were determined to be 4.031 Å (in equilibrium with Ru and C), 4.017 Å (in equilibrium with Rh and C) and 3.994 ± 0.002 Å (in equilibrium with $ScIr_2$ and C), respectively (Holleck 1977).

6.6. *Ternary R–Mn (Tc and Re) carbides*

In the rare-earth–manganese–carbon systems three types of ternary carbides were found: the tetragonal $LaMn_{11}C_{2-x}$ for R = La–Nd (Jeitschko and Block 1985), the rhombohedral $Pr_2Mn_{17}C_{3-x}$ for R = La–Sm (Block and Jeitschko 1986a) and the hexagonal $Tb_2Mn_{17}C_{3-x}$ for R = Gd–Tm, Lu, Y (Block and Jeitschko 1986b, 1987a) structures. These are derived from those of the binary compounds $BaCd_{11}$ (Sanderson and Baenziger 1953). Th_2Zn_{17} (Makarov and Vinogradov 1956) and Th_2Ni_{17} (Florio et al. 1956), respectively. The lattice parameters have been measured.

In the samples of the compounds $R_2Mn_{17}C_{3-x}$ relatively large variations in the lattice parameters were observed. For example, the compound $Gd_2Mn_{17}C_{3-x}$ in the as-cast condition has $a = 8.775(1)$ Å, $c = 8.557(1)$ Å and $v = 570.6(1)$ Å^3. After annealing at 800°C the lattice parameters had decreased to $a = 8.728(2)$ Å, $c = 8.545(3)$ Å and $v = 563.7(3)$ Å^3. Smaller homogeneity ranges were also observed by comparing the lattice parameters of various samples for these compounds. These compounds were presented in the as-cast alloys as well as after annealing, but with an exception; $La_2Mn_{17}C_{3-x}$ was stable only in the as-quenched condition. Annealing of the $La_2Mn_{17}C_{3-x}$ samples at temperatures between 400 and 900°C leads to the formation of $LaMn_{11}C_{2-x}$. No structural transition between the $Pr_2Mn_{17}C_{3-x}$ and the $Tb_2Mn_{17}C_{3-x}$ structures was found. Apparently, the stabilities of both structure types depend on the size of the rare earth elements.

The structures of the three types of compounds have been determined for $LaM_{11}C_{2-x}$ type (Jeitschko and Block 1985), $Pr_2Mn_{17}C_{3-x}$ (Block and Jeitschko 1986a) and $Tb_2Mn_{17}C_{3-x}$ (Block and Jeitschko 1987a) from single-crystal X-ray diffractometer data. The authors also reported the atom parameters and interatomic distances in these compounds.

In $Pr_2Mn_{17}C_{3-x}$, $Tb_2Mn_{17}C_{3-x}$ and $LaMn_{11}C_{2-x}$, the positions of the metal atoms correspond to those of the binary compounds Th_2Zn_{17}, Th_2Ni_{17} and $BaCd_{11}$. The carbon atoms fill voids of approximately octahedral shape formed by four Mn atoms and two R atoms at opposite corners. The R_2Mn_4C octahedra are connected by common R atoms, and the nets of octahedra correspond to the interatomic bonds in the diamond structure, the rhombohedral graphite structure (stacking ABCABC ...) and the hexagonal graphite structure (stacking ABAB ...), respectively, for $LaMn_{11}C_{2-x}$, $Pr_2Mn_{17}C_{3-x}$ and $Tb_2Mn_{17}C_{3-x}$. Thus, like the binary host structures Th_2Zn_{17} and Th_2Ni_{17}, the structures of the corresponding carbides $Pr_2Mn_{17}C_{3-x}$ and $Tb_2Mn_{17}C_{3-x}$ are stacking variants of each other.

The cell volumes of the $Tb_2Mn_{17}C_{3-x}$- and the $Pr_2Mn_{17}C_{3-x}$-type carbides were plotted by Block and Jeitschko (1986a, 1987a). The value for the cerium compound is smaller than that for $Pr_2Mn_{17}C_{3-x}$. This indicates that cerium is at least partially tetravalent in this compound. As shown from a comparison of lattice parameters, this behavior of cerium can also be observed in the $LaMn_{11}C_{2-x}$-type carbides. The cell volume of the lutetium compound is greater than that of the thulium compound. This anomalous behavior is probably the result of a slight deviation from the ideal Lu/Mn ratio. Apparently the cell volume is becoming too small for the lutetium compound to form a thermodynamically stable $Tb_2Mn_{17}C_{3-x}$-type carbide with ideal composition.

The structure refinements showed that in $Tb_2Mn_{17}C_{3-x}$, $Pr_2Mn_{17}C_{3-x}$ and $LaMn_{11}C_{2-x}$ the occupancy of the carbon position was found to be 81 ± 4, 59 ± 3 and $76 \pm 1\%$, respectively, corresponding to the compositions $Tb_2Mn_{17}C_{2.43}$ ($x = 0.57$), $Pr_2Mn_{17}C_{1.77}$ ($x = 1.23$) and $LaMn_{11}C_{0.52}$ ($x = 0.48$). Block and Jeitschko (1987a) examined the homogeneity range of the $Tb_2Mn_{17}C_{3-x}$ structure with respect to the carbon positions in the Sm–Mn–C and Ho–Mn–C systems, indicating that the thermodynamically stable compositions are carbon deficient.

It has been shown that the average Mn–C distances are essentially the same, i.e. 1.92, 1.914 and 1.94 Å in $Tb_2Mn_{17}C_{3-x}$, $LaMn_{11}C_{2-x}$ and $Pr_2Mn_{17}C_{3-x}$, respectively. The average Tb–C distance of 2.52 Å in $Tb_2Mn_{17}C_{3-x}$ compares favorably with the Pr–C distance of 2.566 Å in $Pr_2Mn_{17}C_{3-x}$, considering the lanthanide contraction. However, almost all Mn–Mn distances are larger in $Pr_2Mn_{17}C_{3-x}$ (with average Mn–Mn distance 2.647 Å) than in $Tb_2Mn_{17}C_{3-x}$ (average Mn–Mn distance 2.627 Å), while the average Pr–Mn distance of 3.282 Å is in good agreement with the average Th–Mn distance of 3.229 Å. Thus, a comparison of the coordination polyhedra and interatomic distances of the two stacking variants $Pr_2Mn_{17}C_{3-x}$ and $Tb_2Mn_{17}C_{3-x}$ does not clearly reveal why the $Pr_2Mn_{17}C_{3-x}$ structure is preferred by the large lanthanides and the $Tb_2Mn_{17}C_{3-x}$ structure by the small lanthanides. Block and Jeitschko (1987a, b) discussed the splitting of the uniform coordination of the Pr atom in $Pr_2Mn_{17}C_{3-x}$ to two different coordinations for the Tb atoms in $Tb_2Mn_{17}C_{3-x}$ and suggested that in the latter structure the relatively smaller Tb atoms have a smaller effective coordination number than the Pr atoms.

Block and Jeitschko (1987a, b) also studied the magnetic properties of these manganese compounds, showing that the compounds $LaMn_{11}C_{2-x}$, $CeMn_{11}C_{2-x}$, $Pr_2Mn_{17}C_{3-x}$ and $Tb_2Mn_{17}C_{3-x}$ are paramagnetic with Curie–Weiss behavior at high temperatures and no magnetic order, at least above about 100 K. The Mn atoms carry magnetic moments in all four compounds. The superconductivity measurements for $LaMn_{11}C_{2-x}$, $La_2Mn_{17}C_{3-x}$, $Y_2Mn_{17}C_{3-x}$ or $Lu_2Mn_{17}C_{3-x}$ showed that down to 1.9 K no transition to a superconducting state could be observed.

Finally, it should be noted that a new rhenium compound has been synthesized with a composition $La_{12}Re_5C_{15}$ (Block and Jeitschko 1987b), which in addition to the C_2 pairs also contains carbon atoms that do not form pairs.

6.7. *The rare-earth-metal–(chromium, molybdenum, tungsten)–carbon systems*

6.7.1. *Phase diagram of the Gd–Cr–C system*

Phase equilibria for the isothermal section at 900°C of compound compositions in the ternary gadolinium–chromium–carbon system are shown in fig. 28 (Jeitschko and Behrens 1986). Only one ternary carbide, $Gd_2Cr_2C_3$, exists in this system and it was found to be isotypic with $Ho_2Cr_2C_3$. The variation of the lattice parameters of samples with different compositions indicates that the homogeneity range of $Gd_2Cr_2C_3$ seems to be small.

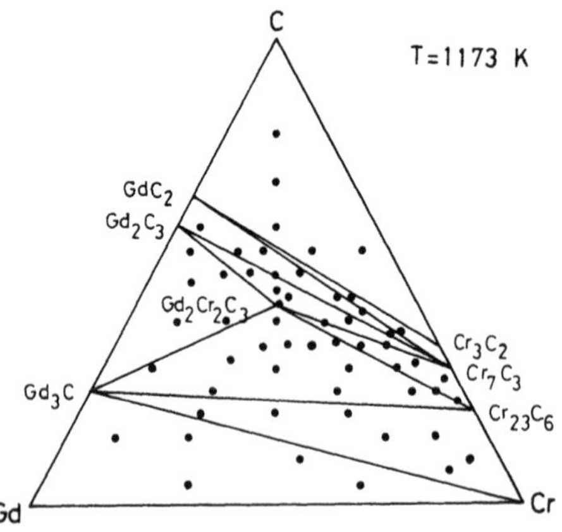

Fig. 28. The isothermal section of the ternary Gd–Cr–C system at 900°C (Jeitschko and Behrens 1986).

6.7.2. *Ternary R–(Cr, Mo, W)–C systems of the $R_2T_2C_3$ and RTC_2 types*

Samples with the composition $R_2T_2C_3$ for combinations of R = Sc, Y, La, Sm, Gd–Lu with T = Cr, Mo, W have been prepared (Behrens and Jeitschko 1984a, b, Jeitschko and Behrens 1986). The $Ho_2Cr_2C_3$-type compounds are formed only for the compositions $R_2Cr_2C_3$ (R = Y, Tb, Dy, Er, Tm, Lu) and $R_2Mo_2C_3$ (R = Ho, Er, Tm, Lu), while for the combinations of R = Sc, La, Sm, Yb with T = Cr, Mo, W, this type of structure was not found. In addition, Jeitschko and Behrens (1986) also found another type of compound with the compositions $RMoC_2$ (R = Y, Gd, Tb, Dy, Ho, Er, Tm) and RWC_2 (R = Y, Tb, Dy, Ho, Er, Tm). The lattice parameters of $R_2Cr_2C_3$ (R = Y, Gd–Tm, Yb), $R_2Mo_2C_3$ (R = Ho, Er, Tm, Lu), $RMoC_2$ (R = Y, Gd–Tm) and RWC_2 (R = Y, Tb–Tm) have been measured (Jeitschko and Behrens 1986). $R_2Cr_2C_3$ for R = Gd, Dy, Ho, Er and $RMoC_2$ for R = Gd–Er were formed after arc melting.

The ternary carbides $R_2T_2C_3$ and RTC_2 have the C-centered monoclinic $Ho_2Cr_2C_3$ (space group C2/m with $Z = 2$ formula units per cell) and the orthorhombic $UMoC_2$-type structure (Cromer et al. 1964), respectively. The structure of $Ho_2Cr_2C_3$ was determined from single-crystal X-ray data and refined to a residual of $R = 0.027$ for 739 independent structure factors and 24 variable parameters. The positional and thermal parameters of $Ho_2Cr_2C_3$ and the interatomic distances in this compound have been determined (Jeitschko and Behrens 1986).

The crystal structures and coordination polyhedra in $UMoC_2$-type and $Ho_2Cr_2C_3$-type compounds are comparable. The metal positions in both structures constitute a distorted body-centered cubic arrangement with a similar order of the two kinds of metal atoms. Nevertheless, in the $UMoC_2$-type structure one third of the (distorted) octahedral voids formed by the metal atoms are filled by carbon atoms, in the $Ho_2Cr_2C_3$ structure only one quarter of the voids are filled.

In both structures the coordination polyhedra of the metal atoms are also similar. However, the carbon distribution around both metal atoms is somewhat on one side in the $Ho_2Cr_2C_3$ structure. This can be considered as an indication of the essentially covalent character of the metal–carbon bonds. In addition, the distance of the C(1) atoms to the two Cr neighbors (1.906 Å) in the $UMoC_2$-type structure is considerably shorter than the corresponding distance (2.292 Å) of the vacant site (which is occupied in $UMoC_2$) in $Ho_2Cr_2C_3$, whereas the corresponding distances to the Ho atoms do not differ that much.

7. Carbides of rare-earth-metal–(oxygen, nitrogen, hydrogen, halogens)

7.1. *Rare earth oxycarbides*

7.1.1. *Preparation and formation of rare earth oxycarbides*

The existence of ternary carbon–oxygen–rare-earth-metal phases, $R_2O_2C_2$ (R = La, Ce, Pr, Nd) (Seiver and Eick 1976, Pialoux 1988, Butherus et al. 1966, Karen and Hájek 1986), $Ce_4O_2C_2$ (Clark and McColm 1972, Anderson et al. 1968), R_4O_3C (R = La, Nd, Gd, Ho, Er) (Butherus and Eick 1968), $CeO_{1.48}(C_2)_{0.07}$ and $CeO_{1.27}(C_2)_{0.34}$ (Clark and McColm 1972), R_2OC (R = Sc, Y, Dy), and their corresponding NaCl-type vacant rare earth oxycarbide phase R_1–(O, C, □) (Karen et al. 1986, Hájek et al. 1984a–d, Brožek et al. 1985, Haschke and Eick 1970b, Haschke and Deline 1980) has been established. The $La_4O_2C_2$ oxycarbide reported by Clark and McColm (1972) has a different structure and different hydrolysis product for the R_2OC oxycarbide, although an identical stoichiometric composition. Thus, it is possible that more than five types of rare earth oxycarbides may be present in ternary rare-earth–oxygen–carbon systems.

The rare earth oxycarbides are commonly prepared by carboreduction of the oxides, and are found to coexist with the rare earth carbides or oxides in the samples. For example, the preparation of β-$Ce_2O_2C_2$ by a progressive carbon-reduction of cerium dioxide was always accompanied by the formation of Ce_2O_3 or β-CeC_2, C, CO (Pialoux 1988). In the reaction of carbon with the rare earth sesquioxide (Sc_2O_3, Y_2O_3 and Dy_2O_3), the products contained not only R_2OC (R = Sc, Y, Dy) but also $Sc_{15}C_{19}$, RC_2(R = Y, Dy) (Hájek et al. 1984b). In some cases, the carboreduction conditions were modified (carbon deficit, lower temperature, high CO pressure) to obtain samples richer in the methanide-type oxycarbide R_2OC (Hájek et al. 1984a, b, d), where R represents the rare earth elements with smaller atomic radii such as Sc, Y, Dy.

Other techniques were also used to prepare these rare earth oxycarbides, such as arc melting under a carbon monoxide atmosphere for the formation of the $R_2O_2C_2$ phase alone (Butherus et al. 1966, Achard 1957) and zone-melting techniques for pure compounds (Butherus and Eick 1968), as well as direct reaction of the rare earth elements with carbon in oxygen at high temperatures (Karen and Hájek 1986). However, in many cases the formation and the yield of the rare earth oxycarbides depend strictly on the starting materials and synthesis conditions (temperature, stoichiometry and CO partial pressure) (Hájek et al. 1984a), e.g. the oxycarbide $Nd_2O_2C_2$ could be prepared only in the presence of a CO partial pressure (Butherus et al. 1966).

7.1.2. *Phase diagrams and thermodynamics*

With respect to the phase relationships of the formation of the rare earth oxycarbides, three R–O–C system phase diagrams have been reported for R = Ce (Pialoux 1988, Clark and McColm 1972), Y (Brožek et al. 1985) (fig. 29) and Sm (Haschke and Deline 1980). The phase diagrams of the cerium and the yttrium systems can be regarded as typical of the light and the heavy rare earth systems, respectively, and that of the samarium system is similar to the latter, showing that only one ternary phase $SmO_{0.5}C_{0.4}$ is present in this system. Comparing the two types of phase diagrams, it is apparent that no $Ce_2O_2C_2$-type ternary phase was found to be present in the yttrium–oxygen–carbon system, while in the cerium–oxygen–carbon system the homogeneity region of $Ce_4O_2C_2$ (corresponding to the Y_2OC phase) seems to be uncertain. The existence of the rare earth oxycarbides, R_4O_3C, was reported for five lanthanide elements, La, Nd, Gd, Ho and Er, and was considered to exist possibly throughout the entire lanthanide series (Butherus and Eick 1968). However, the reported phase diagrams of either the Ce–C–O or Y–C–O system gave no indication of the presence of this phase.

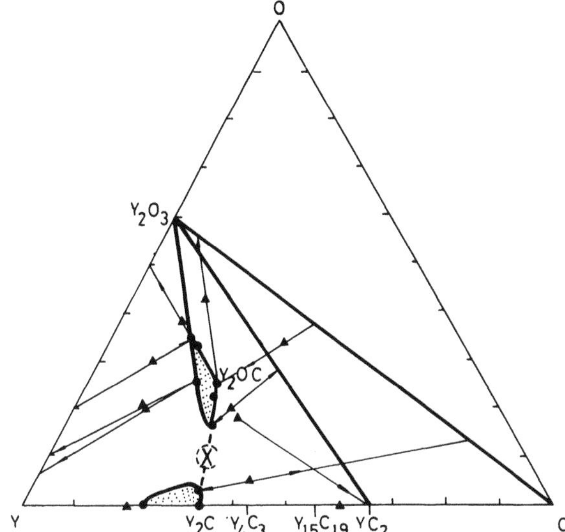

Fig. 29. Regions of homogeneity of the NaCl-type vacant yttrium oxycarbide Y–(O, C, □) showing the redistribution of the initial composition of the pellets among the phases formed after thermal treatment at 900–1200°C (Brožek et al. 1985). ▲, initial elemental composition of the pellet; ●, elemental composition of the Y–(O, C, □) phase; ×, additional phase exhibiting rather complex X-ray diffraction patterns.

Corresponding to the studies on the preparation of the rare earth oxycarbide, the thermodynamics of the formation of the compounds $Nd_2O_2C_2$ (Butherus et al. 1966), $Ce_2O_2C_2$ (Pialoux 1988) and Yb_2OC (Haschke and Eick 1970a, b) have been investigated in detail. For the Ce–C–O system the equilibrium pressures and standard free energies of formation for the various compounds that appear in the course of progressive carboreduction of cerium dioxide were determined from the two large monovariant fields [Ce_2O_3, β-$Ce_2O_2C_2$, C, CO] and [β-$Ce_2O_2C_2$, β-CeC_2, C, O] by high-temperature (1600–2000 K) X-ray diffraction under a controlled carbon monoxide pressure (between 10^{-6} and 1 bar). Corresponding to each equilibrium monovariant,

(i) $Ce_2O_3 + 3C \rightleftharpoons \beta\text{-}Ce_2O_2C_2 + CO,$

(ii) $\beta\text{-}Ce_2O_2C_2 + 4C \rightleftharpoons 2\beta\text{-}CeC_2 + 2CO,$

the equilibrium pressures of CO can be expressed by the following equations:

(i) $\ln P_1(\text{bar}) = -5.62 \times 10^4/T + 26.67 \quad (1610 \leqslant T \leqslant 1993\ \text{K}),$

(ii) $\ln P_2(\text{bar}) = -4.97 \times 10^4/T + 22.35 \quad (1738 \leqslant T \leqslant 2043\ \text{K}).$

The standard free energies of formation were determined to be as follows:

$$\Delta G_f^\circ(\beta\text{-}Ce_2O_2C_2) = -1.217 \times 10^6 + 232.3T - 24.5T \log T \quad (\text{J mol}^{-1}),$$

$$\Delta G_f^\circ(\beta\text{-}CeC_2) = -80 \times 10^3 + 16.9T - 12.5T \log T \quad (\text{J mol}^{-1}),$$

and at 1883 K

$$\Delta G_f^\circ(\beta\text{-}Ce_2O_2C_2) = -930\ \text{kJ mol}^{-1},$$

$$G_f^\circ(\beta\text{-}Ce_2O_{2.1}C_{1.9}) = -963\ \text{kJ mol}^{-1},$$

$$\Delta G_f^\circ(\beta\text{-}CeC_2) = -125\ \text{kJ mol}^{-1},$$

$$G_f^\circ(\beta\text{-}CeC_{1.95}C_{0.5}) = -142\ \text{kJ mol}^{-1}.$$

These data have been used to construct the phase diagram of the Ce–O–C system at about 1800 K (Pialoux 1988).

7.1.3. *Structure and homogeneity region of rare earth oxycarbides*

According to the different hydrolysis products, the rare earth oxycarbides reported so far have usually fallen into one of two categories. The first category is one in which acetylide ions, C_2^{2-}, randomly replace oxygen ions in an oxide lattice, such as in $CeO_{1.48}(C_2)_{0.07}$, $CeO_{1.27}(C_2)_{0.34}$ (Clark and McColm 1972) and $R_2O_2C_2$ (R = La, Ce, Nd) (Butherus et al. 1966, Butherus and Eick 1973, Seiver and Eick 1976, Pialoux 1988), or in an unknown and unlikely oxide, e.g. $Ce_4O_2C_2$ (Clark and McColm 1972, Anderson et al. 1968). The second is that in which the methanide ions, C^{4-}, and oxygen ions are randomly distributed at the anion sites of an NaCl lattice. This category includes the R_4O_3C series (R = La, Nd, Gd, Ho, Er) (Butherus and Eick 1968), the compound R_2OC, R = Yb (Haschke and Eick 1970a, b), Y, Sc, Dy (Brožek et al. 1985, Hájek et al. 1984a–d) and their vacant phases R_1–(O, C, □) (Karen et al. 1986). In addition, an allylene type of carbide was found as an intermediate product of the yttrium oxide carboreduction. This phase is probably a partially oxygen-substituted compound, $Y_{15}(O, C)_{19}$ (Hájek et al. 1984a) and the stability of this phase increases in the direction from Dy, Ho towards Lu (Karen and Hájek 1986).

The cubic oxycarbides generally exhibit a significant stoichiometry range through a simple substitution to parent oxide, although the monoxide, which is the basis for methanide substitution, is probably only stable in the presence of carbon or nitrogen contamination. It has been shown that between the hitherto known cubic phases of the Sc_2OC carbide–oxide (a = 4.5612 Å) and the $Sc_{2-3}C$ carbide there exists a continuous region of mixed crystals of NaCl-type structure (Fm3m, Z = 2). In the oxygen-rich region the homogeneity range extends as far as $ScC_{0.3}O_{0.7}$. This composition is approximately in equilibrium both with the metal and with the Sc_2O_3 at 900°C. The lattice parameter of the oxycarbide is 4.536 Å and the substance can be regarded as a mixed crystal of Sc_2OC and the nonexistent ScO. It is apparent that the homogeneity region of the NaCl-type phase will expand with increasing temperature, at least in the direction towards ScO (Karen et al. 1986). In the carbon-rich range, ScC_xO_y is in equilibrium with the $Sc_{15}C_{19}$ carbide. The homogeneity region gradually extends from $ScC_{0.5}O_{0.5}$ to the binary-defect carbide $Sc_{2-3}C$. This behavior is consistent with the tendency of the Y–(O, C, □) phase (fig. 29) (Brožek et al. 1985). In the Y–O–C ternary diagram at 900–1200°C, there exist two homogeneity regions, one adjacent to the Y_1–$(O, C)_1$ component in the $Y_2O_{1+x}C_{1-x}$ region with x = 0–0.4 and directed towards Y_2C up to $YO_{0.31}C_{0.51}□_{0.18}$, and the other adjacent to the

Y_1–$(C, \square)_1$ component in the region between $YC_{0.3}$ and $YC_{0.5}$ and directed towards Y_2OC up to $YO_{0.04}C_{0.5}\square_{0.46}$.

Corresponding to the deviation in composition from the stoichiometry, the compounds R_2OC (R = Sc, Y, Dy, Sc + Dy, Y + Sc) can contain, in the nonmetal sublattice, as much as 4% vacancies at the expense of carbon (Hájek et al. 1984d) and the $SmO_{0.5}C_{0.4}$ has 10% vacancies (Haschke and Deline 1980). Their behavior as good electrical conductors is consistent with an inherent conduction-band population. On the other hand, the ytterbium compound $YbO_{0.50}C_{0.47}$, closely approaching the ideal ratio and with a low defect-induced conduction population, behaves like a semiconductor (Haschke and Eick 1970a, b). From these facts, it is likely that for the NaCl-type rare earth oxide carbides the apparent increase in the C:R ratio and the accompanying decrease in the defect concentration and conductivity would occur across the lanthanide series.

As reported by Butherus and Eick (1968), the NaCl-type oxide carbides probably exist for all the lanthanides except europium. The compound with the stoichiometry R_4O_3C could also be regarded as a carbon-stabilized monoxide of rare earth elements, like R_2OC. In the europium system, neither R_2OC and R_4O_3C was found. The lattice parameter (5.14 Å) reported for Eu_2OC (Darnell 1977) does not correlate with the parameters of adjacent oxide carbides (5.066 Å for $SmO_{0.5}C_{0.4}$) and is identical with that of EuO (Haschke and Deline 1980). Haschke and Eick (1970a, b) suggested that the greater stability of the monoxide prevents formation of the europium oxide carbide.

In addition to the fcc NaCl-type rare earth oxycarbides with the methanide ions, C^{4-}, the orthorhombic $Ce_4O_2C_2$, hexagonal $CeO_{1.48}(C_2)_{0.07}$, with $a = 3.902$ Å, $c = 6.027$ Å (Anderson et al. 1968) and $CeO_{1.27}(C_2)_{0.34}$, as well as monoclinic and hexagonal $R_2O_2C_2$, definitely belong to the acetylide compound, which have no homogeneity range (for $Ce_4O_2C_2$) or a narrow one.

The crystal structure of $La_2O_2C_2$ determined from three-dimensional X-ray diffraction on a twin crystal is of monoclinic symmetry, space group C/2m (Seiver and Eick 1976). The lattice parameters are $a = 7.069(8)$ Å, $b = 3.985(4)$ Å, $c = 7.310(9)$ Å and $\beta = 95.70(6)°$; the calculated density is 5.41 $g\,cm^{-3}$. In this structure, the lanthanum atom has four oxygen and four carbon atoms situated in a distorted bicapped trigonal prismatic arrangement. Interatomic La–O distances range from 2.392(8) to 2.823(9) Å and La–C distances from 2.86(1) to 3.11(1) Å. The carbon atoms are present as C_2 units with an interatomic C–C distance of 1.21(3) Å. Oxygen atoms are tetrahedrally coordinated, as in the sesquioxide.

In recent work (Pialoux 1988) it has been shown that $Ce_2O_2C_2$ has two modifications, α-$Ce_2O_2C_2$ isotypic with $La_2O_2C_2$ and the high-temperature form β-$Ce_2O_2C_2$. The structure of β-$Ce_2O_2C_2$ ($P\bar{3}12/m$) can be derived from that of α-Ce_2O_3 ($P\bar{3}2/m1$) by substitution of the O^{2-} ions at the octahedral sites of the hexagonal crystalline network by C_2^{2-} pairs. Moreover, there is a close structural relationship between the α and β modifications: their anisotropic behavior becomes more emphasized between 273 and 1333 K and then more so from 1333 up to 1923 K ($7.294 \geqslant c \geqslant 7.279$ Å, $7.029 \leqslant a \leqslant 7.094$ Å, $1.039 \geqslant c/a \geqslant 1.026$). A displacive α monoclinic $\rightleftarrows$ β hexagonal transition occurs at 1333 K.

In summary, the tendency is obvious for the formation of the oxycarbides of the rare earth elements; La, Ce, Pr and Nd form the carbide–oxides $R_2O_2C_2$ or $Ce_4O_2C_2$ of an acetylide nature and with complex crystal structures, while the "smaller" rare earth elements Sm, Dy, Ho, Lu, Y and Sc form the fcc methanide-type carbide–oxides, M–(C, O, □).

7.2. *Rare earth nitride carbides*

The formation of the rare-earth–nitrogen–carbon compounds has been investigated only for the light lanthanide systems with the emphasis on the existence of the "monocarbide" of La, Ce, Pr and Nd, in the presence of nitrogen. It is well known that no evidence for the existence of a face-centered cubic phase, assumed to be the monocarbide, was found in the detailed work on the light-lanthanide–carbon systems (Spedding et al. 1958, Dancy et al. 1962, Anderson et al. 1969).

Anderson et al. (1969), Colquhoun et al. (1975) and McColm et al. (1977) prepared samples containing unstable and highly nonstoichiometric nitride–carbides of cerium, praseodymium and lanthanum with a NaCl structure by high-temperature reactions between RC_2 and RN, RN and C, as well as R and HCN etc. The methods employed to produce alloys in the lanthanum–nitrogen–carbon, cerium–nitrogen–carbon and praseodymium–nitrogen–carbon systems only lead to nonequilibrium phases.

The composition ranges of the nitride–carbide have been determined. For the cerium system and the praseodymium system, two series of fcc nitride–carbide phases were identified.

One series is found in the presence of small amounts of rare earth metals and the total nonmetal content $(x + y)$ is close to and not lower than unity. For CeN_xC_y, e.g. $CeN_{0.74}C_{0.60}$ ($a = 5.111$ Å) or $CeN_{0.65}C_{0.86}$ ($a = 5.032$ Å), owing to hydrolysis products, the Ce compounds could be regarded as the acetylide. Although Colquhoun et al. (1975) found an acetylide fcc praseodymium nitride–carbide $PrN_{0.73}C_{0.69}$ with a large lattice parameter of 5.180 Å and an $x + y$ value much greater than unity, this phase seems particularly unstable for vacuum annealing at 1300°C and rapidly decomposes to a high-carbon methanide with $a = 5.136$ Å. Therefore, it is apparent that Pr has a greater ability to stabilize the methanide phase than Ce. In the lanthanum system, this type of compound is not found (McColm et al. 1977).

The second series of the fcc nitride carbides were found to coexist with the sesquicarbides of the light lanthanides, La, Ce and Pr. The composition ranges of this type of rare earth nitride–carbides are different for different systems, e.g. for LaN_xC_y, this range is very narrow, $x + y = 0.3$–0.4, $x \geqslant 0.04$, in contrast to $x + y \geqslant 0.8$–1.0, $y = 0.08$–0.4, $x = 0.4$–0.7 for CeN_xC_y and $x + y \geqslant 0.53$–0.92, $y = 0.05$–0.55, $x = 0.55$–0.72 for PrN_xC_y. These compounds contain methanide C_1 carbon units and are also nonequilibrium phases. The lattice parameters vary over a wide range depending on the carbon and nitrogen contents in the compounds, e.g. for LaN_xC_y, the lattice parameter $a = 5.295 - 0.171y + 0.186x$ (McColm et al. 1977) and for PrN_xC_y a parameter increases on increasing the nitrogen content from 5.123 Å up to 5.166 Å. Thus, this type of compound might be thought of as a nitrogen-substituted defective monocarbide phase.

McColm et al. (1977) considered that the composition ranges seem to be related to the amounts of R(IV) found in the respective nitrides and carbides (Lorrenzelli et al. 1970, Atoji 1962). In the cerium systems, Ce(IV) is present up to 70%, while in the praseodymium case the amount is less and in the lanthanum system the higher oxidation state is absent. This suggests that Ce(IV) and Pr(IV) do assist in preventing the catenation that leads to acetylide ion formation. However, the reviewers suggest that this factor is probably minor while the size factor of the rare earth atom is essential.

In conclusion, it should again be emphasized that the presence of nitrogen is shown to be essential to the formation of methanide carbide in the light-lanthanide–nitrogen–carbon systems.

7.3. *Ternary rare-earth–hydrogen–carbon compounds*

Studies concerning the preparation and properties of the rare earth carbide hydrides showed that only two or three ternary compounds, hexagonal $RC_{0.5}H$ (R = Y, La, Yb), fcc $RCH_{0.5}$ (R = Yb) and possibly YbC_2H exist in the rare-earth–carbon–hydrogen system; for Yb (Haschke 1975), for La and Y (Peterson and Rexer 1962, Rexer and Peterson 1964). The fcc phase of variable composition with lattice parameters ranging from 4.88 to 4.96 Å obtained by treating ytterbium dihydride with graphite (Lallement 1966) has also been regarded as the carbide hydride $YbCH_{0.5}$ (Haschke and Eick 1970a, b). The phase diagram for the Yb–C–H system at 1173 K and 0.5 atm of hydrogen are shown in Fig. 30 (Haschke 1975).

X-ray diffraction data for hexagonal $YbC_{0.5}H$ prepared under different hydrogen pressures show larger variations in lattice parameters with a ranging from 3.574 to 3.593 Å and c ranging from 5.723 to 5.852 Å. The composition limits for this phase

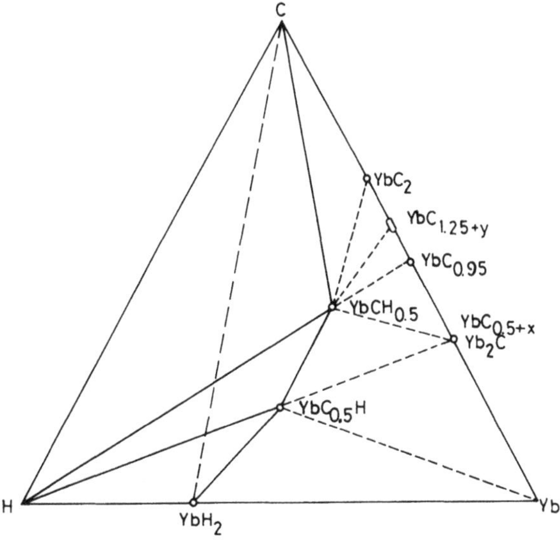

Fig. 30. Phase diagram for the Yb–C–H system at 1173 K and 0.5 atm of hydrogen (Haschke 1975). (Reprinted by permission of the publisher, The American Chemical Society, Inc.)

have not been determined (Haschke 1975). For this type of compound for La or Y, no data on composition were reported and thus these compounds were assigned the RCH_x formula (Samsonov et al. 1970). The fcc $YbCH_{0.5}$ hydrocarbide has a lattice parameter $a = 4.974 \pm 0.001$ Å, and the exact composition limits of this phase are not known.

Reference to fig. 30 shows that $YbC_{0.5}H$ and $YbCH_{0.5}$ lie on lines of constant C:Yb ratio connecting the known carbide composition with hydrogen. These phases relationships suggest the possibility of preparing these and other carbide hydrides by hydrogen substitution into binary carbides. Thus, YbC_2H, an acetylide hydride with the conduction electron bound as a hydride ion would be expected to form by the low-temperature hydrogenation of YbC_2 (Haschke 1985) and structural similarities between the carbide and hydrocarbide phase might be expected, as well as regions of nonstoichiometry might be observed.

Ionic models have been proposed for explaining the observed nonstoichiometry and properties of these hydrocarbides on the basis of magnetic, electrical and hydrolysis data (Haschke 1975). The structure of the hexagonal $YbC_{0.5}H$ phase was considered to be composed of the metal occupying the closest-packed positions with the methanide carbons occupying half of the octahedral interstices, and hydrogens occupying half of the tetrahedral voids. Occupancy of the remaining octahedral and tetragonal voids by hydrogen essentially accounts for the wide composition range of this type of compound. The fcc $YbCH_{0.5}$ phase was proposed as a methanide acetylide hydride, the anion-to-cation ratio is 1.25:1; this phase also has a NaCl-type structure. Carbide ions presumably occupy three-quarters of the octahedral sites, and hydride ions fill one quarter of the tetrahedral holes. This model is similar to that proposed for the hexagonal carbide hydride, and the same processes for nonstoichiometry are expected to be operative. An immeasurably large resistivity was found for the hexagonal phase, and a value of 10^{12} Ω cm was observed for the fcc phase.

It has been shown that in the presence of oxygen, a metastable fcc quaternary phase (with lattice parameters in the range 4.85–4.96 Å) forms between the $YbO_{0.5}C_{0.5}$ and $YbCH_{0.5}$ compositions. This phase is a nonequilibrium one and disproportionates into $YbCH_{0.5}$ and $YbC_{0.5}O_{0.5}$ at equilibrium.

The preparative reaction with europium dihydride failed to produce analogous carbide hydrides. Obviously, the failure is consistent with the high stability of divalent europium. In $YbC_{0.5}H$, $YbCH_{0.5}$ and $YbO_{0.5}C_{0.5}$, ytterbium is clearly trivalent, as shown by magnetic data (Haschke 1975), and for La and Y the hexagonal phases have also been found (Lallement and Veyssie 1968). Therefore, it can be expected that the hexagonal $RC_{0.5}H$ phases probably exist across the lanthanide series, which exhibit a trivalent state as their stable valence state, with the exception of europium.

7.4. *Ternary rare-earth–carbon-halide compounds*

There is another series of cluster-type ternary carbides that contain not only metal and carbon atoms but also the group VIIA elements, Cl, Br and I (Simon et al. 1981, Warkentin et al. 1982, Simon and Warkentin 1983, Schwanitz-Schüller and Simon 1985, see also chapter 100, this volume). Electronic structure calculations have been performed on $Gd_{10}Cl_{18}C_4$ (Satpathy and Anderson 1985), $Gd_{10}Cl_{17}C_4$, $Gd_{12}I_{17}C_6$ (Bullett 1985) and $Gd_2C_2Cl_2$ (Miller et al. 1986). The carbon atoms in these compounds also form pairs. For $Gd_{10}Cl_{18}C_4$, $Gd_{10}Cl_{17}C_4$ and $Gd_{12}I_{17}C_6$ which contain short metal–metal contacts the emphasis has been on the extent of metal–metal bonding. Other chloride carbides, R_2Cl_2C (R = Sc, Y) with the 1 T type structure (Hwu et al. 1986), $Gd_6Cl_5C_3$ (Simon et al. 1988), $Gd_5Cl_9C_2$ (Simon et al. 1981) and the quaternary $Cs_2Lu_7Cl_{18}C$ compound have been synthesized and their structures have been determined. In addition, the Sc compounds $Sc_7X_{12}C$ (X = I, Br), $Sc_6I_{11}C_2$ and $Sc_4I_6C_2$ have also been studied with respect to their synthesis and structure (Dudis et al. 1986, Dudis and Corbett 1987).

8. Physical properties of the rare earth carbides

The magnetic and electrical properties of the rare earth carbides have received considerable attention owing to their unusual crystal structures. Lallement (1968) and Lallement and Veyssie (1968) have examined the magnetic and electrical properties of R_xC, which exists as a solid solution of 25 to 50 at.% C in the temperature range of 4 to 300 K. Vickery et al. (1959) first measured the magnetic susceptibilities of RC_2, which follow the Curie–Weiss law in the range of 60–450 K without showing any magnetic transition. The neutron diffraction studies of Atoji (1967a, b, 1968, 1970, 1971, 1972, 1978) revealed the existence of the magnetic ordering in RC_2 and R_2C_3. Sakai, Adachi and Shiokawa further systematically studied the magnetic and electrical properties of RC_2 and their solid solutions, as well as the ternary RB_2C_2-type layer compounds (Sakai et al. 1979a, b, 1980, 1981a–c, 1982a, b). More recently, owing to the development of rare earth permanent magnet materials, the magnetic properties of the ternary rare-earth-metal–transition-metal (such as Fe, Co and Ni) carbides have also been studied comprehensively.

One of the important properties of the rare earth carbides is their superconductivity at low temperatures. It has been established that yttrium thorium sesquicarbide $(Y_{0.7}Th_{0.3})C_{1.55}$ has a surprisingly high transition temperature T_c of 17.0 K (Krupka et al. 1969b), and YC_2 (Giorgi et al. 1968) and La_2C_3, as well as $La_{0.9}Th_{0.3}C_{1.58}$ (Cort et al. 1984) also appear to be superconducting at low temperatures.

8.1. *Physical properties of binary rare earth carbides and their solid solutions*

8.1.1. *Magnetic structures of the rare earth dicarbides*

Atoji (1967a, b 1968, 1970, 1972) and Atoji and Flowers (1970) have investigated the magnetic structures of CeC_2, PrC_2, NdC_2, TbC_2, DyC_2, HoC_2, ErC_2, TmC_2 and YbC_2 at low temperatures and at 300 K in great detail by means of neutron diffraction. The light rare earth dicarbides CeC_2, PrC_2 and NdC_2 have been shown to become a body-centered first-kind antiferromagnet with Néel temperatures of 33, 15 and 29 K, and with the ordered moments being 81, 44 and 90% of the free-ion values, respectively. The moment direction is parallel to the *c* axis in these dicarbides (Atoji 1967).

The isostructural heavy rare earth dicarbides TbC_2 (Atoji 1967), DyC_2 (Atoji 1968), HoC_2 (Atoji 1967) and ErC_2 (Atoji 1972) become antiferromagnetic with Néel temperatures of 66, 59, 26 and 19 ± 2 K. Below the Néel temperatures, TbC_2 and HoC_2 exhibit an elliptic helical spin alignment propagating along the *a* axis with the repetition being four times that of the atomic spacing, and the root mean square ordered moments are 5.1 and $6.9\mu_B$, respectively. Here the moments lie in the *bc* plane. A modulation of the above-mentioned structure takes place below 40 and 16 K, with the maximum possible additional moments of 0.54 and $1.65\mu_B$ at 2 K in TbC_2 and HoC_2, respectively. In TbC_2, another complex magnetic structure coexists below about 33 K. For DyC_2, below the Néel temperature the spin alignment is of the linear, transverse-wave mode and this static-moment wave propagates along the *a* axis and is polarized along the *c*-axis direction. The root mean square and maximum saturation moment per dysprosium atom are 8.37 and $11.8\mu_B$, respectively, the latter being considerably larger than the order moment of the free Dy^{3+} ion, $10.0\ \mu_B$. The wavelength of the moment wave is about 1.3 times the *a* spacing and is practically independent of temperature. Like TbC_2, an additional, coexisting spin alignment with a small moment also takes place below 31 K (Atoji 1968) for DyC_2.

For ErC_2 below Néel temperature, the moment alignment is governed by several narrowly dispersed incommensurate transverse magnetization waves. The major modulation waves propagate along the *a* axis and are polarized along the *b*-axis direction. The magnetization wavelengths are 5.78, 5.4, 6.7 and 7.1 Å, and the amplitudes are, respectively, 6.55, 2.2, 2.1 and $1.6\mu_B$ at 4.2 K. The ferromagnetic alignment in the plane perpendicular to *a* is slightly altered by the propagation of several long-periodicity modulations along the *b* and *c* axes. A commensurate, antiferromagnetic wave also participates in the moment alignment below 10 ± 3 K. The commensurate wave is represented by setting the wavelength equal to the *a* spacing in the incommensurate wave mode. The commensurate moment is $2.5\mu_B$ at 4.2 K. The root mean square value of all the moment amplitudes is $7.8\mu_B$ at 4.2 K and increases to $7.9\mu_B$ at 2 K. These features are likely to be caused by a strong perturbation between the Rüderman–Kittel–Kasaya–Yoshida (RKKY) coupling and the crystal field effect (Atoji 1972).

Unlike other heavy rare earth dicarbides, RC_2 (R = Tb, Dy, Ho, Er), TmC_2 and YbC_2 exhibit no detectable long-range magnetic ordering in neutron diffraction patterns at temperatures down to 1.7 and 2 K, respectively (Atoji 1970, Atoji and Flowers 1970). However, for YbC_2 experiments at temperatures below 2 K are needed to clarify the modulation of the paramagnetic diffuse scattering near 5 K, which is probably the result of an antiferromagnetic short-range ordering.

The magnetic properties of a rare earth system are essentially determined by crystalline electric field and exchange effects. Wegener et al. (1981) studied the crystalline electric field and molecular field splitting in CeC_2 by the energy spectra of neutron inelastic scattering. In the paramagnetic state, the $\Gamma t_7^{(2)}$ doublet has been determined as the ground state and the Γt_6 doublet at 7 meV and the $\Gamma t_7^{(1)}$ doublet at 18 meV as the excited states. In the magnetically ordered state, the molecular field parameter was determined to be $(2.2 \pm 0.1) \times 10^{25}\ \mathrm{Oe^2\ erg^{-1}}$.

8.1.2. *Magnetic and electrical properties of the rare earth dicarbides and their solid solutions*

As described in previous sections, the rare earth dicarbides RC_2 are composed of a rare earth ion, an acetylide ion C_2^{2-} and a conduction electron in one formula unit. This electron may be responsible for the formation of a conduction band through combinations of $\pi_g 2p$ orbitalds, the C_2^{2-} antibonding orbitals, with the 5d orbitals of the R^{3+} ion (Atoji 1962). This d–π_g2p combination is multicentered and largely delocalized, and hence the electrons partly filling this band can be considered as conduction electrons. This concept has been confirmed by the Hall effect, which indicates 0.6–1.1 conduction electrons per RC_2 (Yupko et al. 1974). Therefore, the unique magnetic and electrical properties of the rare earth dicarbides are considered to be closely associated with the conduction electrons and the bonds in these compounds.

Sakai et al. (1979a, b, 1980, 1981a, b, 1982) experimentally measured the temperature dependences of the magnetic susceptibility χ_m and of the electrical resistivity ρ for RC_2 compounds and their solid solutions. The results are tabulated in tables 20 and 21. When the RC_2 compounds (R = Ce, Nd, Sm, Gd, Tb, Dy, Ho, Er) become antiferromagnetic below their individual Néel temperatures, a rapid decrease in

TABLE 20
Magnetic data of the rare earth dicarbides (Sakai et al. 1981a, b, 1979a, b).

Species	Lattice parameters (Å)	T_{ord} (K)	θ_p (K)	μ_{eff} (μ_B)	μ_{cal} (μ_B)	de Gennes factor, $(g-1)^2 J(J+1)$
LaC_2	$a = 3.939$ $c = 6.579$					
CeC_2	$a = 3.878$ $c = 6.488$	$T_N = 30$	+ 12	2.40	2.54	0.18
PrC_2	$a = 3.851$ $c = 6.430$	$T_c(H \to 0) = 7$	0	3.31	3.58	0.80
NdC_2	$a = 3.818$ $c = 6.397$	$T_N = 24$	− 10	3.29	3.62	1.86
SmC_2	$a = 3.765$ $c = 6.139$	$T_N = 21$[a]			0.84	4.46
EuC_2	$a = 4.118$ $c = 6.756$	$T_c(H \to 0) = 20$	+ 14	8.04	7.94	
GdC_2	$a = 3.722$ $c = 6.295$	$T_N = 42$	− 53	7.68	7.94	15.75
TbC_2	$a = 3.687$ $c = 6.205$	$T_N = 67$ 20	− 150	10.88	9.72	10.50
DyC_2	$a = 3.675$ $c = 6.179$	$T_N = 61$	− 89	10.20	10.63	7.08
HoC_2	$a = 3.659$ $c = 6.154$	$T_N = 25$	− 36	10.75	10.58	4.50
ErC_2	$a = 3.618$ $c = 6.096$	$T_N = 8$	− 17	9.43	9.59	2.55

[a] This value does not obey the Curie–Weiss law.

TABLE 21
Electrical resistivity data of the rare earth dicarbides (Sakai et al. 1981a, b, 1979a, b): ρ_{res}, the residual resistivity; ρ_m, the magnetic resistivity.

Species	Transition temperature (K)		$\rho_{273\,K}$ ($\mu\Omega$ cm)	ρ_{res} ($\mu\Omega$cm)	ρ_m ($\mu\Omega$ cm)	$d\rho/dT$ ($\mu\Omega$ cm K^{-1})
LaC_2	1.6[a,b],	1.44[c]	51.4	4.8		0.24
CeC_2	30,	33[d]	82.1	1.3		0.24
PrC_2	18,	15[d]	41.9	2.3	2.5	0.18
NdC_2	25,	29[d]	43.1	3.3	5.5	0.16
SmC_2	21		115.7	21.4	13.0	0.32
GdC_2	43		55.3	8.0	15.2	0.14
TbC_2	1.67,	66[d]	48.5	10.6	9.1	0.13
	2.20,	< 33[d]				
DyC_2	60,	59[d]	52.5	13.0	5.8	0.17
HoC_2	24,	26[d]	80.3	23.2	4.1	0.25
ErC_2	8,	1.16[d]	98.3	39.4	0.6	0.24
		2.10 ± 3[d]				

[a] Sakai et al. [b] Green et al. (1969). [c] Bowman and Krikorian (1976): superconducting transition temperature. [d] Taken from Atoji (1967a, 1968, 1970, 1971, 1972, 1978).

resistivity occurs. The values of the paramagnetic moments observed are in agreement with the calculated values for the R^{3+} ions, and the susceptibility obeys a Curie–Weiss law in the paramagnetic region with the exception of SmC_2. The compounds PrC_2 and EuC_2 are ferromagnetic below 7 K and 20 K, respectively (Sakai et al. 1981b, 1982), and YC_2 and LaC_2 are Pauli paramagnetic (Sakai et al. 1979b).

Atoji has previously reported that PrC_2 is an antiferromagnet with a Néel temperature of 15 ± 2 K by neutron diffraction techniques at zero field. In order to clarify the discrepancy between the two results, the temperature dependency of the resistivity was examined (Sakai et al. 1981a). The change in the slope of the resistivity–temperature curve occurred at about 18 K, which is close to the Néel temperature reported by Atoji. This finding indicates that PrC_2 possesses a metamagnetic transition in fields below 700 Oe.

The transition temperatures, T_t, on the resistivity–temperature curves were comparable to the value of T_N reported by Atoji (1967a, 1968, 1970, 1971, 1972, 1978). The values for CeC_2, NdC_2, TbC_2, DyC_2 and HoC_2 by Sakai et al. are in good agreement with those obtained by Atoji. However, the T_N value of 8 K for ErC_2 is only compatible with the second Néel temperature (10 ± 3 K) measured by Atoji (1972), and the two transition temperatures for TbC_2, at 67 K and at 20 K, on the ρ–T or the χ_m–T curves can be attributed to the existence of the type-I spin alignment process (T_N = 67 K) and the coexistence of the type-II process with the type-I process below 33 K, respectively, as described by Atoji (1971).

For PrC_2, in addition to the Curie temperature obtained of 7 K in the zero field from the $T_c(H)$ versus H plot, a change in the slope of the ρ–T curve at approximately

18 K, which is close to the Néel temperature of 15 ± 2 K reported by Atoji (1967a, b), was observed (Sakai et al. 1981a). According to Sakai et al., at this temperature, there is a metamagnetic transition.

The Néel temperatures of CeC_2 and GdC_2 are exceptional when taking into consideration the T_N or ρ_m (the magnetic resistivity) versus the de Gennes factor plot. The gadolinium dicarbide has a value only one third of that expected, while the value of T_N for the cerium dicarbide is anomalously larger than that expected from the T_N versus the de Gennes factor relation for all the other dicarbides. This exceptional behavior of GdC_2 can be attributed to the deviation of the crystal structure (the c/a ratio of the lattice parameters is higher than that expected from a plot of c/a versus R^{3+} ionic radii), i.e. the variation of the RKKY function $\Sigma_F(2k_F R)$ (Sakai et al. 1979a). For CeC_2, in addition to the anomalously high T_N value, the slope of the ρ–T curve increases with decreasing temperature in contrast to the relation observed for other rare earth dicarbides, except for SmC_2. According to the assumption of Sakai et al. (1980), the special spread of the 4f orbitals of the Ce^{3+} ion is the largest among the R^{3+} ions; these 4f orbitals may overlap significantly with the C_2^{2-} valency orbitals and, therefore, the effect of the crystal field on the Ce^{3+} ions would become very strong and a superexchange interaction between the 4f orbitals via the C_2^{2-} orbitals would occur as an additional magnetic interaction.

For SmC_2, as mentioned above, the reciprocal susceptibility does not obey the Curie–Weiss law above the Néel temperature of 21 K (Sakai et al. 1981a) in contrast to the results of Vickery et al. (1959). This anomalous magnetic behavior resembles that reported for other samarium compounds and seems to be attributed to the contribution of the first excited multiplet state ($J = \frac{7}{2}$) to the ρ_m value.

It is established that europium dicarbide is a ferromagnet with a zero-field Curie temperature of 20 K. A reciprocal plot of magnetic susceptibility (χ_m^{-1}) against temperature obeys a Curie–Weiss law, giving an effective moment of $8.04\mu_B$ and a paramagnetic Curie temperature (θ_p) of +14 K. The effective moment is in good agreement with that expected for Eu^{2+} (Sakai et al. 1982b) and the ferromagnetic interaction is the result of the f–f interaction between Eu^{2+} nearest neighbors throughout the 5d–4f exchange interaction. Therefore, the exchange mechanism in EuC_2 differs from that of the trivalent rare earth dicarbides, in which the antiferromagnetic interaction is the result of the f–f indirect exchange interaction via conduction electrons.

In order to establish the continuous change in the exchange mechanism with changing composition for the solid solution of EuC_2 and the nonmagnetic trivalent lanthanum dicarbide or the magnetic trivalent gadolinium dicarbide, which has an identical spin configuration with EuC_2, the electrical and magnetic properties have been measured by Sakai et al. (1982b). They have shown that all the effective moments obtained from the χ_m^{-1} versus T linear relation are in good agreement with the calculated values on the assumption that all the Eu ions in the compounds are divalent. The solid solution, $La_xEu_{1-x}C_2$, becomes ferromagnetic at low temperatures over the composition range from $x = 0.08$ to 0.76. The values of T_c (or θ_p) remain almost constant over the composition range from 0 to 0.47 and decrease rapidly with further dilution. In the solid solutions $Gd_xEu_{1-x}C_2$, the kind of magnetic order varies from ferromagnetic to antiferromagnetic with the increase of the concentration of the Gd^{3+} ion. In addition, the electrical resistivities of EuC_2 and its solid solutions decrease with decreasing temperature down to 160 K and have sharp peaks near the Curie temperatures. When a magnetic field is applied, the peaks become broader and are shifted to higher temperatures, confirming the formation of magnetic polarons. According to the magnetic impurity model proposed by Kasuya for the Eu chalcogenides (Yanase and Kasuya 1968), Sakai et al. (1982) discussed their experimental results and suggested that an excess electron with spin $\boldsymbol{S}_i$, which is trapped by a trivalent ion, magnetically polarizes the nearest-neighbor Eu spin, $\boldsymbol{S}_n$, by means of a strong exchange interaction between the impurity electron and the 4f electrons, forming magnetic polarons. Based on this assumption, they explained the above-mentioned electrical behavior.

In addition to the behavior of the solid solutions of the divalent europium dicarbide as compared to the trivalent rare earth dicarbides, the magnetic properties of the solid solutions of mixed trivalent rare earth dicarbides, such as the systems $R_xDy_{1-x}C_2$ (R = Ce, Pr, Nd, Sm, Gd, Tb, Ho, Er, Y) (Sakai et al. 1979a, b, 1980, 1981b) and $Ce_xLa_{1-x}C_2$ (Sakai et al. 1980) have been examined. The Néel temperatures (T_N) and magnetic resistivities (ρ_m) for the solid solutions in combination with the data of the individual rare earth dicarbides have been discussed in terms of the RKKY theory. The results suggest that the T_N values for the light rare earth dicarbides vary widely with the de Gennes factor G, $(g-1)^2 J(J+1)$, whereas those of the heavy rare-earth dicarbides and their solid solutions vary continuously although they indicate a gradual departure from the RKKY theoretical expectation. The paramagnetic Curie temperatures (θ_p) of the RC_2 compounds, except for GdC_2, increase continuously with increasing G values. The ρ_m values of the heavy rare earth dicarbides are linearly correlated with the G values, obeying the theoretical equation for the magnetic resistivity ρ_m. On the basis of the above-mentioned discussion, Sakai et al. (1981b) proposed two models to explain the anomalous Néel temperatures of CeC_2 and GdC_2 and the ferromagnetic behavior of PrC_2 at low temperatures.

The RKKY theory has also been employed to explain the magnetism of the solid solutions $Gd_xDy_{1-x}C_2$ and $Y_xDy_{1-x}C_2$ in which GdC_2 and YC_2 are antiferromagnetic and Pauli paramagnetic, respectively (Sakai et al. 1979a, b). For the two solid-solution systems, the susceptibilities follow a Curie–Weiss law in the paramagnetic region. In the $Y_xDy_{1-x}C_2$ system with increasing yttrium ion concentration, the values of the paramagnetic Curie temperatures (θ_p) and the Néel temperatures (T_N) decrease linearly against the G values, suggesting that the RKKY theory is applicable with respect to the magnetic interaction in this system. In the $Gd_xDy_{1-x}C_2$ system, two Néel points T_{N_1} and T_{N_2} as well as a second transition point T_F were observed on the susceptibility versus temperature curves. Applying fields from 3.7 to 7.3 kOe had no influence on these transition temperatures. The T_{N_1} temperatures decrease with an increase in the concentration of Gd ions whereas T_{N_2} increases and T_F remains unaffected. These results have been explained successfully on the basis of the RKKY theory, suggesting that at T_{N_1}, the spins of the Dy^{3+} ions order antiferromagnetically while those of the Gd^{3+} ions remain paramagnetic, and at T_{N_2} the Gd^{3+} ions also undergo spin ordering.

8.1.3. *Magnetic structure of* R_2C_3

Neutron diffraction measurements have shown that Pr_2C_3, Nd_2C_3, Dy_2C_3 (Atoji 1978), Tb_2C_3 (Atoji 1971) and Ho_2C_3 (Atoji and Tsunoda 1971) become antiferromagnetic below 8.24, 22, 33 ± 4 and 19 ± 1 K, respectively, but no detectable coherent magnetic peak was found in the Ce_2C_3 patterns between 300 and 4 K. The four sesquicarbides, Pr_2C_3, Nd_2C_3, Tb_2C_3 and Dy_2C_3, exhibit a similar magnetic structure. In the uniaxial moment model having two antiferromagnetic and two paramagnetic body diagonals (Atoji 1971), the saturation order moments per metal atom are 1.3, 3.0, 9.0 and $9.5\mu_B$, respectively, being 41, 92, more than 98, and 95% of the respective free-ion values.

Although Ho_2C_3 also exhibits an antiferromagnetic structure, this structure is markedly different from the ordered spin structure of the isostructural Tb_2C_3. Two out of four body-diagonal-linked arrays of Ho atoms become the ferromagnetically ordered chains and the remaining two arrays show one-dimensional antiferromagnetic alignment, while in the magnetic structure of Tb_2C_3 two become antiferromagnetic linear chains and the other two exhibit no ordered moment. Among three nearest Ho–Ho neighbors bridging between the different body-diagonal arrays are two antiferromagnetic pairs and one ferromagnetic pair instead of only ferromagnetic pairs as in Tb_2C_3. Furthermore, the moment direction is most likely parallel to one of the ferromagnetic linear arrays in Ho_2C_3. However, for Tb_2C_3 it is along the face diagonal which lies in the plane formed by two body diagonals parallel to the ordered spin arrays, and another probable moment direction lies longitudinally along any one of the two ordered arrays. The maximum spontaneous moment is $7.3 \pm 0.2\mu_B$, which is considerably smaller than the free Ho^{3+} value of $10\mu_B$. As a common point, Tb_2C_3 and Ho_2C_3 exhibit a sizable magnetic diffuse scattering having the characteristics of short-range order in the magnetically long-range ordered region (Atoji 1971, Atoji and Tsunoda 1971).

It can be seen from the comparison of the known Néel temperatures between R_2C_3 and RC_2 (Atoji 1978) that in the light rare earth compounds the crystal field effect is often predominant, as exemplified by the fact that praseodymium carbides have exceptionally low values of T_N and strongly suppressed ordered moments ($1.14\mu_B$ in PrC_2). In the heavy lanthanide compounds, the exchange interaction and anisotropy energy become the major factors in the magnetization.

According to the molecular field equation, $T_N = AG + B$, where G is the de Gennes factor, $G = (g-1)^2 J(J+1)$ (Elliott 1972), the first term represents the RKKY bilinear scalar exchange interaction and the second term signifies mainly the anisotropy energy. For the heavy R_2C_3 phases, T_N can be expressed as $T_N = 2.38G + 7.15$, indicating a relatively large anisotropy term. For the heavy RC_2 phases, a least-squares linear fit gives $T_N = 6.98G - 1.70$, which is in rather poor agreement with the experimental data. Moreover, the anisotropy term is negative. In RC_2, the anisotropic and/or higher-order exchange interaction may have to be considered and the crystal field effect is very different from that in R_2C_3.

8.1.4. *Physical properties of* R_2C_3

The Debye characteristic temperatures θ_D (K) of R_2C_3 have been given, based on the neutron scattering temperature factor coefficients of R_2C_3 at 296 K, to be 247 K

(Atoji et al. 1973), 344 K (Carter et al. 1974), 181, 181 K (Atoji and Tsunoda 1971), 177 K (Atoji 1965), and 176 K (Atoji 1967a, b) for R = La, Ce, Pr, Nd, Dy, Ho, respectively. Yupko et al. (1974) also reported the θ_D values for R = La, Ce, Nd to be 189, 199 and 203 K, respectively, which are, except for Ce_2C_3, insignificantly different from Atoji's data. These θ_D values for R_2C_3 are comparable to those of the rare earth metals (Elliot 1972) but are roughly 100 K lower than those for RC_2 (Atoji 1967a, b).

Yupko et al. (1974) have also measured the Hall coefficient and the electrical resistivity of La_2C_3, Ce_2C_3, Nd_2C_3 and Y_2C_3. The resulting electron carrier concentration data are, however, mutually inconsistent.

In fact, the above-mentioned physical properties of R_2C_3 are closely related to the conduction band and bond characters. For RC_2, the conduction band wavefunctions are composed only of the metal d orbitals and the C_2 $\pi_g 2p$ antibonding orbitals. However, in R_2C_3 the C_2 $\pi_g 2p$ contribution to the conduction band is smaller than in RC_2 (Atoji 1962). Because the C–C bond distances are all significantly longer than the acetylene C–C distance of 1.20 Å, it suggests that although the C–C antibonding orbitals ($\pi_g 2p$) also participate in the valency bond hydridization with the R orbitals (Atoji 1967a, b), the contribution is relatively smaller. In addition, the bond distances between the R atoms R_0–$2R_2$ and R_0–$3R_1$ are about the same as the nearest neighbor distance and considerably shorter than in the R metal. These bond distances are of high s character with a minor d contribution, and of high d character, respectively. Furthermore, these R–R bonds are closely conjugated with the R–C bonds through the multiple-centered or asymmetric s–p–d hybridization. The RKKY, crystal field and related magnetic interaction in R_2C_3 should hence have considerable anisotropy, as demonstrated in the order moment configuration.

8.1.5. *Magnetic structure of the rare earth hypocarbides*

8.1.5.1. *Magnetic structure of the cubic R_3C compounds.* Several investigators have reported the ferromagnetic transition at low temperatures for $ErC_{0.45}$ (Lallement 1966), Dy_3C (Aoki and Williams 1979), $GdC_{0.33-0.5}$ and $ErC_{0.33-0.5}$ (Takaki and Mekata 1969). Atoji (1981b) studied the magnetic structure of the cubic NaCl-type $Er_{1.67}C$ compound by neutron diffraction at low temperatures, showing that $Er_{1.67}C$ is paramagnetic above 90 K exhibiting a free Er^{3+} ion moment of 9.4 (2)μ_B, and below this temperature $Er_{1.67}C$ becomes a ferromagnet with a saturation moment of 2.5μ_B (only 28% of the maximum free-ion moment), indicating a large crystal field effect. The direction of the ferromagnetic moment in crystals of cubic symmetry has been determined by inducing a preferential crystallite orientation with an applied magnetic field (Atoji et al. 1981) to be parallel to the $\langle 100 \rangle$ axis. They also found that the paramagnetic scattering curve exhibits a small hump near $\sin\theta = 0$, which becomes more pronounced at lower temperatures down to 1.6 K. This implied that a ferromagnetic short-range ordering coexists with the ferromagnetic long-range ordering (Trammell 1963).

8.1.5.2. *Magnetic structure of the trigonal R_2C compounds.* All but the yttrium hypocarbides become ferromagnetic at low temperature (Atoji 1969, 1981a, b, c, Atoji and Kikuchi 1969, Lallement 1966). The Curie temperatures are 266 ± 2 K for Tb_2C or $Tb_{2.1}C$, 168 K for Dy_2C and 100 K for Ho_2C. The saturation moments in the ferromagnetic phase are 7.6, 8.6, 6.8 and 7.16μ_B, corresponding to 84, 96, 68 and 72% of the free-ion moments, respectively, for Tb_2C, $Tb_{2.1}C$, Dy_2C and Ho_2C, indicating a considerable crystal field effect for Dy_2C and Ho_2C. The preferential crystallite orientation induced by the applied magnetic field has shown that the ordered moment directions are aligned parallel to the *c* axis in Tb_2C and the [104] axis for Dy_2C and Ho_2C, which corresponds to the [100] axis of the high-temperature cubic modification of Dy_2C and Ho_2C. Like the cubic $Er_{1.67}C$ compound (Atoji 1981b), Tb_2C and Ho_2C also exhibit a ferromagnetic short-range order superposing on the ferromagnetic long-range order at 4 K.

For Tb_2C, an unusual magnetostriction along *c* was found by determining the linear thermal expansion coefficient between 4 and 298 K. This may be correlated with the fact that the direction of the ferromagnetic moment is parallel to the *c* axis in Tb_2C.

For Ho_2C, Lallement (1966) has reported a neutron diffraction pattern at 4 K and postulated its ordered magnetic structure as an HoN-type retarded ferromagnet; also determined were the Curie temperatures of 100, 90, 90, and 100 K by the temperature dependency of the magnetization using neutron data, magnetic susceptibility, electrical resistivity and thermoelectric power, respectively. However, the result of Atoji (1981a) that the coherent reflections in the ferromagnetic phase showed no broadening in their peak profiles is contrary to the HoN-type magnetic ordering (Lallement 1966, Child et al. 1963).

8.1.6. *Physical properties of the rare earth hypocarbides*

In addition to the magnetic susceptibility, the electrical resistivity and the thermoelectric power at low temperatures for the RC_x- and R_2C-type carbides have been measured (Lallement 1968). The resultant magnetic and electrical data are shown in table 22.

TABLE 22
Magnetic and electrical data of several rare earth hypocarbides (Lallement 1968).

Compound	T_c (K)	θ_p (K)	μ (in μ_B)		ρ_m (μΩ cm)	ρ_{res} (μΩ cm)
			Experimental	Theoretical		
SmC_x + traces of Sm_2C_3	35	− 10	Sm^{3+}	Sm^{3+}	45	85
GdC_x + traces of metal	400	373			80–110	36–110
Gd_2C	400	245			110	80
$DyC_{0.35}$	160				90	40
Dy_2C	165	175	11.3	10.6	75	75
$HoC_{0.4}$		90	10.9	10.6		90
Ho_2C	90	85–105	10.5		45	80
$ErC_{0.4-0.45}$	50	0	8.5	9.6	50	65
Er_2C	70	60	8.5		45	
YbC + Yb_2C_3		−4 to −5	3–4	4.56		

The results show that the magnetic susceptibility of these carbides agrees well with a Curie–Weiss law of the form $X = X_0 + (C/T - \theta_p)$ above the temperature T_c. The only exception is samarium hypocarbide, for which a deviation from this law has been observed. This type of deviation also occurs in SmC_2 and has been attributed to Van Vleck paramagnetism.

Takaki and Mekata (1969) pointed out that in the composition range $0.33 < x < 0.5$, the ferromagnetic Curie temperatures of GdC_x were more or less constant and were about 20% higher than that of hcp Gd metal, while the magnetic moment per Gd atom was increased by up to 4% relative to Gd metal. For ErC_x, the T_C temperatures were up to 14% higher than the highest magnetic transition temperature observed in the Er metal, but the magnetic moment per Er atom was down by about 5% relative to hcp Er metal. Takaki and Mekata ascribed this reduction in magnetic moment to the effects of magnetocrystalline anisotropy. Their results are also in agreement with the relationship of the variation of the magnetic order–disorder temperatures of the metals and the hypocarbides of the heavy rare earths with the de Gennes factor (Atoji 1981c). According to this relationship, the difference, $\Delta T_t = (T_t$ of the R metal$) - (T_t$ of the $R_2C)$, is progressively decreased in the order from gadolinium to holmium. The ΔT_t values are 36, -8 and -30 K for R = Tb, Dy, Ho, respectively. The differences ΔT_t can be also correlated to the strengths of the R(I)–R(III) bonds, which are 5%, 4% and 3% shorter than the bond distances in the respective R metals. Therefore, the above relations can be interpreted as follows: in the gadolinium hypocarbide, both the exchange interaction and the metal–metal bonds are considerably stronger than those in the Gd metal, and these differences become smaller in the heavier rare earths until Ho and then the relation is reversed in Ho.

In the rare earth hypocarbides, the carbon atoms provide a considerable number of bonding electrons to the s–d bond orbitals among the rare earth atoms, and thus strengthen the metal–metal bonds, as demonstrated by the fact that the lattice parameter of RC_x decreases as the carbon content increases (Atoji 1981c). However, Aoki and Williams (1979) found that in the composition range of $DyC_{0.43}$–$DyC_{0.53}$, the ferromagnetic Curie temperature seems to decrease slightly as the C content is increased, while at $x = 0.33$ the Curie temperature shows a maximum value of 216 K which is 36 K higher than that of Dy metal, and the lattice parameter corresponds to a minimum. Therefore, it is apparent that the relation found by Atoji seems to be applicable only for a limited composition range, although Aoki and Williams assigned this slight decrease in T_C to variations in the radius of the Fermi surface of conduction electrons.

8.1.7. *Magnetic properties of* EuC_6

Suematsu et al. (1981) have observed a complex magnetic structure in EuC_6 depending on the magnetic field; below 40 K EuC_6 exhibits antiferromagnetism while above 40 K EuC_6 is paramagnetic. The magnetization for $H \perp c$ axis shows a field-induced phase transition from a state with a small magnetic moment ($\approx 0.6\mu_B$) at lower fields to a larger moment (2.2–$2.5\mu_B$) above 25 kG. The transition field depends on temperature and is found to vary from 5 to 20 kG. The magnetization for $H \parallel c$ has only a sublinear field dependence. Above 40 K the magnetic susceptibility obeys the Curie–Weiss law with $\theta = +1.3$ K and the molar Curie constant $C_m = 6.78$, which suggests that the Eu ion is divalent in EuC_6. The high-field magnetization measurements of Suematsu et al. (1981) (up to 400 kG below 40 K) of EuC_6 have revealed that the spins of Eu^{2+} ions exhibit an XY magnetic anisotropy, i.e., they are inclined to be along the layers. On the origin of the XY magnetic anisotropy observed in EuC_6, a new mechanism was proposed by Akera and Kamimura (1983), who suggested that it is the result of the combined action of the π–f exchange and the spin–orbit interaction in antiferromagnetic EuC_6.

The behavior of the first-stage intercalation compound EuC_6 under high fields up to 400 kG and below 40 K is very interesting. According to the work of Suematsu et al. (1981), in the magnetization curve at 4.2 K four regions have been found which correspond to different magnetization processes. For $H \perp c$, when $H < 22$ kG, the magnetization increases steeply with the field and shows an apparent moment of about $0.6\mu_B$. In the region from 22 to 82 kG, the magnetization shows a slight increase from 2.2 to $2.7\mu_B$ per Eu ion, which corresponds to one third of the full moment of the Eu^{2+} ion. In the region of 82–205 kG, there is a linear field dependence of the magnetization and above 205 kG the saturation moment is found to be $6.2\mu_B$. The four regions of the magnetization processes are probably related to a phase sequence of: antiferromagnet (I)–ferrimagnet (II)–fan-state (III)–ferromagnet (IV). For $H \parallel c$, the magnetization increases monotonically and approaches a saturation value corresponding to $6\mu_B$ above 240 kG.

8.2. *Superconductivity of the rare earth carbides and their solid solutions*

8.2.1. *Superconductivity of the rare earth sesquicarbides and the Th-containing solid solutions*

The superconductivity of the rare earth carbides was first found by the Krupka–Giorgi–Krikorian group (Krupka et al. 1969a, Giorgi et al. 1969). Y_2C_3, which was synthesized by a high-pressure, high-temperature technique, exhibited a transition temperature of 11.5 K and La_2C_3 of 11 K.

The phase composition has a significant effect on the superconducting transition temperature over the entire homogeneity range. The superconducting transition temperature varies from 6 K to 11 or 11.5 K, and the highest observed transition temperature occurs at the intermediate composition of $LaC_{1.41}$ and $YC_{1.38-1.48}$ rather than at the highest carbon concentration of the homogeneity range.

Addition of thorium enhances the superconducting temperature T_c of Y_2C_3 to 17.1 K (Krupka et al. 1969b) and a smaller T_c increase (about 3 K) also occurs in the lanthanum thorium sesquicarbides (Giorgi et al. 1970). These high-temperature superconductors, e.g. ytttrium–thorium sesquicarbides, were also prepared by high-pressure, high-temperature techniques. High-temperature and ambient-pressure annealing results in the destruction of both the bcc structure and the high-temperature superconductivity. In addition, in the yttrium–thorium–carbon system a ternary bct phase also exists over a wide ternary field but is not superconducting.

Incorporation of elements other than thorium into yttrium sesquicarbide also leads to a resultant enhancement of the superconducting temperature (Krupka et al. 1969b). These elements include the group IV and VI transition metals, as well as gold, germanium and silicon. But, in these cases, it is necessary to ensure the solubility of these elements in the parent sesquicarbide lattice.

In the studies on the superconductivity of the sesquicarbide phase in the lanthanum–thorium–carbon system (Giorgi et al. 1970), it has been found that the transition temperature is relatively insensitive to large variations in the lanthanum-to-thorium ratio, in sharp contrast to the results obtained in the yttrium–thorium–carbon system (Krupka et al. 1969b). The high-temperature, high-pressure $La_xTh_{1-x}C_{1.4\pm0.1}$ phases, except for the phase with an La:Th ratio between 8:2 and 5:5, exhibit superconductivity with the maximum transition temperature 14.1 and 14.2 K, respectively, at the compositions with an La:Th ratio of 8:2 and 5:5. Of course, the superconducting compositions also include bcc Pu_2C_3-type Th_2C_3 (Krupka 1970) and the Th_2C_3-based solutions with an La:Th ratio of less than 5:5.

Other superconducting Th_2C_3-based rare earth carbides $(Th, R)_2C_3$ (R = Ho, Er, Lu, Sc) have also been explored by using high-temperature, high-pressure synthesis (Krupka et al. 1973). Transition temperatures are dependent upon the rare earth element and the thorium:rare earth atomic ratio, and vary from 4.1 to 11.7 K. The maximum transition temperature was observed in the region of the Th:R atomic ratio of 8:2 to 7:3. The occurrence of superconductivity in these systems is limited by the thermodynamic stability of the superconducting phase and the conversion from the superconducting state to a paramagnetic state. Light rare earth elements, such as Ce, Pr and Nd, also form a superconducting Th_2C_3-based solid solution but with depressed transition temperatures (<4 K). The intermediate rare earth elements Gd, Tb, and Dy form magnetic solid solutions at the concentration levels of the Th:R atomic ratio from 8:2 to 9:1.

Summing up what has been described above, the maximum transition temperatures observed for the additive elements Y and La occur at a valence electron per atom of 3.7, and for Ho, Er, Lu and Sc this is 3.9. Cooper et al. (1970) have demonstrated that a new peak of the superconducting transition temperatures exists at the general valence electron per atom concentration of 3.7 to 3.9 for a number of crystal structures of which the Pu_2C_3 structure is one.

8.2.2. *Superconductivity of the rare earth dicarbides and mixed dicarbides*

LaC_2, YC_2 and LuC_2, which crystallized in the body-centered tetragonal C_{11a} (CaC_2-type) structure, were also found to be superconducting with transition temperatures of 1.61, 3.75 and 3.33 K, respectively (Green et al. 1969). For YC_2, another value of the transition temperature was reported to be 3.88 K (Giorgi et al. 1968).

The superconducting transition temperatures of the solid solutions $La_xY_{1-x}C_2$, were a little higher than those expected from the bct dicarbides, although the changes of the crystal structure (bct → fcc) occur in the neighborhood of 50 mol% LaC_2 (Sakai et al. 1981a, b). These results can be explained reasonably in terms of the BCS theory.

8.2.3. *Specific heats and critical magnetic field curves of the rare earth and the rare-earth–thorium sesquicarbides*

The specific heats of the superconductors $LaC_{1.35}$ (11.0 K), $La_{0.9}Th_{0.1}C_{1.6}$, $YC_{1.35}$ (Cort et al. 1984) and $Y_{0.7}Th_{0.3}C_{1.58}$ (17.0 K) (Stewart et al. 1978) have been measured. The Debye temperatures, θ_D, are 318, 274, 557, and 346 K, respectively. The linear-term coefficients, γ, of the specific heat have the values, 4.8, 2.3, 2.8 and 4.7 mJ mol^{-1} K^{-2}. Thus, the electronic density of states $N(0)$, which is proportional to γ, is quite low for the four compounds. Specific heat measurements have shown neither the arc-melted lanthanum nor the lanthanum–thorium sesquicarbide to be a bulk superconductor. However, yttrium and yttrium–thorium sesquicarbide exhibit a bulk specific heat anomaly at $T_c = 10.5$ K. Therefore, the addition of thorium does not necessarily induce a bulk transition. From comparison of γ, it is apparent that the occurrence of bulk superconductivity is not correlated with the electronic density of states in this system.

The critical magnetic field curve measurements of lanthanum and yttrium–thorium sesquicarbide (Francavilla et al. 1976, Francavilla and Carter 1976) show that the initial slopes of the critical magnetic field curve are $(2.4 \pm 0.2) \times 10^4$ Oe K^{-1} and 1.8×10^4 Oe K^{-1}, as well as the maximum critical magnetic fields are $H_{c_2}(0)$, 280 ± 24 kOe and $H_{c_2}(4.2)$, 95 kOe, respectively, for $Y_{0.7}Th_{0.3}C_{1.5}$ and La_2C_3. Other superconducting parameters such as the electron–phonon coupling constant and the band structure density of states at the Fermi energy as well as the upper and lower critical fields, the electronic coefficient of specific heat (γ) and the Ginzburg–Landau k value were obtained for La_2C_3 (Francavilla and Carter 1976). However, the γ value of $6.10 + 1.3$ mJ mol^{-1} K^{-2} for La_2C_3 based upon critical-field measurements is larger than that obtained by Cort et al. (1984) from specific heat measurements, and this material used in critical-field measurements also seems unlikely to be a bulk superconductor. The upper and lower critical fields, H_{c_2} (4.2 K), 95 kOe, and H_{c_1} (4.2 K), 22 ± 4 Oe, are also different from the values 53 kOe and 750 Oe, respectively, obtained in the previous work (Carter et al. 1975) of the same research group. Despite these contradictions, it has been established that La_2C_3 is a type-II superconductor.

Carter et al. (1975) and Cadieu et al. (1980) took into consideration the C_2 bonding in the Pu_2C_3 structure type, and regarded the structure of the yttrium sesquicarbide system as being of the anti-Th_3P_4 type with pairs of carbon atoms replacing the thorium atoms and wrote the compound as $Y_4(C_2)_3$, and then computed the valence electron per atom for seven atoms in the structure to give a valence electron per atom of 4.3. Based on this valence electron number per atom and the C–C bonding character, they explained the occurrence of the maximum value of T_c in the system corresponding to an intermediate value of the lattice parameter and the effect of adding a third element on T_c.

8.3. *Physical properties of ternary rare earth carbides*

8.3.1. *Magnetic and electrical properties of RB_2C_2*

As has been established (Bauer and Bars 1980), the crystal structure of RB_2C_2 consists of alternating rare earth metals and covalently bonded B–C sheets. The B–C network contains four- and eight-membered rings. Owing to such a layer structure, these compounds are expected to be new materials with interesting physical properties that embody both those of the rare earth borides and the carbides.

The RB_2C_2 (R = Ce, Nd, Sm, Gd, Tb, Er, Tm) compounds are antiferromagnetic (Sakai et al. 1981c). Their Néel temperatures vary from 7 K of CeB_2C_2 to 47.5 K of GdB_2C_2 and the paramagnetic Curie temperatures from about 55 K of CeB_2C_2 to 5 K of ErB_2C_2. The Curie–Weiss fits to the susceptibilities lead to effective moments that are in agreement with those expected for R^{3+} ions. In addition, CeB_2C_2 and TbB_2C_2 also exhibit metamagnetic transitions at low fields. The compounds DyB_2C_2 and HoB_2C_2 are ferromagnetic with values of T_C of 16.4 and 7 K, respectively. The spontaneous magnetizations below 5 K are considerably smaller than the theoretical values, suggesting the existence of complicated ferromagnetic structures. An exception is PrB_2C_2, which becomes a Van Vleck paramagnet at low temperatures.

Since the B–C bonds in RB_2C_2 do not require any electron transfer from rare earth metals, all the rare earth diborodicarbides are expected to have three conduction electrons per formula unit. Therefore, it is reasonable to assume that the main magnetic interaction in RB_2C_2 is the f–f indirect exchange via conduction electrons (the RKKY interaction), as was subsequently verified by the relation of the ρ_m values with the de Gennes factors (Sakai et al. 1982a).

The temperature dependences of resistivity for RB_2C_2 (R = Ce, Nd, Gd, Tb, Dy, Ho, Er, Tm) also show a sharp change in slope at the magnetic ordering temperatures found from the magnetic susceptibility measurements (Sakai et al. 1982a, b). For TmB_2C_2, the rapid increase in resistivity below the Néel temperature of 16.5 K was observed and may be ascribed to a transition in the crystal. YB_2C_2 and LuB_2C_2 exhibit superconducting transitions at 3.6 K and 2.4 K, respectively, while the compound LaB_2C_2 remains in the normal state down to 1.8 K.

In conclusion, it can be said that the rare earth diborodicarbides have a variety of peculiar physical properties because of their characteristic crystal or band structures and may be exploited as a new type of material.

8.3.2. *Superconductivity of $La_5B_2C_6$ and other lanthanum–boron carbides*

The borocarbide LaBC and the alloy of nominal composition $La_5B_2C_8$ were determined to be not superconducting down to 4.2 K, but $La_5B_2C_6$ becomes superconducting at $T_c = 6.9$ K (Bauer and Politis 1982, Bauer and Bars 1983).

8.3.3. *Electrical and magnetic properties of $EuB_{6-x}C_x$*

Kasaya et al. (1978) studied the electrical properties of single crystals of EuB_6 and $EuB_{5.95}C_{0.05}$. Both single crystals exhibit metallic behavior between 30 and 300 K and their resistivities at 300 K are of the order of 10^{-4} Ω cm. However, each one shows a rather different temperature dependence of resistivity at low temperatures. For EuB_6, the resistivity shows a small peak at 18 K and then decreases down to 25 μΩ cm at 4.2 K while for $EuB_{5.95}C_{0.05}$ the resistivity goes through a minimum and then increases rapidly below 20 K, showing a sharp maximum at 7 K and decreasing again down to 4.2 K. At room temperature, the resistivity of EuB_6 is larger than that of $EuB_{5.95}C_{0.05}$, but the value of $d\rho/dT$ is smaller. This may be connected with the increase in the number of conduction electrons and the higher phonon scattering.

In EuB_6 or YbB_6, the carbon substitution yields n-type metallic conductors (Tarascon et al. 1980) and the conduction electrons would modify the magnetic interactions and induce a sign change of the paramagnetic Curie temperature θ_p. In fact, for $EuB_{6-x}C_x$, θ_p changes sign at $x \simeq 0.13$ (Tarascon et al. 1981), but the magnetic ordering temperature passes through a minimum around $x \simeq 0.06$. The ordering temperature was defined by the appearance of either magnetic hyperfine splitting in the ^{151}Eu Mössbauer spectrum (Coey et al. 1979) or of a peak in the magnetization measured in a constant small field (Kasaya et al. 1979).

The results of the Mössbauer study revealed a temperature-dependent broadening of the magnetic hyperfine lines for $x = 0.05$, implying that the exchange interactions are inhomogeneous on a microscopic scale (Coey et al. 1979). This result, together with the magnetization curve at 4.2 K, shows that 90% of the saturation is achieved in fields of less than 5 kOe for EuB_6, but that 100 kOe is required to produce the same degree of saturation in the $x = 0.21$ sample (Kasaya et al. 1979). The authors suggest that the magnetic order evolves from ferromagnetism ($x = 0$) to antiferromagnetism ($x = 0.21$), via an intermediate micromagnetic phase. Neutron diffraction investigation (Tarascon et al. 1981) further determined the magnetic structure of $EuB_{6-x}C_x$ ($x = 0$ to 0.20) and showed that EuB_6 is a simple ferromagnet, whereas the $x = 0.20$ compound has an incommensurate spiral structure with propagation vector $\boldsymbol{\tau} = (0.08, 0, 0)$. For a magnetically inhomogeneous intermediate composition $x = 0.05$, there is a mixture of ferromagnetic and helimagnetic domains with $\boldsymbol{\tau} = (0, 0, 0.104)$. The helimagnetism in $EuB_{6-x}C_x$ arises from a competition between ferromagnetic nearest-neighbor exchange and antiferromagnetic interactions resulting from conduction electrons.

8.3.4. *Magnetic properties of the $RNiC_2$ compounds*

Recently, the magnetic properties of the rare-earth-metal–transition-metal carbides, in particular, Fe, Co and Ni carbides, have received significant attention. As a part of this study, the magnetic properties of the series of compounds $RNiC_2$ (R = Pr, Nd, Gd, Tb, Dy, Ho, Er, Tm, Y) have been investigated. The compounds for Nd, Gd, Tb, Dy, Er and Tm show ordering temperatures between 7 K (Nd) and 25 K (Tb), while $PrNiC_2$ and $HoNiC_2$ do not show any magnetic ordering down to 4.2 K. For $YNiC_2$, a temperature-independent susceptibility occurs. None of the compounds follows the Curie–Weiss law in the temperature region 4.2–250 K (Kotsanidis et al. 1989). The magnetization values are much smaller than the free R^{3+} ion values. The ordering temperatures of the compounds studied do not follow the de Gennes $T_0 = c(g_J - 1)^2 J(J+1)$ law (de Gennes 1962).

Based on the absence of any magnetic moment for $YNiC_2$, it can be seen that the magnetic contribution of nickel is zero. This may be the result of the transfer of 5d electrons of the rare earth atoms to the 3d nickel band. Consequently, only the rare earth ions contribute to the magnetic interaction, which is commonly of the RKKY type and occurs via polarization of the conduction electrons. The breakdown of the de Gennes relationship suggests that the situation is less simple in $RNiC_2$.

Below the ordering temperature, the field dependence of the magnetization exhibits a metamagnetic transition in low fields (3–4 KOe) for the compounds with R ≡ Gd, Tb, Er, Tm.

The crystal field interaction is probably relatively high in $RNiC_2$ compounds because the rare earth ions occupy a low-symmetry site (C_{2v}) (Gignoux et al. 1982). This may cause the deviations from Curie–Weiss behavior observed for this series of compounds.

8.3.5. *Physical properties of* $R_2Fe_{14}C$

8.3.5.1. *Magnetic properties of* $R_2Fe_{14}C$. As has been described in sect. 6.4.3, the compounds $R_2Fe_{14}C$ have the same crystal structure as $R_2Fe_{14}B$ and the magnetic properties of the $R_2Fe_{14}C$ compounds are similar to those of the $R_2Fe_{14}B$ series. This has led to interest in these materials, not only from the viewpoint of fundamental physics, but also with a view to their potential application as permanent-magnet materials (Denissen et al. 1988a).

^{57}Fe Mössbauer spectra were taken of the compounds $Lu_2Fe_{14}C$ and $Gd_2Fe_{14}C$ (Denissen et al. 1988a) at 300 K and 10 K, $Ce_2Fe_{14-x}Mn_{0.3}C$ at 300 K and 20 K (Jacobs et al. 1989), $Nd_2Fe_{14}C$ at room temperature (Buschow et al. 1988a) and 20, 125, 205, 300 K (Denissen et al. 1988b). Analysis of the spectra showed that the average iron moment in $R_2Fe_{14}C$ is approximately the same as in $R_2Fe_{14}B_3$, e.g. the moments of $Gd_2Fe_{14}C$ and $Lu_2Fe_{14}C$ are equal to $31.9\mu_B$ and $27.5\mu_B$, respectively. For $Nd_2Fe_{14}C$, it is $30.9\mu_B$ per formula unit and the average iron moment is equal to $2.2\mu_B$/(Fe atom) at low temperatures while the corresponding value for $Nd_2Fe_{14}B$ is $2.26\mu_B$/(Fe atom) (Onodera et al. 1987a, b). This indicates that the iron moments in the carbides are only slightly lower (3%) than in the borides. When decomposing these Mössbauer spectra into six subspectra according to the six different iron sites (denoted in the Wyckoff notation by the symbols $8j_2$, $16k_1$, $16k_2$, $8j_1$, 4e and 4c) and deriving the hyperfine parameters from fits to the ^{57}Fe Mössbauer spectra of $Gd_2Fe_{14}C$, $Lu_2Fe_{14}C$ and $Nd_2Fe_{14}B$, it can be seen that in both series, $R_2Fe_{14}B$ and $R_2Fe_{14}C$, there are large differences between the moments of the six crystallographically different iron sites (Denissen et al. 1988a, b).

The shifts in the hyperfine fields at iron sites of $Gd_2Fe_{14}C$ to values smaller than those determined for $Gd_2Fe_{14}B$, indicate a lowering in the average iron magnetic moment, which is probably due to a stronger electron transfer from the carbon atom to the iron d band or to pd hybridization. The hyperfine field at the carbon nuclei was found to be 2.38 T at the maximum spin echo amplitude, which has the same value as that determined for the hyperfine field at the boron nuclei in $Gd_2Fe_{14}B$ (Erdmann et al. 1988, 1989) (table 23). Most of the compounds $R_2Fe_{14}C$ are ferromagnetic with their preferred magnetization direction along the tetragonal *c* axis. At low temperatures, the rare earth (R) sublattice magnetization prevails and the competition with the iron sublattice anisotropy leads to spin reorientations near room temperature for $Tm_2Fe_{14}C$ and $Er_2Fe_{14}C$ (de Boer et al. 1988a). A spin reorientation of a different type also occurs in $Nd_2Fe_{14}C$ at $T_{SR} = 140$ K (Denissen et al. 1988b) and in $Ho_2Fe_{14}C$ at $T_{SR} = 35$ K (de Boer et al. 1988b). These data may be compared with

TABLE 23

Hyperfine field (HF), quadrupole splitting (QS) and isomer shifts (IS) relative to α-Fe at the six different iron sites 16k, 4c and at the metalloid site 4g occupied by carbon or boron determined from Mössbauer and nuclear magnetic resonance (NMR) spectra (Erdmann et al. 1988, 1989). All these data were determined at 4.2 K.

	$Gd_2Fe_{14}C$				$Gd_2Fe_{14}B$	
	Mössbauer			NMR HF	Mössbauer HF (T)	NMR HF (T)
	HF (T)	QS ($mm\,s^{-1}$)	IS ($mm\,s^{-1}$)			
$16k_1$ (iron)	32.4	0.09	0.15	32.3	34.1	33.3
$16k_2$ (iron)	35.3	0.10	0.00	34.9	35.9	35.9
$8j_1$ (iron)	32.3	–	–	31.4	32.9	–
$8j_2$ (iron)	39.3	0.32	0.23	39.0	40.1	40.1
4e (iron)	30.8	–	–	–	32.1	–
4c (iron)	27.9	–	–	–	29.7	–
4g (C, B)				2.38		2.40

those for the $R_2Fe_{14}B$ compounds (Coey 1986). For instance, the spin reorientation temperature for $Nd_2Fe_{14}B$ is at 148 K based on an anomalous behavior of the isomer shift derived from the fits to the ^{57}Fe Mössbauer spectra (Onodera et al. 1987a, b) and is only slightly higher than that of $Nd_2Fe_{14}C$.

From the above-described results, it can be seen that the behavior of the $R_2Fe_{14}B$ compounds is followed almost completely by the $R_2Fe_{14}C$ compounds, but with the only difference that the Curie temperatures of the former are slightly higher than those of the latter (Pedziwiatr et al. 1986, Buschow 1987, de Boer et al. 1988a, Gueramian et al. 1987). Values for the Curie temperatures, T_C, together with the saturation magnetizations, M_s, are summarized in table 24.

Gueramian et al. (1987) made a comparison of Curie temperatures between both the $R_2Fe_{14}B$ and $R_2Fe_{14}C$ series for the rare earths from gadolinium through lutetium, relating to the de Gennes factor $(g_J - 1)^2 J(J+1)$, and indicated that the Curie temperatures of the carbide series are 40–50 K lower, on the average, than those of the boride series. The values of the saturation magnetizations for the carbide series are also generally lower by several per cent than those of the corresponding borides (Pedziwiatr et al. 1986, Bolzoni et al. 1985, Gueramian et al. 1987).

The magnetic coupling between iron spins and rare earth spins in $R_2Fe_{14}C$ is antiferromagnetic, similar to that observed in $R_2Fe_{14}B$ (Sinnema et al. 1984). Consequently, at low temperatures, $M_s(T)$ is expected to decrease with increasing temperatures for the light rare earth compounds, and to increase for the heavy rare earth compounds (table 24).

In addition, it is known that the coercivity and anisotropy are higher in some of the $R_2Fe_{14}C$ compounds (Liu and Stadelmaier 1986, Bolzoni et al. 1985). In particular, because these carbides do not crystallize from the melt, the high coercivity-yielding microstructure must be obtained through annealing the cast materials, such as

TABLE 24

Curie temperature, T_C, and saturation magnetization, M_s, of the $R_2Fe_{14}C$ compounds in a field $B = 8\,T$ (Gueramian et al. 1987).

Rare earth	T_c (K)	$M_s (A^2/kg)$ $T = 250K$	$T = 5\,K$	μ_s (μ_B/formula unit)	M_{Fe} (μ_B/Fe)	M_R (μ_B/R)	gJ
La							
Ce	345[a]	124[a]		23.9[a]			
Pr							
Nd	535[b]			30.9[b]	2.2[b] (3.1[d])	3.3[b]	–
Sm	371	120	125	24.5	1.65	0.0	0.7
Gd	613, 643[f]	80	90	17.9 (31.9[c])	2.28	6.5	7.0
Tb	572	40	45	9.0	1.93	10.9	9.0
Dy	534, 553[e]	45	42	8.4, 9.7[e] (4.2 K)	2.03, 2.12[e]	11.2	10.0
Ho	516	62	52	10.5	2.18	10.2	10.0
Er	500, 501[e]	90	60	12.1, 11.3[e] (4.2 K)	2.15, 2.10[e]	9.4	9.0
Tm	495	110	83	16.8	2.20	7.0	7.0
Lu	493	120	133	27.3 (27.5[c])	1.95 (1.97[c])	0.0	0.0

[a] Jacobs et al. (1989).
[b] de Mooij and Buschow (1988).
[c] Denissen et al. (1988a).
[d] Derived from the 4.2 K neutron diffraction data (Helmholdt and Buschow 1988).
[e] Pedziwiatr et al. (1986).
[f] Abache and Oesterriecher (1985).

$Dy_2Fe_{14}C$ and $Dy_6Nd_9Fe_{14}B_{0.1}C_{0.9}$ (Liu and Stadelmaier 1986, Liu et al. 1987). Therefore, it could be expected that these compounds will be good candidates for permanent-magnet applications.

8.3.5.2. *Substitution effect in* $R_2Fe_{14}C$: (i) *substitution of manganese for iron.* As described above, the successful preparation of $Ce_2Fe_{14}C$ was achieved by substituting 2% manganese for iron in this compound (Jacobs et al. 1989). The magnetic data of $Ce_2Fe_{14-x}Mn_xC$ compounds ($x = 0.3$ and 0.8) have been measured, showing that the Curie temperatures are 315 and 265 K, the magnetizations at 4.2 K in $H = 1500\,kA\,m^{-1}$ are 109 and 83.5 $A\,m^2\,kg^{-1}$, and the moments per formula unit of $Ce_2Fe_{14-x}Mn_xC$ are 21.0 and 16.1μ_B at 4.2 K for $x = 0.3$ and 0.8, respectively. It is apparent that the substitution of manganese for iron could give rise to the reduction in the Curie temperature and the magnetization as well as the moment per formula unit in $Ce_2Fe_{14}C$.

In the $Nd_2Fe_{14}C$ compound the substitution of manganese for iron also exhibits the same effect as in $Ce_2Fe_{14}C$ (Buschow et al. 1988b). Pertinent data derived from the magnetic measurements showed that the substitution of manganese leads to a strong reduction in the saturation magnetization, the Curie temperature and the reorientation temperature, and when the manganese concentration is in excess of $x = 2$, the Curie temperature falls below room temperature. Furthermore, corresponding to the decrease in the saturation magnetization, a moment reduction of about 4μ_B per manganese atom added was calculated, which on the basis of comparing ^{57}Fe Mössbauer spectra of two $Nd_2Fe_{14-x}Mn_xC$ compounds with those of

$Nd_2Fe_{14}C$ can be interpreted as being due to the occurrence of a highly disordered magnetic structure comprising large differences in the orientations of the magnetic moments surrounding a given iron atom. However, as in $Ce_2Fe_{14}C$, the manganese substitution can also promote the formation of the tetragonal $Nd_2Fe_{14}C$ compound, and the single-phase samples can be easily prepared after annealing at 800°C as well as at 900°C, while for the pure $Nd_2Fe_{14}C$ compound only through annealing in a fairly narrow temperature range around 850°C can the single-phase sample be obtained (Buschow et al. 1988a).

Therefore, taking into consideration the practical permanent-magnet applications of this material, the recommendation is made that the manganese substitution of only small amounts ($x \approx 0.3$ in $Nd_2Fe_{14-x}Mn_xC$) is reasonable (Buschow et al. 1988b)

8.3.5.3. *Substitution effect in* $R_2Fe_{14}C$: (ii) *Boron and Carbon in the* $R_2Fe_{14}X$ *phase.* Deppe et al. (1988) reported the results of a Mössbauer spectroscopy study for two alloys with the major phases $Dy_2Fe_{14}C$ and $Nd_2Fe_{14}B_{0.1}C_{0.9}$, which were obtained by annealing the arc-melted specimens of composition $R_{15}Fe_{77}X_8$ at 900°C. It is shown that the hyperfine fields of the iron sites Fe(j_2) (33.2 T and 33.5 T), Fe(k_2) (29.8 T and 29.2 T) and Fe(k_1) (27.4 T and 27.6 T) for $Nd_2Fe_{14}B_{0.1}C_{0.9}$ and $Dy_2Fe_{14}C$ lie significantly below those of the corresponding borides $R_2Fe_{14}B$. For dysprosium, the field reduction amounts to 7.2%, for neodymium it is 4.3%. They suggested that the observed reductions principally result from an increased transfer of electrons into the 3d band of iron when carbon replaces boron.

Nevertheless, in a permanent-magnet application, the compounds $R_2Fe_{14}B_xC_{1-x}$ have been exploited. Liu and Stadelmaier (1986) demonstrated that bulk alloys with the composition $R_{15}Fe_{77}(B, C)_8$, can be heat-treated to produce high values of iH_c; in the cases of $Dy_{15}Fe_{77}C_8$ and $Nd_9Dy_6Fe_{77}B_{0.8}C_{7.2}$ the values are as high as 12.5 kOe. Subsequently, Liu et al. (1987) prepared a series of alloys with the compositions $R_{15}Fe_{77}X_9$, $R = Nd_{0.6}Dy_{0.4}$ and $X = B_{0.1}C_{0.9}$, and determined the coercivities of the samples after heat treatment. For the alloys $R_{15}Fe_{76}X_9$, $R_{15}Fe_{77}X_8$ and $R_{14.5}Fe_{77.5}X_8$ annealed at 900°C for 6 h, the $\mu_0 iH_c$ are 0.95 T, 0.85 T and 0.7 T, respectively. The temperature dependence of iH_c for $Nd_9Dy_6Fe_{77}B_{0.9}C_{7.2}$ is practically linear. The slope is $-0.042\ \mathrm{kOe\,K^{-1}}$ and the coercivity loss between 20 and 100°C is 39%, which is comparable to the 42.5% loss for a typical sintered boride magnet (Li et al. 1985). The high coercivity is related to a cellular microstructure of $R_2Fe_{14}X$ in which the cell size is approximately 1 μm. The cell structure, which originates from a peritectoid-like transformation from primary $Fe_{17}R_2$,

$$4(R_2Fe_{17}X_{0.75}) + 2RFeX \longrightarrow 5R_2Fe_{14}X,$$

is quite stable and does not change during prolonged annealing.

Liu et al. (1987) have studied the phase region of stable $R_2Fe_{14}B_xC_{1-x}$ compounds and the observed phases at 800°C, indicating that the miscibility of the $R_2Fe_{14}C$ carbides with the $R_2Fe_{14}B$ borides is reported for the lighter rare earth metals. They suggested that like the borides, the carbides are magnetically hard, but unlike them, the carbides do not normally crystallize from the melt, and this property can be

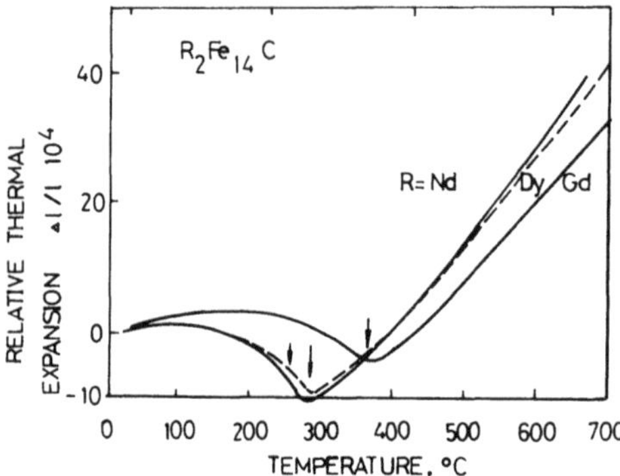

Fig. 31. Temperature dependence of the relative thermal expansion of $R_2Fe_{14}C$ (Buschow 1988).

exploited to produce intrinsic coercivities above 12 kOe in cast materials without the additional special sintering processing and melt spinning.

8.3.5.4. *Magnetovolume effects in* $R_2Fe_{14}C$. Recently, it was shown that huge Invar-type anomalies are also present in ternary compounds of the type $R_2Fe_{14}B$ (Buschow and Grössinger 1987) and these compounds are virtually so-called line compounds in which none of the component elements has regions of solid solubility (Buschow et al. 1986). The investigation has also been extended to the series of $R_2Fe_{14}C$ compounds for R = Nd, Gd, Tb, Dy, Ho, Er, Lu (Buschow 1988) and the experimental results obtained for various members of the $R_2Fe_{14}C$ series are shown in fig. 31. The corresponding Curie temperatures are also indicated in the figure by means of small arrows. It can be seen from fig. 31 that $\Delta l/l$ in all these compounds tends towards linear behavior at high temperatures. A highly anomalous thermal expansion behavior is manifest at low temperatures below the Curie temperature. However, no clear correspondence was found between the magnitude of the anomaly and the size of the iron moments, although it is true that the anomaly is most pronounced in $R_2Fe_{14}B$ ($\mu_{Fe} \approx 2.2\mu_B$/Fe atom) followed by $R_2Fe_{14}C$ ($\mu_{Fe} \approx 2.1\mu_B$/Fe atom).

The results obtained for the compounds of the series $R_2Fe_{14}C$ confirmed the results on the isotypic $R_2Fe_{14}B$ compounds, showing that the spontaneous volume magnetostriction extends to temperatures substantially above T_C in this type of ternary compounds. This observation lends credence to the view that the local iron moments in iron-based metallic materials do not disappear above T_C (Holden et al. 1984).

To conclude, several of the studied compounds, e.g. $Nd_2Fe_{14}C$, have an almost vanishing thermal expansion in the range 20–150°C, and these properties make them good candidates for applications as Invar-type alloys.

8.3.6. *Magnetic properties of a series of compounds* $RFe_{10}SiC_{0.5}$ (R = *Ce, Pr, Nd, Sm*)

Magnetic measurements performed between 1.5 K and 300 K showed that at room temperature ferromagnetic or ferrimagnetic behavior is observed for all the $RFe_{10}SiC_{0.5}$ compounds, and in low fields a large susceptibility is found correspond-

TABLE 25
The Curie temperature T_C, spontaneous magnetization M_s at 4.2 and 300 K, theoretical moment for the rare earth ion R^{3+} and value of the iron moment of the compound $RFe_{10}SiC_{0.5}$.

Compound	T_C (K)	M_s (4.2 K) [μ_B/formula unit]	M_s (300 K) [μ_B/formula unit]	$g_J J$	M_{Fe} [μ_B/Fe atom]
$CeFe_{10}SiC_{0.5}$	390	13.2	6.5	2.14	1.32
$PrFe_{10}SiC_{0.5}$	430	17.1	9.2	3.20	1.39
$NdFe_{10}SiC_{0.5}$	410	17.5	8.4	3.27	1.42
$SmFe_{10}SiC_{0.5}$	460	14.8	11.1	0.71	1.41

ing to the expected slope of the demagnetizing fields. In higher fields, the magnetization tends to saturate. The spontaneous magnetization values deduced from these data in the different compounds are given in table 25 (Le Roy et al. 1987). At low temperatures the compounds with a magnetic R atom (R = Pr, Nd, Sm) have a much weaker susceptibility than $CeFe_{10}SiC_{0.5}$. Above a critical field of the order of 1 kOe, the magnetization increases rapidly and tends to saturate above 5 kOe. Measurements of hysteresis loops have confirmed that this phenomenon results from intrinsic coercivity, which occurs at low temperatures in highly anisotropic compounds (Barbara 1973).

Acknowledgement

The authors wish to express their thanks to Miss Mikiko Miyamoto for her sincere help.

References

Abache, C., and H. Oesterreicher, 1985, J. Appl. Phys. **57**, 4112.

Achard, J.C., 1957, C. R. Acad. Sci. **245**, 1064.

Adachi, G., H. Kotani, N. Yoshida and J. Shiokawa, 1970, J. Less-Common Met. **22**, 517.

Adachi, G., T. Nishihata and J. Shiokawa, 1973, J. Less-Common Met. **32**, 301.

Adachi, G., K. Ueno and J. Shiokawa, 1974, J. Less-Common Met. **37**, 313.

Adachi, G., Y. Shibata, K. Ueno and J. Shiokawa, 1976, J. Inorg. & Nucl. Chem. **38**, 1023.

Adachi, G., R. Tonomura, Y. Shibata and J. Shiokawa, 1978, J. Inorg. & Nucl. Chem. **40**, 489.

Akera, H., and H. Kamimura, 1983, Solid State Commun. **48**, 467.

Al-Shahery, G.Y.H., D.W. Jones, I.J. McColm and R. Steadman, 1982a, J. Less-Common Met. **85**, 233.

Al-Shahery, G.Y.H., D.W. Jones, I.J. McColm and R. Steadman, 1982b, J. Less-Common Met. **87**, 99.

Al-Shahery, G.Y.H., R. Steadman and I.J. McColm, 1983, J. Less-Common Met. **92**, 329.

Anderson, J.S., and A.N. Bagshaw, 1970, in: Les Elements des Terres Rares, Vol. 1 (Centre National de la Recherche Scientifique, Paris) p. 397.

Anderson, J.S., and A.N. Bagshaw, 1972, Rev. Chim. Miner. **9**, 115.

Anderson, J.S., N.J. Clark and I.J. McColm, 1968, J. Inorg. & Nucl. Chem. **30**, 105.

Anderson, J.S., N.J. Clark and I.J. McColm, 1969, J. Inorg. & Nucl. Chem. **31**, 1621.

Aoki, Y., and D.E. Williams, 1979, J. Less-Common Met. **65**, 35.

Arbuckle, J., and E. Parthé, 1962, Acta Crystallogr. **15**, 1205.

Aronsson, B., 1958, Acta Chem. Scand. **12**, 31.

Aronsson, B., 1960, Acta Chem. Scand. **14**, 1414.

Atoji, M., 1961, J. Chem. Phys. **35**, 1950.

Atoji, M., 1962, J. Phys. Soc. Jpn., Suppl. B-II **17**, 395.

Atoji, M., 1965, Nucl. Instrum. Methods **35**, 13.

Atoji, M., 1967a, J. Chem. Phys. **46**, 1891.

Atoji, M., 1967b, J. Chem. Phys. **46**, 4148.

Atoji, M., 1968, J. Chem. Phys. **48**, 3384.

Atoji, M., 1969, J. Chem. Phys. **51**, 3872.

Atoji, M., 1970, J. Chem. Phys. **52**, 6431.

Atoji, M., 1971, J. Chem. Phys. **54**, 3504.

Atoji, M., 1972, J. Chem. Phys. **57**, 2410.

Atoji, M., 1978, J. Solid State Chem. **26**, 51.

Atoji, M., 1980, J. Chem. Phys. **73**, 5796.

Atoji, M., 1981a, J. Chem. Phys. **74**, 1893.

Atoji, M., 1981b, J. Chem. Phys. **74**, 1898.

Atoji, M., 1981c, J. Chem. Phys. **75**, 1434.

Atoji, M., and R.H. Flowers, 1970, J. Chem. Phys. **52**, 6430.

Atoji, M., and M. Kikuchi, 1968, in: Crystal Structures of Cubic and Trigonal Yttrium Hypocarbides: a Dimorphically Interphased Single-Crystal Study, Rep. ANL-7441 (Chemistry Division, Argonne, National Laboratory, IL).

Atoji, M., and M. Kikuchi, 1969, J. Chem. Phys. **51**, 3863.

Atoji, M., and R.C. Medrud, 1959, J. Chem. Phys. **31**, 332.

Atoji, M., and Y. Tsunoda, 1971, J. Chem. Phys. **54**, 3510.

Atoji, M., and D.E. Williams, 1961, J. Chem. Phys. **35**, 1960.

Atoji, M., K.A. Gschneidner Jr, A.H. Daane, R.E. Rundle and F.H. Spedding, 1958, J. Am. Chem. Soc. **80**, 1804.

Atoji, M., I. Atoji, C. Do-Dink and W.E. Wallace, 1973, J. Appl. Phys. **44**, 5096.

Auer-Welsbach, H., and H. Nowotny, 1961, Monatsh. Chem. **92**, 198.

Avery, D.F., J. Cuthbert and C. Silk, 1967, Brit. J. Appl. Phys. **18**, 1133.

Bacchella, G.L., P. Meriel, M. Pinot and R. Lallement, 1966, Bull. Soc. Fr. Mineral. & Cristallogr. **89**, 226.

Baker, F.G., E.J. Huber Jr, C.E. Holley Jr and N.H. Krikorian, 1971, J. Chem. Thermodyn. **3**, 77.

Balducci, G., A. Capalbi, G. de Maria and M. Guido, 1965, J. Chem. Phys. **43**, 2136.

Balducci, G., A. Capaldi, G. de Maria and M. Guido, 1967, AFML-TR-67-335, October (National Technical Information Service, Springfield, VA 22151).

Balducci, G., A. Capaldi, G. de Maria and M. Guido, 1968, J. Chem. Phys. **48**, 5275.

Balducci, G., A. Capaldi, G. de Maria and M. Guido, 1969a, J. Chem. Phys. **50**, 1969.

Balducci, G., A. Capaldi, G. de Maria and M. Guido, 1969b, J. Chem. Phys. **51**, 2871.

Balducci, G., A. Capaldi, G. de Maria and M. Guido, 1969c, in: Proc. 1st Int. Conf. on Calorimetry and Thermodynamics, ed. H.V. Kehiahan (Polish Scientific Publishers, Warsaw) p. 415.

Balducci, G., G. de Maria and M. Guido, 1969d, J. Chem. Phys. **51**, 2876.

Balducci, G., G. de Maria and M. Guido, 1970, Thermodynamics of Rare Earth Carbide Systems, in; Proc. 1st Int. Conf. on Calorimetry and Thermodynamics (Polish Scientific Publishers, Warsaw, 1970).

Balducci, G., G. de Maria and M. Guido, 1972, J. Chem. Phys. **56**, 1431.

Barbara, B., 1973, J. Phys. (Paris) **34**, 1039.

Bärnighausen, H., and G. Schiller, 1985, J. Less-Common Met. **110**, 385.

Bauer, J., 1974, J. Less-Common Met. **37**, 161.

Bauer, J., and D. Ansel, 1985, J. Less-Common Met. **109**, L9.

Bauer, J., and O. Bars, 1980, Acta Crystallogr. B **36**, 1540.

Bauer, J., and O. Bars, 1982, J. Less-Common Met. **83**, 17.

Bauer, J., and O. Bars, 1983, J. Less-Common Met. **95**, 267.

Bauer, J., and H. Bienvenu, 1980, C.R. Acad. Sci. Ser. C **290**, 387.

Bauer, J., and J. Debuigne, 1972, C.R.Acad. Sci. Ser. C **274**, 1271.

Bauer, J., and J. Debuigne, 1975, J. Inorg. & Nucl. Chem. **37**, 2473.

Bauer, J., and H. Nowotny, 1971, Monatsh. Chem. **102**, 1129.

Bauer, J., and C. Politis, 1982, J. Less-Common Met. **88**, L1.

Bauer, J., P. Venneques and J.L. Vergneau, 1985, J. Less-Common Met. **110**, 295.

Behrens, R.K., and W. Jeitschko, 1984, Acta Crystallogr. A **40** (Supplement), C-238.

Bezruk, E.T., and L.Ya. Markovskii, 1972, Zh. Prikl. Khim. **45**, 3.

Binder, I., 1956, Powder Metall. Bull. **7**, 74.

Binder, I., 1960, J. Am. Ceram. Soc. **43**, 287.

Block, G., and W. Jeitschko, 1986a, Inorg. Chem. **25**, 279.

Block, G., and W. Jeitschko, 1986b, Z. Kristallogr. **174**, 19.

Block, G., and W. Jeitschko, 1987a, J. Solid State Chem. **70**, 271.

Block, G., and W. Jeitschko, 1987b, Z. Kristallogr. **178**, 25.

Bodak, O.I., and E.P. Marusin, 1979, Dokl. Akad. Nauk Ukr. SSR, Ser. A **12**, 1048.

Bodak, O.I., E.P. Marusin and V.A. Bruskov, 1980, Kristallografiya **25**, 617.

Bodak, O.I., E.P. Marusin and V.S. Fundamenskii, 1982, Kristallografiya **27**(6), 1098.

Bolzoni, F., F. Leccabue, L. Pareti and J.L. Sanchez, 1985, J. Phys. (Paris) **46**, 305.

Börner, G., T. Gorgenyi and P. Venet, 1972, Prakt. Metallogr. **9**, 431.

Bowman, A.L., and N.H. Krikorian, 1976, in; Treatise on Solid State Chemistry, Vol. 3, eds N.B. Hanny (Plenum, New York) p. 253.

Bowman, A.L., T.C. Wallace, J.L. Yarnell, R.G. Wenzel and E.K. Storm, 1965, Acta Crystallogr. **19**, 6.

Bowman, A.L., G.P. Arnold, T.C. Wallace and N.G. Nereson, 1966, Acta Crystallogr. **21**, 670.

Bowman, A.L., N.H. Krikorian, G.P. Arnold, T.C. Wallace and N.G. Nereson, 1967, in: High-temperature Neutron Diffraction Study of LaC_2 and YC_2 (U.S. Atomic Energy Commission) LA-DC-8451, CONF-670502-1, CESTI-7.

Bowman, A.L., N.H. Krikorian, G.P. Arnold, T.C. Wallace and N.G. Nereson, 1968, Acta Crystallogr. B **24**, 459.

Breant, T., D. Pensec, J. Bauer and J. Debuigne, 1978, C.R. Acad. Sci. Ser. C **287**, 261.

Bredig, M.A., 1960, J. Am. Ceram. Soc. **43**, 493.

Brewer, L., and O. Krikorian, 1956, J. Electrochem. Soc. **103**, 38.

Brožek, V., M. Popl and B. Hájek, 1970, Collect. Czech. Chem. Commun. **35**, 1832.

Brožek, V., P. Karen and B. Hájek, 1985, J. Less-Common Met. **107**, 295.

Bullet, D.W., 1985, Inorg. Chem. **24**, 3319.

Buschow, K.H.J., 1987, in: Proc. 9th Int. Workshop on Rare Earth Magnets, Bad Soden, August, eds C. Herget and R. Porschke (Deutsche Physikalische Gesellschaft, Bad Honnef, FRG) p. 459.

Buschow, K.H.J., 1988, J. Less-Common Met. **144**, 65.

Buschow, K.H.J., and R. Grössinger, 1987, J. Less-Common Met. **135**, 39.

Buschow, K.H.J., D.B. de Mooij, J.L.C. Daams and H.M. van Noort, 1986, J. Less-Common Met. **115**, 357.

Buschow, K.H.J., D.B. de Mooij and C.J.M. Denissen, 1988a, J. Less-Common Met. **143**, L15.

Buschow, K.H.J., D.B. de Mooij and C.J.M. Denissen, 1988b, J. Less-Common Met. **142**, L13.

Butherus, A.D., and H.A. Eick, 1968, J. Am. Chem. Soc. **90**, 1715.

Butherus, A.D., and H.A. Eick, 1973, J. Inorg. & Nucl. Chem. **35**, 1925.

Butherus, A.D., R.B. Leonard, G.L. Buchel and H.A. Eick, 1966, Inorg. Chem. **5**, 1567.

Button, T.W., and I.J. McColm, 1984, J. Less-Common Met. **97**, 237.

Cadieu, F.J., J.J. Cuomo and R.J. Gambino, 1980, in: The Rare Earths in Modern Science and Technology, Vol. 2, eds G.J. McCarthy, J.J. Rhyne and H.B. Silber (Plenum, New York) p. 173.

Carlson, O.N., and W.M. Paulson, 1968, Trans. AIME **242**, 846.

Carlson, O.N., R.R. Lichtenberg and J.C. Warner, 1974, J. Less-Common Met. **35**, 275.

Carter, F.L., T.L. Francavilla and R.A. Hein, 1975, in: Proc. Eleventh Rare Earth Research Conference, Vol. 1, eds J.M. Haschke and H.A. Eick, Traverse City, MI, p. 36.

Chang, R., 1961, Acta Crystallogr. **14**, 1097.

Child, H.R., M.K. Wilkinson, J.W. Cable, W.C. Koehler and E.O. Wollan, 1963, Phys. Rev. **131**, 922.

Chupka, W.A., J. Berkowitz, C.F. Giese and M.G. Inghram, 1958, J. Phys. Chem. **62**, 611.

Clark, N.J., and I.J. McColm, 1972, J. Inorg. & Nucl. Chem. **34**, 117.

Coey, J.M.D., 1986, J. Less-Common Met. **126**, 21.

Coey, J.M.D., O. Massenet, M. Kasaya and J. Etourneau, 1979, J. Phys. (Paris) **40**, 333.

Colquhoun, I., N.H. Greenwood, I.J. McColm and G.E. Turner, 1972, J. Chem. Soc. Dalton Trans., 1337.

Colquhoun, I., I.J. McColm and W. Batey, 1975, J. Inorg. & Nucl. Chem. **37**, 1705.

Cooper, A.S., E. Corenzwit, L.D. Longinotti, B.T. Matthias and W.H. Zachariasen, 1970, Proc. Nat. Acad. Sci. **67**, 313.

Cort, B., G.R. Stewart and A.L. Giorgi, 1984, J. Low. Temp. Phys. **54**, 149.

Cromer, D.T., A.C. Larson and R.B. Roof Jr, 1984, Acta Crystallogr. **17**, 272.

Cuthbert, J., R.L. Faircloth, R.H. Flowers and F.C.W. Pummery, 1967, Proc. Brit. Ceram. Soc. **8**, 155.

Dancy, E., L.E. Everett and C.L. McCabe, 1962, Trans AIME **224**, 1095.

Darnell, A.J., 1977, in: Proc. Thirteenth Rare Earth Research Conference, eds G.J. McCarthy and J.J. Rhyne (Plenum, New York) p. 297.

de Boer, F.R., Huang Ying-kai, Zhang Zhi-dong, D.B. de Mooij and K.H.J. Buschow, 1988a, J. Magn. & Magn. Mater. **72**, 167.

de Boer, F.R., R. Verhoef, Z.-D. Zhang, D.B. de Mooij and K.H.J. Buschow, 1988b, J. Magn. & Magn. Mater. **73**, 263.

de Gennes, P.G., 1962, J. Phys. **23**, 510.

de Maria, G., and G. Balducci, 1972, Int. Rev. Sci. Phys. Chem. Ser. I, **10**, 209.

de Maria, G., M. Guido, L. Malaspina and B. Pesce, 1965, J. Chem. Phys. **43**, 4449.

de Maria, G., G. Balducci, A. Capalbi and M. Guido, 1967, Proc. Brit. Ceram. Soc. **8**, 127.

de Mooij, D.B., and K.H.J. Buschow, 1988, J. Less-Common Met. **142**, 349.

De Novion, C., J.P. Krebs and P. Meriel, 1966, C.R. Acad. Sci. Ser. B **263**, 457.

Denissen, C.J.M., D.B. de Mooij and K.H.J. Buschow, 1988a, J. Less-Common Met. **139**, 291.

Denissen, C.J.M., D.B. de Mooij and K.H.J. Buschow, 1988b, J. Less-Common Met. **142**, 195.

Deppe, P., M. Rosenberg and H.H. Stadelmaier, 1988, J. Less-Common Met. **143**, 77.

Drowart, J., and P. Goldfinger, 1967, Angew. Chem. Int. Ed. Engl. **6**, 581.

Dudis, D.S., and J.D. Corbett, 1987, Inorg. Chem. **26**, 1933.

Dudis, D.S., J.D. Corbett and S.-J. Hwu, 1986, Inorg. Chem. **25**, 3434.

El-Makrini, M., D. Guérard, P. Lagrange and A. Hérold, 1980, Physica B **99**, 481.

El-Masry, N.A., H.H. Stadelmaier, C.J. Shanwan and L.T. Jordan, 1983, Z. Metallk. **74**, 33.

Elliott, R.J., ed., 1972, Magnetic Properties of Rare Earth Metals (Plenum, New York).

Erdmann, K., M. Rosenberg and K.H.J. Buschow, 1988, J. Appl. Phys. **63**, 4116.

Erdmann, K., T. Sinnemann, M. Rosenberg and K.H.J. Buschow, 1989, J. Less-Common Met. **146**, 59.

Faircloth, R.L., R.H. Flowers and F.C.W. Pummery, 1968, J. Inorg. & Nucl. Phys. **30**, 499.

Filby, E.E., and L.L. Ames, 1971a, J. Phys. Chem. **75**, 848.

Filby, E.E., and L.L. Ames, 1971b, High Temp. Sci. **3**, 41.

Filby, E.E., and L.L. Ames, 1972, Inorg. Nucl. Chem. Lett. **8**, 855.

Fishel, N.A., and H.A. Eick, 1969, J. Inorg. & Nucl. Chem. **31**, 891.

Flahaut, J., 1969, in: Les éléments des Terres Rares (Massonet Cie Ed., Paris).

Florio, J.V., N.C. Baenziger and R.E. Rundle, 1956, Acta Crystallogr. **9**, 367.

Francavilla, T.L., and F.L. Carter, 1976, Phys. Rev. **14**, 128.

Francavilla, T.L., F.L. Carter and A.W. Webb, 1976, Phys. Lett. A **59**, 388.

Gebelt, R.E., and H.A. Eick, 1964, Inorg. Chem. **3**, 335.

Gebelt, R.E., and H.A. Eick, 1966a, in: Thermodynamics, Proc. Symp. Thermodynamics with Emphasis on Nuclear Materials and Atomic Transport in Solids, Vol. 1, Vienna, 22–27 July, 1965 (International Atomic Energy Agency, Vienna) p. 291.

Gebelt, R.E., and H.A. Eick, 1966b, J. Chem. Phys. **44**, 2872.

Gerss, M.H., and W. Jeitschko, 1986, Z. Naturforsch. B **41**, 946.

Gerss, M.H., W. Jeitschko, L. Boonk, J. Nientiedt, J. Grobe, E. Mörsen and A. Leson, 1987, J. Solid State Chem. **70**, 19.

Gignoux, D., B. Hennion and A.N. Saada, 1982, in: Crystalline Electric Field Effects in f-electron Magnetism, eds R.P. Guertin, W. Suski and Z. Zolnierek (Plenum, New York) p. 485.

Gingerich, K.A., 1969, J. Chem. Phys. **50**, 2255.

Gingerich, K.A., 1974, J. Chem. Phys. **60**, 3703.

Gingerich, K.A., 1980, in: Current Topic in Materials Science, Vol. 6, ed. E. Kaldis (North-Holland, Amsterdam) p. 345.

Gingerich, K.A., 1985, J. Less-Common Met. **110**, 41.

Gingerich, K.A., and D.L. Cocke, 1979, Inorg. Chim. Acta **33**, L107.

Gingerich, K.A., and R.Haque, 1980, J. Chem. Soc., Faraday Trans. II, **76**, 101.

Gingerich, K.A., D.L. Cocke and J.E. Kingcade, 1976a, Inorg. Chim. Acta **17**, L1.

Gingerich, K.A., D.L. Cocke and J.E. Kingcade, 1976b, in: Annu. Conf. Mass Spectrometry and Allied Topics, San Diego, CA, May 9–14, 1976.

Gingerich, K.A., B.M. Nappi, R. Haque and M.

Pelino, 1981a, in: Proc. 29th Annu. Conference Mass Spectrometry and Allied Topics (Am. Soc. for Mass Spectrometry, Minneapolis, MN) p. 89.

Gingerich, K.A., M. Pelino and R. Haque, 1981b, High Temp. Sci. **14**, 137.

Gingerich, K.A., R. Haque and M. Pelino, 1982, J. Chem. Soc., Faraday Trans. I, **78**, 341.

Giorgi, A.L., E.G. Szklarz, M.C. Krupka, T.C. Wallace and N.H. Krikorian, 1968, J. Less-Common Met. **14**, 247.

Giorgi, A.L., E.G. Szklarz, M.C. Krupka and N.H. Krikorian, 1969, J. Less-Common Met. **17**, 121.

Giorgi, A.L., E.G. Szklarz, N.H. Krikorian and M.C. Krupka, 1970, J. Less-Common Met. **22**, 131.

Green, R.W., E.O. Thorland, J. Croat and S. Legvold, 1969, J. Appl. Phys. **40**, 3161.

Gschneidner Jr, K.A., 1961, Rare Earth Alloys (van Nostrand, Princeton, NJ).

Gschneidner Jr, K.A., and F.W. Calderwood, 1986, Bull. Alloy Phase Diagrams **7**, 564.

Gueramian, M., A. Bezinge, K. Yvon and J. Muller, 1987, Solid State Commun. **64**, 639.

Guérard, D., and A. Hérold, 1975, C.R. Acad. Sci. Ser. C **281**, 929.

Guido, M., and G. Gigli, 1973, J. Chem. Phys. **59**, 3437.

Guido, M., G. Balducci and G. de Maria, 1972, J. Chem. Phys. **57**, 1475.

Hackstein, K., H. Nickel and P. Venet, 1971, in: Production, Characterization and Hydrolysis of Samarium Carbide, EUR-4697, Nuklear-Chemie und Metallurgie, GmbH, Wolfgang bei Hanau, FRG, p. 78.

Haines, H.K., and P.E. Potter, 1970, Rep. AERE R 6513.

Hájek, B., V. Brožek and M. Popl, 1971, Collect. Czech. Chem. Commun. **36**, 1537.

Hájek, B., P. Karen and V. Brožek, 1984a, Collect. Czech. Chem. Commun. **49**, 936.

Hájek, B., P. Karen and V. Brožek, 1984b, Collect. Czech. Chem. Commun. **49**, 944.

Hájek, B., P. Karen and V. Brožek, 1984c, J. Less-Common Met. **96**, 35.

Hájek, B., P. Karen and V. Brožek, 1984d, J. Less-Common Met. **98**, 245.

Hájek, B., P. Karen and V. Brožek, 1986, Monatsh. Chem. **117**, 1271.

Halliday, A.K., G. Hughes and S.M. Walker, 1973, in: Comprehensive Inorganic Chemistry, Vol. 1, eds J.C. Bailor Jr, H.J. Emeléus, S.R. Nyholm and A.F.T. Dickenson (Pergamon, Oxford) p. 591.

Haque, R., and K.A. Gingerich, 1979, J. Chem. Soc., Faraday Trans. I, **75**, 985.

Haque, R., and K.A. Gingerich, 1981, J. Chem. Phys. **74**, 6407.

Harris, T.R., and M. Norman, 1967, J. Less-Common Met. **13**, 629.

Haschke, H., H. Nowotny and F. Benesovsky, 1966a, Monatsh. Chem. **97**, 716.

Haschke, H., H. Nowotny and F. Benesovsky, 1966b, Monatsh. Chem. **97**, 1045.

Haschke, J.M., 1975, Inorg. Chem. **14**, 779.

Haschke, J.M., and T.A. Deline, 1980, Inorg. Chem. **19**, 527.

Haschke, J.M., and T.A. Deline, 1982, J. Chem. Thermodyn. **14**, 1019.

Haschke, J.M., and H.A. Eick, 1968, J. Phys. Chem. **72**, 1697.

Haschke, J.M., and H.A. Eick, 1970a, J. Am. Chem. Soc. **92**, 1526.

Haschke, J.M., and H.A. Eick, 1970b, Inorg. Chem. **9**, 851.

Helmholdt, R.B., and K.H.J. Buschow, 1988, J. Less-Common Met. **144**, L33.

Herbst, J.F., J.J. Croat, F.E. Pinkerton and W.B. Yelon, 1984, Phys. Rev. B **29**, 4176.

Hoenig, C.L., N.D. Stout and P.C. Nordine, 1967, J. Am. Ceram. Soc. **50**, 385.

Hoffmann, R.-D., and W. Jeitschko, 1987, Z. Kristallogr. **178**, 110.

Hoffmann, R.-D., and W. Jeitschko, 1988, Z. Kristallogr. **182**, 137.

Hoffmann, R.-D., W. Jeitschko, M. Reehuis and S. Lee, 1989, Inorg. Chem. **28**, 934.

Holden, A.J., V. Heine and J.H. Samson, 1984, J. Phys. F **14**, 1005.

Holleck, H., 1972, J. Nucl. Mater. **42**, 278.

Holleck, H., 1975, Ber. Bunsenges. Phys. Chem. **79**, 975.

Holleck, H., 1977, J. Less-Common Met. **52**, 167.

Huber Jr, E.J., C.E. Holley Jr and N.H. Krikorian, 1973, in: The Enthalpies of Formation of Some Gadolinium Carbides, LA-UR-73-694 (Los Alamos Scientific Laboratory, University of California, Los Alamos, NM 87544).

Hultgren, R., P.D. Desai, D.T. Hawkins, M. Gleiser, K.K. Kelley and D.D. Wagman, 1973, in: Selected Values of the Thermodynamic Properties of the Elements (American Society for Metals, Metal Park, OH).

Hwu, S.-J., R.P. Ziebarth, S. von Winbush, J.E. Ford and J.D. Corbett, 1986, Inorg. Chem. **25**, 283.

Ivashchenko, V.I., A.A. Lisenko and E.A. Zhurakovski, 1984, Phys. Status Solidi b **121**, 583.

Jackson, D.D., R.G. Bedford and G.W. Barton Jr, 1963, UCRL-7362-T (Lawrence Livermore Radiation Laboratory, University of California, CA).

Jacobs, T.H., C.J.M. Denissen and K.H.J. Buschow, 1989, J. Less-Common Met. **153**, L5.

Jedlicka, H., H. Nowotny and F. Benesovsky, 1971, Monatsh. Chem. **102**, 389.

Jeitschko, W., and R.K. Behrens, 1986, Z. Metallk. **77**, 788.

Jeitschko, W., and G. Block, 1985, Z. Anorg. & Allg. Chem. **528**, 61.

Jeitschko, W., and M.H. Gerss, 1986, J. Less-Common Met. **116**, 147.

Jeitschko, W., H. Nowotny and F. Benesovsky, 1963, Monatsh. Chem. **94**, 565.

Jeitschko, W., H. Nowotny and F. Benesovsky, 1964, Monatsh. Chem. **95**, 1040.

Johnson, R.W., and A.H. Daane, 1961, J. Phys. Chem. **65**, 909.

Jones, D.W., I.J. McColm, R. Steadman and J. Yerkess, 1984, J. Solid State Chem. **53**, 367.

Jones, D.W., I.J. McColm, R. Steadman and J. Yerkess, 1986, J. Solid State Chem. **62**, 172.

Karen, P., and B. Hájek, 1986, Collect. Czech. Chem. Commun. **51**, 1411.

Karen, P., B. Hájek and V. Valvoda, 1986, J. Less-Common Met. **120**, 337.

Kasaya, M., J.M. Tarascon, J. Etourneau and P. Hagenmüller, 1978, Mater. Res. Bull. **13**, 751.

Kasaya, M., J.M. Tarascon, J. Etourneau, P. Hagenmüller and J.M.D. Coey,. 1979, J. Phys. (Paris) **40**, C5-393.

Kingcade, J.E., D.L. Cocke and K.A. Gingerich, 1976, Inorg. Chem. Acta **17**, L1

Kingcade, J.E., D.L. Cocke and K.A. Gingerich, 1983, High Temp. Sci. **16**, 89.

Kingcade, J.E., D.L. Cocke and K.A. Gingerich, 1984, Inorg. Chem. **23**, 1334.

Kohl, F.J., and C.A. Stearns, 1970, J. Chem. Phys. **52**, 6310.

Kohl, F.J., and C.A. Stearns, 1971a, NASA-TN-D-7039 (National Aeronautic and Space Administration, Washington, DC 20546).

Kohl, F.J., and C.A. Stearns, 1971b, J. Chem. Phys. **54**, 1414.

Kosolapova, T.Ya., and G.N. Makarenko, 1964, Ukr. Khim. Zh. **30**, 784.

Kosolapova, T.Ya., G.N. Makarenko and L.T. Domasevich, 1971, Zh. Prikl. Khim. (Leningrad) **44**, 953.

Kotroczo, V., I.J. McColm and N.J. Clark, 1987, J. Less-Common Met. **132**, 1.

Kotsanidis, P., J.K. Yakinthos and E. Gamari-Seale, 1989, J. Less-Common Met. **152**, 287.

Krikorian, N.H., 1971, J. Less-Common Met. **23**, 271.

Krikorian, N.H., T.C. Wallace and M.G. Bowman, 1967, Phase relationships of the high-carbon portion of lanthanide carbon system, in: Proprieties Thermodynamiques Physiques et Structurales des Dérivés Semi-Metalliques, Orsay, 28 Sept.–1 Oct. 1965 (Centre National Recherche Scientifique, Paris) p. 489.

Krikorian, N.H., A.L. Bowman, M.C. Krupka and G.P. Arnold, 1969a, High Temp. Sci. **1**, 360.

Krikorian, N.H., A.L. Giorgi, E.G. Szklarz, M.C. Krupka and B.T. Matthias, 1969b, J. Less-Common Met. **19**, 253.

Krupka, M.C., 1970, J. Less-Common Met. **20**, 135.

Krupka, M.C., and N.H. Krikorian, 1970, High-pressure synthesis of new heavy rare earth carbides, in: Proc. Eight Rare Earth Research Conference, Vol. 2, eds T.A. Henrie and R. E. Lindstrom (National Technical Information Service, Springfield, VA) p. 382.

Krupka, M.C., N.H. Krikorian and T.C. Wallace, 1968, Polymorphism in heavy rare-earth dicarbides at high pressures and temperatures, in: Proc. Seventh Rare-Earth Research Conference, 28–30 Oct., Coronado, CA, p. 197.

Krupka, M.C., A.L. Giorgi, N.H. Krikorian and E.G. Szklarz, 1969a, J. Less-Common Met. **17**, 91.

Krupka, M.C., A.L. Giorgi, N.H. Krikorian and E.G. Szklarz, 1969b, J. Less-Common Met. **19**, 113.

Krupka, M.C., A.L. Giorgi and E.G. Szklarz, 1973, J. Less-Common Met. **30**, 217.

Lallement, R., 1966, Centre d'Etudes Nucléaires de Fonteney-aux-Roses Rapport, CEA-R 3043.

Lallement, R., 1968, Proc. Brit. Ceram. Soc. **10**, 115.

Lallement, R., and J.J. Veyssie, 1968, in: Progress in the Science and Technology of the Rare Earths, Vol. 3, ed. L. Eyring (Pergamon, Oxford) pp. 284–342.

Langer, S., 1963, General Atomics Report, GA-4450 Sept. 5, 1963.

Laplace, A., and N. Lorenzelli, 1970, in: Les Elements des Terres Rares, Vol. 1, ed. M.F. Trombe (Centre National de la Recherche Scientifique, Paris) p. 385.

Le Roy, L., J.M. Moreau, C. Bertrand and M.A. Fremy, 1987, J. Less-Common Met. **136**, 19.

Lee, S., W. Jeitschko and R.-D. Hoffmann, 1989, Inorg. Chem. **28**, 4094.

Li, D., H.F. Mildrum and K.J. Strnat, 1985, J. Appl. Phys. **57**, 4140.

Li, J., and R. Hoffman, 1987, J. Am. Chem. Soc. **109**, 6600.

Li, J., and R. Hoffman, 1989, Chem. Mater. **1**, 83.

Liu, N.C., and H.H. Stadelmaier, 1986, Mater. Lett. **4**, 377.

Liu, N.C., H.H. Stadelmaier and G. Schneider, 1987, J. Appl. Phys. **61**, 3574.

Loe, I.R., I.J. McColm and T.A. Quigley, 1976, J. Less-Common Met. **46**, 217.

Lorenzelli, N., and J.P. Marcon, 1972, J. Less-Common Met. **26**, 71.

Lorenzelli, N., J. Melamed and J.P. Marcon, 1970, Colloq. Int. CNRS **180**(1), 375.

Makarenko, G.N., L.T. Pustovoit, V.L. Yopko and B.M. Rud, 1965, Izv. Akad. Nauk SSSR Neorg. Mater. **1**, 1787.

Makarov, E.S., and S.I. Vinogradov, 1956, Sov. Phys. Crystallogr. **1**, 499.

Markovskii, L.Ya., N.V. Vekshina and G. Pron, 1965, Zh. Prikl. Khim. (Leningrad) **38**, 245.

Markovskii, L.Ya., N.V. Vekshina and Yu.D. Kondrashev, 1972, Zh. Prikl. Khim. (Leningrad) **45**, 1183.

Markovskii, L.Ya., N.V. Vekshina and G.F. Pron, 1986, Zh. Prikl. Khim. (Moscow) **38**, 245.

Marusin, E.P., O.I. Bodak, A.O. Tsokol and M.G. Baivel'men, 1985a, Kristallografiya **30**(3), 584.

Marusin, E.P., O.I. Bodak, A.O. Tsokol and V.S. Fundamenskii, 1985b, Kristallografiya **30**(3), 581.

Mayer, I., and I. Shidlovsky, 1968, J. Appl. Crystallogr. **1**, 194.

Mayer, I., and I. Shidlovsky, 1969, Inorg. Chem. **8**, 1240.

McColm, I.J., 1981, J. Less-Common Met. **78**, 287.

McColm, I.J., N.J. Clark and B. Mortimer, 1971, J. Inorg. & Nucl. Chem. **33**, 49.

McColm, I.J., I. Colquhoun and N.J. Clark, 1972, J. Inorg. & Nucl. Chem. **34**, 3809.

McColm, I.J., T.A. Quigley and N.J. Clark, 1973, J. Inorg. & Nucl. Chem. **35**, 1931.

McColm, I.J., I. Colquhoun, A. Vendl and G. Dufek, 1977, J. Inorg. & Nucl. Chem. **39**, 1981.

McColm, I.J., V. Kotroczo, T.W. Button, N.J. Clark and B. Bruer, 1986, J. Less-Common Met. **115**, 113.

Meyer, G., and T. Schleid, 1987, Inorg. Chem. **26**, 217.

Miedema, A.R., P.F. de Châtel and F.R. de Boer, 1980, Physica B **100**, 1.

Miller, G.J., J.K. Burdett, C. Schwarz and A. Simon, 1986, Inorg. Chem. **25**, 4437.

Moissan, H., 1896a, C. R. Acad. Sci. **122**, 573.

Moissan, H., 1896b, Bull. Soc. Chim. Fr. **15**, 1293.

Moissan, H., 1896c, C. R. Acad. Sci. **122**, 357.

Moissan, H., 1900a, C. R. Acad. Sci. **131**, 595.

Moissan, H., 1900b, C. R. Acad. Sci. **131**, 924.

Nasonov, A.V., 1981, in: Phase Equilibriums in the Al–R–C system, Khim. Fak. Mosk. Gos. Univ. Deposited Doc. VINITI 575-82, 602-5 (Avail, VINITI).

Nasonov, A.V., L.S. Guzei, L.M. Azieva and E.M. Sokolovskaya, 1988, Vestn. Mosk. Univ. Ser. II, Khim. **29**(5), 532.

Niessen, A.K., and F.R. de Boer, 1981, J. Less-Common Met. **82**, 75.

Nordine, P.C., G.S. Smith and Q. Johnson, 1964, Report UCRL-12205 (Lawrence Livermore Radiation Laboratory, University of California, CA).

Novion, C., J.P. Krebs and P. Meriel, 1966, C. R. Acad. Sci. Ser. B **263**, 457.

Novokshonov, V.I., 1980, Zh. Neorg. Khim. **25**, 684.

Nowotny, H., 1968, Proc. Seventh Rare Earth Research Conf., p. 309.

Nowotny, H., and H. Auer-Welsbach, 1961, Monatsh. Chem. **92**, 789.

Onodera, H., A. Fujita, H. Yamamoto, M. Sagawa and S. Hirosawa, 1987a, J. Magn. & Magn. Mater. **68**, 6.

Onodera, H., A. Fujita, H. Yamamoto, M. Sagawa and S. Hirosawa, 1987b, J. Magn. & Magn. Mater. **68**, 15.

Paccard, L., and D. Paccard, 1988, J. Less-Common Met. **136**, 297.

Paccard, L., D. Paccard and C. Bertrand, 1987, J. Less-Common Met. **135**, L5.

Paderno, Yu.B., V.L. Yupko, B.M. Rud and G.N. Makarenko, 1966, Izv. Akad. Nauk SSSR Neorg. Mater. **2**, 626.

Paderno, Yu.B., V.L. Yupko and G.N. Makarenko, 1969a, Izv. Akad. Nauk SSSR Neorg. Mater. **5**, 889.

Paderno, Yu.B., V.L. Yupko and G.N. Makarenko, 1969b, Izv. Akad. Nauk SSSR Neorg. Mater. **5**, 386.

Palmer, P.E., H.R. Burkholder, B.J. Beaudry and K.A. Gschneidner Jr, 1982, J. Less-Common Met. **87**, 135.

Panek, P., and R. Hoppe, 1972, Z. Anorg. & Allg. Chem. **393**, 13.

Park, H.Y., H.H. Stadelmaier and L.T. Jordan, 1982, Z. Metallk. **73**, 399.

Parnell, D.G., N.H. Brett and P.E. Potter, 1985, J. Less-Common Met. **114**, 161.

Parthé, E., 1960, Acta Crystallogr. **13**, 868.

Pasto, A.E., and C.S. Morgan, 1976, ORNL/TM-5244, Contract No. VV-7405-eng-26.

Pauling, L., 1960, The Nature of the Chemical Bond, 3rd Ed. (Cornell University, Ithaca, New York).

Pedziwaitr, A.T., W.E. Wallace and E. Burzo, 1986, J. Magn. & Magn. Mater. **59**, L179.

Pelino, M., and K.A. Gingerich, 1989, J. Phys. Chem. **93**(4), 1581.

Pelino, M., K.A. Gingerich, B. Nappi and R. Haque, 1984, J. Chem. Phys. **80**(9), 4478.

Pelino, M., K.A. Gingerich, R. Haque and J.E. Kingcade, 1985, J. Phys. Chem. **89**, 4257.

Pelino, M., K.A. Gingerich, R. Haque and L. Bencivenni, 1986, J. Phys. Chem. **90**, 4358.

Pelino, M., K.A. Gingerich, R. Haque and L. Bencivenni, 1988a, High Temp. & High Pressures **20**(4), 413.

Pelino, M., R. Haque, L. Bencivenni and K.A. Gingerich, 1988b, J. Chem. Phys. **88**, 6534.

Peterson, D.T., and J. Rexer, 1962, J. Inorg. & Nucl. Chem. **24**, 519.

Petterson, O., 1895, Chem. Ber. **28**, 2419.

Pialoux, A., 1988, J. Less-Common Met. **143**, 219.

Post, B., D. Moskowitz and F.W. Glaser, 1956, J. Am. Chem. Soc. **78**, 1800.

Putyatin, A.A., 1987, Vestn. Mosk. Univ. Ser. II, Khim. **28**(3), 294.

Putyatin, A.A., and L.G. Sevastyanova, 1987, Vestn. Mosk. Univ. Ser. II, Khim. **28**, 199.

Rassaerts, H., H. Nowotny, G. Vinck and F. Benesovsky, 1967, Monatsh. Chem. **98**, 460.

Rexer, J., and D.T. Peterson, 1964, in: Compounds of Interest in Nuclear Reactor Technology, U.S. Atomic Energy Commission Report IS-875, AIME Michigan (Iowa State University, Ames, Iowa) Apr.

Richards, S.M., and J.S. Kasper, 1969, Acta Crystallogr. B **25**, 1237.

Rogl, P., 1978, J. Nucl. Mater. **73**, 198.

Rogl, P., 1979, J. Nucl. Mater. **79**, 154.

Rosen, S., and P.G. Sprang, 1966, Adv. X-ray Anal. **9**, 131.

Ross, M., H.E. Dewitt and W.B. Hubbard, 1981, Phys. Rev. A **24**, 1016.

Sagawa, M., S. Hirosawa, H. Yamamoto, S. Fujimura and Y. Matsuura, 1987, Jpn. J. Appl. Phys. **26**, 785.

Sakai, T., G. Adachi and J. Shiokawa, 1979a, J. Appl. Phys. **50**, 3592.

Sakai, T., G. Adachi and J. Shiokawa, 1979b, Mater. Res. Bull. **14**, 791.

Sakai, T., G. Adachi and J. Shiokawa, 1980, Mater. Res. Bull. **15**, 1001.

Sakai, T., G. Adachi, T. Yoshida and J. Shiokawa, 1981a, J. Chem. Phys. **75**, 3027.

Sakai, T., G. Adachi, T. Yoshida and J. Shiokawa, 1981b, J. Less-Common Met. **81**, 91.

Sakai, T., G. Adachi and J. Shiokawa, 1981c, Solid State Commun. **40**, 445.

Sakai, T., G. Adachi and J. Shiokawa, 1982a, J. Less-Common Met. **84**, 107.

Sakai, T., G. Adachi, T. Yoshida, S. Ueno and J. Shiokawa, 1982b, Bull. Chem. Soc. Jpn. **55**, 699.

Samsonov, G.V., G.N. Makarenko and T.Ya. Kosolapova, 1961, Zh. Prikl. Khim. (Leningrad) **34**, 1444.

Samsonov, G.V., M.M. Antonova and V.V. Morozov, 1970, Poroshk. Metall. **4**, 66.

Sanderson, M.J., and N.C. Baenziger, 1953, Acta Crystallogr. **6**, 627.

Satpathy, S., and D.K. Anderson, 1985, Inorg. Chem. **24**, 2604.

Schwanitz-Schüller, U., and A. Simon, 1985, Z. Naturforsch. B **40**, 710.

Schwarz, K., P. Weinberger and A. Neckel, 1969, Theor. Chim. Acta **15**, 149.

Schwetz, K.A., M. Hoerle and J. Bauer, 1979, Ceramurqia Int. **5**, 105.

Seiver, R.L., and H.A. Eick, 1971, High Temp. Sci. **3**, 292.

Seiver, R.L., and H.A. Eick, 1976, J. Less-Common Met. **44**, 1.

Semonenko, K.N., A.A. Putyatin, I.V. Nikol'skaya and V.V. Burnasheva, 1983, Russ. J. Inorg. Chem. **28**, 943.

Sheng, Luo, Lui Zhinhui, Zhang Guowei, Pei Xiedi and Jiang Wei, 1987, IEEE Trans. Magn. **M23**, 3095.

Simon, A., and E. Warkentin, 1983, Z. Anorg. & Allg. Chem. **497**, 79.

Simon, A., E. Warkentin and R. Masse, 1981, Angew. Chem. Int. Ed. Engl. **20**, 1013.

Simon, A., C. Schwarz and W. Bauhofer, 1988, J. Less-Common Met. **137**, 343.

Sinnema, S., R.J. Radwanski, J.J.M. Franse, D.B. de Mooij and K.H.J. Buschow, 1984, J. Magn. & Magn. Mater. **44**, 333.

Smith, G.S., Q. Johnson and P.C. Nordine, 1965, Acta Crystallogr. **19**, 668.

Smith, P.K., 1964, PhD. Thesis (University of Kansas, Lawrence, KS).

Smith, P.K., and P.W. Gilles, 1967, J. Inorg. & Nucl. Chem. **29**, 375.

Spedding, F.H., K.A. Gschneidner Jr and A.H. Daane, 1958, J. Am. Chem. Soc. **80**, 4499.

Spedding, F.H., K.A. Gschneidner Jr and A.H. Daane, 1959, Trans. AIME **215**, 192.

Stadelmaier, H.H., and N.A. El-Masry, 1985, in: Proc. Eighth Int. Workshop on Rare Earth Magnets and Their Application, ed. K.J. Strnat (University of Dayton, Dayton, OH) p. 613.

Stadelmaier, H.H., and S.B. Kim, 1984, Z. Metallk. **75**, 381.

Stadelmaier, H.H., and N.-C. Liu, 1985, Z. Metallk. **76**, 585.

Stadelmaier, H.H., and H.K. Park, 1981, Z. Metallk. **72**, 417.

Stearns, C.A., and F.J. Kohl, 1971, J. Chem. Phys. **54**, 5180.

Stecher, P., A. Neckel, F. Benesovsky and H. Nowotny, 1964, Plansee Ber. Pulvermetall. **12**, 181.

Stewart, G.R., A.L. Giorgi and M.C. Krupka, 1978, Solid State Commun. **27**, 413.

Storms, E.K., 1967, in: Refractory Carbides (Academic Press, New York).

Storms, E.K., 1971, High Temp. Sci. **3**, 99.

Stout, N.D., C.L. Hoenig and P.C. Nordine, 1969, J. Am. Ceram. Soc. **52**, 145.

Suematsu, H., K. Ohmatsu, K. Sugiyama, T. Sakakibara, M. Motokawa and M. Date, 1981a, Solid State Commun. **40**, 241.

Suematsu, H., K. Ohmatsu and R. Yoshizaki, 1981b, Solid State Commun. **38**, 1103.

Takaki, H., and M. Mekata, 1969, in: Structures and Physical Properties of Non-Stoichiometric Compounds, Meeting of Japanese Physical Society, cited by Y. Aoki and D.E. Williams, 1979, J. Less-Common Met. **65**, 35.

Tarascon, J.M., J. Etourneau, P. Dordor, P. Hagenmüller, M. Kasaya and J.M.D. Coey, 1980, J. Appl. Phys. **51**, 574.

Tarascon, J.M., J.L. Soubeyroux, J. Etourneau, R. Georges, J.M.D. Coey and O. Massenet, 1981, Solid State Commun. **37**, 133.

Toth, T., H. Nowotny, F. Benesovsky and E. Rudy, 1961, Monatsh. Chem. **94**, 565.

Trammell, G.T., 1963, Phys. Rev. **131**, 932.

Tsokol, A.O., O.I. Bodak, E.P. Marusin and M.G. Baivel'man, 1986, Kristallografiya **31**(4), 791.

Tsokol, A.O., O.I. Bodak and E.P. Marusin, 1986a, Kristallografiya **31**(1), 73.

Tsokol, A.O., O.I. Bodak and E.P. Marusin, 1986b, Kristallografiya **31**(4), 788.

Vereshchagin, L.F., V.I. Novokshonov and V.V. Evdokimova, 1978, Fiz. Tver. Tela **20**, 3109.

Vereshchagin, L.F., V.I. Novokshonov, V.V. Evdokimova and E.P. Khlybov, 1979, Fiz. Tverd. Tela. **21**, 569.

Verhagen, G., S. Smoes and J. Drowart, 1965, WADD-TR-60-782, Pt. 28 AD-464599; cited by F.J. Kohl and C.A. Stearns, 1971, J. Chem. Phys. **54**, 1414.

Verhoef, R., F.R. de Boer, Yang Fu-ming, Zhang Zhi-dong, D.B. de Mooij and K.H.J. Buschow, 1989, J. Magn. & Magn. Mater. **80**, 37.

Vickery, R.C., R. Sedlacek and A. Ruben, 1959, J. Chem. Soc. **104**, 503.

von Stackelberg, M., 1930, Z. Phys. Chem. **89**, 437.

Wakefield, G.F., A.H. Daane and F.H. Spedding, 1965, in: Rare Earth Research Conf. III, ed. L. Eyring (Gordon and Breach, New York) p. 469.

Wallace, T.C., N.H. Krikorian and P.L. Stone, 1964, J. Electrochem. Soc. **111**, 1404.

Warf, J.C., 1955, Status report, U.S. Dept of Ordnance, Research Project 683, Contract No. DA-04-495-Ord-421, 20 July.

Warkentin, E., R. Masse and A. Simon, 1982, Z. Anorg. & Allg. Chem. **491**, 323.

Wegener, W., A. Furrer, W. Bührer and S. Hautecler, 1981, J. Phys. C: **14**, 2387.

Winchell, P., and N.L. Baldwin, 1967, J. Phys. Chem. **71**, 4476.

Yanase, A., and T. Kasuya, 1968a, J. Appl. Phys. **39**, 430.

Yanase, A., and T. Kasuya, 1968b, J. Phys. Soc. Jpn. **25**, 1025.

Yupko, V.L., G.N. Makarenko and Yu.B. Paderno, 1974, Physical properties of carbides of rare earth metals, in: Refractory Carbides, ed G.V. Samsonov (Consultants Bureau, New York) p. 251.

Yvon, K., W. Jeitschko and E. Parthé, 1977, J. Appl. Crystallogr. **10**, 73.

Zhukov, V.P., V.A. Gubanov, O. Jepsen, N.E. Christensen and O.K. Andersen, 1987, Phys. Status Solidi **143**, 173.

Handbook on the Physics and Chemistry of Rare Earths, Vol. 15
edited by K.A. Gschneidner, Jr. and L. Eyring

Chapter 100

METAL-RICH HALIDES

Structure, Bonding and Properties

A. SIMON, HJ. MATTAUSCH, G. J. MILLER, W. BAUHOFER, and R. K. KREMER
Max-Planck-Institut für Festkörperforschung, Heisenbergstrasse 1, W-7000 Stuttgart 80, FRG

Contents

1. Introduction

The chemistry of metal-rich halides of the rare earth metals with halogen-to-metal ratios $X/R < 2$ has become a rapidly growing field of research. The structures follow a general pattern: with hardly any exception, they are derived from transition metal M_6X_8 or M_6X_{12} units, which as discrete clusters contain octahedral M_6 cores surrounded by eight X atoms above the faces (M_6X_8) or twelve X atoms above the edges (M_6X_{12}), respectively (fig. 1). Such units occur as discrete entities or condensed into extended arrays (Simon 1976, 1981, 1985, 1988, Corbett 1981, Corbett and McCarley 1986, Meyer, 1988) as with 4d and 5d metals, Nb, Ta, Mo, W and Re in particular. Similar to silicates that have been systematized in terms of the connection of the SiO_4 tetrahedra as characteristic building units (Liebau 1985), the rare earth

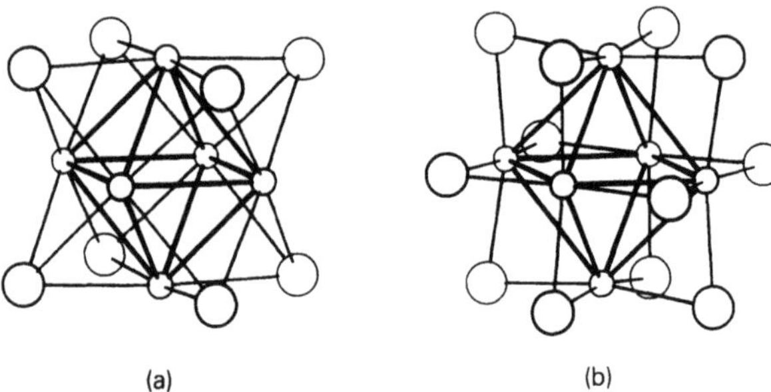

Fig. 1. M_6X_8- and M_6X_{12}-type clusters: the X atoms (X^i-type) center faces and edges of the M_6 octahedron. Up to six additional X atoms of the type X^a can be located above the apices of the octahedron. Any bridging functions of these atoms between clusters is denoted as X^{i-i}, X^{i-a}, X^{a-i} or X^{a-a} (Schäfer and Schnering 1964).

metal compounds of interest here are derived from R_6 octahedra, isolated or condensed into twin octahedra, chains (fig. 2), twin chains, planar or puckered layers and, finally, three-dimensional frameworks.

In contrast to the 4d and 5d metals, where the clusters generally exhibit empty M_6 cores, Zr and Th being distinct exceptions, the octahedral voids with rare earth metal compounds as a rule are occupied by interstitial atoms. A variety of elements throughout the periodic table can be incorporated in the octahedral cavities. Thus, except for a few examples with exclusively metal–metal bonded cluster species, the chemical bonding ranges from metal–metal bonds mediated by strong heteropolar bonds with interstitial atoms to units with only metal–interstitial bonds and no metal–metal bonding left. The latter compounds can be discussed in terms of simple salts, although they still have the atomic arrangement in common with cluster compounds. In fact, the $R_6X_{12}Z$ cluster with an interstitial atom Z in the center of the octahedron frequently occurs with rare earth metals. The X atoms display a cuboctahedron centered by the Z atom, i.e. the 13 atoms are part of a cubic close-packing (ccp). Packing of these clusters therefore leads to an extended ccp and the R atoms occupy the octahedral voids around the Z atoms as in a rocksalt type structure with ordered defects.

The specific role of the interstitial atoms is intimately related to the valence electron deficiency of the clusters formed from rare earth metals. The interstitial atoms Z in a formal sense add electrons to the cluster, yet they actually remove electrons from (weakly) metal–metal bonding bands and build up strong R–Z bonds. Structures with empty clusters seem to be confined to the sesquichlorides Gd_2Cl_3, Tb_2Cl_3, Tm_2Cl_3, Y_2Cl_3, and the analogous bromides which are derived from chains of transedge sharing clusters, as well as Sc_7Cl_{10} with twin chains of metal octahedra. Normally carbon (single atoms or C_2 units), boron, silicon, and transition metal atoms occupy the octahedral voids between R atoms. Nitrogen atoms frequently prefer tetrahedral voids as hydrogen atoms do. In the compounds RXH_2 the hydrogen atoms are found in both the tetrahedral and (pairwise) in octahedral voids.

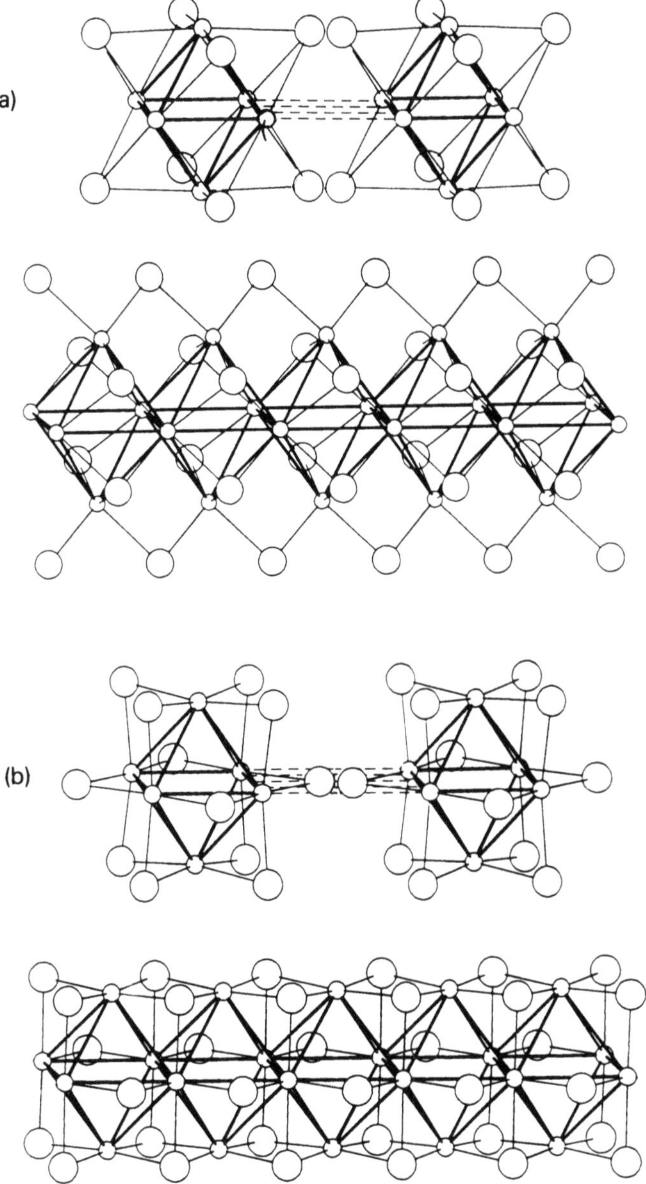

Fig. 2. Principle of cluster condensation for (a) M_6X_8-type clusters (as in Gd_2Cl_3) and (b) M_6X_{12}-type clusters (as in $NaMo_4O_6$ or including interstitial C atoms in e.g. Gd_4I_5C).

In what follows, the existing metal-rich halides are discussed first along the line of their structural chemistry. As an ordering scheme, the increasing degree of cluster condensation is used whether the Z atoms are present or not, i.e. the kind of R/X framework alone is used as a guideline. The arguments used to understand the chemical bonding in these compounds are based on a simplified picture of ionic bonding between R and X atoms, which yields the valence electron concentration available for metal–metal bonding.

In the second section this electron counting scheme is rationalized in terms of recent band structure calculations. The limitations of the ionic model as well as a variety of structural and electronic relationships for metal-rich halides are also discussed using these theoretical treatments.

The close contacts between R atoms in the structures and the low dimensionality of the metal–metal bonded regions in most of the structures lead to interesting electrical and magnetic phenomena. These are described in the last section.

This review was intended to be written several years ago. Fortunately, it was postponed year after year and it is only now that the chemical understanding of the systems is such as to be predictable in terms of general structural features of new compounds. Furthermore, our physical measurements (most of which still have to be published in detail) are frequently revealing interesting properties and great potential for surprising results in the future.

2. Compounds and structures

2.1. *Discrete clusters*

2.1.1. *Octahedral units*

Discrete R_6X_{12} units (all centered by interstitial atoms) have been found in a number of phases.

$R_7X_{12}Z$. The structure can be written as $R_6X_{12}Z \cdot R$. Single R atoms occur between the R_6X_{12} units. The phases having this structure are summarized in table 1. Figure 3 presents a slab of the structure between $\frac{1}{6} \leqslant z \leqslant \frac{1}{2}$ which contains a close packed, slightly buckled layer of X atoms in $z \approx \frac{1}{3}$ with one atom replaced by the interstitial atom Z at $(\frac{2}{3}, \frac{1}{3}, \frac{1}{6})$ and R_6 antiprisms (octahedra) around it. The isolated R atoms lie in $(0, 0, \frac{1}{2})$. The X atoms coordinate the R_6 unit above all edges. The six X atoms around the waist of each R_6 octahedron simultaneously occupy positions above corners of adjacent M_6 octahedra. Using the established notation (Schäfer and Schnering 1964) the structure is described as $R_6ZX^{i}_{6}X^{i-a}_{6/2}X^{a-i}_{6/2}$. The additional isolated R atoms are in an octahedral environment of those X atoms which do not link clusters (X^i). In terms of chemical bonding these R atoms act as electron donors for the clusters and remain as isolated R^{3+} ions.

At the beginning, structural investigations were performed on low-yield single-crystalline material thought to represent binary phases R_7X_{12} (Corbett et al. 1978, Berroth 1980, Corbett et al. 1982). All refinements suffered from excess electron density in various parts of the structures, in particular, at the cluster centers, which amounted to approximately 5 electrons for "Sc_7Cl_{12}", 13 for "La_7I_{12}" and even 23 for "Er_7I_{12}". This excess density could be well accounted for by disorder and twin models which refined to residual factors near 5% in the last two cases. Later, it became clear that this structure type only exists with interstitial atoms.

TABLE 1
Crystallographic data for $R_7X_{12}Z$ compounds R: cluster atom; R′: isolated atom.

Compound	Space group; lattice constants (Å)	d_{R-R} (Å)	d_{R-X} (Å)	d_{R-Z} (Å)	$d_{R'-X}$ (Å)	Ref.[a]
$Sc_7I_{12}C$	R3; $a = 14.717$, $c = 9.847$	3.226–3.324	2.884–2.940	2.32	2.851–2.983	[1]
$Gd_7I_{12}C$	$R\bar{3}$; $a = 15.288$, $c = 10.291$	3.765–3.795	3.054–3.228	2.673	3.035	[2]
$Sc_7Br_{12}C$	R3; $a = 13.628$, $c = 9.203$	3.220–3.301	2.657–2.998	2.28	2.635–2.769	[1]
$Sc_7Cl_{12}N$	$R\bar{3}$; $a = 12.990$, $c = 8.835$	3.237–3.256	2.550–2.744	2.296	2.550	[3]
$Sc_7Cl_{12}B$	$R\bar{3}$; $a = 13.014$, $c = 8.899$	3.281–3.293	2.557–2.730	2.324	2.547	[1]
$Sc_7I_{12}Co$	$R\bar{3}$; $a = 14.800$, $c = 10.202$	3.386–3.489	2.927–3.310	2.431	2.874	[4]
$Y_7I_{12}Fe$	$R\bar{3}$; $a = 15.351$, $c = 10.661$	3.665–3.747	3.085–3.317	2.621	3.022	[4]
$Ca_{0.65}Pr_{0.35}Pr_6I_{12}Co$	$R\bar{3}$; $a = 15.777$, $c = 10.925$	3.906–3.928	3.178–3.350	2.770	3.128	[4]
"Sc_7I_{12}"	$R\bar{3}$; $a = 12.959$, $c = 8.825$	3.201–3.229	2.554–2.747	–	2.561	[5]
"La_7I_{12}"	$R\bar{3}$; $a = 15.930$, $c = 10.768$	3.760–3.761	3.245–3.480	–	3.200	[6]
"Ce_7I_{12}"	$R\bar{3}$; $a = 15.766$, $c = 10.610$	3.729–3.735	3.200–3.439	–	3.166	[6]
"Pr_7I_{12}"	$R\bar{3}$; $a = 15.644$, $c = 10.552$	3.706–3.707	3.177–3.415	–	3.144	[6]
"Gd_7I_{12}"	$R\bar{3}$; $a = 15.507$, $c = 10.494$	3.832–3.849	3.102–3.280	–	3.084	[6]
"Tb_7I_{12}"	$R\bar{3}$; $a = 15.438$, $c = 10.358$	3.794–3.838	3.109–3.117	–	3.061	[6]
"Er_7I_{12}"	$R\bar{3}$; $a = 15.267$, $c = 10.609$	3.601–3.696	3.112–3.157	–	3.007	[6]

[a] References: [1] Dudis et al. (1986); [2] Simon and Warkentin (1982); [3] Hwu and Corbett (1986); [4] Hughbanks and Corbett (1988); [5] Corbett et al. (1982); [6] Berroth (1980).

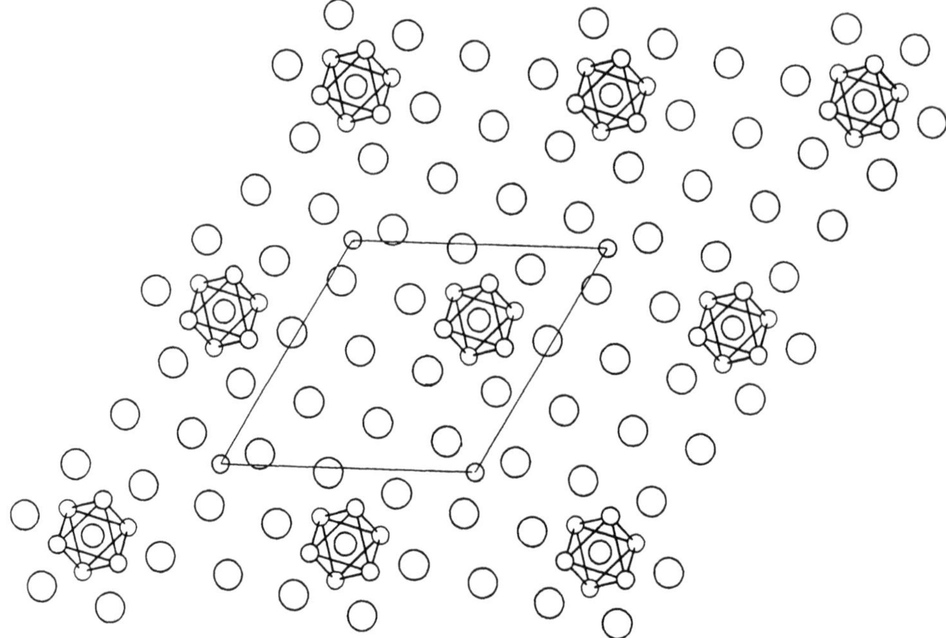

Fig. 3. One slab (around $z = \frac{1}{3}$) of the structure of $Sc_7I_{12}C$ projected along [001]. The Sc_6 octahedra centered at C atoms and the unit cell are outlined (Hwu and Corbett 1986).

Even with such (chemically) rather well-defined compounds, this description of the structure is an idealized one. The real structure as determined from single crystals (table 1) exhibits characteristic deviations from the idealized structure: (i) the single R atom frequently shows unusually elongated "thermal ellipsoids", and (ii) excess electron density is found in all octahedral voids surrounded by X and R atoms. These deviations are elaborated in the following

(i) Characteristic tensor ratios $B_{33}/B_{11} = 3$ to 9 are found for the single R atoms in "Sc_7Cl_{12}" (Corbett et al. 1982) and "La_7I_{12}" (Berroth 1980). Refinement with a split-position for the single atom leads to significant improvement in the latter case. The structures of $Sc_7Br_{12}C$ and $Sc_7I_{12}C$ (Dudis et al. 1986) have been refined in an acentric space group. The distances of the single Sc atom to the Sc atoms of the two adjacent Sc_6 octahedra are then different by approximately 0.5 Å. $Sc_7I_{12}C$ represents one of the rare cases with an ordered displacement. Normally, crystals consist of antiphase domains (with displacements of R atoms in opposite directions), as, e.g. in the cases of $Sc_7Cl_{12}B$ and $Sc_7Cl_{12}N$ (Berroth 1980; Hwu and Corbett, 1986) which apparently crystallize with space group $R\bar{3}$. Only this space group seems to be real when the interstitial is a transition metal atom (with much less polar bonding), as in $Sc_7I_{12}Co$, $Y_7I_{12}Fe$, $Ca_{0.65}Pr_{0.35}Pr_6I_{12}Co$ and $Gd_7I_{12}Fe$ (Hughbanks et al. 1986, Hughbanks and Corbett 1988). In all these cases no significant anisotropy of the electron density distribution around the single R atom, which takes a central position in the X_6 octahedron, is detected.

(ii) Frequently scattering density is observed in the centers of all X_6 octahedra due to disorder phenomena other than the antiphase domains (Berroth 1980). This is due to the conventional reverse–obverse twinning as well as a disorder which can be described by a 180° rotation about the [140] axis of the hexagonal cell. As the structure represents a ccp arrangement of X atoms with every 13th atom substituted by an R_6 unit, such twinning creates additional scattering density from X atoms in all cluster centers and from R atoms in the centers or X atom octahedra. This kind of disorder is not possible if only complete $R_6X_{12}Z$ units are present, and its occurrence gives strong evidence for a limited chemical intergrowth of $R_7X_{12}Z$ layers with RX_2 layers of the $CdCl_2$- (Pauling and Hoard 1930) or PrI_2(V)-type (Warkentin and Bärnighausen 1979) arrangement.

It is certainly clear from the preceding discussion that considerable further work is necessary to elucidate the nature of all of these imperfections. Such effort seems worthwhile in view of their interesting properties, e.g. the field-dependent magnetic behavior found for "Tb_7I_{12}" in preliminary measurements (Simon 1982).

Before moving on to other phases a brief discussion of chemical bonding in R_7X_{12}-type compounds should be helpful. Taking into account the ideal structure of, e.g. $Sc_7I_{12}C$ together with the characteristic displacements of the C atom and the single R atoms, the compound has a defect rocksalt structure with strong ionic bonding present. X^- and C^{4-} ions are cubic close-packed. Six R^{3+} ions gather in octahedral voids around the highly charged anion. According to this description, $R_7X_{12}C$ compounds may be formulated as $(R^{3+})_7(X^-)_{12}C^{4-}(e^-)_5$. Covalent contributions will reduce the ionic charges but will not change the number of valence electrons available for metal–metal bonding, which is essentially confined to the R_6 unit. The empty cluster in a binary compound $R^{3+}(R_6X_{12})^{3-}$ would have 9 electrons available, significantly less than 14–16 which is the normal range for M_6X_{12}-type clusters. With rare earth metals the stabilization by the interstitial C atom is needed which formally adds its 4 valence electrons to the cluster. The resulting count of 13 electrons is due to the fact that the insertion of the C atom into the empty cluster does not add any more bonding MOs (molecular orbitals) but only additional electrons to the cluster. A more detailed discussion of bonding including the bonding of d metals in the cluster cavity will be given in sect. 3.

$R_6X_{11}Z$. The structure of $Sc_6I_{11}C_2$ (table 2) (Dudis and Corbett 1987) consists of discrete M_6I_{12}-type octahedral clusters that are elongated along a pseudo-four-fold axis to accommodate a C_2 unit with the orientation, as illustrated in fig. 4. The clusters are pairwise linked by four iodine atoms, two of the kind I^{i-i} and two of I^{i-a}, respectively. The remaining intercluster bonds involve five of I^{i-a} type. So, the structure may be represented as $Sc_6(C_2)I^{i}_4I^{i-i}_{2/2}I^{i-a}_{6/2}I^{a-i}_{6/2}$. The basis of the Sc_6 octahedron exhibits fairly uniform Sc–Sc distances of 3.14–3.17 Å which are shorter than the distances $d_{Sc-Sc} = 3.46$–3.65 Å involving the apical atoms. This kind of differentiation is due to the spacial requirements of the C_2 unit, and is therefore always found with interstitial C_2 units. Short R–C distances to the apical R atoms are also quite

TABLE 2
Crystallographic data for rare earth metal compounds with discrete clusters.

Compound	Space group; lattice constants (Å); angle (deg)	d_{R-R} (Å)	d_{R-X} (Å)	d_{R-Z} (Å)	Ref.[a]
$Sc_6I_{11}C_2$	$P\bar{1}$; $a = 10.046$, $b = 14.152$ $c = 9.030$; $\alpha = 104.36$ $\beta = 14.152$, $\gamma = 89.27$	3.143–3.647	2.790–3.577	2.07–2.37	[1]
$Cs_2Lu_7Cl_{18}C$	Not published	3.577–3.615	2.583–2.691	2.543	[2]
$Y_6I_{10}Ru$	$P\bar{1}$; $a = 9.456$, $b = 9.643$ $c = 7.629$; $\alpha = 97.20$ $\beta = 105.04$, $\gamma = 107.79$	3.688–3.969	2.983–3.244	2.556–2.772	[3]
$Gd_{10}Cl_{18}C_4$	$P2_1/c$; $a = 9.182$, $b = 16.120$, $c = 12.886$; $\beta = 119.86$	3.212–4.086	2.612–3.260	2.21–2.66	[4]
$Gd_{10}Cl_{17}C_4$	$P\bar{1}$; $a = 8.498$, $b = 9.174$ $c = 11.462$; $\alpha = 104.56$ $\beta = 91.98$, $\gamma = 110.35$	3.121–4.014	2.633–2.965	2.20–2.63	[4]
$Gd_{10}I_{16}C_4$	$P\bar{1}$; $a = 10.463$, $b = 16.945$ $c = 11.220$; $\alpha = 99.15$ $\beta = 92.68$, $\gamma = 88.06$	3.278–4.000	2.920–3.232	2.20–2.69	[5]

[a] References: [1] Dudis and Corbett (1987); [2] Meyer and Schleid (1987); [3] Hughbanks and Corbett (1989); [4]Warkentin et al. (1982); [5] Simon and Warkentin (1982).

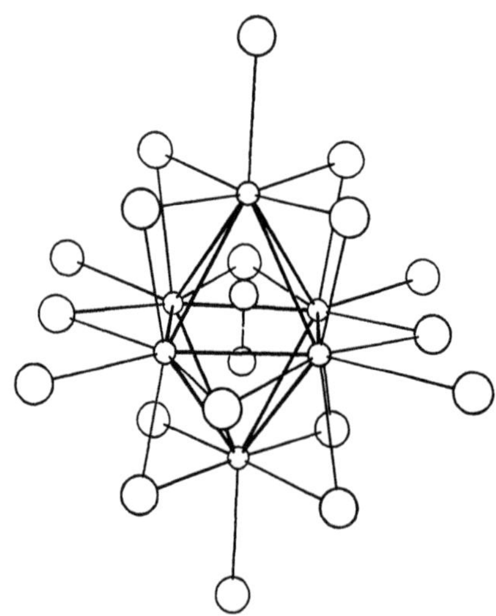

Fig. 4. $Sc_6I_{18}C_2$ cluster in $Sc_6I_{11}C_2$ (Dudis and Corbett 1987).

characteristic. In $Sc_6I_{11}C_2$ they are 2.08 Å, i.e. 0.26 Å shorter than those to the basis atoms. The Sc–I^i distances range from 2.79 to 2.99 Å. The Sc–I^{i-a} distances are more varied (3.06–3.58 Å), obviously in order to achieve an efficient packing of the clusters. A carbon-free framework Sc_6I_{11} would provide 18–11 = 7 electrons per C_2, which is sufficient to fill all MOs of the C_2 unit with the exception of the highest antibonding one and leave one electron in a metal–metal bonding state. Therefore, one would expect a C_2 entity (formally, an ethane analogue C_2^{6-}) with a slightly shortened single bond distance. The C–C distance in $Sc_6I_{11}C_2$ is rather short, 1.39 Å. This shortening is more obvious here than for a C_2 unit in a Gd_6 cage, and the interesting question arises whether this difference is only due to the different cage sizes or to some extent due to stronger charge transfer from occupied C_2 π^* orbitals into empty Sc d states.

Anyhow, according to the formulation $(Sc^{3+})_6(I^-)_{11}C_2^{6-}e^-$ the compound lies very near to the borderline of simple salts. There is only one electron in a metal–metal bonding state left. This fact should not be confused with the other fact that according to the formal way of counting cluster bonding electrons, the C_2 unit adds $8 - 2 = 6$ electrons, thus leading to the same total electron count (13) for $Sc_6I_{11}C_2$ as for $Sc_7I_{12}C$.

$Cs_2R_7X_{18}Z$. The single-crystal investigation of the phase $Cs_2Lu_7Cl_{18}C$ (Meyer and Schleid 1987) revealed another structure based on discrete M_6X_{12}-type clusters (table 2), which is isostructural with the zirconium compound $K_2Zr_7Cl_{18}H$ (Imoto et al. 1981). Each Lu_6Cl_{12} unit is coordinated by six Cl atoms in X^a positions which belong to $LuCl_6^{3-}$ anions. Evidence for an interstitial C atom comes only from the X-ray investigation. The compound must be considered as rather unusual: the ionic limit $(Cs^+)_2(Lu^{3+})_7(Cl^-)_{18}C^{4-}e^-$ shows that there is one electron available for metal–metal bonding, rather similar to the situation for $Sc_6I_{11}C_2$. On the other hand, the formal electron count – the C atom adding its valence electrons to the Lu_6Cl_{12} cluster – is extremely low, only 9. It corresponds to that in the pure binary compounds M_7X_{12}, which obviously cannot be prepared.

$R_6X_{10}Z$. Attempts to prepare $Y_7I_{12}Ru$ yielded the new cluster compound $Y_6I_{10}Ru$ (Hughbanks and Corbett 1989) with the novel feature of a heavy transition metal atom acting as interstitial (table 2). We find an extension of the $R_6X_{11}Z$ structure: infinite chains of X^{i-i}-linked discrete clusters occur. Upon inspecting the central cluster in fig. 5, it is obvious that four of its inner X atoms are shared with adjacent clusters in the X^{i-i} mode. The complete interconnection pattern can be formulated as $Y_6RuI_2^iI_{4/2}^{i-i}I_{6/2}^{i-a}I_{6/2}^{a-i}$. With respect to structural details the construction of chains from the clusters through I^{i-i} connections accompanied by the I^{i-a} bonding between chains is quite similar to that in the $Sc_6I_{11}C_2$ structure. The different stoichiometry is due to coupling only pairs of clusters via X^{i-i} bridges in the latter compound. The cluster exhibits a tetragonal compression (in contrast to the cluster in $Sc_6I_{11}C_2$) to produce two short Y–Ru distances of 2.56 Å compared to four of average length 2.76 Å involving the octahedral basis. The question remains whether this distortion of

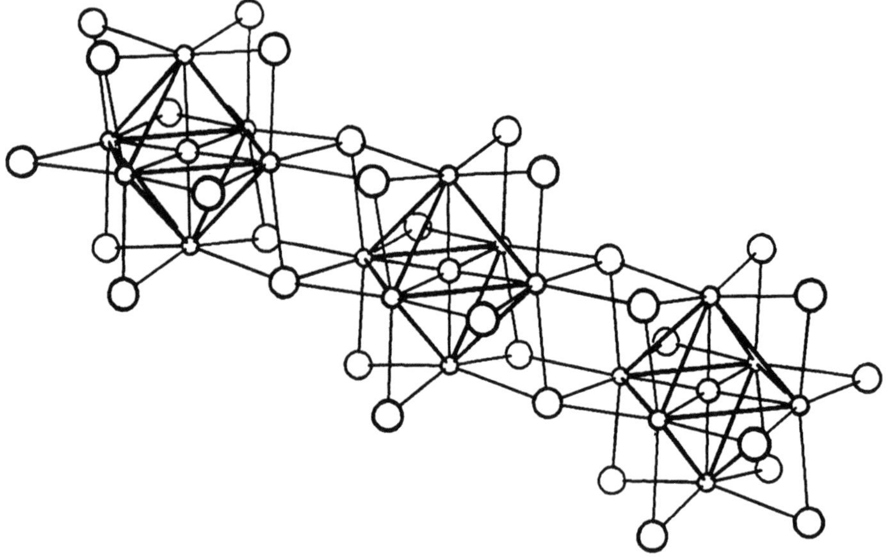

Fig. 5. Interconnection of $Y_6I_{12}Ru$ cluster in the structure of $Y_6I_{11}Ru$ (Hughbanks and Corbett 1989).

the cluster is due to the special type of cluster linkage or whether it has reasons related to the electronic configuration.

Finally, it is worthwhile to mention that a very similar interconnection of (empty!) clusters is found in the structure of $BaMo_6O_{10}$ (Lii et al. 1988) which contains discrete Mo_6O_{12} clusters linked via intercluster metal–metal bonds into zigzag chains.

2.1.2. *Bioctahedral units*

The first step of condensation of M_6X_{12}-type clusters is achieved in the compounds (see table 2) $Gd_{10}Cl_{18}C_4$, $Gd_{10}Cl_{17}C_4$, $Gd_{10}Br_{17}C_4$ and $Gd_{10}I_{16}C_4$ (Simon et al. 1981, Masse and Simon 1981, Warkentin et al. 1982, Simon 1985): the structures are so closely related that they will not be discussed separately. The metal atoms form a Gd_{10} unit by connection of two Gd_6 octahedra via a common edge. Both octahedra are centered at C_2 groups and the free edges are coordinated by X atoms as in the M_6X_{12}-type cluster. The different compound compositions arise from a different interconnectivity according to

$$Gd_{10}(C_2)_2Cl^{i}_{10}Cl^{i-a}_{8/2}Cl^{a-i}_{8/2},$$

$$Gd_{10}(C_2)_2Cl^{i}_{8}Cl^{i-i}_{2/2}Cl^{i-a}_{8/2}Cl^{a-i}_{8/2},$$

$$Gd_{10}(C_2)I^{i}_{6}I^{i-i}_{4/2}I^{i-a}_{8/2}I^{a-i}_{8/2}.$$

The quasi-molecular units are packed in a way to provide an additional ligand from adjacent units for all but the two central M atoms to coordinate them in X^a positions. Figure 6, together with the formal descriptions of the structures, indicates that the

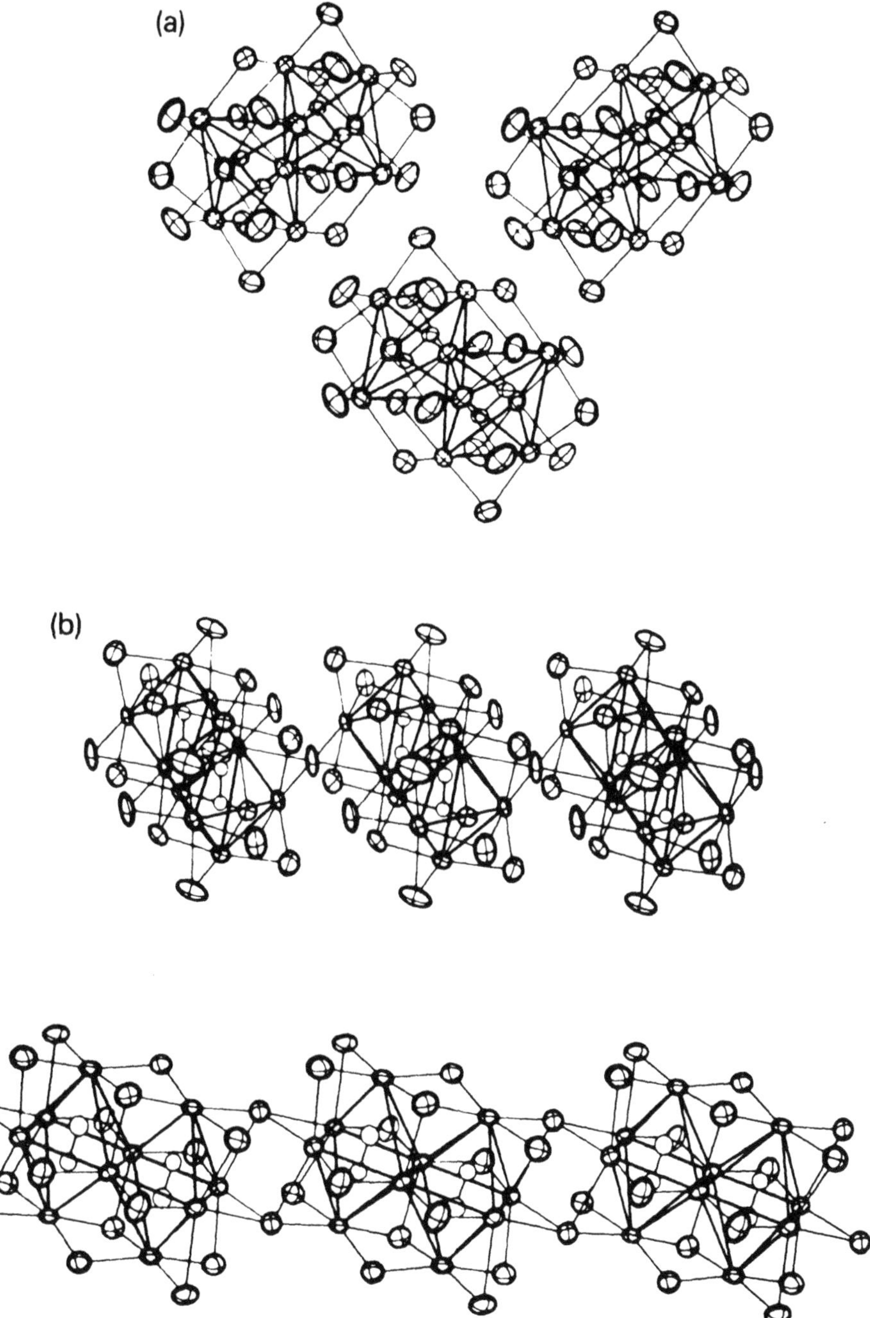

Fig. 6. Arrangement of $Gd_{10}X_{18}C_4$ units in (a) $Gd_{10}Cl_{18}C_4$, (b) $Gd_{10}Cl_{17}C_4$ and (c) $Gd_{10}I_{16}C_4$ (Simon et al. 1981).

units are isolated (except for the X^{i-a} contacts) in $Gd_{10}Cl_{18}C_4$, whereas neighboring clusters are linked via X^{i-i} bridges in $Gd_{10}Cl_{17}C_4$ and $Gd_{10}I_{16}C_4$. The similar manner of interconnection between the single and bioctahedral units in $Y_6I_{10}Ru$ and $Gd_{10}I_{16}C_4$, respectively, is evident from a comparison of figs. 5 and 6c. Again the structural analogue with empty bioctahedral clusters is well known from $La_2Mo_{10}O_{16}$ (Hibble et al. 1988) and $Pb_2Mo_{10}O_{16}$ (Dronskowski and Simon 1989). In $La_2Mo_{10}O_{16}$ ($Pb_2Mo_{10}O_{16}$) the number of electrons in metal–metal bonding states is 34 (32). With each C_2 entity adding 6 electrons to the bioctahedral unit in the Gd compounds, one arrives at $30 - 18 + 12 = 24$ electrons for $Gd_{10}Cl_{18}C_4$ and 26 for $Gd_{10}I_{16}C_4$. A large valence electron concentration of the M atom allows the formation of empty clusters with only metal–metal bonds, whereas electron deficiency of metal atoms calls for a stabilization of the cluster arrangement by interstitial atoms or groups.

In the case of the bioctahedral Gd clusters a critical borderline is crossed which becomes evident from the ionic descriptions $(Gd^{3+})_{10}(Cl^-)_{18}(C_2{}^{6-})_2$, $(Gd^{3+})_{10}(Cl^-)_{17}(C_2^{6-})_2e^-$ and $(Gd^{3+})_{10}(I^-)_{16}(C_2^{6-})_2(e^-)_2$. Bonding of the interstitial C_2 unit in $Gd_{10}Cl_{18}C_4$ removes all electrons from metal–metal bonding states and, therefore, actually destroys the cluster (if one associates metal–metal bonding with the term "cluster"). The alternative description used already with M_7X_{12}-type compounds therefore seems more adequate (fig. 7). Starting from a cubic close packing of X^- and C_2^{6-} anions in the given stoichiometric ratios, the Gd^{3+} ions occupy all octahedral voids around pairs of C_2^{6-} ions leading to electroneutrality in the case of $Gd_{10}Cl_{18}C_4$, but leaving 1 and 2 electrons in a metal–metal bonding level in $Gd_{10}Cl_{17}C_4$ and $Gd_{10}I_{16}C_4$, respectively. $Gd_{10}Cl_{18}C_4$ is a purely salt-like carbide halide. The mean of all Gd–Cl distances is close to the sum of ionic radii with only minor differentiations ($d_{Gd-Cl^i} = 2.70$ and 2.75 Å, $d_{Gd-Cl^{i-a}} = 2.85$ and 2.81 Å for $Gd_{10}Cl_{18}C_4$ and $Gd_{10}Cl_{17}C_4$, respectively). The additional metal–metal bonding in $Gd_{10}Cl_{17}C_4$ is not really obvious when the average Gd–Gd distances (3.74 Å in $Gd_{10}Cl_{18}C_4$ and 3.71 Å in $Gd_{10}Cl_{17}C_4$) are compared. But the difference is striking for the shared edge in the bioctahedron which shortens from 3.21 to 3.12 Å with the additional electron which, as MO calculations show (Satpathy and Andersen 1985), is rather localized in the central metal–metal bond. (The larger value of 3.28 Å in the structure of $Gd_{10}I_{16}C_4$ may be explained in terms of a matrix effect due to the larger I atoms.)

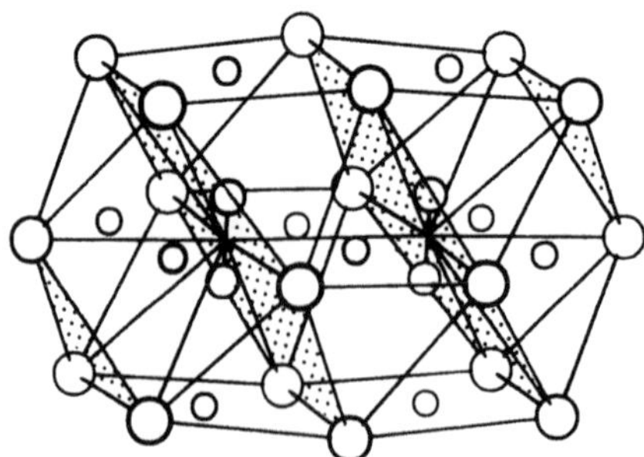

Fig. 7. Indication of the ccp arrangement of anions in the $Gd_{10}Cl_{18}C_4$ unit.

The C–C distances in all three compounds have comparable lengths ($d_{C-C} = 1.45 \pm 0.02$ Å). The value corresponds to a Pauling bond order of 1.3 (Pauling 1960). Of course, one may argue that the single-bond distance in a hydrocarbon is not a suitable reference, yet the assumption of a slightly shortened single bond due to backbonding effects is in agreement with the MO calculations of Satpathy and Andersen (1985). Hydrolysis experiments, too, reveal the compounds as ethanides: at room temperature $Gd_{10}Cl_{18}C_4$ reacts with water or dilute H_2SO_4 to form 90–95% C_2H_6, besides small amounts of C_2H_4 and C_2H_2. The reduced state of $Gd_{10}Cl_{17}C_4$ is reflected in the presence of approximately 10% H_2 in the gaseous reaction products (Schwarz 1987).

2.2. *Cluster chains*

2.2.1. *Single chains*

The extension of the structural principle found in the bioctahedron of $Gd_{10}Cl_{18}C_4$ leads to a series of general composition $R_{4n+2}X_{6n+6}C_x$, n denoting the number of R_6 octahedra with $x = n$ or $2n$ depending on whether single C atoms or C_2 units fill the octahedra. Structures containing such oligomeric units with $n > 2$ have not been verified experimentally so far. Only the final member, namely, the infinite chain has been realized in a number of compound compositions which differ due to different types of interconnections between the chains (table 3). By comparison with the 4d metal Mo (Mattausch et al. 1986a, Simon et al. 1986, Hibble et al. 1988, Lii et al. 1988, Dronskowski and Simon 1989), where discrete (empty) clusters with $n = 1$–6 have been realized, one could imagine that similar units are accessible with rare earth metals, too.

$Sc_4I_6C_2$. The compound (Dudis and Corbett 1987) may be described as a "mixed cluster compound" with a remarkably complex chain structure (in contrast to $Sc_4Cl_6C_2$). As illustrated in fig. 8, two kinds of C_2-centered Sc_6 octahedra are linked via Sc_2 units in a way to result in strongly distorted trigonal prisms centered as well at C_2 entities. The octahedra have approximate D_{2h} symmetry and exhibit the same pattern of Sc–Sc distances as the Gd–Gd distances in $Gd_{10}Cl_{18}C_4$. The common edges are shortened to 3.02 and 3.05 Å, while the other basal distances are 3.34 and 3.49 Å, respectively. The distances involving the apical atoms lie in the range

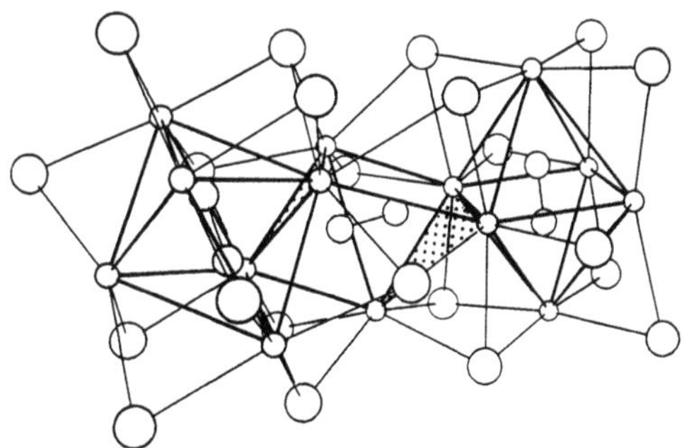

Fig. 8. Arrangement of clusters in $Sc_4I_6C_2$: direct metal–metal bonds between clusters lead to distorted trigonal prisms whose trigonal faces are emphasized (Dudis and Corbett 1987).

TABLE 3
Crystallographic data of metal-rich rare earth halides with one-dimensional infinite chains.

Compound	Space group; lattice constants (Å); angle (deg)	d_{R-R} (Å)	d_{R-X} (Å)	d_{R-Z} (Å)	Ref.[a]
$Sc_4I_6C_2$	$P\bar{1}$; $a = 10.803$, $b = 13.959$ $c = 10.793$; $\alpha = 106.96$ $\beta = 119.20$, $\gamma = 87.80$	3.02–3.89	2.765–3.878	2.01–2.47	[1]
$Gd_{12}Br_{17}C_6$	C2/c; $a = 18.253$, $b = 11.716$ $c = 17.795$; $\beta = 90.22$	3.160–4.140	2.741–3.131	2.202–2.663	[2]
$Gd_{12}I_{17}C_6$	C2/c; $a = 19.297$, $b = 12.201$ $c = 18.635$; $\beta = 90.37$	3.187–4.269	3.049–3.268	2.21–2.79	[3]
Y_4I_5C	C2/m; $a = 18.4797$, $b = 3.947$ $c = 8.472$; $\beta = 103.22$	3.248–3.513	3.036–3.503	2.410–2.556	[4]
Gd_4I_5C	C2/m; $a = 18.587$, $b = 3.978$ $c = 8.561$; $\beta = 103.3$	3.331–3.978	3.078–3.499	2.460–2.600	[3]
"Er_4I_5"	C2/m; $a = 18.521$, $b = 4.015$ $c = 8.478$; $\beta = 103.07$	3.353–4.015	3.022–3.411	–	[5]
Gd_4I_5Si	C2/m; $a = 18.9175$, $b = 4.1966$ $c = 8.710$; $\beta = 130.30$	3.733–4.197	3.098–3.318	2.727–2.808	[6]
Sc_5Cl_8C	C2/m; $a = 17.80$, $b = 3.5259$ $c = 12.052$; $\beta = 130.11$	3.041–3.526	2.541–2.791	2.238–2.328	[7]
Sc_5Cl_8N	C2/m: $a = 17.85$, $b = 3.5505$ $c = 12.090$; $\beta = 130.13$	3.090–3.551	2.548–2.777	2.273–2.353	[7]
"Sc_5Cl_8"	C2/m; $a = 17.78$, $b = 3.523$ $c = 12.04$; $\beta = 130.10$	3.021–3.523	2.530–2.799	–	[8]
"Gd_5Br_8"	C2/m; $a = 20.997$, $b = 3.884$ $c = 13.553$; $\beta = 133.20$	3.369–3.884	2.860–3.145	–	[9]
"Tb_5Br_8"	C2/m; $a = 20.705$, $b = 3.859$ $c = 13.367$; $\beta = 133.07$	3.325–3.859	2.828–3.106	–	[9]
Sc_4Cl_6B	Pbam; $a = 11.741$, $b = 12.187$ $c = 3.5988$	3.197–3.599	2.570–2.664	2.329–2.407	[10]
Sc_4Cl_6N	Pbam; $a = 11.634$, $b = 12.144$ $c = 3.550$	3.086–3.550	2.548–2.690	2.279–2.350	[10]

[a] References: [1] Dudis and Corbett (1987); [2] Schwarz (1987); [3] Simon and Warkentin (1983); [4] Kauzlarich et al. (1988); [5] Berroth and Simon (1980); [6] Nagaki et al. (1989); [7] Hwu et al. (1987); [8] Poeppelmeier and Corbett (1978); [9] Mattausch et al. (1980a); [10] Hwu and Corbett (1986).

3.46–3.64 Å. The C_2 entities lie roughly in line with two apex atoms in the Sc_6 octahedra and point towards the connecting edges in the trigonal prism. The C–C distances vary considerably from $d_{C-C} = 1.48$ Å in one octahedron to $d_{C-C} = 1.24$ Å in the other (and 1.40 Å in the trigonal prism). It would be interesting to know about the origin of this differentiation. The simple electron count in the ionic limit according to $(Sc^{3+})_4(I^-)_6C_2^{6-}$ yields C–C single bonds for the C_2 entities in the compound. The distance of 1.48 Å in fact corresponds to the single-bond distance in $Gd_{10}Cl_{18}C_4$. The distance of 1.40 Å has been observed also in the structure of $Sc_6I_{11}C_2$ and is discussed there. The distance $d_{C-C} = 1.24$ Å, however, contradicts the above formulation of the electron count and causes difficulties in the understanding of the chemical bonding in this compound.

$Gd_{12}X_{17}C_6$. The structure of the compound $Gd_{12}I_{17}C_6$ (Simon and Warkentin, 1983), which also occurs with the bromide (Schwarz 1987), is closely related to that of the preceding Sc compound. As fig. 9 illustrates, $Gd_{10}(C_2)_2$ bioctahedra (instead of Sc_6C_2 octahedra) are linked via M_2C_2 entities in a way to yield a trigonal antiprism (instead of a prism) between them. The structure of $Gd_{12}I_{17}C_6$ is therefore entirely made up from edge-sharing Gd_6 octahedra. The alternation between the condensation via *cis*- and *trans*-positioned edges leads to a folded chain of octahedra which are

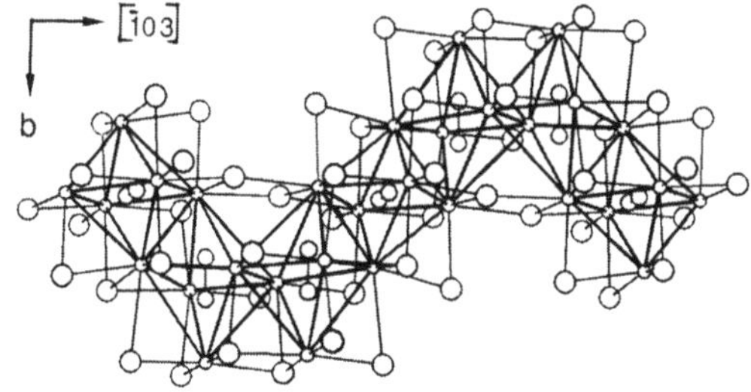

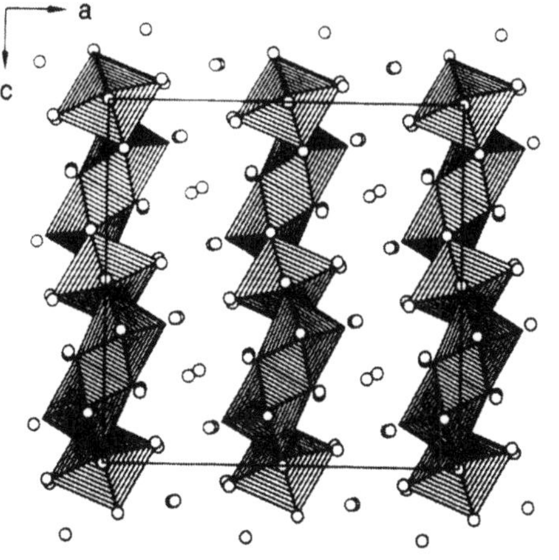

Fig. 9. Folded chains of Gd_6 octahedra in the structure of $Gd_{12}I_{17}C_6$ (Simon and Warkentin 1983).

surrounded by I atoms above all free edges as in the M_6X_{12} cluster. Such chains run parallel to the crystallographic *c* axis and are interconnected via bridges of the type X^{i-i} and X^{i-a}. All Gd atoms except those in shared edges are surrounded by five X atoms as in the discrete $M_6X^i_{12}X^a_6$ cluster. The Gd–Gd distances show a large scatter covering the range $3.19 \leqslant d_{Gd-Gd} \leqslant 4.27$ Å. The shared edges in the chain are again shortest (3.19 and 3.36 Å). The C–C distances in the C_2 entities are 1.41 and 1.45 Å, which correspond to shortened single-bond distances expected from the electron count with one electron per formula unit remaining in a metal–metal bonding band. As in the case of $Sc_4I_6C_2$, the C–C distances in the structure of $Gd_{12}Br_{17}C_6$ are shorter than in the isotypic iodide, 1.36 and 1.44, respectively. Tentative explanations for the observation of such shortened C–C distances have been given before, yet one possible explanation has not been discussed: any undetected hydrogen in the compounds which could be located in tetrahedral voids changes the electron count in a way that less electrons are donated to the C_2 entity. Thus, the C–C bond distances might be probes for additional undetected interstitial atoms.

In the structure of $Gd_{12}I_{17}C_6$, as in other structures with interstitial C_2 units, the latter are directed towards those corners of the octahedra which are not involved in the condensation, with each octahedron elongated parallel to this direction and contracted in the equatorial plane. To prevent very short I–I contacts this special orientation of the C_2 groups needs a folded chain of clusters which fits very well to the close-packed iodine matrix. In fact, the iodine matrix is only slightly distorted from an ideal ccp structure of the anions.

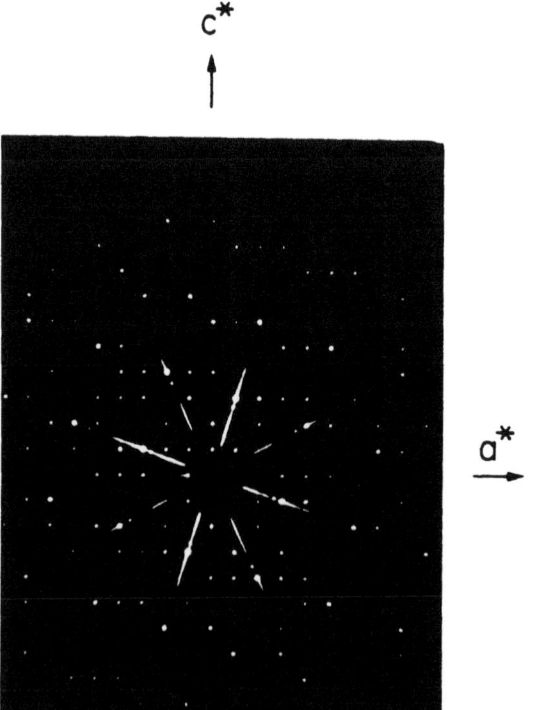

Fig. 10. Precession photograph of the *h0l* plane of $Gd_{12}Br_{17}C_6$: the axes of the “NaCl subcell” are rotated by approximately 25°.

It has been mentioned in the introduction that the "condensed cluster" halides of the rare earth metals based on the M_6X_{12}-type cluster with an interstitial atom (or molecular unit) generally exhibit a defect rocksalt structure. Figure 10 provides clear evidence for this remark. The "NaCl" subcell in the structure of $Gd_{12}I_{17}C_6$, marked by strong streaks is only weakly distorted ($a = 6.07$, $b = 6.10$, $c = 5.92$ Å, $\alpha = \gamma = 90°$, $\beta = 91°$) by the ordering of I atoms and C_2 units and occupation of all voids around the C_2 units by Gd atoms.

R_4X_5Z. This type of compound is realized in Y_4I_5C (Kauzlarich et al. 1988), Gd_4I_5C and Er_4I_5C (Simon and Warkentin 1982). The structure is also formed with interstitial Si in Gd_4I_5Si (Nagaki et al. 1989). What was once described as "Er_4I_5" (Berroth 1980) does not exist as a binary compound but contains interstitial carbon. As illustrated in fig. 11 the anions again form a ccp structure like in $Gd_{12}I_{17}C_6$. In all close-packed planes parallel to (101), every sixth iodine atom is substituted by a carbon atom. The R atoms occupy the octahedral voids around the C atoms. Talking in terms of R^{3+} ions, these are shifted towards the highly charged carbide anions.

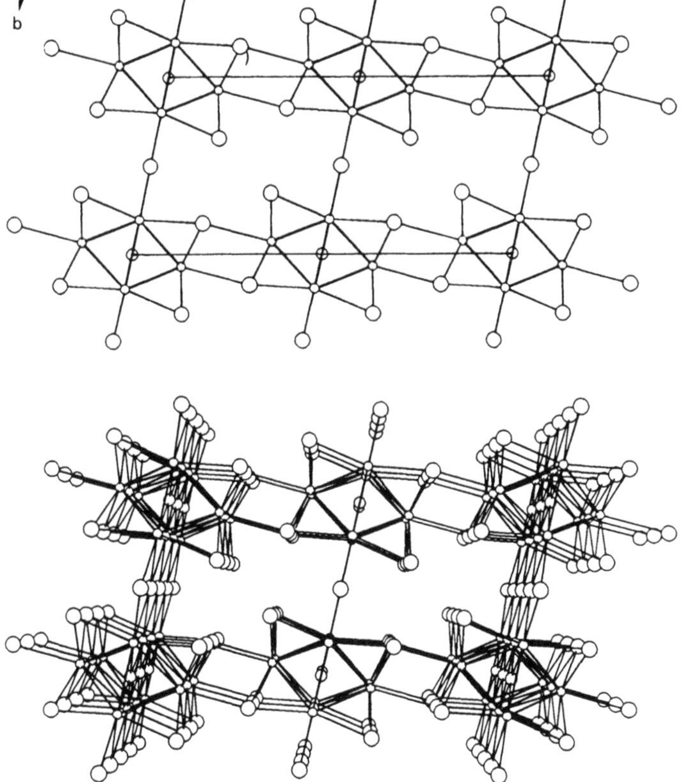

Fig. 11. Projections of the structure of Er_4I_5C along [010]: the central projection emphasizes cluster chains (Simon and Warkentin 1982).

The special arrangement of ions generates metal octahedra that share *trans*-edges to form infinite linear chains (fig. 2b). The three-dimensional structure can be represented by the formula $R_2R_{4/2}CI^{i}_{2}I^{i-i}_{2/2}I^{i-a}_{2/2}I^{a-i}_{2/2}$. It follows from this description that two iodine atoms coordinate only one octahedron, two connect neighboring octahedra in X^{i} positions along [001], and four iodine atoms in X^{i-a} positions bridge neighboring octahedra along [100] above edges and corners.

The metal octahedra are characteristically elongated, 3.98 Å in the chain direction versus 3.33 Å for the shared edge in Gd_4I_5C. The long repeat distance in the chain direction seems mainly related to the large ionic size of the I atom. The average value of the Gd–Gd distance in Gd_4I_5C is 3.62 Å, and thus significantly shorter than the average for $Gd_{10}I_{16}C_4$ and $Gd_{12}I_{17}C_6$, 3.75 and 3.81 Å, respectively. Of course, a larger average is expected for the C_2 interstitial. On the other hand, the electron count for Gd_4I_5C [according to $(Gd^{3+})_4(I^-)_5C^{4-}(e^-)_3$] is comparatively high and metal–metal bonding should therefore be strong in this compound.

R_5X_8Z. The previously described binary compounds "Sc_5Cl_8" (Poeppelmeier and Corbett 1978), "Tb_5Br_8" and "Gd_5Br_8" (Mattausch et al. 1980a) are all prepared in very low yield and are definitely ternary compounds with, perhaps, carbon as an interstitial atom. Single-crystal investigations have been performed on Sc_5Cl_8Z with Z=C, N (Hwu et al. 1987). The distances determined for Sc_5Cl_8C and "Sc_5Cl_8" are identical within three standard deviations.

The structure of Sc_5Cl_8C is shown in fig. 12. As in the structure of R_4I_5C, linear chains of edge-sharing metal octahedra are the characteristic feature. In addition (similar to R_7X_{12}), single Sc atoms occupy octahedral voids of the Cl sublattice. They obviously act as electron donors to the metal–metal bonded chains leading to the same balance of three electrons per repeat unit R_4X_8C in metal–metal bonding states as for R_4I_5C. In contrast to R_4I_5C, where neighboring chains of octahedra are linked via X^{i-i}-type bridges, the chlorine atoms in the structure of Sc_5Cl_8C belong only to one cluster chain. The chains are connected by the chlorine atoms (denoted X′) of the octahedra around the isolated scandium atoms according to $Sc_2Sc_{4/2}CCl^{i}_{4}Cl'^{i}_{2}Cl'^{a}_{2}\cdot Sc$. A rather remarkable feature of this structure is the very homogeneous packing of the cluster chains and the isolated Sc atoms. All Cl–Cl contacts around, between and along the chains as well as within the $ScCl_6$ units fall in a narrow range of 3.52–3.59 Å, with an average 3.54 Å in the carbide (3.50–3.60 Å, with an average 3.55 in the nitride). Besides, the Sc_6 octahedra in the chains of Sc_5Cl_8C exhibit the same characteristic

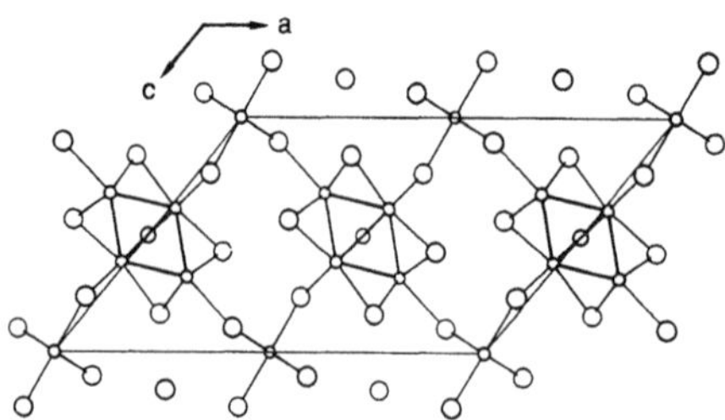

Fig. 12. Projection of the structure of Sc_5Cl_8C along [010] (Hwu et al. 1987).

geometrical details of condensed clusters as discussed before. The shortest Sc–Sc distance is found in the shared edges (3.04 Å) and the longest in the chain direction (3.53 Å). The average Sc–Sc distance is slightly longer when N is the interstitial, 3.29 compared to 3.25 Å for the carbide. The difference can be explained in terms of less coulombic attraction between Sc^{3+} and N^{3-} than C^{4-}.

Sc_4Cl_6Z. Very unusual structures based on the chain of *trans*-edge-sharing M_6 octahedra (clusters of the M_6X_{12}-type) have been found for Sc_4Cl_6N and Sc_4Cl_6B (Hwu and Corbett 1986). A similar heavy-atom arrangement occurred in a "Tb_2Br_3" modification (Mattausch and Simon 1979), whose structure has never been published in detail because of severe lack of understanding ("Heißluft" modification; Corbett 1981). In the structure of Sc_4Cl_6Z the cluster chains are linked according to $Sc_2Sc_{4/2}ZCl^{i}_{4}Cl^{i-a}_{2/2}Cl^{a-i}_{2/2}$. Formally, the structure (fig. 13) is related to rutile. Yet, the similarity with the structure of $NaMo_4O_6$ (Torardi and McCarley 1979) is more obvious. There are differences between the two structures: the Sc_4Cl_6N structure is orthorhombic, crystallizing in a maximal subgroup (Pbam) of the tetragonal space group P4/mbm of $NaMo_4O_6$, with each M_6 octahedron filled by an interstitial atom. The major difference, however, lies in the fact that the channels between the cluster chains are filled by large cations in $NaMo_4O_6$, whereas they are empty in Sc_4Cl_6N. The tilting of the chains in space group Pbam (Sc–Cl–Sc=167°) closes the channels somewhat as in the case of $BaMo_4O_6$; (Torardi and McCarley 1986), yet the size of the cavities is amazing. In difference Fourier maps, no trace of an atom can be located in the channel. So, what are the forces that keep the structure open in the absence of strong directional bonding?

2.2.2. *Gd_2Cl_3 family*

The structure of Gd_2Cl_3 contains linear chains of trans-edge-sharing metal octahedra. However, there are distinct differences which require a separate discussion of this structure. First, the compounds R_2X_3 (together with Sc_7Cl_{10}) seem to be the only binary metal-rich ($X/R < 2$) halides of the rare earth metals. Second, the chains in the structure are formally derived from the M_6X_8-type cluster. Last, but not least, incorporation of interstitial atoms leads to a number of phases, whose structures are closely related to that of Gd_2Cl_3. The structural family is summarized in table 4.

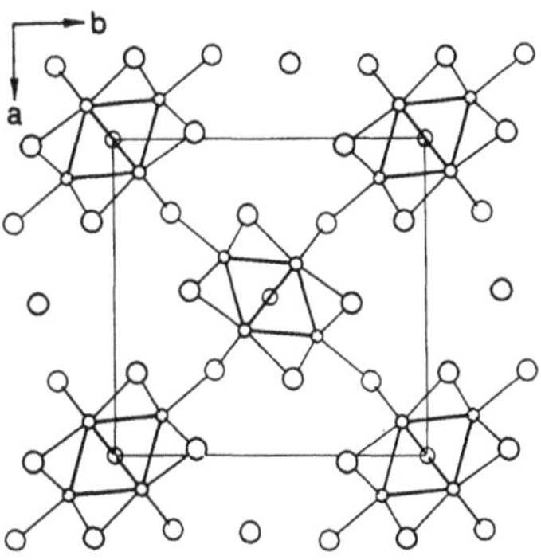

Fig. 13. Projection of the structure of Sc_4Cl_6C along [001] (Hwu and Corbett 1986).

TABLE 4
Crystallographic data of metal-rich rare earth halides with structures derived from Gd_2Cl_3. Calculated with the atomic parameters of (a) Gd_2Cl_3, (b) α-Gd_2Cl_3N, (c) β-$Y_2Cl_3N_x$.

Compound	Space group; lattice constants (Å); angle (deg)	d_{R-R} (Å)	d_{R-X} (Å)	d_{R-Z} (Å)	Ref.[a]
Y_2Cl_3	C2/m; $a = 15.144$, $b = 3.825$ $c = 10.077$; $\beta = 118.24$	3.266–3.825	2.699–3.130	–	[1]
Y_2Br_3	C2/m; $a = 16.463$, $b = 3.898$ $c = 10.695$; $\beta = 120.08$	3.358–3.898	2.890–3.539	–	[1]
Gd_2Cl_3	C2/m; $a = 15.237$, $b = 3.896$ $c = 10.179$, $\beta = 117.66$	3.371–3.896	2.729–3.096	–	[2]
Gd_2Br_3	C2/m; $a = 16.418$, $b = 3.983$ $c = 10.753$; $\beta = 119.27$	3.325–3.983	2.868–3.506	–	[2]
Tb_2Cl_3 (a)	C2/m; $a = 15.184$, $b = 3.869$ $c = 10.135$; $\beta = 118.07$	3.365–3.869	2.726–3.088	–	[3]
Tm_2Cl_3 (a)	C2/m; $a = 15.197$, $b = 3.881$ $c = 10.147$; $\beta = 117.65$	3.363–3.881	2.730–3.087	–	[3]
Lu_2Cl_3 (a)	C2/m; $a = 15.176$, $b = 3.871$ $c = 10.116$; $\beta = 117.89$	3.360–3.871	2.726–3.084	–	[3]
La_2Cl_3 (a)	C2/m; $a = 15.890$, $b = 4.404$ $c = 10.231$; $\beta = 119.14$	3.509–4.404	2.839–3.012	–	[4]
α-Y_2Cl_3N (b)	Pbcn; $a = 12.761$, $b = 6.676$ $c = 6.100$	3.303–3.922	2.739–2.945	2.238–2.262	[5]
β-$Y_2Cl_3N_x$	C2/m; $a = 15.238$, $b = 3.8535$ $c = 10.156$; $\beta = 118.38$	3.290–3.854	2.713–2.839	2.240–2.253	[5]
α-Gd_2Cl_3N	Pbcn; $a = 13.027$, $b = 6.7310$ $c = 6.1403$	3.350–3.967	2.773–2.973	2.262–2.285	[6]
β-$Gd_2Cl_3N_x$(c)	C2/m; $a = 15.290$, $b = 3.912$ $c = 10.209$; $\beta = 117.79$	3.295–3.912	2.725–3.160	2.250–2.286	[5]
Gd_3Cl_6N	$P\bar{1}$; $a = 7.107$, $b = 8.156$ $c = 9.707$; $\alpha = 75.37$ $\beta = 109.10$, $\gamma = 114.65$	3.448–3.918	2.667–3.255	2.238–2.313	[7]

[a] References: [1] Mattausch et al. (1980); [2] Simon et al. (1979); [3] Simon (1981); [4] Araujo and Corbett (1981); [5] Meyer et al. (1989); [6] Schwanitz-Schüller and Simon (1985a); [7] Simon and Koehler (1986).

R_2X_3. Definitely Gd_2X_3, Tb_2X_3 (X = Cl, Br) (Mee and Corbett 1965, Lokken and Corbett 1970, 1973, Simon et al. 1979) and Y_2Cl_3 (Mattausch et al. 1980b) exist as binary compounds with the Gd_2Cl_3 structure. Y_2Br_3 (Mattausch et al. 1980b), Tm_2Cl_3 and Lu_2Cl_3 (Simon 1981) have been described, but as they are only accessible in low yields, these phases might be impurity stabilized. In particular, the unusual lattice constants of "La_2Cl_3" (Araujo and Corbett 1981) ($b = 4.40$ Å compared to 3.90 Å for Gd) raise doubts about the binary nature of this compound. As could be shown by tracer co-crystallization experiments, Tb (but not Ce, Nd, Eu, Dy, Tm or Yb) can substitute for Gd in Gd_2Cl_3 (Simon et al. 1987a, Mikheev et al. 1988). Caloric measurements on Gd_2Cl_3 indicate only a weak stability of the compound against disproportionation into $GdCl_3$ and Gd (Morss et al. 1987). The structure of Gd_2Cl_3 is shown in fig. 14. The chain of edge-linked Gd_6 octahedra together with their characteristic environment of Cl atoms can be derived schematically by condensing M_6X_8-type clusters as indicated in fig. 2a (Simon 1976). As in the chains formally derived from the M_6X_{12}-type cluster, the octahedral basis is elongated in the chain direction (3.90 versus 3.37 Å); the distances to the apex atoms lie in an intermediate range (3.71–3.78 Å). In Gd_2Br_3 the repeat distance in the chain is increased to 3.98 Å according to the larger anion matrix. This elongation is compensated by a slight contraction of the shared edges to 3.32 Å.

The unusual electron count of 1.5 e^- per Gd atom raises the question whether intercalation reactions are possible. Between the cluster chains the packing of Cl atoms offers approximately octahedral voids (edge lengths 3.82, 3.86 and 4.19 Å, respectively). All experiments aimed at an occupation of these voids by cations of suitable size, e.g. Li^+ or Mg^{2+}, have to date been unsuccessful (Corbett and Simon 1979). Surprisingly, voids in the partial lattice of the metal atoms can be occupied by interstitial atoms, although not in a reversible topochemical reaction. In contrast to the preceding structures with interstitial atoms in octahedral voids, in the nitride halides tetrahedral voids play a dominant role.

β-$R_2Cl_3N_x$. Both β-$Gd_2Cl_3N_x$ and β-$Y_2Cl_3N_x$ ($x \approx 0.8$), (Meyer et al. 1989) are isostructural with the corresponding binary chlorides, except for the presence of N atoms in the metal atom substructures. Crystals of the β-phase form in the presence of

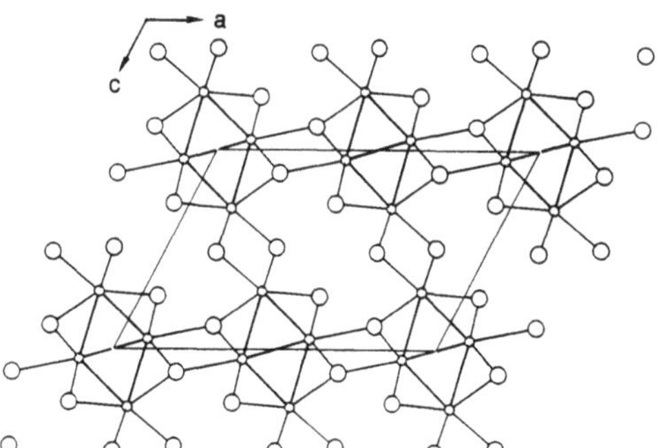

Fig. 14. Projection of the structure of Gd_2Cl_3 along [010] (Simon et al. 1979).

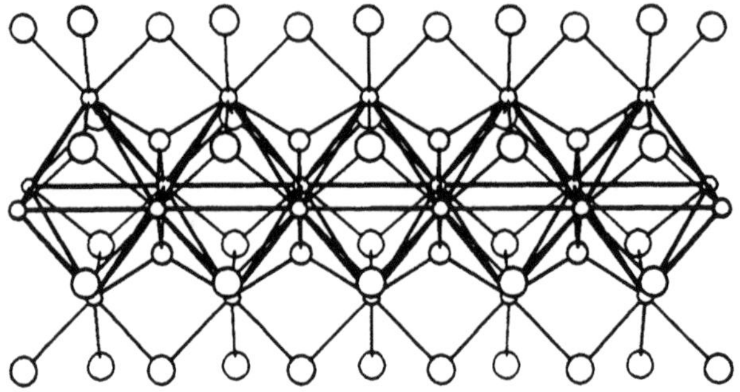

Fig. 15. $Gd_2Cl_3N_x$ chain with N atoms tetrahedrally coordinated by Gd atoms in the structure of β-$Gd_2Cl_3N_x$ (Meyer et al. 1989).

an excess of metal. They are black in color and in accordance with this observation, the structure refinement reveals a significant N atom deficiency which amounts to approximately 20% in the case of β-$Y_2Cl_3N_x$. The ionic description for such a compound shows that the metal–metal bonding bands should still be partly filled $[(Y^{3+})_2(Cl^-)_3(N^{3-})_{0.8}(e^-)_{0.6}]$ resulting in metallic properties. As fig. 15 illustrates, the N atoms are tetrahedrally coordinated by metal atoms.

The differences between the structures of Gd_2Cl_3 and β-Gd_2Cl_3N are surprisingly small. The lattice constants (Gd_2Cl_3: $a = 15.237$ Å, $b = 3.896$ Å, $c = 10.179$ Å, $\beta = 117.6°$; β-Gd_2Cl_3N: $a = 15.290$ Å, $b = 3.912$ Å, $c = 10.209$ Å, $\beta = 117.79°$) are nearly the same and the molar volume of Gd_2Cl_3 (80.63 cm^3/mol) differs by less than 1 cm^3 from the molar volume of Gd_2Cl_3N [α-Gd_2Cl_3: 81.05 cm^3/mol; β-Gd_2Cl_3N: 81.34 cm^3/mol]! We meet the interesting effect that the shrinkage of the Gd atoms upon charge transfer to the N atom just compensates for the volume increment of N^{3-} (~19 cm^3; Biltz 1934). Such effects are well known with the electropositive metals and may even lead to an overall contraction as in the oxidation reaction of alkali metals (Simon 1979). Needless to say that any analysis of metal–metal distances in terms of bond orders with these electropositive metals must be rejected (Simon 1988).

There is a difficulty in understanding the relationship between Gd_2Cl_3 and β-$Gd_2Cl_3N_x$. Both phases are isostructural (with respect to Gd and Cl) and have nearly identical lattice constants. Moreover, the tetrahedral voids are only partially occupied by N atoms. So, one would expect a range of homogeneity between $Gd_2Cl_3N_{0.0}$ and $Gd_2Cl_3N_{0.8}$, but the reproducibility of various physical properties (e.g. the semiconducting behavior) of differently prepared Gd_2Cl_3 does not indicate an accidental incorporation of N.

α-R_2Cl_3N. From stoichiometric amounts of $GdCl_3$ and GdN the phase α-Gd_2Cl_3N is formed (Schwanitz-Schüller and Simon 1985a). The isotypic compound α-Y_2Cl_3N also exists (Meyer et al. 1989). The stoichiometric compounds are colorless in agreement with their ionic description as $(R^{3+})_2(Cl^-)_3N^{3-}$. α-Gd_2Cl_3N crystallizes in a new structure type with the N atoms tetrahedrally coordinated by Gd atoms and the tetrahedra connected via common edges into infinite chains (fig. 16). One may play the game of occupying each octahedron in a chain of octahedra by one (anionic)

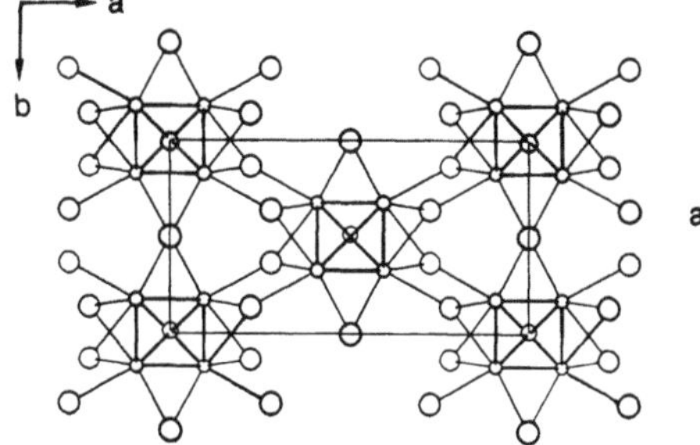

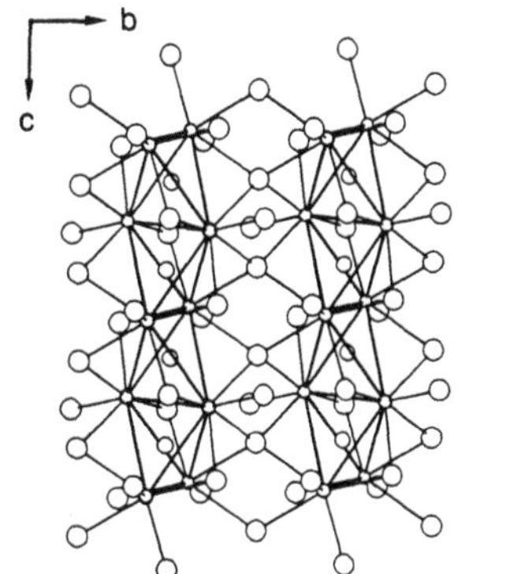

Fig. 16. (a) Projection of the structure of α-Gd_2Cl_3N along [001] and (b) Gd_2Cl_3N chains in this structure (Schwanitz-Schüller and Simon 1985a).

interstitial as in the case of Sc_4Cl_6C or by two as in the case of Gd_2Cl_3N. In the latter case the Coulombic repulsion between the nitride anions leads to a deformation of the octahedron into edge-sharing tetrahedra. Thus, a relation between the structures of Gd_2Cl_3 and α-Gd_2Cl_3N is constructed, but this is a formal one as α-Gd_2Cl_3N is a salt-like compound with the typical tricapped trigonal prismatic coordination of the Gd^{3+} ion as in $GdCl_3$ itself. Gd_2Cl_3N, of course, is no metal-rich compound in the sense of metal–metal bonding being present. The outlined tetrahedra in fig. 16 just represent the topology of the structure. The same holds for the next compound, Gd_3Cl_6N. Yet, both structures nicely illustrate the borderline between metal–metal bonded systems and simple salts.

Gd_3Cl_6N. The structure of the colorless compound $(Gd^{3+})_3(Cl^-)_6N^{3-}$ contains Gd_6N_2 bitetrahedra as a characteristic unit (fig. 17) (Simon and Koehler 1986). This unit is surrounded by Cl atoms in a way to yield rather distorted tricapped trigonal prisms as typical coordination polyhedra for Gd^{3+}. The close relation between the atomic arrangements in α-Gd_2Cl_3N and Gd_3Cl_6N becomes evident from fig. 18. In fact, the bitetrahedron is the characteristic building unit in both the structures of α- and β-Gd_2Cl_3N.

Figure 19 illustrates the linkage of the bitetrahedra via opposite edges into chains of the SiS_2 type as they occur in α-Gd_2Cl_3N and the linkage via four edges in a "parallel

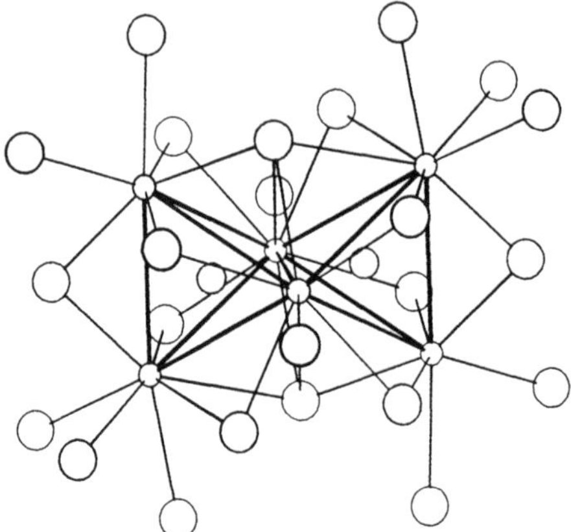

Fig. 17. Unit of two edge-sharing Gd_4N tetrahedra in the structure of Gd_3Cl_6N (Simon and Koehler 1986).

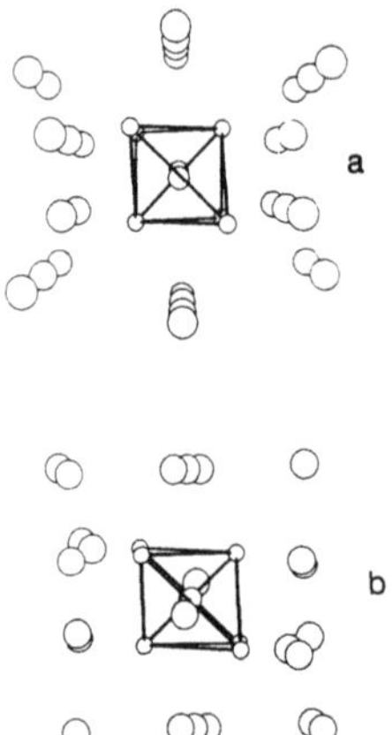

Fig. 18. Comparison of the atomic arrangements in the structures of (a) α-Gd_2Cl_3N and (b) Gd_3Cl_6N (projections along the N–N axes) (Simon and Koehler 1986).

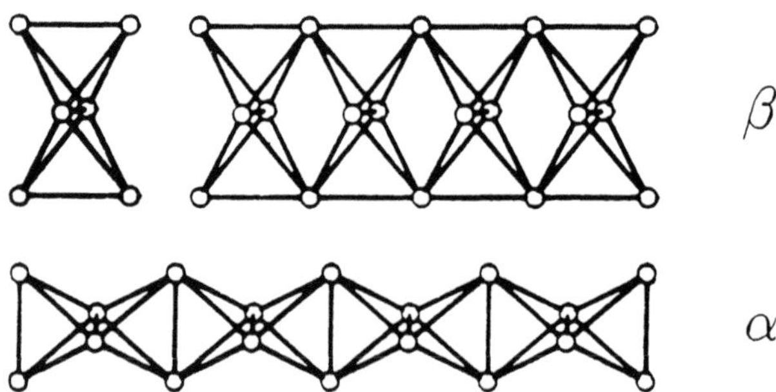

Fig. 19. Schematic representation of the bitetrahedra of Gd atoms in Gd_3Cl_6N and linked in β-$Gd_2Cl_3N_x$ and α-Gd_2Cl_3N.

mode" as in β-$Gd_2Cl_3N_x$. The geometrical details are remarkably similar in all three compounds (distances in Å):

	$\bar{d}_{Gd-Gd}$	$\bar{d}_{Gd-Cl}$	$\bar{d}_{Gd-N}$	$\bar{d}_{Cl-N}$	$\bar{d}_{N-N}$
Gd_3Cl_6N	3.72	2.88	2.27	3.27	3.04
α-Gd_2Cl_3N	3.77	2.88	2.27	3.42	3.07
β-$Gd_2Cl_3N_x$	3.73	2.84	2.27	3.24	3.08

2.2.3. *Twin chains*

The next step of condensation is reached when chains are fused together. There exist a number of different structure types based on the characteristic building unit of the twin chain, e.g. in $R_6X_7C_2$, $R_7X_{10}C_2$, Sc_7Cl_{10} and Gd_3I_3C. With the exception of Sc_7Cl_{10} the presence of interstitial atoms is well established in all of these structures.

TABLE 5
Crystallographic data of metal-rich rare earth halides with twin chains.

Compound	Space group; lattice constants (Å); angle (deg)	d_{R-R} (Å)	d_{R-X} (Å)	d_{R-Z} (Å)	Ref.[a]
$Y_6I_7C_2$	C2/m; $a = 21.557$, $b = 3.9093$ $c = 12.374$; $\beta = 123.55$	3.329–3.909	2.973–3.592	2.360–2.651	[1]
$Gd_6Br_7C_2$	C2/m; $a = 20.748$, $b = 3.819$ $c = 11.885$; $\beta = 124.73$	3.402–3.819	2.820–3.193	2.40–2.62	[2]
"Tb_6Br_7"	C2/m; $a = 20.571$, $b = 3.814$ $c = 11.800$; $\beta = 124.59$	3.384–3.814	2.815–3.188	–	[3]
"Er_6I_7"	C2/m; $a = 21.375$, $b = 3.869$ $c = 12.319$; $\beta = 123.50$	3.431–3.869	2.948–3.653	–	[3]
$Gd_6I_7C_2$	C2/m; $a = 21.767$, $b = 3.946$ $c = 12.459$; $\beta = 123.6$	3.394–3.946	3.078–3.499	2.46–2.60	[4]
Gd_3I_3C	C2/m; $a = 11.735$, $b = 3.926$ $c = 8.658$; $\beta = 92.26$	3.316–3.926	3.089–3.516	2.425–2.680	[5]
Sc_7Cl_{10}	C2/m; $a = 18.620$, $b = 3.5366$ $c = 12.250$; $\beta = 91.98$	3.147–3.537	2.443–3.208	–	[6]
$Sc_7Cl_{10}C_2$	C2/m; $a = 18.620$, $b = 3.4975$ $c = 11.810$; $\beta = 99.81$	3.072–3.498	2.499–2.799	2.185–2.339	[7]

[a] References: [1] Kauzlarich et al. (1988); [2] Schwanitz-Schüller (1984); [3] Berroth et al. (1980); [4] Simon and Warkentin (1982); [5] Mattausch et al. (1987); [6] Poeppelmeier and Corbett (1977); [7] Hwu et al. (1985).

In particular, the previously described phases "Tb_6Br_7" and "Er_6I_7" (Berroth et al. 1980) are carbide halides. The structurally characterized compounds are summarized in table 5.

First, the structures are introduced. Later, it is shown that these structures can be systematized in terms of different series which comprise other known condensed cluster structures and suggest that further related structures should exist.

Sc_7Cl_{10} and $R_7X_{10}C_2$. Comparison of the structures of Sc_7Cl_{10} (Poeppelmeier and Corbett 1977a) and $Sc_7Cl_{10}C_2$ (Hwu et al. 1985) projected down the short monoclinic *b* axis in fig. 20 reveals no big difference between them. Both contain twin chains of edge-sharing Sc_6 octahedra. A one-dimensional unit is shown in fig. 21. The atomic arrangement can be decomposed into two chains of *trans*-edge-sharing octahedra that are fused by sharing two *cis*-edges per octahedron. Each Sc_6 octahedron therefore has four edges in common with adjacent ones. The carbon atoms are located near the centers of the octahedra. As in the structure of R_5X_8Z, additional single R atoms occur in the structures which are octahedrally coordinated by Cl atoms. The orientations of these $ScCl_6$ octahedra are different in the two structures. A more profound difference is that the Cl atoms nearest to the twin chain cap the octahedra above faces in Sc_7Cl_{10} and above edges in $Sc_7Cl_{10}C_2$. The difference finds a simple explanation in terms of electrostatic interactions between the anions: the occupation of the cluster centers of Sc_7Cl_{10} by the highly charged carbide anions results in short C–Cl

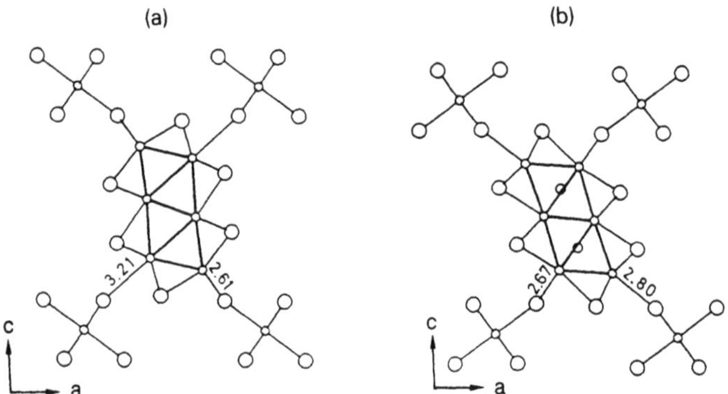

Fig. 20. Comparison of the structures of (a) Sc_7Cl_{10} and (b) $Sc_7Cl_{10}C_2$ projected along [010]. Distances in Å units. (Hwu et al. 1985).

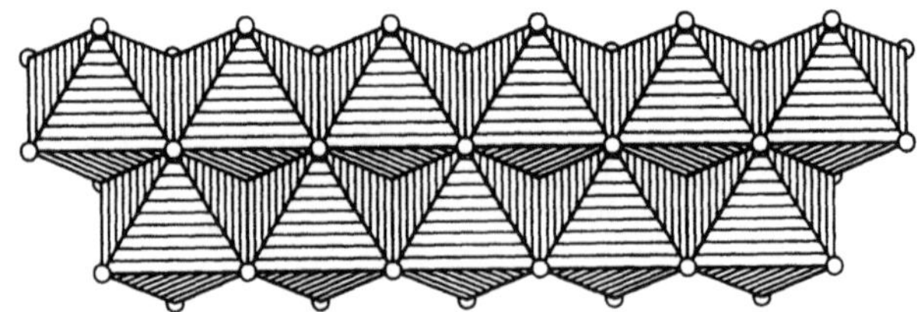

Fig. 21. Twin chain of edge-sharing Sc_6 octahedra in the structures of Sc_7Cl_{10}.

distances (2.88 Å) and strong repulsion between these ions. As the Cl^- ions are shifted above octahedral edges in $Sc_7Cl_{10}C_2$, the C–Cl distances are increased to 3.48 Å. In fact, the M_6X_8-type arrangement in Sc_7Cl_{10} witnesses that there are no interstitial atoms present in the octahedral voids. If at all, they are located in tetrahedral voids.

As in the case of Gd_2Cl_3 and Gd_2Cl_3N, the comparison of the unit cell volumes of Sc_7Cl_{10} and $Sc_7Cl_{10}C_2$ leads to a remarkable result. The ternary compound has a nearly 7% smaller cell volume. Above all, this overcompensation for the volume of the additional C atoms is due to a more effective space-filling with the ccp arrangement of the anions in the ternary compound. The rearrangement of the Cl atoms in both compounds leads to drastic changes in the Sc–Cl distances from the twin chains (fig. 20). Whereas one distance to the $ScCl_6$ unit is slightly longer in $Sc_7Cl_{10}Cl_2$ (2.80 versus 2.61 Å), the other one is shorter by more than 0.5 Å.

$R_6X_7C_2$. This structure type is realized with $Y_6I_7C_2$ (Kauzlarich et al. 1988), $Gd_6X_7C_2$ (X = Br I). The structures first reported as "Tb_6Br_7" and "Er_6I_7" refine as $Tb_6Br_7C_{2.0\pm0.4}$ and $Er_6I_7C_{1.5\pm0.4}$ when using the original data sets (Simon and Warkentin 1982). Figure 22 shows the twin chains of metal octahedra coordinated by halogen atoms above all the free edges. Such twin chains form a close packing of rods along [010] and are interconnected according to $(RR_{1/3}R_{2/2}R_{2/3}C)_2X^i_4X^{i-i}_{2/2}X^{i-a}_{2/2}X^{a-i}_{2/2}$. The formula means that four halogen atoms coordinate only one twin chain, two halogen atoms are located above the edges of neighboring twin chains (X^{i-i}), and two halogen atoms bridge via edge and corner positions (X^{i-a}, X^{a-i}). The small number of interchain links explains the very fibrous nature of the crystals, which easily splice along [010]. As indicated in fig. 23, the structure of $R_6X_7C_2$ can be derived from a ccp

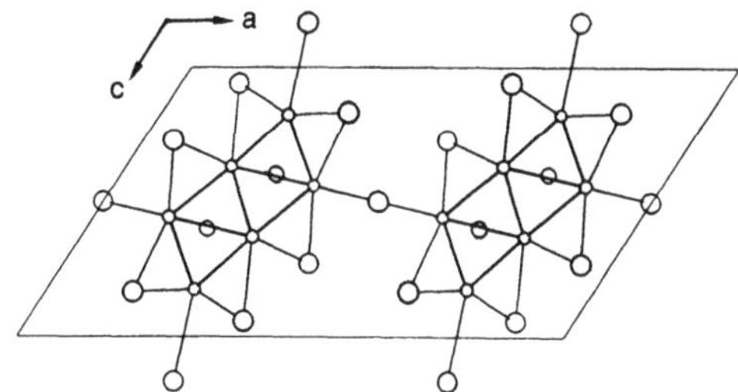

Fig. 22. Projection of the structure of $Gd_6I_7C_2$ along [010] (Simon 1988).

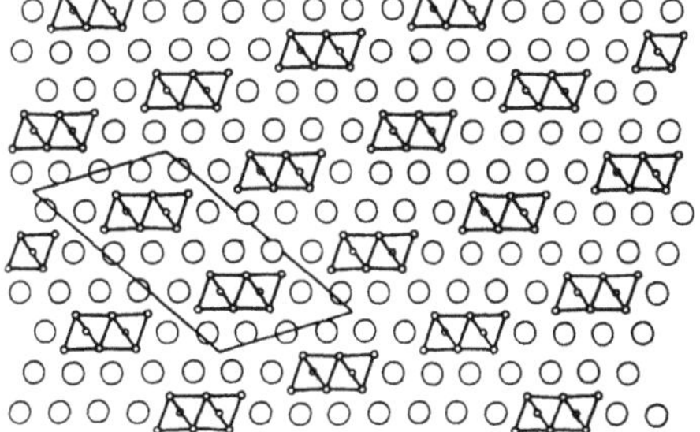

Fig. 23. Projection of the structure of $Gd_6I_7C_2$ indicating the ccp arrangement of anions.

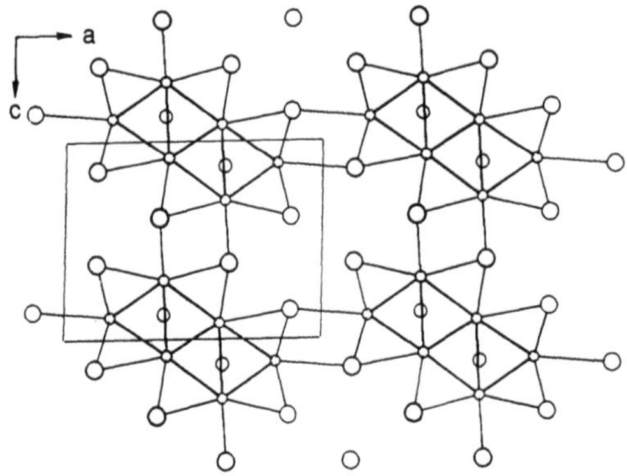

Fig. 24. Projection of the structure of Gd_3I_3C along [010] (Mattausch et al. 1987).

arrangement of X atoms (stacking vector, e.g. [304]) with $\frac{2}{9}$ of the X atoms substituted by C atoms. The R atoms take the positions of the octahedral voids around all C atoms and they repeat along the stacking vector [304] after $18 \times 3 = 54$ layers. The R_6 octahedra are rather distorted with the shortest R–R distance along the shared edges ($Y_6I_7C_2$: 3.33, 3.44 Å), compared with the largest separation along the repeat unit (3.91 Å) and the waist–apex distances ranging from 3.52 to 3.68 Å.

Gd_3I_3C. The structure of $Gd_3I_3C \mathrel{\hat{=}} Gd_6I_6C_2$ (Mattausch et al. 1987) is closely related to the structure of the compounds $R_6X_7C_2$. It also contains twin chains of C-centered Gd_6 octahedra surrounded by I atoms above all the free edges. These one-dimensional units are oriented along [010] and linked according to $(GdGd_{1/3}Gd_{2/2}Gd_{2/3}C)_2I^{i}_{2}I^{i-i}_{4/2}I^{i-a}_{2/2}I^{a-i}_{2/2}$. The lower X/R ratio compared to $R_6X_7C_2$ results in more X atom bridges. Comparison of figs. 22 and 24 indicates close similarities, but also distinct differences in their patterns of void filling. In both the $R_6X_7C_2$ and Gd_3I_3C structures close-packed planes of X and C atoms exist which are stacked in a ccp sequence along [304] and [101], respectively. The main difference lies in the fact that in the structure of Gd_3I_3C pure I atom layers alternate with I- and C-containing layers, whereas in the structure of $R_6X_7C_2$ all layers are X–C mixed. This difference can be systematized in terms of different structural series derived by fusing chains of condensed R_6 octahedra.

2.2.4. *Chain condensation*

The characteristic features which make the structures of Gd_3I_3C and $R_6X_7C_2$ different are both simultaneously present in the single-chain structure of R_4X_5C. Figure 25 shows the projection of the structure along [010] with two different types of layering indicated by arrows. Both clearly reveal the ccp anionic sublattice, with the R_6C octahedron replacing an X atom. The mode of replacement is different with respect to the horizontally oriented layers in the two presentations.

(a) In a layer parallel to (101) every sixth row of X atoms is replaced by a chain of R_6C octahedra. The repeat distance corresponds to $3 \times 6 = 18$ layers.

(b) Parallel to ($\bar{2}01$) one third of the X atom rows is replaced by a chain in every second layer. The repeat distance is ten layers.

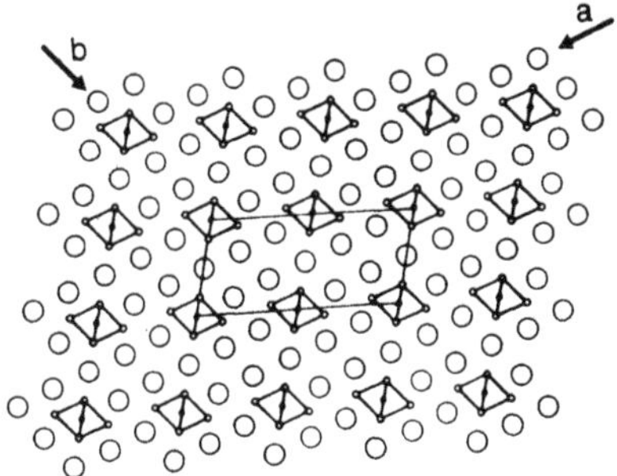

Fig. 25. Projection of the structure of Gd_4I_5C with two kinds of layering indicated by stacking directions: (a) layers parallel to (101) and (b) layers parallel to ($\bar{2}01$).

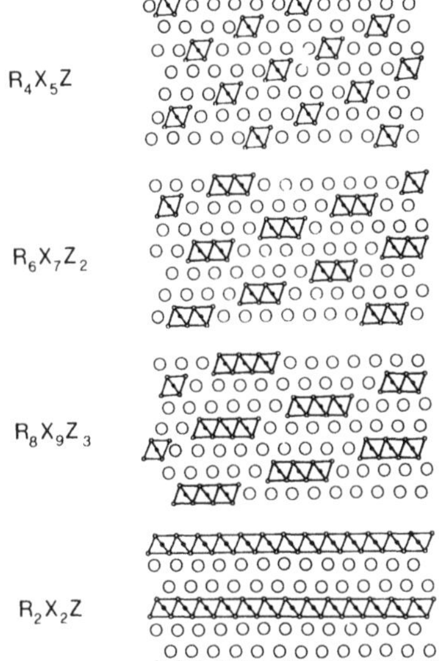

Fig. 26. Structural series $R_{2n+2}X_{2n+3}Z_n$ derived from fig. 25a by increasing the cluster content within the (101) layers.

Starting from the two settings and replacing adjacent rows of X atoms by chains of R_6Z octahedra fragments in a linear fashion lead to two different structure series. Starting from (a) results in compounds which follow the general formula $R_{2n+2}X^{i}_{2}X^{i}_{2n-2}X^{i-i}_{2/2}X^{i-a}_{2/2}X^{a-i}_{2/2}Z_n$ where n denotes the number of condensed chains and Z indicates that interstitials other than C, e.g. Si, may occur. Every chain unit is coordinated by $2n$ X^{i} above the edges of the octahedra, two X^{i-i} above edges of neighboring chains, and two anions X^{i-a} and X^{a-i} between edges and corners of two chains. The series of compounds $R_{2n+2}X_{2n+3}Z_n$ is schematically shown in fig. 26. So far, the compounds R_4X_5Z with $\frac{1}{6}$ of the X atoms replaced and $R_6X_7C_2$ with $\frac{2}{9}$ replaced are known. The series ends in a final member R_2X_2Z which contains van der Waals bonded slabs with a stacking sequence AγbαC (positions of R: α, γ; X: A, C; Z: b). Polymorphs of this single-slab structure are known. Starting from case (b) the general structural formula $R_{2n+2}X^{i}_{2}X^{i-i}_{(2n-2)/2}X^{i-i}_{2/2}X^{i-a}_{2/2}X^{a-i}_{2/2}Z_n$ results. In this series of compounds only two halogen atoms coordinate exclusively one unit above the edges of the octahedron; the others link adjacent units via edges and corners as indicated.

Figure 27 shows the series of compounds with a general composition $R_{2n+2}X_{n+4}Z_n$. In a ccp arrangement of X atoms, $n/(2n+4)$ of these are replaced by Z atoms. The final member of this series has the composition R_2XZ, and consists of twin layers of R atoms centered at Z atoms and sandwiched by the halide atoms above the edges of the

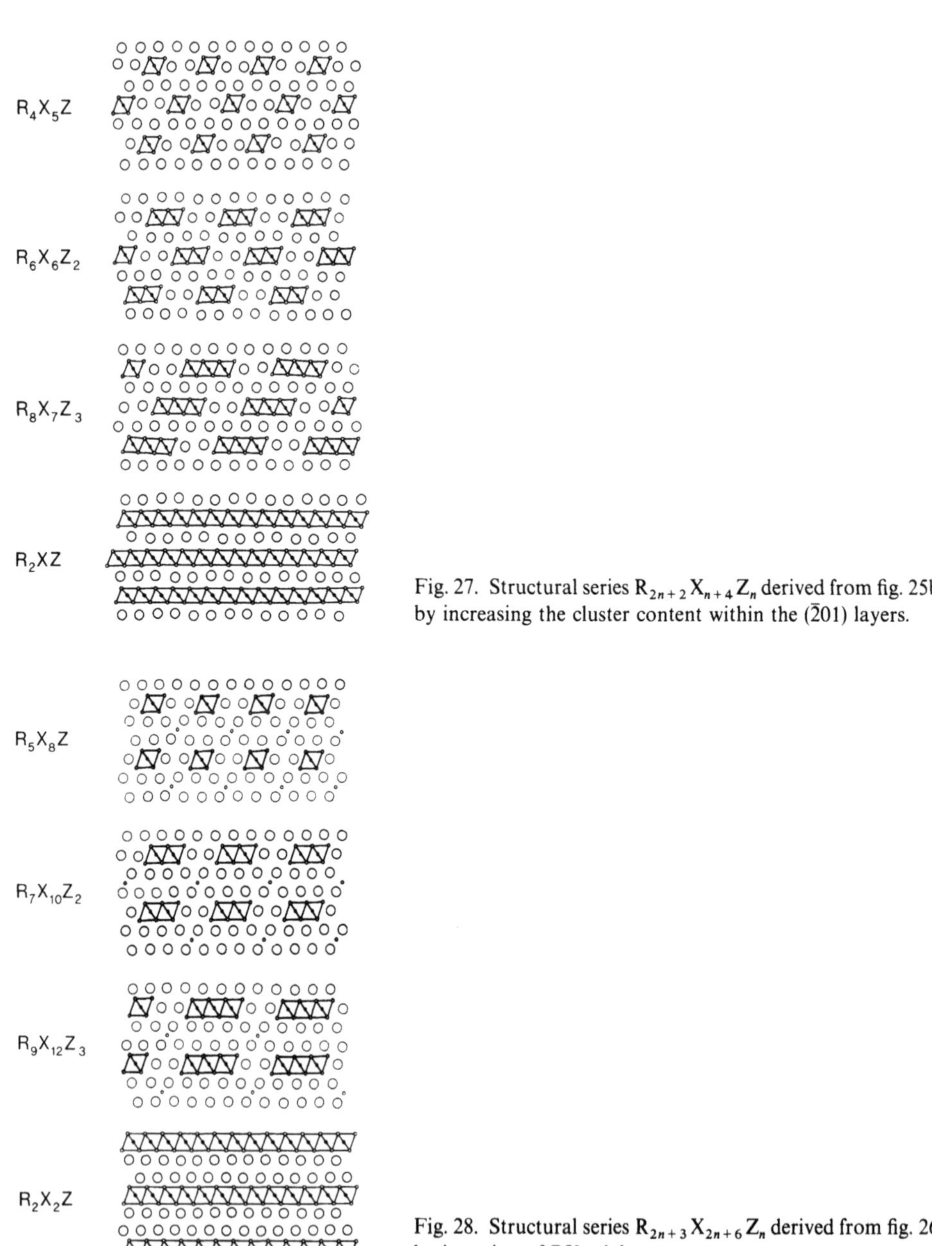

Fig. 27. Structural series $R_{2n+2}X_{n+4}Z_n$ derived from fig. 25b by increasing the cluster content within the ($\bar{2}01$) layers.

Fig. 28. Structural series $R_{2n+3}X_{2n+6}Z_n$ derived from fig. 26 by insertion of RX_3 slabs.

metal octahedra. These structures are realized with the polytypes of Gd_2XC (X = Cl, Br, I).

Last, but not least, the principle realized in the $R_7X_{10}C_2$-type compounds relates closely to the case (b) series. In a formal step RX_3 layers are inserted between the layers containing the clusters. Thus, the anionic substructures are characterized by the alternation of two pure X atom layers and one X and C mixed layer as illustrated in fig. 28. The stacking sequence is changed from pure cubic (c) into a chh sequence. If one omits the $n = 0$ member (RX_2, $CdCl_2$ type structure, realized in one of the PrI_2 modifications), R_5X_8Z and $R_7X_{10}Z$ (Berroth and Simon 1980) are the first members of this series which has the general composition $R_{2n+3}X_{2n+6}Z_n$.

2.3. *Layers and networks*

The fusion of chains of condensed R_6 octahedra led to different series of structures discussed in the last subsection and finally ended up in layered structures. These divide into structures with planar four-layer slabs (. . . XRRX . . .), structures with planar three-layer slabs (. . . XRR . . .) and structures with puckered layers. Compounds based on three-dimensional networks of R_6 octahedra are rare and are therefore discussed together with the layered compounds.

In all cases the tetrahedral and/or octahedral voids in the metal sublattices are filled with interstitial atoms. To cover the wealth of existing compounds, a subdivision into carbide halides and hydride halides has been made in the following discussion.

2.3.1. *Carbide halides*

In all structures the carbido species, which might be single C atoms or C_2 units, occupy octahedral voids. The compounds are summarized in table 6.

R_2X_2C and $R_2X_2C_2$. Figure 29 shows the different structures of rare earth metal carbide halides based on planar four-layer slabs.

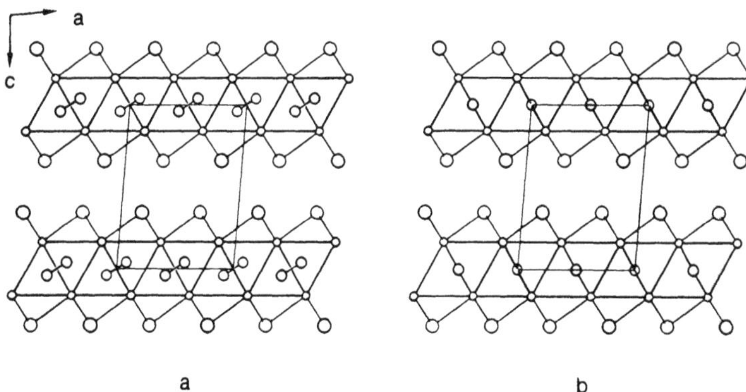

Fig. 29. Projections of the structures of (a) $Gd_2Br_2C_2$ and (b) Gd_2Br_2C along [010] (Schwanitz-Schüller and Simon 1985b).

TABLE 6
Crystallographic data of metal-rich rare earth halides with metal atom bilayers.

Compound	Space group; lattice constants (Å); angle (deg)	d_{R-R} (Å)	d_{R-X} (Å)	d_{R-Z} (Å)	Ref.[a]
Sc_2Cl_2C(1T)	$P\bar{3}m1$; $a = 3.399$, $c = 8.858$	3.123–3.399	2.573	2.308	[1]
Sc_2Cl_2C(3R)	$R\bar{3}m$; $a = 3.4354$, $c = 26.604$	3.214–3.435	2.605	2.352	[1]
Sc_2Cl_2N(1T)	$P\bar{3}m1$; $a = 3.349$, $c = 8.808$	3.057–3.349	2.556	2.267	[1]
Y_2Cl_2C(1T)	$P\bar{3}m1$; $a = 3.705$, $c = 9.183$	3.304–3.705	2.748	2.482	[1]
$Gd_2Cl_2C_2$	C2/m; $a = 6.954$, $b = 3.783$ $c = 9.361$, $\beta = 95.23$	3.465–3.965	2.764–2.923	2.344–2.679	[2]
$Gd_2Br_2C_2$	C2/m; $a = 7.025$, $b = 3.8361$ $c = 9.868$; $\beta = 94.47$	3.451–4.002	2.908–3.030	2.37–2.66	[3]
Gd_2Br_2C(1T)	$P\bar{3}m1$; $a = 3.8209$, $c = 9.8240$	3.432–3.821	2.946	–	[3]
Lu_2Cl_2C(1T)	$P\bar{3}m1$; $a = 3.5972$, $c = 9.0925$	3.245–3.597	2.689	2.424	[4]
Lu_2Cl_2C(3R)	$R\bar{3}m$; $a = 3.6017$, $c = 27.160$	3.315–3.602	2.702	2.448	[4]
Gd_2ClC	$R\bar{3}m$; $a = 3.6902$, $c = 20.308$	3.499–3.690	2.920	2.543	[2]
Gd_2BrC	$P6_3/mmc$; $a = 3.7858$, $c = 14.209$	3.300–3.786	3.185	2.510	[2]
Gd_2IC	$P6_3/mmc$; $a = 3.8010$, $c = 14.794$	3.385–3.801	3.259	2.545	[5]
$Gd_6Cl_5C_3$	C2/m; $a = 16.688$, $b = 3.6969$ $c = 12.824$; $\beta = 128.26$	3.399–3.889	2.755–2.903	2.442–2.757	[6]
Gd_3Cl_3C	$I4_132$; $a = 10.734$	3.294–3.676	2.800–2.876	2.513	[7]
Gd_3Cl_3B	$I4_132$; $a = 10.960$	3.363–3.753	2.859–2.936	2.565	[8]
Gd_3I_3Si	$I4_132$; $a = 12.052$	3.697–4.121	3.144–3.233	2.814	[9]

[a] References: [1] Hwu et al. (1986); [2] Schwarz (1987); [3] Schwanitz-Schüller and Simon (1985b); [4] Schleid and Meyer (1987); [5] Mattausch et al. (1987); [6] Simon et al. (1987a); [7] Warkentin and Simon (1983); [8] Mattausch et al. (1984); [9] Nagaki et al. (1989).

The similarity with the structures of the binary compounds ZrX is evident. Close-packed bilayers of metal atoms are sandwiched by layers of halogen atoms. Such slabs are bonded to adjacent ones via van der Waals interactions. In space group $R\bar{3}m$ for both ZrCl and ZrBr, identity along [001] occurs after three slabs with the stacking sequence ... AbcABcaBCabC ... (ZrCl) (Izmailovich et al. 1974, Adolphson and Corbett 1976) and ... AcbABacBCbaC ... (ZrBr) (Daake and Corbett 1977). The same space group holds for the so-called 3R form of R_2X_2C, which has a stacking sequence ... AbaBCacABcbC .... For the 1T form (Marek et al. 1983) the translational period corresponds to one slab (space group $P\bar{3}ml$) as it does for the structure of $Gd_2X_2C_2$ (Schwanitz-Schüller and Simon 1985b, Schwarz 1987) (space group C 2/m). The ordering of the layers is centrosymmetric in both cases (AbaB, in contrast to the noncentrosymmetric sequence AbaC).

Formally, all these structures are related in the following way: if the distinction between metal and halogen atoms is neglected, the ZrCl structure is cubic close-packed (c), the arrangement in (1T-)R_2X_2C and $Gd_2X_2C_2$ corresponds to hexagonal close-packing (h), and that in ZrBr to a mixture (cch). The differentiation between metal and halogen atoms for (c) reduces Fm3m to R$\bar{3}$m (3R-R_2X_2C), and for (h) reduces $P6_3$/mmc to P$\bar{3}$ml (1T-R_2X_2C). Through the introduction of the C_2 unit at an inclined position to the stacking direction, the symmetry is further decreased to C2/m for $Gd_2X_2C_2$.

So far, the similarities between these layered carbide halides and the ZrX structures have been emphasized. The main difference between them is due to the different stacking of layers within one slab. In the ZrX structures the X atoms cap triangular faces which enclose the octahedral voids in the metal atom bilayers (related to the arrangement in the discrete M_6X_8 cluster), whereas in the carbide halides the triangular faces above tetrahedral voids are capped. The latter means an edge-capping to the condensed octahedra which relates to the arrangement in the discrete M_6X_{12}-type cluster. The difference clearly has reasons based on electrostatics as the occupation of the octahedral void by negatively charged interstitials makes the X atom position above octahedral faces unfavourable.

For both polytypes of R_2X_2C the R–R distances in the layers are longer than the distances across adjacent layers. The occupation of the octahedral voids by the C_2 unit in $Gd_2X_2C_2$ lengthens the octahedra in a direction which is parallel to the C–C bond, yet the shortest Gd–Gd distances are still those between adjacent layers.

The C–C distance $d_{C-C} = 1.27$ Å in $Gd_2Br_2C_2$ (Schwanitz-Schüller and Simon 1985b) is particularly short. The finding is easily explained in terms of the special electron count for this compound. According to $(Gd^{3+})_2(Br^-)_2C^{4-}$ four electrons from the framework enter the MOs of the C_2 unit. Two antibonding MOs are unoccupied corresponding to a C=C double bond. The electron count for R_2X_2C agrees with the observation of single C atoms.

Sc_2Cl_2N. (Hwu et al. 1986) is the only nitride compound in this series. It has the 1T structure and – similar to Ta_2S_2C (Beckmann et al. 1970) which occurs with the 1T and 3R structures – has an excess of electrons which occupy extended metal–metal bonding states according to $(Sc^{3+})_2(Cl^-)_2N^{3-}(e^-)_2$. The bonding in Sc_2Cl_2N indicates that strong electron donors might be intercalated into the corresponding R_2X_2C compounds generating metal–metal bonds. Yet, neither Sc_2Cl_2C (Hwu et al. 1986) nor Lu_2Cl_2C (Schleid and Meyer 1987) could be intercalated, although 1T-$M_xY_2Cl_2C$ (Meyer et al. 1986) phases are known.

Gd_2XC. As discussed in the preceding section, the compounds R_2XC are the final members of a series which starts from R_4X_5C and $R_6X_6C_2$ and has the general formula $R_{2n+2}X_{n+4}C_n$. The compounds Gd_2XC could be prepared and characterized (X = Cl, Br, I) (Mattausch et al. 1987, Schwarz 1987). Figure 30 shows perspective views of the structure of (a) Gd_2ClC and (b) Gd_2XC (X = Br, I). Both structures

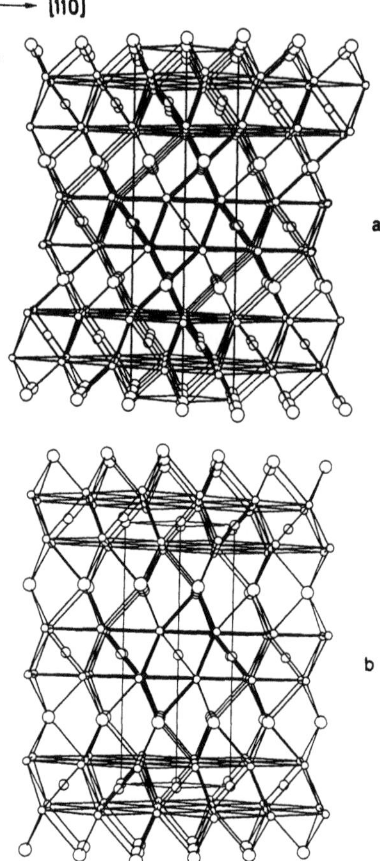

Fig. 30. Central projections of the structures of (a) Gd_2ClC and (b) Gd_2IC along [110] (Mattausch et al. 1987).

contain planar layers of edge-sharing Gd_6C octahedra. Such layers are linked via single layers of halogen atoms which lie above the edges of the octahedra. The difference between the two structure types is concerned with the X atom coordination. The Cl atom is coordinated by a trigonal antiprism of Gd atoms, whereas the Br and I atoms lie in centers of trigonal prisms. In the Gd_2ClC structure three GdGdCl slabs are stacked in a single translational periods, whereas in the Gd_2(Br, I)C structure only two.

In a generalized view, Gd_2ClC has a rock salt structure with two kinds of atoms forming the anionic substructure. Alternatively, the structure is described as a filled $CdCl_2$-type structure. This view bears some interest as it relates the metal–metal bonded compound $((Gd^{3+})_2Cl^-C^{4-}e^-)$ to Ag_2F (Argay and Naray-Szabo 1966) which has the simple $CdCl_2$-type structure stabilized by metal–metal bonding without the need of interstitial atoms $[(Ag^+)_2F^-e^-]$. The stacking sequence of the anions in Gd_2BrC and Gd_2IC is ... ABACA ... With the occupation of all octahedral voids a layered intergrowth of NaCl- and NiAs-type fragments results, which is already known from the "H"-phase structure, e.g. Ti_2SC (Reiss 1971).

$Gd_6Cl_5C_{3+x}$. Figure 31 presents a projection of the structure of $Gd_6Cl_5C_3$ (Simon et al. 1988) along [010]. Edge-sharing Gd_6C octahedra form undulating layers with Cl atoms capping all free edges of the octahedra.

The anions together form a ccp arrangement with close-packed layers in the (101) and ($\bar{1}01$) planes. The substitution pattern in the ($\bar{1}01$) plane is particularly interesting as it closely relates the structures of $Gd_6Cl_5C_3$ and Gd_3I_3C. In both, an identical number of adjacent rows of X atoms in a layer are substituted by C atoms. Whereas in $Gd_3I_3C \hat{=} Gd_6I_6C_2$ pure X atom layers are between such substituted ones according to $(Gd_6C_2X_2)(X_4)$, in $Gd_6Cl_5C_3$ these intermediate layers are also substituted according to $(Gd_6C_2X_2)(CX_3)$, and the layers are shifted in a way to yield octahedral metal atom coordination for all C atoms. Strong electrostatic repulsions between the additional C atoms and the C atoms (anions) in the twin chains can be inferred from the imbalance of the C–Gd distances for these atoms (2.59 and 2.44 Å, respectively, to the atoms in the basis of the Gd_6 octahedron).

$Gd_6Cl_5C_3$ exhibits a range of homogeneity indicated by significant changes of the lattice constants and physical properties. These changes are discussed in terms of a partial substitution of the single C atoms by C_2 units. Following an ionic description as $(Gd^{3+})_6(Cl^-)_5(C^{4-})_3e^-$, one electron has to enter a delocalized band having essentially Gd 5d character. A partial substitution of the C^{4-} ions is possible up to a composition $Gd_6Cl_5C_{3.5}$ when, according to $(Gd^{3+})_6(Cl^-)_5(C^{4-})_{2.5}(C_2^{6-})_{0.5}$, all electrons are removed from the metal–metal bonding band and localized in C–C bonds. An observed volume increase by 9.7 cm^3/mol and a transition from metallic to semiconducting behavior is in agreement with this interpretation, which could not be derived directly from the diffraction data. Such simultaneous presence of different carbon entities in rare earth metal carbide halides might be observed more frequently, although it should be limited to C_2^{6-} occurring together with C^{4-} or C_2^{4-} but not C^{4-} together with C_2^{4-}. Simultaneously present C^{4-} and C_2^{4-} would undergo a symproportionation reaction. Thus, the missing homogeneity range between Gd_2X_2C and $Gd_2X_2C_2$ agrees with expectation.

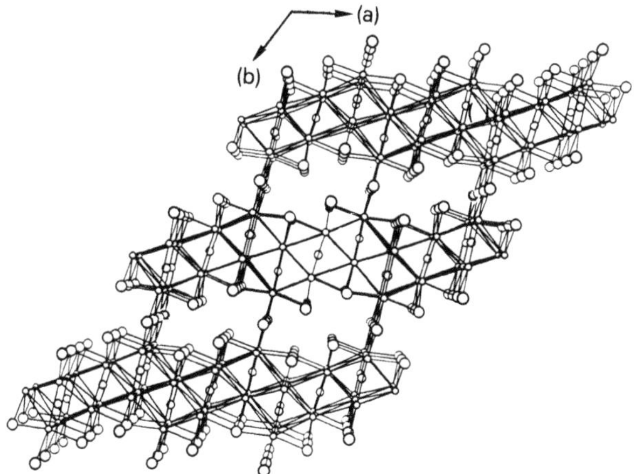

Fig. 31. Central projection of the structure of $Gd_6Cl_5C_3$ along [010] (Simon et al. 1988).

Gd_3Cl_3C. The structural principle met with $Gd_6Cl_5C_3$ offers a way to build up three-dimensional, condensed cluster structures by substituting C atoms for halogen atoms. Examples of this kind are not known. To our knowledge the Gd_3Cl_3C-type structure (Warkentin and Simon 1983) is the only one containing R_6Z octahedra condensed to give a three-dimensional network. The structure is also found for Gd_3Cl_3B (Mattausch et al. 1984) and Gd_3I_3Si (Nagaki et al. 1989). The interconnection pattern is illustrated in fig. 32. Each octahedron shares edges with three others in a plane. As one would expect, the repulsion between the highly charged interstitial atoms results in an elongation of all Gd–Gd distances in that plane (3.68 Å), whereas the shortest distances occur in the shared edges (3.29 Å). The Cl atoms cap free edges as in the M_6X_{12}-type cluster. The repetition of the interconnection pattern for the peripheral clusters in fig. 32 leads to a cubic structure.

The formulation as $Gd_{6/2}CCl^{i-i}_{9/3}$ corresponds to a description in terms of condensed clusters. On the other hand, the structure can be described as a defect rock salt structure again: the Cl atoms form a ccp arrangement with $\frac{1}{4}$ of them substituted by C atoms. Only the octahedral voids around these C atoms are occupied by the Gd atoms. In fact, Gd_3Cl_3C is isotypic with Ca_3PI_3 (Hamon et al. 1974) which is a colorless, salt-like compound. The difference between the two compounds lies in different electron counts which shift Gd_3Cl_3C towards the class of cluster compounds. According to $(Gd^{3+})_3(Cl^-)_3C^{4-}(e^-)_2$ two electrons per formula unit occupy delocalized metal–metal bonds. In agreement with this simplified description of the bonding, Gd_3Cl_3C is metallic.

At the end of the description of the rare earth metal carbide halides it seems worthwhile to summarize some facts. These compounds contain single C atoms, or C_2 entities with C–C single and double bonds. The kind of species seems entirely related to the number of residual valence electrons at the metal site. As we are dealing with electropositive metals, these electrons will be transferred to MOs of the C_2 unit and it is the number of vacant antibonding MOs which determines the kind of carbido species. Thus, the ideas of Atoji (1961) concerning binary carbides can be extended to the rare earth metal carbide halides. A more detailed discussion of the bonding will be given in sect. 3.

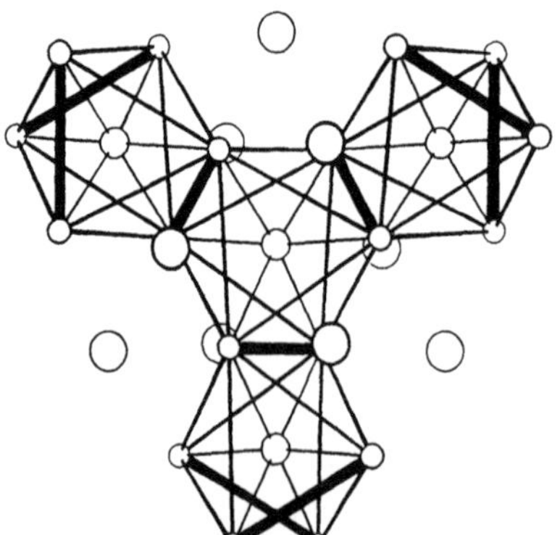

Fig. 32. Building unit of the edge-sharing Gd_6C octahedra with Cl atoms above all the free edges in the structure of Gd_3Cl_3C: all shared edges are emphasized by bold lines (Warkentin and Simon 1983).

2.3.2. *Hydride halides*

In contrast to all other rare earth metal compounds discussed so far, the hydride halides contain mobile interstitial atoms that can be added or (partly) removed at will in a topochemical reaction. These compounds therefore offer a unique possibility to study the delocalization of electrons in extended metal–metal bonds versus the localization at interstitial atoms. The known compounds are summarized in table 7.

RXH_x. The hydride halides RXH of the divalent rare earth metals have been known for a long time. All of them, EuXH, YbXH with X = Cl, Br, I, and SmBrH (Beck and Limmer 1982) crystallize in the PbFCl-type structure, which is also adopted by the hydride halides of the alkaline earth metals MXH (Ehrlich et al. 1956), by the mixed halides RXX′ of divalent lanthanides, and many oxyhalides ROX of the trivalent metals. The colorless compounds RXH of R = Sm, Eu, Yb therefore have to be addressed as "normal" salts. The hydrogen content of these compounds is strictly stoichiometric.

The hydride halides RXH_x (Mattausch et al. 1985a) of the trivalent metals are distinctly different as they look like graphite and are metallic. They all exhibit a range of homogeneity which ends at an upper value $x = 1.0$. There is still some discrepancy concerning the lowest possible hydrogen content in special systems. Definitely, the "monohalides" (Simon et al. 1976, Poeppelmeier and Corbett 1977b, Mattausch et al. 1980c) first described as binary compounds are hydride halides (Simon et al. 1987b). Obviously these hydride halides can be prepared from all trivalent rare earth metals. Up to now the following phases have been prepared and characterized: RClH (R = Y, La, Ce, Pr, Gd, Tb), RBrH (R = Y, La, Ce, Pr, Nd, Gd, Tb) and GdIH (Mattausch et al. 1985a, Michaelis and Simon 1986, Müller-Käfer 1988). The heavy-atom substructures that have been found with RXH_x are drawn in fig. 33. In all cases close-packed bilayers of metal atoms (layers of edge-sharing octahedra) are sandwiched

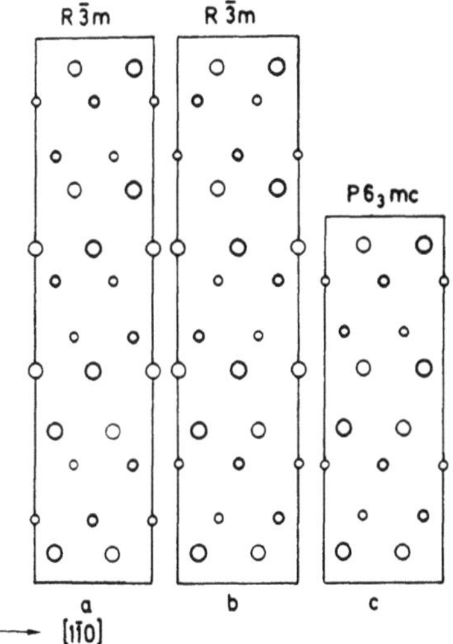

Fig. 33. The different stacking variants for the heavy-atom arrangements in hydride halides RXH_x projected along [110] of the hexagonal unit cells which are outlined: (a) ZrCl type, (b) ZrBr type and (c) "2s" type.

TABLE 7

Crystallographic data of rare earth hydride halides RXH_x ($x \leqslant 1.00$) and RXH_2; D positions have been determined only for $TbClD_{0.78}$, $TbBrD_{0.88}$ and $TbBrD_2$,

Compound	Space group; structure type; lattice constants (Å)	d_{R-R} (Å)	d_{R-X} (Å)	d_{R-Z} (Å)	Ref.[a]
$ScClH_{0.38}$	$R\bar{3}m$; ZrBr;				
to	$a = 3.4760$, $c = 26.710$	3.354–3.476	2.617	2.12	[1]
$ScClH_{0.70}$	$a = 3.4766$, $c = 26.622$				
$ScClH_{0.70}$	$R\bar{3}m$; ZrCl;				
to	$a = 3.4766$, $c = 26.622$	3.222–3.478	2.584	2.06	[1]
$ScClH_{1.06}$	$a = 3.4785$, $c = 26.531$				
$YClH_{0.69}$	$R\bar{3}m$; ZrBr;				
to	$a = 3.7518$, $c = 27.516$	3.515–3.752	2.773	2.23	[2]
$YClH_{0.81}$	$a = 3.7521$, $c = 27.460$				
$YClH_{0.82}$	$R\bar{3}m$; ZrCl;				
to	$a = 3.7526$, $c = 27.387$	3.377–3.753	2.743	2.17	[2]
$YClH_{0.91}$	$a = 3.7534$, $c = 27.375$				
$LaClH_{0.82}$	$R\bar{3}m$; ZrBr;				
to	$a = 4.0991$, $c = 27.585$	3.648–4.099	2.935	2.33	[1]
$LaClH_{0.98}$	$a = 4.0983$, $c = 27.568$				
$CeClH_{0.67}$	$R\bar{3}m$; ZrBr;				
to	$a = 4.036$, $c = 27.614$	3.592–4.036	2.890	2.30	[3]
$CeClH_{0.97}$	$a = 4.034$, $c = 27.580$				
$CeBrH_{0.70}$	$R\bar{3}m$; ZrCl;				
to	$a = 4.064$, $c = 29.437$				
$CeBrH_{0.97}$	$a = 4.068$, $c = 29.414$	3.754–4.068	3.015	2.41	[3]
$PrClH_{0.69}$	$R\bar{3}m$; ZrBr;				
to	$a = 3.992$, $c = 27.608$	3.610–3.995	2.886	2.30	[3]
$PrClH_{0.95}$	$a = 3.995$, $c = 27.561$				
$PrBrH_{0.70}$	$R\bar{3}m$; ZrCl;				
to	$a = 4.0250$, $c = 29.386$	3.737–4.025	2.994	2.38	[3]
$PrBrH_{0.94}$	$a = 4.0299$, $c = 29.348$				
$NdBrH_{0.89}$	$R\bar{3}m$; ZrCl;				
to	$a = 3.9913$, $c = 29.277$	3.716–3.991	2.975	2.36	[4]
$NdBrH_{1.00}$	$a = 3.991$, $c = 29.276$				
$GdClH_{0.73}$	$R\bar{3}m$; ZrBr;				
to	$a = 3.8232$, $c = 27.557$	3.544–3.823	2.81	2.25	[5]
$GdClH_{1.00}$	$a = 3.8261$, $c = 27.471$				
$GdBrH_{0.69}$	$R\bar{3}m$; Zr;				
to	$a = 3.8694$, $c = 29.150$	3.548–3.869	2.924	2.27	[5]
$GdBrH_{0.93}$	$a = 3.8261$, $c = 29.041$				
$GdIH_{0.80}$	$R\bar{3}m$; ZrCl;				
	$a = 3.9297$, $c = 31.003$	3.901–3.982	3.028	2.46	[5]
$GdIH_{0.80}$	$R\bar{3}m$; ZrBr;				
	$a = 3.9821$, $c = 31.326$	3.404–3.930	3.245	2.28	[5]
$GdIH_{0.80}$	$P6_3mc$; 2s				
	$a = 3.9825$, $c = 20.832$	3.592–3.9825	3.166	2.33	[5]
$TbBrD_{0.69}$	$R\bar{3}m$; ZrCl;				
to	$a = 3.8346$, $c = 29.057$	3.528–3.837	2.905	2.253–2.268	[6]
$TbBrD_{0.88}$	$a = 3.8371$, $c = 28.986$				

TABLE 7 (cont'd)

Compound	Space group; structure type; lattice constants (Å)	d_{R-R} (Å)	d_{R-X} (Å)	d_{R-Z} (Å)	Ref.[a]
$TbClD_{0.78}$	$R\bar{3}m$; ZrBr; $a = 3.7800$, $c = 27.494$	3.524–3.780	2.786	2.230–2.247.	[7]
$TbClD_{0.86}$	$R\bar{3}m$; ZrCl; $a = 3.784$, $c = 27.418$	3.391–3.784	2.759	2.23	[2]
$ErBrH_{0.70}$	$R\bar{3}m$; ZrCl; $a = 3.7685$, $c = 28.729$	3.481–3.769	2.862	2.23	[1]
$LuClH_x$	$R\bar{3}m$; ZrCl; $a = 3.6383$, $c = 27.102$	3.364–3.638	2.685	2.15	[8]
$YClH_2$	$R\bar{3}m$; $a = 3.712$, $c = 29.346$	3.712–3.805	2.766	2.15–2.32	[2]
$CeClH_2$	$R\bar{3}m$; $a = 3.963$, $c = 29.840$	3.931–3.963	2.900	2.29–2.46	[3]
$CeBrH_2$	$R\bar{3}m$; $a = 4.000$, $c = 31.513$	4.000–4.090	2.986	2.324–2.50	[3]
$PrClH_2$	$R\bar{3}m$; $a = 3.915$, $c = 29.650$	3.898–3.915	2.870	2.27–2.43	[3]
$PrBrH_2$	$R\bar{3}m$; $a = 3.953$, $c = 31.397$	3.953–4.065	2.951	2.29–2.47	[3]
$NdBrH_2$	$R\bar{3}m$; $a = 3.923$, $c = 31.320$	3.923–4.048	2.934	2.27–2.45	[3]
$GdClH_2$	$R\bar{3}m$; $a = 3.777$, $c = 29.467$	3.777–3.836	2.799	2.19–2.35	[2]
$GdBrH_2$	$R\bar{3}m$; $a = 3.819$, $c = 31.003$	3.819–3.987	2.876	2.21–2.39	[2]
$GdID_2$	$R\bar{3}m$; $a = 3.899$, $c = 33.192$	3.899–4.209	2.996	2.26–2.46	[2]
$TbClD_2$	$R\bar{3}m$; $a = 3.738$, $c = 29.349$	3.738–3.950	2.836	2.16–2.38	[2]
$TbBrD_2$	$R\bar{3}m$; $a = 3.781$, $c = 30.883$	3.781–3.965	2.855	2.189–2.378	[5]

[a] References: [1] Meyer et al. (1986); [2] Mattausch et al. (1984); [3] Müller-Käfer (1988); [4] Michaelis and Simon (1986); [5] Mattausch et al. (1985); [6] Cockcroft et al. (1989a); [7] Ueno et al. (1984); [8] Schleid and Meyer (1987).

by layers of halogen atoms. The X atoms cap the faces of the R_6 octahedra as in the equivalent structures of the binary phases ZrX. Thus, the structures can be derived from condensed M_6X_8-type clusters. The heavy-atom substructures (Mattausch et al. 1985a) exhibit the stacking sequences . . . AbcABcaBCabC . . . (ZrCl), . . . AcbABacBCbaC . . . (ZrBr) and . . . AbcABacB . . . ("2s" in analogy to NbS_2; Jellinek 1962) where A, B and C denote the positions of the X atom layers. In these structures all atoms are close-packed. Only in the stacking pattern . . . AbcAAcbA . . . ("2H" as it corresponds to the stacking in 2H–NbS_2 (Huster and Franzen 1985)) the XRRX slabs are arranged in a way to yield trigonal prismatic voids between them. This stacking was never observed with hydride halides, but was

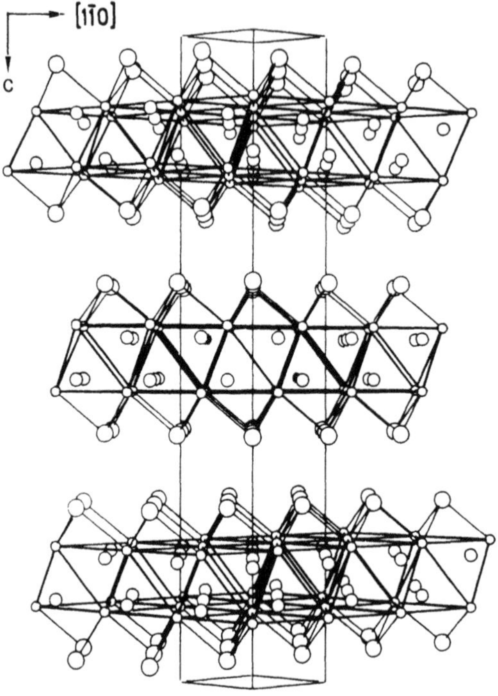

Fig. 34. Central projection of the structure of $TbClD_{0.8}$ along [110] of the hexagonal unit cell: the D atoms statistically occupy the tetrahedral voids within metal atom bilayers (Ueno et al. 1984).

described for intercalation compounds Li_xYClO (Ford et al. 1983, Ford and Corbett 1985).

Before the occurrence of the different stackings with RXH_x is discussed, the content and arrangement of the hydrogen atoms need to be analyzed. In the metal atom bilayers ($x \leqslant 1.0$) only tetrahedral voids are occupied by H atoms (fig. 34), as has been shown by neutron diffraction of $TbClD_{0.8}$ (Ueno et al. 1984). The arrangement of the X atoms above the octahedral faces is therefore easily understood in terms of electrostatics. Filling all tetrahedral voids leads to the upper limit $RXH_{1.00}$. The experimental finding for the lower limit in a number of systems corresponds to $RXH_{0.67}$, with an overall statistical distribution of the H atoms in $\frac{2}{3}$ of the available voids. A way to reach this lower limit is to heat single-phase RXH_x in a closed evacuated Ta capsule which is permeable to hydrogen (Simon et al. 1987b). The Ta capsule is sealed in vacuo in a Pt tube and the whole assembly is heated in a stream of O_2 to keep the H_2 pressure as low as possible. Single-phase products analyzed as $RXH_{0.67\pm0.03}$ for $YClH_x$ (Mattausch et al. 1986b), $CeXH_x$ (X = Cl, Br), $PrXH_x$ (X = Cl, Br) (Müller-Käfer 1988), and $GdXH_x$ (X = Cl, Br) (Simon et al. 1987b). This value was confirmed independently for $TbBrD_{0.69}$ (Cockcroft et al. 1989a) via neutron scattering at low temperatures. Single-phase products were never observed for $x < 0.67$.

The lattice constants of RXH_x depend in a remarkable way on the H content. As shown for the Gd compounds in fig. 35, the length of the *a* axis increases, whereas that of the *c* axis decreases with rising *x* (Simon et al. 1987b). The volume changes by an

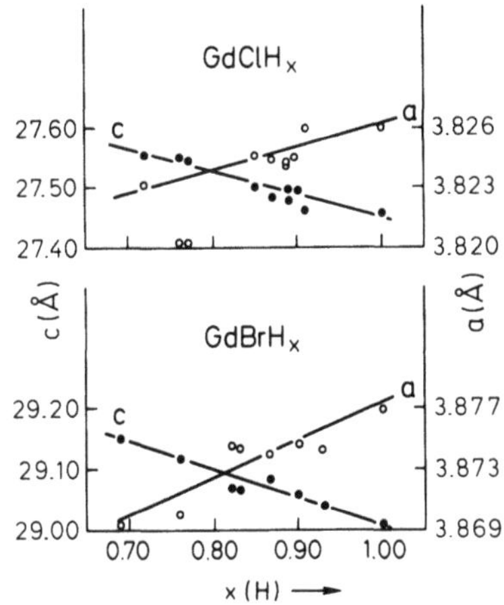

Fig. 35. Lattice constants of $GdXH_x$ (X = Cl, Br) as a function of x (Simon et al. 1987b).

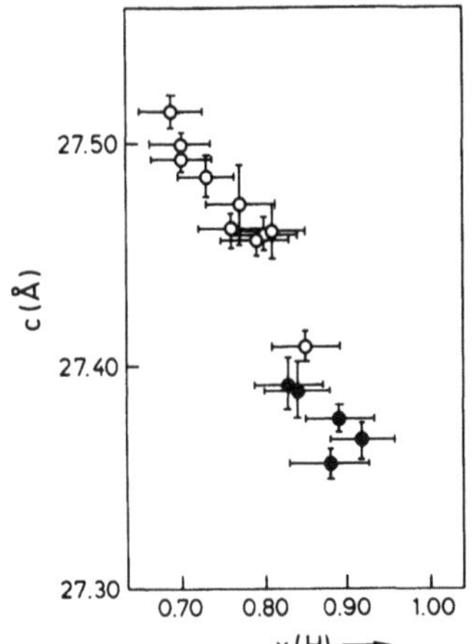

Fig. 36. Lattice constants of $YClH_x$ as a function of x: open circles denote ZrBr-type and solid circles ZrCl-type structure.

insignificant value of 0.1 cm^3/mol between $x = 0.7$ and 1.0. We meet the case that the volume increase by insertion of additional atoms is counterbalanced by the shrinkage of the metal atom size, as it was already discussed with Gd_2Cl_3N (and is well known for metallic hydrides). Similar monotonous changes of the lattice constants are found with the compounds of Ce and Pr.

Whereas $GdClH_x$ has the ZrBr-type structure, $GdBrH_x$ (as do all other $RBrH_x$) adopts the ZrCl-type structure. $GdBrH_{0.69}$ has also been observed with the "2s"-type structure. $TbClH_x$ and $YClH_x$ undergo changes from ZrCl- to ZrBr-type structures as a function of the hydrogen content. According to X-ray and neutron diffraction, $TbClD_x$ has the ZrBr-type structure up to $x = 0.8$, after which it adopts the ZrCl-type arrangement. In the $YClH_x$ system the change from the ZrBr-type ($x \lesssim 0.8$) to the ZrCl-type (Mattausch et al. 1986b) structure shows up particularly in the change of the c axis length as a function of the H content (fig. 36). For the special composition

$GdIH_{0.8}$ both the ZrCl- and "2s"-type structures have been observed (Mattausch et al. 1985a).

Clearly, the knowledge about the RXH_x phases is still rather fragmentary. The relative stabilities of the different polymorphs are by no means clear. Moreover, the amount of hydrogen needed to stabilize RXH_x phases is still a matter of dispute. Hydrogen absorption/desorption experiments on $ScClH_x$ yield $0.38 \leqslant x \leqslant 1.06$ which seems a very large range compared to the lanthanide compounds (Corbett 1986). Possibly intercalation of alkali metals can stabilize the phases RXH_x at lower H contents. Small amounts of Li or Na enter the antiprismatic voids between the XRRX slabs and phases like $Li_{0.1}GdClH_{0.45}$ have been described (Meyer et al. 1986). This phase obviously contains significantly less hydrogen than is possible with the alkali-metal-free hydride halides.

RXH_2. Heating the metallic phases RXH_x ($x \leqslant 1$) in hydrogen to 400°C produces transparent, nonconducting RXH_2 (Simon et al. 1987b). The phases characterized so far are summarized in table 7. They are isotypic and the structure has been solved for $TbBrD_2$ via neutron diffraction (Mattausch et al. 1985b). The result is shown in fig. 37. At first sight the structural principle of RXH_x ($x \leqslant 1$) appears to be preserved with the hydrogenation reaction: the topochemical reaction leaves the close-packed twin layers of metal atoms which are sandwiched by halogen atoms. Yet there are significant differences. For example, whereas the Tb–Tb distances within the layers of $TbClD_{0.8}$ and $TbBrD_2$ are the same (3.78 Å), the Tb–Tb distances between atoms in adjacent layers increase considerably from 3.58 to 3.96 Å when the additional hydrogen (deuterium) is incorporated. An even more dramatic change occurs with the X atoms which layerwise shift from positions above octahedral faces to positions above the triangular faces belonging to tetrahedral sites in the metal twin layers. Thus, the stacking sequence changes from AcbA to BcbC within one XRRX slab, which is the characteristic sequence for the carbide halides $R_2X_2C_x$.

The cause for the martensitic-type transition that occurs when RXH_x phases are hydrogenated to RXH_2 is easily understood along the lines of arguments presented for the layered carbide halides. In the structure of RXH_x ($x \leqslant 1$) only tetrahedral voids between the metal atom layers are occupied by H atoms. Therefore, positions near the octahedral voids are electrostatically favorable for the X atoms. In RXH_2 both the tetrahedral voids and the octahedral voids are occupied by H atoms. As the number of

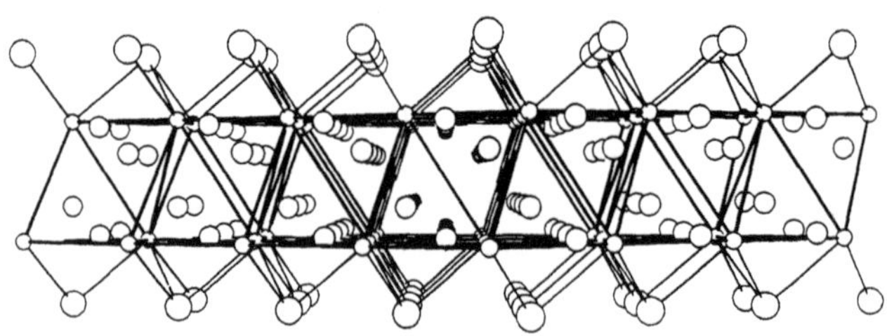

Fig. 37. Single slab in the structure of $TbBrD_2$ with D atoms in tetrahedral and trigonal planar coordination (Mattausch et al. 1985b).

octahedral voids is only half the number of tetrahedral voids, two H atoms have to enter each octahedral void. The insulating properties of RXH_2 show that all electrons are localized in heteropolar bonds according to an approximate description as $R^{3+}X^-(H^-)_2$. The strong repulsive interactions between the H^- ions lead to the significant dilatation of the metal atom bilayers in the [001] direction and fix the H^- ions in the trigonal faces of the R_6 trigonal antiprisms ("octahedra"). The positions above these faces therefore become less favorable for the X atoms than the positions near the (also filled) tetrahedral voids.

3. Chemical bonding and electronic structure

As illustrated in the previous section, the metal-rich rare earth metal halides and their interstitial derivatives provide a vast collection of compounds that transcends the structural chemistry of both molecules and extended solids. On the one hand, these substances can be considered as connected or condensed clusters of the M_6X_{12}- or M_6X_8-type, which may contain interstitial species. On the other hand, many of them can be derived from the structures of simple salts, e.g. NaCl or La_2O_2S.

These descriptions influence the way in which the chemical bonding and electronic structure of these materials are discussed. From the point of view of metal octahedra, the binary rare earth metal halides are more electron deficient than the analogous group 5 and group 6 chalcogenides, chalcogenide halides and halides. To overcome this deficiency, the group 3 and group 4 halides incorporate a variety of atomic or small molecular species at the center of the metal clusters. These compounds are stabilized relative to structures constructed from empty clusters by introducing strong interactions between the metal atoms and the interstitials at the expense of weakening the metal–metal interactions. When these materials are viewed as derivatives of simple salts, the halide and interstitial species together form some kind of close-packing, e.g. frequently ccp, and the rare earth metals occupy the octahedral holes. Arguments based on electrostatics corroborate the observation that the formally trivalent rare earth metals encapsulate the highly charged interstitial atoms. However, with respect to the common salts like NaCl, the metal-rich halides under consideration are electron-rich; the additional electrons contribute to some degree of clustering and metal–metal bonding between the cations. If the interstitial atom is a transition metal or a relatively electropositive main group atom, e.g. Be, Al, or B, then stability criteria for the formation of intermetallic phases become more relevant for the discussion of their structures than simple electrostatic interactions.

Several methods are available to analyze the electronic structure and discuss the chemical bonding of molecules and solids in a quantitative as well as qualitative way. Various levels of sophistication exist. From the chemist's perspective, electron counting schemes, such as the Zintl–Klemm (Klemm 1958; Schäfer and Eisenmann 1981, von Schnering 1985), Mooser–Pearson (Mooser and Pearson 1956, 1960), or the $(8 - N)$ rule (Kjekshus and Rakke 1974), which were frequently applied in the preceding section, are useful tools to understand relationships between composition and structure, structure and bonding, bonding and properties. Molecular orbital methods provide even more quantitative information, and even justify many of these simple electron counting rules (Albright et al. 1985). The most common, and, perhaps, the most useful approximation involves the formation of molecular orbitals by a linear superposition of orbitals on each of the component atoms in the structure, the LCAO approximation (or, the tight-binding approximation in the vernacular of the solid state physicist) (Ashcroft and Mermin 1976). Other computational methods use pseudopotentials, chemical pseudopotentials, linear muffin-tin orbitals, and augmented plane waves in order to explain and understand electronic and physical properties. Although these methods are able to provide excellent agreement with experimental observations in relatively simple structures, e.g. Si, ZnS, NaCl and CsCl, many are cumbersome when trying to examine the somewhat complex interstitial rare earth metal halides.

In the following section, we shall review the attempts to understand the electronic structure and chemical bonding in the binary and interstitial rare earth metal halides. The computational results are usually presented in density of states (DOS) diagrams, which can illustrate how the various constituents contribute to the total DOS (in terms of projections of the total DOS). Integration of each DOS projection up to the Fermi level determines the number of electrons assigned to the specified component, whether it is an individual atom or a small molecular fragment. In addition, Hoffmann and co-workers have developed a DOS scheme weighted by the overlap population between two constituent atoms in order to examine the nature of the orbital overlap between these two sites as a function of energy, the so-called COOP curves (Hughbanks and Hoffmann 1983, Hoffmann 1989). Regardless of the presentation of the computational results, the theoretical model should predict or confirm (or both!) various physical and structural characteristics of the specific compound or class of compounds examined in order to be accepted. A frequently used indicator is to compare how well the calculated total DOS corresponds with the photoelectron spectrum, although in most cases only qualitative agreement occurs.

For the rare earth metal halides and their interstitial derivatives, a simple ionic model related to the aforementioned electron counting rules has been successful to rationalize a number of structural and physical observations. This model, which has been repeatedly used in the preceding section, assigns to each halide an oxidation state of -1, and to the highly electropositive rare earth metal, usually its maximum oxidation state, which in most cases is trivalent. If interstitial species occurs, these accept any excess electrons to the extent of becoming closed shell ions. Any additional electrons will then enter the metal-centered band. The following examples will address the validity and the limitations of such a treatment.

3.1. *Discrete cluster units*

Empty clusters of the M_6X_8- and M_6X_{12}-types form a significant part of the structural chemistry of the groups 4 through 7 transition metals (Schäfer and Schnering 1964, Simon 1981, 1988). M_6X_8 clusters are favored for higher d electron concentrations (24 valence electrons per M_6 unit is the optimal electron count) than the M_6X_{12} clusters, which occur for 14 to 16 cluster valence electrons. Molecular orbital theory has been rather successful in elucidating electronics-based reasons for these two different "magic numbers" (M_6X_8: Cotton and Haas 1964, Mattheiss and Fong 1977, Bullett 1977, Nohl et al. 1982; M_6X_{12}: Cotton and Haas 1964, Robbins and Thomson 1972). In the structure of each isolated cluster the metal atoms sit in an approximately square planar ligand field of anions (the local site symmetry of each metal is 4 mm in the idealized case). For a convenient molecular orbital description, a local coordinate system at each metal atom is chosen with the z-axis perpendicular to the plane of the four coordinating anions. The corresponding x- and y-axes are oriented such that the local xz and yz planes each contain four metal atoms of the octahedron (see fig. 38). The energies of the metal d orbitals in this ligand field are split into one high-lying level (in our coordinate system convention, d_{xy} for M_6X_8 and $d_{x^2-y^2}$ for M_6X_{12}), i.e. metal–anion σ antibonding and four low-lying levels [d_{xz}, d_{yz}, d_{z^2}, and either $d_{x^2-y^2}$ (M_6X_8), or d_{xy} (M_6X_{12})] that are essentially metal–anion nonbonding (with π donors like O^{2-}, Cl^-, S^{2-}, etc.; these are metal–anion π antibonding).

Condensation of these local fragments into the complete cluster [$M_6X_8 = (MX_{4/3})_6$ and $M_6X_{12} = (MX_{4/2})_6$] could reduce or even eliminate the ligand field splitting of the d manifold as a consequence of both symmetry (through bond coupling) and increased electron delocalization (orbital dilution). However, the topology of both M_6X_8 and M_6X_{12} clusters keeps the six metal–anion σ antibonding orbitals well separated from the remaining 24 d orbitals, which interact with each other to produce the splitting patterns for these clusters, as shown in fig. 39. The metal–metal bonding energy levels are labelled according to irreducible representations of the point group m3m. Analysis of the nodal characteristics of each cluster orbital indicates that there are 12 metal–metal bonding levels in the M_6X_8 cluster and clearly 7 metal–metal bonding orbitals in M_6X_{12}. The a_{2u} orbital for the M_6X_{12} cluster is also metal–metal bonding, but has significant metal–anion π antibonding character that it can be regarded as approximately overall nonbonding.

In structures containing these clusters, each metal atom is capped by an additional anion (in the notation $M_6X^i_8X^a_6$, these are the X^a atoms). These capping anions may

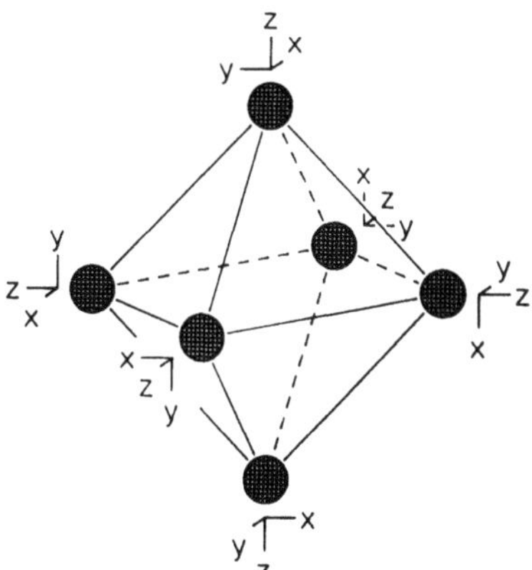

Fig. 38. Local coordinate system at each metal center of the M_6 octahedron.

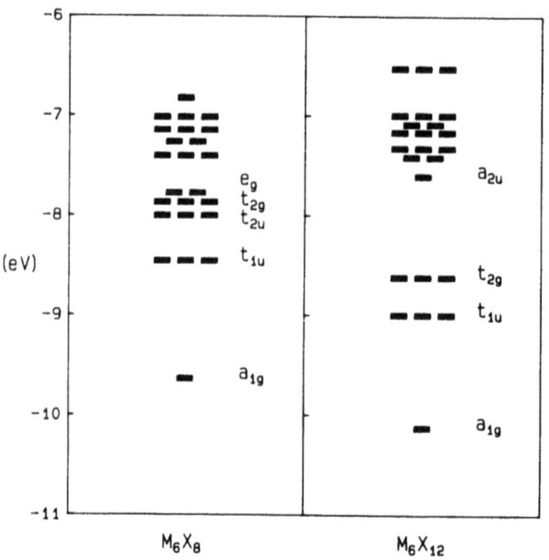

Fig. 39. Energy level diagrams for the metal-centered orbitals of isolated M_6X_8-type (left) and M_6X_{12}-type (right) clusters. Metal–metal bonding orbitals are M_6X_{12} labelled.

assume X^i positions for a neighboring cluster, as in $PbMo_6S_8$ (Chevrel et al. 1971, Matthias et al. 1972), bridge two clusters, e.g. I^a in Nb_6I_{11} ($=Nb_6I^i_8I^{a-a}_{6/2}$) (Simon et al. 1967), or simply act as terminating ligands, e.g. Cl^a in $MoCl_2$ ($=Mo_6Cl^i_8Cl^a_2Cl^{a-a}_{4/2}$) (Schäfer et al. 1967). The orbitals of these anions, X^a, will most strongly interact with the radial orbitals of the M_6 unit, i.e. d_{z^2}, s and p_z. However, the distances $d(M–X^a)$ are significantly longer than the distances $d(M–X^i)$. As fig. 40 shows, these X^a atoms increase the HOMO–LUMO (highest occupied–lowest unoccupied) gaps in both M_6X_8 and M_6X_{12} clusters with no significant alterations in the metal–metal bonding manifolds of the two clusters. In addition, the anion bands become stabilized due to interactions with the metal cluster orbitals. Therefore, these model calculations suggest an important structure determining interaction in compounds containing either M_6X_8 or M_6X_{12} clusters: since the LUMOs of the metal d orbital manifold are strongly perturbed by the orbitals of the capping anions, X^a, there exists a strong donor–acceptor interaction that stabilizes the total electronic energy of the individual clusters. For certain stoichiometries, these donor–acceptor bonds will lead to lattice symmetries lower than cubic. The energy level schemes in figs. 39 and 40 are consistent with the bulk of experimental data on compounds with these fundamental units, but certainly do not elucidate all details. Especially interesting are compounds in which not all of the metal–metal bonding cluster orbitals are occupied, e.g. Nb_6I_{11} (Finley et al. 1981a, b, Imoto and Simon 1982, Brown et al. 1988).

When the metal atoms are rare earth elements R and the anions are halogens X, both cluster types are highly electron deficient. One avenue for increasing the valence electron concentration is through condensation of these units via common vertices, edges or faces. Of course, the relative stoichiometry, i.e. the X/R ratio, necessarily changes. Another way is to incorporate interstitial species Z within the octahedral cluster. However, strong R–Z interactions now occur at the expense of the metal–metal bonds. The discussion begins with compounds with interstitials in “isolated” clusters and then proceed towards ever increasing levels of cluster condensation.

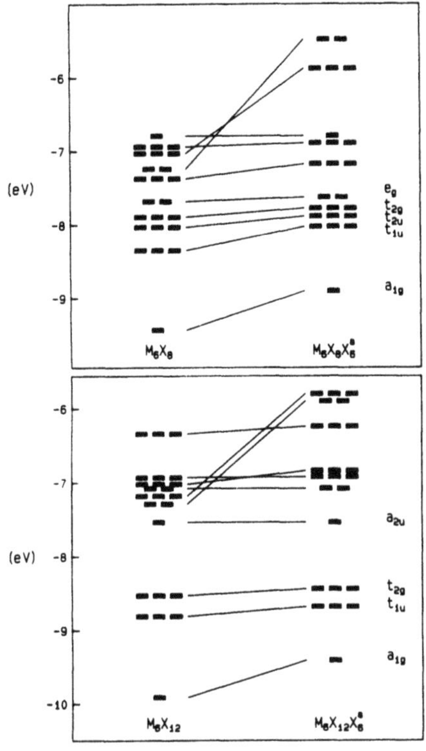

Fig. 40. The effects on the metal-centered orbitals of M_6X_8-type (top) and M_6X_{12}-type (bottom) clusters by capping each metal atom with an X^a atom. The metal–metal bonding orbitals are labelled.

3.2 *Interstitial species*

From the molecular orbital point of view, the simplest interstitial species is the H atom. In the octahedral cavity the H 1s orbital transforms as a_{1g} and will overlap with the a_{1g} levels of either the M_6X_8- or M_6X_{12}-units to form an M_6–H bonding and antibonding pair. Since the cluster a_{1g} orbital is metal–metal bonding and the M_6–H antibonding level is destabilized above the remaining metal–metal bonding manifold, the number of cluster bonding states remains unchanged, and only the number of electrons in these states is increased by the electron of the H atom. The stabilization of the cluster arises primarily from the low energy of the a_{1g} level, which is essentially localized on the H atom. Except for $Nb_6I_{11}H$ (Simon 1967), $CsNb_6I_{11}H$ (Imoto and Corbett 1980) and some Zr halide cluster compounds (Chu et al. 1988), H atoms seldom occupy these octahedral centers, perhaps due to the size of the cavity.

The most frequently encountered interstitial species are the second period main group elements C and N. From the main group atoms in the periodic table, B and Si have also been reported to occupy interstitial sites in the reduced rare earth halides. Their valence atomic orbitals, s, p_x, p_y and p_z, transform as $a_{1g} + t_{1u}$ in an octahedral field (Bursten et al. 1980). In addition to the a_{1g} orbital, the metal–metal bonding levels of the clusters also contain a t_{1u} orbital which is well suited to overlap strongly with the orbitals on the interstitial. Figure 41 illustrates the molecular orbital diagram

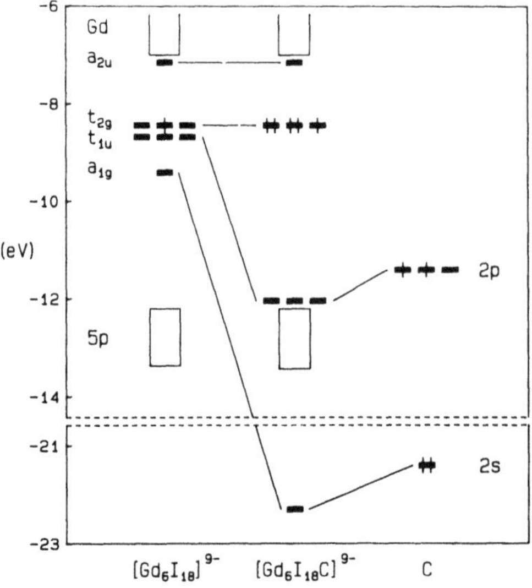

Fig. 41. Molecular orbital interaction diagram between a C atom and the $(Gd_6I_{12}I_6)^{9-}$ octahedral fragment to form $(Gd_6I_{18}C)^{9-}$: orbital occupation is appropriate for $R_7X_{12}C$.

for an interstitial C atom in a $(Gd_6I_{12}I_6)^{9-}$ cluster, which was calculated using the extended Hückel method (EH; Hoffmann 1963). These results are consistent with those of Hwu and Corbett (1986) on $(Sc_6Cl_{18}B)^{9-}$, and represent a typical scheme for $(R_6X_{18}Z)^{9-}$ units. The interaction between the cluster valence orbitals and those on the interstitial species produce four R–Z bonding levels, which are occupied, and four R–Z antibonding levels, which occur at high energy. Since C, N and perhaps B are more electronegative than the rare earth elements, the bonding orbitals are mostly centered on the interstitial. Therefore, we can conclude that the M_6 octahedron becomes oxidized upon introduction of the interstitial, i.e. charge flows from the cluster to the interstitial. Furthermore, since the a_{1g} and t_{1u} metal–metal bonding levels are utilized in the orbital overlap with the interstitial, the strength of the direct metal–metal bond decreases, although the number of cluster bonding states again remains unchanged. The t_{2g} metal–metal bonding orbital remains unperturbed because there exists no appropriate set of orbitals on the interstitial with which to interact. Likewise, the a_{2u} cluster orbital is unaffected, but due to the R–R separations imposed by the R–Z distances, this cluster level remains high in energy through the greater significance of the R–X π^* interactions. Therefore, the optimal valence electron concentration is six electrons per cluster when the t_{2g} orbital is completely occupied.

These discrete clusters are found in the $R_7X_{12}Z$ structures (Z = B, C, N), from which formal charges may be assigned as $R^{3+}(R_6X_{12}Z)^{3-}$. EH molecular orbital calculations on the complete structure of $Sc_7I_{12}C$ confirm this electron counting scheme as the d orbitals on the single Sc^{3+} ion remain well above the Fermi level. In addition, the cluster bonding t_{2g} level becomes stabilized through the symmetry allowed mixing with the t_{2g} orbitals on the single Sc^{3+} (note: in $Zr_6I_{12}C$, which does not contain this additional cation, the enhanced electronic stability of the cluster arises from occupation of the a_{2u} level and concomitant shortening of the Zr–Zr distance; Smith and Corbett 1985). The optimal valence electron concentration t_{2g}^6 occurs when Z = N, as in $Sc_7Cl_{12}N$ (Hwu and Corbett 1986). Fewer valence electrons are possible, as for $Sc_7Cl_{12}B$ (t_{2g}^4; Hwu and Corbett 1986), which shows a rather complex temperature dependence of its magnetic susceptibility, and $Sc_7I_{12}C$ (t_{2g}^5; Dudis et al. 1986). Since the t_{2g} orbital is only partially occupied in the boride, we can also expect slightly larger Sc–Sc distances in $Sc_7Cl_{12}B$ (3.281 and 3.293 Å) than in $Sc_7Cl_{12}N$ (3.237 and 3.256 Å), although a difference in the sizes of B and N can account for these different distances as well. Higher valence electron concentrations are as yet unobserved. For example, oxygen cannot be substituted into the Sc_7Cl_{12} framework due in large part to the high stability of ScOCl and the need to populate the high-lying a_{2u} cluster level.

Given the degenerate ground states for both $Sc_7Cl_{12}B$ (t_{2g}^4: $^3T_{1g}$) and $R_7X_{12}C$ (t_{2g}^5: $^2T_{2g}$), first-order Jahn–Teller arguments predict some type of distortion to remove this degeneracy. The rhombohedral symmetry of these systems is certainly sufficient, which produces two symmetry inequivalent R–R distances within the cluster. The predicted distortion for these two electronic configurations, compression of the octahedron along the three-fold axis, is observed to a slight extent in every case. In $Sc_7I_{12}C$, the displacement of the isolated Sc atom polarizes the cluster to give higher electron density on the face closer to this atom. The capacity for metal–metal bonding increases in this face, decreases in the opposite face, and the cluster distorts as described in sect. 2.1. The shift of the cluster with respect to the interstitial C atom is also in accord with this description (Dudis et al. 1986).

The chemical bonding in compounds like $Gd_7I_{12}Fe$ with a transition metal acting as an interstitial atom takes on an entirely different aspect (Hughbanks et al. 1986, Hughbanks and Corbett 1988). Figure 42 illustrates the molecular orbital diagram for the cluster fragment $[(Gd_6I_{12}Fe)I_6]^{4-}$. What are the important differences between this structural class and the materials with main group interstitials? Although the Fe atoms are also more electronegative than the cluster metals (indicated partly from the larger work function of the later transition metals over the earlier metals), their valence orbitals include the 4s and 4p levels in addition to the 3d functions, which lie respectively higher and lower in energy than the d orbitals of the cluster atoms. These orbitals transform respectively as a_{1g}, t_{1u} and $t_{2g} + e_g$ in an octahedral field. The 4s and 4p levels serve to stabilize the a_{1g} and t_{1u} cluster orbitals with most of the charge

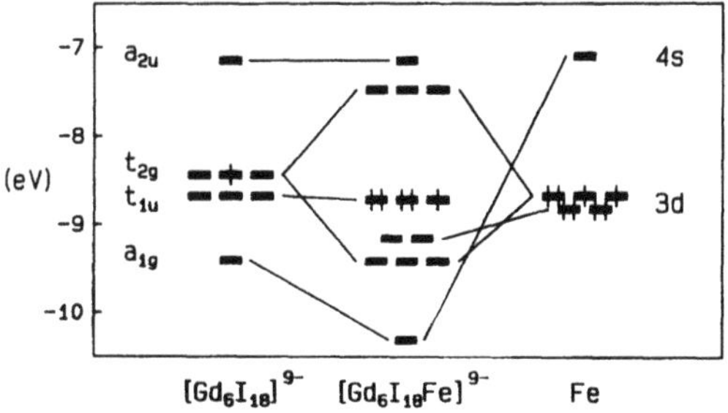

Fig. 42. Molecular orbital interaction diagram between an Fe atom and the $(Gd_6I_{12}I_6)^{9-}$ octahedral fragment to form $(Gd_6I_{18}Fe)^{9-}$: only the cluster-centered orbitals are illustrated; orbital occupation is appropriate for $R_7X_{12}Fe$.

in this case remaining on the clusters in these metal–metal bonding orbitals. In fact, due to the large energy separation between the Fe 4p and the cluster t_{1u} orbitals, the cluster t_{1u} level is only slightly stabilized. The Fe e_g orbitals are stabilized by the e_g-type LUMOs of the cluster and remain mostly Fe-centered. The interaction between the two t_{2g} orbital sets involves nearly isoenergetic levels, which results in a nearly equal partitioning of charge.

Occupation of the nine energy levels ($a_{1g} + t_{2g} + e_g + t_{1u}$) produces the magic number for maximum electronic stability at 18 valence electrons per cluster, which neatly separates R–Z bonding and antibonding states. $R_7X_{12}Co$ satisfies this criterion and should show diamagnetic properties. Electron counts higher than 18 would involve occupation of either the nonbonding a_{2u} cluster orbital or the R–Z t_{2g} antibonding component. The relative ordering of these two levels is rather sensitive to both the R–Z distance and the energy separation between the cluster a_{2u} orbital and the transition metal d orbitals. Until now, efforts to introduce Ni into the cluster have failed. Although by no means a proof, this suggests the nature of the LUMO to have strong R–Z antibonding character. However, this may also reflect the tendency to occupy states up to a significant energy gap. Unfortunately, the nature of the HOMO as well cannot be clearly assessed, although EHMO calculations consistently give the t_{1u} level as the HOMO. As the valence electron concentration is reduced from 18, i.e. by incorporation of Cr, Mn or Fe into the octahedral cluster, both the increased size of the interstitial as well as the reduced occupation of R–Z bonding levels tend toward increased distances within the cluster.

Although the electronic requirements of these transition metal systems differ from those of the carbides, the clusters are also trigonally compressed, but to a much greater extent, cf. sect. 2.1. This is consistent with the analysis of overlap populations for the two inequivalent edges of the octahedron in the trigonal field of the complete $Y_7I_{12}Fe$ structure, and originates from Y–Y interactions due to a negligible contribution of the Fe 4p orbitals in the HOMO.

An example containing a 4d transition metal in the octahedral cavity is $Y_6I_{10}Ru$ with 16 electrons in cluster valence orbitals (Hughbanks and Corbett 1989). One notable structural feature is the 0.21 Å tetragonal compression of the octahedron. Given t_{1u}^4 as the ground state configuration, the observed distortion splits the HOMO into a two-below-one pattern. Since the HOMO contains little Ru 4p character, this energy level scheme also originates with Y–Y interactions (note: in ML_6 units with t_{1u}^4 configuration, the reverse distortion mode is expected through the M–L overlap).

At this point, we note that except for H in $Nb_6I_{11}H$ (Simon et al. 1967) and $CsNb_6I_{11}H$ (Imoto and Corbett 1980), no other kind of interstitial species has been observed at the center of the metal octahedron in M_6X_8 clusters. Possible explanations include the restrictive size of the interstitial site as well as X–Z repulsions through the faces of the octahedron. In any case, since no examples exist in the rare earth halide family and no detailed theoretical treatment has been undertaken, we shall not discuss this specific example further.

How can we partition charge when B, Si, or even a transition element is an interstitial? Formally, we have the same problem here that occurs in the binary intermetallic phases between the rare earth elements and B, Si, Mn, or Fe. Experiments, which include photoelectron spectroscopy, NMR and magnetic measurements on the binary alloys, all suggest some degree of charge transfer from the group 3 metal to the other constituent. The molecular orbital calculations on the $(Gd_6I_{18}Fe)^{4-}$ cluster also assigns higher electron density to the Fe atom to produce a formal charge between Fe^0 and Fe^{2-}. For B and Si, the description as closed-shell anions, B^{5-} and Si^{4-}, is rather formal. Yet when B or Si are only partially reduced, they tend to catenate as observed in FeB (Pearson 1972), Gd_2B_5 (Schwarz and Simon 1987), GdSi (Nagaki and Simon 1990), and $ThSi_2$ (Pearson 1972), to name only a few. Therefore, these compounds with interstitial B and Si are excellent candidates for NMR investigations in order to probe the nature of the metal–boron or metal–silicon interaction.

The previous structures, which contain distinct R_6 octahedral clusters, necessarily have $R/Z > 6$. When $R/Z > 4$, we can find a finite number of these clusters condensed to give larger molecular units. For rare earth halide systems, only edge-sharing octahedra have been observed, as delineated in sect. 2, in which only two octahedra are involved and the interstitial moiety is a C_2 dimer. Satpathy and Andersen (1985) applied the LMTO-ASA method to produce the energy DOS illustrated in fig. 43. Their model examined the interaction between an isolated $Gd_{10}Cl_{18}$ cluster with two

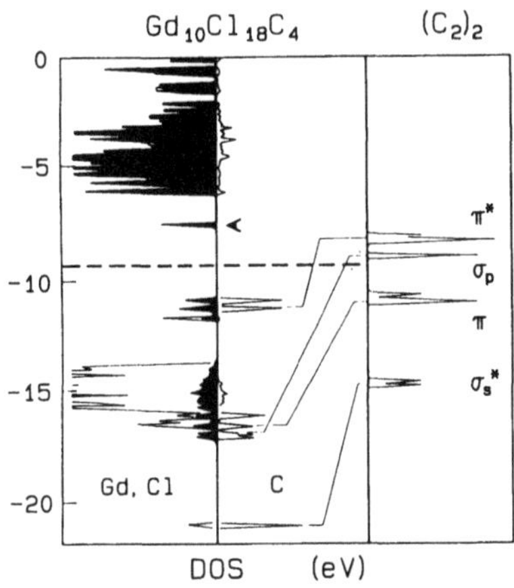

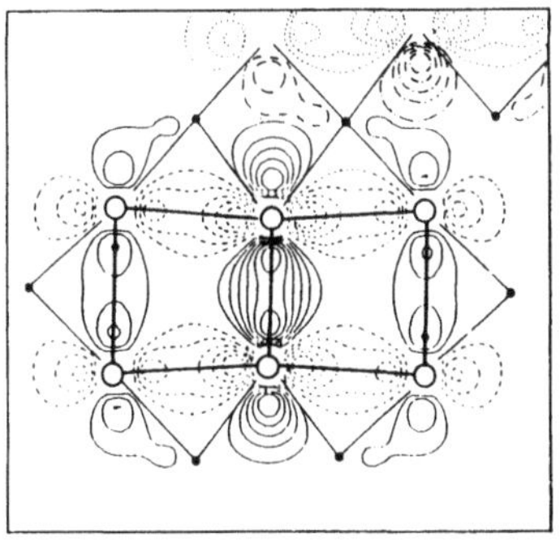

Fig. 43. (Top) Projected DOS for the $Gd_{10}Cl_{18}C_4$ unit with the Gd projection shaded (left) and C_2 projection (middle); the energy scheme for two free C_2 units is also presented (right): the Fermi level is indicated by the dotted line. (Bottom) Representation of the Gd–Gd bonding orbital above the Fermi level marked by the arrow: the orbital is drawn in the plane of the Gd_{10} bioctahedral base (Gd: circles; Cl: points).

C_2 units using the observed distance, d(C–C) = 1.45 Å found in $Gd_{10}Cl_{18}C_4$ (Simon et al. 1981). Although the arrangement of Cl and C_2 units taken together corresponds to a ccp arrangement, interactions between cluster units are sufficiently weak so as not to affect the energy levels near the Fermi level for these models. The condensation of clusters will mostly alter the band dispersion of the Cl component. Their results indicate that for $Gd_{10}Cl_{18}C_4$ an appropriate description of the C_2 electronic configuration is complete occupation of the C_2 π^* levels. The σ-type orbitals of the C_2 dimer are stabilized by 8.2–9.5 eV and the π-type levels by 2.7–5.4 eV, primarily through interactions with the apical Gd atomic orbitals. The resulting π^* levels have about 30% Gd contribution. These two results certainly suggest a strong Gd–C interaction, especially for the apical Gd atoms of the cluster, although the authors conclude that rather weak or essentially no covalent Gd–C bonds occur. Moreover, from a population analysis, they find no support for a preferentially strong Gd–C bond along the apical direction. However, the significant dispersion of the π^* component of C_2 in the Gd d manifold indicates a certain degree of Gd–C orbital overlap, although it does not preclude any significant amount of charge transfer. Therefore, the Gd–C interaction may exhibit appreciable ionic character, with relatively small electron density localized along the apical Gd–C vector. EHMO results on both a Gd–C–C–Gd fragment from $Gd_2Cl_2C_2$ (Miller et al. 1986) and $Sc_6I_{11}C_2$ (Dudis and Corbett 1987) indicate strong dπ interaction in this direction. The results of Bullett (1985) and of Xu and Ren (1987) on $Gd_{10}Cl_{18}C_4$ parallel the results of Satpathy and Andersen, with very similar decompositions. However, Xu and Ren, who performed INDO calculations on $Gd_{10}Cl_{18}C_4$, conclude that the "ionic" model which attributes the formation of the compound to the electrostatic attraction between ions seems incorrect because the atomic net charges of the compound are far from the values estimated by the model. In truth, the real situation inevitably lies somewhere between the ionic and the covalent picture.

From the model cluster calculation on $Gd_{10}Cl_{18}C_4$, we can also extract some information on $Gd_{10}Cl_{17}C_4$. With one less halide anion, $Gd_{10}Cl_{17}C_4$ has one extra electron with which to occupy the Gd orbital manifold, i.e. the LUMO of $Gd_{10}Cl_{18}C_4$. The d block in $Gd_{10}Cl_{18}C_4$ shows one low-lying orbital, which is isolated from the remainder of the d levels, and has strong Gd–Gd bonding character between the metal atoms in the basal plane, especially along the shared edge, shown in fig. 43. There is no C_2 contribution to the LUMO from purely symmetry arguments. Geometrical comparisons between $Gd_{10}Cl_{18}C_4$ and $Gd_{10}Cl_{17}C_4$ nicely confirm the nature of this orbital as the Gd–Gd distance of the shared edge decreases from 3.21 to 3.12 Å with no observable change in the C–C distance (Warkentin et al. 1982).

Earlier, we questioned how to partition the charge between the octahedral cluster and the interstitial species, especially for those like B, Si or Fe. With the C_2 units, a chemical and structural check exists to monitor the extent of charge transfer, as well as the number of electrons that remain in cluster valence levels, i.e. electrons that contribute to metal–metal bonding. In fact, a simple ionic treatment in conjunction with the qualitative molecular orbital scheme for the C_2 dimer can reasonably predict the expected C–C distance. For the rare earth elements, these donate as many of their valence electrons as the nonmetallic species can accommodate. For

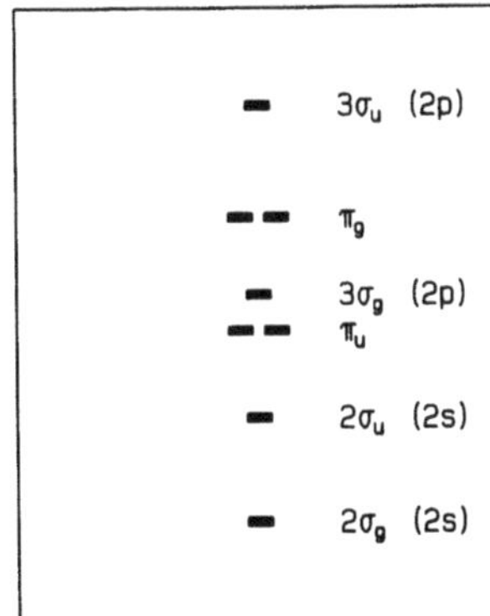

Fig. 44. Qualitative molecular orbital scheme for a C_2^{x-} dimer. The electronic configurations for certain values of x are – C_2^{2-}: $(2\sigma_g)^2(2\sigma_u)^2(\pi_u)^4(3\sigma_g)^2$; C_2^{4-}: $(2\sigma_g)^2(2\sigma_u)^2(\pi_u)^4(3\sigma_g)^2(\pi_g)^2$; C_2^{6-}: $(2\sigma_g)^2(2\sigma_u)^2(\pi_u)^4(3\sigma_g)^2(\pi_g)^4$.

example, $Gd_{10}Cl_{18}C_4$ can be formulated as $(Gd^{3+})_{10}(Cl^-)_{18}(C_2^{6-})_2$. According to the qualitative molecular orbital scheme for a homonuclear diatomic in fig. 44, the 14 electron C_2^{6-} species completely fills the π_g level, in agreement with the computational results of both Andersen and Bullett. In addition, the ionic formulation leaves no excess electrons to occupy metal–metal bonding levels. In $Gd_{10}I_{16}C_4$ (Warkentin and Simon 1982) and $Sc_6I_{11}C_2$ (Dudis and Corbett 1987), an effective negative charge of -7 is not assigned to each C_2 group, since this would necessitate half occupancy of the $3\sigma_u$ orbital and an extreme lengthening in the C–C bond distance. Instead, these compounds are formulated as $(Gd^{3+})_{10}(I^-)_{16}(C_2^{6-})_2(e^-)_2$ and $(Sc^{3+})_6(I^-)_{11}C_2^{6-}e^-$, in which two and one extra electron(s), respectively, occupy the low-lying metal–metal bonding band. In agreement with this formulation, the C–C distance does not significantly change from $Gd_{10}Cl_{18}C_4$ to $Gd_{10}I_{16}C_4$. In $Gd_{10}Cl_{17}C_4$ a single unpaired electron occupies the metal-centered orbital. However, with ten Gd atoms each with an effective configuration of $4f^7$ in the cluster unit, it is unlikely that magnetic measurements would be illuminating.

In all known examples, the maximum formal charge assigned to the C_2 unit is -6. Certainly, when the formal charge becomes -8, the dimer is no longer stable with respect to homolytic dissociation into two C^{4-} ions, as in the case of He_2, which is unstable with respect to two He atoms due to closed-shell repulsions. On the other hand, when the formal charge on C_2 becomes less than -6, we expect shorter C–C distances. The binary and ternary carbides listed in table 8 confirm this expectation.

3.3. *Extended structures*

There are many similarities between the metal dicarbides and the $Gd_2X_2C_2$ (X = Cl, Br) structures. Both have an octahedron of metal atoms surrounding each C_2 unit with the dimer parallel or nearly parallel to one of the tetragonal or pseudotetragonal axes of the octahedron. Dicarbides of Y and the lanthanides as well as the $Gd_2X_2C_2$ compounds show metallic behavior, while CaC_2, SrC_2 and BaC_2 are insulators.

TABLE 8
Comparison of C–C distances in various binary and ternary carbides. n is the formal negative charge per C_2 unit.

Compound	n	d(C–C)(Å)	Ref.[a]
CaC_2	2	1.19	[1]
TbC_2	3	1.29	[1]
ThC_2	4	1.33	[2]
$Gd_2Br_2C_2$	4	1.28	[3]
$Gd_{10}Cl_{18}C_4$	6	1.47	[4]
$Gd_{12}I_{17}C_6$	6	1.44	[5]
$Gd_{10}Cl_{17}C_4$	6	1.47	[6]
$Gd_{10}I_{16}C_4$	6	1.43	[7]
Gd_4I_5C	8	–	[7]
$Gd_6I_7C_2$	8	–	[8]
Gd_3Cl_3C	8	–	[9]
Gd_3I_3C	8	–	[10]

[a] References: [1] Atoji (1961); [2] Bowman et al. (1968); [3] Schwanitz-Schüller and Simon (1985); [4] Simon et al. (1981); [5] Simon and Warkentin (1983); [6] Warkentin et al. (1982); [7] Simon and Warkentin (1982); [8] Simon (1988); [9] Warkentin and Simon (1983); [10] Mattausch et al. (1987).

In this case the simple ionic model has only a limited ability to predict electronic properties. For the alkaline earth dicarbides, each C_2^{2-} dimer has all σ and π bonding levels filled (isoelectronic to N_2, place 10 electrons in the molecular orbital scheme of fig. 44) with HOMO and LUMO separated by a significant energy gap. The C–C distances are rather short, approximately 1.21 Å, which is similar to the C–C distance in acetylene. In the lanthanide dicarbides, the additional electron could occupy either a free-electron-type conduction band (a situation proposed for the rare earth hexaborides) or the π_g level of the dimer. Within the simple ionic formulation, the second possibility at first suggests nonmetallic properties, in contradiction with observation, and yet, the C–C bond distances increases to ca 1.28 Å. Band structure calculations on LaC_2 suggest strong interactions between the C_2 π_g orbital and the symmetry adapted linear combination of La d orbitals, which have a large dispersion in reciprocal space. The bottom of the conduction band has significant contributions from both the C_2 π_g orbitals and Gd d orbitals, which is consistent with the observed C–C distance and the electrical conductivity. In $Gd_2Cl_2C_2$ and $Gd_2Br_2C_2$ the formal assignment of C_2^{4-} leaves no electrons available for either free-electron-type or metal-centered conduction bands. However, as in the case of the lanthanide dicarbides, it is common knowledge that filling energy bands gives no clue as to how these bands may be separated or to their dispersion in reciprocal space.

The photoelectron spectrum of $Gd_2Cl_2C_2$, shown at the top of fig. 45, exhibits nonzero density of states at the Fermi level. The dramatic shift of the Gd 4f band to higher energy and the results from the rare earth metal dicarbides suggest that backbonding from the occupied C_2 π_g states into the empty Gd d states contributes primarily to their metallic behavior. The analysis of the electronic structure of $Gd_2Cl_2C_2$ using extended Hückel calculations verifies this interpretation (Miller et al. 1986). The calculated density of states (omitting the Gd 4f levels) and the projected DOS for Gd 5d orbitals are plotted in fig. 46. The COOP diagrams (see also fig. 46) for

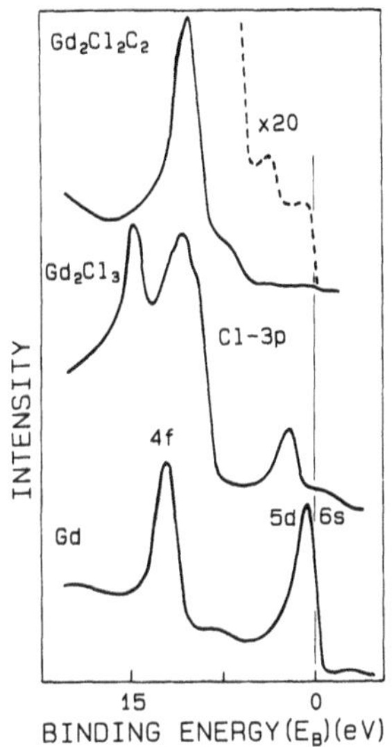

Fig. 45. Photoelectron spectra of Gd and Gd_2Cl_3 taken with He(II) radiation (40.8 eV; Ebbinghaus et al. 1982), and of $Gd_2Cl_2C_2$ (He(I), 21.2 eV; Miller et al. 1986): the narrow f band is marked; structures above the Fermi level ($E_B = 0$) in the spectra of Gd and Gd_2Cl_3 arise from excitations of electrons from the 4f band by the 50.3 eV satellite line.

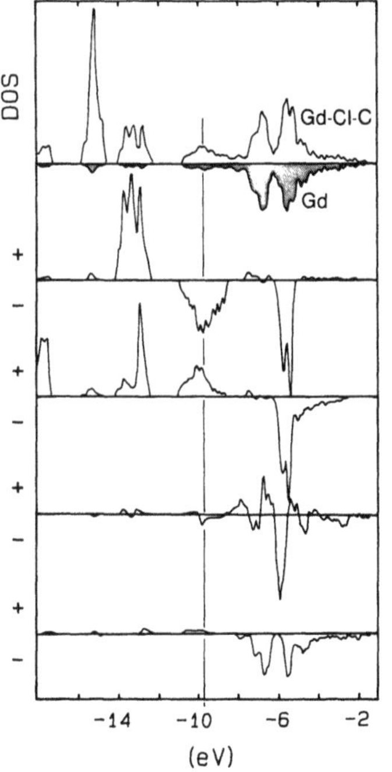

Fig. 46. Results of EH calculations for $Gd_2Cl_2C_2$ (Miller et al. 1986): (from top to bottom) total DOS and projected DOS for Gd 5d (shaded); COOP curves for C–C, C–Gd_{axial}, Gd_{basal}–Gd_{basal}, and Gd_{axial}–Gd_{basal} interactions, respectively; the calculated Fermi level is marked by the vertical line.

C–C, Gd–C, and two different Gd–Gd interactions effectively characterize the states at the Fermi level. As expected, they are C–C antibonding (π^*), but bonding with respect to Gd–C interactions. The stronger interactions involve the Gd atoms that are nearly collinear with the C_2 dimer, but the band width arises primarily via the coupling with Gd atoms in the basal positions. The occupation of these states is a compromise between the weakening of the C–C interaction and formation of π bonds between C and Gd. It is also evident from fig. 46 that the large Gd 5d component near the Fermi level is due primarily to the covalent Gd–C bonds and contributes very little (essentially nothing) to direct metal–metal bonding. It is possible to intercalate $Gd_2Cl_2C_2$ electrochemically with Li to yield a limiting composition $Li_{0.9}Gd_2Cl_2C_2$ (Schwarz 1987). Unfortunately, structural characterizations have been hindered by the disordering of the layers during the reaction. The electronic structure analysis under a rigid-band approximation predicts an increase in the C–C bond distance via chemical reduction. Given the presence of heavy atoms, e.g. Gd and the halides, in these carbides, the C–C bond distances cannot be so exactly determined from an X-ray experiment. Raman measurements on a variety of these compounds resulted in C–C stretching frequencies in accord with theoretical expectations. Some of these results are listed in table 9 (Schwarz 1987).

The discussion of the interstitial C_2 moieties has carried our review into structures which are constructed from R_6X_{12} clusters condensed into an infinitely extending two-dimensional layer. However, since we have begun with molecular units, the next logical step is the formation of infinite chains. Gd_2Cl_3 crystallizes with chains of edge-sharing octahedra of Gd atoms and Cl atoms capping the triangular faces, i.e. a linear condensation of "Gd_6Cl_8" clusters, as described in sect. 1. The electronic density of states is shown in fig. 47, and agrees with the description of Bullett (1980, 1985), who utilized a chemical pseudopotential approach. The Gd d manifold splits to give three low-lying bands per chain, which are separated by 0.7 eV from the remainder of the d block. Examination of the dispersion of these energy bands along the axis in reciprocal space parallel to the chain axis shows that this energy gap arises via an avoided crossing. The photoelectron spectrum for Gd_2Cl_3 in fig. 45 nicely confirms

TABLE 9

Results from Raman spectroscopy on various ternary carbides (Schwarz 1987).

Compound	C species	ν(C–C) (cm^{-1})	Peak shape	f(C–C) (mdyn/Å)	d(C–C) (Å)
$Gd_{10}Cl_{18}C_4$	C_2^{6-}	1158	Sharp	4.74	1.47
$Gd_{10}Cl_{17}C_4$	C_2^{6-}	1175	Sharp	4.91	1.47
$Gd_{12}Br_{17}C_6$	C_2^{6-}	1175	Broad[a]	4.91	1.41[b]
$Gd_2Cl_2C_2$	C_2^{4-}	1578	Broad	8.91	1.35
$KLiC_2$[c]	C_2^{2-}	1872	Broad	12.40	NCD[d]

[a] Contains two independent C_2 units.
[b] Average between 1.44 (2x) and 1.36 (1x).
[c] Nesper (1987).
[d] d(C–C) not crystallographically determined.

these predictions (Ebbinghaus et al. 1982), as there exists a small contribution to the DOS nearly 0.5 eV below the Fermi level (suggesting an energy gap of the order of 1 eV).

The nature of the chemical bonding within the metal network can be seen from a close examination of the four energy bands in fig. 48 which were calculated for a single $^{1}_{\infty}[Gd_4Cl_4]$ chain. Each band can be classified according to how it transforms with

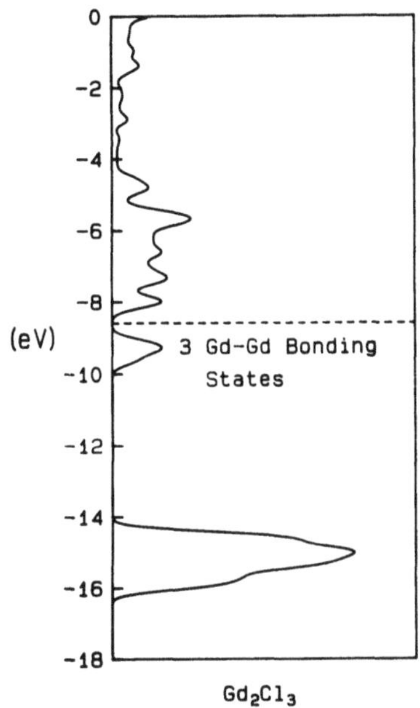

Fig. 47. Total DOS of Gd_2Cl_3: the Fermi level is indicated by the dashed line; the peak between −10 and −9 eV contains three metal–metal bonding states for two formula units.

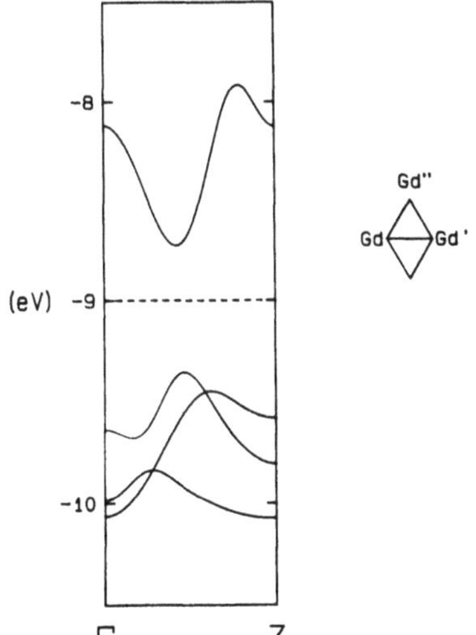

Fig. 48. Dispersion of four lower d bands calculated for a single $^{1}_{\infty}[(Gd_4Cl_4)^{2+}]$ chain. Gd atoms in the chain are labelled as described in the text. The Fermi level for this model chain is indicated by the dashed line.

respect to reflection in each of the two perpendicular mirror planes parallel to the chain axis (note: in the complete structure of Gd_2Cl_3, these two mirror planes do not exist). Due to the translational symmetry of the infinite one-dimensional chain, orbitals on translationally equivalent sites are in phase at the zone center Γ and out of phase at the zone edge Z. Therefore, for bonds parallel to the chain axis bonding interactions occur only within a restricted region of reciprocal space, whereas for bonds perpendicular to the chain axis bonding contributions occur at all k for a given energy band. In Gd_2Cl_3, the two lowest metal d bands are σ and π bonding, respectively, between the two basal Gd atoms (Gd and Gd′), and have σ- and δ-type overlap between metal atoms along the chain (Gd–Gd and Gd′–Gd′) at Γ. The third band at Γ is primarily δ bonding between Gd and Gd′, and has σ overlap along the chain axis. As k increases, this band rises in energy. The fourth band at Γ is π, bonding perpendicular to the chain, and π^* parallel to the chain. As k approaches Z, this band decreases in energy. Both bands have identical symmetry characteristics except at Γ and Z. Therefore, for a general point in reciprocal space, these bands may not cross, and a gap occurs. The electron count for Gd_2Cl_3 is appropriate to completely occupy the three bands below the energy gap. Although bond orders for the individual bonds cannot be strictly assigned from this band picture, it is seen that translational symmetry controls to some extent the structural tendencies in Gd_2Cl_3. These arguments predict the following order of increasing Gd–Gd distances as it is found: Gd–Gd′ (3.37 Å), (Gd, Gd′)–Gd″ (3.73 Å), and then Gd–Gd, Gd′–Gd′, Gd″–Gd″ (3.90 Å).

An interstitial "derivative" of the Gd_2Cl_3 structure is the nitride, α-Gd_2Cl_3N. According to the ionic formulation $(Gd_3^+)_2(Cl^-)_3N^{3-}$, no electrons remain in metal–metal bonding states, but are involved in strong Gd–N interactions. Bullett (1985) performed band structure calculations for this compound and arrived at the population analysis of $(Gd^{2+})_2(Cl^{0.7-})_3N^{1.8-}$, which differs from the extreme ionic picture due to the orbital mixing between the Gd atoms and the anions, i.e. significant covalency. However, the essential point remains that the Gd d band is unoccupied so as to preclude the existence of Gd–Gd bonds in the structure. Nevertheless, the Gd–Gd distance of the shared edges in the chain (3.35 Å) are comparable with the bond length of the shared edges in Gd_2Cl_3 (3.371 Å), in which strong Gd–Gd interactions are important. The origin of this short distance comes primarily from constraints placed by the Gd–N interactions with only a small amount of direct, through space, Gd–Gd d orbital overlap. If we use the purely ionic formulation, then the minimum ionic energy of such a chain occurs at a Gd–N–Gd angle of 90°, in which the two Gd atoms refer to the shared edge (in this case the calculated d(Gd–Gd) is 3.21 Å for d(Gd–N) = 2.27 Å). A molecular orbital treatment predicts an angle ca. 100° (less than the tetrahedral angle, 109.5°) due to closed-shell repulsions between the nitrogen atoms. The observed angle of 94.9° represents a compromise between minimization of the ionic and covalent energy terms in the total energy of the system.

Chains of edge-sharing octahedra using M_6X_{12} clusters also occur for the rare earth elements, in which the metal octahedra contain an interstitial atom. Of the three observed structure types described in sect. 2.2, extended Hückel calculations have been performed on the model chains, $(Sc_4Cl_8B)^{2-}$ and $(Y_4I_8C)^{3-}$, in order to account for the UPS (ultraviolet photoelectron spectroscopy) valence spectra of Sc_4Cl_6B (Hwu and Corbett 1986) and Y_4I_5C (Kauzlarich et al. 1988), respectively, and to qualitatively discuss the chemical bonding in such phases. Similar results on Gd_4I_5Si give typical DOS for systems with these weakly interacting chains as shown in fig. 49. Due to the relatively long separations between the main structural units, the general features for their DOS plots differ very little from one compound to the next. Only the relative electronegativity differences of the components will affect the position of the bands. At lower energy values we find the *s* and *p* bands of the halide and interstitial, which in the case of carbides are well separated from the predominantly metal-centered conduction band. In Gd_4I_5Si, closed-shell repulsions between the anions, strong Gd–Si interactions, and the relative electropositive nature of Si as compared to C result in much smaller separations. In any event, electron counting places three electrons into the conduction band which has metal–metal bonding character near the bottom. The relative strength of the Gd–Si as compared to the Gd–Gd interactions is clearly revealed by the COOP curves in fig. 49 (only the Gd–Gd contribution along the shared edges of the octahedra are represented). A significant contribution to the Gd–Gd overlap population occurs as a component of the Si *p* band, which arises from the interaction between the Si p orbitals and the "t_{1u}"-type orbital of the condensed cluster. Since the average Gd–Gd overlap population is 0.085 in the complete

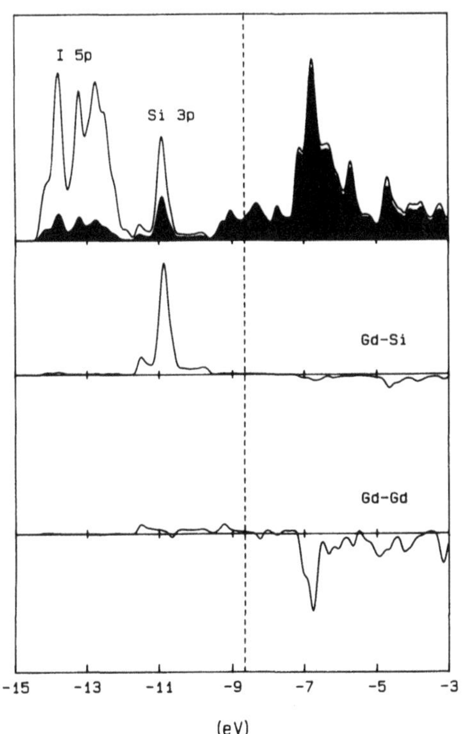

Fig. 49. (Top) Total DOS for Gd_4I_5Si with the Gd projection shaded, (middle) COOP curve for Gd–Si interactions and (bottom) COOP curve for the Gd–Gd interactions along the shared edges of the octahedra; the Fermi level is marked by the dashed line.

structure, metal–metal bonding in these systems certainly contributes to their electronic stability.

Condensation of two chains of edge-sharing octahedra is observed in the compound $Gd_6Br_7C_2$. Again, three electrons are in the conduction band. In fact, for the series of compounds $Gd_{2n+2}X_{2n+3}C_n$, regardless of the number of Gd atoms, there are three conduction electrons per formula unit. When n reaches infinity, however, no conduction electrons remain and a lamellar structure occurs, e.g. Gd_2Cl_2C. The ionic model readily shows the closed shell nature of this material as the positive charge assigned to the metals is exactly compensated by the anionic charge. The semiconducting properties of these compounds are also predicted from band calculations on a two-dimensional layer, Gd_2Cl_2C, in which the Cl and C bands are completely occupied and the Gd d band is empty (see fig. 50; Ziebarth et al. 1986, Burdett and Miller 1987). What the calculated DOS does show that is not revealed from the ionic treatment is that the lower part of the conduction band can accommodate up to three electrons in metal–metal bonding states. These states become occupied in Zr_2Cl_2C and Zr_2Cl_2N (Hwu et al. 1986).

The structures within the homologous series $Gd_{2n+2}X_{2n+3}C_n$ have been described in sect. 2 as a ccp arrangement of X and C atoms with Gd atoms in $\frac{2}{3}$ of the octahedral holes, i.e. as ordered defect derivatives of rock salt. That closed-shell situations do not always maximize stability is excellently demonstrated for the $n = 1$ and $n = 2$ members of the series because removal of one Gd atom per formula unit would produce closed-shell compounds in each case, "Gd_3X_5C" and "$Gd_5X_7C_2$". Both metal–metal bonding interactions introduced by the three conduction electrons as well as the requirements of the highly charged interstitial atom to have a "spherical shell" of

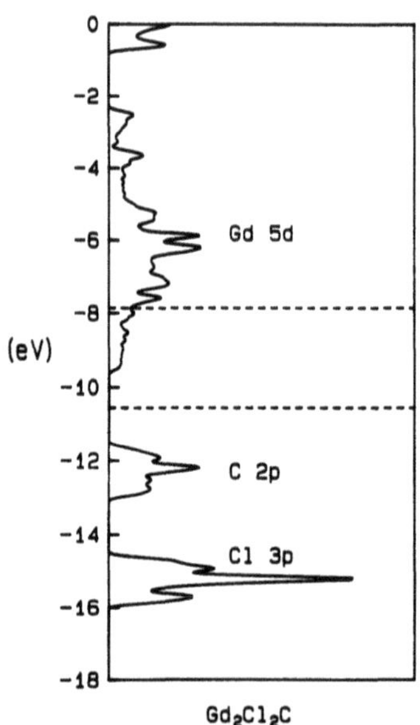

Fig. 50. Total DOS for Gd_2Cl_2C (Burdett and Miller 1987): the lower dashed line marks the Fermi level for this compound and the upper line indicates where the conduction band changes from metal–metal bonding to antibonding character.

positively charged cations contribute to the stability of the members of the $Gd_{2n+2}X_{2n+3}C_n$ series. For n greater than two, the structural building principle will necessitate crystallographically inequivalent (and, therefore, electronically inequivalent) carbon atom sites. For these cases, it is possible that thermodynamic reasons preferentially lead to mixed products of Gd_2X_2C with either $Gd_6X_7C_2$ or Gd_4X_5C rather than a single component of a higher analogue, although $Gd_6Cl_5C_3$ is one example that does contain such inequivalent carbon atoms.

In compounds described by the formula $R_{2n+3}X_{2n+6}C_n$, there are three conduction band electrons as well, which are now formally supplied by the trivalent metal. These additional metal atoms connect the metal carbohalide units together and effect a change in the close-packing sequence of anions from ccp (. . . ccc. . .) to (. . . hhc. . .) in order to minimize electrostatic repulsions between cations. Again, removal of the single metal atom would produce a closed-shell electronic situation. However, these metal atoms contribute to the stability of the structure not only as a bridge between cluster units, but also as an electron source. Indeed, these two series of compounds, $R_{2n+2}X_{2n+3}C_n$ and $R_{2n+3}X_{2n+6}C_n$, confirm the importance of occupation of metal–metal bonding states for the stability of the condensed cluster units.

A number of other gadolinium carbide halides which were described in sect. 2 as defect rock salt derivatives have only one electron in the conduction band. $Gd_6Cl_5C_3$ exhibits a range of homogeneity of $Gd_6Cl_5C_{3.0-3.5}$, which can be rationalized by a partial substitution of C^{4-} by C_2^{6-}. At the point $Gd_{12}Cl_{10}C_5(C_2)$ $(=Gd_6Cl_5C_{3.5})$, in which $\frac{1}{6}$ of the interstitial sites are occupied by C_2^{6-} units, all metal–metal bonding states are formally depleted (Simon et al. 1988). The change from metallic to semiconducting character is consistent with this model. However, why is the C^{4-} ion not replaced by equally charged C_2^{4-} units? The additional Madelung term in the total energy from the C_2^{6-} species and the energy gap that opens between occupied and unoccupied states as the $C_2\pi^*$ band is pushed to lower energy both contribute to this phase width.

In Gd_2ClC EH tight-binding calculations indicate a single metal–metal bonding band which could accommodate at most two electrons (fig. 51). A similar situation occurs in Gd_2IC, although it adopts an H-phase structure type rather than a rock salt derivative. Metal–metal bonding states occur up to the minimum in the total DOS just below -8 eV for both compounds. However, the difference in structural preference for Gd_2ClC and Gd_2IC are not yet understood.

As the final set of examples of the metal-rich rare earth halides, the chemical bonding and electronic stability of the hydride halides, RXH_x, are presented. Their structures have been discussed in sect. 2.3.

In general, the character of the metal–hydrogen interaction, as to whether it is predominantly metallic or ionic, is usually a controversial point and blurs the distinction between “hydrogen solid solutions” and “hydrides”. The discussion of interstitial hydrogen atoms in sect. 3.1 indicated that only a restricted number of cluster or crystal orbitals of the metal halide framework can interact with the totally symmetric H 1s orbital. Furthermore, the R–H bonding combinations have largely hydrogen character, but are still metal–metal bonding. Therefore, the electronic structure of hydrogen in a metal or metal halide may not be accurately understood

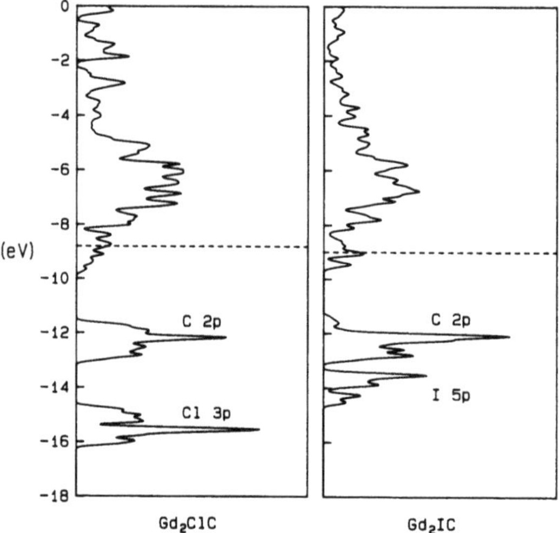

Fig. 51. Total DOS for Gd_2ClC (left) and Gd_2IC (right); the Fermi levels for the two compounds are labelled.

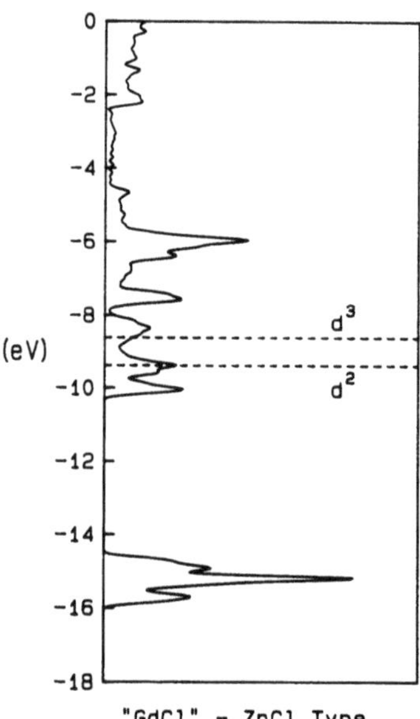

Fig. 52. Total DOS for hypothetical "GdCl" in the ZrCl structure; the Fermi levels for d^2 and d^3 systems are labelled.

from a simple rigid-band treatment of the host system. The interaction between hydrogen and the "substrate" produce significant changes in the electronic density of states near the Fermi level of the metal part of the host, so that one cannot simply add electrons to the DOS of the host substructure. In the examples of RXH_x that follow, further confirmation of this conclusion will be illustrated.

Figure 52 illustrates the total DOS for a two-dimensional model of GdCl in the ZrCl structure (Bullett 1980, Burdett and Miller 1987). The Fermi levels for d^2 ("GdCl") and d^3 (ZrCl) systems indicate a possible instability associated with the binary d^2 system. Careful examination of the lower part of the conduction band reveals three metal–metal bonding energy bands per two formula units. Two of these levels have large metal–metal σ and π overlap between the adjacent sheets, and the third level is a mixture of σ and π interactions between metals within a single layer. For d^3 metals like Zr in ZrCl, these bands are occupied with the Fermi level falling just below an intersheet metal–metal π* band. The high $N(E_F)$ for d^2 systems arises primarily from an avoided crossing. However, this large $N(E_F)$ also indicates a certain degree of electronic instability which may be alleviated by either a geometrical distortion (similar to a first-order Jahn–Teller effect) or by introducing interstitial species into the structure.

Burdett and Miller (1987) have examined the problem of placing hydrogen into the various interstitial sites in "GdCl"; their conclusions are summarized, using a simplified model, in fig. 53. Only the following stoichiometries are considered: GdCl, $GdClH_{0.5}$, GdClH, $GdClH_{1.5}$ and $GdClH_2$. For the host framework, there are three metal–metal bonding bands per two metal atoms in the cell, which are indicated by the three bars under the major Gd d component (these should not be considered as localized states, but only serve as a means of counting relevant energy bands). These five compounds introduce, respectively, zero, one, two, three and four hydrogen bands to the DOS slightly above the X p bands. The three metal–metal bonding bands have the correct symmetry characteristics with which to interact with the H 1s bands. Therefore, with the addition of each H atom to the structure, one metal–metal bonding band each forms a bonding/antibonding combination with the H 1s function. The Gd–H bonding orbital has mostly hydrogen character, while the antibonding combination is pushed well above the bottom of the conduction band. Thus, in $GdClH_{0.5}$ there are two low-lying metal–metal bonding bands, and in GdClH, there is one such band. From fig. 53, GdClH represents an electronically stable situation since metal–metal bonding is maximized. Qualitative confirmation of this tight-binding extended Hückel treatment of GdClH comes from comparison between its photoelectron spectrum and its calculated total DOS in fig. 54. Hydrogen contribution to the total DOS lies primarily under the Cl 3p peak, while the peak at 1.0 eV is due to the single Gd–Gd bonding band. A subtle result from the work of Burdett indicates

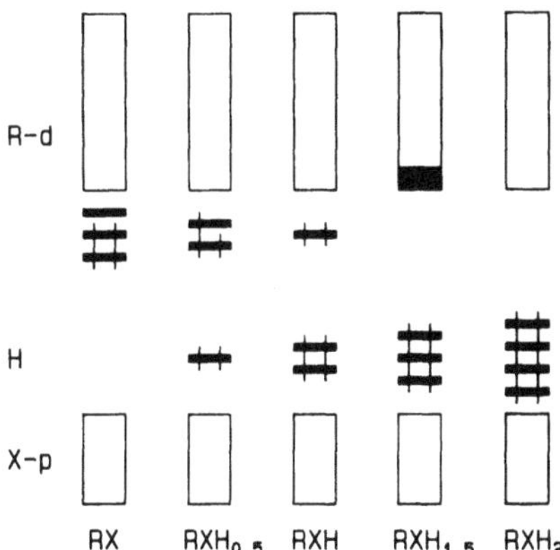

Fig. 53. Schematic diagram of the DOS for various RXH_x compositions ($x = 0.0, 0.5, 1.0, 1.5, 2.0$): bars above the p band of X (X – p) represent H levels and those below the d band of R (R – d), R–R bonding bands; these bars do not represent localized states; the shaded portion in "$RXH_{1.5}$" must necessarily be occupied.

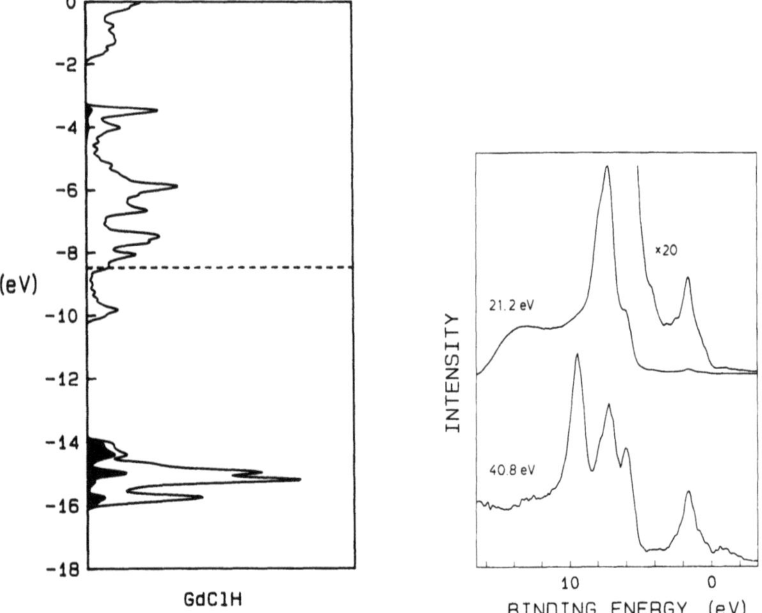

Fig. 54. Calculated total DOS (Burdett and Miller 1987) and the photoelectron spectrum for GdClH (Schwarz 1987).

that the instability associated with a large $N(E_F)$ is not quite alleviated in the hemihydrate. However, for stoichiometries slightly above H/Gd = 0.5, electron occupations proceed into regions of low total DOS. Dispersion of these metal–metal bonding bands in reciprocal space indicate metallic behavior for these systems.

For $GdClH_{1.5}$ there now remain no metal–metal bonding levels at the bottom of the conduction band, and yet to maintain charge neutrality, one electron must occupy the high-lying Gd d states (or form one Gd^{2+} per two formula unit. This highly unstable situation (in terms of electronic energy) is eliminated when an additional H atom is added to the structure to form $GdClH_2$. The fourth H 1s band allows this electron to adopt a stable state which is Gd–H bonding, and results in insulating electrical transport properties for the material. Regarding the absence of RXH_x compounds with $1 < x < 2$, these theoretical arguments strongly suggest that rather than occupying an electronically unstable metal d state to form a single-phase product, a mixture of RXH_x ($x \leqslant 1$) and RXH_2 is preferred.

To summarize, the tight-binding approximation has been extremely successful in assessing the variety of structural information and electronic properties in the metal rich rare earth metal halide systems. Of particular importance is the observation that the interstitial species provides increased stability through the R–Z overlap while leaving the number of cluster bonding states unchanged. The additional electrons supplied by the Z atoms fill metal–metal bonding levels at the bottom of the conduction band. However, all of these treatments have neglected the effects due to the f states of the lanthanide metals, which contribute to the magnetic and transport properties these materials have. In the following section we discuss the experiments and physical models that have been used to examine these particular problems.

4. Electrical and magnetic properties

Investigations of the electrical and magnetic properties of the metal-rich rare earth halides have focussed on the Gd halide hydrides (deuterides) and carbides, and the Tb halide hydrides (deuterides). Table 10 summarizes some significant electrical and magnetic data of Gd, Tb, Sc and Y compounds. The binary compounds, the carbide halides with cluster chains or planes, and the hydride halides are discussed in detail.

The ionic model for the metal-rich R halides assuming closed-shell anionic species, e.g. X^-, H^-, C^{4-}, C_2^{6-}, and trivalent cations R^{3+} has been shown to be very effective in explaining details of the crystal structure and bonding. However, this simple picture is unable to generally predict the magnitude of the electrical conductivities. Gd_2Cl_3 $[=(Gd^{3+})_2(Cl^-)_3(e^-)_3]$ contains three electrons per formula unit in metal–metal bonding states and should have metallic properties. With the same argument, $Gd_2Cl_2C_2$ $[=(Gd^{3+})_2(Cl^-)_2C_2^{4-}]$, on the other hand, should be a semiconductor. In reality, Gd_2Cl_3 is semiconducting while $Gd_2Cl_2C_2$ is metallic. Detailed band structure analyses lead to the same results (Bullett 1980, 1985, Miller et al. 1986).

From their crystal structures the metal-rich R halides can be classified with respect to their electrical properties as follows (see table 10):

- Compounds with either isolated clusters or chains condensed from single octahedra are semiconductors;
- Compounds containing either twin chains or layers of octahedra are metallic, at least at room temperature (disregarding the compounds RXH_2 and $R_2X_2C_2$).

Among the compounds with metallic conductivity at room temperature, only Gd_2XC compounds show a decreasing resistivity towards lower temperatures. In most other cases a slight upturn of the resistivity with decreasing temperature is observed. These results were obtained on powder compacts and, therefore, give an average of the different components of the resistivity tensor. The single-crystal investigation of the layer compound $Gd_2Br_2C_2$ indicates an in-plane resistivity with metallic temperature behavior and with two orders of magnitude smaller values than for the direction perpendicular to the layers. Indeed, a pronounced anisotropy of the electrical properties should be expected for compounds containing one-dimensional (1D) or two-dimensional (2D) structural elements.

For a small number of the investigated compounds a very strong increase of the resistivity upon lowering the temperature is observed. Such behavior has been found for $GdXH_{0.67}$ ($X = Br, I$), Gd_3I_3C and $Gd_6Cl_5C_{3.5}$. Explanations for these metal-to-semiconductor transitions that take place in a rather broad temperature range still remain speculative and will be outlined below.

Many of the Gd and Tb compounds show features in their temperature-dependent resistivities which are characteristic of magnetic phase transitions. The magnetic

TABLE 10

Electrical and magnetic properties[a] of metal-rich halides of the rare earth metals. T_N(K): Néel temperature; T_C(K): Curie temperature; θ(K): paramagnetic Curie temperature; MI: data not accessible because of magnetic impurity phases; SC: semiconductor; M: metal; ?: data not available.

Compound	Condensation of R octahedra	[c]Interstitial (X)	T_N (K)	θ (K)		μ_{eff} (μ_B) per R atom	Electrical properties	Ref.[b]
Gd_2Cl_3	Chains	–	26	−180		7.7	SC; $E_g=0.85$ eV	[1]
Tb_2Cl_3	Chains	–	41	−91, −116		9.7	SC; $E_g=1.1$ eV	
$GdClH_y$	Layers	Tetrahedral (1X)	?		?			
$y=0.72$			47		?		Weak M→SC	
$y=0.97$							M	
$GdBrH_y$	Layers	Tetrahedral (1X)	36	+63		7.7	Very strong	
$y=0.7$			45	−44		7.6	M→SC ($T<20$ K)	
$y=0.8$							Weak M→SC	
$y=1.0$			53		MI		M	
$GdIH_y$	Layers	Tetrahedral (1X)	?		?		Very strong,	
$y=0.67$								
$y=0.92$			≈50		?		M→SC ($T<30$ K)	
							M	
$GdBrH_2$	Layers	Tetrahedral (1X)	2.5	−8.5		7.9	SC; $E_g=2.66$ eV	
		Octahedral (2X)						
$TbBrD_y$	Layers	Tetrahedral (1X)	<1.5	+23		9.5	Weak,	[2]
$y=0.69$							M→SC ($T<50$ K)	
$y=0.8$			20 (1)	−47		9.5	Very weak	
$y=0.9$			22.5 (0.5)	−66		9.5	M	
$TbBrD_2$	Layers	Tetrahedral (1X)	5.5	−4		9.4	$E_g=2.66$ eV	
		Octahedral (2X)						

$Gd_{10}Cl_{18}C_4$	Double octahedra	C_2^{6-}	<2	−32	8.1	SC	
$Gd_{10}Cl_{17}C_4$	Double octahedra	C_2^{6-}	25	MI	MI	SC	
$Gd_{12}Br_{17}C_6$	Zigzag chains	C_2^{6-}	≈30	MI	MI	SC	
Gd_4I_5C	Chains	C^{4-}	<2	MI	MI	SC	
$Gd_6Br_7C_2$	Twin chains	C^{4-}	12; 30	+41	7.9	M	
Gd_3I_3C	Twin chains	C^{4-}	25; 100	–	–	M→SC (<100 K)	
$Gd_2Cl_2C_2$	Layers	C_2^{4-}	<2	−21	7.9	M	
$Gd_2Br_2C_2$	Layers	C_2^{4-}	≈35	−37	7.9	M	
Gd_2ClC	Layers	C_2^{4-}	32	+66	7.9	M	
Gd_2BrC	Layers	C_2^{4-}	$T_c=110$	+123	7.9	M	
Gd_2IC	Layers	C_2^{4-}	$T_c=182$	+206	7.6	M	
$Gd_6Cl_5C_{3+x}$	Buckled layers	C^{4-} and C_2^{6-}	? –	MI	MI	$x=0$: M $x=0.5$: M→SC ($T<50$ K)	[3]
$Sc_7Cl_{12}B$	Single octahedra	B	?	MI	MI	?	[4]
$Sc_7Cl_{12}C$	Single octahedra	C	15	−36	1.7[e]	?	[5]
$Sc_6I_{11}C_2$	Single octahedra	C_2	?	?	1.8[e]	M→SC ($T<50$ K)	[6]
Sc_7Cl_{10}	Twin chains	–	–	1.4	0.9[e]	ρ(300 K) $<5\cdot10^{-2}\ \Omega$cm	[7]
$Y_6I_7C_2$	Twin chains	C	TFIP[d]	TFIP[d]	–	M	[8]

[a] The majority of the data of the Gd and Tb compounds has been published in Bauhofer et al. (1988).

[b] References: [1] Kremer (1985); [2] Kremer et al. (1990); [3] Simon et al. (1988); [4] Hwu and Corbett (1986); [5] Dudis et al. (1986); [6] Dudis and Corbett (1987); [7] DiSalvo et al. (1985); [8] Kauzlarich et al. (1988).

[c] H sites for hydrides and deuterides; C species for carbides.

[d] Temperature- and field-independent paramagnetism.

[e] μ_{eff} (μ_B) per formula unit.

properties of these compounds have been investigated in more detail by measuring the susceptibility, by neutron diffraction, and by Mössbauer spectroscopy. The susceptibilities in the paramagnetic temperature region generally follow a Curie–Weiss law with $\mu_{eff} \approx 7.9\mu_B$ and $\approx 9.6\mu_B$ corresponding to free Gd^{3+} and Tb^{3+} ions, respectively. A negative paramagnetic Curie temperature Θ is observed for the majority of the compounds indicating predominant antiferromagnetic (AF) coupling between the magnetic moments. A positive Θ, indicative for predominant ferromagnetic coupling, occurs for the C^{4-}-containing carbides Gd_2BrC, Gd_2IC and $Gd_6Br_7C_2$. Since Gd is a strong thermal neutron absorber, the determination of magnetic structures by neutron diffraction necessitates the use of a hot neutron source, which is available in the D4B powder diffractometer of the Institut-Laue-Langevin in Grenoble.

Magnetic ordering is a common property of the reduced Gd and Tb halides. With two exceptions, Gd_2BrC and Gd_2IC, antiferromagnetism is found. No ordering down to 2 K has been observed for $Gd_{10}Cl_{18}C_4$, Gd_4I_5C, $Gd_2Cl_2C_2$ and $TbBrD_{0.67}$. The latter compound represents a remarkable example of a magnetic system whose spin glass behavior is induced by diluting its nonmagnetic substructure. It should be pointed out that the isostructural compound $GdBrD_{0.67}$ behaves completely differently since it orders antiferromagnetically at $T_N = 35$ K. Different magnetic properties for isostructural Gd and Tb compounds are expected because the Tb^{3+} ion with its orbital angular momentum experiences much stronger crystal field effects than Gd^{3+} in its S ground state.

The frequent occurrence of metal atom chains and layers in the crystal structures of reduced rare earth metal halides suggests that these compounds could serve as model systems for low-dimensional magnetism. Three-dimensional (3D) magnetic ordering for Gd_2Cl_3 and Tb_2Cl_3 as determined by neutron diffraction, together with tiny features superimposed on a broad background as found in susceptibility and specific heat at the transition temperature, underlines this conjecture.

4.1. *Binary chain compounds*

The binary halides R_2X_3 and Sc_7Cl_{10} have chain structures. Early measurements of the electrical conductivity of Gd_2Cl_3 by Mee and Corbett (1965) already indicated the semiconducting character of the sesquihalides. More detailed investigations on powder samples and on single crystals using a contactless microwave method revealed a temperature-activated conductivity for both Gd_2Cl_3 and Tb_2Cl_3 (fig. 55, Bauhofer and Simon 1982). From the slopes of log ρ versus $1/T$ energy gaps of 0.85 eV for Gd_2Cl_3 and 1.1 eV for Tb_2Cl_3 could be derived. The value for Gd_2Cl_3 is in agreement with band structure calculations (Bullett 1980, 1985) and UV photoemission spectra (fig. 45, Ebbinghaus et al. 1982), while for Tb_2Cl_3 some electron density of states at the Fermi level was interpreted as an excess of Tb metal for the investigated sample. The resistivities of the single crystals along the crystallographic b direction are in both cases considerably higher than the values obtained with pellets which represent an average over all lattice directions. Presuming that these differences are not due to the different measurement methods, an anisotropic conductivity must be inferred. As

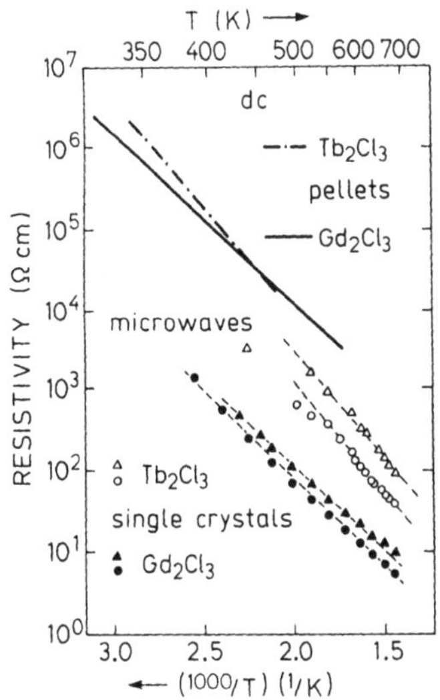

Fig. 55. Resistivity of Tb_2Cl_3 and Gd_2Cl_3 as a function of temperature; the microwave measurements shown in the lower part were obtained on two different single crystals (Bauhofer and Simon 1982).

compared with the other directions the conductivity along the chains is higher by two orders of magnitude for Gd_2Cl_3 and by a factor of 5 for Tb_2Cl_3. Anisotropic behavior is also observed for the optical reflectivity of Tb_2Cl_3 in the range 1.1–2.5 eV.

For Sc_7Cl_{10} a room-temperature resistivity of $\rho < 0.05\ \Omega\,cm$ has been reported by DiSalvo et al. (1985). This result is in agreement with the general statement that all R rich halides containing twin chains show metallic conductivity at room temperature. The magnetic susceptibility of Sc_7Cl_{10} shows a Curie–Weiss behavior indicating the existence of local moments (DiSalvo et al. 1985). The small value of the Curie constant cannot be simply understood on the basis of the structure of Sc_7Cl_{10}. In addition, the lack of a resolved hyperfine structure of ^{45}Sc in the EPR (electron paramagnetic resonance) spectra, which suggests rapid spin diffusion, complicates possible interpretations of the source of the local moments. The situation seems to be clearer for the sesquihalides investigated so far, although some discrepancies remain between the results of different measurement methods. Y_2Cl_3 is diamagnetic at room temperature with $\chi = -31 \times 10^{-6}$ emu/mol. Gd_2Cl_3 has been thoroughly investigated by ESR, NMR and magnetic susceptibility measurements (Kremer 1985). With these methods an unambiguous evidence for magnetic ordering could not be found. Figure 56 shows typical results of susceptibility measurements. The broad maximum around 25 K suggests the case of a 1 D magnetic system. The influence of spurious amounts of Gd metal can be suppressed with a high enough magnetic field.

Neutron diffraction experiments on single crystals of Gd_2Cl_3 clearly proved the occurrence of a 3D magnetic phase transition at 26.8 K. Magnetic structure calculations of the ordered state led to a model for the arrangement of the magnetic moments as depicted in fig. 57. The moments of the Gd atoms in the octahedra bases (Gd1) are ferromagnetically coupled in the chain direction. The two columns of Gd1 moments of the same chain have antiparallel orientations. Refinement of the moment magnitudes at 14.5 K yields $5.5(5)\mu_B$ for Gd 1. The extrapolation to $T = 0$ K is in best

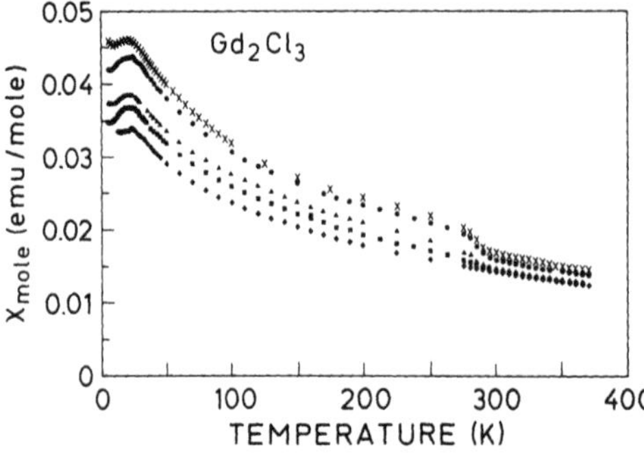

Fig. 56. Powder molar susceptibility of Gd_2Cl_3 for different applied magnetic fields: 0.01, 0.03, 0.3, 1 and 5 T from top to bottom. The step in the range of 300 K originates from a spurious contamination of ferromagnetic Gd metal which is saturated with increasing magnetic field (Kremer 1985).

agreement with the theoretical value of $7\mu_B$ for the saturation moment. For the apex atoms (Gd2), on the other hand, only a small fraction of the expected magnetic moment is ordered, reflecting their frustration in a triangular arrangement. Finally, adjacent chains are coupled antiferromagnetically.

The specific heat of Gd_2Cl_3 (fig. 58a) displays a tiny anomaly around 23 K which indicates the 3D ordering transition. The difference between the ordering temperatures obtained from the neutron and the specific heat measurements remains unclear. Comparison with nonmagnetic Y_2Cl_3 reveals that the magnetic contributions to the specific heat for Gd_2Cl_3 extend up to 100 K. Figure 58b shows a comparison of the magnetic part of the specific heat with the theoretical temperature dependence for a 1D Heisenberg chain with spin $S = \frac{7}{2}$ and an exchange coupling $J = -2.6$ K. Obviously, only a small part of the specific heat is connected with the 3D ordering. It must be concluded that strong 1D magnetic correlations lead to appreciable short-range order well above the 3D ordering temperature.

The single-crystal susceptibilities of Tb_2Cl_3 show a distinct splitting between $\chi \parallel b$ (in-chain direction) and $\chi \perp b$ at temperatures below 60 K (fig. 59). However, clear evidence for long-range magnetic order cannot be drawn from these measurements. The specific heat C_p, on the other hand, exhibits a small anomaly at 46 K which obviously marks the onset of 3D order (fig. 60). The entropy involved in this anomaly is only 0.4 J mol^{-1} K^{-1}. This value is even far from $R \ln 2$ expected for the ordering of a $S = \frac{1}{2}$ system. Additionally, a comparison between C_p of Tb_2Cl_3 and Y_2Cl_3 reveals a broad supplementary contribution for Tb_2Cl_3 which extends to temperatures higher

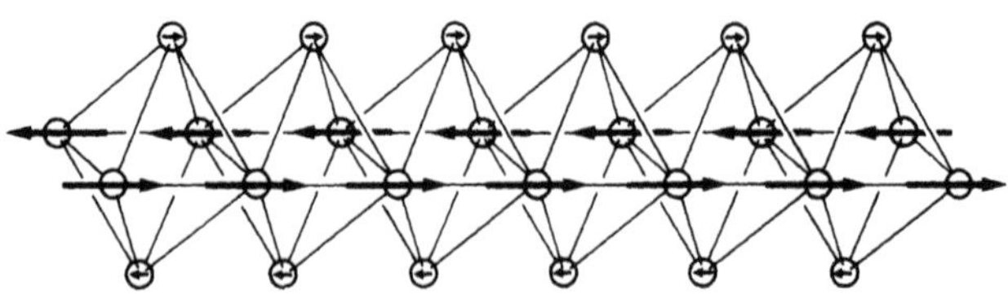

Fig. 57. Magnetic structure of Gd_2Cl_3: the lengths of the arrows correspond to the magnitude of the ordered moments.

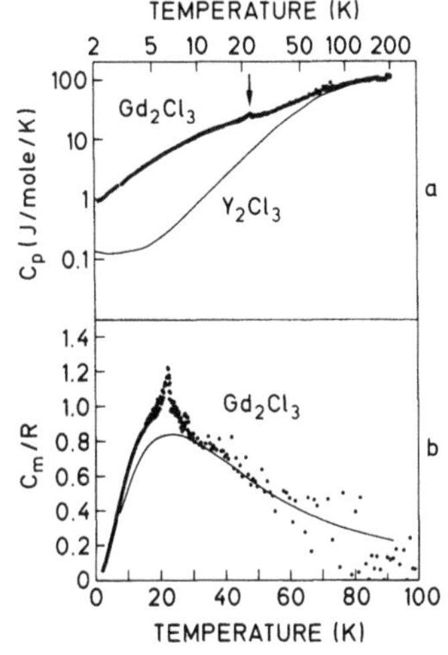

Fig. 58. (a) Specific heats of Gd_2Cl_3 and Y_2Cl_3 (full line): the arrow indicates the 3D ordering transition. (b) Magnetic part of the specific heat of Gd_2Cl_3 (per formula unit $GdCl_{1.5}$): the full line is the specific heat of an $S = \frac{7}{2}$ Heisenberg chain with exchange constant $J = -2.6$ K (after Kremer 1985).

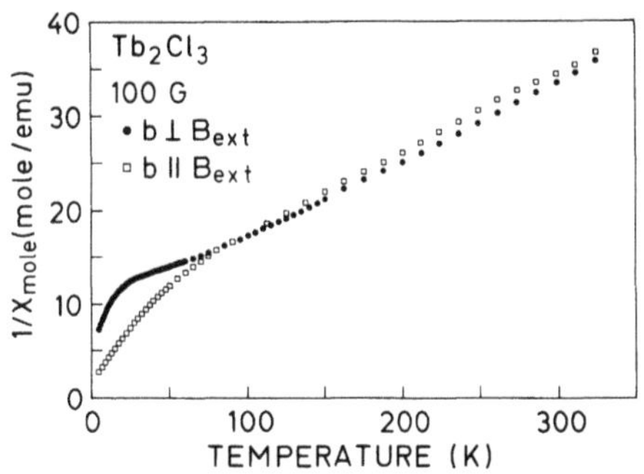

Fig. 59. Single-crystal susceptibilities of Tb_2Cl_3 with external magnetic field applied parallel and perpendicular to the *b*-axis (per formula unit $TbCl_{1.5}$).

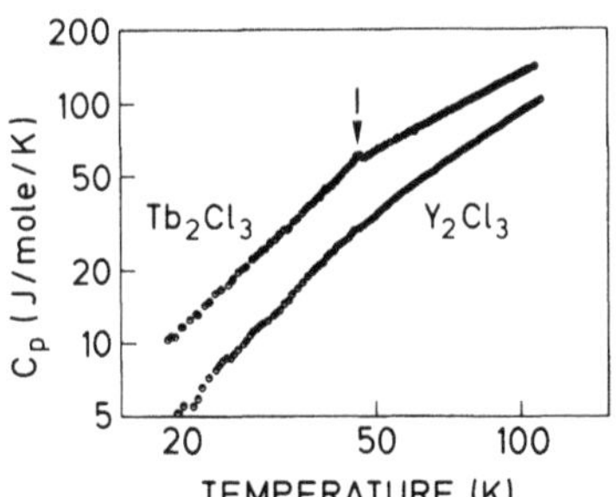

Fig. 60. Specific heats of Tb_2Cl_3 and Y_2Cl_3: the arrow indicates the 3D ordering transition.

than 100 K. This extra contribution might be due to crystal field excitations, but short-range ordering due to low-dimensional magnetism would have the same effect.

Three-dimensional antiferromagnetic ordering of Tb_2Cl_3 has been proved unequivocally by neutron diffraction carried out by Schmid et al. (1988). A powder pattern taken at 5 K shows strong additional Bragg reflections when compared with a

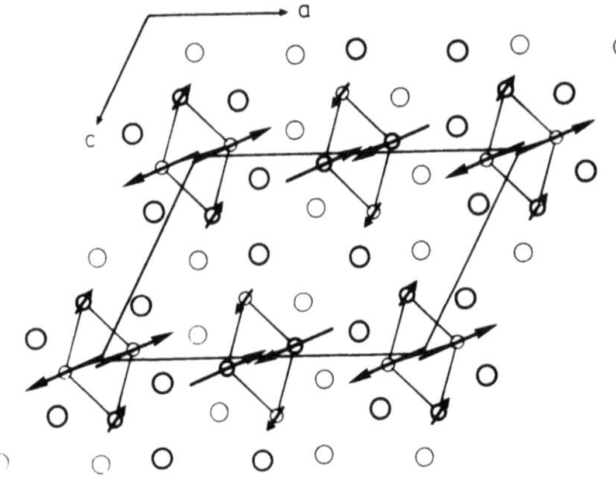

Fig. 61. Projection of the magnetic structure of Tb_2Cl_3 along [010] (Tb: small circles, Cl: large circles): the lengths of the arrows correspond to the magnitude of the ordered moments (after Schmid et al. 1988).

diagram in the paramagnetic state at 50 K. Rietveld refinement of the magnetic structure leads to the model depicted in fig. 61. This model is closely related to the magnetic order of Gd_2Cl_3 with the differences that the size of the magnetic cell is identical to the size of the crystallographic cell and that the major components of the magnetic moments are perpendicular to the *b* axis in the case of Tb_2Cl_3. Similar to Gd_2Cl_3 the magnetic moments on the basal atoms (Tb1) are a factor of three larger than the moments of the apex atoms. This difference could be due to varying crystal field splittings of the Tb^{3+} ions according to their coordination geometries. However, the analogy to the Heisenberg magnet Gd_2Cl_3 also suggests a frustration of the Tb2 moments. The moments of Tb1 are close to the maximal possible value of $9\mu_B$ for Tb^{3+}. From the neutron measurements a Néel temperature $T_N = 40.8 \pm 0.2$ K is derived. This value is 5 K lower than that found from C_p, which might be another indication of low-dimensional properties in Tb_2Cl_3.

4.2. *Carbide halides*

Carbide halides form a wide variety of different structure types ranging from isolated C-filled R_6 octahedra contained in $Gd_7I_{12}C$ to the three-dimensional R_6 network of Gd_3Cl_3C. The carbon species, either C or C_2, are situated in octahedral holes of the R substructure, and no partial filling is observed. $Gd_6Cl_5C_{3+x}$ is the only example so far where C and C_2 units occur in the same compound. The great variety of crystal structures is reflected in the large number of different electrical and magnetic properties of the carbide halides. Four groups of compounds with characteristically different structural features have been selected for further investigations and will be presented in more detail: chain structures, layer structures with interleaving X atom bilayers, layer structures with interleaving single X atom layers, and buckled layer structures. Information on the physical properties of other Gd, Sc and Y carbide halides can be found in table 10.

Chain carbides of the types R_4X_5C, R_3X_3C and $R_6X_7C_2$ have been investigated by physical measurements. Considering the crystal structures, R_4X_5C is closely related to Gd_2Cl_3. Both compounds contain straight chains of edge-sharing R_6 octahedra which, in the case of R_4X_5C, are filled with single carbon atoms. The semiconducting behavior of Gd_4I_5C (fig. 62) cannot be rationalized with the same arguments used for Gd_2Cl_3 (Gd–Gd triple bond at shared edges) since an odd number of electrons, $(Gd^{3+})_4(I^-)_5C^{4-}(e^-)_3$, cannot be accommodated in the bonding of one Gd–Gd pair. The low room-temperature resistivity value of Gd_4I_5C indicates a distinctly smaller energy gap than 0.85 eV found for Gd_2Cl_3. A small room-temperature resistivity has also been reported for Y_4I_5C by Kauzlarich et al. (1988). Another interesting chain carbide, Sc_5Cl_8C, has been predicted to be metallic (Hwu et al. 1987). This would represent the first example of a metallic single octahedra chain carbide. The magnetic susceptibility of Gd_4I_5C shows no indication for a magnetic phase transition down to 2 K.

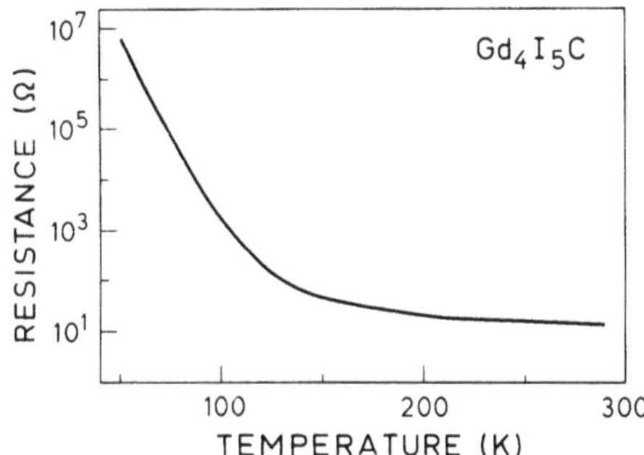

Fig. 62. Temperature-dependence of resistance for Gd_4I_5C.

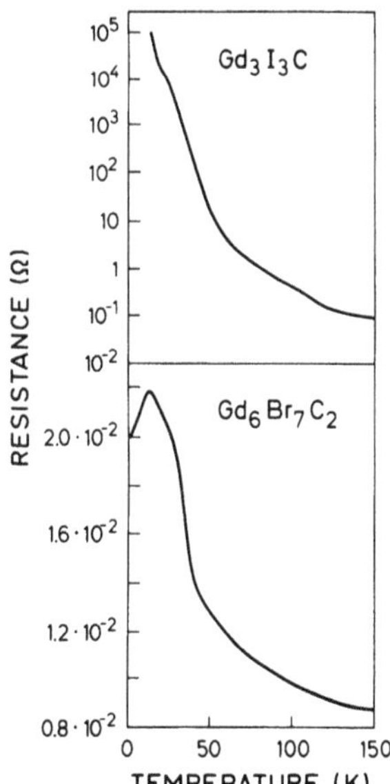

Fig. 63. Temperature dependence of resistance for Gd_3I_3C and $Gd_6Br_7C_2$.

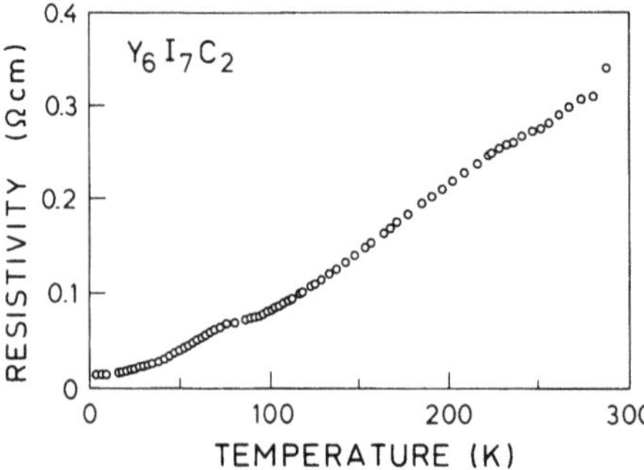

Fig. 64. Temperature dependent resistivity for $Y_6I_7C_2$ (Kauzlarich et al. 1988).

Twin chains are contained in the structures of $R_7X_{10}C_2$, Gd_3I_3C and $R_6X_7C_2$. The electrical resistivities of some of these compounds are depicted in figs. 63 and 64. Gd_3I_3C becomes semiconducting at low temperatures, whereas $Gd_6Br_7C_2$ and $Y_6I_7C_2$ stay metallic. This difference might be connected with the even number of electrons per formula unit in Gd_3I_3C $[=(Gd^{3+})_3(I^-)_3C^{4-}(e^-)_2]$ in contrast to the odd electron number in $Gd_6Br_7C_2$ $[=(Gd^{3+})_6(Br^-)_7(C^{4-})_2(e^-)_3]$. It should be mentioned that the resistivity measurements on $Y_6I_7C_2$ were performed on single crystals (Kauzlarich et al. 1988). Both resistivity curves of fig. 63 show two magnetic anomalies which can easily be identified in the magnetic susceptibilities as local maxima at 12 K and 30 K for Gd_3I_3C and at 25 K and 100 K for $Gd_6Br_7C_2$. Antiferromagnetic ordering transitions can be assumed at these temperatures. This assumption has been confirmed for the 25 K transition in Gd_3I_3C by neutron diffraction where additional Bragg reflections are observed at low temperatures. Fitting the inverse susceptibility of $Gd_6Br_7C_2$ to a Curie–Weiss law leads to $\mu_{eff} = 7.92\mu_B$ and $\theta = +41$ K. A high positive θ is rather surprising for an antiferromagnetic material with $T_N = 30$ K and clearly proves the existence of strong ferromagnetic coupling. Impurity phases prevent such an analysis for Gd_3I_3C.

$Gd_2X_2C_2$ with X = Cl and Br are two examples for layered carbides where the R bilayers are interleaved with bilayers of X atoms. Each Gd octahedron is filled by a C_2 unit. Figure 65 shows the resistivity of a $Gd_2Cl_2C_2$ powder pellet (Schwarz 1987) in comparison with a contactless measurement on a $Gd_2Br_2C_2$ single crystal. The difference of two orders of magnitude reflects the difference between the in-plane resistivity of the single crystal and the averaged resistivity of the powder sample. In addition, the sharp drop at 30 K in the resistivity of $Gd_2Br_2C_2$ hints at a magnetic phase transition. The minimum of the inverse susceptibility at the same temperature confirms the occurrence of an antiferromagnetic transition (Schwanitz-Schüller 1984). Interestingly, no such transition is observed for $Gd_2Cl_2C_2$ down to 1.5 K. This might be related to the fact that $\theta = -37$ K is deduced for $Gd_2Br_2C_2$ in contrast to the smaller value of $\theta = -21$ K for $Gd_2Cl_2C_2$.

The most metal-rich carbide halides have the formula R_2XC and are hitherto only known for R = Gd. Their crystal structures can either be described as condensed layers of edge-sharing octahedra of Gd atoms separated by X atom layers, or they can be derived from the Gd_2C structure by interleaving neighboring Gd atom bilayers

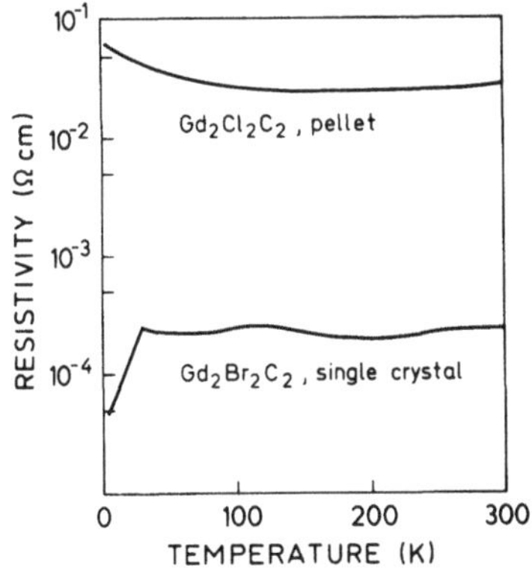

Fig. 65. Temperature dependent resistivity for a $Gd_2Cl_2C_2$ powder pellet and a $Gd_2Br_2C_2$ single crystal.

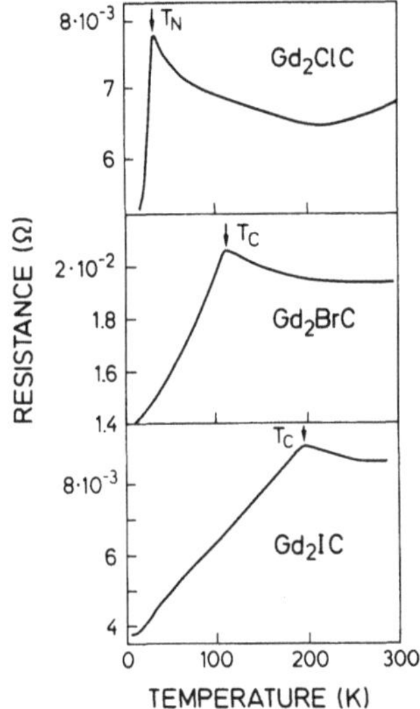

Fig. 66. Temperature dependent resistivity for Gd_2XC (X = Cl, Br, I): the magnetic transition temperatures are indicated by arrows.

with a single X atom layer. All three compounds (X = Cl, Br, I) have a low-temperature resistivity smaller than at room temperature as expected for metallic materials (fig. 66). Phase transitions are clearly visible in the resistivity curves. From susceptibility measurements Gd_2ClC is found to order antiferromagnetically whereas Gd_2BrC and Gd_2IC are ferromagnets (fig. 67). This behavior can be explained with respect to the different coordination of the X atoms. In Gd_2ClC, straight Gd–Cl–Gd connections lead to antiferromagnetic coupling whereas the different stacking in Gd_2XC (X = Br, I) favors Gd–X–Gd angles smaller than 180° and ferromagnetic coupling (Schwarz 1987). The ferromagnetism of Gd_2XC should be viewed in context with the close structural relationship to Gd_2C, for which a T_C of 400 K has been reported (Lallement 1968).

$Gd_6Cl_5C_{3+x}$ seems to represent the only example for a R carbide halide containing C and C_2 units simultaneously. This feature might explain the dependence of the

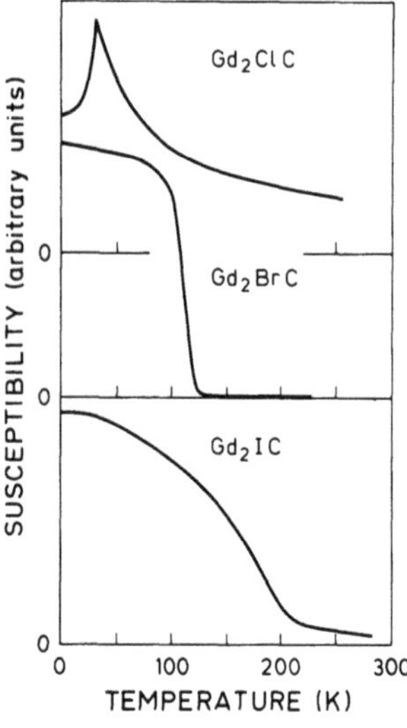

Fig. 67. Temperature dependent susceptibilities for Gd_2XC (X = Cl, Br, I).

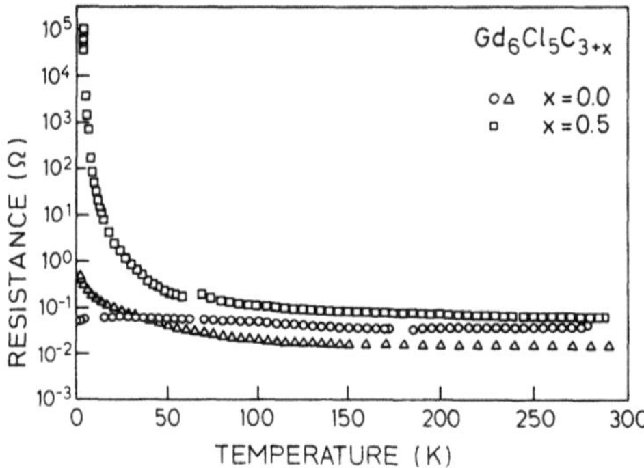

Fig. 68. Temperature dependence of resistance for $Gd_6Cl_5C_{3+x}$ (Simon et al. 1988).

resistivity on x. The electrical resistivities of $Gd_6Cl_5C_{3+x}$ are shown in fig. 68 for different values of x. For $x = 0$, equivalent to $(Gd^{3+})_6(Cl^-)_5(C^{4-})_3e^-$, a temperature-independent metallic-like resistivity is observed. The $x = 0.5$ compound, which can be described as $(Gd^{3+})_{12}(Cl^-)_{10}(C^{4-})_5C_2^{6-}$, exhibits a strong increase of the resistivity towards lower temperatures according to localization of the charge carriers (Simon et al. 1988).

4.3. *Hydride halides*

Our investigations of the systems $GdClH_x$, $GdBrH_x$ and $YClH_x$ have shown that the existence range for x is limited to $0.67 \leqslant x \leqslant 1$ and $x = 2$. The dihydrides are salt-like compounds with typically yellow to greenish color. From measurements of the diffuse reflectance optical band gaps in the range from 2.5–3 eV have been derived (Müller-Käfer 1988). The exponential decrease of the electrical resistivity with rising temperature corresponds to an activation energy of 0.55 eV.

Magnetic susceptibility measurements have been performed on $CeBrD_2$, $GdBrD_2$ and $TbBrD_2$. A Curie–Weiss law is obeyed in all cases at higher temperatures with a Curie constant corresponding to the magnetic moments of the free trivalent metal ions. The paramagnetic Curie temperatures are found to be negative, $\theta_{Ce} = -3$ K, $\theta_{Gd} = -8.5$ K, $\theta_{Tb} = -4$ K, pointing to antiferromagnetic interactions. Indeed, minima in $\chi^{-1}(T)$ for $GdBrH_2$ at 2.5 K and for $TbBrD_2$ at 5.5 K indicate antiferromagnetic ordering temperatures (fig. 69). $GdID_2$ has been further investigated by powder neutron diffraction and broad additional features characteristic for short-range magnetic order have been observed up to 10 K (fig. 70, Cockcroft et al. 1989b).

All RXH_x compounds with $x \leqslant 1$ are metallic conductors at room temperature (fig. 71). The change of the complex dielectric constant at the metal-to-semiconductor transition during the reversible reaction $GdBrH_x + \frac{1}{2}(2 - x)/H_2 = GdBrH_2$ (at 700 K) has been measured in situ using a contactless microwave method (Bauhofer et al. 1989). The homogeneity range $0.67 \leqslant x \leqslant 1$ suggests an interesting influence of the hydrogen concentration x on the physical properties. Measurements of $GdXH_x$ first pointed out the existence of a strong correlation between the hydrogen concentration and the temperature dependence of the electrical resistivity. Generally, the

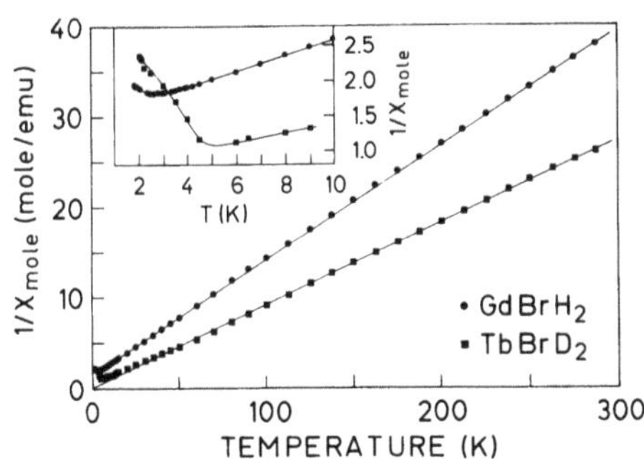

Fig. 69. Reciprocal molar susceptibility of $GdBrD_2$ and $TbBrD_2$.

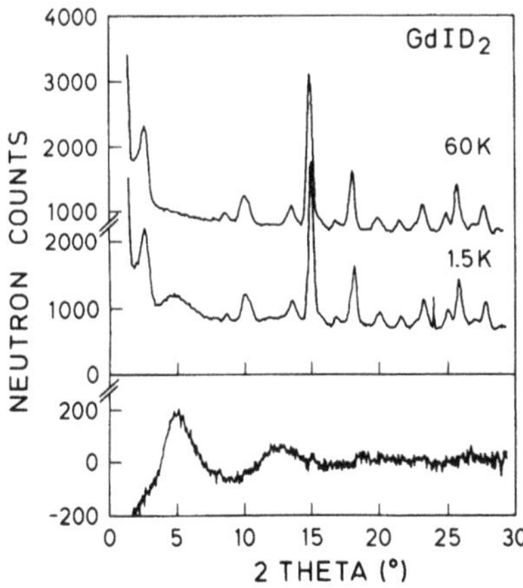

Fig. 70. Powder neutron diffraction patterns (top) of $GdID_2$ at 60 K and 1.5 K and difference pattern (bottom); neutron wavelength $\lambda = 0.5$ Å (Cockcroft et al. 1989b).

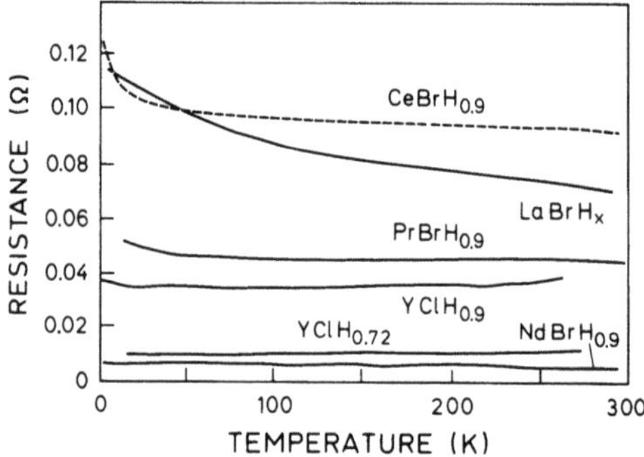

Fig. 71. Temperature dependence of resistance for different RXH_x with $x \leqslant 1$.

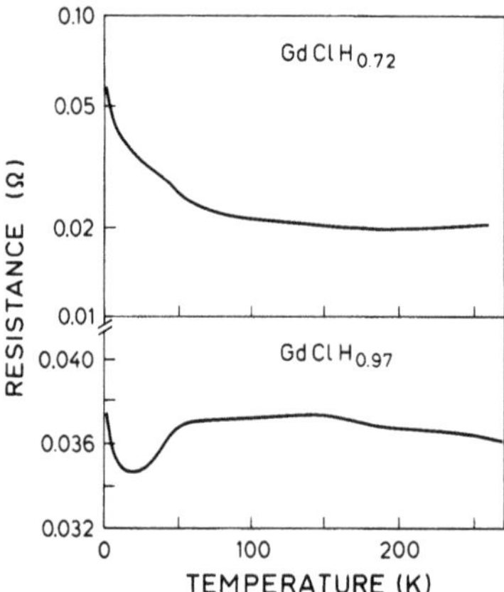

Fig. 72. Temperature dependence of resistance for $GdClH_{0.72}$ and $GdClH_{0.97}$.

resistivity of $GdXH_x$ increases at low temperatures. This increase becomes more pronounced as x approaches its lower limit $\sim\frac{2}{3}$. Figure 72 shows this behavior for the chloride system. For $x = 0.97$, the resistivity value at 1.5 K is only slightly larger than at room temperature, whereas for $x = 0.72$, an increase by a factor of nearly 3 is observed.

The low-temperature increase of the resistivity at low x is much stronger for the bromide and the iodide compounds (figs. 73 and 74). For $GdIH_{0.67}$ an increase by four orders of magnitude is found. For $GdBrH_{0.69}$ this increase reaches the record value of six orders of magnitude. In addition, the results for the bromide system show a strictly monotonic growth of the resistivity increase with decreasing x. Deuterated samples prepared for neutron diffraction behave exactly in the same way. What causes the resistivity to rise so strongly at low temperatures when x approaches its lower limit $\sim\frac{2}{3}$? It is appealing to correlate this behavior with an ordering of the hydrogen atoms which might be possible for $x = \frac{2}{3}$. Such an ordering produces a slight reconstruction of the lattice which could result in a localization of the charge carriers at low temperatures. However, the absence of any nuclear superstructure reflections in X-ray and neutron diffractograms excludes a 3D ordering of the hydrogen atoms. A 2D ordering occurring independently within different Gd bilayers would account for

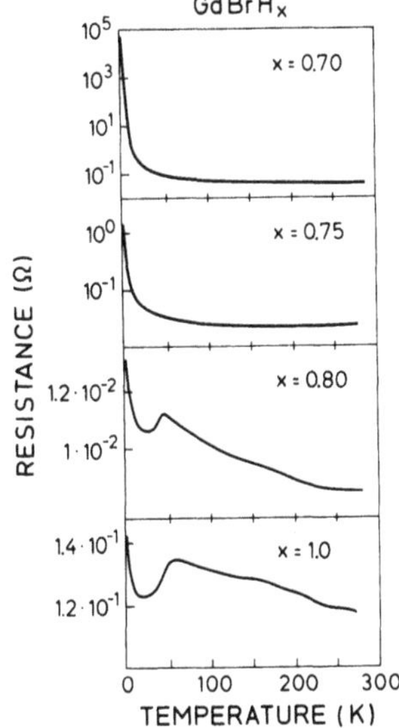

Fig. 73. Temperature dependence of resistance for $GdBrH_x$ for different values of x.

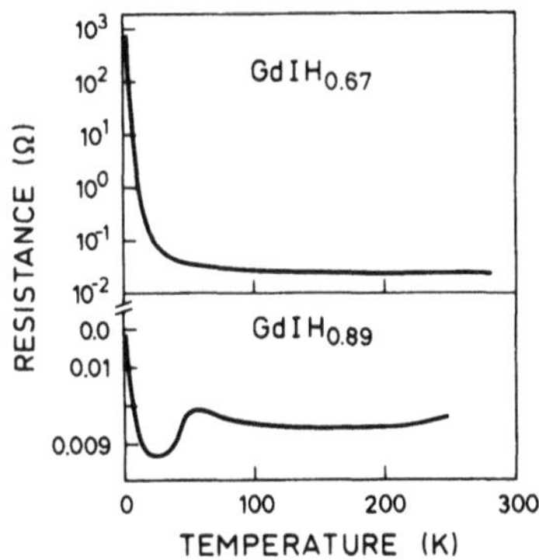

Fig. 74. Temperature dependence of resistance for $GdIH_{0.67}$ and $GdIH_{0.89}$.

the experimental fact that no abrupt phase transition but rather a sluggish change from metallic to semiconducting behavior is observed. In the most symmetric ordering of the hydrogen atoms (conserving the hexagonal symmetry) islands of four hydrogen filled tetrahedra are formed leading to a 2D unit cell of $a\sqrt{3} \times a\sqrt{3}$. Mössbauer spectroscopy on ^{155}Gd suggests a deviation from hexagonal symmetry and favors a chain-like structure with $3a \times a\sqrt{3}$ (Czjzek 1989). The free electrons could localize in bonds between Gd atoms. A similar model has been derived from neutron diffraction for $ZrBrD_{0.5}$ by Wijeyesekera and Corbett (1986).

Some important questions remain to be answered in this context:

– Why is the low-temperature resistivity increase much stronger for $GdBrH_{0.7}$ and $GdIH_{0.7}$ than for the chloride?

– Why is a strong low-temperature resistivity increase observed only for Gd hydride halides and not for other rare earth metal hydride halides?
– Is there a connection between antiferromagnetic ordering and the low-temperature resistivity increase?

Many resistivity curves of $GdXH_x$ show anomalies around 50 K in their temperature dependences which typically occur at magnetic phase transitions. These anomalies become more pronounced when x approaches its upper limit 1.0. The electrical resistivity drops at a magnetic ordering transition due to the reduction of spin-disorder scattering. For the $GdXH_x$ compounds this decrease is superimposed by the general increase of the resistivity towards lower temperatures, leading to a characteristic "scoop-like" shape for $T \leqslant 50$ K. In the bromide and the iodide system no magnetic anomaly can be observed at low values of x due to the strong resistivity increase for low temperatures. In these cases an antiferromagnetic ordering can be inferred with the necessary reservation (see below the result for $TbBrD_{0.7}$) from measurements of the magnetic susceptibility (fig. 75a). Figure 75b displays the dependence of the Néel temperature T_N on the hydrogen concentration x for $GdBrH_x$: T_N drops from 53 K for $x = 1.0$ to 35 K for $x = 0.67$. The strong negative magnetoresistivity found for $GdBrD_{0.7}$ at low temperatures supports the idea of a connection between the low-temperature resistivity increase and antiferromagnetic order.

Neutron powder measurements are shown in fig. 76 for $GdBrD_{0.7}$ and $GdID_{1.0}$ (Cockcroft et al. 1989b). In both cases the low-temperature diffraction patterns contain structures in addition to the nuclear peaks at higher temperatures. The extra peaks at small diffraction angles can be indexed as 009 on the basis of the nuclear unit cell with a doubled c axis. It should be emphasized that at low temperatures $GdBrD_{0.7}$ exhibits antiferromagnetic ordering together with electrically insulating behavior. Thus, the RKKY interaction cannot be the dominant mechanism for the magnetic coupling to produce magnetic order.

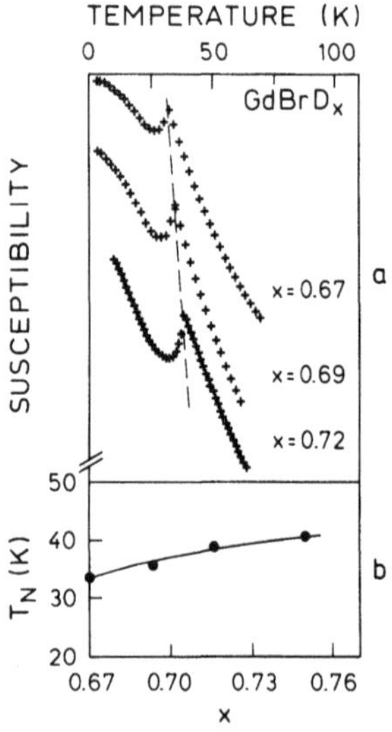

Fig. 75. (a) Magnetic susceptibility of $GdBrD_x$ in the range of the Néel temperature. (b) Shift of the Néel temperature with x.

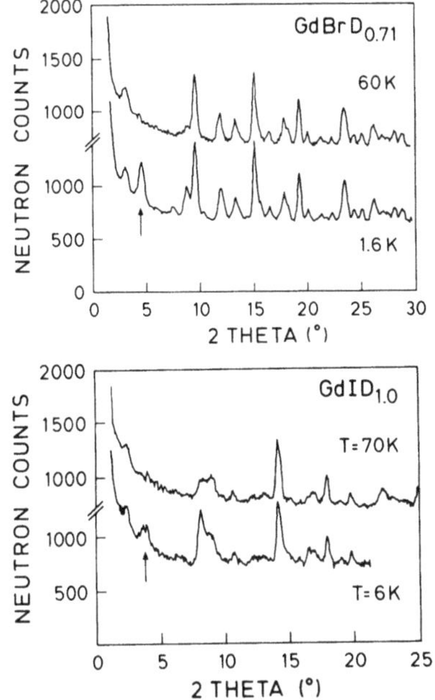

Fig. 76. Powder neutron diffraction patterns of $GdBrD_{0.71}$ and $GdID_{1.0}$ at a neutron wavelength $\lambda = 0.5$ Å. The arrows indicate the 009 additional reflection with respect to a nuclear cell with doubled c axis (Cockcroft et al. 1989b).

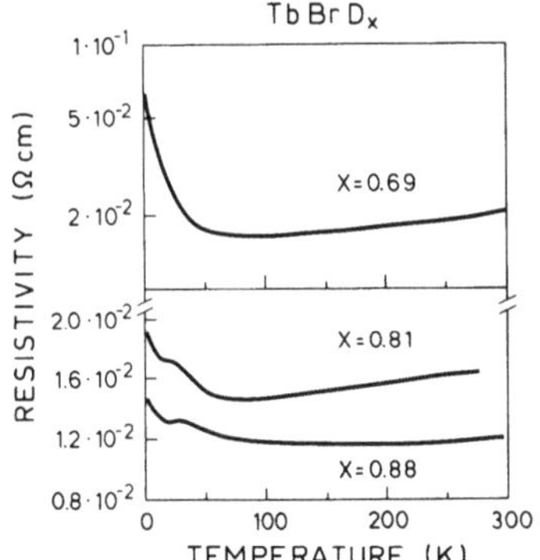

Fig. 77. Temperature dependence of resistivities for $TbBrD_x$ for different values of x (Cockcroft et al. 1989a).

Extensive neutron diffraction investigations have been performed on the analogous $TbBrD_x$ system to obtain some experimental evidence for hydrogen ordering and to refine the magnetic structure, since with Tb problems due to neutron absorption are avoided (Cockcroft et al. 1989a). However, as will be shown, the physical properties of $TbBrD_x$ and $GdBrD_x$ are not at all as similar as the equivalent crystal structures might suggest. In particular, the increase of the resistivity towards lower temperatures is much less pronounced for $TbBrD_{0.69}$ than in the case of Gd (fig. 77). The resistivity grows only by a factor of three. The temperatures of the magnetic anomalies are shifted (in comparison with the corresponding Gd compounds) down to 20 and 22 K for $x = 0.81$ and $x = 0.88$, respectively.

Low-temperature neutron diffraction has been carried out on $TbBrD_x$ for $x = 0.88$, 0.81 and 0.69. Magnetic diffraction peaks are clearly observed for $x = 0.88$ and $x = 0.81$. No additional reflections could be detected for the $x = 0.69$ sample. This result excludes any 3D superstructure for the low hydrogen concentration limit. For the higher concentrations the magnetic structure was determined and refined using the Rietveld method. The magnetic order is such that ferromagnetic alignment is found in a single metal atom layer with the moment vector confined to the (001) plane while antiferromagnetic coupling dominates between two metal atom layers of the same slab BrTbDDTbBr. Neighboring bilayers are coupled ferromagnetically resulting in the scheme $+ - \cdot\cdot - + \cdot\cdot + -$. The magnetic structure of $TbBrD_x$ $(0.8 = x < 1.0)$ can be partially traced back to the ordering in metallic antiferromagnetic $GdD_{1.93}$ (Arons and Schweizer 1982).

The Gd moments in $GdD_{1.93}$ have collinear order of the antiferromagnetic type II. Figure 78 clearly shows that the ferromagnetic order within (111) planes and the antiferromagnetic order between adjacent (111) planes of $GdD_{1.93}$ corresponds to the magnetic structure of the Tb atom bilayers in $TbBrD_x$ $(0.8 < x < 1.0)$. The ordered components of the Tb moments were refined to $4.3\mu_B$ for $x = 0.88$ and $2.6\mu_B$ for $x = 0.81$. Apart from x-dependent crystal field effects, these low values of the magnetic moments (compared to the saturation moment of $9\mu_B$ for the free ion) could as well be due to a reduction of the magnetic correlation with decreasing x. This interpretation is supported by the fact that the paramagnetic Curie temperature θ changes sign with decreasing x, ranging from $\theta = -66$ K for $x = 0.9$ to $\theta = +23$ K for $x = 0.7$ which indicates competition between ferromagnetic and antiferromagnetic interactions (fig. 79). Competing interactions together with disorder are an essential ingredient for the occurrence of spin glass behavior in magnetic systems. Disorder might be introduced in $TbBrD_{0.7}$ by random insertion of vacancies in the D positions. This is in

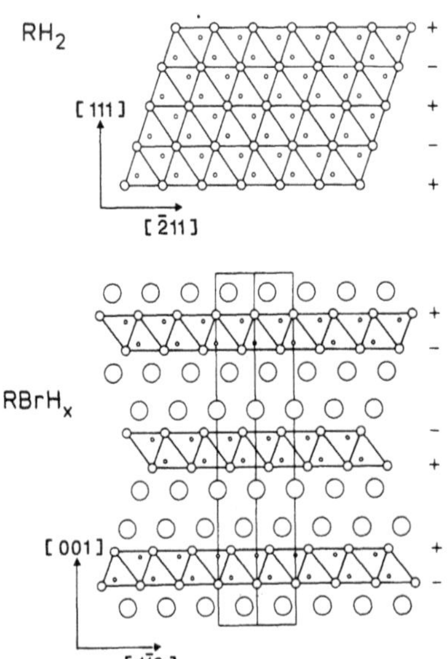

Fig. 78. Projections of the crystal structure of RH_2 and $RBrH_x$ (ZrCl-type): small circles, H; medium circles, R; large circles, X; the signs on the right-hand side represent the magnetic ordering of $GdD_{1.93}$ and $TbBrD_x$ $(x \geqslant 0.81)$, respectively. The moments are parallel within metal atom planes (111) for $GdD_{1.93}$ and within (001) for $TbBrD_x$ $(x \geqslant 0.81)$.

agreement with the weak resistivity increase at low temperatures, which favors a disordered D distribution as well. Indeed, $TbBrD_{0.7}$ displays all the properties characteristic for spin glass behavior (Kremer et al. 1990): a thermal hysteresis of the DC-field-cooled and zero-field-cooled susceptibilities (fig. 80a), a time-dependent increase of the zero-field-cooled susceptibility at low temperatures, a frequency-dependent cusp in the real part of the AC susceptibility at the freezing temperature T_f and a steep increase of the imaginary part at T_f (fig. 80b), and a broad maximum of the magnetic part of the specific heat centered at about $\frac{3}{2}T_f$.

It has been suggested that the spin glass behavior of $TbBrD_{0.7}$ is due to random magnetic anisotropies which might arise from the locally different crystal field splittings that each Tb^{3+} experiences due to a randomly occupied D substructure. This interpretation also explains the absence of spin glass behavior for the isotypic $GdBrD_{0.7}$ since Gd^{3+}, to a very good approximation, has a spin-only moment for which crystal field effects and, hence, anisotropies can be neglected. $TbBrD_{0.7}$ is a

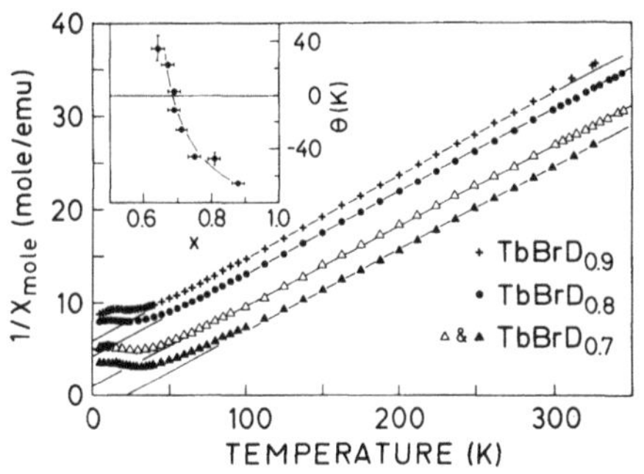

Fig. 79. Reciprocal molar susceptibilities of $TbBrD_x$ for different values of x. The inset shows the paramagnetic Curie temperatures Θ as a function of x.

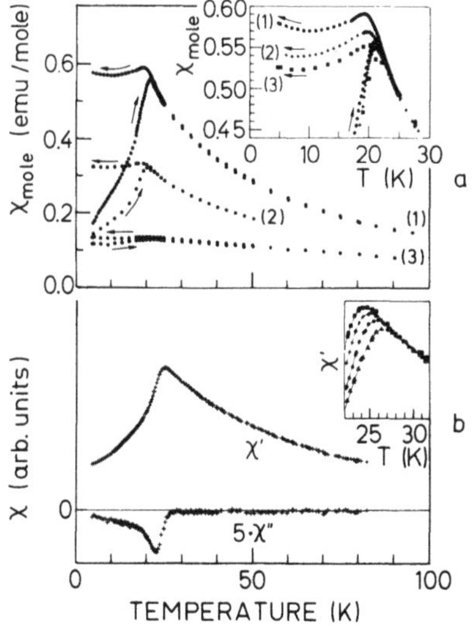

Fig. 80. (a) Field-cooled (left arrow) and zero-field-cooled (right arrow) magnetic susceptibilities of three different $TbBrD_x$ samples: (1) $x = 0.7$ at 1.6 G; (2) $x = 0.7$ at 100 G; (3) $x = 0.8$ at 100 G; the inset shows field dependence for $TbBrD_{0.7}$: (1) 1.6 G; (2) 26 G; (3) 100 G. (b) Real and imaginary parts χ' and χ'' of the AC magnetic susceptibility at 11.7 Hz of $TbBrD_{0.7}$; the inset shows shift of the χ' "cusp" with frequency: (from left to right) 3.51; 11.7; 117 Hz; 1.17 kHz (full lines are guides for the eye) (Kremer et al. 1990).

remarkable new spin glass example because its spin glass behavior arises from disorder in a nonmagnetic substructure, whereas the sublattice carrying the magnetic moments remains well ordered.

4.4. *Magnetic exchange interactions*

The identification of the nature of magnetic exchange interactions is in general a difficult problem even for well-known magnetic materials. For the rare earth metals and a few groups of their compounds the dominant exchange mechanisms have been elucidated, i.e. RKKY interaction for the magnetic metals, cation–anion–cation superexchange for the Eu chalcogenides, dipolar and exchange interaction for the trihalides. In the case of a novel class of compounds (as the metal-rich halides) the magnetic exchange couplings are accessible only by a comparison of structural features and electrical properties of a variety of compounds with the paramagnetic Curie temperatures and the magnetic ordering temperatures. The metal-rich halides can be regarded as intermediates between the trihalides RX_3 and the metals R. We restrict our discussion to the Gd compounds where the amount of available data is by far the largest. Removal of the halide ions X^- from Gd_2XC leads to Gd_2C, and from GdXD leads to GdD_2 when all tetrahedral sites are occupied by the hydrogen atoms. Table 11 summarizes all available data on magnetic ordering and paramagnetic Curie temperatures which are of interest in this context.

The upper part of table 11 shows a series of compounds starting with $GdCl_3$ and proceeding to elemental Gd via the carbide halides with single C interstitials. The first step $GdCl_3 \rightarrow Gd_2Cl_3$ clearly shows the enhancement of the magnetic couplings

TABLE 11

Magnetic ordering and paramagnetic Curie temperatures of selected Gd compounds (notations as in table 10).

Compound	Magnetic ordering temperature	Paramagnetic θ (K)	Reference
$GdCl_3$	$T_C = 2.2$ K	+2.6	Wolf et al. (1961)
Gd_2Cl_3	$T_N = 26$ K	−180	
Gd_4I_5C	<2 K	MI	
Gd_3I_3C	$T_{N1} = 100$ K, $T_{N2} = 25$ K	MI	
$Gd_6Br_7C_2$	$T_{N1} = 30$ K, $T_{N2} = 12$ K	+37	
Gd_2Br_2C	?	?	
Gd_2IC	$T_C = 182$ K	+206	
Gd_2C	$T_C = 400$ K	>400	Lallement (1968)
Gd	$T_C = 292$ K	>292	Elliot et al. (1953)
$GdBrD_{0.7}$	$T_N = 36$ K	+63	
$GdBrD_{0.8}$	$T_N = 45$ K	−44	
$GdBrH_2$	"T_N" = 2.5 K	−8.5	
$GdD_{1.93}$	$T_N = 20$ K	?	Arons and Schweizer (1982)
GdD_2	$T_N = 15.5$ K	?	

through clustering of the R atoms. The magnetic structure of Gd_2Cl_3 suggests ferromagnetic 90° cation–anion–cation superexchange along the chains of octahedra and direct exchange via fd-hybridization for the Gd pairs perpendicular to the chain direction. Inserting C atoms into the octahedral holes of the Gd chains in Gd_2Cl_3 leads to the C-filled Gd chains contained in the Gd_4I_5C structure. Gd_4I_5C shows no magnetic ordering down to 2 K which suggests that C interstitials introduce ferromagnetic coupling and destroy the antiferromagnetic ordering of single-chain halides. Condensation to twin chains restores the antiferromagnetism. The positive value of $\theta = 37$ K for $Gd_6Br_7C_2$ shows, however, that strong ferromagnetic exchange is still present. Finally, ferromagnetism becomes dominant for the layer compounds Gd_2XC, with the highest Curie temperature of 200 K for Gd_2IC, while the antiferromagnetic ordering of Gd_2ClC is probably due to a different stacking of ferromagnetic Gd atom bilayers. The decrease of the Curie temperature from Gd_2C to Gd can consistently be explained by the ferromagnetic coupling introduced by C interstitials in octahedral holes of a close-packed structure of Gd atoms.

The Gd hydride (deuteride) halides all crystallize in a layered structure containing alternating Gd and X atom and the ordering temperatures for varying D content (table 11, lower part) the following conclusions can be drawn. Insertion of deuterium in tetrahedral sites of a close-packed Gd structure induces strong antiferromagnetic coupling. Situated in octahedral holes, the D atoms reduce the AF coupling as substantiated by the pair $GdD_{1.93} \rightarrow GdD_2$ since in GdD_{2-x} octahedral sites are occupied for $x \leqslant 0.05$. A similar observation holds for the pair $GdBrD \rightarrow GdBrD_2$ although this case is complicated by substantial structural changes. Since $GdBrD_{0.7}$ with $T_N = 36$ K is a semiconductor at low temperatures, the RKKY interaction cannot be the essential exchange mechanism. We rather assume a direct connection between T_N and the deuterium content. As in the carbide halides the ferromagnetic coupling for Gd atoms within the same layer should be mediated via 90° cation–halide–cation exchange. The Tb compounds show a qualitatively analogous behavior in many respects. However, the spin glass properties of $TbBrD_{0.7}$ represent a marked difference to the isostructural Gd compound. They are due to the distinct single site anisotropy of Tb^{3+} in contrast to the spin-only ground state of Gd^{3+}.

Acknowledgement

We gratefully acknowledge the help of R. Eger, B. Krauter and I. Remon with drawing and typing.

Appendix

Controlled preparation of compounds has to precede any measurement of their physical properties (Schäfer 1971). In particular, this remark holds for the research on metal-rich halides of rare earth elements which is marked by trial and error in the early stages. Often, phases could only be prepared as a minority in multicomponent

TABLE 12
Synthetic details and *d* values for various metal-rich rare earth halides.

Compound	Starting materials	T (K)/Time	Yield	Remarks	*d* values (Å) of the 10 strongest reflections; the number in parentheses give the relative intensity	Ref[a]
Gd_2Cl_3	1.00 mmol $GdCl_3$, 1.00 mmol Gd	890/3 w	++	–	3.309 (100); 7.251 (89); 9.014 (63); 2.713 (50); 3.081 (50); 2.382 (45); 2.467 (34); 3.373 (31); 2.479 (29); 1.985 (26)	[1]
Gd_2Br_3	1.00 mmol $GdBr_3$, 1.00 mmol Gd	1100/4 w	++	–	7.830 (100); 3.399 (84); 9.382 (60); 2.769 (48); 3.181 (46); 2.519 (43); 2.643 (35); 2.641 (30); 3.581 (25); 2.370 (22)	[1]
Tb_2Cl_3	1.00 mmol $TbCl_3$, 1.00 mmol Tb	900/4 w	++	–	3.284 (100); 7.242 (91); 8.943 (63); 2.691 (50); 3.062 (50); 2.374 (46); 2.456 (34); 3.351 (30); 2.473 (29); 1.970 (26)	[1]
Y_2Cl_3	1.00 mmol YCl_3, 2.40 mmol Y	1000/4 w	++	–	3.251 (100); 7.220 (93); 8.878 (64); 2.666 (50); 3.032 (50); 2.359 (46); 2.447 (34); 3.350 (32); 2.461 (29); 1.956 (26)	[2]
Y_2Br_3	1.00 mmol YBr_3, 5.00 mmol Y	1020/5 w	0	–	7.862 (100); 3.333 (78); 9.255 (58); 2.746 (45); 3.128 (44); 2.503 (40); 2.640 (35); 3.562 (25); 2.635 (25); 2.047 (20)	[2]
Sc_7Cl_{10}	3.90 mmol $ScCl_3$, 4.45 mmol Sc	1150–1170/6 w	–	–	2.765 (100); 2.345 (25); 7.287 (20); 2.217 (20); 9.304 (20); 7.534 (20); 2.878 (20); 2.435 (17); 1.768 (17); 1.764 (15)	[3]
$Gd_{10}Cl_{18}C_4$	5.69 mmol $GdCl_3$, 3.79 mmol Gd, 3.83 mmol C	1030/7 d	++	–	2.778 (100); 7.800 (73); 7.963 (50); 2.674 (45); 9.184 (30); 4.362 (27); 1.954 (26); 1.889 (25); 7.139 (22); 4.081 (21)	[4]
$Gd_{10}Cl_{17}C_4$	3.79 mmol $GdCl_3$, 3.90 mmol Gd, 2.68 mmol C	1170/7 d	++	–	2.842 (100); 10.829 (92); 2.678 (70); 2.754 (20); 7.311 (67); 8.117 (63); 7.452 (45); 8.135 (40); 5.415 (30); 2.084 (30)	[4]
Sc_5Cl_8C	8.00 mmol $ScCl_3$, 3.00 mmol Sc, 3.00 mmol C	1130/12 d	++	–	2.649 (100); 6.021 (45); 1.781 (45); 2.246 (42); 2.681 (37); 2.385 (35); 2.269 (30); 1.763 (25); 3.016 (25); 2.414 (25)	[5]
$Sc_7Cl_{10}C_2$	10.00 mmol $ScCl_3$, 11.00 mmol Sc, 6.00 mmol C	1130/7 d	+	–	2.603 (100); 2.298 (45); 1.781 (40); 6.673 (37); 9.174 (35); 2.314 (35); 2.629 (32); 3.001 (30); 2.492 (27); 1.749 (25)	[6]

Lu_2Cl_2C	16.00 mmol $LuCl_3$, 5.00 mmol Lu, 3.00 mmol C, 6.00 mmol CsCl	1. 1220/4 d 2. 970/2 d	–	–	9.053 (100); 2.705 (50); 2.834 (35); 3.009 (30); 2.297 (30); 1.801 (20); 2.431 (15); 1.736 (15); 3.018 (10); 2.263 (10)	[7]
Gd_2ClC	1.52 mmol $GdCl_3$, 7.59 mmol Gd, 4.55 mmol C	1370–1470/7 w	++	Only powder	2.705 (100); 3.048 (85); 1.845 (35); 3.385 (25); 1.988 (25); 2.512 (20); 1.524 (15); 2.148 (12); 6.769 (10); 2.356 (10)	[8]
Gd_3Cl_3C	4.00 mmol $GdCl_3$, 8.00 mmol Gd, 4.00 mmol C	1150/5 w	+	0.8 g $GdCl_3$ + 1.1 g Gd + 0.04 g C; 1050 K; 5 d; 2 mm crystals	2.684 (100); 7.590 (75); 3.099 (70); 1.898 (60); 2.869 (50); 2.289 (25); 1.618 (22); 1.200 (20); 1.549 (17); 1.363 (15)	[9]
$Gd_6Cl_5C_3$	5.50 mmol $GdCl_3$, 6.50 mmol Gd, 4.50 mmol C	1170–1370/7 d	++	1270 K; 7 d, excess $GdCl_3$; large crystals	10.069 (100); 2.677 (78); 3.053 (50); 2.707 (45); 1.901 (25); 1.905 (25); 3.150 (22); 3.158 (22); 3.033 (20); 2.822 (20)	[10]
$Gd_2Cl_2C_2$	7.19 mmol $GdCl_3$, 15.18 mmol Gd, 22.76 mmol C	1230/7 d	++	1230 K; 7 d, excess $GdCl_3$; large crystals	9.322 (100); 2.650 (60); 2.909 (35); 3.320 (30); 3.172 (30); 1.948 (20); 4.661 (20); 1.971 (20); 3.347 (15); 3.107 (15)	[8]
Y_2Cl_2C	2.00 mmol YCl_3, 4.00 mmol Y, 3.00 mmol C	1220/10 d	++	Small crystals	2.630 (100); 9.183 (90); 3.209 (30); 1.853 (25); 1.867 (17); 3.029 (15); 2.215 (15); 1.515 (10); 2.296 (8); 4.592 (5)	[11]
Sc_2Cl_2C	2.00 mmol $ScCl_3$, 4.00 mmol Sc, 3.00 mmol C	1220/5 d	++	Small crystals	2.474 (100); 8.916 (45); 2.973 (25); 1.717 (22); 1.783 (18); 4.458 (12); 1.410 (10); 2.102 (7); 2.229 (6); 1.236 (5)	[11]
$Sc_7Br_{12}C$	7.00 mmol $ScBr_3$, 30.00 mmol Sc	1151–1173/24 d	+	–	2.667 (100); 3.084 (50); 1.890 (30); 1.882 (30); 3.068 (15); 1.542 (14); 1.609 (14); 1.191 (10); 1.334 (8); 1.604 (7)	[12]
$Gd_{12}Br_{17}C_6$	2.01 mmol $GdBr_3$, 2.25 mmol Gd, 2.12 mmol C	1320/3 w	++	Quenching	2.929 (100); 2.882 (90); 2.817 (85); 8.617 (75); 9.860 (65); 9.126 (60); 2.031 (55); 2.054 (50); 6.599 (45); 2.027 (25)	[8]
$Gd_6Br_7C_2$	7.00 mmol $GdBr_3$, 11.00 mmol Gd, 6.00 mmol C	1220/6 w	++	–	2.765 (100); 8.526 (55); 2.809 (45); 1.979 (27); 1.962 (25); 2.799 (25); 3.156 (22); 9.768 (20); 1.910 (17); 2.619 (20)	[13]
$Gd_2Br_2C_2$	1.26 mmol $GdBr_3$, 2.52 mmol Gd	1320/20 d	++	Graphite tube in Ta; vacuum sealed	2.728 (100); 9.838 (77); 2.964 (63); 3.361 (48); 2.021 (37); 1.944 (33); 2.825 (33); 2.753 (24); 4.919 (22); 2.460 (19)	[14]

TABLE 12 (cont'd)

Compound	Starting materials	T (K)/Time	Yield	Remarks	d values (Å) of the 10 strongest reflections; the number in parentheses give the relative intensity	Ref[a]
Gd_2Br_2C	1.26 mmol $GdBr_3$, 2.52 mmol Gd	1220/3 w	0	Graphite tube in Ta; vacuum sealed	2.744 (100); 9.824 (60); 2.744 (55); 3.309 (50); 1.910 (35); 1.972 (25); 2.456 (17); 1.568 (15); 3.136 (14); 2.328 (12)	[14]
Gd_2BrC	1.00 mmol $GdBr_3$, 4.25 mmol Gd 3.00 mmol C	1370/4 w	+	Powder	2.696 (100); 3.279 (25); 1.839 (23); 1.479 (18); 1.549 (17); 1.199 (8)	[8]
$Sc_7I_{12}C$	4.00 mmol ScI_3, 12.00 mmol Sc 1.04 mmol C	1120/3 w	+	Sc metal as foil	2.818 (100); 3.327 (76); 1.733 (33); 2.041 (31); 2.020 (30); 3.282 (25); 1.664 (15); 1.720 (12); 1.740 (12); 1.279 (10)	[12]
Y_4I_5C	0.43 mmol YI_3, 5.95 mmol Y 0.27 mmol C	1120/1 w	+	Y metal as foil	2.888 (100); 2.998 (50); 2.080 (27); 2.080 (27); 1.974 (17); 2.937 (17); 3.203 (16); 3.297 (16); 2.659 (15); 1.649 (14)	[15]
$Sc_6I_{11}C_2$	0.46 mmol ScI_3, 0.30 mmol Sc 0.26 mmol C	1120/2 w	+ +	–	2.950 (100); 2.875 (88); 2.853 (85); 3.349 (75); 3.347 (72); 3.331 (72); 3.331 (72); 3.330 (70); 2.058 (70); 2.058 (30)	[16]
$Sc_4I_6C_2$	0.50 mmol ScI_3, 1.50 mmol C	1120/2 w	+	–	3.312 (100); 3.182 (90); 3.149 (80); 3.109 (70); 3.103 (67); 3.099 (67); 3.125 (65); 2.062 (40); 2.034 (40); 2.049 (40)	[16]
$Y_6I_7C_2$	0.43 mmol YI_3, 6.68 mmol Y 0.36 mmol C	1220/1 w	+ +	Y metal as foil	2.858 (100); 2.947 (45); 2.078 (27); 2.027 (25); 3.242 (20); 2.910 (18); 1.955 (17); 2.719 (16); 3.258 (13); 1.629 (13)	[15]
Gd_3I_3C	4.00 mmol GdI_3, 8.00 mmol Gd, 4.00 mmol C	1250/3 w	+ +	Quenching	2.912 (100); 2.936 (45); 2.774 (25); 2.096 (23); 2.040 (22); 2.035 (20); 2.664 (20); 1.966 (18); 2.552 (18); 11.743 (18)	[17]
Gd_2IC	1.00 mmol GdI_3, 5.00 mmol Gd, 3.00 mmol C	1270/3 w	+ +	Large crystals	2.738 (100); 3.292 (25); 1.901 (22); 1.505 (19); 1.561 (16); 2.466 (15); 1.206 (15); 1.471 (13); 1.973 (13); 1.163 (7)	[17]
$Sc_7Cl_{12}N$	1.30 mmol $ScCl_3$, 1.00 mmol Sc	1130/1 w	+	Nb open; sealed in fused silica; 1 atm N_2	2.549 (100); 6.495 (30); 1.801 (27); 1.803 (27); 2.942 (13); 1.471 (12); 1.394 (10); 1.140 (10); 6.948 (10); 4.745 (10)	[18]

Gd_2Cl_3N	3.00 mmol $GdCl_3$, 2.00 mmol $GdN_{0.978}$	1120/5 d	++	Ta open; sealed in fused silica; 1 atm N_2	3.722 (100); 6.509 (38); 5.980 (32); 2.644 (30); 2.878 (30); 2.440 (18); 2.211 (18); 2.731 (17); 1.772 (12); 1.875 (12)	[14]
Sc_4Cl_6N	1.30 mmol $ScCl_3$, 1.30 mmol Sc	1210/37 d	+	Nb open; sealed in fused silica; 1 atm N_2	8.401 (100); 2.405 (31); 2.711 (19); 2.391 (15); 1.758 (11); 2.426 (11); 1.940 (9); 1.775 (7); 1.853 (6); 1.425 (4)	[18]
Sc_5Cl_8N	8.00 mmol $ScCl_3$, 7.00 mmol Sc	1130–1230/5 w	0	Nb open; sealed in fused silica; 1 atm N_2	2.663 (100); 6.039 (44); 1.788 (44); 2.256 (42); 2.687 (38); 2.396 (32); 2.275 (29); 1.775 (24); 3.033 (23); 2.425 (23)	[5]
Gd_3Cl_6N	0.56 mmol $GdCl_3$, 0.28 mmol $GdN_{0.978}$	885/3 d	++	Ta open; sealed in fused silica; 1 atm N_2	9.088 (100); 2.630 (78); 7.345 (76); 2.634 (69); 2.559 (67); 6.251 (66); 4.544 (62); 2.544 (59); 4.081 (57); 4.439 (55)	[19]
Sc_2Cl_2N	2.00 mmol $ScCl_3$, 4.00 mmol Sc, 3.00 mmol NaN_3	1000–1130/5 w	++	–	2.423 (100); 8.080 (37); 2.901 (23); 1.675 (22); 1.754 (18); 1.378 (13); 2.064 (10); 4.404 (9); 1.214 (8); 1.333 (6)	[11]
$Sc_7I_{12}B$	5.00 mmol $ScCl_3$, 10.00 mmol Sc 1.00 mmol B	1220/2 w	0	Powder; contaminated with Sc and unidentified ScI_x	Isotypic with $Sc_7I_{12}C$; no lattice constants	[12]
$Sc_7Cl_{12}B$	4.00 mmol $ScCl_3$, 3.00 mmol Sc, 1.00 mmol B	1130/2 w	++	Crystals; 1170–1210 K; Sc on the hot ampoule end	2.558 (100); 6.507 (28); 1.813 (28); 1.805 (28); 1.142 (16); 1.475 (12); 4.761 (12); 2.949 (11); 1.146 (10); 2.699 (10)	[18]
Sc_4Cl_6B	2.00 mmol $ScCl_3$, 2.00 mmol Sc, 1.00 mmol B	1120/2 w	+	–	8.455 (100); 2.429 (34); 2.740 (21); 2.404 (15); 2.448 (13); 1.769 (12); 2.975 (8); 1.799 (8); 1.872 (6); 1.702 (5)	[18]
Gd_3Cl_3B	1.90 mmol $GdCl_3$, 1.90 mmol Gd, 0.60 mmol B	1280/3 w	++	Powder	2.739 (100); 7.747 (75); 3.163 (70); 1.937 (60); 2.928 (50); 2.336 (27); 1.652 (23); 1.225 (20); 1.491 (18); 1.581 (18)	[20]
$Sc_7I_{12}Co$	1.00 mmol ScI_3, 1.10 mmol Sc, 0.30 mmol CoI_2	1120/3 w	++	–	2.917 (100); 3.357 (73); 2.052 (32); 2.072 (32); 3.401 (26); 3.678 (15); 1.299 (12); 1.309 (12); 1.770 (12); 1.757 (10)	[21]
$Sc_7I_{12}Ni$	1.00 mmol ScI_3, 1.10 mmol Sc, 0.30 mmol NiI_2	1120/3 w	++	–	2.910 (100); 3.357 (75); 2.054 (32); 2.061 (31); 3.372 (25); 1.754 (22); 1.678 (15); 1.300 (12); 1.759 (10); 1.752 (10)	[21]

TABLE 12 (cont'd)

Compound	Starting materials	T (K)/Time	Yield	Remarks	d values (Å) of the 10 strongest reflections; the number in parentheses give the relative intensity	Ref [a]
$Pr_7I_{12}M$ (M = Mn, Fe, Co, Ni)	1.00 mmol PrI_3, 1.10 mmol Pr, 0.30 mmol NI_2	1120/3 w	M = Fe, Co, Ni: + +; M = Mn: +	Powder	M = Fe: 3.1065 (100); 2.194 (32); 7.911 (31); 2.199 (30); 3.584 (17); 1.792 (14); 8.476 (10); 1.388 (10); 1.553 (10); 1.391 (9)	[21]
$Gd_7I_{12}M$ (M=Mn, Fe, Co,)	1.00 mmol GdI_3, 1.10 mmol Gd, 0.30 mmol NI_2	1120/3 w	M = Fe, Co: ++; M=Mn: +	Powder	M = Mn: 3.052 (100); 7.751 (35); 2.150 (32); 2.167 (30); 1.758 (14); 3.515 (24); 8.346 (13); 1.361 (10); 3.229 (10); 1.526 (10)	[21]
$Y_7I_{12}M$ (M = Fe, Co)	1.00 mmol YI_3, 1.10 mmol Y, 0.30 mmol NI_2	1120/3 w	M = Fe: ++; M = Co: +	Crystals: M = Fe	M = Fe: 3.032 (100); 3.485 (34); 2.160 (33); 2.129 (32); 1.742 (16); 1.384 (12); 3.554 (12); 1.364 (11); 1.826 (10); 1.516 (10)	[21]
$Ca_{0.65}Pr_{0.35}$ $(Pr_6I_{12})Co$	3.00 mmol PrI_3, 3.00 mmol Pr, 1.00 mmol CoI_2 1.00 mmol Ca	1120/3 w	–	–	3.114 (100); 2.216 (33); 2.188 (33); 3.580 (17); 1.790 (15); 8.533 (15); 1.386 (12); 1.400 (12); 7.889 (10); 1.269 (8)	[21]
$YClH_x$ $0.81 \leqslant x \leqslant 1.00$	2.60 mmol YCl_3, 5.30 mmol YH_2	1160/3 d	+ +	$YClN_x$ $0.81 \leqslant x \leqslant 1.00$ ZrCl type	$YClH_{0.9}$: 2.936 (100); 9.125 (46); 2.357 (40); 1.877 (25); 4.563 (17); 3.228 (14); 2.094 (13); 1.581 (13); 1.209 (11); 2.795 (11)	[22]
$YClH_x$ $0.69 \leqslant x \leqslant 0.80$	2.60 mmol YCl_3, 5.30 mmol YH_2	1220/3 d	+ +	$YClH_x$ $0.69 \leqslant x \leqslant 0.80$ ZrBr type	$YClH_{0.7}$: 2.798 (100); 3.162 (70); 2.504 (56); 1.876 (36); 4.586 (21); 1.452 (17); 2.100 (15); 1.558 (14); 1.982 (14); 2.293 (14)	[22]
$CeClH_x$ $0.69 \leqslant x \leqslant 1.00$	2.60 mmol $CeCl_3$, 5.30 mmol CeH_2	970/1 w	+ +	ZrBr type	$CeClH_{0.69}$: 2.953 (100); 9.205 (95); 3.388 (93); 2.616 (63); 2.018 (40); 1.517 (17); 4.602 (17); 2.039 (15); 1.666 (15); 3.118 (15)	[23]

$CeBrH_x$ $0.69 \leqslant x \leqslant 1.00$	2.60 mmol $CeBr_3$, 5.30 mmol CeH_2	950/1 w	++	ZrCl type	$CeBrH_{0.97}$: 3.177 (100); 2.544 (44); 2.034 (27); 9.805 (20); 1.713 (14); 1.565 (13); 1.310 (12); 4.902 (10); 2.258 (10); 2.451 (10)	[23]
$PrClH_x$ $0.69 \leqslant x \leqslant 1.00$	2.60 mmol $PrCl_3$, 5.30 mmol PrH_2	1210/1 w	++	ZrBr type	$PrClH_{0.97}$: 2.930 (100); 9.187 (97); 3.356 (94); 2.599 (65); 1.998 (40); 1.507 (18); 4.594 (17); 3.092 (16); 2.029 (16); 1.651 (15)	[23]
$PrBrH_x$ $0.69 \leqslant x \leqslant 1.00$	2.60 mmol $PrBr_3$, 5.30 mmol PrH_2	900/1 w	++	ZrCl type	$PrBrH_{0.94}$: 3.152 (100); 2.529 (44); 2.015 (26); 9.783 (20); 1.698 (14); 1.555 (13); 1.298 (12); 4.891 (12); 2.446 (10); 2.246 (10)	[23]
$NdBrH_{0.9}$	2.60 mmol $NdBr_3$, 5.30 mmol NdH_2	1065/3 d	++	ZrCl type	$NdBrH_{0.90}$: 3.125 (100); 2.512 (44); 1.995 (26); 9.757 (22); 1.682 (14); 1.544 (13); 4.879 (12); 1.286 (12); 2.439 (10); 2.234 (10)	[24]
$GdClH_x$ $0.69 \leqslant x \leqslant 1.00$	2.60 mmol $GdCl_3$, 5.30 mmol GdH_2	1170/3 d	++	ZrBr type	$GdClH_{0.89}$: 9.165 (100); 2.837 (88); 3.220 (85); 2.532 (60); 1.912 (34); 2.984 (17); 4.582 (17); 3.055 (16); 1.486 (16); 1.995 (15)	[25]
$GdBrH_x$ $0.69 \leqslant x \leqslant 1.00$	2.60 mmol $GdBr_3$, 5.30 mmol GdH_2	1170/3 d	++	ZrCl type	$GdBrH_{0.87}$: 3.047 (100); 2.446 (45); 1.937 (26); 9.695 (25); 4.848 (14); 1.635 (14); 1.513 (13); 1.249 (12); 2.198 (11); 2.424 (10)	[25]
$GdBrH_2$	$GdBrH_x$ ($x \leqslant 1.0$)	670/24 h	++	Mo boat in flowing H_2	$GdBrH_2$: 2.893 (100); 5.147 (38); 1.891 (35); 3.256 (33); 10.294 (30); 1.922 (28); 2.497 (26); 2.629 (25); 3.015 (20); 2.247 (18)	[26]

[a] References: [1] Simon et al. (1979); [2] Mattausch et al. (1980); [3] Poeppelmeier and Corbett (1977a); [4] Warkentin et al. (1982); [5] Hwu et al. (1987); [6] Hwu et al. (1985); [7] Schleid and Meyer (1987); [8] Schwarz (1987); [9] Warkentin and Simon (1983); [10] Simon et al. (1988); [11] Hwu et al. (1986); [12] Dudis et al. (1986); [13] Schwanitz-Schüller (1985); [14] Schwanitz-Schüller and Simon (1985b); [15] Kauzlarich et al. (1988); [16] Dudis and Corbett (1987); [17] Mattausch et al. (1987); [18] Hwu and Corbett (1986); [19] Simon and Koehler (1986); [20] Mattausch et al. (1984); [21] Hughbanks and Corbett (1988); [22] Mattausch et al. (1986b); [23] Müller-Käfer (1988); [24] Michaelis and Simon (1986); [25] Mattausch et al. (1985a); [26] Simon et al. (1987b).

systems. The term "yield", in contrast to, e.g. reactions between gases, does not have a thermodynamic meaning with most solid state reactions as entropy effects are of minor importance, and a temperature-dependent equilibrium is missing. As a rule, any deviation from a quantitative yield must be caused by (a) kinetic effects or (b) accidental stabilization of compounds by impurities. Most of the early "binary" compounds turned out to be ternaries and can now be prepared by purposely adding the "contaminants" which occur as interstitial atoms. There are hints as to the presence of different interstitials in different environments. We therefore add a list of detailed preparations and characterizations (table 12). If not stated otherwise, the reactants are sealed in arc-welded Ta tubes under a reduced pressure of Ar. A "quantitative yield" is denoted by " + + ", the presence of a compound as a majority phase (60–90%) by "+", as a minority phase (10–30%) by "0", and as a few selected crystals by "–".

References

Adolphson, D.G., and J.D. Corbett, 1976, Inorg. Chem. **15**, 1820.

Albright, T.A., J.K. Burdett, M.-H. Whangbo, 1985, Orbital Interactions in Chemistry (Wiley, New York).

Araujo, R.E., and J.D. Corbett, 1981, Inorg. Chem. **20**, 3082.

Argay, G., and I. Nary-Szabo, 1966, Acta Chim. Acad. Sci. Hung. **49**, 329.

Arons, R.R., and J. Schweizer, 1982, J. Appl. Phys. **53**, 2645.

Ashcroft, N.W., and N.D. Mermin, 1976, Solid State Physics (Holt, Rinehart and Winston, New York).

Atoji, M., 1961, J. Chem. Phys. **35**, 1950.

Bauhofer, W., and A. Simon, 1982, Z. Naturforsch. A **37**, 568.

Bauhofer, W., J.K. Cockcroft, R.K. Kremer, Hj. Mattausch, C. Schwarz and A. Simon, 1988, J. Phys. (Paris) **49**, C8–893.

Bauhofer, W., Hj. Mattausch and A. Simon, 1989, Z. Naturforsch. A **44**, 625.

Beck, H.P., and A.Z. Limmer, 1982, Z. Naturforsch. B **37**, 574.

Beckmann, O., H. Boller and H. Nowotny, 1970, Monatsh. Chem. **101**, 945.

Berroth, K., 1980, Thesis (Stuttgart, FRG).

Berroth, K., and A. Simon, 1980, J. Less-Common Met. **76**, 41.

Berroth, K., Hj. Mattausch and A. Simon, 1980, Z. Naturforsch. B **35**, 626.

Biltz, W., 1934, Raumchemie der festen Stoffe (Verlag Leopold Voss, Leipzig).

Bowman, A.L., N.H. Krikorian, G.P. Arnold, T.C. Wallace and N.G. Nereson, 1968, Acta Crystallogr. B **24**, 1121.

Brown, P.J., K.R.A. Ziebeck, A. Simon and M. Sägebarth, 1988, J. Chem. Soc. Dalton Trans., 111.

Bullet, D.W., 1977, Phys. Rev. Lett. **39**, 664.

Bullet, D.W., 1980, Inorg. Chem. **19**, 1780.

Bullet, D.W., 1985, Inorg. Chem. **24**, 3319.

Burdett, J.K., and G.J. Miller, 1987, J. Am. Chem. Soc. **109**, 4092.

Bursten, B.E., F.A. Cotton and G.G. Stanley, 1980, Isr. J. Chem. **19**, 132.

Chevrel, R., M. Sergent and J. Prigent, 1971, J. Solid State Chem. **3**, 515.

Chu, P.J., R.P. Ziebarth, J.D. Corbett and B.C. Gerstein, 1988, J. Am. Chem. Soc. **110**, 5324.

Cockcroft, J.K., W. Bauhofer, Hj. Mattausch and A. Simon, 1989a, J. Less-Common Met. **152**, 227.

Cockcroft, J.K., W. Bauhofer, Hj. Mattausch, A. Simon and R.R. Arons, 1989b, unpublished.

Corbett, J.D., 1981, Acc. Chem. Res. **14**, 239.

Corbett, J.D., 1986, private communication.

Corbett, J.D., and R.E. McCarley, 1986, in: Crystal Chemistry and Properties of Materials with Quasi-One-Dimensional Structures (Reidel, Dordrecht) p. 179.

Corbett, J.D., and A. Simon, 1979, unpublished.

Corbett, J.D., R.L. Daake, K.R. Poeppelmeie and D.H. Guthrie, 1978, J. Am. Chem. Soc. **100**, 652.

Corbett, J.D., K.R. Poeppelmeier and R.L. Daake, 1982, Z. Anorg. & Allg. Chem. **491**, 51.

Cotton, F.A., and T.E. Haas, 1964, Inorg. Chem. **3**, 10.

Czjzek, G., 1989, private communication.

Daake, R.L., and J.D. Corbett, 1977, Inorg. Chem. **16**, 2029.

DiSalvo, F.J., J.V. Wasczak, W.M. Walsh Jr, L.W. Rupp and J.D. Corbett, 1985, Inorg. Chem. **24**, 4624.

Dronskowski, R., and A. Simon, 1989, Angew. Chem. **101**, 775; Angew. Chem. Int. Ed. Engl. **28**, 758.

Dudis, D.S., and J.D. Corbett, 1987, Inorg. Chem. **26**, 1933.

Dudis, D.S., J.D. Corbett and S.-J. Hwu, 1986, Inorg. Chem. **25**, 3434.

Ebbinghaus, G., A. Simon and A. Griffith, 1982, Z. Naturforsch. A **37**, 564.

Ehrlich, P., B. Alt and L. Gentsch, 1956, Z. Anorg. & Allg. Chem. **283**, 58; and **288**, 148 and 156.

Elliott, J.F., S. Legvold and F.H. Spedding, 1953, Phys. Rev. **91**, 28.

Finley, J.J., R.E. Camley, E.E. Vogel, V. Zevin and E. Gmelin, 1981, Phys. Rev. B **24**, 1023.

Finley, J.J., H. Nohl, E.E. Vogel, H. Imoto, R.E. Camley, V. Zevin, O.K. Andersen and A. Simon, 1981, Phys. Rev. Lett. **46**, 1472.

Ford, J.E., and J.D. Corbett, 1985, Inorg. Chem. **24**, 4120.

Ford, J.E., J.D. Corbett and S.-J. Hwu, 1983, Inorg. Chem. **22**, 2789.

Greiner, J.D., J.F. Smith, J.D. Corbett and F.J. Jelinek, 1966, J. Inorg. & Nucl. Chem. **28**, 971.

Hamon, C., R. Marchand, Y. Laurent and J. Lang, 1974, Bull. Soc. Fr. Mineral. & Crystallogr. **97**, 6.

Hibble, S.J., A.K. Cheetham, A.R. Bogle, H.R. Wakerley and D.E. Cox, 1988, J. Am. Chem. Soc. **110**, 3295.

Hoffmann, R., 1963, J. Chem. Phys. **39**, 1397.

Hoffmann, R., 1989, Solids and Surfaces: A Chemist's View of Bonding in Extended Structures (VCH, New York).

Hughbanks, T., and J.D. Corbett, 1988, Inorg. Chem. **27**, 2022.

Hughbanks, T., and J.D. Corbett, 1989, Inorg. Chem. **28**, 631.

Hughbanks, T., and R. Hoffmann, 1983, J. Am. Soc. Chem. **105**, 3528.

Hughbanks, T., G. Rosenthal and J.D. Corbett, 1986, J. Am. Chem. Soc. **108**, 8289.

Huster, J., and H.F. Franzen, 1985, J. Less-Common Met. **113**, 119.

Hwu, S.-J., and J.D. Corbett, 1986, J. Solid State Chem. **64**, 331.

Hwu, S.-J., J.D. Corbett and K.R. Poeppelmeier, 1985, J. Solid State Chem. **57**, 43.

Hwu, S.-J., R.P. Ziebarth, J.V. Winbush, J.E. Ford and J.D. Corbett, 1986, Inorg. Chem. **25**, 283.

Hwu, S.-J., D.S. Dudis and J.D. Corbett, 1987, Inorg. Chem. **26**, 469.

Imoto, H., and J.D. Corbett, 1980, Inorg. Chem. **19**, 1241.

Imoto, H., and A. Simon, 1982, Inorg. Chem. **21**, 308.

Imoto, H., J.D. Corbett and A. Cisar, 1981, Inorg. Chem. **20**, 145.

Izmailovich, A.S., S.I. Troyanov and V.I. Tsirelnikov, 1974, Russ. J. Inorg. Chem. **19**, 1597.

Jellinek, F., 1962, J. Less-Common Met. **4**, 9.

Kauzlarich, S.M., T. Hughbanks, J.D. Corbett, P. Klavins and R.N. Shelton, 1988, Inorg. Chem. **27**, 1791.

Kjekshus, A., and T. Rakke, 1974, Struct. & Bonding **19**, 45.

Klemm, W., 1958, Proc. Chem. Soc. (London) p. 329.

Kremer, R.K., 1985, Thesis (Darmstadt, FRG).

Kremer, R.K., W. Bauhofer, Hj. Mattausch and A. Simon, 1990, Solid State Commun. **73**, 281.

Lallement, R., 1968, Proc. Brit. Ceram. Soc. **10**, 115.

Liebau, F., 1985, Structural Chemistry of Silicates (Springer, Berlin).

Lii, K.H., C.C. Wang and S.L. Wang, 1988, J. Solid State Chem. **77**, 407.

Lokken, D.A., and J.D. Corbett, 1970, J. Am. Chem. Soc. **92**, 1799.

Lokken, D.A., and J.D. Corbett, 1973, Inorg. Chem. **12**, 556.

Marek, H.S., J.D. Corbett and R.L. Daake, 1983, J. Less-Common Met. **89**, 243.

Masse, R., and A. Simon, 1981, Mater. Res. Bull. **16**, 1007.

Mattausch, Hj., and A. Simon, 1979, unpublished.

Mattausch, Hj., A. Simon and R. Eger, 1980a, Rev. Chim. Miner. **17**, 516.

Mattausch, Hj., J.B. Hendricks, R. Eger, J.D. Corbett and A. Simon, 1980b, Inorg. Chem. **19**, 2128.

Mattausch, Hj., A. Simon, N. Holzer and R. Eger, 1980c, Z. Anorg. & Allg. Chem. **466**, 7.

Mattausch, Hj., A. Simon and R. Eger, 1984, unpublished.

Mattausch, Hj., W. Schramm, R. Eger and A. Simon, 1985a, Z. Anorg. & Allg. Chem. **530**, 43.

Mattausch, Hj., A. Simon and K.R.A. Ziebeck, 1985b, J. Less-Common Met. **113**, 149.

Mattausch, Hj., A. Simon and E.-M. Peters, 1986a, Inorg. Chem. **25**, 3428.

Mattausch, Hj., A. Simon and R. Eger, 1986b, unpublished.

Mattausch, Hj., C. Schwarz and A. Simon, 1987, Z. Kristallagor **178**, 156.

Mattheiss, L.F., and C.Y. Fong, 1977, Phys. Rev. B **15**, 1760.

Matthias, B.T., M. Marezio, E. Corenzwit, A.S. Cooper and H.E. Barz, 1972, Science **175**, 1465.

Mee, J.E., and J.D. Corbett, 1965, Inorg. Chem. **4**, 88.

Meyer, G., 1988, Chem. Rev. **88**, 93.

Meyer, G., and Th. Schleid, 1987, Inorg. Chem. **26**, 217.

Meyer, G., S.-J. Hwu, S.D. Wijeyesekera and J.D. Corbett, 1986, Inorg. Chem. **25**, 4811.

Meyer, H.-J., N.L. Jones and J.D. Corbett, 1989, Inorg. Chem. **28**, 2635.

Michaelis, C., and A. Simon, 1986, unpublished.

Mikheev, N.B., A. Simon and Hj. Mattausch, 1988, Radiokhimiya **30**, 314.

Miller, G.J., J.K. Burdett, C. Schwarz and A. Simon, 1986, Inorg. Chem. **25**, 4437.

Mooser, E., and W.B. Pearson, 1956, Phys. Rev. **101**, 1608.

Mooser, E., and W.B. Pearson, 1960, in: Progress in Semiconductors, Vol. 5 (Wiley, New York) p. 103.

Morss, L.R., Hj. Mattausch, R.K. Kremer, A. Simon and J.D. Corbett, 1987, Inorg. Chim. Acta **140**, 107.

Müller-Käfer, R., 1988, Thesis (Stuttgart, FRG).

Nagaki, D.A., and A. Simon, 1990, Acta Crystallogr. C **46**, 1197.

Nagaki, D.A., A. Simon and H. Borrmann, 1989, J. Less-Common Met. **156**, 193.

Nesper, R., 1987, personal communication.

Nohl, H., W. Klose and O.K. Andersen, 1982, in: Superconductivity in Ternary Compounds I (Springer, Berlin) p. 165.

Pauling, L., 1960, The Nature of the Chemical Bond, 3rd Ed. (Cornell University Press, Ithaca, NY).

Pauling, L., and J.L. Hoard, 1930, Z. Crystallogr. **74**, 546.

Pearson, W.B., 1972, Crystal Chemistry and Physics of Metals and Alloys (Wiley-Interscience, New York).

Poeppelmeier, K.R., and J.D. Corbett, 1977a, Inorg. Chem. **16**, 1107.

Poeppelmeier, K.R., and J.D. Corbett, 1977b, Inorg. Chem. **16**, 294.

Poeppelmeier, K.R., and J.D. Corbett, 1978, J. Am. Chem. Soc. **100**, 5039.

Reiss, H., 1971, Progress in Solid State Chemistry, Vol. 5 (Pergamon, Oxford).

Robbins, D.J., and A.J. Thomson, 1972, J. Chem. Soc. Dalton Trans., p. 2350.

Satpathy, S., and O.K. Andersen, 1985, Inorg. Chem. **24**, 2604.

Schäfer, H., 1971, Angew. Chem. **83**, 35; Angew. Chem. Int. Ed. Engl. **10**, 43.

Schäfer, H., and B. Eisenmann, 1981, Rev. Inorg. Chem. **3**, 29.

Schäfer, H., and H.G. Schnering, 1964, Angew. Chem. **76**, 833; Angew. Chem. Int. Ed. Engl. **3**, 117.

Schäfer, H., H.G. von Schnering, J.V. Tillack, F. Kuhnen, H. Wöhrle and H. Baumann, 1967, Z. Anorg. & Allg. Chem. **353**, 281.

Schleid, Th., and G. Meyer, 1987, Z. Anorg. & Allg. Chem. **552**, 90.

Schmid, B., P. Fischer, R.K. Kremer, A. Simon and A.W. Hewat, 1988, J. Phys. (Paris) **49**, C8–841.

Schwanitz-Schüller, U., 1984, Thesis (Stuttgart, FRG).

Schwanitz-Schüller, U., and A. Simon, 1985a, Z. Naturforsch. B **40**, 705.

Schwanitz-Schüller, U., and A. Simon, 1985b, Z. Naturforsch. B **40**, 710.

Schwarz, C., 1987, Thesis (Stuttgart, FRG).

Schwarz, C., and A. Simon, 1987, Z. Naturforsch. B **42**, 935.

Simon, A., 1967, Z. Anorg. & Allg. Chem. **355**, 295.

Simon, A., 1976, Chemie in unserer Zeit **10**, 1.

Simon, A., 1979, Struct. & Bonding **36**, 81.

Simon, A., 1981, Angew. Chem. **93**, 23; Angew. Chem. Int. Ed. Eng. **20**, 1.

Simon, A., 1982, in: Crystalline Electric Field Effects in f-Electron Magnetism (Plenum, New York) p. 443.

Simon, A., 1985, J. Solid State Chem. **57**, 2.

Simon, A., 1988, Angew. Chem. **100**, 164; Angew. Chem. Int. Ed. Engl. **27**, 160.

Simon, A., and T. Koehler, 1986, J. Less-Common Met. **116**, 279.

Simon, A., and E. Warkentin, 1982, unpublished.

Simon, A., and E. Warkentin, 1983, Z. Anorg. & Allg. Chem. **497**, 79.

Simon, A., H.G. von Schnering and H. Schäfer, 1967, Z. Anorg. & Allg. Chem. **355**, 295.

Simon, A., Hj. Mattausch and N. Holzer, 1976,

Angew. Chem. **88**, 685; Angew. Chem. Int. Ed. Engl. **15**, 624.

Simon, A., N. Holzer and Hj. Mattausch, 1979, Z. Anorg. & Allg. Chem. **456**, 207.

Simon, A., E. Warkentin and R. Masse, 1981, Angew. Chem. **93**, 1071; Angew. Chem. Int. Ed. Engl. **20**. 1013.

Simon, A., W. Mertin, Hj. Mattausch and R. Gruehn, 1986, Angew. Chem. **98**, 831; Angew. Chem. Int. Ed. Engl. **25**, 845.

Simon, A., Hj. Mattausch, N.B. Mikheev and C. Keller, 1987a, Z. Naturforsch. B **42**, 666.

Simon, A., Hj. Mattausch and R. Eger, 1987b, Z. Anorg. & Allg. Chem. **550**, 50.

Simon, A., C. Schwarz and W. Bauhofer, 1988, J. Less-Common Met. **137**, 343.

Smith, J.D., and J.D. Corbett, 1985, J. Am. Chem. Soc. **107**, 5704.

Torardi, C.C., and R.E. McCarley, 1979, J. Am. Chem. Soc. **101**, 3963.

Torardi, C.C., and R.E. McCarley, 1986, J. Less-Common Met. **116**, 169.

Ueno, F., K.R.A. Ziebeck, Hj. Mattausch and A. Simon, 1984, Rev. Chim. Miner. **21**, 804.

von Schnering, H.G., 1985, Nova Acta Leopold. **59**, 165, Neue Folge Nr. 26.

Warkentin, E., and H. Bärnighausen, 1979, Z. Anorg. & Allg. Chem. **459**, 187.

Warkentin, E., and A. Simon, 1983, Rev. Chim. Miner. **20**, 488.

Warkentin, E., R. Masse and A. Simon, 1982, Z. Anorg. & Allg. Chem. **491**, 323.

Wijeyesekera, S.D., and J.D. Corbett, 1986, Inorg. Chem. **25**, 4709.

Wolf, W.P., M.J.M. Leask, B. Magnum and A.F.G. Wyatt, 1961, J. Phys. Soc. Jpn. Suppl. BI **17**, 487.

Xu, G.-X., and J. Ren, 1987, Lanthanide & Actinide Res. **2**, 67.

Ziebarth, R.P., S.-J. Hwu and J.D. Corbett, 1986, J. Am. Chem. Soc. **108**, 2594.

Handbook on the Physics and Chemistry of Rare Earths, Vol. 15
edited by K.A. Gschneidner, Jr. and L. Eyring

Chapter 101

FLUORIDE GLASSES

RUI M. ALMEIDA
Centro de Física Molecular and Departamento de Engenharia de Materiais, Instituto Superior Técnico, Av. Rovisco Pais, 1000 Lisboa, Portugal

Contents

Symbols and abbreviations

a	half-length of Griffith crack
$\mathscr{A}$	pre-exponential term of viscosity
A	wavelength-independent material attenuation parameter
AC	alternating current
AS	antisymmetric stretch of F_b
AS(C)	antisymmetric stretch of F_b with cation motion
$\mathscr{B}$	Fulcher activation energy for viscosity
B	wavelength-independent material attenuation parameter
BT	BaF_2–ThF_4 based (glasses)
c	speed of light
C	wavelength-independent material attenuation parameter
CVD	chemical vapor deposition
DC	direct current
DR	Raman depolarization ratio
DSC	differential scanning calorimetry
DTA	differential thermal analysis
e	charge of the electron
E	Young's modulus
$\mathscr{E}$	activation energy for crystal growth
E_G	electronic energy gap
ESR	electron spin resonance
EXAFS	extended X-ray absorption fine structure
E_σ	activation energy for electrical conduction
f	oscillator strength
F_b	bridging fluorine atom
F_{nb}	nonbridging fluorine atom
FH	fluorohafnate
FWHM	full width at half maximum
FZ	fluorozirconate
g	Landé factor
GCFN	generalized central force network model
h	Planck's constant
HMFG	heavy metal fluoride glasses
IR	infrared
k	crystallization rate constant
k_i	ith vibrational force constant

K	Kelvin
K	bulk modulus
K_I	stress intensity factor
K_{IC}	critical stress intensity factor
m	mass of the electron
M	metal element
MD	molecular dynamics
$M(\lambda)$	material dispersion
n	stress corrosion susceptibility parameter
n_2	nonlinear refractive index
n_1, n_2	core, cladding refractive indices
n_D, n_F, n_C	refractive indices for the D, F and C discharge lines
$\boldsymbol{n}$	Avrami exponent
N	number of ions per unit volume of glass
NMR	nuclear magnetic resonance
$N(\gamma, T)$	Bose–Einstein population factor
r	atomic radial distance
R	gas constant
$\mathcal{R}$	Fresnel reflectivity
RAP	reactive atmosphere processing
RDF	radial distribution function
RE	rare earth
RF	radio frequency
SS	symmetric stretch of F_b
SS(C)	symmetric stretch of F_b with cation motion
SS(F_{nb})	symmetric stretch of F_{nb}
t	time
T	absolute temperature
T_0	Fulcher limiting glass transition temperature
T_{fr}	spin-glass transition temperature
T_g	glass transition temperature
T_ℓ	nuclear magnetic resonance spin-lattice relaxation time
T_m	melting temperature
T_x	temperature of onset of crystallization
TM	transition metal
TMFG	transition metal fluoride glasses
u	crystal growth velocity
V	vitreous
v	crack velocity
VS	vibrational spectroscopy
x	optical pathlength
$\mathscr{x}$	volume fraction crystallized after a time t
XANES	X-ray absorption near-edge structure
XRD	X-ray diffraction
Y	crack geometry factor
α	optical absorption coefficient
α_0	infrared absorption coefficient at 0 K
α_{min}	minimum optical loss coefficient
α_t	total optical attenuation coefficient
α_{MP}	multiphonon absorption coefficient
α_{RS}	Rayleigh scattering coefficient
α_T	linear thermal expansion coefficient
γ	glass cooling rate
γ_c	glass critical cooling rate
ΔH	nuclear magnetic resonance chemical shift
ΔT_c	thermal shock resistance
$\Delta\lambda$	wavelength bandwidth
η	viscosity
θ	network bridging angle
θ_D	Debye temperature
Θ	Curie temperature
λ	wavelength of light
λ_0	zero material dispersion wavelength
λ_c	UV cut-off wavelength
λ_{min}	wavelength of minimum attenuation
μ	reduced mass
ν	frequency of light
ν_0	average optical phonon frequency
ν_c	frequency factor for crystal growth
ν_C	cage-like vibration of cations
ν_D	Abbé number
ν_{LF}	low frequency Raman vibration
ν_P	Poisson's ratio
ρ	density
σ	radiative emission cross section
σ_0	electrical conductivity pre-exponential factor
σ_e	electrical conductivity
σ_A	applied stress
σ_F	fracture strength
χ_m	magnetic susceptibility
ω	angular frequency
Ω_i	ith Judd–Ofelt parameter

1. Introduction

The fluoride glass designation includes two main families of vitreous materials: (1) the beryllium fluoride based (fluoroberyllate) glasses and (2) the heavy-metal fluoride glasses (HMFG), where the metal cations present are considerably heavier than Be. Fluoroberyllate glasses have been extensively investigated for optical and laser applications, particularly those doped with Nd^{3+}, in view of their small nonlinear refractive index (Weber et al. 1978). A comprehensive review of the preparation, structure and properties of BeF_2-based glasses was published by Baldwin et al. (1981). Since then, although the structural role of lanthanide cations such as Nd^{3+} has continued to be investigated, e.g. with X-ray absorption methods (EXAFS and XANES) by Weber and Wong (1987), few significant developments have been reported, with the exception of some advances in CVD processing for fiber optics (Sarhangi 1987). Therefore, only the second type of glasses, the HMFG, will be discussed in the present review.

HMFG include (1) glasses based on ZrF_4 or HfF_4, which may contain significant amounts of rare earth (RE) compounds like ScF_3, LaF_3, NdF_3, GdF_3, etc., (2) glasses based on BaF_2 and ThF_4, which also contain YbF_3 or AlF_3 and ZnF_2, (3) glasses based on ZnF_2 and/or CdF_2, usually containing also BaF_2, LaF_3 and/or AlF_3 and PbF_2, (4) glasses based on AlF_3, but also including some ZrF_4, LaF_3 and/or YbF_3, or sometimes PbF_2, CaF_2, BaF_2, ThF_4, etc., and (5) the transition metal fluoride glasses, which usually contain a significant amount of PbF_2 plus one or more transition metal fluorides and may also include small amounts of AlF_3 and/or YF_3. In addition to the book edited by Almeida (1987a), which presents detailed accounts of these different fluoride glass systems, a few other reviews devoted to HMFG are available, namely, those by Drexhage (1985) and Lucas (1985). The former is a comprehensive review of general character up to the beginning of 1984, with 215 references, while the latter is a shorter review dealing specifically with RE in fluoride glasses. The review by Baldwin et al. (1981) deals also extensively with early developments in HMFG during the first five years after their discovery. Finally, the recent book edited by Comyns (1989), covers some of the latest developments in the field.

In this chapter, particular attention will be devoted to fluorozirconate (FZ) glasses. These have a large transparency domain extending from the near UV (~ 0.2 μm) to the middle IR (~ 7 μm), a reasonably good glass forming ability and also acceptable chemical durability and temperature resistance, which are a set of characteristics not presently found in any other glass system outside the HMFG family. Thus, FZ glasses have an enormous potential as ultralow loss middle infrared (IR) fiber optics materials and in other applications in fiber form (sensors, active optical fibers) or in bulk form (windows, light pipes, prisms or other active components). The present chapter will attempt to present a general review of the most important aspects concerning the preparation, structure and properties of HMFG, with particular emphasis on FZ systems and on glasses where RE compounds play a significant role, either from the composition or application viewpoints. A separate section for applications is not included, but those most relevant will be briefly discussed together with the particular property to which they relate.

2. Glass formation

Glass formation in HMFG systems was first reported for the pair AlF_3–PbF_2 by Sun (1949). One should note that, although aluminum cannot be considered a truly heavy metal, it may be taken as being heavy when compared with beryllium; the fluoroaluminate glasses often include heavy-metal fluorides such as PbF_2. Based on the known glass-forming ability of the Al_2O_3–CaO pair, Baldwin et al. (1981) reported the preparation of AlF_3–CaF_2 and AlF_3–BaF_2 binary glasses which could be further stabilized by ThF_4 additions yielding pieces up to 1–2 mm thick without severe quenching. Videau et al. (1979) studied a glass with $40AlF_3$–$40CaF_2$–$20BaF_2$ (all glass compositions are given in mol%, unless specified otherwise) and they also prepared other fluoroaluminate glasses containing MgF_2, NaF and YF_3. More recently, Izumitani et al. (1987) reported a series of multicomponent glasses with AlF_3, CaF_2, SrF_2, BaF_2, ZrF_4 and YF_3 as main constituents. Pure AlF_3 sublimes under atmospheric pressure and it cannot be melted into a glass, as opposed to pure BeF_2. In general, it is difficult to prepare thick samples of AlF_3-based glasses without a certain degree of crystallization. Table 1 lists some of the most important fluoroaluminate glass compositions developed so far.

Glass formation in ZrF_4-based systems was first reported 15 years ago by Poulain et al. (1975) for a ternary glass containing BaF_2 and NaF. The first indications concerning possible glass formation had become obvious during an attempt to prepare a quaternary crystal which included NdF_3 and the three previous compounds. Since then, many other FZ glasses have been prepared and studied, including binary ZrF_4–BaF_2 and ZrF_4–SrF_2 glasses prepared by Almeida and Mackenzie (1981) in thicknesses up to 2–3 mm, mostly for structural studies. Other binary glass systems were reported shortly thereafter, such as ZrF_4–ThF_4 (Almeida and Mackenzie 1983a) and ZrF_4–PbF_2 (Almeida and Mackenzie 1985). The most stable FZ glass compositions presently available, for which the smallest dimension in bulk pieces may reach ~3 cm, are usually multicomponent systems containing four to six compounds. A typical example is the ZBLAN composition (ZrF_4–BaF_2–LaF_3–AlF_3–NaF), widely used in bulk glass and fiber optics applications. Compared to oxide or chalcogenide glasses, fluorozirconate glass-forming regions are usually quite small, as illustrated in fig. 1a for the ZrF_4–BaF_2–LaF_3 diagram. Therefore, compositional variations within a particular system tend to be limited. In most FZ glasses, ZrF_4 is present in excess of 50 mol% and it plays the role of glass network former. Pure zirconium tetrafluoride is a highly volatile solid which sublimes at 904°C under atmospheric pressure (Sense et al. 1954) and, like AlF_3, it cannot be melted into a glass. Table 1 lists some of the most representative FZ glass compositions and their corresponding acronyms. These will sometimes be used in the text to designate other glasses in the same systems even if with somewhat different compositions.

HfF_4 is chemically similar to ZrF_4 and its room-temperature stable crystalline phase is isomorphic with β-ZrF_4. Therefore, it can replace ZrF_4, partially or totally, in any FZ glass. The first report of glass formation in fluorohafnate (FH) systems was a patent by Lucas et al. (1977). HfF_4 has not been extensively used to replace ZrF_4 in

TABLE 1
Compositions of some typical heavy-metal fluoride glasses.

Composition (in mol%)	Acronym	Reference
$64ZrF_4$–$36BaF_2$	ZB	Almeida and Mackenzie (1981)
$66HfF_4$–$34BaF_2$	HB	Almeida and Mackenzie (1983b)
$70ZrF_4$–$30SrF_2$	ZS	Almeida and Mackenzie (1981)
$43HfF_4$–$57ThF_4$	HT	Almeida and Mackenzie (1984)
$70ZrF_4$–$30PbF_2$	ZP	Almeida and Mackenzie (1985)
$57ZrF_4$–$34BaF_2$–$9ThF_4$	ZBT	Poulain and Lucas (1978)
$62ZrF_4$–$30BaF_2$–$8LaF_3$	ZBL	Lecoq and Poulain (1979)
$60ZrF_4$–$34BaF_2$–$6NdF_3$	ZBNd	Poulain and Lucas (1978)
$63ZrF_4$–$33BaF_2$–$4GdF_3$	ZBGd	Mitachi et al. (1981)
$55ZrF_4$–$30BaF_2$–$15UF_4$	ZBU	Aliaga et al. (1978)
$60ZrF_4$–$33ThF_4$–$7LaF_3$	ZTL	Matecki et al. (1978)
$57ZrF_4$–$36BaF_2$–$3LaF_3$–$4AlF_3$	ZBLA	Lecoq and Poulain (1980)
$57ZrF_4$–$18BaF_2$–$3LaF_3$–$5AlF_3$–$17NaF$	ZBLAN	Grilo et al. (1987)
$56ZrF_4$–$14BaF_2$–$6LaF_3$–$4AlF_3$–$20KF$	ZBLAK	Grilo et al. (1987)
$56ZrF_4$–$14BaF_2$–$6LaF_3$–$4AlF_3$–$10NaF$–$10KF$	ZBLANK	Grilo et al. (1987)
$30AlF_3$–$10ZrF_4$–$8YF_3$–$4MgF_2$–$20CaF_2$–$13SrF_2$–$11BaF_2$–$4NaF$	AZYMgCSBN	Izumitani et al. (1987)
$26CdF_2$–$10LiF$–$31AlF_3$–$33PbF_2$	CLAP	Tick (1985)
$50AlF_3$–$50CaF_2$	AC	Baldwin et al. (1981)
$40AlF_3$–$22CaF_2$–$22BaF_2$–$16YF_3$	ACBY	Kanamori et al. (1981)
$30ZnF_2$–$30CdF_2$–$40BaF_2$	ZCdB	Matecki et al. (1982b)
$19BaF_2$–$27ThF_4$–$27YbF_3$–$27ZnF_2$	BTYbZn	Slim (1981)
$19BaF_2$–$27ThF_4$–$27LuF_3$–$27ZnF_2$	BTLuZn	Drexhage et al. (1982)
$30BaF_2$–$10ThF_4$–$10YbF_3$–$20ZnF_2$–$30InF_3$	BTYbZnI	Bouaggad et al. (1987)
$43PbF_2$–$17GaF_3$–$17InF_3$–$19ZnF_2$–$4LaF_3$	PGIZnL	Nishii et al. (1989)
$50PbF_2$–$25MnF_2$–$25FeF_3$	PMFe	Miranday et al. (1979)

HMFG, since the improvement in properties is often insufficient to justify the large increase in price.

Within the HMFG family, the Zr-free glasses are generally less stable against devitrification than the FZ materials. The BaF_2–ThF_4-based (BT) fluorides represent another type of glasses whose preparation requires relatively fast quenching. However, Slim (1981) reported the fabrication of samples up to 3–4 mm thick for glass compositions near $19BaF_2$–$27ThF_4$–$27YbF_3$–$27ZnF_2$ and Drexhage et al. (1982) were able to replace YbF_3 with LuF_3. When InF_3 is added to the base quaternary composition, e.g. in a $30BaF_2$–$10ThF_4$–$10YbF_3$–$20ZnF_2$–$30InF_3$ glass, the crystallization tendency is reduced and samples over 1 cm thick can be prepared (Bouaggad et al. 1987).

A series of glasses were reported by Matecki et al. (1982b) in the ZnF_2–CdF_2–BaF_2 system, to which other fluorides can be added such as LaF_3, AlF_3 and NaF. These

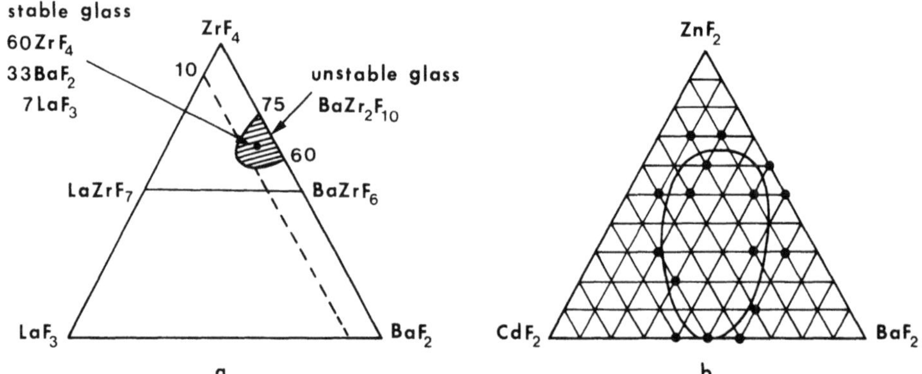

Fig. 1. Glass forming regions of (a) ZrF_4–BaF_2–LaF_3 system (Lucas 1987) and (b) ZnF_2–CdF_2–BaF_2 system (Matecki et al. 1982b).

stabilize the crystallization-prone ternary compositions. Figure 1b shows the glass-forming region of the ZnF_2–CdF_2–BaF_2 ternary system. Tick (1985) reported a new family of quaternary glasses based on CdF_2, AlF_3 and PbF_2, but also including LiF, designated by the acronym CLAP (for CdLiAlPb). These glasses have recently been found by El-Bayoumi et al. (1987) and Randall et al. (1988) to exhibit liquid–liquid phase separation which enhanced crystallization. Tick (1988a) has argued that an optimum amount of oxygen is needed to prevent crystallization or phase separation. Boehm et al. (1988) have also reported phase separation in PbF_2-containing FZ glasses. Nishii et al. (1989) have recently prepared glasses in the system PbF_2–GaF_3–InF_3–ZnF_2–LaF_3, whose thermal stability allowed the preparation of samples more than 15 mm thick.

Finally, the transition metal fluoride glasses (TMFG), discovered by Miranday et al. (1979), consist of one or more trivalent or divalent transition metal fluorides, combined with a substantial concentration of PbF_2 and eventually other minor additions, in order to improve stability. A typical composition is $50PbF_2$–$25MnF_2$–$25FeF_3$. The TMFG are generally less stable than the FZ compositions and the addition of ZrF_4 as a minor constituent may improve glass stability (Jian-Feng and Ji-Jian 1989a), although the same authors have later reported liquid–liquid phase separation in TMFG with ZrF_4 (Jian-Feng and Ji-Jian 1989b).

The previous listing of glass-forming systems represents the most unique and important compositions which have so far been developed, but it is not exhaustive. It is now clear that the number of glass-forming mixtures is endless, in particular when glasses with five or more components are considered. In addition, other glasses have been prepared which combine heavy-metal fluorides with BeF_2, e.g. in the PbF_2–BeF_2–AlF_3 system. Therefore, given the previous considerations and also the frequent absence of a compound which can clearly be singled out as a network former, an accurate classification of the HMFG families is rather difficult at this point and it will probably become almost impossible in the future.

The subject of glass-forming criteria, to be analyzed in more detail in sect. 4, should also be mentioned in connection with fluoride glasses. Although Zachariasen's rules (Zachariasen 1932) cannot be used with such systems, the kinetic theory of glass formation (Turnbull 1952, Turnbull and Cormia 1961, Klein et al. 1977) is certainly applicable to the HMFG, but it has the same drawbacks as when applied, e.g. to oxide systems; namely, the need for data which are usually lacking and are often collected only after glass formation has been detected. Also, due to the low viscosities of HMFG melts, near ~0.01 Pa s at the liquidus temperatures, crystallization velocities tend to be substantially high and the critical cooling rates are quite large when compared to most oxide glass systems. There is some evidence that glass formation in HMFG is associated with compositions lying in the vicinity of deep eutectics, but the usefulness of this criterion is hindered by the lack of experimental phase diagrams for virtually all the systems of interest from the glass-forming viewpoint. Finally, two of the earlier criteria developed for explaining glass formation in oxide systems have been adapted to fluoride glasses with mixed success. Thus, Baldwin and Mackenzie (1979) used the bond energy approach (Sun 1947) to predict which compounds would exhibit network forming, intermediate or network modifying character, based on the M–F (M = metal) single-bond energy. In order of decreasing glass-forming ability, TiF_4, ScF_3, BeF_2, HfF_4, ZrF_4 and AlF_3 would be classified as network formers, which agrees to some extent with the experimental observations. However, this criterion is of limited usefulness in the many examples of multicomponent HMFG where a clear network former cannot be identified. Finally, Poulain (1982) proposed an adaptation of the well-known cation field strength criterion (Dietzel 1942) to HMFG, in general agreement with previously known glass formation trends.

3. Preparation and handling

The preparation of HMFG involves a number of experimental difficulties. Some are characteristic of all fluoride glasses, including fluoroberyllates, namely, their toxicity (particularly when actinide or very heavy RE elements are present), the hygroscopicity of the raw and vitreous materials and the reactivity of the melts with most crucibles available. Others are typical of HMFG, most notably the high crystallization velocities in supercooled melts, usually related to low melt viscosities. Therefore, both preparation and handling involve several precautions rarely necessary in the fabrication of oxide glasses.

The best method to prepare most HMFG (although not the cheapest) starts with anhydrous metallic fluorides MF, MF_2, MF_3 and MF_4. Fluorides of high purity are available commercially, especially monovalent and divalent ones, but for applications where ultra high purity is required, including the ppb or sub-ppb range as in the case of ultra low loss fiber optics, further purification is mandatory. For highly volatile compounds like ZrF_4 and AlF_3, sublimation (Almeida and Mackenzie 1981) is one of the preferred techniques, whereas this technique may also be used for some of the other nonvolatile fluorides, but with utilization of the residue rather than the sublimate. Chemical vapor purification (Folweiler and Guenther 1985), where Zr metal is converted to $ZrCl_4$ by gaseous Cl_2 and the zirconium chloride is then fluorinated by SF_6 or HF into ZrF_4, has also been successful in reducing the transition metal (TM) ion content to very low levels. These and other purification methods were thoroughly discussed by Robinson (1987). Another important aspect related to the preparation of ultra high purity raw materials and also pertinent to the final glasses is the need for very sophisticated analytical tools for the detection of sub-ppb levels of TM, RE, ammonium, hydroxide or other ions. Spark-source mass spectrography, atomic absorption spectroscopy, plasma emission spectroscopy, neutron activation analysis and ion exchange chromatography are some of the available methods, as discussed by Klein (1987). The ubiquitous hydroxide impurity, in particular, is very difficult to measure quantitatively with accuracy.

An alternative method of HMFG preparation (Poulain and Lucas 1978) starts with less expensive metallic oxides which are converted to fluorides, usually with ammonium bifluoride $NH_4F \cdot HF$, as in the reactions

$$2ZrO_2\,(s) + 7NH_4F \cdot HF\,(s) = 2(NH_4)_3ZrF_7\,(s) + NH_3\,(g) + 4H_2O\,(g), \tag{1}$$

$$(NH_4)_3ZrF_7\,(s) = ZrF_4\,(s) + 3NH_4F\,(g), \tag{2}$$

which may be carried out to completion in the melting crucible near 400°C, in the presence of excess ammonium bifluoride. These reactions release large concentrations of ammonia and water that may contaminate the fluorides produced, as well as the melt and the glass prepared afterwards. In addition, the highly corrosive nature of NH_3 gas prevents the reactions to be carried out under a clean environment such as a glove box. Despite those disadvantages, even in the preparation of ultrapure HMFG it is common practice to use small quantities of $NH_4F \cdot HF$ in a premelting step, in order to completely fluorinate any residual oxides present in the anhydrous fluoride batch mixture.

Because of the high reactivity of HMFG melts with most crucible materials, platinum or vitreous carbon are usually preferred, although even these choices are not trouble free. On the other hand, given the considerable hygroscopicity of most fluorides of interest for glass making, not only must the raw materials be carefully dehydrated under vacuum prior to the melting operation, but this must also be carried out under an extremely dry atmosphere, usually in an inert-gas filled glove box. Although the detrimental effects of oxide species in the glasses and the reactivity of fluoride melts with oxygen at high temperatures are not yet absolutely clear, the presence of oxygen in the melting atmosphere is usually also avoided. Commercial glove box systems are capable of maintaining H_2O and O_2 concentrations at ppm or sub-ppm levels in inert atmospheres such as Ar or N_2.

Most HMFG are melted at temperatures reaching maximum values near 700–1000°C, in resistance or radio frequency induction heated furnaces. Because of the high vapor pressure of some of the fluorides involved (e.g. ZrF_4 sublimes at 904°C and its vapor pressure becomes appreciable above 600°C), preferential volatilization of certain species is a problem. Therefore, melting times tend to be kept to a minimum. One usually tries to operate in tightly capped crucibles and sometimes an excess is added of a given compound in order to account for its preferential loss from the melt. In addition to the previous precautions, the ultimate reduction of oxide and particularly hydroxide species in the final glasses is usually accomplished via reactive atmosphere processing (RAP) techniques (Robinson et al. 1980, Robinson 1987), often employed during the melting procedure in silica reaction vessels. Several different atmospheres have been used to date, including CCl_4, CF_4, HF, SF_6 and NF_3, with the first and last ones being frequently used at present. For example, in the case of CCl_4 the gaseous chloride is transported to the reaction vessel by bubbling an inert carrier gas through the liquid and the removal of hydroxide species may be accomplished via the following reactions,

$$CCl_4\,(g) + 2OH^-\,(melt) = CO_2\,(g) + 2Cl^-\,(melt) + 2HCl\,(g), \tag{3}$$

$$CCl_4\,(g) + 2H_2O\,(melt) = CO_2\,(g) + 4HCl\,(g), \tag{4}$$

in which large quantities of gaseous CO_2 and Cl^- ions are produced. The presence of residual CO_2 molecules in the glasses may lead to a significant resonant absorption in the infrared region near 4.24 μm (France et al. 1987, Almeida et al. 1988), whereas the incorporation of chloride species in the fluoride melt may eventually increase the crystallization tendency during its supercooling (Neilson et al. 1985). Also, platinum crucibles cannot be used with CCl_4 RAP, due to attack of the noble metal by Cl^- species. On the other hand, however, these Cl^- species and those introduced by Cl_2 as a result of the decomposition of CCl_4 at high temperatures may compensate for fluorine deficiencies in the melt. In the case of FZ systems, these are perhaps attributable to reduced ZrF_3 or ZrF_2 dark-colored phases (Robinson et al. 1980, Robinson et al. 1981), or to Zr^{3+} centers (Carter et al. 1987).

After the fining operation, consisting of the removal of gaseous bubbles from the liquid which is facilitated in the present case by the low melt viscosities, the compositions are cast into graphite, brass or stainless steel molds, eventually preheated. Gold-coated aluminum may also be used. Finally, the glasses are annealed near the glass transition temperature (T_g). Good glass-forming compositions such as ZBLAN, e.g. can be left cooling inside the crucible without visible crystallization, as long as the smallest dimension is not much larger than ~3 cm, whereas many others, including binary systems such as ZrF_4–BaF_2 or ternary systems such as AlF_3–CaF_2–BaF_2 and most TMFG, need relatively large quenching rates into molds. Although most HMFG are hygroscopic, they are generally stable under atmospheric air of average humidity and storing in a desiccator is convenient but not mandatory.

All the above considerations apply to the preparation of glasses in bulk form.

Optical fibers, mostly of FZ glasses, are usually drawn from preforms and subsequently coated with a polymer. At present, preforms are almost always multimode and they are prepared by melting, with the core composition being poured into a previously cast hollow cladding. The cladding is either cast into a mold which is immediately upset (built-in-casting method), or the heated mold is positioned horizontally right after pouring of the cladding composition and it is very rapidly rotated, leading to a cladding of very uniform thickness and also to a smooth core/clad interface. The latter method, due to Tran et al. (1982), is known as rotational casting. A problem often encountered in the preparation of HMFG, particularly with FZ samples, is the occurrence of bubbles, especially when the melts are allowed to cool down inside the crucibles or when rod-like shapes are fabricated. These include optical fiber preforms, where bubbles may lead to unacceptable scattering levels in the fibers. The nature of such bubbles has been discussed by Mcnamara and MacFarlane (1987) in the case of a ZBLAN glass and it was speculated that these might either be under vacuum or contain some gas, probably with the same composition as the melting atmosphere. Tick and Mitachi (1989) found certain bubbles which contained only CO_2 and others which contained the melting atmosphere. Thin films have also been prepared from some glass compositions. For example, Poignant et al. (1985) were able to deposit films from ZBLA and ZBLAN melts, resistance-heated and boiling under vacuum, and Poignant (1987) and Jacoboni et al. (1987) succeeded in depositing 10–20 μm films of TMFG such as PbF_2–MnF_3–GaF_3 onto ZBLA substrates, with a final composition near $50PbF_2$–$50GaF_3$. More recently, Bruce (1989) has deposited FZ glass films onto a variety of substrates by electron beam evaporation.

4. Structure

The knowledge of the structures of HMFG is still very incomplete, being in fact limited to a few fluoroaluminate glasses and to several FZ systems. However, even in the latter case, a great deal of controversy is still apparent.

Compared to vitreous BeF_2, whose structure is essentially a weakened version of the structure of vitreous silica (Baldwin et al. 1981) (with the same analogy holding approximately between fluoroberyllates and silicates), all HMFG appear to exhibit very different structures. Namely, a clear glass network former is absent in many cases. The cation coordination numbers, in particular for the higher valence ones, are considerably larger than four and there is some evidence concerning the occurrence of edge sharing between neighboring fluorinated polyhedra in certain glasses.

In HMFG, whenever a clear network forming cation, present in a concentration of at least 50 mol %, does not exist, it is useful to consider the higher-valence, more electronegative cations as the network forming species, whereas the lower-valence fluorides may be taken as network modifiers and the intermediate valence fluorides, if any, may be considered network intermediates. In this fashion, one may still talk about modified glasses in the Zachariasen sense, where the structure of the network former is disrupted by nondirectional ionic bonds between the modifying cations and the anions, although Zachariasen's rules themselves (Zachariasen 1932) are obviously not applicable in the present case. Even if a unified structural framework is perhaps impossible to achieve for all the highly diverse HMFG structural types, Almeida (1987) tried to identify the main structural features which appeared to be common to the most stable halide glass forming compositions. The main conclusions were the following: (1) the network forming cation coordination is hardly a major factor, (2) the degree of covalency of the glass network is not particularly important, (3) network modification always increases the glass-forming ability of a network-forming halide and (4) chain-like structures, where orientational freedom is highest, appear to be the most favorable configuration for halide systems.

The role of the RE elements in HMFG glasses was discussed in detail by Lucas (1985). In the case of FZ systems, he concluded that the lighter lanthanide elements, from La to Gd, given their suitable ionic size, occupy a small fraction of the nine-fold coordinated and a part of the eight-fold coordinated sites in the glasses, stabilizing them against devitrification, despite being present only in small amounts. This corresponds to intermediate character in the Zachariasen–Sun sense for oxide glasses. It should be mentioned that, in FZ systems, BaF_2 is always the prime network modifier, presumably because the Ba^{2+} ion radius ($\sim$0.135 nm) is similar to that of F^- ($\sim$0.136 nm); although PbF_2 is also a good replacement for BaF_2 in terms of glass-forming ability, glasses containing SrF_2 instead of BaF_2, e.g., were found to be considerably less stable (Almeida and Mackenzie 1981). In the case of zirconium-free HMFG, larger amounts of the heavier and more covalent lanthanides from Tb to Lu are usually present and Lucas (1985) argued that they play a more fundamental role in glass formation, with the smaller, more covalent elements (generally eight-fold coordinated in crystalline fluorides) leading to the most stable glasses.

The structure of water-quenched AlF_3–CaF_2–BaF_2 glasses was discussed by Videau et al. (1979), based on Raman scattering spectroscopy and X-ray diffraction studies of the crystallized glasses. For the composition $40AlF_3$–$40CaF_2$–$20BaF_2$, the high-frequency region of the Raman spectrum above 450 cm^{-1} was highly polarized and it was proposed that most Al atoms were six-fold coordinated, probably forming chain-like structures, with Ca playing a network-forming role together with Al, and Ba behaving as a modifier. Some of the Al atoms might also occur in four-fold coordinated environments.

The structure of FZ and also FH glasses has already been the subject of relatively extensive studies. Since all the available evidence strongly indicates that these two glass families are structurally isomorphous (Etherington et al. 1984a), only the first one will be discussed in detail. The main constituents of FZ glasses are ZrF_4 and BaF_2. Zirconium tetrafluoride has a low-temperature β phase where Zr is eight-fold coordinated at the center of a square Archimedian antiprism and each F is coordinated by two Zr atoms (Burbank and Bensey 1956), thus being of the bridging type (F_b); the antiprisms are joined together by sharing all eight corners, forming a three-dimensional network. ZrF_4 has also a high-temperature α phase which is a three-dimensional array of triangular ZrF_8 dodecahedra where all fluorine atoms are bridging type and both corners and edges are shared (Papiernik et al. 1982). Barium fluoride has the fluorite structure, with eight-fold coordinated Ba atoms (Wyckoff 1963). On the other hand, metazirconate and dizirconate compounds have been reported in the relevant system ZrF_4–BaF_2 by Laval et al. (1979). For the metazirconate composition $BaZrF_6$, the low-temperature α form, stable below 544°C, contains seven-fold coordinated Zr atoms forming edge-shared $Zr_2F_{12}^{4-}$ dimers (Laval et al. 1978), whereas the high-temperature β form consists of infinite chains of eight-coordinated ZrF_8 dodecahedra sharing two edges each (Melhorn and Hoppe 1976). For the dizirconate composition $BaZr_2F_{10}$, the high-temperature β form, stable above 460°C, consists of distorted ZrF_7 pentagonal bipyramids sharing an edge and two corners each (Laval et al. 1988).

Early studies of FZ glass structure, such as those performed on the systems ZrF_4–BaF_2–ThF_4 (Poulain et al. 1977), ZrF_4–ThF_4–RF_3 (R = La, Nd) (Matecki et al. 1978) and ZrF_4–BaF_2–MF_n (M = Na, Ca, Nd, Th) (Poulain et al. 1979), were usually based on a presumed analogy between the structures of the glasses and those of crystalline phases of relatively similar compositions, or based on the behavior of a selected physical property. Visible spectroscopy was also used to examine the structure of a ZrF_4–BaF_2–LaF_3–NdF_3 glass (Lucas et al. 1978) and the ZrF_4–BaF_2–UF_4 system (Aliaga et al. 1978). In the former glass, Nd^{3+} was used as a local probe and, based on its absorption and fluorescence spectra in the glass and related crystals, it was proposed that both eight- and nine-coordinated Nd atoms were present in the vitreous material, together with ZrF_6, ZrF_7, ZrF_8 and LaF_8 polyhedra, forming a three-dimensional array through corner sharing.

Starting with Almeida and Mackenzie (1981), it was realized that the rather complex structures of FZ glasses should preferably be studied beginning with binary systems, namely with the simplest possible ones – ZrF_4–MF_2 (M = Ba, Sr, Pb) – consisting of the network former and a network modifier MF_2, since pure ZrF_4 could not be melted and cast as a glass. Despite a few other studies which have been done on multicomponent FZ glasses, the rest of this discussion on experimental studies of structure will concentrate almost exclusively on the binary glasses. The first structural model, proposed by Almeida and Mackenzie (1981) on the basis of IR absorption and polarized Raman scattering spectroscopy studies, was built around the most stable dizirconate composition ($2ZrF_4$–BaF_2) and it consists of zigzag chains of elongated ZrF_6 octahedra sharing two corners each, cross-linked by Ba–F ionic bonds, plus a small amount of eight- and nine-membered rings, with an estimated average Zr–F–Zr bridging angle near 136°. This structure is shown in fig. 2a. The two other structural prototypes were the metazirconate glass (ZrF_4–BaF_2), which was proposed to contain seven-fold coordinated Zr atoms similar to α-$BaZrF_6$ and the trizirconate composition ($3ZrF_4$–BaF_2), which included mostly six- but also some five-fold coordinated Zr atoms, in a more two-dimensional type of structure. Vibrational spectroscopy was also used by Almeida and Mackenzie (1983b) to study the structure of binary FH glasses. Here the isomorphic substitution of Hf for Zr allowed a more definitive vibrational assignment to be made and led to a refinement of the previous structural model, valid for FH and FZ glasses. It was again suggested that six-fold coordination predominated for the network forming cations around the 2:1

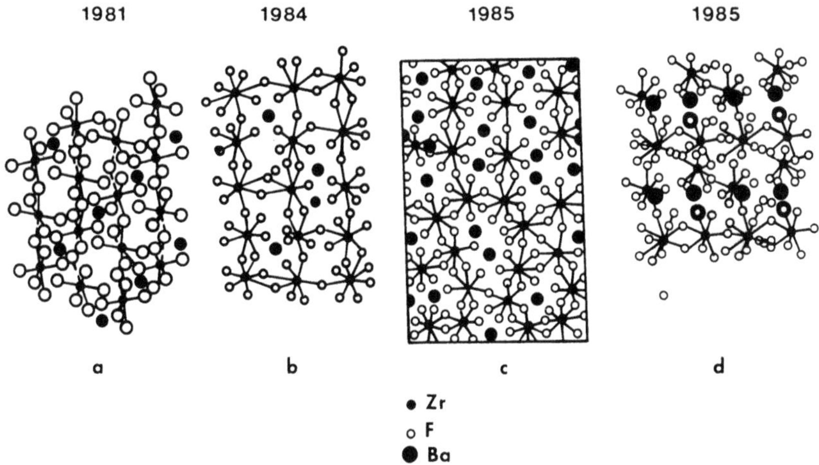

Fig. 2. Structural models for barium FZ compositions: (a) dizirconate glass (Almeida and Mackenzie 1981); (b) dizirconate glass (Lucas et al. 1984); (c) metazirconate glass (Kawamoto et al. 1985); (d) $3ZrF_4$–$2BaF_2$ glass (Yasui and Inoue 1985).

composition, with the degree of bridging increasing with the Hf (Zr) concentration. Substitution of Ba by Sr did not appear to affect the glass structure and, due to the large Ba (Sr):F ratio near 10, it was suggested that neighboring chain overlap could lead to sharing of a single nonbridging fluorine (F_{nb}) atom by two Hf (Zr) atoms belonging to neighboring chains, increasing the apparent Zr coordination when certain structural tools such as X-ray diffraction (XRD) were used. In fact, in addition to vibrational spectroscopy, XRD has also been used in experimental studies of FZ glass structure. For example, Coupé et al. (1983) studied the structure of a series of barium FZ glass compositions by X-ray radial distribution function (RDF) analysis and they found that the Zr coordination increased with the ZrF_4 content between 7.08–8.35, lying at 7.35 for the dizirconate composition, with an average Zr–F–Zr bridging angle near 150°–160°. Etherington et al. (1984a) studied the structure of barium dizirconate and dihafnate glasses also by Fourier analysis of XRD data, with the isomorphic substitution of Zr by Hf again proving useful in the assignment of the different peaks in the total correlation function, which was used here instead of the RDF. A Zr coordination number of $\sim$7.4 was calculated for the dizirconate glass and it was concluded that the apparent disagreement with the previous Raman results of Almeida and Mackenzie was probably due to a substantial interaction between chains as discussed by Almeida and Mackenzie (1983b). The average Zr–F–Zr bridging angle was estimated at $\sim$170°. No evidence was found at this point for edge-sharing between different anionic structural units.

Fourier analysis of neutron diffraction data (time-of-flight method) was also employed by Etherington et al. (1984b) to probe the structure of the barium dizirconate glass and the combination of the X-ray and neutron methods allowed an assignment of the different peaks in the total correlation function with a good degree of confidence. The average Zr coordination was found to be near ~6.8 and the F–F coordination number was ~5, in general agreement with the chain-like structure schematically depicted in fig. 2a. Le Bail et al. (1987) used neutron diffraction in the variable 2θ scanning configuration, in order to study the structure of barium FZ glasses with 60–75 mol % ZrF_4 and they discussed a quasicrystalline model for the dizirconate composition, derived from the β-$BaZr_2F_{10}$ structure, including both seven- and eight-fold coordinated Zr atoms. Finally, Wagner et al. (1987) obtained the total reduced atomic distribution function for a lead dizirconate glass by time-of-flight neutron diffraction, determining a Zr coordination number of 7.1. Figure 3 shows a comparison of this function for the different glasses analyzed by Wagner et al. (1987).

Extended X-ray absorption fine structure (EXAFS) has also been used as a structural tool with some success. For example, Ma et al. (1988) examined the structure of barium FZ glasses and they concluded that the most probable coordination number for Zr in the 60–75 mol % ZrF_4 composition range was always six. Boulard (1989), on the other hand, has recently utilized EXAFS in the study of barium FZ glasses and the Zr coordination number was found to be 7.3 ± 0.8 for the dizirconate composition. Given the poor precision associated with the EXAFS-derived coordination numbers ($\pm 15\%$), the author preferred to use the EXAFS-derived interatomic distances (less precise also than those obtained from XRD data) and their correlation with known crystalline coordinations. This approach might

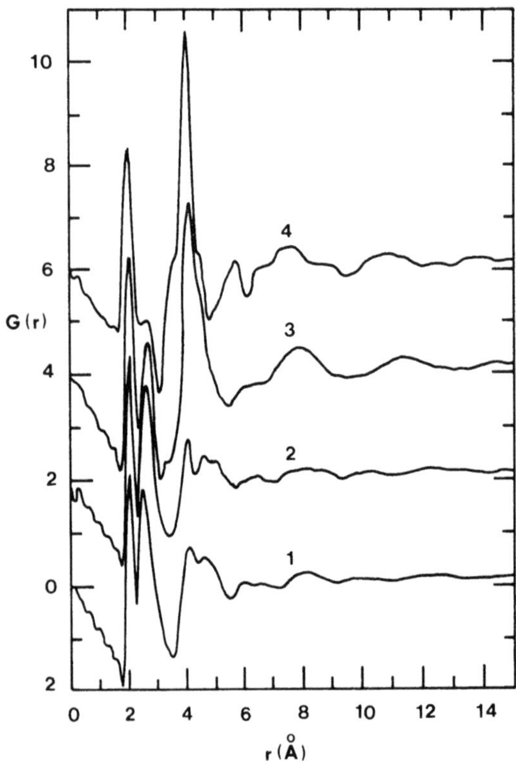

Fig. 3. Reduced atomic distribution $G(r) = 4\pi r[\rho(r) - \rho_0]$ for HMFG as a function of the radial distance r:
(1) $2ZrF_4$–PbF_2 (neutron);
(2) $2ZrF_4$–BaF_2 (neutron);
(3) $2ZrF_4$–BaF_2 (X-ray);
(4) $2HfF_4$–BaF_2 (X-ray).
(Wagner et al. 1987)

eventually lead to erroneous conclusions. For example, in the case of crystalline fluorozirconates, there are cases where the Zr–F interatomic distances are actually larger for six- and seven-fold coordinations than for seven- and eight-fold coordinations, respectively. Another uncertainty associated with the EXAFS technique stems from its large sensitivity to the degree of disorder in the structure, as pointed out by Boulard (1989).

Other experimental methods which have been used to study the structure of FZ glasses, such as ESR (Cases et al. 1985b), NMR (Bray et al. 1983) and XPS (Almeida et al. 1984), have met with less success. For example, Almeida et al. (1984) were unable to differentiate between bridging and nonbridging fluorine species on the basis of XPS, although MacFarlane et al. (1989) recently identified multiple fluoride ion sites in HMFG using NMR. Modeling of glass structure has also been attempted, in particular through molecular dynamics (MD) simulations. A good review of this computer technique has been written by Parker (1987).

Lucas et al. (1984) carried out a reinterpretation of the previous X-ray RDF results of Coupé et al. (1983) for the barium dizirconate glass composition ($BaZr_2F_{10}$) using MD. Here the aperiodic three-dimensional network was proposed to have the $Zr_2F_{13}^{5-}$ edge-shared dimer (with seven- and eight-fold coordinated Zr atoms) as the basic motif and these units were assumed to be bonded together via corner sharing. The occurrence of edge sharing was postulated as a result of the computer simulations and the experimental evidence for it was believed to lie in a small peak at 0.36 nm in the radial distribution functions of Coupé et al. (1983), which was not apparent in the Fourier transformations of Etherington et al. (1984a, b). The structural model of Lucas et al. (1984), which included both seven- and eight-fold coordinated Zr atoms, is shown in fig. 2b. Kawamoto and Horisaka (1983) and Kawamoto et al. (1985) studied the structure of barium, strontium and lead metazirconate glasses, based on RDF analysis and MD simulations of the glasses plus β-$BaZrF_6$ and $PbZrF_6$ crystals, suggesting a structural model composed of corner- or edge-shared chains of ZrF_8 dodecahedra similar to the crystalline phases as shown in fig. 2c. In the work of Yasui and Inoue (1985) and Inoue et al. (1985), the structure of $3ZrF_4 \cdot 2BaF_2$ and $3ZrF_4 \cdot BaF_2$ glasses was also studied with a combination of Fourier analysis of XRD and MD simulations and the coordination number of Zr was found to lie in the 7.3–8.0 interval. The corresponding structure, depicted in fig. 2d for the former composition, consisted essentially of ZrF_8 polyhedra sharing corners or edges. A common feature of the structures shown in figs. 2b–2d is the occurrence of a mixture of seven- and eight-fold coordinated Zr atoms and both corner- and edge-sharing, as opposed to the model of fig. 2a, from which all the other models appear to derive graphically. In any case, the existence in FZ systems of network-forming cation coordinations significantly higher than in oxide glasses is now well established. Hamill and Parker (1985a, b) have also performed MD simulations on barium FZ glass compositions and they found a coordination number of 8 for Zr and $\sim$10–11 for Ba, as well as nearly linear Zr–F–Zr bridges for corner-shared sequences.

Among the techniques which have been used in the structural analysis of FZ glasses, IR and Raman spectroscopy and X-ray and neutron diffraction are probably the most direct ones. These are likely to provide a structural picture closest to the atomic level reality, despite the disagreements still existing, quite typical of the problems always found in the structural characterization of the vitreous state. MD computer simulations are a very useful technique in spite of certain drawbacks and the considerable CPU times usually needed.

Very little has been done concerning the structure of Al- and Zr-free glasses. There have been suggestions of more ionic structures in the case of BT glasses (Lucas et al. 1981), disordered fluorite-type structures for CdF_2-containing glasses (Matecki et al. 1982b) and chains of corner-sharing octahedra in TMFG, where the TM ions were found to be six-fold coordinated (Miranday et al. 1979, 1981).

More work is needed concerning the structural study of other glass compositions, e.g. by a more extensive use of X-ray and neutron diffraction techniques and also of inelastic neutron scattering, in conjunction with IR and Raman spectroscopy.

5. Properties

5.1. *Density*

As one might expect, the densities (ρ) of HMFG are large compared to common silicate or fluoroberyllate glasses (for which $\rho \sim 2.5\ g/cm^3$), ranging from $\sim$4.0–6.5 g/cm^3 depending on the composition. Typical values for the most representative glass systems are given in table 2, where it may be noticed that the density is lowest for fluoroaluminate glasses, closely followed by the fluorozirconates; the values increase considerably upon substitution of HfF_4 for ZrF_4 and they are largest for nonmodified FH glasses and for BaF_2–ThF_4-based compositions with YbF_3 or LuF_3. In general, rare earth fluorides increase the density of most HMFG. Another relevant aspect is the fact that most HMFG have packing densities $\gtrsim 0.6$, a value which is considerably larger than that observed in typical silicate glass systems ($\lesssim 0.5$). In fact, there is usually little change in density between crystalline FZ and the corresponding glass compositions. For example, the barium dizirconate glass has a density of 4.64 g/cm^3 (Etherington et al. 1984a, b) which is higher than that of α-$BaZr_2F_{10}$ and β-$BaZr_2F_{10}$ at 4.35 g/cm^3 (Laval et al. 1979). Of course, the packing density for a random close packing of identical spheres (0.637) is larger than 0.6; and HMFG, whose structures might be considered as a random close packing of F^- and Ba^{2+} ions with the Zr^{4+} cations occupying the interstices, could in principle have a packing density even higher than 0.637. But the above points clearly show that the ionicity of FZ glasses is higher than that of their silicate counterparts. Finally, some studies have attempted to correlate density changes with compositional variations within a particular HMFG system (Lecoq and Poulain 1979, 1980a, b), but the observed changes are usually small due to the reduced extent of compositional variations which are possible within the characteristically limited glass-forming ranges available. Grilo et al. (1987) have studied the influence of melting conditions on the density of a series of FZ glasses and it was generally concluded that the density tended to increase with the melting temperature, probably due to higher ZrF_4 losses. It was also observed that the substitution of KF for NaF substantially decreased the glass density, although the

TABLE 2
Density, thermal expansion coefficient and characteristic temperatures of some representative HMFG.

Composition (in mol%)	ρ (g/cm^3)	$\alpha_T \times 10^7$ (K^{-1})	T_g (°C)	T_x (°C)	T_m (°C)	T_L (°C)
$64ZrF_4$–$36BaF_2$	4.65	186	300	352	525	589
$66HfF_4$–$34BaF_2$	6.10	–	308	360	552	607
$70ZrF_4$–$30PbF_2$	5.21	–	255	305	–	–
$43HfF_4$–$57ThF_4$	–	–	500	583	–	–
$57ZrF_4$–$34BaF_2$–$9ThF_4$	4.80	180	320	400	560	628
$62ZrF_4$–$33BaF_2$–$5LaF_3$	4.79	170	307	380	547	610
$60ZrF_4$–$34BaF_2$–$6NdF_3$	4.56	–	320	380	–	–
$63ZrF_4$–$33BaF_2$–$4GdF_3$	–	175	310	390	575	630
$60ZrF_4$–$30ThF_4$–$10LaF_3$	–	–	450	540	–	810
$57ZrF_4$–$36BaF_2$–$3LaF_3$–$4AlF_3$	4.63	187	310	390	562	–
$56ZrF_4$–$14BaF_2$–$6LaF_3$–$4AlF_3$–$20NaF$	–	–	270	–	472	558
$57ZrF_4$–$23BaF_2$–$3LaF_3$–$5AlF_3$–$12NaF$	–	–	280	372	445	–
$30AlF_3$–$10ZrF_4$–$8YF_3$–$4MgF_2$–$20CaF_2$–$13SrF_2$–$11BaF_2$–$4NaF$	3.85	152	392	~490	~660	~720
$26CdF_2$–$10LiF$–$31AlF_3$–$33PbF_2$	5.89	–	245	345	–	–
$40AlF_3$–$22CaF_2$–$22BaF_2$–$16YF_3$	4.00	–	430	–	–	–
$30ZnF_2$–$30CdF_2$–$40BaF_2$	5.53	–	290	330	–	–
$19BaF_2$–$27ThF_4$–$27LuF_3$–$27ZnF_2$	6.45	–	353	452	–	–
$30BaF_2$–$10ThF_4$–$10YbF_3$–$20ZnF_2$–$30InF_3$	5.56	171	324	447	–	650
$43PbF_2$–$17GaF_3$–$17InF_3$–$19ZnF_2$–$4LaF_3$	6.32	–	236	327	460	–
$50PbF_2$–$25MnF_2$–$25CrF_3$	5.90	–	261	310	–	–

density of crystalline KF is only slightly lower than that of NaF. Given the usual differences between the nominal glass compositions and the final ones (sect. 3), small density variations reported in the literature may sometimes not be significant.

5.2. *Glass transition*

The glass transition temperature (T_g) may be defined as the temperature at which the solid–liquid transition occurs in a system involving a noncrystalline solid and it is generally assumed to be an approximately isoviscous point corresponding to a viscosity of $\sim 10^{12}$ Pa s. This temperature range, which is usually determined by differential scanning calorimetry (DSC), differential thermal analysis (DTA) or by dilatometry, is found to vary somewhat with the heating or cooling rate during the measurement and with the thermal history of the system. T_g marks the useful upper temperature limit for the practical utilization of the glass, although certain aging phenomena have been noticed for HMFG at temperatures as low as 100 °C below a T_g of $\sim$300 °C (Moynihan et al. 1984) or perhaps even at room temperature (Lehman et al. 1989). The graphical determination of the T_g value from a DSC plot is illustrated in fig. 4 for a ZBLAN glass prepared and analyzed in our laboratory in Lisbon. Table 2

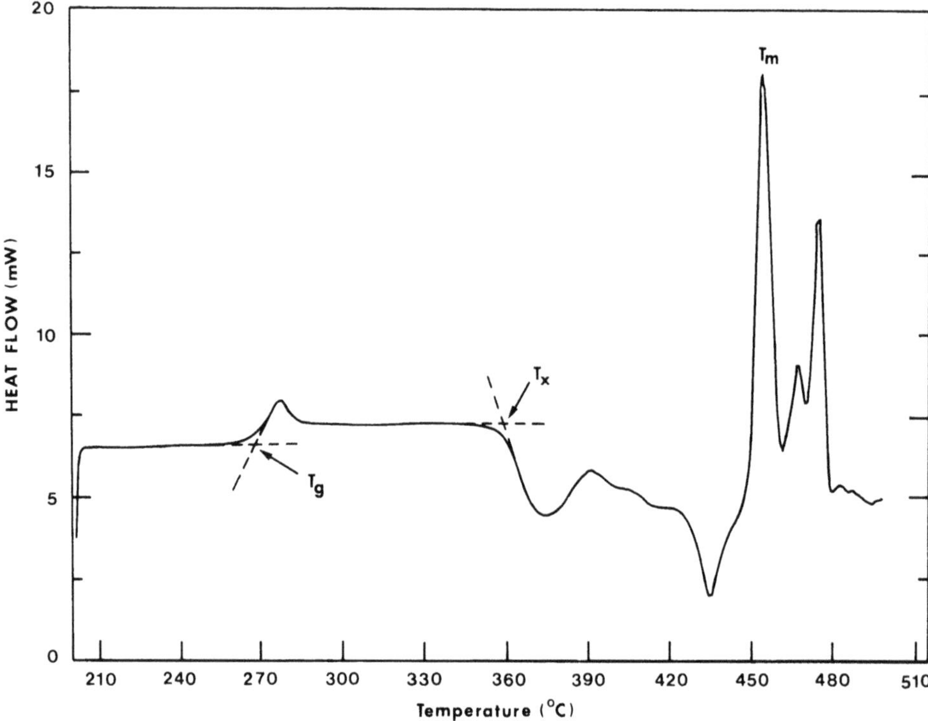

Fig. 4. DSC scan of ZBLAN glass (10 K/min) showing the graphic procedure for determining T_g, T_x and T_m.

lists glass transition temperatures for a series of representative HMFG. It may be observed that T_g has similar values for most alkaline-earth modified FZ or FH glasses, whereas it is lower for alkali or lead fluoride containing glasses and it is substantially higher for nonmodified glasses and for the fluoroaluminate systems. Table 2 also lists some values for the linear thermal expansion coefficient α_T, which again appears to vary little among the different glasses but has in general a very high value (as expected from the low glass transition temperatures), about double that for window glass and similar to that of stainless steel. This fact renders the annealing operation a very critical and somewhat difficult one. The normal tendency, for most glasses, of α_T decreasing with an increase in T_g is not clearly observed in the present case.

5.3. *Viscosity*

Viscosity is one of the properties which more directly controls the glass-forming ability of a given system. The viscosity of most HMFG-forming melts at the liquidus temperature is only $\leqslant 0.1$ Pa s, which accounts for their well-known tendency towards devitrification. The corresponding viscosity of, e.g. SiO_2 is $\sim 10^6$ Pa s and that of BeF_2 is $>10^5$ Pa s. The very low viscosity, on the other hand, has some advantages; e.g. fining poses no problems and the possibility of using rotational casting for preform fabrication exists as mentioned in sect. 3.

The activation energy for viscous flow in HMFG systems varies considerably with temperature (Mackenzie et al. 1987), which means that the temperature dependence of viscosity is highly non-Arrhenian, as illustrated in fig. 5 for a ZBLA glass in comparison with other glass-forming systems. The activation energy is very small near the liquidus temperature, close to 100 kJ/mol for typical FZ systems, compared to ~500 kJ/mol for BeF_2 (Mackenzie et al. 1987), but it becomes very high near T_g, corresponding to a "short" glass with a very limited working range. This behavior makes it very difficult to draw fibers from the melt by the double crucible method. Angell (1985) has characterized this type of behavior as that of a fragile liquid, where the structural relaxation times decrease extremely rapidly with increasing temperature above T_g.

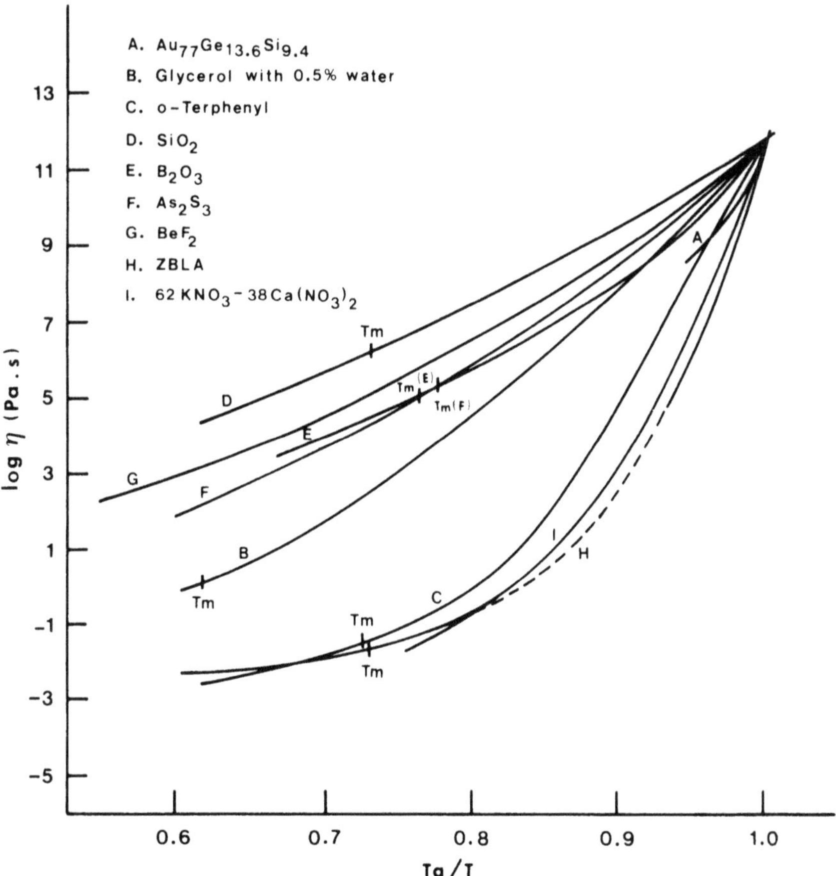

Fig. 5. Plots of log viscosity versus reduced inverse temperature for various glass forming liquids (Mackenzie et al. 1987).

The effect of compositional changes on the viscosity of molten fluorozirconates was studied by Hu and Mackenzie (1983), who concluded that above the liquidus temperature the viscosity of a barium dizirconate base melt was lowered by 10 mol % additions of modifying fluorides such as those of Li, Na, K, Ca, Zn and Pb, whereas it was raised by the intermediate fluorides of Al, Y, La and Th. Although quantitative studies are lacking at present, rare earth fluorides are generally expected to increase the viscosity of modified HMFG melts. Hu and Mackenzie (1983) noticed also that above the liquidus the viscosity (η) versus absolute temperature (T) relationship was well described by the Vogel–Fulcher–Tammam equation:

$$\eta = \mathscr{A} \exp[\mathscr{B}/(T - T_0)], \tag{5}$$

where $\mathscr{A}$, $\mathscr{B}$ and T_0 are constants. Since T_0 was found to be larger than T_g in all cases, whereas it should be lower, this was interpreted as signifying that eq. (5) was not applicable to the supercooled melts at low temperatures near T_g. Tran et al. (1982) have also shown that the addition of 20 mol % LiF of a ZBLA-based glass composition not only significantly decreased the viscosity near T_g, but it also decreased the activation energy by $\sim$138 kJ/mol. This effect was especially pronounced when 5 mol % PbF_2 was also added, as shown in fig. 6, reducing the activation energy by $\sim$485 kJ/mol relative to the ZBLA composition and leading to a much larger working range for the glass, thus improving its viscosity behavior for fiber drawing from the melt or from a preform rod. This activation energy referred to short temperature intervals, where the viscosity obeyed the Arrhenius equation (Tran et al. 1982).

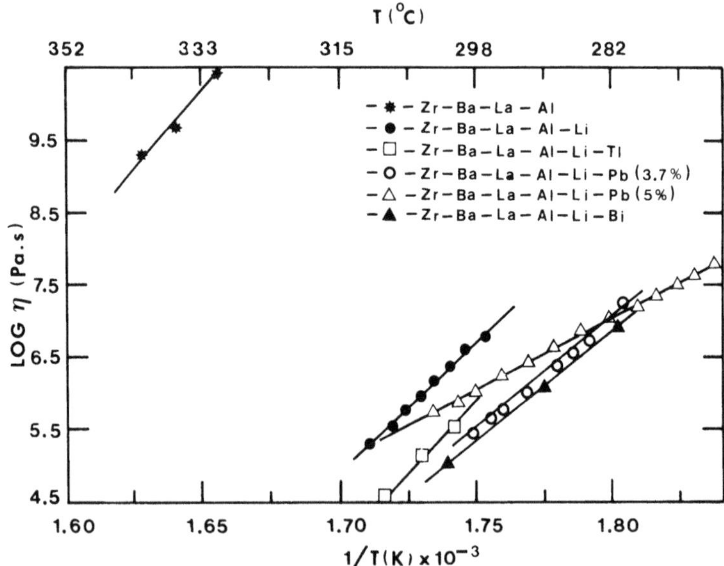

Fig. 6. Arrhenius plots of viscosity for selected fluorozirconate systems over limited temperature ranges (Tran et al. 1982).

5.4. *Crystallization behavior*

The crystallization temperature T_x is commonly defined as the temperature corresponding to the onset of formation of the first crystalline phase which appears within a homogeneous glassy phase and it is usually determined from DTA or DSC plots as shown in fig. 4 for a ZBLAN glass. T_x is also the upper temperature limit for processing the melt or drawing fibers from preforms. The crystallization tendency of HMFG is high compared to most oxide systems and this represents one of the most serious obstacles to the technological development of the former glasses. It also explains, on the other hand, why this phenomenon has received so much attention in recent years. Table 2 lists T_x values for some of the most representative HMFG compositions and the difference $(T_x - T_g)$, which may be taken as a measure of the glass thermal stability, is found to be close to or only slightly larger than 100°C for the best compositions, namely, the fluorozirco–aluminate glass AZYMgCSBN (98°C), BTLuZ (99°C), CLAP (100°C), ZBLAN (106°C) and BTYbZI (123°C), but it is $\ll$ 100°C for all ternary or binary systems (for the acronyms, see table 1). For FZ-type glasses, Grilo et al. (1987) found that the reduced parameter $(T_x - T_g)/T_g$ offered a somewhat better measure of the glass thermal stability. The addition of RE fluorides to modified HMFG, within their short solubility limits, usually increases this parameter and improves glass stability.

The crystallization behavior of a given glass, which may involve the nucleation and growth of several different crystalline phases from the glassy phase remaining at a given time, corresponding to exothermic processes on a DSC plot, may also involve exothermic transformations of metastable crystalline phases in stable ones and endothermic allotropic transformations. The existence of more than one crystalline phase at the solidus temperature may lead to more than one endothermic melting peak on the DSC scan. Table 2 lists values for the temperature corresponding to the peak of the first major melting endotherm (T_m) and the liquidus temperature T_L. The so-called "two-thirds rule", i.e. $T_g/T_L = \frac{2}{3}$ (T_g, T_L in Kelvin) (Sakka and Mackenzie 1971), is largely valid for HMFG. The determination of T_m is illustrated in fig. 4. The crystallization behavior is also a function of the aggregation state of the sample, as illustrated in fig. 7 by the differences observed between powdered and bulk glass DSC scans, due to a predominance of heterogeneous nucleation in the case of the powdered specimen.

Although there are a large number of experimental studies devoted to the crystallization behavior of HMFG, dealing almost exclusively with FZ glasses, most of these are of a qualitative nature. Several of them have attempted to identify the first phases which crystallize out of barium FZ glasses, using DSC, XRD and both optical and electron microscopy (Bansal and Doremus 1983, Weinberg et al. 1983, Bansal et al. 1984, Neilson et al. 1984, 1985, Parker et al. 1985, Miniscalco et al. 1985, Parker et al. 1986). There is now general agreement that, when such compositions are reheated slightly above T_x, the metastable high-temperature forms β-$BaZrF_6$ and β-$BaZr_2F_{10}$ are the first major phases to precipitate out of the glass, whereas at longer heating times or higher temperatures, phases such as α-$BaZrF_6$, α-$BaZr_2F_{10}$ and sodium or sodium–barium FZ compounds may form (Almeida and Mackenzie 1981, Moynihan

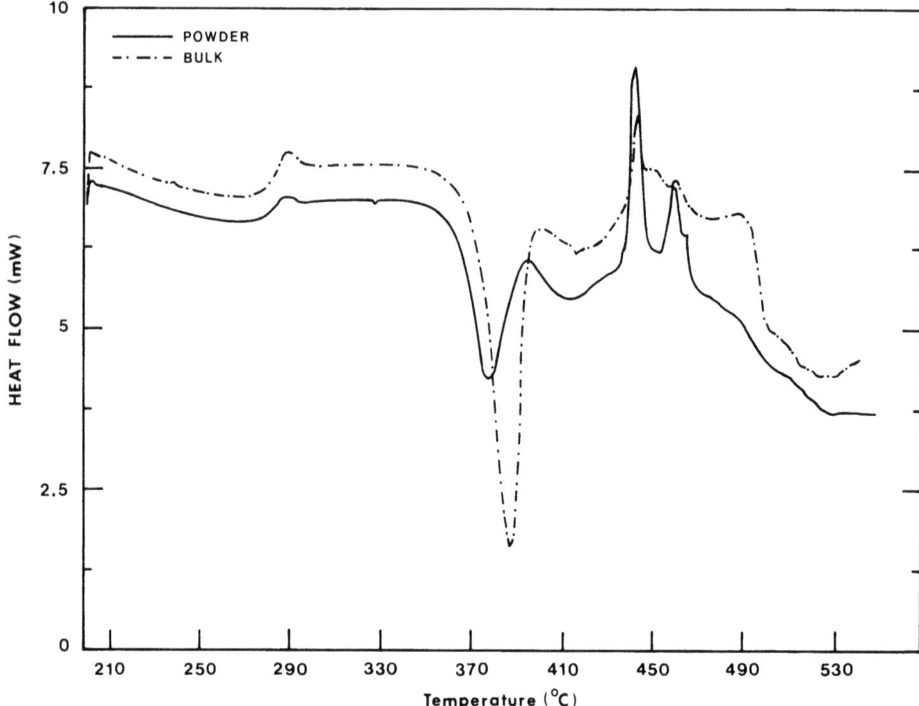

Fig. 7. DSC scans for powdered and bulk ZBLAN glasses at 10 K/min scanning rate.

1987). There is also evidence that the first minor phases to crystallize upon heating or cooling the supercooled liquid may include ZrF_4, AlF_3 and LaF_3 (Moynihan 1987) or even NaF (Sanghera 1989). The metastable crystalline phases may be responsible for some of the excess light scattering which has been observed in these glasses and they may also serve as nuclei for crystallization of the major phases indicated above.

So far the crystallization kinetics of HMFG have been less studied. Both continuous heating and isothermal experiments have been performed using DTA or DSC techniques. Bansal et al. (1983) studied the crystallization kinetics of a ZBL glass by means of isothermal and nonisothermal DSC heating above T_g. Under isothermal conditions, the volume fraction x crystallized after a time t was found to follow the Avrami equation,

$$x = 1 - \exp[-(kt)^n], \tag{6}$$

where the Avrami exponent n depends on the nucleation rate and the geometry of crystal growth and k is a rate constant. The value obtained for n (~ 3) indicated three-dimensional crystal growth from a constant number of nuclei and k was found to follow the Arrhenius equation

$$k = \nu_c \exp[-\mathscr{E}/RT] \tag{7}$$

where ν_c is a frequency factor and $\mathscr{E}$ is the activation energy for crystal growth. Bansal et al. (1983) determined $\mathscr{E} \sim 300$ kJ/mol. This value is similar to the Arrhenius activation energy for viscous flow which one would approximately expect for this composition in the temperature range analyzed (~50–100°C above T_g), since this activation energy was found by Moynihan et al. (1983) to be near 1460 kJ/mol at the glass transition of a similar composition and it was determined by Hu and Mackenzie (1983) at ~125 kJ/mol for a ZBL melt at ~200°C above T_g. In general, the compositions with the lowest values of the activation energy for viscous flow are the ones with the lowest crystallization tendency. *T–T–T* curves of temperature versus time versus % transformation (crystallized fraction) have also been constructed by certain authors, e.g. Bansal et al. (1985), Esnault-Grosdemouge et al. (1985) and Busse et al. (1985), based on isothermal DSC studies of crystallization phenomena. In the latter study, critical cooling rates of 78 K/min and 8 K/min were determined for ZBLALi and ZBLAN compositions, respectively, showing the latter glass to be less prone to devitrification. Critical cooling rates have also been determined in a less accurate fashion via DTA/DSC cooling of melts from above the liquidus and recording the lowest cooling rate for which the crystallization exotherms were suppressed, e.g. by Kanamori and Takahashi (1985) and several other authors more recently. A more accurate method was utilized by Kanamori (1987), where the cooling rate r in a DSC scan was plotted as a function of supercooling $\Delta T = T_m - T_x$ in the form ln r versus $1/(\Delta T)^2$ (approximately straight lines) and the critical cooling rate r_c was obtained by extrapolating the curves to $1/(\Delta T)^2 = 0$. Figure 8 shows a series of isocritical ZBLAN compositions and the most stable ones have r_c as low as ~1 K/min.

The determination of nucleation and crystallization rates and the nature of the nucleation phenomena in HMFG systems are topics which have not been addressed

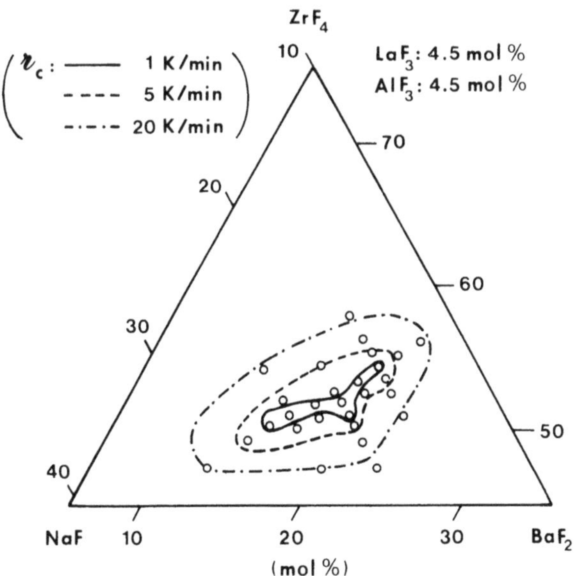

Fig. 8. Composition dependence of critical cooling rates r_c for glass formation in ZBLAN system (Kanamori 1987).

very often. MacFarlane and Moore (1987) studied a series of barium FZ compositions by DSC and they found that nucleation started to occur at temperatures just above T_g, appearing to be of the homogeneous type. Other authors, however, seem to favor a heterogeneous nucleation mechanism instead, due, e.g. to small Pt particles from the melting crucibles or to undissolved oxide crystals (Lu and Bradley 1985). It is not clear at present whether nucleation is homogeneous or heterogeneous in the different HMFG systems, although the excess light scattering observed in FZ glasses appears to be associated preferentially with bulk rather than surface crystallization (Moynihan 1987). If nucleation were in fact homogeneous, the extrinsic light scattering would be extremely difficult to eliminate.

Parker et al. (1987) estimated crystal growth velocities for β-$BaZrF_6$ in a ZBLAN glass ($u \sim 26$ μm/h at ~ 10°C below T_x) and the corresponding activation energy was computed at ~ 370 kJ/mol. Drehman (1987), on the other hand, measured nucleation rates as a function of temperature in another similar ZBLAN composition. He found that the nucleation rate for two different types of crystals formed was maximum only a few Kelvin above T_g and also that it was proportional to time, indicating that it was homogeneous. Finally, Hart et al. (1988) simultaneously measured the nucleation and crystal growth rates in the ZBLAN composition developed by Tokiwa et al. (1985). The nucleation rates exhibited large scatter, from 10^{-3}–1 nuclei $cm^{-3}\,s^{-1}$ [some 4–7 orders of magnitude lower than the values determined by Drehman (1987)] and heterogeneous nucleation appeared to dominate; the crystallized phase was always $BaZr_2F_{10}$ and the growth rate u was found to be linear with time, but the maximum in the curve of u versus T (located above 400°C) could not be determined. The nucleation and crystallization rates are shown as a function of temperature in figs. 9a and 9b, respectively.

Miniscalco et al. (1985) used the photoluminescence of Ce^{3+}, Nd^{3+}, Eu^{3+}, Cr^{3+}, Fe^{3+} and Mn^{2+} in ZBLA-doped glasses to investigate the crystallization process, to which Cr^{3+} was particularly sensitive. Ferrari et al. (1988) have also used the

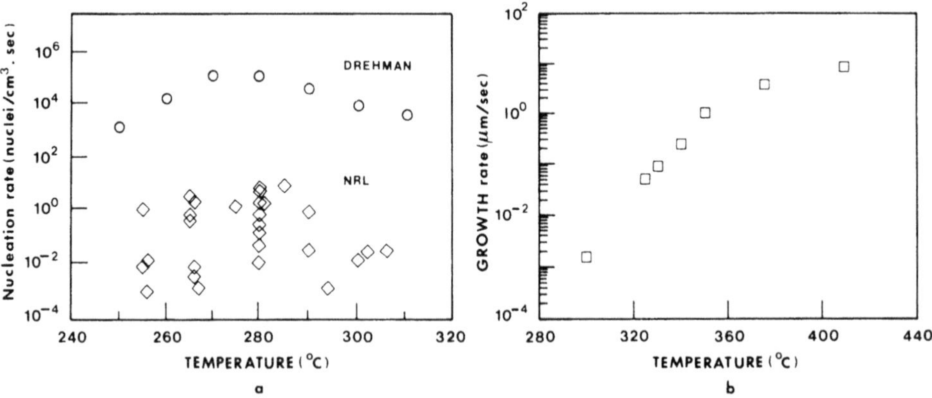

Fig. 9. (a) Nucleation rates for ZBLAN glass: NRL (◇) and Drehman (1987) data (○); (b) crystal growth rate for ZBLAN glass (Hart et al. 1988).

fluorescence spectra of Pr^{3+} in several FZ glasses to follow the early stages of crystallization.

Among the different HMFG systems, the crystallization of FZ-type glasses has overwhelmingly been studied compared to the Zr-free compositions. The isothermal crystallization kinetics study of a $35AlF_3$–$16YF_2$–$16MgF_3$–$33SrF_2$ glass by Kanamori (1983) is an example of the latter type of composition, where the activation energy $\mathscr{E}$ was found to be ~ 570 kJ/mol at 30–50°C above T_g (= 429°C) and the Avrami exponent was $n \sim 1.5$, suggesting one- or two-dimensional crystal growth from a constant number of nuclei.

The crystallization behavior is certainly an area of paramount importance for the practical use of HMFG and much work needs to be done, especially the measurement of nucleation and crystallization rates and the differentiation between homogeneous and heterogeneous nucleation. It is likely that crystallization will always remain a problem during the fabrication of HMFG.

5.5. *Chemical durability*

Despite some early optimism considering the chemical resistance of HMFG, e.g. the claims that ZBT glasses were stable in wet atmospheres up to 350°C (Poulain et al. 1977) and that FZ glasses were very resistant to fluorinating agents such as F_2, HF and UF_6 (Poulain and Lucas 1978), it soon became evident that the chemical durability was actually poor, especially for the FZ systems in water or acidic environments. Nevertheless, HMFG are stable under humid atmospheres provided water vapor condensation on their surface does not occur and the fluorinating agent resistance is probably better than that of fluoroberyllate or oxide glasses, although quantitative tests are lacking at present.

Baldwin et al. (1981) reported that the dissolution rate of a $57ZrF_4$–$34BaF_2$–$9ThF_4$ glass in boiling water was $\sim 4.7 \times 10^{-5}$ g cm^{-2} min^{-1}, about a thousand times faster than a typical silicate glass. Simmons and Simmons (1986) studied the reaction of FZ glasses with unbuffered static water and they found that (1) corrosion proceeded primarily by matrix dissolution in fluoride form, (2) LiF, NaF and AlF_3 dissolved faster than ZrF_4, BaF_2 and especially LaF_3, leading to a La-rich, porous, hydrated corrosion surface layer which did not act as protective, and (3) the leaching solution pH drifted into acidic values with time, due to a F^-/OH^- ion exchange, raising the solubility of ZrF_4 and accelerating the glass dissolution. Ravaine and Perera (1986), on the other hand, reported congruent dissolution for a ZBT glass in buffered liquid water, although the occurrence of F^-/OH^- ion exchange was confirmed. In a later publication, Simmons (1987b) found a strong effect of pH on the chemical durability of HMFG. For a ZBLALi glass, this was maximum in buffered solutions with a pH in the range 6.0–8.0 (in which the FZ glass was only ~ 50 times less resistant than Pyrex), whereas the dissolution became very rapid and congruent for lower pH values. The formation of zirconium and lanthanum hydroxide species on the attacked glass surfaces has been reported by Le Toullec et al. (1988) and Pantano and Brow (1988), with the help of FTIR and XPS, respectively. The addition of RE fluorides to modified FZ glasses tends to improve their corrosion resistance.

The chemical durability of Zr-free HMFG has generally been found to be superior to the FZ systems, which has prompted a relatively large number of studies on the corrosion resistance of the Zr-free glasses, most of them including a comparison between the characteristics of different types of HMFG. Thus Tregoat et al. (1986), Ravaine and Perera (1986), Simmons (1987a, b), Tregoat (1987) and Le Toullec et al. (1988) have all studied the chemical durability of BaF_2–ThF_4-based glasses in aqueous environments. These materials generally showed an approximate 100-fold improvement in water resistance compared to FZ systems without a hydrated layer forming on the surface but with matrix dissolution of the component fluorides also appearing to dominate the leaching process (Simmons 1987a). Figure 10 shows a comparison of leach rates of some HMFG and silicate glasses; it is worth noticing that while the leach rates of FZ systems continuously decrease with time, similar to the silicate glasses, those of the BT glasses appear to show a minimum. Guery et al. (1988b) and Ravaine and Perera (1986) have also studied the chemical durability of UF_4-containing glasses in aqueous solutions. Guery et al. (1988b) concluded that the leaching characteristics of the UF_4-based glasses were similar to the other HMFG families, namely that the leach rate of the former was minimum for pH 7 and it continuously decreased with time, with values intermediate between those of the FZ and BT systems. The water durability of CLAP glasses is also comparable to that of the BT materials and hermetic coatings may not be needed for the former in certain environments (Tick 1988b). At present, hermetic coatings are absolutely mandatory whenever HMFG are to be used in contact with aqueous liquid environments.

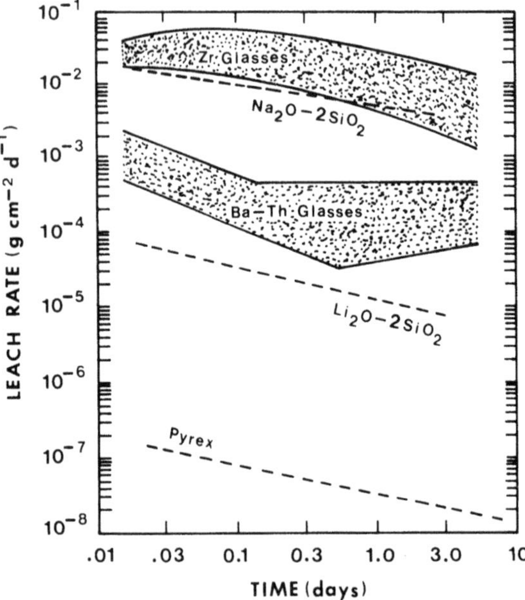

Fig. 10. Comparison of leach rates of BaF_2–ThF_4-based glasses with those measured for FZ glasses, Pyrex and nondurable silicates. (Simmons 1987a).

5.6. *Mechanical properties*

Mechanical properties are perhaps the least documented topic in the already extensive HMFG literature data base. Therefore, only a short account of this subject will be given here.

Table 3 contains values for some of the relevant mechanical properties of FZ and other Zr-free fluoride glasses. It may be observed that a typical HMFG has mechanical properties which are slightly better than a representative chalcogenide glass, but which are worse than those of a common silicate glass and much worse than those of fused silica. The Young's modulus E has values in the range between 55–65 GPa, similar to modified oxide glasses and much higher than for chalcogenides. However, the bulk glass fracture strengths (~20 MPa) are much lower than for fused silica (~70 MPa) or most silicates, the Vicker's hardness values are typically low (~250–350 kg/mm^2) compared to SiO_2 (~635 kg/mm^2) and the Poisson's ratios of HMFG are relatively high. The thermal expansion coefficients are also quite high (sect. 5.2), leading to low thermal shock resistances. The approximate formula (Kingery et al. 1976)

$$\Delta T_c \sim (1 - \nu_P) \frac{\sigma_F}{E\alpha_T}, \tag{8}$$

where ΔT_c is the instantaneous temperature rise at which failure occurs, ν_P is the Poisson's ratio and σ_F is the fracture strength of the glass, may be used in order to compute the thermal shock resistance of HMFG. The value for a HBLA glass, e.g. was 60°C compared to 80°C for a modified silicate (Tran et al. 1984). The fracture toughness K_{IC} (which is the critical stress intensity factor) of bulk HMFG specimens was found to range between 0.25–0.40 MPa m$^{1/2}$, compared with ~0.75 MPa m$^{1/2}$

TABLE 3

Selected mechanical properties of HMFG systems and other nonhalide glasses.

Composition (in mol%)	E (GPa)	K	ν_p	Vicker's hardness (kg/mm^2)	K_{IC} (MPa m$^{1/2}$)	n
$57ZrF_4$–$34BaF_2$–$9ThF_4$	55.1	43.0	0.287	–	–	–
$61ZrF_4$–$34BaF_2$–$5LaF_3$	44.0	–	0.305	~228	0.25	–
$60ZrF_4$–$33ThF_4$–$7LaF_3$	82.4	54.0	0.247	–	–	–
$57ZrF_4$–$36BaF_2$–$3LaF_3$–$4AlF_3$	56.0	48.0	0.30	267	0.31	~11
$57HfF_4$–$36BaF_2$–$3LaF_3$–$4AlF_3$	55.0	48.0	0.30	271	0.31	~11
$56ZrF_4$–$24BaF_2$–$6LaF_3$–$4AlF_3$–$10NaF$	58.3	35.3	0.23	221	0.34	–
$30AlF_3$–$10ZrF_4$–$8YF_3$–$4MgF_2$–$20CaF_2$–$13SrF_2$–$11BaF_2$–$4NaF$	63.8	55.9	0.31	–	–	–
$40AlF_3$–$22CaF_2$–$22BaF_2$–$16YF_3$	63.8	53.2	0.30	360	0.38	–
$16BaF_2$–$28ThF_4$–$28YbF_3$–$28ZnF_2$	–	–	–	276	0.32	–
Borosilicate (Pyrex)	61.0	36.0	0.22	–	–	~31
Fusel Silica (SiO_2)	74.0	36.0	0.16	635	0.73	~36
$Ge_{10}As_{40}Se_{50}$	15.9	~14	~0.29	173	0.23	–
$Ge_{30}As_{10}Se_{60}$	18.6	–	–	236	0.24	–

for fused silica. This parameter should be a measure of the intrinsic strength, but it has been shown to depend somewhat on the test speed and environment (Freiman 1980).

The static fatigue of HMFG systems has been studied to some extent. Pantano (1987) gives an account of these phenomena for FZ glasses. His surface studies, related to the subcritical slow crack growth phenomena, showed that aqueous environments were especially aggressive. Figure 11 displays a series of crack velocity versus stress intensity diagrams for a ZBLA glass in several different environments, determined by the constant moment double cantilever beam technique (Pantano 1987). In liquid water, a threshold and a velocity plateau were observed, whereas both features were absent for nonaqueous environments. It was proposed that the stress intensity thresholds and velocity plateaus could be due to stress-enhanced diffusion of molecular water ahead of the crack tip. The crack velocity v is usually related to the stress intensity factor K_I ($K_I = Y\sigma_A\sqrt{a}$, where Y is a crack geometry related factor, σ_A is the applied stress and a is the half-length of the crack), in region I (where crack propagation is controlled by the rate of a stress-induced chemical reaction at the crack tip), by the empirical power law

$$v = AK_I^n, \tag{9}$$

where A is a constant and n is the stress corrosion susceptibility parameter, whose value is related to the slope of the $\log v$ versus K_I curve. A large value of n such as in liquid methanol or in formamide ($n \sim 77$) is associated with a lower susceptibility to slow crack growth (or delayed failure) compared with liquid water or humid gas,

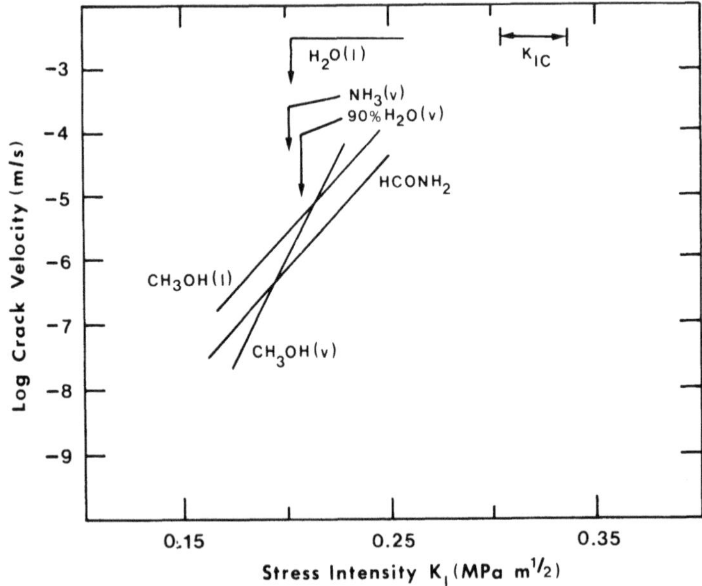

Fig. 11. Crack velocity versus stress intensity factor for a ZBLA glass in different environments (Pantano 1987).

although the absence of a threshold in the former case indicates that the nonaqueous environments are also active in promoting slow crack growth. In liquid water, despite the existence of a threshold, the cracks run at a high velocity above it, indicating a complex crack growth behavior compared to common silicate glasses ($n \sim 15$–20) or bulk vitreous SiO_2 ($n \sim 36$). Mecholsky et al. (1985a) noticed that the actual value of n appears to depend on the testing technique, being larger for dynamic fatigue or delayed failure tests than for crack growth measurements. Mecholsky et al. (1985b) have also argued that the n value might not be a unique way of identifying stress corrosion susceptibility in HMFG as it is in silicate glasses, in agreement with Freiman (1980). These facts must be kept in mind when using n for obtaining lifetime estimates from proof testing. The large stress corrosion susceptibility in water is in good agreement with the chemical durability characteristics discussed in sect. 5.5 and this may present a serious problem in the case of fibers. For example, Nakata et al. (1983) found a very significant reduction of Teflon-jacketed FZ glass fiber strength with time in humid atmospheres. Pantano (1987) observed that the crack velocity was insensitive to the pH value for ZBLA glass, concluding that the stress corrosion mechanism in FZ glasses might be different from the chemical corrosion mechanisms themselves. A crack healing phenomenon recently reported for an InF_3-containing ZBLAN glass (Lehman et al. 1989) could also be a consequence of the above type of behavior.

It should be emphasized that the existing volume of mechanical property data is small and, due to the variations in preparation methods (i.e. purity, use of RAP or not, etc.), storing conditions of these moisture-sensitive glasses and also in the final chemical compositions versus the nominal ones, the mechanical characterization of HMFG should be considered as only of approximate nature. The wide scatter observed for most data is also in part a result of this fact.

With respect to compositional effects on the values of the mechanical properties of HMFG, Mecholsky et al. (1985a) noted that K_{IC} appeared to increase with the AlF_3 content in FZ glasses, whereas K_{IC}, the Vicker's hardness and the Young's modulus tended to increase with the BaF_2 concentration in the BaF_2–ThF_4–YbF_3–ZnF_2 system. However, these findings offer little help, since the AlF_3 amount can hardly be increased in ZBLA-based glasses without crystallization problems. The same is approximately true for BaF_2 in the latter system. The effects of rare earth compounds on the mechanical properties of HMFG are not well documented yet.

Due to the very important application of HMFG as optical waveguide materials, the mechanical properties of fibers have also been tested. This topic was recently reviewed by Schneider (1988), who pointed out that the two major problems which need to be solved in order to improve the strength of ZBLA-based fibers are (1) the suppression of surface (heterogeneous) crystallization to form ZrO_2, which may be achieved by introducing NF_3 in the furnace atmosphere during the drawing process, and (2) preventing bulk (homogeneous) crystallization when casting the preforms and also at the drawing temperature during the time needed for fiber fabrication. It was found that drawing in the NF_3-doped atmosphere suppressed surface hydrolysis and led to defect-free fibers; for ZBLA uncladded fibers, the mean (two-point) bending strength was doubled from ~700–1400 MPa and for uncladded ZBLAN fibers it was raised from ~550–850 MPa, with smaller additional improvements in strength being obtained by a final fiber etching step in a $ZrOCl_2$-based solution. Schneider (1988) showed also that etched ZBLAN fibers exhibited strength values close to the intrinsically expected ones as calculated with the help of a crystal growth model which assumed spherical growth of a single phase at a constant linear growth rate.

From the previous discussion, it becomes obvious that coatings are mandatory for the practical use of HMFG: in bulk form, to prevent chemical and optical deterioration, and in fiber form, to preserve the inherent glass strength. Thus Al-Jumaily et al. (1985) tested an ion-assisted deposition technique to deposit MgF_2 submicron coatings on a HBLA glass, which provided antireflectivity and abrasion resistance without deteriorating the UV cut-off (MgF_2 itself is approximately as water durable as BTYbZ glass). More recently, Schultz et al. (1987) have reported the application of a series of coatings ~27–100 nm thick to ZBLAN bulk glasses, by RF sputtering. It was found that thermal expansion match was an important factor and MgO, MgF_2 and ThF_4 appeared to give good results. A low-temperature on-line technique was also being investigated for coating HMFG fibers during drawing. The great tendency of these glasses to crystallize during fiber drawing led to the development of fiber coatings, a problem that has been addressed in different ways. Thus, while the early solution was the jacketing of HMFG preforms with Teflon-FEP, Nakata et al. (1983) showed that the strength of Teflon-FEP-coated HMFG fibers degraded rapidly in wet atmospheres due to water permeation through the polymer, which leads to surface

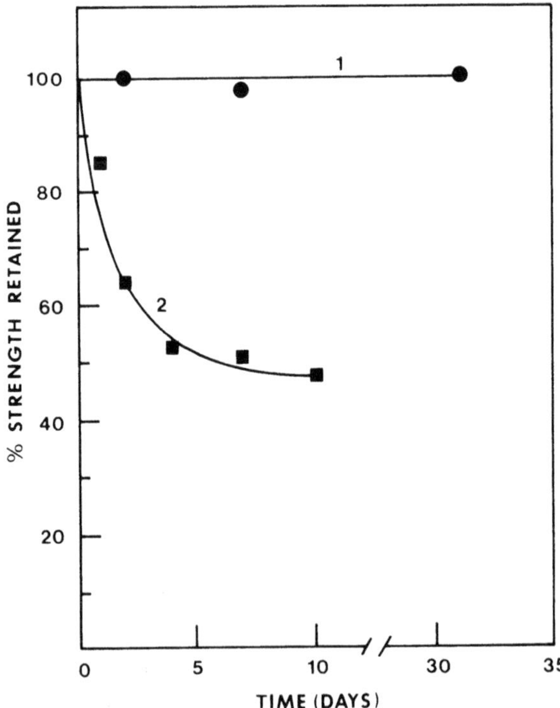

Fig. 12. Comparison of strengths of FZ glass fibers with different jackets: (1) chalcogenide glass jacket, and (2) Teflon-FEP jacket. (Nakata et al. 1985).

attack on the fibers. These authors later proposed the use of a high thermal expansion, water-resistant chalcogenide glass jacket for the protection of FZ glass fibers instead of Teflon-FEP (Nakata et al. 1985). Figure 12 shows that the strength of chalcogenide-jacketed ZBLALiPb fibers did not change after 31 days in an atmosphere of 100% relative humidity, whereas that of Teflon-FEP-jacketed fibers was reduced to less than half after 10 days only. The full strength of the former fibers was actually maintained even after 1 year in the same atmosphere.

As an alternative to the above solutions, oxide glass overcladding with a phosphate jacket has been proposed for the so-called FLOX fibers (Vacha et al. 1987). Although diamond-like carbon (Stein et al. 1983, Tran et al. 1984) and other types of ceramic or metallic hermetic coatings have also been investigated, UV-curable acrylate coatings appear to be currently preferred for HMFG fibers such as those produced by Infrared Fiber Systems Inc., U.S.A. (IFS).

5.7. *Optical properties*

The relatively unusual optical properties of HMFG constitute the main justification for the technological interest which they have aroused, despite less favorable chemical and mechanical properties. Namely, their unusual near-UV to middle IR transparency and the fact that these glasses constitute a suitable atomic matrix for optically active element doping have attracted a good deal of interest. On the other hand, the atomic structure of these glasses is also somewhat unusual, most notably due to the large cation coordination numbers and the possibility of a range of different nearest-neighbor geometries and this has justified the large scientific curiosity on such vitreous materials. Here also the optical and spectroscopic properties are of paramount importance. Therefore, the present section will cover in detail the most relevant optical parameters of these glasses.

5.7.1. *Refractive index and dispersion*

The refractive index and its variation with the wavelength (dispersion) are very important from the viewpoint of applications, both in bulk optical components such as lenses, prisms, light pipes, etc., and in fiber optics, as well as from the viewpoint of structure.

The refractive index of most HMFG for the sodium D-line n_D ($\lambda = 589.3$ nm) falls in the range of ~ 1.48–1.54, with those glasses containing large concentrations of AlF_3 or NaF having the lowest values. On the other hand, the transition metal fluoride glasses, due in part to the presence of large amounts of PbF_2, have the largest refractive indices: as high as 1.663 for the $50PbF_2$–$25MnF_2$–$25CrF_3$ composition (Miranday et al. 1981). Although these particular glasses have also high dispersion, most HMFG exhibit low dispersion, corresponding to high Abbé numbers in the range of ~ 70–90, defined as $\nu_D = (n_D - 1)/(n_F - n_C)$, where F and C refer to the blue and red wavelengths of 486.1 nm and 656.3 nm, respectively. Therefore, on a plot of refractive index n_D versus reciprocal dispersion ν_D, the HMFG fall near the fluorophosphate glasses (Baldwin et al. 1981), although the Abbé number of the former is still higher (lower dispersion).

It is important that a glass has a low refractive index when it is to be crossed by high-power laser beams since it will also have a low nonlinear refractive index n_2, reducing the possibility of damage due to self-focusing. In this respect, fluoroberyllate glasses are the best materials and vitreous BeF_2 itself has the lowest n_2 value at 0.23×10^{-13} esu (Baldwin et al. 1981). The nonlinear index has been estimated for ZBT and ZBNd glasses at ~ 0.8–0.9×10^{-13} esu by Lucas et al. (1978), slightly lower than the value for fused silica (0.95×10^{-13} esu). The relatively low refractive index values of most HMFG can be understood in terms of the Lorentz–Lorenz equation (Kingery et al. 1976). In fact, the average molar refractivity (weighted by the ionic fractions) will be relatively low for a given glass despite the significant cationic refractions of the high atomic number heavy metals, because of the dominant contribution from the weakly polarizable, highly electronegative fluoride anions.

Mitachi and Miyashita (1983) measured the refractive indices of a series of pure and doped ZBGdA glasses between 404.7 nm and 5.304 μm and they found that 4 mol% of YF_3, CsF, CdF_2, LiF, SnF_2 and PbF_2 dopants raised the refractive index of the ZBGdA-based glass, with the magnitude of the effect increasing in the same order. They also noted that the refractive index could be expressed as the following function of the wavelength λ,

$$n(\lambda) = \frac{a}{\lambda^4} + \frac{b}{\lambda^2} + c + d\lambda^2 + e\lambda^4, \tag{10}$$

similar to other dispersion equations such as the Cauchy or Sellmeier expressions. Figure 13 shows the dispersion curves for two ZBGdA glasses, where increasing amounts of AlF_3 are found to decrease the refractive index. At short wavelengths, the

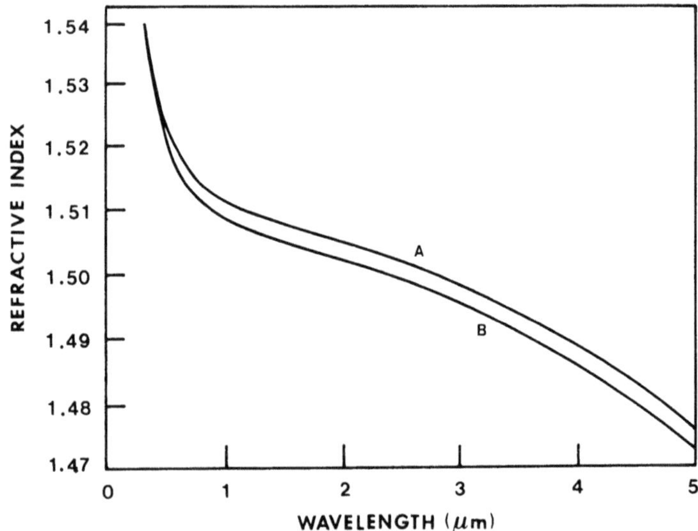

Fig. 13. Refractive index dispersion for AlF_3-doped ZrF_4–BaF_2–GdF_3 glasses: (A) 2 mol% AlF_3 and (B) 6 mol% AlF_3. (Mitachi and Miyashita 1983).

anomalous dispersion associated with the electronic absorption may also be observed. These dispersion curves are generally flatter than that of v-SiO_2. With respect to the effect of LiF on the refractive index, the results of Mitachi and Miyashita (1983) are in apparent disagreement with those of Takahashi et al. (1981), who found that LiF, in concentrations up to ~24 mol%, decreases the refractive index of ZBGd glasses. It should be noted that the former data contained only one doped glass composition with 4 mol% LiF. Lecoq and Poulain (1979) studied also the influence of 0–30 mol% concentrations of LiF and NaF on the refractive index of ZBL glasses, concluding that both compounds decrease the index of the base composition, with NaF having the strongest effect. They also observed that the LiF-containing compositions exhibited a discontinuity in the slope of the curve of n_D versus mol% LiF at ~ 18 mol%, which was attributed to a change in the Li coordination from four- to six-fold, corresponding to a simultaneous change in character of the Li^+ cations from network former to network modifier. In a recent work of Zhao and Sakka (1988), although LiF, NaF and KF (in concentrations up to 55, 35 and 30 mol%, respectively) were all found to decrease the index of a ZBA-based composition, with KF having the strongest effect, no such discontinuities were found in any case in spite of the curvature of the n_D versus concentration plot being more pronounced for the LiF-containing glasses. Those authors suggested also that the three different alkali cations were all six-fold coordinated over the whole glass composition ranges studied and they further suggested that the discontinuity observed by Lecoq and Poulain (1979) was probably a result of the fact that the concentration of BaF_2 was also varied in that work, in addition to the alkali concentration.

In general, there is now strong evidence that the refractive index of FZ glasses increases with PbF_2 or BiF_3 additions, decreasing with LiF, NaF, KF, AlF_3 or HfF_4. The indices n_1 and n_2 of the core and cladding, respectively ($n_1 > n_2$), together with the numerical aperture NA $= \sqrt{(n_1^2 - n_2^2)}$, are often controlled in ZBLAN glass cladded optical fibers by the NaF concentration (with the total concentration of NaF + BaF_2 remaining constant), or by replacing some of the ZrF_4 with HfF_4 (Drexhage 1987). Mansfield (1983) found that the use of CCl_4 RAP significantly increased the refractive index of a series of FZ and FH compositions, probably due to Cl incorporation into the glass. Finally, Matecki et al. (1982a) tabulated a series of refractive index values (for the sodium D-line) for nonmodified ZrF_4–ThF_4–MF_3 glasses (M = Al, Y, Sc, Lu), which fell in the range of 1.53–1.56 and corresponded to values intermediate between the fluoroaluminate and the transition metal fluoride glasses.

The refractive index of Zr-free HMFG has thus far been little investigated. Videau et al. (1979) reported refractive indices for AlF_3-based glasses containing MgF_2, CaF_2, SrF_2 or BaF_2 plus YF_3 or NaF, which varied between 1.40 and 1.45 and therefore represented the low-index end of the HMFG families. Drexhage et al. (1982) measured the refractive index for BT fluoride glasses containing Yb or Lu, which exhibited values near 1.53, similar to the nonmodified FZ glasses. The dispersion curve of a glass with 19BaF_2–27ThF_4–27YbF_3–27ZnF_2 was also given, showing slopes similar to those found for FZ and FH glasses. Finally, the transition metal fluoride glasses exhibit the largest refractive index values, from n_D = 1.547 for a glass with $36PbF_2$–$24MnF_2$–$35GaF_3$–$5YF_3$–$2AlF_3$ (Jacoboni 1987) to 1.663 for a $50PbF_2$–$25MnF_2$–$25CrF_3$ glass (Miranday et al. 1981). The lead fluoride concentration is again here a major factor.

For fiber optics applications, the information carrying capacity (bandwidth) is often limited by the material dispersion, which is proportional to the parameter $M(\lambda) = -(\lambda/c)(d^2n/d\lambda^2)$. Ideally, therefore, one would like to use the optical fibers at the wavelength λ_0, where $M(\lambda) = 0$, provided this coincides with the wavelength of minimum attenuation λ_{min}. This, however, is usually not the case and most HMFG have λ_0 in the neighborhood of 1.6 μm ($\lambda_0 \sim 1.3$ μm for vitreous silica), whereas $\lambda_{min} \geqslant 2.5$ μm (Drexhage 1985). Nevertheless, the dispersion curves measured for HMFG are flatter than the curve of v-SiO_2 (Miyashita and Manabe 1982), so that operation near λ_{min}, coupled with dispersion shifting designs, should not significantly sacrifice the intrinsic bandwidth. Even single-mode fibers coupled to laser emitters will exhibit a certain degree of material dispersion, given the finite line width $\Delta\lambda$ of the semiconductor lasers utilized.

The temperature dependence of the refractive index is also important in high-power laser applications for predicting possible damage which may occur in the glassy materials. Mitachi and Miyashita (1983) measured dn/dT in a ZBGA glass and they found $dn/dT = -1.1 \times 10^{-5}$°$C^{-1}$, a value of the same magnitude as that of v-SiO_2 but of opposite sign, probably due in part to the smaller polarizability of the F^- ions compared to O^{2-} ions.

5.7.2. *Transparency*

As previously pointed out, the HMFG owe much of their current interest to the unusually wide optical transparency from the near UV to the middle IR. Figure 14

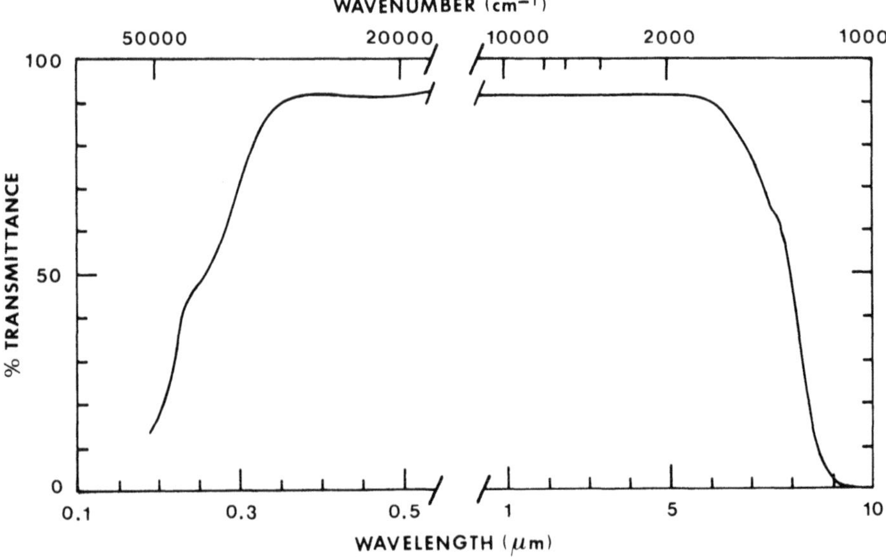

Fig. 14. UV, visible and IR absorption spectrum of a barium dizirconate glass, 1.5 mm thick.

illustrates this point in the case of a bulk binary barium dizirconate glass, where small shoulders on the high- and low-frequency sides are tentatively attributed to iron and oxide impurities, respectively. The UV cut-off wavelength was about 180 nm (for 10% transmittance), whereas the IR cut-off occurred at about 8.6 μm. This transparency range is typical of most FZ glasses. On the other hand, glass compositions such as BTYbZ, PGIZL or the TMFG transmit typically ~0.5–2.0 μm farther into the IR, whereas most glasses in the latter family are virtually opaque in the UV–visible region and the Yb-containing glasses exhibit a relatively narrow but strong absorption band near 0.98 μm, which has been attributed to the $^2F_{7/2} \rightarrow {}^2F_{5/2}$ transition of the Yb^{3+} ion (Drexhage et al. 1982).

In the case of glass optical fibers, the ultimate intrinsic transparency is determined by a superposition of the electronic absorption (or Urbach) edge at shorter wavelengths (in the UV–visible for most systems of interest), the multiphonon absorption edge at longer wavelengths (IR) and the Rayleigh scattering losses in the intermediate frequency regime. For the HMFG of current interest, which are fully transparent in the visible and whose vacuum wavelength of minimum attenuation λ_{min} falls in the middle IR, the contribution of the exponential Urbach edge to the total attenuation is negligible at wavelengths near λ_{min}. Therefore, the total attenuation coefficient α_t is the sum of a $1/\lambda^4$ Rayleigh scattering (RS) term and the exponential multiphonon (MP) term (Shibata et al. 1981):

$$\alpha_t = \alpha_{RS} + \alpha_{MP} = \frac{A}{\lambda^4} + \frac{B}{\exp(C/\lambda)}, \tag{11}$$

where α_t is usually given in units of dB/km (1 dB/km = 2.3×10^{-6} cm^{-1}) and A, B, and C are material parameters approximately wavelength-independent. The Rayleigh light scattering results from refractive index fluctuations in the material due to density or composition fluctuations on a scale $\ll \lambda$ (Schroeder 1977), whereas the multiphonon absorption results from overtones and other higher-order features of the fundamental (or first-order) vibrational modes of the glasses, which have peaks at $\lambda > 15$ μm for most known HMFG. The minimum values of λ and α_t are easily found by differentiation of eq. (11) with respect to λ. Figure 15 shows the results of a recent calculation of intrinsic attenuation for three different HMFG, where a PGIZL glass was estimated to have $\lambda_{min} = 4$ μm and $\alpha_{min} = 5 \times 10^{-4}$ dB/km (Nishii et al. 1989), while the estimates for the ZBGd and ACBY glasses were $\alpha_{min} \sim 10^{-3}$ dB/km at $\lambda \sim 3.2$ μm and $\alpha_{min} \sim 10^{-2}$ dB/km at $\lambda \sim 2.6$ μm, respectively (Shibata et al. 1981). The attenuation curves exhibit a typical V shape. Longer λ_{min} values usually lead to lower intrinsic attenuations and hence the ongoing search for glass compositions with heavier cations, including cations heavier than Zr in the Zr-free glasses.

Less optimistic predictions have been advanced by France et al. (1987), who estimated $\alpha_{min} \sim 3.5 \times 10^{-2}$ dB/km for ZrF_4-based glass fibers at $\lambda \sim 2.55$ μm, by taking into account the predicted effects of extrinsic absorption such as that due to transition metal ions, certain lanthanide ions and OH ions. The intrinsic value of attenuation has already been achieved in practice for v-SiO_2 at ~0.16 dB/km for $\lambda = 1.55$ μm (Miya et al. 1979). Figure 16a shows an actual measurement for a fluoride glass fiber prepared at the Naval Research Laboratory (NRL), Washington, DC, USA

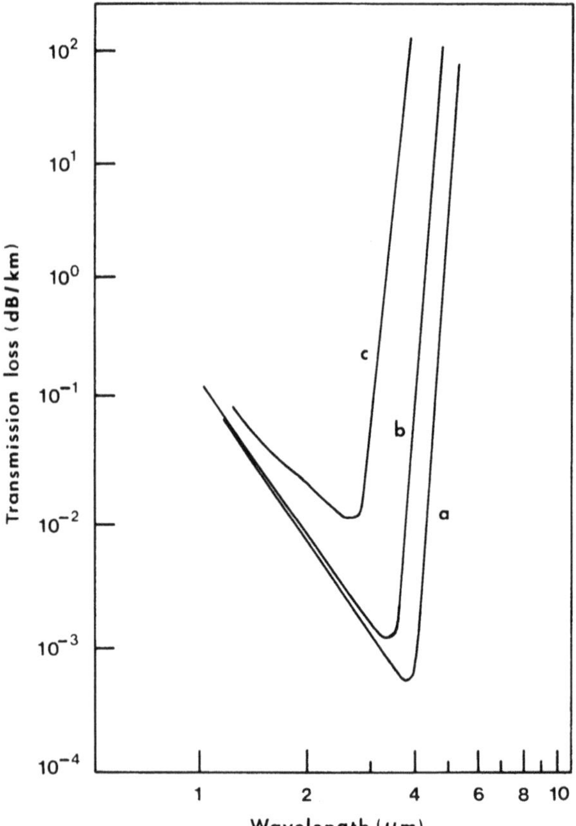

Fig. 15. Calculated loss spectra for (a) ZIGPL glass, (b) ZBGd glass and (c) ACBY glass. (Nishii et al. 1989).

(Tran et al. 1986) and fig. 16b shows two other attenuation curves for glass-cladded HMFG fibers commercially supplied by Infrared Fiber Systems Inc. (Silver Spring, MD, USA). For the NRL fiber, which is representative of the current state-of-the-art [together with those prepared at Nippon Telegraph and Telephone Telecommunications Laboratory (Ibaraki, Japan) and British Telecom Research Laboratories (Ipswich, UK)], α_{min} was 0.9 dB/km at 2.55 μm, corresponding to an absorption loss of 0.7 dB/km and a scattering loss of 0.2 dB/km. Here, most of the absorption was believed to be due to impurities such as Ni, whereas most of the scattering was λ-independent (non-Rayleigh). This latter observation and also the fact that absorption at λ_{min} was higher than scattering show that attenuation in HMFG is still dominated by extrinsic contributions. Moreover, although the OH peak corresponded to only $\sim$30 dB/km on top of the other contributions at $\sim$2.9 μm, reaching a presumable intrinsic α_{min} at wavelengths near the low-frequency edge of the OH fundamental stretching peak (see, e.g. Shibata et al. 1981) appears impossible at present. Nevertheless, France et al. (1987) predicted that the OH-related loss in FZ glasses should be less than 10^{-3} dB/km near the low-loss window of 2.55 μm and

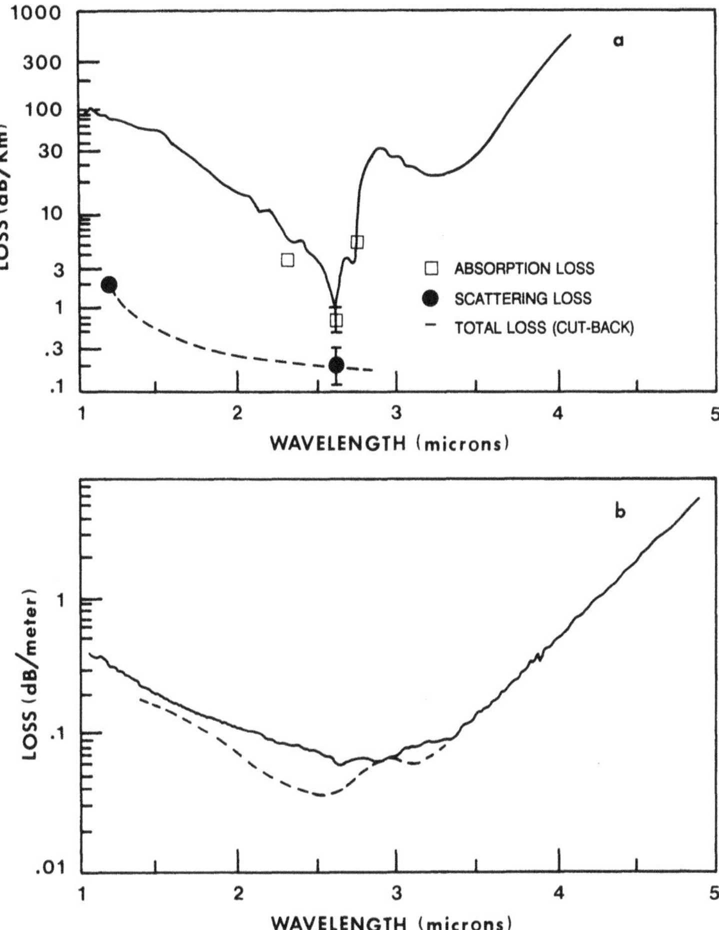

Fig. 16. (a) Optical loss measurement for a FZ glass fiber prepared at NRL (Tran et al. 1986); (b) cut-back loss for HMFG fiber supplied by Infrared Fiber Systems Inc. (USA) – solid line corresponds to reagent grade raw materials and broken line corresponds to purer materials.

preliminary results of France (1985) and Loehr et al. (1988) have shown that deuteration is a possible route to shifting the OH peak to longer wavelengths.

The multiphonon edge, although somewhat difficult to measure accurately, is not always strictly exponential, probably due to intrinsic compositional characteristics. In addition, non-Rayleigh scattering has often been measured, e.g. exhibiting $1/\lambda^2$ or $1/\lambda^0$ dependence, perhaps due to microcrystals, bubbles, etc., and residual transition metal ion, lanthanide ion and OH group absorptions will always be present. In conclusion, the estimates for intrinsic minimum loss in HMFG systems appear optimistic but, ultimately, the actual values at ultralow loss levels will be determined mostly by extrinsic factors.

5.7.3. *UV–visible absorption and luminescence*

There is already a considerable body of literature concerning the absorption and luminescence of HMFG in the ultraviolet and visible regions of the electromagnetic spectrum. The reviews by Lucas (1985), Drexhage (1985, 1987) and Reisfeld (1987) are recommended for more details or for certain topics which will not be covered here, such as radiation effects (Drexhage 1987). The review by Reisfeld and Jørgensen (1987) deals extensively with absorption, luminescence and energy transfer in oxide glasses and it refers also briefly to fluoride glasses.

Figure 14 shows an example of the UV cut-off in HMFG, which may be taken as the wavelength λ_C corresponding to 10% transmittance for a glass thickness of 2 mm, the transmittance being approximately defined as

$$T = \frac{(1-\mathscr{R})^2 \exp(-\alpha x)}{1-\mathscr{R}^2 \exp(-2\alpha x)}, \tag{12}$$

where T is the percentage transmittance for a glass slab of thickness x (in cm), α is the absorption coefficient in cm^{-1} and $\mathscr{R}$ is the percentage reflectivity for one glass/air interface. Alternatively, the UV cut-off may be defined as $\lambda_C(\mu m) = 1.24/E_G(eV)$, where E_G is the average electronic energy gap of the glassy material. This parameter can be determined based on optical transmission or reflectivity measurements, but not from electrical conductivity data, since most HMFG are almost exclusively anionic conductors as discussed in sect. 5.8. For typical HMFG with and without Zr, $\lambda_c \sim 0.20$–0.25 μm corresponding to $E_G \sim 5$–6 eV. Although this transparency is not as good as that of fused silica, for which $\lambda_C \sim 0.16$ μm ($E_G \sim 8.0$ eV) (Sigel 1977), similar to v-BeF_2 ($\lambda_C \sim 0.15$ μm) (Dumbaugh and Morgan 1980), it is still quite good and this has been attributed to the high electronegativity of the fluoride anions which reduces the electronic transitions responsible for the UV absorption (Poulain and Lucas 1978). The effects of composition on the UV absorption edge of HMFG have not yet been studied in detail, but in general it appears that λ_c is red-shifted when the extent of glass modification increases, e.g. by BaF_2 additions (Matecki et al. 1982a, Drexhage et al. 1983) and Zr-free glasses containing electronically dense heavy elements such as Th and Yb appear to have lower UV transparency than the other HMFG (Drexhage 1987). Most impurities present in the glasses will also adversely affect their UV-visible transmission, including chlorine when a CCl_4 RAP is used (Brown et al. 1982), which also leads to radiation-induced defects (Friebele and Tran 1985).

The electronic transitions which are responsible for the Urbach edge have been investigated by Izumitani and Hirota (1985) and Izumitani et al. (1986) by UV reflection spectroscopy with synchrotron radiation in ZrF_4–BaF_2 and AlF_3–CaF_2–P_2O_5 glasses. In the FZ glasses, in addition to weaker features at 7.3, 9.3, 13.2 and 19.2 eV, two strong reflection bands were recorded at 11.6 and 23.2 eV, which were also present in the fluoroaluminate glass. The longer wavelength band, which was dominant, was attributed to the F_{nb} anions, while the shorter wavelength feature at 23.2 eV was tentatively suggested as due to the F_b species.

The visible absorption and luminescence spectra of HMFG have been studied by a number of researchers, namely R. Reisfeld, C.K. Jørgensen, R. Alcala and co-workers,

J. Lucas and co-workers and W.A. Sibley and co-workers. Due to the large volume of published data, only a brief account will be given here. Optical absorption of Fe, Co, Ni, Ti, V, Mn, Cr and Cu ions were studied by Ohishi et al. (1983) in ZBGd, ZBGdA and ZBGdAN fluorozirconate glasses and their absorption coefficients in the vitreous fluoride matrices were determined. With the exception of iron, all the ions appeared to be in six-fold coordination, leading to lower absorption coefficients than in silicate glasses, where coordination is usually tetrahedral for TM ions in their highest oxidation state. Also, the absorption bands in HMFG were red-shifted relative to oxide glass matrices, indicating that the fluoride ligand field was weaker than the oxide field. Reisfeld (1987) reported absorption spectra for Cr^{3+} and Ni^{2+} in ZBLA glass and it was concluded that these ions had octahedral coordination and that the Cr–F and Ni–F distances in the glass were ~ 0.003 nm longer than in crystalline compounds, although this has not yet been confirmed by other techniques. The absorption and fluorescence spectra of Cr^{3+}, Mn^{2+}, Fe^{2+}, Cu^{+}, Pb^{2+} and Sn^{2+} were also investigated by Hollis et al. (1985).

The electric dipole forbidden f → f transitions of the lanthanide ions are crystal field induced by admixing of f–f wavefunctions with odd components of the crystal field potential (Reisfeld 1987) and their intensities in glass may be treated by the Judd–Ofelt method (Krupke 1974). As an example, a recent study of the optical properties of Tm^{3+} ions in In-based fluoride glasses was carried out by Guery et al. (1988a). The absorption spectrum of Tm^{3+} in a $30BaF_2$–$10ThF_4$–$(10-x)YF_3$–$20ZnF_2$–$30InF_3$–$x\,TmF_3$ is shown in fig. 17a at two different temperatures, with little change in peak positions. The emission spectra are shown in fig. 17b. Lifetimes were also measured for the $^1G_4$ and $^3H_4$ levels. The oscillator strengths depend on the three phenomenological Judd–Ofelt parameters Ω_2, Ω_4 and Ω_6. These are obtained to within $\pm 10\%$ from a least-squares fit of the absorption line strengths calculated by means of the Judd–Ofelt theory (Krupke 1974) to the measured oscillator strengths,

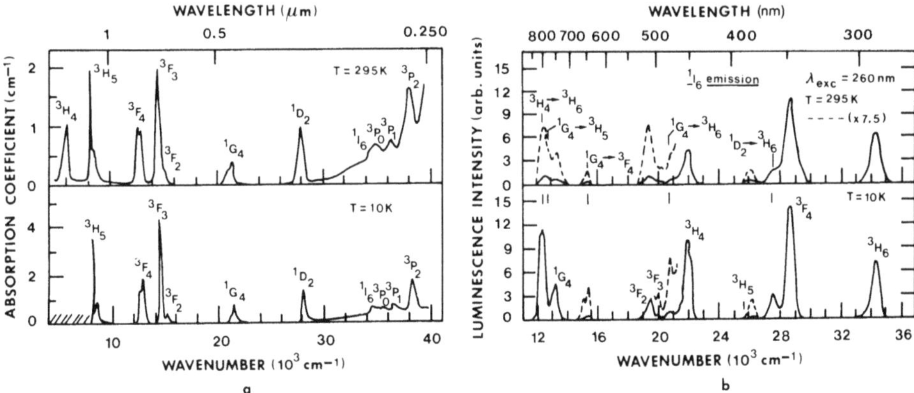

Fig. 17. Optical spectra of Tm^{3+} in $30BaF_2$–$10ThF_4$–$(10-x)YF_3$–$20ZnF_2$–$30InF_3$–$x\,TmF_3$ at 10 K and 295 K: (a) absorption and (b) emission. (Guery et al. 1988a).

given by, respectively:

$$f_{\text{calc}} = \frac{8\pi^2 m\nu}{3h}\frac{1}{2J+1}\frac{(n^2+2)^2}{9n}\sum_{t=2,4,6}\Omega_t(\langle SLJ\|U^{(t)}\|S'L'J'\rangle)^2 \tag{13}$$

$$f_{\text{meas}} = \frac{mc}{\pi e^2 N}\int \alpha(\nu)\,\mathrm{d}\nu, \tag{14}$$

for an electric dipole transition $SLJ \rightarrow S'L'J'$ of average frequency ν, between two J states, where n is the refractive index at this frequency, m is the mass and e the charge of the electron, the quantities in brackets are the matrix elements of the tensor operators U^t, which do not depend significantly on the nature of the vitreous matrix, N is the number of optically active Tm^{3+} ions per unit volume determined by the glass composition and $\alpha(\nu)$ is the measured absorption coefficient at frequency ν. Values for both f_{meas} and f_{calc} are given in table 4 and they are found to be $\ll 1$, due to the forbidden character of the transitions. The radiative transition probabilities for the electric-dipole emissions are then calculated, together with the spontaneous emission probabilities, radiative lifetimes and branching ratios. The occurrence of cross-relaxation between Tm^{3+} ions was also observed for the $^1G_4$ and $^3H_4$ levels, at TmF_3 concentrations as low as 0.5 mol%. Sanz et al. (1987) have studied the optical properties of Tm^{3+} in a ZBLALi glass.

Nd was the first lanthanide element whose absorption and fluorescence spectra were studied in HMFG (Poulain et al. 1975), and Nd-doped glasses are possible candidates for high-power laser materials. Lucas et al. (1978) investigated the structure of a Nd-doped ZBL glass by absorption and fluorescence spectroscopy. The $^4I_{9/2} \rightarrow {}^2P_{1/2}$ absorption transition of Nd^{3+} from the lowest Stark-split level had a maximum at 426.8 nm at 4 K, indicating a nephelauxetic shift between that of $NdZrF_7$ (427.8 nm) and NdF_3 (426.1 nm), where Nd^{3+} ions are eight- and nine-fold coordinated, respectively. This suggested that both eight- and nine-fold coordinated

TABLE 4
Oscillator strengths of Tm^{3+} ions in $30BaF_2$–$10ThF_4$–$(10-x)YF_3$–$20ZnF_2$–$30InF_3$–$xTmF_3$ glass

Transition	λ (nm)	f_{meas} ($\times 10^8$)	f_{calc} ($\times 10^8$)
$^3H_6 \rightarrow {}^3F_4$	1709	161	171
$^3H_4$	786	185	178
$^3F_2$, $^3F_3$	685	289	285
$^1G_4$	468	64	52
$^1D_2$	358	195	192
$^1I_6$, $^3P_0$	287	80	75
$^3P_1$	275	30	59
$^3P_2$	262	196	199

$\Omega_2 = 2.05\ \text{pm}^2$
$\Omega_4 = 1.58\ \text{pm}^2$
$\Omega_6 = 1.12\ \text{pm}^2$

Nd ions were present in the glass, with the former being predominant. The Judd–Ofelt parameters were also obtained, spontaneous emission cross sections were determined ($\sigma \sim 2.95 \times 10^{-20}$ cm^2 for the $^4F_{3/2} \rightarrow {}^4I_{11/2}$ emission at 1049 cm^{-1}) similar to silicate glasses and the corresponding radiative lifetimes were calculated at 433–450 μs, with a quantum efficiency near 1. At higher Nd concentrations, these relatively long fluorescence lifetimes decreased due to nonradiative processes. Cross-relaxation between Nd^{3+} ions and energy transfer between Mn^{2+} and Nd^{3+} in $56ZrF_4$–$34BaF_2$–$4LaF_3$–$4AlF_3$–$1MnF_2$–$1NdF_3$ and $36PbF_2$–$24MnF_2$–$35GaF_3$–$2.8LaF_3$–$2AlF_3$–$0.2NdF_3$ glasses have been discussed by Reisfeld (1987) and Reisfeld et al. (1986), respectively. Lucas et al. (1978) found that although the $^4I_{9/2} \rightarrow {}^2P_{1/2}$ absorption of Nd^{3+} in FZ glass was broad compared to crystalline NdF_3 and $NdZrF_7$, the two main fluorescence transitions, $^4F_{3/2} \rightarrow {}^4I_{11/2}$ and $^4F_{3/2} \rightarrow {}^4I_{9/2}$, had bandwidths similar in the glass and crystalline $NdZrF_7$, probably because Nd^{3+} played a network-forming role in FZ glasses as opposed to a modifying role in oxide glasses. In the latter case, the Nd fluorescence peaks are broader due to large site-to-site variations in the local ligand field. Therefore, laser action in minilasers or even high-power lasers (although the nonlinear refractive index is not conveniently low in all HMFG presently known) is one of the potential applications of the Nd-doped fluoride glasses. Reisfeld (1987) argued that the near-IR laser characteristics of Nd^{3+} in HMFG were at least as good as in a Li–Ca alumino–silicate glass. In fact, Nd is a convenient active ion for four-level solid state lasers, since at room temperature the population of the final lasing level $^4I_{11/2}$ (at $\sim$2000 cm^{-1}) is only $\sim 5 \times 10^{-5}$ of that of the ground state, facilitating population inversion with the initial lasing level $^4F_{3/2}$ and allowing laser operation at low threshold values (Singh 1988). The above properties coupled with the possibility of making optical fibers from HMFG has generated interest in the development of fiber lasers. Thus, a Nd FZ multimode glass fiber laser was demonstrated by Brierley and France (1987) at 1.05 μm. Although the output power with Nd is always highest at 1.05 μm, it is very important to obtain fiber lasers or amplifiers for the technologically relevant telecommunications window at 1.3 μm. Some success in that area has already been reported for ZBLANI (Miniscalco et al. 1988) and ZBLANP (Brierley and Millar 1988) multimode fibers, with the $^4F_{3/2} \rightarrow {}^4I_{13/2}$ transition of Nd^{3+} in the 1.33–1.35 μm range.

Another lanthanide ion whose laser action has been explored in HMFG systems is Er^{3+}. Shinn et al. (1983) noticed that the optical behavior of Er^{3+} in FZ glasses was similar to crystalline $RbMgF_3$ and that the fluoride glasses were at least as good laser hosts as the oxides. Allain et al. (1989) reported three-level lasing of Er^{3+} in quasi-single-mode ZBLAN fibers at 1.00 μm, with powers in the 1 mW range and a gain near 100, corresponding to the $^4I_{11/2} \rightarrow {}^4I_{15/2}$ transition. Pollack and Robinson (1988) have also reported Er^{3+} laser action in a ZBLAN bulk glass parallelpiped at 2.78 μm, arising from the $^4I_{11/2} \rightarrow {}^4I_{13/2}$ transition, where the OH absorption had been reduced to a level low enough not to inhibit the laser emission. Finally, Moore (1989) reported CW lasing at 2.7 μm in an Er^{3+}-doped FZ glass fiber, where the fiber laser served as source in a fluoride fiber system operating at 2.7 μm, not too distant from the 2.55 μm attenuation minimum although $\sim$1 μm beyond the zero dispersion wavelength λ_0. As pointed out by Pollack and Robinson (1988), one of the main difficulties in fabricating good lasers from HMFG bulk glasses and fibers is to obtain high optical quality materials, free of any bubbles, precipitates or undissolved particles. The purity requirements (Singh 1988), on the other hand, should not be difficult to meet in view of the developments already achieved in the optical fiber technology of HMFG. Also, although energy loss by transfer between ions may be a problem, nonradiative vibrational relaxation is less severe than in oxide systems, due to the characteristically low phonon frequencies of the HMFG (Almeida 1987).

Ho^{3+} has also been evaluated as a potential active ion for laser emission in FZ glasses (Tanimura et al. 1984, Sibley 1985, Adam and Sibley 1985, Haixing and Fuxi 1985). Adam et al. (1988) characterized the IR transitions of Ho^{3+} in FZ and BT glasses, in particular the $^5I_7 \rightarrow {}^5I_8$ three-level lasing emission at 2.08 μm. Eyal et al. (1987) studied the laser properties of Ho^{3+} and Er^{3+} in $9BaF_2$–$28ThF_4$–$22YF_3$–$27ZnF_2$–$6ZrF_4$–3LiF–5NaF glass. Miniscalco and Andrews (1988) presented a short review of HMFG fiber-optic lasers and amplifiers.

Another important topic related to fluorescence spectra of lanthanide ions in HMFG is the phenomenon of up-conversion, where one or two lanthanide ions are excited with IR radiation and transfer it immediately to another neighboring lanthanide ion, which is excited twice in sequence to a high-lying level and then emits radiation of a shorter wavelength, e.g. in the visible. The occurrence of this phenomenon in HMFG was discussed in detail by Quimby (1988). Quimby et al. (1987) and Yeh et al. (1987) studied the IR to visible light conversion of Er^{3+} only and Yb^{3+} to Er^{3+} in BT fluoride glasses, while Chamarro and Cases (1988a) have studied the up-conversion of Yb^{3+} to Ho^{3+} and Tm^{3+}. Okada et al. (1988) reported near IR to visible conversion of Er^{3+} in fluorozircoaluminate glass.

Other energy transfer phenomena have also been investigated, e.g. between Yb^{3+} ions and Ho^{3+} or Tm^{3+} (Chamarro and Cases 1988b) and between Yb^{3+} and Pr^{3+}, Sm^{3+} or Dy^{3+} (Chamarro and Cases 1989). The optical properties (absorption and fluorescence) of other lanthanide ions have also been studied in FZ glasses, e.g. Pr^{3+} (Adam and Sibley 1985b, Ferrari et al. 1988), Sm^{3+} (Canalejo et al. 1988), Gd^{3+} (Alonso et al. 1988), Dy^{3+} (Orera et al. 1988, Adam et al. 1988) and Yb^{3+} and Lu^{3+} (Drexhage et al. 1982). The optical absorption of Eu^{2+} was characterized for ZBT glasses in terms of structure (Shafer and Perry 1979) and the $^7F_1 \rightarrow {}^7F_6$ absorption of Eu^{3+} at 2.2 μm, which is modulated by temperature, was proposed by Ohishi and Takahashi (1986) as the basis for a thermo-optical low-temperature HMFG fiber sensor. Miniscalco et al. (1987) showed that photoluminescence was a suitable technique for the analysis of impurities such as Fe^{3+}, Pr^{3+} and Nd^{3+} in ZBLAN glass fibers, at sub-ppb or even sub-ppt levels.

5.7.4. *IR absorption and Raman scattering*

Vibrational spectroscopy (VS) is very important in the context of HMFG, since it is one of the most powerful techniques for the study of glass structure and also because the main applications of these glasses derive from their IR transparency. The principal objective of this section is to discuss the nature of the fundamental (first-order or one-phonon) vibrational modes of representative HMFG and some of their structural implications.

Although VS does not usually provide direct structural information, it is very useful for probing nonbridging or weakly coupled species in terms of short-range order and it may also be compared with model calculations to yield more direct short-range or even intermediate-range structural data. Among the different experimental methods available at present, only IR absorption or reflection and Raman scattering have been used for HMFG.

IR transparency depends on the absolute value of the absorption coefficient given in the multiphonon regime by eq. (11) for α_{MP}, whose temperature dependence may be accounted for in cases where eq. (11) holds by the following expression (Bendow 1977)

$$\alpha_{MP}(\nu, T) = \alpha_0[N(\nu_0, T) + 1]^{\nu/\nu_0}[N(\nu, T) + 1]^{-1}\exp(-C\nu), \tag{15}$$

where $N(\nu, T)$ is the Bose–Einstein thermal population factor for frequency ν and temperature T, ν_0 is an average optical phonon frequency and the coupling constant α_0 is the absorption coefficient at $0\ cm^{-1}$ and 0 K. The values of C, α_0 and ν_0 are obtained by curve fitting. Absolute values of the absorption coefficient may be obtained in the fundamental regime via IR reflectivity on thick samples (Bendow et al. 1981), but this involves elaborated data manipulation, e.g. through Kramers–Kronig analysis (Moss 1959) and, for the purpose of only structural studies qualitative IR spectra from powdered glass pellets are often sufficient. In the case of Raman spectra, given the intrinsic weakness of the inelastic light scattering process in the order of 10^{-6} of the incident laser light intensity, and also due to the small size of the scattering volume crossed by the focused laser beam and the weak dependence of Raman intensity on this volume, only the fundamental spectrum is recorded under normal conditions for bulk samples and use of thin films or pellets is not necessary. Perhaps as a result of the very active search for ultralow loss in HMFG, their fundamental vibrational spectra have not received much attention and have been studied almost exclusively in simple FZ systems plus a few BT glasses. Short reviews, not dealing exclusively with fluoride glasses, were written by Almeida (1987, 1988).

Figure 18 shows a series of first-order transmission spectra of modified binary FZ and FH glasses, where data were recorded for both KBr pellets ($\nu \geqslant 300\ cm^{-1}$) and polyethylene pellets ($\nu \leqslant 300\ cm^{-1}$). All glasses showed strong high- and intermediate-frequency bands near $500\ cm^{-1}$ and $250\ cm^{-1}$, respectively, plus a weak band near $110\ cm^{-1}$, in agreement with reflectivity data presented later in this section. Figure 19 shows the polarized Raman spectra of a lead dizirconate glass, where the depolarization ratio DR represents the ratio of the depolarized or perpendicular spectrum to the polarized or parallel spectrum as defined by Almeida (1988). In the absence of a theory which allows the prediction of the DR values, one usually makes an approximate interpretation based on the molecular model namely, the assumption that values near zero correspond to totally symmetric vibrations, whereas large values (DR $\geqslant 0.5$) correspond to antisymmetric vibrations. In fig. 18, which displays the effects of isomorphic substitutions of Hf for Zr and Sr or Pb for Ba in the most stable 2:1 compositions, it may be observed that on going from Zr → Hf, the frequency of the high- and intermediate-frequency peaks decreased slightly but more markedly in the latter case. However, when going from Sr → Ba → Pb-containing glass samples, the frequency of the lowest frequency band (or shoulder) decreased roughly as the

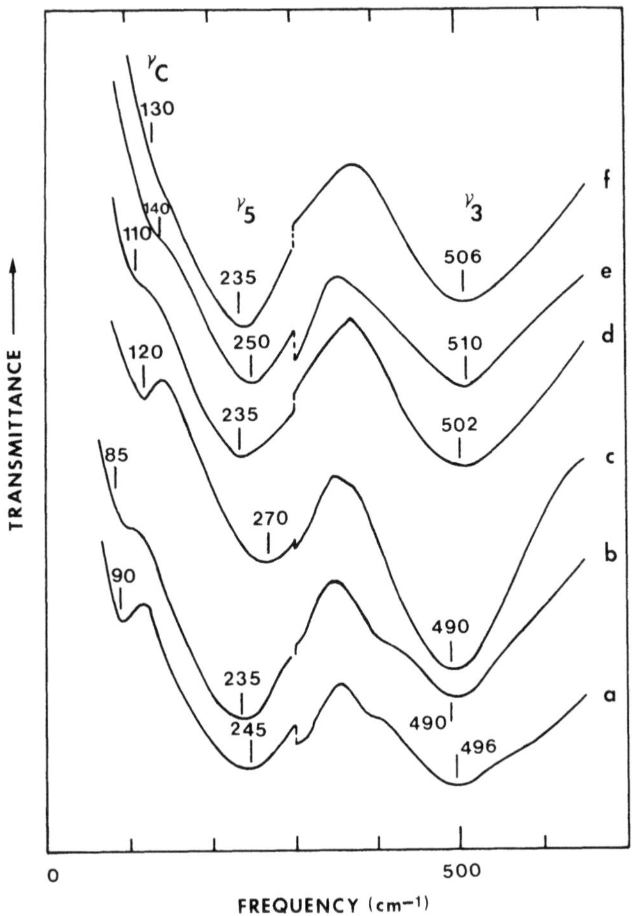

Fig. 18. Infrared absorption spectra of HMFG: (a) $70ZrF_4$–$30PbF_2$, (b) $70HfF_4$–$30PbF_2$, (c) $64ZrF_4$–$36BaF_2$, (d) $66HfF_4$–$34BaF_2$, (e) $70ZrF_4$–$30SrF_2$, (f) $70HfF_4$–$30SrF_2$.

square root of the modifying cation mass and its intensity increased in the same direction. Figure 19 shows that the Raman spectrum of a typical FZ glass contains two very strong bands, one almost completely polarized near 577 cm^{-1} (DR = 0.08 and FWHM = 74 cm^{-1}) and the other depolarized near 40 cm^{-1} (DR = 0.6), plus a weakly polarized peak at 477 cm^{-1} and weakly depolarized bands near 390 cm^{-1} and 185 cm^{-1}. It has been shown that the position of the high-frequency Raman peak is insensitive to the replacement of Zr by Hf and the frequency of the band near 200 cm^{-1} in the Raman spectrum does not depend on the type of modifier cation (Sr, Ba or Pb) (Almeida and Mackenzie 1981, 1983b, Almeida 1987).

The main IR- and Raman-active vibrational modes may be assigned in the following way (Almeida 1987), where ν_i designates each particular peak as shown in figs. 18 and 19: ν_7 is the symmetric stretch of nonbridging fluorines without Zr(Hf)

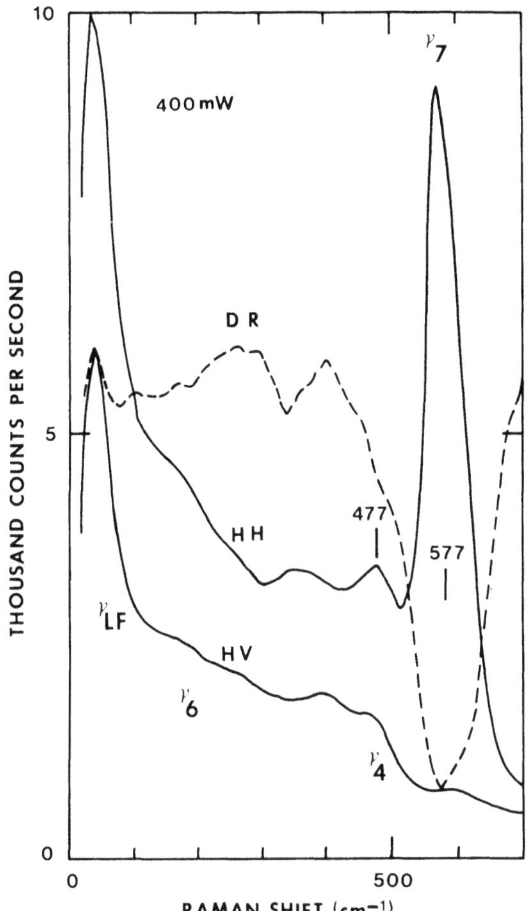

Fig. 19. Polarized Raman spectra of lead dizirconate glass in the polarized (HH) and depolarized (HV) configurations. (DR varies linearly between 0 and 1 when the number of counts varies from zero to ten thousand.)

cation motion, SS(F_{nb}); ν_4 the antisymmetric stretch of bridging fluorines without Zr(Hf) motion, AS; ν_6 the symmetric stretch of bridging fluorines without Zr(Hf) motion, SS; ν_{LF} the low-frequency Raman mode due to a combination of the Bose–Einstein thermal population effect with vibrational modes of optic and acoustic character; ν_3 the antisymmetric stretch of bridging fluorines with simultaneous Zr(Hf) cation motion, AS(C); ν_5 the symmetric stretch of bridging fluorines with Zr(Hf) motion, SS(C); and ν_C the cage-like vibration of the modifying cations in their network sites. Using a generalization of the central force network model of Sen and Thorpe (1977), which includes a nonbridging stretching force constant k_1, a bridging stretching force constant k_2 and a bridging bending (noncentral) force constant k_3, and representing by m_F and θ the mass of the fluorine atom and the F–Zr(Hf)–F bridging angle, respectively, the squares of the angular frequencies of the above modes may be given by the following equations (Almeida and Mackenzie 1983, Almeida 1987)

$$\omega^2_{SS(F_{nb})} = \frac{k_1}{m_F}, \tag{16}$$

$$\omega^2_{AS} = \frac{k_2}{m_F}(1 - \cos\theta), \tag{17}$$

$$\omega^2_{SS} = \frac{1}{m_F}[k_2(1 + \cos\theta) + 2k_3(1 - \cos\theta)], \tag{18}$$

$$\omega^2_{AS(C)} = \frac{k_2}{\mu}(1 - \cos\theta), \tag{19}$$

$$\omega^2_{SS(C)} = \frac{1}{\mu}[k_2(1 + \cos\theta) + 4k_3(1 - \cos\theta)], \tag{20}$$

where μ is a reduced mass of the F atom and a $Zr(Hf)F_y$ particle. The parameter y represents the number of F_{nb} atoms around the central network forming cation; when y varies between 2–5, the reduced mass μ_{F-ZrF_y} will vary from 16.56–17.24. The model just described allows the simultaneous calculation of force constants and bridging angles from a given set of vibrational spectra. The values of the force constants which have been calculated to date (Almeida and Mackenzie 1981, 1983b, Almeida 1985, 1987), $k_1 \sim 360$–$390\ Nm^{-1}$, $k_2 \sim 120$–$160\ Nm^{-1}$ and $k_3 \sim 4$–$8\ Nm^{-1}$, are interchangeable between different FZ and FH systems. Average bridging angles have been calculated for such glasses between 140°–180°, being larger than the Si–O–Si angle in v-SiO_2.

The value of y, which has only a small effect on the reduced mass, is, however, important from the structural viewpoint, being 4 for the dizirconate structure of fig. 2a where Zr atoms are six-fold coordinated, being 5 for the seven-coordinated metazirconate structure (Almeida and Mackenzie 1981) and being <4 for the dizirconate structures of figs. 2b–2d where seven and eight-fold coordinated Zr atoms are present. As argued by Almeida (1988), although the latter models are in better agreement with existing XRD data, serious difficulties are found when trying to interpret the known vibrational spectra on their basis. In fact, Toth et al. (1973) showed that in the absence of bridging there is an inverse correlation between the $\omega_{SS(F_{nb})}$ Raman frequency and the coordination number of Zr in the complex ions present in the structure of several crystalline compounds. Also, Walrafen et al. (1988) have suggested that bridging effects in FZ glasses, if present, do not influence significantly the symmetric stretching Raman frequencies. Therefore, the occurrence of a single, sharp and completely polarized Raman line involving no Zr cation motion near $580\ cm^{-1}$ – the same frequency of similar dominant Raman bands in octahedrally coordinated Li_2ZrF_6 and Cs_2ZrF_6 crystals (Toth et al. 1973) – is in maximum agreement with highly symmetrical F atom environments about Zr, predominantly six-fold coordinated with a significant degree of covalence. Walrafen et al. (1988) have shown that when the F/Zr ratio varied from $\sim$4.6–8.0 in FZ glasses the symmetric stretching Raman frequency decreased from $600\ cm^{-1}$ to $555\ cm^{-1}$. This was interpreted as due to an increase in the average CN of Zr from roughly 5 to 8 and a simultaneous decrease in the values of the symmetric stretching force constants (unspecified whether bridging and/or nonbridging). It is interesting to notice that the results of Walrafen et al. (1988) for the glasses with large F/Zr ratio and those of Walrafen et al. (1985) for FZ melts show that when the degree of Zr–F–Zr bridging decreases drastically, the Raman bands near $\sim 490\ cm^{-1}$ and $\sim 200\ cm^{-1}$ completely disappear in agreement with the above assignment of these bands to the bridging anion modes AS (ν_4) and SS (ν_6).

The above generalized central force network model (GCFN) for the vibrational modes of both modified and nonmodified glasses shows also that the intermediate frequency Raman band of FZ-type glasses near $200\ cm^{-1}$ and the intermediate frequency IR band near $250\ cm^{-1}$, whose positions are not sensitive to the mass of the modifying cation, do not involve vibrations of these cations contrary to the suggestions of Bendow et al. (1981a). The model explains also why the position of the IR absorption edge of FH glasses is only slightly shifted to lower frequencies relative to FZ glasses, although the atomic weight of Hf (178.49) is almost double that of Zr (91.22). In fact, since the k_2 values are almost identical for the two glass families (Almeida and Mackenzie 1983b) an AS(C) frequency of $502\ cm^{-1}$ for barium dihafnate glass would lead to an estimate of $511\ cm^{-1}$ for the same mode in barium dizirconate glass via eq. (19) (provided the average bridging angle θ were assumed to be the same), rendering the IR absorption edges almost coincident. (The experimental value for the FZ sample is $490\ cm^{-1}$, probably due to different k_2 and/or θ values.) The small estimated frequency difference stems from the small difference in reduced mass between F–HfF_4 (17.68) and F–ZrF_4 (17.06), taking the six-coordinated case as an example.

Almeida (1985) put forward a pair of tentative empirical selection rules for the VS of modified HMFG: (1) the IR spectrum is dominated by stretching vibrations of bridging F atoms accompanied by a small amount of network forming cation motion; (2) the Raman spectrum is dominated by high-frequency symmetric stretching vibrations of nonbridging F atoms around fixed network-forming cations. Such rules should be useful as a zeroth-order approach to the empirical assignment of VS in unknown HMFG.

The fundamental VS of multicomponent FZ-type glasses such as ZBLA or ZBLAN have not been extensively investigated, but it is known that minor components such as LaF_3, AlF_3 and even NaF, when present in small concentrations, have virtually no effect in the Raman spectra and little effect, if any, in the IR spectra. As an example, fig. 20a is a comparison of far-FTIR reflectivity spectra of FZ glasses with and without NaF and fig. 20b shows the polarized components of the Raman spectra for two different FZ glasses. The reflectivity bands near $540\ cm^{-1}$ and $260\ cm^{-1}$ correspond to IR-active AS(C) and SS(C) modes, respectively, while the shoulder near $600\ cm^{-1}$, which has also been observed in transmission for binary FZ glasses without AlF_3 (see fig. 18) and in ZrF_4–BaF_2–LaF_3 glasses (Almeida and Mackenzie 1983c), is probably due to residual oxide species and it is not associated with Al as previously suggested by Bendow et al. (1983).

Inoue and Yasui (1987) have calculated the Raman spectrum of barium dihafnate glass from the time history of the polarizability tensor in a MD simulation between

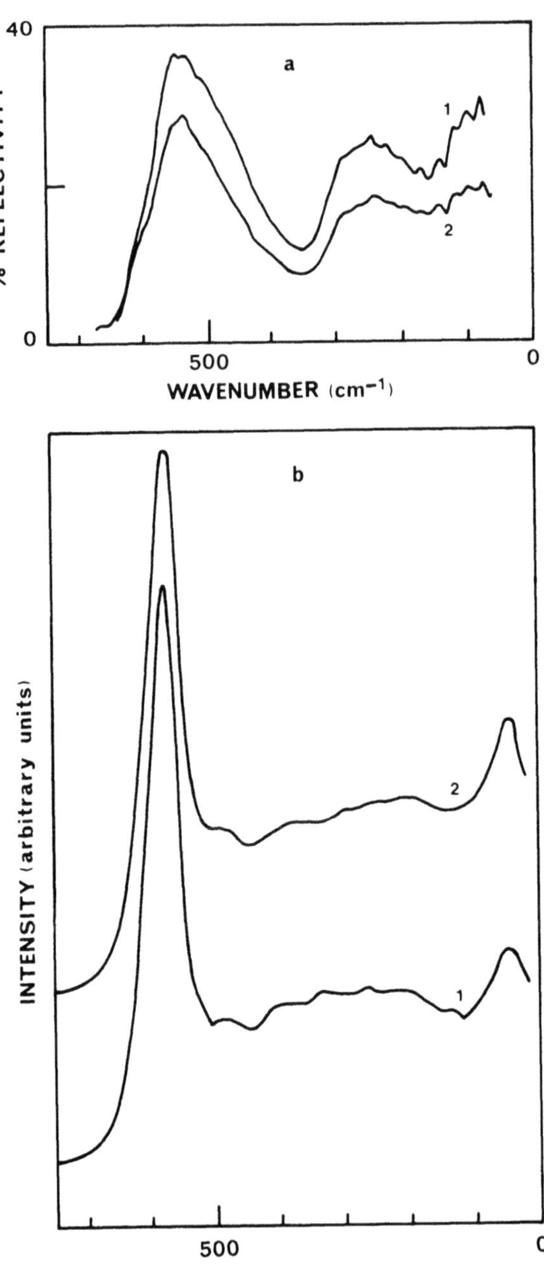

Fig. 20. (a) Far-FTIR specular reflectivity of (1) ZBLA and (2) ZBLAN glasses; (b) polarized (HH) components of the Raman spectra of (1) ZBLA and (2) ZBLALiP glasses.

400–700 cm^{-1}. Although the agreement was generally poor, the dominant band of the calculated spectrum, of similar shape but higher frequency, was attributed to the symmetric stretching of HfF_8 polyhedra. Also, stimulated Raman scattering and Raman amplification have been reported in slightly multimode ZBLAN fibers (Durteste et al. 1985).

The influence of the various FZ and FH glass constituents on their IR absorption edge was semi-empirically assessed by Moynihan et al. (1981) and later received additional experimental support (Chung and Moynihan 1987). The main trends follow the observed fundamental and multiphonon spectra of the individual crystalline fluorides and it was generally concluded that the IR edge of typical multicomponent FZ and FH glasses (containing ~50–60 mol% of network former) was unshifted by monovalent fluorides not lighter than NaF, divalent fluorides not lighter than CaF_2, trivalent fluorides not lighter than LaF_3 and tetravalent fluorides not lighter than ThF_4.

The fundamental IR and Raman spectra of binary thorium FZ glasses (Almeida and Mackenzie 1983a) and binary thorium FH glasses (Almeida and Mackenzie 1984) have also been studied. The first-order VS of Zr-free glasses are less documented at present. Videau et al. (1979) studied the Raman spectrum of a $40AlF_3$–$40CaF_2$–$20BaF_2$ glass; a polarized peak at 570 cm^{-1} and a polarized shoulder at ~620 cm^{-1} were interpreted as due to six- and four-fold coordinated Al complexes, respectively. Fonteneau et al. (1980) studied the bulk IR transmission spectra of a series of RE-containing RF_3–BaF_2–ZnF_2 glasses (R = Y, La, Ce, Pr, Nd, Sm, Eu, Gd, Tb, Dy, Ho, Er, Tm, Yb) with ~20 mol% RF_3, having observed, in addition to the multiphonon edges, the near-IR bands characteristic of the $f \rightarrow f$ electronic transitions of the RE ions. The vibrations associated to the R–F bonding environments were not characterized. Jia and Zhang (1988), on the other hand, have tabulated stretching force constants for the different lanthanide trifluoride molecules (C_{3v} point group), which could in principle be extrapolated for the solid state case. Because of the lanthanide contraction, the vibrational frequency values remain quite similar throughout the whole elemental series, with the atomic masses and the force constants increasing simultaneously. The IR reflectivity (Bendow et al. 1983) and polarized Raman spectra (Bendow et al. 1985) have been recorded for BaF_2–ThF_4-based glasses. Most Raman spectra were relatively featureless. In the case of glasses which contained YF_3 and AlF_3, this was probably due to simple fluorescence interference as evidenced by depolarization ratios almost constant and near one (perhaps associated with metallic impurities in those two raw materials), whereas the Raman spectra of the other BT glasses had a more credible appearance. However, no definite vibrational mode assignments were attempted. Figure 21 shows the HH and HV spectra of a ZnF_2–CdF_2-based glass recorded in our laboratory from samples prepared by R. Alcala and co-workers at the University of Zaragoza (Spain). The composition of the glasses was similar to a ZBLALi glass with Zr replaced by approximately equal amounts of Zn and Cd and where fluorescence interference was identified above ~600 cm^{-1} by using two different laser lines, but was almost nonexistent below that frequency. The spectra are very different from the FZ-type spectra of fig. 20a exhibiting mostly depolarized features except for a band near 370 cm^{-1}. The former features were found to be strongly IR-active, while the polarized mode was only weakly active in the infrared (absorption and reflectivity). It is possible that the weak band near 560 cm^{-1} (whose DR might be artificially large due to the onset of the fluorescence interference) is related to six-fold coordinated Al and the ~370 cm^{-1} mode is associated with octahedrally coordinated Zn, whereas

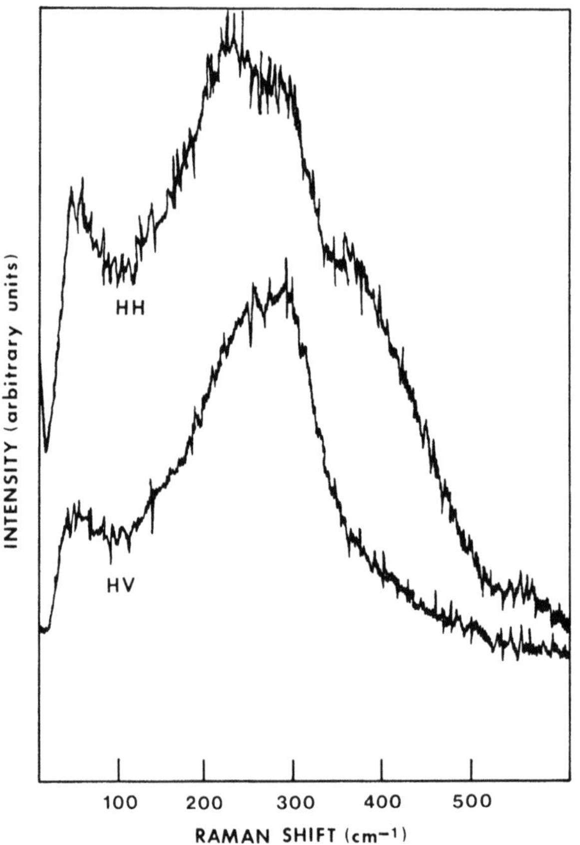

Fig. 21. Polarized Raman spectra of $32ZnF_2$–$28CdF_2$–$20BaF_2$–$4LaF_3$–$5AlF_3$–$11LiF$ glass in the polarized (HH) and depolarized (HV) configurations.

the depolarized region below $\sim 300\ cm^{-1}$ may contain contributions from both Cd and Ba. These cations could play an intermediate role in these glasses as opposed to the case of the more covalent FZ systems. For more ionic glasses like the ZnF_2–CdF_2-based ones, the previous selection rules do not strictly apply.

5.8. *Electrical properties*

Unlike most oxide glasses (which are cationic conductors with Na^+, Li^+ or K^+ as typical charge-carrying species), but similar to fluoroberyllate glasses and several crystalline fluorides such as BaF_2 and PbF_2, the HMFG studied to date (almost exclusively of FZ-type) are anionic conductors. F^- ions have been identified as the conducting species in FZ glasses by the Tubandt test using Pb/PbF_2 electrodes (Leroy and Ravaine 1978) and e.m.f. measurements of sodium amalgam cells gave a mean ionic transport number of 0.999 ± 0.004. The enhanced mobility of F^- ions compared to O^{2-} ions may be due to the larger strength of the M–O bonds compared to M–F bonds. The blocking electrode method showed that in such glasses the electronic transport number did not exceed 0.0068 (Ravaine and Leroy 1980). The electrical conductivity of FZ glasses is several orders of magnitude larger than that of fluoroberyllates. This has aroused some interest in the former glasses as solid state electrolyte materials.

Leroy et al. (1978) showed that the conductivity of ternary barium FZ glasses containing ThF_4, LaF_3, PrF_3 or NdF_3 increased substantially with the BaF_2 content and Chandrashekhar and Shafer (1980) found that the conductivity of ZrF_4–BaF_2–ThF_4 glasses decreased when BaF_2 was replaced by NaF, showing that Na^+ ions were indeed not mobile and also that the mobility of the F_{nb} species was larger than that of the F_b species. Almeida and Mackenzie (1982) assumed that in FZ glasses only the F_{nb} species had appreciable mobility under an applied electric field and the same conclusion was later reached by Matusita et al. (1987).

Most studies of the electrical properties of HMFG report conductivity versus temperature measurements, either DC (see, e.g. Almeida and Mackenzie 1982) or AC with use of complex impedance plots extrapolated to zero frequency (see, e.g. Ravaine 1985). Arrhenius behavior was always observed for the electrical conductivity σ_e below T_g

$$\sigma_e = \sigma_0 \exp(-E_\sigma/RT), \tag{21}$$

where σ_0 is a pre-exponential factor and E_σ is the activation energy for conduction. Table 5 gives conductivity and E_σ values for a series of representative HMFG, mostly

TABLE 5
Selected electrical properties of HMFG systems and other nonhalide glasses.

Composition (mol%)	σ_e(200°C) ($\times 10^4$ S m^{-1})	Activation energy (eV)	ε_r
$64ZrF_4$–$36BaF_2$	2.83	0.70	13.9
$74ZrF_4$–$26BaF_2$	1.00	0.70	–
$50ZrF_4$–$35BaF_2$–15CsF	19.95	0.66	–
$55ZrF_4$–$37BaF_2$–$8ThF_4$	5.10	0.77	11.8
$55ZrF_4$–$31BaF_2$–$14ThF_4$	3.30	0.78	–
$61ZrF_4$–$30BaF_2$–$9ThF_4$	2.55	0.83	7.4
$62ZrF_4$–$30BaF_2$–$8LaF_3$	3.20	0.79	7.0
$62ZrF_4$–$30BaF_2$–$8NdF_3$	3.50	0.78	–
$62ZrF_4$–$30BaF_2$–$8PrF_3$	4.30	0.75	–
$56ZrF_4$–$14BaF_2$–$6LaF_3$–$4AlF_3$–20NaF	0.84	0.87	11.2
$60ZrF_4$–$33ThF_4$–$7LaF_3$	0.37	0.82	–
$40AlF_3$–$22CaF_2$–$22BaF_2$–$16YF_3$	2.53	0.83	–
$16BaF_2$–$28ThF_4$–$28YbF_3$–$28ZnF_2$	0.65	0.83	–
$40ScF_3$–$20YF_3$–$40BaF_2$	13.5	0.70	–
$50PbF_2$–$20MnF_2$–$30GaF_3$	54.95	0.61	–
$35PbF_2$–$30MnF_2$–$35InF_3$	2.81×10^2	0.56	–
$40PbF_2$–$10MnF_2$–$50FeF_3$	1.00	–	–
$58PbF_2$–$21MnF_2$–$21FeF_3$	2.40×10^2	–	–
$80BeF_2$–20CsF	4.8×10^{-5}	1.12	–
$63SnF_2$–$14PbF_2$–$23P_2O_5$	9.2	0.75	–
$74SiO_2$–$26Na_2O$	7.16×10^{-6}	0.70	–

FZ-type, together with some available values of relative dielectric constant. In general, the electrical conductivity was not very sensitive to the composition of the glasses and the activation energy values were similar to those measured for modified silicate glasses and lower than those of fluoroberyllate glasses. Assuming that the charge carriers are only the F_{nb} species, some general trends may be found for the data of table 5. Namely, ThF_4 plays neither a network forming nor a network modifying role in FZ glasses, exhibiting an intermediate character between ZrF_4 and BaF_2; this is either due to structural changes introduced by ThF_4, or it is because the Th atoms will be surrounded by fluoride anions with a somewhat intermediate character between F_b and F_{nb}. On the other hand, the lanthanide ions, although also intermediate in character, may behave slightly more as modifiers than will Th. The case of the TMFG is special in the sense that, for FeF_3-containing glasses, the conductivity is both anionic and electronic, the transport number of electrons increasing with the Fe content and the overall conductivity increasing with the PbF_2 content (Kawamoto et al. 1987).

The compositional dependence of conductivity and, in particular, of the activation energy, was interpreted in FZ glasses via the classical Anderson and Stuart model (Anderson and Stuart 1954), which has been applied successfully to silicate and fluoroberyllate glasses. Based on this model, Almeida and Mackenzie (1982) concluded that in binary barium FZ glasses the electrostatic energy term, which is the electrostatic energy needed to remove an F_{nb} ion from a Zr cation, predominated over the strain energy term, which is the energy needed to open up the glass network enough to allow the diffusional motion of the F_{nb} anions. The activation energy was the sum of these two terms.

Fast fluoride ion conduction was explored by Kawamoto and Nohara (1987a) in FZ glasses such as $50ZrF_4$–$35BaF_2$–15CsF, for which $\sigma_{200°C} = 2 \times 10^{-3}\ S\,m^{-1}$ and $E_\sigma = 0.66$ eV, as well as in Zr-free glasses like $35PbF_2$–$30SnF_2$–$35InF_3$, for which $\sigma_{200°C} = 0.25\ S\,m^{-1}$ and $E_\sigma = 0.49$ eV (Kawamoto and Nohara 1987b). For the highly conductive crystalline fluoride $PbSnF_4$, $\sigma_{200°C} > 1\ S\,m^{-1}$.

Although there have been no reports of mixed alkali effect in fluoroberyllate glasses in the literature (Baldwin et al. 1981) and despite the accumulated experimental evidence showing FZ glasses to be exclusively anionic conductors, some authors have investigated the possibility of occurrence of a mixed alkali effect in FZ glasses (Perazzo et al. 1985, Xiujian et al. 1987, Tatsumisago et al. 1988). The mixed alkali effect consists in large deviations from linearity in dynamic properties such as the electrical conductivity as a function of composition, when an alkali oxide is progressively substituted for another in alkali cation conducting oxide glasses. In FZ glasses, there are some obvious differences in the conductivity behavior compared to silicates. For example, in $40ZrF_4$–$20PbF_2$–$10AlF_3$–xMF–$(30 - x)$M′F glasses (M, M′ = Li, Na, K) (Xiujian et al. 1987), pure Li and K compositions are more conductive than Na glasses and the observed DC conductivity minima are much shallower, being more pronounced for mixed Li–Na than mixed Li–K glasses. Nevertheless, both the previous authors and Tatsumisago et al. (1988) have suggested that these FZ glasses are alkali rather than fluoride ion conductors, at least for larger concentrations of alkali fluorides.

Finally, Hahn et al. (1987) studied the decrease in the electrical conductivity of FZ glasses upon crystallizing and some authors have investigated the relationship between certain kinds of localized F^- ion motions and the electrical conductivity of HMFG using ^{19}F nuclear magnetic resonance (NMR) spectroscopy (Bray et al. 1983, Ji-Jian and Jian-Feng 1988). The short-range motions had activation energies less than half of those corresponding to the long-range diffusive motions responsible for the electrical conduction processes (Bray et al. 1983).

Although the electrical conductivity of HMFG is lower than that of a typical fast ion conducting glass, the fact that the conductivity is anionic and relatively high in the former case raises the possibility of interesting applications.

5.9. *Magnetic properties*

A short review of the magnetic properties of HMFG was published by Dupas (1985). These materials offer the possibility of incorporating large amounts of transition metal or lanthanide magnetic fluorides in a vitreous matrix up to ~50 mol%, allowing a detailed evaluation of magnetic properties in nondilute systems.

The TMFG have been the most studied so far, followed by the BT glasses, with magnetic ions such as Mn^{2+} (d^5), Fe^{3+} (d^5), Fe^{2+} (d^6), Dy^{3+} (f^9), Ho^{3+} (f^{10}), Gd^{3+} (f^7) and Eu^{2+} (f^7). The AC magnetic susceptibility χ_m of these glasses was found by Dupas (1985) to follow a Curie–Weiss law over a large temperature range, with a Curie temperature $\theta \sim 100$ K, indicating mainly antiferromagnetic interactions. Well below θ, the behavior of the susceptibility gradually changed to the Curie law, and at temperatures of a few Kelvin a maximum in χ_m clearly showed spin glass behavior very similar to the spin glass characteristics exhibited by metallic glasses. The spin freezing temperature T_{fr} corresponding to the spin glass transition was only about 1% of θ, indicating a high degree of topological disorder in the HMFG tested, namely those corresponding to compositions $PbMnFeF_7$ and Pb_2MnFeF_9. The value of T_{fr} was found to be dependent upon the frequency of the electromagnetic field, particularly in glasses containing ions with local anisotropy such as Dy^{3+} and Ho^{3+}.

Mössbauer spectroscopy is another technique which has also been used to characterize the magnetic behavior of HMFG. Thus Renard et al. (1980) recorded the Mössbauer spectra of ^{57}Fe nuclei in $PbMnFeF_7$ and Pb_2MnFeF_9 glasses, whose temperature dependence suggested a progressive freezing of the magnetic moments with a decrease in temperature. Coey et al. (1981) studied the ^{151}Eu Mössbauer spectra in $61ZrF_4$–$(32 - x)BaF_2$–$7ThF_4$–$xEuF_2$ glasses and their results indicated that most of the europium was present as Eu^{2+}, with a coordination number between 8 and 12 and an effective Debye temperature θ_D of 145 K, corresponding to network modifying behavior. A small fraction of the lanthanide element (~10%) was present as Eu^{3+} in well-defined sites with an effective $\theta_D \sim 261$ K, corresponding to network forming behavior, considered by the authors as representative of the other trivalent RE elements in FZ glasses. Recent Mössbauer data have also been obtained by Ji-Jian and Jian-Feng (1988) in a structural study of ZrF_4–PbF_2–YF_3–FeF_3 glasses, with isomer chemical shifts in the ranges of ~0.5 to ~1.1 mm/s, which are very close to those found for octahedrally coordinated Fe^{3+} and Fe^{2+}, respectively, indicating six-

fold coordination for iron ions in the FZ glasses. The same authors measured the magnetic susceptibility of the glasses, and spin glass behavior was characterized by T_{fr} values near 6.5 K, irrespective of the Zr and Fe content of the glasses. Finally, ^{19}F NMR and electron spin resonance (ESR) spectra were also recorded. The NMR chemical shift ΔH was close to 6.5 G and did not vary much with the composition, indicating that the topology of the fluorine nuclei in the ZrF_7 and FeF_6 polyhedra remained unchanged. The ESR spectra of iron in the glasses furnished values of the Landé factor g at 4.3 and 2.0, the first being attributed to distorted octahedra and the second to Fe^{3+}–Fe^{3+} spin interactions. ^{19}F NMR was also recorded by Bray et al. (1983) in FZ glasses with measurement of spin–lattice relaxation times T_ℓ, from which activation energies were calculated for short-range fluorine motions, as mentioned in sect. 5.8. The ^7Li NMR was obtained as well and a distribution of activation energies was invoked in this instance. Recently, MacFarlane et al. (1989) studied the ^{19}F NMR spectra of ZBL glass and they identified three different fluoride ion sites – one bridging and two nonbridging. The ESR spectra of paramagnetic ions in HMFG have been examined in several cases, often in connection with radiation effects. Yoneda et al. (1983) studied the X-band ESR spectra of Gd^{3+} (0.05 mol%) in BaF_2–ZnF_2–RF_3 (R = Y, La, Nd) glasses, which were characterized by a long T_ℓ value and very broad absorptions attributed to large site-to-site variations. Cases et al. (1985b) examined radiation effects in FZ and FH glasses after irradiation with X-rays or γ-rays. A series of defects could be identified which were associated with F_2^- molecular ions and F^0 atoms (originating from low-temperature irradiation of F_{nb} species) and also with Zr^{3+} or Hf^{3+} and Pb^{3+} in lead-containing compositions. No defects could be associated with La, Al or Li atoms in the glasses. In BT glasses containing YbF_3 or YF_3 and ZnF_2, Cases et al. (1985a) found results basically similar to the FZ case. glasses, only after irradiation did ESR signals appear, attributed to F_2^- defects and perhaps to F^0 atoms as well. The occurrence of F_2^- centers could not be observed for
glasses, only after irradiation did ESR signals appear, attributed to F_2^- defects and perhaps to F^0 atoms as well. The occurrence of F_2^- centers could not be observed for
YbF_3-containing glasses, probably due to the broadening of the ESR signals referred above. Finally, Harris (1987a,b) has published detailed studies on Ti^{3+} and iron-group transition metal ions in FZ glasses, respectively, by ESR. In ZBLANP glass with ~0.1 mol% TiF_4 added, a broad asymmetric absorption at low temperatures was interpreted as due to Ti^{3+} ions in a single type of site with distorted octahedral symmetry. In the same base glass composition, Cr^{3+}, Mn^{2+}, Fe^{3+}, Co^{2+}, Ni^{2+} and Cu^{2+} ions were also found most likely to be in octahedral coordination.

6. Conclusions

Heavy-metal fluoride glasses hold a large potential for optical applications in the middle infrared, namely in fiber form as optical waveguides, fiber lasers and sensors, but also in bulk form as prisms, lenses, light pipes or optical windows. This potential has stimulated extensive research into their structure and properties, but present knowledge is still largely incomplete due to unusual structural characteristics such as high coordinations and also as a result of the large number of chemical elements of the most important systems as well as the endless number of different compositions which have been synthesized. Large amounts of rare earth elements can be incorporated into most HMFG, and their optical and magnetic properties, which have already been investigated quite extensively, appear very promising in future applications. Nevertheless, most HMFG compositions developed so far still exhibit poor thermal and chemical stability, whereas their optical transparency, of paramount importance for ultralow loss fiber optics, remains dominated by extrinsic factors. Significant research and development work remains to be done so that the full potential usefulness of these glasses may be realized.

Acknowledgements

I am indebted to the Instituto de Engenharia de Sistemas e Computadores, for supporting the present work and I acknowledge the financial support of the Junta Nacional de Investigação Científica e Tecnológica and the Instituto Nacional de Investigação Científica, in Lisboa. I am also grateful to Prof. M.A. Fortes for reviewing the manuscript.

References

Adam, J.L., and W.A. Sibley, 1985a, Mater. Sci. Forum **6**, 641.
Adam, J.L., and W.A. Sibley, 1985b, J. Non-Crystal. Solids **76**, 267.
Adam, J.L., A.D. Docq and J. Lucas, 1988, J. Solid State Chem. **75**, 403.
Al-Jumaily, G.A., J.R. McNeil, B. Bendow and D.J. Martin, 1985, Mater. Sci. Forum **6**, 721.
Aliaga, N., G. Fonteneau and J. Lucas, 1978, Ann. Chim. Fr. **3**, 51.
Allain, J.Y., M. Monerie and H. Poignant, 1989, Electron Lett. **25**, 318.
Almeida, R.M., 1985, Mater. Sci. Forum **6**, 427.
Almeida, R.M., ed., 1987a, Halide Glasses for Infrared Fiberoptics (Martinus Nijhoff, Dordrecht).
Almeida, R.M., 1987b, Vibrational spectroscopy studies of halide glass structure, in: Halide Glasses for Infrared Fiberoptics, ed. R.M. Almeida (Martinus Nijhoff, Dordrecht) pp. 57–72.
Almeida, R.M., 1988, J. Non-Crystal. Solids **106**, 347.
Almeida, R.M., and J.D. Mackenzie, 1981, J. Chem. Phys. **74**, 5954.
Almeida, R.M., and J.D. Mackenzie, 1982, J. Mater. Sci. **17**, 2533.
Almeida, R.M., and J.D. Mackenzie, 1983a, Glastech. Berichte **56**, K850.
Almeida, R.M., and J.D. Mackenzie, 1983b, J. Chem. Phys. **78**, 6502.
Almeida, R.M., and J.D. Mackenzie, 1983c, J. Non-Crystal. Solids **56**, 63.
Almeida, R.M., and J.D. Mackenzie, 1984, J. Non-Crystal. Solids **68**, 203.
Almeida, R.M., and J.D. Mackenzie, 1985, J. Phys. Colloq. (Suppl. 12) **46**, C8–75.
Almeida, R.M., J. Lau and J.D. Mackenzie, 1984, J. Non-Crystal. Solids **69**, 161.
Almeida, R.M., M.C. Conçalves and J.L. Grilo, 1988, Mater. Sci. Forum **32–33**, 427.
Alonso, P.J., V.M. Orera, R. Cases and R. Alcala, 1988, J. Luminescence **39**, 275.
Anderson, O.L., and D.A. Stuart, 1954, J. Am. Ceram. Soc. **37**, 573.
Angell, C.A., 1985, J. Non-Crystal. Solids **73**, 1.
Baldwin, C.M., and J.D. Mackenzie, 1979, J. Am. Ceram. Soc. **62**, 573.
Baldwin, C.M., R.M. Almeida and J.D. Mackenzie, 1981, J. Non-Crystal. Solids **43**, 309.

Bansal, N.P., and R.H. Doremus, 1983, J. Am. Ceram. Soc. **66**, C132.

Bansal, N.P., R.H. Doremus, A.J. Bruce and C.T. Moynihan, 1983, J. Am. Ceram. Soc. **66**, 233.

Bansal, N.P., R.H. Doremus, A.J. Bruce and C.T. Moynihan, 1984, Mater. Res. Bull. **19**, 577.

Bansal, N.P., A.J. Bruce, R.H. Doremus and C.T. Moynihan, 1985, J. Non-Crystal. Solids **70**, 379.

Bendow, B., 1977, Annu. Rev. Mater. Sci. **7**, 23.

Bendow, B., M.G. Drexhage, P.K. Banerjee, J. Goltman, S.S. Mitra and C.T. Moynihan, 1981, Solid State Commun. **37**, 485.

Bendow, B., P.K. Banerjee, M.G. Drexhage, O.H. El-Bayoumi, S.S. Mitra, C.T. Moynihan, D.L. Gavin, G. Fonteneau, J. Lucas and M. Poulain, 1983, J. Am. Ceram. Soc. **66**, C-64.

Bendow, B., P.K. Banerjee, J. Lucas, G. Fonteneau and M.G. Drexhage, 1985, J. Am. Ceram. Soc. **68**, C-92.

Boehm, L., K.H. Chung and C.T. Moynihan, 1988, J. Non-Crystal. Solids **102**, 150.

Bouaggad, A., G. Fonteneau and J. Lucas, 1987, Mater. Res. Bull. **22**, 685.

Boulard, B., 1989, Thesis (Université du Maine, Le Mans, France) unpublished.

Bray, P.J., D.E. Hintenlang, R.B. Mulkern, S.G. Greenbaum, D.C. Tran and M.G. Drexhage, 1983, J. Non-Crystal. Solids **56**, 27.

Brierley, M.C., and P.W. France, 1987, Electron. Lett. **23**, 815.

Brierley, M.C., and C.A. Millar, 1988, Electron. Lett. **24**, 438.

Brown, R.N., B. Bendow, M.G. Drexhage and C.T. Moynihan, 1982, Appl. Opt. **21**, 361.

Bruce, A.J., 1989, private communication.

Burbank, R.D., and F.N. Bensey, 1956, Union Carbide Nuclear Division, Report K-1280.

Busse, L.E., G. Lu, D.C. Tran and G.H. Sigel Jr, 1985, Mater. Sci. Forum **5**, 219.

Canalejo, M., R. Cases and R. Alcala, 1988, Phys. & Chem. Glasses **29**, 187.

Carter, S.F., P.W. France, M.W. Moore and E.A. Harris, 1987, Phys. & Chem. Glasses **28**, 22.

Cases, R., R. Alcala, J. Lucas and G. Fonteneau, 1985a, Mater. Sci. Forum **6**, 749.

Cases, R., D.L. Griscom and D.C. Tran, 1985b, J. Non-Crystal. Solids **72**, 51.

Chamarro, M.A., and R. Cases, 1988a, J. Luminescense **42**, 267.

Chamarro, M.A., and R. Cases, 1988b, Solid State Commun. **68**, 953.

Chamarro, M.A., and R. Cases, 1989, J. Non-Crystal. Solids **107**, 178.

Chandrashekhar, G.V., and M.W. Shafer, 1980, Mater. Res. Bull. **15**, 221.

Chung, K.H., and C.T. Moynihan, 1987, Mater. Sci. Forum **19–20**, 615.

Coey, J.M.D., A. McEvoy and M.W. Shafer, 1981, J. Non-Crystal. Solids **43**, 387.

Comyns, A.E., ed., 1989, Fluoride Glasses (John Wiley, Chichester).

Coupé, R., D. Louer, J. Lucas and A.J. Leonard, 1983, J. Am. Ceram. Soc. **66**, 523.

Dietzel, A., 1942, Z. Elektrochem. **48**, 9.

Drehman, A.J., 1987, Mater. Sci. Forum **19–20**, 483.

Drexhage, M.G., 1985, Heavy metal fluoride glasses, in: Treatise on Materials Science and Technology, Vol. 26, Glass IV, eds M. Tomozawa and R. Doremus (Academic Press, New York) pp. 151–243.

Drexhage, M.G., 1987, Optical properties of fluoride glasses, in: Halide Glasses for Infrared Fiberoptics, ed. R.M. Almeida (Martinus Nijhoff, Dordrecht) pp. 219–234.

Drexhage, M.G., O.H. El-Bayoumi, C.T. Moynihan, A.J. Bruce, K.H. Chung, D.L. Gavin and T.J. Loretz, 1982, J. Am. Ceram. Soc. **65**, C-168.

Drexhâge, M.G., O.H. El-Bayoumi, H. Lipson, C.T. Moynihan, A.J. Bruce, J. Lucas and G. Fonteneau, 1983, J. Non-Crystal. Solids **56**, 51.

Dumbaugh, W.H., and D.W. Morgan, 1980, J. Non-Crystal. Solids **38–39**, 211.

Dupas, C., 1985, Mater. Sci. Forum **6**, 731.

Durteste, Y., M. Monerie and P. Lamouler, 1985, Electron. Lett. **2**, 723.

El-Bayoumi, O.H., C.J. Simmons, E.M. Clausen Jr, M.J. Suscavage and J.H. Simmons, 1987, J. Am. Ceram. Soc. **70**, 510.

Esnault-Grosdemouge, M.A., M. Matecki and M. Poulain, 1985, Mater. Sci. Forum **5**, 241.

Etherington, G., L. Keller, A. Lee, C.N.J. Wagner and R.M. Almeida, 1984a, J. Non-Crystal. Solids **69**, 69.

Etherington, G., C.N.J. Wagner, R.M. Almeida and J. Faber Jr, 1984b, Rep. Hahn-Meitner Institute B **411**, 64.

Eyal, M., R. Reisfeld, C.K. Jørgensen and B. Bendow, 1987, Chem. Phys. Lett. **139**, 395.

Ferrari, M., E. Duval, A. Boyrivent, A. Boukenter and J.L. Adam, 1988, J. Non-Crystal. Solids **99**, 210.

Folweiler, R.C., and D.E. Guenther, 1985, Mater. Sci. Forum **5**, 43.

Fonteneau, G., H. Slim, F. Lahaie and J. Lucas, 1980, Mater. Res. Bull. **15**, 1425.

France, P.W., 1985, U.K. Pat. Application No. GB 2148276A, May.

France, P.W., S.F. Carter, M.W. Moore and J.R.

Williams, 1987, Loss mechanisms in ZrF_4 based IR fibres, in: Halide Glasses for Infrared Fiberoptics, ed. R.M. Almeida (Martinus Nijhoff, Dordrecht) pp. 253–262.

Freiman, S.W., 1980, in: Glass: Science and Technology, Vol. 5, eds D.R. Uhlmann and N.J. Kreidl (Academic Press, New York) pp. 21–75.

Friebele, E.J., and D.C. Tran, 1985, J. Am. Ceram. Soc. **68**, C-279.

Grilo, J.L., M.C. Conçalves and R.M. Almeida, 1987, Mater. Sci. Forum **19–20**, 299.

Guery, J., J.L. Adam and J. Lucas, 1988a, J. Luminescence **42**, 181.

Guery, J., D.J. Chen, C.J. Simmons, J.H. Simmons and C. Jacoboni, 1988b, Phys. & Chem. Glasses **29**, 30.

Hahn, T.-S., W.-J. Cho, S.-S. Choi and S.-J. Park, 1987, J. Non-Crystal. Solids **95–96**, 929.

Haixing, Z., and G. Fuxi, 1985, Mater. Sci. Forum **6**, 617.

Hamill, L.T., and J.M. Parker, 1985a, Phys. & Chem. Glasses **26**, 52.

Hamill, L.T., and J.M. Parker, 1985b, Mater. Sci. Forum **6**, 437.

Harris, E.A., 1987a, Phys. & Chem. Glasses **28**, 112.

Harris, E.A., 1987b, Phys. & Chem. Glasses **28**, 196.

Hart, P., G. Lu and I. Aggerwal, 1988, Mater. Sci. Forum **32–33**, 179.

Hollis, D.B., S. Parke and E.A. Harris, 1985, Mater. Sci. Forum **6**, 621.

Hu, H., and J.D. Mackenzie, 1983, J. Non-Crystal. Solids **54**, 241.

Inoue, H., and I. Yasui, 1987, Phys. & Chem. Glasses **28**, 143.

Inoue, H., H. Hasegawa and I. Yasui, 1985, Phys. & Chem. Glasses **26**, 74.

Izumitani, T., and S. Hirota, 1985, Mater. Sci. Forum **6**, 645.

Izumitani, T., S. Hirota, K. Tanaka and H. Onuki, 1986, J. Non-Crystal. Solids **86**, 361.

Izumitani, T., T. Yamashita, M. Tokida, K. Miura and H. Tajima, 1987, Physical, chemical properties and crystallization tendency of the new fluoroaluminate glasses, in: Halide Glasses for Infrared Fiberoptics, ed. R.M. Almeida (Martinus Nijhoff, Dordrecht) pp. 187–196.

Jacoboni, C., B. Boulard and P. Baniel, 1987, Mater. Sci. Forum **19–20**, 253.

Ji-Jian, C., and H. Jian-Feng, 1988, Mater. Sci. Forum **32–33**, 623.

Jia, Y.-Q., and S.G. Zhang, 1988, Inorg. Chim. Acta. **143**, 137.

Jian-Feng, H., and C. Ji-Jian, 1989a, J. Non-Crystal. Solids **109**, 64.

Jian-Feng, H., and C. Ji-Jian, 1989b, Mater. Lett. **8**, 12.

Kanamori, T., 1983, J. Non-Crystal. Solids **57**, 443.

Kanamori, T., 1987, Mater. Sci. Forum **19–20**, 363.

Kanamori, T., and S. Takahashi, 1985, in: Colloq. Fluoride Glass for Optical Fibers, NTT Corp., Ibaraki, Japan, August 1985.

Kanamori, T., K. Oikawa, S. Shibata and T. Manabe, 1981, Jpn. J. Appl. Phys. **20**, L326.

Kawamoto, Y., and T. Horisaka, 1983, J. Non-Crystal. Solids **56**, 39.

Kawamoto, Y., and I. Nohara, 1987a, Solid State Ionics **22**, 207.

Kawamoto, Y., and I. Nohara, 1987b, Solid State Ionics **24**, 327.

Kawamoto, Y., T. Horisaka, K. Hirao and N. Soga, 1985, J. Chem. Phys. **83**, 2398.

Kawamoto, Y., J. Fujiwara, K. Hirao and N. Soga, 1987, J. Non-Crystal. Solids **95–96**, 921.

Kingery, W.D., H.K. Bowen and D.R. Uhlmann, 1976, Introduction to Ceramics, 2nd Ed. (Wiley, New York).

Klein, L.C., C.A. Handwerker and D.R. Uhlmann, 1977, J. Cryst. Growth **42**, 47.

Klein, P., 1987, Chemical analysis of trace impurities, in: Halide Glasses for Infrared Fiberoptics, ed. R.M. Almeida (Martinus Nijhoff, Dordrecht) pp. 27–33.

Krupke, W.F., 1974, J. Quant. Electron. **QE-10**, 450.

Laval, J.P., R. Papiernik and B. Frit, 1978, Acta Crystallogr. B **34**, 1070.

Laval, J.P., B. Brit and B. Gaudreau, 1979, Rev. Chim. Miner. **16**, 509.

Laval, J.P., B. Brit and J. Lucas, 1988, J. Solid State Chem. **72**, 181.

Le Bail, A., B. Boulard and C. Jacoboni, 1987, Mater. Sci. Forum **19–20**, 127.

Le Toullec, M., C.J. Simmons and J.H. Simmons, 1988, J. Am. Ceram. Soc. **71**, 219.

Lecoq, A., and M. Poulain, 1979, J. Non-Crystal. Solids **34**, 101.

Lecoq, A., and M. Poulain, 1980a, J. Non-Crystal. Solids **41**, 209.

Lecoq, A., and M. Poulain, 1980b, Verres Refract. **34**, 333.

Lehman, L.R., R.E. Hill Jr and G.H. Sigel Jr, 1989, J. Am. Ceram. Soc. **72**, 474.

Leroy, D., and D. Ravaine, 1978, C.R. Acad. Sci. Paris, Ser. C **286**, 413.

Leroy, D., J. Lucas, M. Poulain and D. Ravaine, 1978, Mater. Res. Bull. **13**, 1125.

Loehr, S.R., K.H. Chung and C.T. Moynihan, 1988, J. Am. Ceram. Soc. **71**, C-46.

Lu, G., and J. Bradley, 1985, Mater. Sci. Forum **6**, 551.

Lucas, J., 1985, J. Less-Common Met. **112**, 27.

Lucas, J., M. Poulain and M. Poulain, 1977, German Patent 2 726 170.

Lucas, J., M. Chanthanasinh, M. Poulain, M. Poulain, P. Brun and M.J. Weber, 1978, J. Non-Crystal. Solids **27**, 273.

Lucas, J., H. Slim and G. Fonteneau, 1981, J. Non-Crystal. Solids **44**, 31.

Lucas, J., C.A. Angell and S. Tamaddon, 1984, Mater. Res. Bull. **19**, 945.

Ma, F., Z. Zhen, L. Ye and M. Zhang, 1988, J. Non-Crystal. Solids **99**, 387.

MacFarlane, D.R., and L.J. Moore, 1987, Mater. Sci. Forum **19–20**, 447.

MacFarlane, D.R., J.O. Browne, T.J. Bastow and G.W. West, 1989, J. Non-Crystal. Solids **108**, 289.

Mackenzie, J.D., H. Nasu and J. Sanghera, 1987, Viscosity behavior of halide glasses and melts, in: Halide Glasses for Infrared Fiberoptics, ed. R.M. Almeida (Martinus Nijhoff, Dordrecht) pp. 139–146.

Mansfield, J., 1983, Evaluation of Multicomponent Fluoride Glass Candidates for Double Crucible Drawn Optical Fibers, Final Report No. Rome Air Development Center RADC-TR-83-131.

Matecki, M., M. Poulain, M. Poulain and J. Lucas, 1978, Mater. Res. Bull. **13**, 1039.

Matecki, M., M. Poulain and M. Poulain, 1982a, Mater. Res. Bull. **17**, 1035.

Matecki, M., M. Poulain and M. Poulain, 1982b, Mater. Res. Bull. **17**, 1275.

Matusita, K., T. Komatsu and K. Aizawa, 1987, J. Non-Crystal. Solids **95–96**, 945.

Mcnamara, P., and D.R. MacFarlane, 1987, J. Non-Crystal. Solids **95–96**, 625.

Mecholsky, J.J., A.C. Gonzalez, C.G. Pantano, B. Bendow and S.W. Freiman, 1985a, Mater. Sci. Forum **6**, 703.

Mecholsky, J.J., C.G. Pantano and A.C. Gonzalez, 1985b, Mater. Sci. Forum **6**, 699.

Mehlhorn, B., and R. Hoppe, 1976, Z. Anorg. & Allg. Chem. **425**, 180.

Miniscalco, W.J., and L.J. Andrews, 1988, Mater. Sci. Forum **32–33**, 501.

Miniscalco, W.J., L.J. Andrews, B.T. Hall and D.E. Guenther, 1985, Mater. Sci. Forum **5**, 279.

Miniscalco, W.J., B.A. Thompson and C.F. Fisher, 1987, Mater. Sci. Forum **19–20**, 221.

Miniscalco, W.J., L.J. Andrews, B.A. Thompson, R.S. Quimby, L.J.B. Vacha and M.G. Drexhage, 1988, Electron. Lett. **24**, 28.

Miranday, J.P., C. Jacoboni and R. de Pape, 1979, Rev. Chim. Miner. **16**, 277.

Miranday, J.P., C. Jacoboni and R. de Pape, 1981, J. Non-Crystal. Solids **43**, 393.

Mitachi, S., and T. Miyashita, 1983, Appl. Opt. **22**, 2419.

Mitachi, S., S. Shibata and T. Manabe, 1981, Jpn. J. Appl. Phys. **20**, L337.

Miya, T., T. Terunuma, T. Hosaka and T. Miyashita, 1979, Electron. Lett. **15**, 106.

Miyashita, T., and T. Manabe, 1982, IEEE J. Quant. Electron. **QE-18**, 1432.

Moore, M.W., 1989, Non-Oxide Glass Society News, No. 15, 1.

Moss, T.S., 1959, Optical Properties of Semiconductors (Academic Press, New York) ch. 2.

Moynihan, C.T., 1987, Crystallization behavior of fluorozirconate glasses, in: Halide Glasses for Infrared Fiberoptics, ed. R.M. Almeida (Martinus Nijhoff, Dordrecht) pp. 179–184.

Moynihan, C.T., M.G. Drexhage, B. Bendow, M.S. Boulos, K.P. Quinlan, K.H. Chung and E. Gbogi, 1981, Mater. Res. Bull. **16**, 25.

Moynihan, C.T., D.L. Gavin, K.H. Cung, A.J. Bruce, M.G. Drexhage and O.H. El-Bayoumi, 1983, Glastech. Ber. **56**, K 862.

Moynihan, C.T., A.J. Bruce, D.L. Gavin, S.R. Loehr, S.M. Opalka and M.G. Drexhage, 1984, Polym. & Eng. Sci. **24**, 1117.

Nakata, A.M., J. Lau and J.D. Mackenzie, 1983, in: Environmental Effects on the Strength of Fluoride Glass Fibers, Second Int. Symp. on Halide Glasses, Troy, NY, USA, paper no. 34.

Nakata, A.M., J. Lau and J.D. Mackenzie, 1985, Mater. Sci. Forum **6**, 717.

Neilson, G.F., G.L. Smith and M.C. Weinberg, 1984, Mater. Res. Bull. **19**, 279.

Neilson, G.F., G.L. Smith and M.C. Weinberg, 1985, J. Am. Ceram. Soc. **68**, 629.

Nishii, J., Y. Kaite and T. Yamagishi, 1989, Phys. & Chem. Glasses **30**, 55.

Ohishi, Y., and S. Takahashi, 1986, Appl. Opt. **25**, 720.

Ohishi, Y., S. Mitachi, T. Kanamori and T. Manabe, 1983, Phys. & Chem. Glasses **24**, 135.

Okada, K., K. Miura, I. Masuda and T. Yamashita, 1988, Mater. Sci. Forum **32–33**, 523.

Orera, V.M., P.J. Alonso, R. Cases and R. Alcala, 1988, Phys. & Chem. Glasses **29**, 59.

Pantano, C.G., 1987, Surface chemistry and slow crack-growth behavior of fluorozirconate glasses, in: Halide Glasses for Infrared Fibroptics, ed. R.M. Almeida (Martinus Nijhoff, Dordrecht) pp. 199–217.

Pantano, C.G., and R.K. Brow, 1988, J. Am. Ceram. Soc. **71**, 577.

Papiernik, R., D. Mercurio and B. Frit, 1982, Acta Crystallogr. B **38**, 2347.

Parker, J.M., 1987, Molecular dynamics simulations of fluorozirconate glass structure, in: Halide Glasses for Infrared Fiberoptics, ed. R.M. Almeida (Martinus Nijhoff, Dordrecht) pp. 119–136.

Parker, J.M., A.G. Clare and A.B. Seddon, 1985, Mater. Sci. Forum **5**, 257.

Parker, J.M., G.N. Ainsworth, A.B. Seddon and A.G. Clare, 1986, Phys. & Chem. Glasses **27**, 219.

Parker, J.M., A.B. Seddon and A.G. Clare, 1987, Phys. & Chem. Glasses **28**, 4.

Perazzo, N.L., R. Mossadegh and C.T. Moynihan, 1985, Mater. Sci. Forum **6**, 793.

Poignant, H., 1987, Fluoride glass optical fibers in France, in: Halide Glasses for Infrared Fiberoptics, ed. R.M. Almeida (Martinus Nijhoff, Dordrecht) pp. 265–280.

Poignant, H., J. Le Mellot and Y. Bossis, 1985, Mater. Sci. Forum **5**, 79.

Pollack, S.A., and M. Robinson, 1988, Electron. Lett. **24**, 320.

Poulain, M., 1982, Nature **293**, 279.

Poulain, M., and J. Lucas, 1978, Verres Refract. **32**, 505.

Poulain, M., M. Poulain and J. Lucas, 1975, Mater. Res. Bull. **10**, 243.

Poulain, M., M. Chanthanasinh and J. Lucas, 1977, Mater. Res. Bull. **12**, 151.

Poulain, M., M. Poulain and J. Lucas, 1979, Rev. Chim. Miner. **16**, 267.

Quimby, R.S., 1988, Mater. Sci. Forum **32–33**, 551.

Quimby, R.S., M.G. Drexhage and M.J. Suscavage, 1987, Electron. Lett. **23**, 32.

Randall, M.S., J.H. Simmons and O.H. Al-Bayoumi, 1988, J. Am. Ceram. Soc. **71**, 1134.

Ravaine, D., 1985, Mater. Sci. Forum **6**, 761.

Ravaine, D., and D. Leroy, 1980, J. Non-Crystal. Solids **38–39**, 575.

Ravaine, D., and G. Perera, 1986, J. Am. Ceram. Soc. **69**, 852.

Reisfeld, R., 1987, Optical properties of rare earths and transition metal ions in fluoride glasses, in: Halide Glasses for Infrared Fiberoptics, ed. R.M. Almeida (Martinus Nijhoff, Dordrecht) pp. 237–251.

Reisfeld, R., and C.K. Jørgensen, 1987, Excited State Phenomena in Vitreous Materials, in: Handbook on the Physics and Chemistry of Rare Earths, Vol. 9, eds K.A. Gschneidner Jr and L. Eyring (North-Holland, Amsterdam) pp. 1–90.

Reisfeld, R., M. Eyal, C.K. Jørgensen and C. Jacoboni, 1986, Chem. Phys. Lett. **129**, 392.

Renard, J.P., J.P. Miranday and F. Varret, 1980, Solid State Commun. **35**, 41.

Robinson, M., 1987, High purity components for fluorozirconate glass optical fibers, in: Halide Glasses for Infrared Fiberoptics, ed. R.M. Almeida (Martinus Nijhoff, Dordrecht) pp. 11–24.

Robinson, M., R.C. Pastor, R.R. Turk, D.P. Devor, M. Braunstein and R. Braunstein, 1980, Mater. Res. Bull. **15**, 735.

Robinson, M., R.C. Pastor, R.R. Turk, D.P. Devor and M. Braunstein, 1981, Proc. SPIE **266**, 78.

Sakka, S., and J.D. Mackenzie, 1971, J. Non-Crystal. Solids **6**, 145.

Sanghera, J., 1989, private communication.

Sanz, J., R. Cases and R. Alcala, 1987, J. Non-Crystal. Solids **93**, 377.

Sarhangi, A., 1987, Vapor deposition of fluoride glasses, in: Halide Glasses for Infrared Fiberoptics, ed. R.M. Almeida (Martinus Nijhoff, Dordrecht) pp. 293–301.

Schneider, H.W., 1988, Mater. Sci. Forum **32–33**, 561.

Schroeder, J., 1977, Light scattering of glass, in: Treatise on Materials Science and Technology, Vol. 12, Glass I, eds M. Tomozawa and R.H. Doremus (Academic Press, New York) pp. 157–222.

Schultz, P.C., L.J.B. Vacha, C.T. Moynihan, B.B. Harbison, K. Cadien and R. Mossadegh, 1987, Mater. Sci. Forum **19–20**, 343.

Sen, P.N., and M.F. Thorpe, 1977, Phys. Rev. B **15**, 4030.

Sense, K.A., M.J. Snyder and R.B. Filbert, 1954, J. Phys. Chem. **58**, 995.

Shafer, M.W., and P. Perry, 1979, Mater. Res. Bull. **14**, 899.

Shibata, S., M. Horiguchi, K. Jinguji, S. Mitachi, T. Kanamori and T. Manabe, 1981, Electron. Lett. **17**, 775.

Shinn, M.D., W.A. Sibley, M.G. Drexhage and R.N. Brown, 1983, Phys. Rev. B **27**, 6635.

Sibley, W.A., 1985, Mater. Sci. Forum **6**, 611.

Sigel Jr, G.H., 1977, Optical absorption of glasses, in: Treatise on Materials Science and Technology, Vol. 12, Glass I: Interaction with Electromagnetic Radiation, eds M. Tomozawa and R.H. Doremus (Academic Press, New York) pp. 5–89.

Simmons, C.J., 1987a, J. Am. Ceram. Soc. **70**, 295.

Simmons, C.J., 1987b, J. Am. Ceram. Soc. **70**, 654.

Simmons, C.J., and J.H. Simmons, 1986, J. Am. Ceram. Soc. **69**, 661.

Singh, S.S., 1988, Mater. Sci. Forum **30**, 155.

Slim, H., 1981, Thesis (Université de Rennes, France) unpublished.

Stein, M.L., A. Green, B. Bendow, O.H. El-Bayoumi and M.G. Drexhage, 1983, in: Diamond-Like Carbon Coatings for Protection of Halide Glass Surfaces, Second Int. Symp. on Halide Glasses, Troy, NY, paper No. 48.

Sum, K.H., 1949, U.S. Patent 2 466 509.

Sun, K.H., 1947, J. Am. Ceram. Soc. **30**, 277.

Takahashi, S., S. Shibata, T. Kanamori, S. Mitachi and T. Manabe, 1981, New fluoride glasses for infrared transmission, in: Physics of Fiber Optics, Advances in Ceramics, Vol. II, eds B. Bensow and S.S. Mitra (Am. Ceram. Soc., Columbus, OH) pp. 74–83.

Tanimura, K., M.D. Shinn, W.A. Sibley, M.G. Drexhage and R.N. Brown, 1984, Phys. Rev. B **30**, 2429.

Tatsumisago, M., Y. Akamatsu and T. Minami, 1988, Mater. Sci. Forum **32–33**, 617.

Tick, P.A., 1985, Mater. Sci. Forum **5**, 165.

Tick, P.A., 1988a, Proc. SPIE **843**, 34.

Tick, P.A., 1988b, Mater. Sci. Forum **32–33**, 115.

Tick, P.A., and S. Mitachi, 1989, J. Mater. Sci. Lett. **8**, 436.

Tokiwa, H., Y. Mimura, O. Shinbori and T. Nakai, 1985, J. Lightwave Technol. **3**, 569.

Toth, L.M., A.S. Quist and G.E. Boyd, 1973, J. Phys. Chem. **77**, 1384.

Tran, D.C., R.J. Ginther and G.H. Sigel Jr, 1982, Mater. Res. Bull. **17**, 1177.

Tran, D.C., G.H. Sigel Jr and B. Bendow, 1984, J. Lightwave Technol. **2**, 566.

Tran, D.C., K.H. Levin, M.J. Burk, C.F. Fisher and D. Brower, 1986, Proc. SPIE **618**, 48.

Tregoat, D., 1987, Chemical durability of BaF_2–ThF_4 based glass, in: Halide Glasses for Infrared Fiberoptics, ed. R.M. Almeida (Martinus Nijhoff, Dordrecht) pp. 331–338.

Tregoat, D., M.J. Liepman, G. Fonteneau, J. Lucas and J.D. Mackenzie, 1986, J. Non-Crystal. Solids **83**, 282.

Turnbull, D., 1952, J. Chem. Phys. **20**, 411.

Turnbull, D., and R. Cormia, 1961, J. Chem. Phys. **34**, 820.

Vacha, L.J.B., P.C. Schultz, C.T. Moynihan and S.N. Crichton, 1987, Mater. Sci. Forum **19–20**, 419.

Videau, J.J., J. Portier and B. Piriou, 1979, Rev. Chim. Miner. **16**, 393.

Wagner, C.N.J., S.B. Jost, G. Etherington, M.S. Boldrick, R.M. Almeida, J. Faber Jr and K. Volin, 1987, Mater. Sci. Forum **19–20**, 137.

Walrafen, G.E., M.S. Hokmabadi, S. Guha, P.N. Krishnan and D.C. Tran, 1985, J. Chem. Phys. **83**, 4427.

Walrafen, G.E., M.S. Hokmabadi, M.R. Shariari, G.H. Sigel Jr, R. Pafchek, A. Elyamani and M. Poulain, 1988, Infrared Optical Materials VI, Proc. SPIE **929**, 133.

Weber, M.J., and J. Wong, 1987, Mater. Sci. Forum **19–20**, 141.

Weber, M.J., C.F. Cline, W.L. Smith, D. Milam and R.W. Hellwarth, 1978, Appl. Phys. Lett. **32**, 403.

Weinberg, M.C., G.F. Neilson and G.L. Smith, 1983, J. Non-Crystal. Solids **56**, 45.

Wyckoff, R.W.G., 1963, Crystal Structures, Vol. 1, 2nd Ed. (Interscience, New York) pp. 239–241.

Xiujian, Z., T. Kokubo and S. Sakka, 1987, J. Mater. Sci. Lett. **6**, 143.

Yasui, I., and H. Inoue, 1985, J. Non-Crystal. Solids **71**, 39.

Yeh, D.C., W.A. Sibley, M.J. Suscavage and M.G. Drexhage, 1987, J. Appl. Phys. **62**, 266.

Yoneda, Y., H. Kawazoe, T. Kanazawa and H. Toratani, 1983, J. Non-Crystal. Solids **56**, 33.

Zachariasen, W.H., 1932, J. Am. Chem. Soc. **54**, 3841.

Zhao, X., and S. Sakka, 1988, J. Mater. Sci. **23**, 3455.

Handbook on the Physics and Chemistry of Rare Earths, Vol. 15
edited by K.A. Gschneidner, Jr. and L. Eyring

Chapter 102

KINETICS OF COMPLEXATION AND REDOX REACTIONS OF THE LANTHANIDES IN AQUEOUS SOLUTIONS*

K.L. NASH AND J.C. SULLIVAN
Chemistry Division, Argonne National Laboratory, 9700 S. Cass Ave. Argonne, IL 60439–4831, USA

Contents

* Work performed under the auspices of the Office of Basic Energy Sciences, Division of Chemical Sciences, U.S. Department of Energy, under contract number W-31-109-ENG-38.

1. Introduction

With the electronic configuration $4f^n5s^25p^6$, the lanthanide series provide an opportunity to study the effect of shrinking ionic radius on the electrostatic and ion–dipole interactions of hard-sphere cations in aqueous solutions, uncomplicated by crystal field stabilization effects. The shrinking radii of the lanthanide series is a direct result of the relatively poor shielding of the charge of the nucleus by the deeply submerged f-electrons. In an ideal world, smooth variation of rate parameters might be expected in such a situation. Fortunately, the solution chemistry of the lanthanides displays more interesting variation than simple linear correlation of rate and/or thermodynamic parameters with shrinking cation radius.

Mainly as a result of the inaccessibility of the f-electrons, directed bonding typically observed in the transition elements is not seen in the lanthanides. In many respects, the lanthanides behave more like the alkaline earth elements than like the transition elements. The lanthanides behave as hard-acid cations in solution, preferring interaction with hard-base donors like oxygen and fluoride to that with sulfur or heavy halide donors. Significant lanthanide interaction with nitrogen donors is observed only when steric factors force the interaction (as in aminopolycarboxylic acid complexes).

The hydration number of the trivalent lanthanide ions in solution is generally believed to be eight to nine, with a possible shift from nine to eight coordination near the middle of the series. Water exchange rates are rapid (although less than diffusion-controlled ones), being more labile than the most labile of the first-row transition metal ions (Ti^{3+}, Mn^{3+}), Al^{3+} and Mg^{2+} (Lincoln 1986). Until recent development of high-field NMR (nuclear magnetic resonance) instruments, determination of the rates of water exchange proved to be a moderately difficult task. These aspects of lanthanide solution chemistry have recently been discussed in a splendid review by Lincoln (1986), to which the reader is referred for a detailed discussion. We will confine our discussion of lanthanide complexation and solvation rates to those data which have appeared in the four years since Lincoln's publication.

The trivalent oxidation state is dominant for the lanthanides in aqueous solution. The two redox-active species most commonly known and widely studied are Ce(IV) and Eu(II). Oxidized and reduced forms of neighboring lanthanide elements (e.g. Pr(IV), Sm(II)) have been reported to exist, but these species are formed only in relatively exotic solutions like concentrated carbonate media or as transients in the radiolysis of trivalent lanthanide aqueous solutions. As their stability is very limited, few significant studies of their redox chemistry exist. Our discussion of lanthanide redox reaction rates will deal almost exclusively with the reactions of Ce(IV) and Eu(II). Existing reviews of the literature relating to the redox kinetics of the lanthanides cover the period up to 1979 (Benson 1976, McAuley 1981), so we will confine our discussion primarily to the results of the last 10–12 years.

2. Solvent and ligand exchange kinetics

The rates of solvent and ligand exchange of the lanthanides are in general quite rapid, being accessible only through rapid-kinetic techniques like stopped-flow, temperature- or pressure-jump methods and NMR relaxation. As is true for ligand exchange reactions for the d-transition elements, the rate of water exchange is central to the determination of ligand exchange rates. Until recently, these rates have only been estimated. In the following discussion we will summarize recent developments in water exchange rates and continuing investigations of lanthanide ligand exchange rates.

2.1. *Hydration numbers and water exchange rates*

The composition of the primary hydration sphere of the lanthanide ions continues to be a matter of intense interest from both experimental and theoretical points of view. Lincoln (1986) discussed both hydration numbers and water exchange rates for the lanthanides, ending with the conclusion that the most probable hydrated lanthanide ion is $R(H_2O)_9^{3+}$ for the entire series, with the qualifier that new data could alter this opinion.

This model was used by Cossy et al. (1988) in a variable-temperature ^{17}O-NMR study that determined the rates of water exchange for Tb^{3+} to Yb^{3+}. The derived rates decrease with increasing atomic number and were in good agreement with rate parameters previously reported by Purdie and Farrow (1973) and Fay and Purdie (1969, 1970); the latter were derived from R^{3+}–SO_4^{2-} complexation. In contrast to the observations of Purdie, the results of Cossy et al. appear to indicate that the rate of water exchange continues to increase with lower atomic number (larger ionic radius) for the lanthanide ions lighter than Gd.

The results of calculations by Miyakawa et al. (1988) showed that the difference between free energies of hydration for eight- and nine-coordinated metal ions would energetically favor the nine-coordinated configuration for the lighter members of the series and eight coordination for the heavier lanthanides. A recent neutron scattering study by Cossy et al. (1989a) in dilute perchlorate solutions establishes with great certainty a coordination number of eight for Dy^{3+} and Yb^{3+}, seemingly confirming that a change in hydration number does occur across the lanthanide series.

The results of a subsequent variable-pressure ^{17}O-NMR study by the same authors (Cossy et al. 1989b) were used to suggest that the mechanisms for water exchange of

TABLE 1
Rate and activation parameters for H_2O exchange on $R(H_2O)_8^{3+}$ from Cossy et al. (1989b).

Metal ion	k ($\times 10^7$ s^{-1})	ΔH^* (kJ/mol)	ΔS^* (J mol^{-1} K^{-1})	ΔV^* (cm^3/mol)
Tb	55.8 ($\pm$1.3)	12.1 ($\pm$0.5)	−36.9 ($\pm$1.6)	−5.7 ($\pm$0.5)
Dy	43.4 ($\pm$1.0)	16.6 ($\pm$0.5)	−24.0 ($\pm$1.5)	−6.0 ($\pm$0.4)
Ho	21.4 ($\pm$0.4)	16.4 ($\pm$0.4)	−30.5 ($\pm$1.3)	−6.6 ($\pm$0.4)
Er	13.3 ($\pm$0.2)	18.4 ($\pm$0.3)	−27.8 ($\pm$1.1)	−6.9 ($\pm$0.4)
Tm	9.1 ($\pm$0.2)	22.7 ($\pm$0.6)	−16.4 ($\pm$1.9)	−6.0 ($\pm$0.8)
Yb	4.7 ($\pm$0.2)	23.3 ($\pm$0.9)	−21.0 ($\pm$0.3)	–

Numbers in parentheses represent 1σ uncertainty limit.

the octa-aquolanthanide(III) ions from Tb^{3+} to Tm^{3+} were identical. The activation volumes and entropies for the water exchange reactions are consistent with the interpretation of a concerted associative mechanism for the exchange reaction. The values of the rate and activation parameters from Cossy et al. (1989b) are reproduced in table 1.

2.2. *Lanthanide complexation kinetics*

In addition to a detailed discussion of hydration dynamics of the lanthanides, Lincoln's review also covers the kinetics of solvation in nonaqueous media and complexation kinetics. As our focus is on the aqueous chemistry of the lanthanides, we will not discuss recent developments in nonaqueous lanthanide solution chemistry. The intervening four years have seen the publication of a handful of studies of the kinetics of lanthanide chelate dissociation kinetics. In the following section we will discuss the best of these results.

Recent reports on the rates of lanthanide complexation and dissociation deal exclusively with polydentate ligand complexes. Experimentally, these systems have been investigated by either metal ion or ligand exchange reactions, using mainly spectroscopic methods. The polydentate ligands are aminopolycarboxylic acid complexants with one report of a diaza-crown ether complex. Consistent with Lincoln's review, both the dissociative and interchange mechanisms operate in these systems, often simultaneously. Although recent results do not break any new ground mechanistically, there are a number of complete, well-conceived and properly interpreted studies among them.

Breen et al. (1986) report the results of a stopped-flow investigation of metal ion exchange kinetics between Tb(III) and $Ca(EDTA)^{2-}$. The progress of the reaction was monitored by Tb(III) luminescence. These authors interpret the data to indicate that metal ion exchange occurs via acid-catalyzed dissociation of the $Ca(EDTA)^{2-}$ complex with rapid formation of the Tb(III)–EDTA complex, and through a parallel process involving a dinuclear intermediate. The acid-catalyzed dissociation pathway is the preferred mode.

Brücher and co-workers have long investigated the kinetics of metal ion and ligand exchange for the lanthanide–aminopolycarboxylate complexes. Laurenczy and Brücher (1984) derive the rates of formation of Nd^{3+}, Gd^{3+}, Er^{3+} and Y^{3+} complexes with $H(EDTA)^{3-}$ and $H_2(EDTA)^{2-}$ by analysis of the $Ce(EDTA)^{-}$–R^{3+} exchange rate. The interpretation of their data follows that of Nyssen and Margerum (1970) for the *trans*-(1,2-diaminocyclohexane)tetraacetic acid complexes. By comparing the derived formation rate parameters with order-of-magnitude estimates of the lanthanide hydration rates, the authors conclude that the rate-determining step of the complex formation between R(III) and $H(EDTA)^{3-}$ must be a ring closure reaction, in agreement with most previous data interpretations. For complex formation reactions between R(III) and $H_2(EDTA)^{2-}$, the resolved rates are slower, indicating that proton transfer (from a ligand species in which both nitrogens are protonated) to convert the ligand to a kinetically more active form becomes the rate-limiting process.

Brücher et al. (1984) reported the rate of metal ion and ligand exchange for ethyleneglycol bis(2-aminoethylether)tetraacetic acid (EGTA). As this ligand is potentially octadentate, coordinative saturation of the trivalent lanthanide cation is possible. $Ce(EGTA)^-$ complex dissociation in the presence of Y^{3+} proceeds via a dinuclear intermediate (with and without acid catalysis) and via the acid catalyzed dissociation pathway, with the former being preferred. $Ce(EGTA)^-$ ligand exchange reactions, on the other hand, proceed by the dissociation pathway. A mixed ligand EGTA–Ce(III)–DTPA intermediate is postulated at $pH > 6$. The authors observe that $Ce(EGTA)^- – Y^{3+}$ metal ion exchange is 140 times faster than that for $Ce(EDTA)^-$, while the acid catalyzed ligand exchange reaction is 10^3 times slower. This observation appears to suggest "coordinative saturation" in the $Ce(EGTA)^-$ complex.

Nikitenko et al. (1984a, b, 1985) also observe the same mechanistic pathways for lanthanide EDTA and DTPA complexes. Nikitenko et al. (1984a) observe that the $Nd(DTPA)^{2-} – Yb^{3+}$ exchange mechanism changes as a function of pH. At pH 4–4.5 the reaction proceeds via a dissociative mechanism with a diprotonated intermediate (RH_2L^+); at pH 4.5–5, via a monoprotonated intermediate (RHL); and at pH 5–6, via an associative mechanism and a dinuclear intermediate. At $pH > 5$, the dissociation rate is independent of pH. In Nikitenko et al. (1984b) the authors conclude that both of the exchanging metal ions ($Nd(EDTA)^- + Er^{3+}$, Gd^{3+}, Eu^{3+}, Pr^{3+}) are coordinated in the intermediate.

Finally, we quote the work of Chang et al. (1988) who have reported on the dissociation kinetics of lanthanide (La, Eu, Lu) diaza-crown ether carboxylate complexes, K22MA, K22MP, K22DP (general structure below)

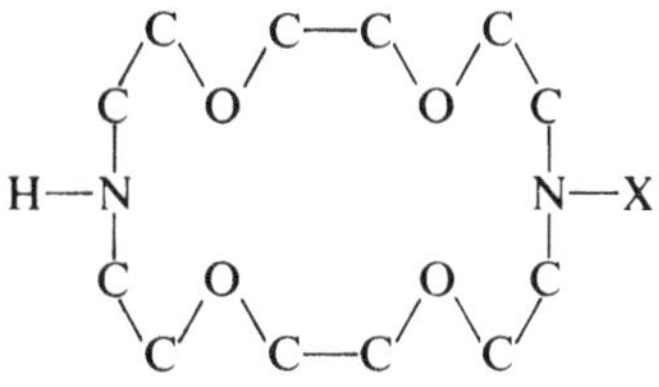

```
       C     C—C     C
      / \   /   \   / \
     C   O       O     C
     |                 |
  H—N                  N—X          (1)
     |                 |
     C   O       O     C
      \ / \     / \   /
       C   C—C     C
```

where X is an acetate group in K22MA and a propionate group in K22MP; a second propionate group replaces H in K22DP. The metal dissociation reaction is in the stopped-flow time regime with Cu(II) used as the exchanging metal ion. No complex is formed with the monopropionate. The dissociation reaction occurs simply via acid-dependent and acid-independent pathways (zeroth order in Cu(II)). The rate of dissociation of K22MA is faster than that for K22DP for all three metal ions. The relative order for the dissociation of the three metal ions (with either complexant) is Lu(III) > La(III) > Eu(III), indicating no simple trend of reaction rate with the cation radius. This pattern is not typically observed (cf. Lincoln 1986, fig. 5, p. 244) although it does correlate inversely with the rate of water exchange inferred from sulfate complexation (i.e. Eu(III) > La(III) > Lu(III)) as reported in Lincoln (1986, fig. 3, p. 241).

3. Oxidation by Ce(IV) in acidic aqueous solutions

In the following section we discuss the kinetics of cerium(IV) oxidation of all classes of organic compounds, nonmetal substrates and metal complexes. The rates of these oxidation reactions vary from extremely slow to extremely rapid. Many different reaction pathways have been postulated.

3.1. *General characteristics of Ce(IV) oxidation reactions*

Cerium(IV) is a powerful oxidant having numerous analytical and preparative applications. The kinetics of oxidation of a variety of substrates have been reported. Interpretation of studies of the kinetics of ceric oxidations is complicated by uncertainty regarding Ce(IV) speciation in aqueous solutions. In perchloric acid, Ce(IV) is at its most powerful, characterized by $E^0 = 1.70$ V (in 1 M acid, Morss 1985). As this potential is above the oxidation potential of water, its existence in aqueous solutions represents a metastable condition. The speciation of Ce(IV) in perchloric acid is dominated by hydrolytic species, even in moderately strong acid. The various ceric species in $HClO_4$ include Ce^{4+} (aquo ion), $Ce(OH)^{3+}$, $Ce(OH)_2^{2+}$, various Ce(IV)–O–Ce(IV) dimers, and an apparent hydrolytic polymer. The latter two polynuclear species are not observed in ceric perchlorate solutions carefully prepared by electrolysis, leading one to suspect that they are one and the same thing, and perhaps are formed irreversibly in aqueous solutions.

In nitric acid solutions the potential is somewhat lowered ($E^0 = 1.61$ V in 1.0 M HNO_3), suggesting a partial complexation of the Ce^{4+} aquo ion by nitrate. The stability and stoichiometry of soluble ceric nitrate complexes is not well established, but certainly must include at least $CeNO_3^{3+}$ and probably $Ce(NO_3)_2^{2+}$ or mixed Ce(IV)–OH–NO_3 species. Since nitrate complexes are intermediate in stability between those of perchlorate (barely perceptible) and sulfate (moderately strong), cerium(IV) speciation in nitric acid probably also includes hydrolytic species. Ceric ammonium nitrate, $(NH_4)_2Ce(NO_3)_6$, is a preferred ceric reagent, because it appears to minimize irreversible polymerization of Ce(IV).

In sulfuric acid solutions the standard potential is further reduced to 1.44 V, indicative of moderately strong complexes. The most widely accepted equilibria for the formation of Ce(IV)–SO_4^{2-} complexes in $[H^+] = 1\,M$ and $2\,M$ ionic strength medium (25 °C) have been reported by Hardwick and Robertson (1951). These are:

$$Ce^{4+} + HSO_4^- \rightleftharpoons CeSO_4^{2+} + H^+, \qquad K_1 = 3500, \qquad (2)$$

$$CeSO_4^{2+} + HSO_4^- \rightleftharpoons Ce(SO_4)_2 + H^+, \qquad K_2 = 200, \qquad (3)$$

$$Ce(SO_4)_2 + HSO_4^- \rightleftharpoons Ce(SO_4)_3^{2-} + H^+, \qquad K_3 = 20. \qquad (4)$$

Various kinetic studies postulate a profusion of mixed cerium(IV)–sulfate, –bisulfate and –hydroxy complexes to attempt explanation of acid and sulfate concentration dependences in cerium(IV) redox studies. The dual effect of sulfate complexation on redox rates (and mechanisms) is to lower the thermodynamic driving force for the reaction (E^0), and to occupy inner-sphere coordination sites around the Ce(IV) cation.

As a result, rates of ceric oxidation in sulfuric acid solutions are normally slower than the corresponding rates in nitric or perchloric acids.

Typically, reactions studied in sulfuric acid conform to simple second-order kinetics (i.e. first order in both ceric ion and substrate), while reactions in $HClO_4$ or in HNO_3 exhibit nonlinear substrate dependence, indicating the formation of a precursor complex. Other indicators of precursor complex formation are the rapid development of color in the solution, and linear plots of $1/k_{obs}$ versus 1/[limiting reagent] (called Michaelis–Menten plots).

A wide variety of oxidizable species have been subjected to oxidation by ceric ion in aqueous (or mixed aqueous–organic) solutions. Partly as a result of this variety, no single set of conditions of temperature and ionic strength/medium can be considered as the representative standard state. For the literature discussed herein, temperatures range from 0–65°C, media include $HClO_4$, H_2SO_4, HNO_3 (including mixtures of these) all in water or in mixed acetic acid/water and at ionic strength (I) from 0.2 M to 6.0 M. Generally, ceric ion is the limiting reagent, although there are a few in which kinetics were studied under second-order conditions, or with the substrate as the limiting reagent.

In the oxidation of organic species, terminal oxidation products are CO_2, HCOOH (formic acid) and H_2O. As Ce(IV) is typically the limiting reagent, intermediate products are often identified to confirm reaction pathways. Since cerium(IV) is a one-electron oxidant and the organic substrates are two-electron donors, the chemistry of the oxidized substrate involves subsequent reactions of free radical species. The presence of free radicals is verified experimentally by the initiation of polymerization upon addition of methacrylic acid, or other easily polymerized organic species. In some cases the radicals are stable enough to have been studied by ESR (electron spin resonance) methods. There are a few examples of reactions which exhibit an inverse dependence on [Ce(III)], which can be interpreted in terms of hydrolytic dimers or reversible reactions of the type

$$\mathrm{Ce(IV) + Subst \rightleftharpoons Ce(III) + Subst^{\bullet}} \qquad (5)$$

in which the organic radical reoxidizes the Ce(III) product. Many of these aspects of ceric oxidations have been discussed in general terms by Hanna et al. (1976).

The kinetics of ceric ion oxidations have been the subject of a number of reviews, though the most recent one covers the literature up to 1979 (Mc Auley 1981). It is the intention of this chapter to discuss the literature of the intervening years, including a few reports previously covered. We will restrict ourselves to the results most completely describing the kinetics of relevant systems, and will deal solely with reaction kinetics in homogeneous aqueous (or mostly aqueous) solutions. We will not include reaction kinetics involving lanthanide–organometallic complexes, lanthanides as catalysts, any heterogeneous or solid-state reactions, or reactions in which the metal ion serves only as an analytical probe for redox chemistry of a complexed ligand species (e.g. NMR shift reagent).

In the following discussion, the results will be grouped according to the general character of the substrate. We will begin with the organic species, proceeding from the least oxidized ones (hydrocarbons) to the most oxidized ones (carboxylic acids), followed by nonmetal reductants, and finally, metal ions and their complexes.

3.2. *Oxidation of hydrocarbons by Ce(IV)*

The primary motivation for these studies is the analysis of the reactivity patterns of organic compounds, when Ce(IV) is used as an oxidant. These patterns are determined for the most part by product analysis of selected series of organic compounds. The results obtained in two studies that bear more directly on the chemical behavior of Ce(IV) as an oxidant for hydrocarbons have been interpreted to indicate different mechanistic behavior of Ce(IV). In a product study of the oxidation of isodurene (1,2,3,5-tetramethyl benzene) by ceric ammonium nitrate compared to anodic oxidation, Eberson and Oberrauch (1979) concluded that the oxidation by Ce(IV) occurs via a H atom transfer from the alkylaromatic compound to Ce(IV). Baciocchi et al. (1980) measured the variation of second-order rate constants for the oxidation of a series of alkylaromatic compounds with added Ce(III). These results along with those from the determination of kinetic deuterium isotope effect were cited to support a mechanism involving radical cations. The Ce(IV)/Ce(III) functions as an electron acceptor/donor in such a mechanism.

3.3. *Oxidation of aminoacids and amines by Ce(IV)*

The electron transfer reactions of these compounds with Ce(IV) provide a fascinating area for the elucidation of the reaction mechanisms of organic compounds. The Ce(IV) is, however, in the majority of these studies used as a one-equivalent acceptor with the obvious result that the bulk of the mechanistic considerations are concerned with the chemical properties of the parent organic compound, subsequent organic radical intermediates, and/or product distribution analyses.

For example, the oxidation of 1-amino-3-arylguanidines by Ce(IV) was studied by Schelenz et al. (1982) in perchloric acid solutions for a homologous series of 17 substituted compounds at four different temperatures. The predominant products of this eight-equivalent oxidation are nitrogen and 3,6-diarylamino-1,2,3,5-terrazine compounds. For such well-defined systems, the resulting isokinetic and Hammet correlation relations calculated from the kinetic data present significant mechanistic information.

There is information from other studies which probe the question of delineation of the electronic spectra and kinetic stability of organic radical intermediates produced by Ce(IV) oxidations. For example, Kemp et al. (1980) determined such spectra for the radical-cation and neutral radical produced in the first two kinetic steps when phenothiazine and phenoxazine are the reductants. The combination of stopped-flow and rapid-scan spectrophotometry were used to evaluate kinetic parameters for these steps.

Gasco and Carlotti (1979) report the results of rate studies of ten dimetacrine, imipramine and phenothiazine drugs. These compounds share a common base structure (6) with varied substituents (X, Y, Z), and have pharmacological activity as tranquilizers, antihistaminics, or anti-depressants. The oxidation reactions in 1.0 *M* H_2SO_4 occur on the stopped-flow time scale and exhibit second-order kinetics implying no complex formation. The initial organic product is a cation radical. The rate constants correlate with the redox potentials of the substrate, where they are known, in accord with the Marcus formalism for outer-sphere electron transfer reactions (table 2),

```
     C   X   C
   C   C   C   C
   |   |   |   |
   C   C   C   C
     C   N   C   Z
         |
         Y
```
(6)

TABLE 2
Ceric oxidation of diarylamino drugs ($[H^+] = 1.0\ M$, $I = 1.0\ M$, $T = 25.0°C$, from Gasco and Carlotti 1979).

Substrate[a]	Second-order rate constant	E^0 (V)
Hydroxypromazine: X = S Y = OH $Z = (CH_2)_3N(CH_3)_2$	$3.5(\pm 0.6) \times 10^6\ M^{-1}s^{-1}$	0.65
Perazine: X = S Y = H $Z = (CH_2)_3N(C_2H_4)_2NCH_3$	$2.0(\pm 0.6) \times 10^6\ M^{-1}s^{-1}$	0.70
Promazine: X = S Y = H $Z = (CH_2)_3N(CH_3)_2$	$1.6(\pm 0.3) \times 10^6\ M^{-1}s^{-1}$	0.72
Levomepromazine: X = S $Y = OCH_3$ $Z = (CH_2)_3N(CH_3)_2$	$1.2(\pm 0.2) \times 10^6\ M^{-1}s^{-1}$	0.725
Chlorpromazine: X = S Y = Cl $Z = (CH_2)_3N(CH_3)_2$	$8.3(\pm 0.3) \times 10^5\ M^{-1}s^{-1}$	0.78
Dietazine: X = S Y = H $Z = CH_2CH(CH_3)N(CH_3)_2$	$3.0(\pm 0.3) \times 10^5\ M^{-1}s^{-1}$	0.83

TABLE 2 (cont'd)

Substrate[a]	Second-order rate constant	E^0 (V)
Promethazine: X = S Y = H Z = $CH_2CH(CH_3)N(CH_3)_2$	$2.7(\pm 0.2) \times 10^5\ M^{-1}s^{-1}$	0.865
Isopromethazine: X = S Y = H Z = $CH(CH_3)CH_2N(CH_3)_2$	$2.3(\pm 0.2) \times 10^5\ M^{-1}s^{-1}$	0.885
Dimetacrine: X = $C(CH_3)_2$ Y = H Z = $(CH_2)_3N(CH_3)_2$	$1.5(\pm 0.1) \times 10^5\ M^{-1}s^{-1}$	–
Imipramine: X = $(CH_2)_2$ Y = H Z = $(CH_2)_3N(CH_3)_2$	$1.3(\pm 0.1) \times 10^5\ M^{-1}s^{-1}$	–

[a] General structure:

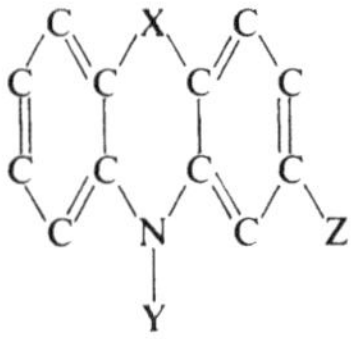

The oxidation of aminopolycarboxylic acids by Ce(IV) produces numerous species which can be used to infer mechanistic pathways. A more well-defined aspect of these systems is revealed in the initial steps of this class of reactions. Prior to the oxidation steps, Ce(IV) forms complexes with these compounds as demonstrated in the studies of Trubacheva and Pechurova (1981) and Hanna and Moehlenkamp (1983). The experimental constraints in the former study with ethylenediaminedisuccinic acid as the substrate resulted in a metastable complex formulated as $Ce(OH)(EDDS)^-$ with $\log K_{stab} = 16.46$. The subsequent redox reaction is extremely slow and exhibits a complex hydrogen ion concentration dependence. The latter system, wherein the substrate is *N*-benzyliminodiacetic acid (BIDA), was described in terms of the sequential reaction:

$$\text{BIDA} + \text{Ce(IV)} \rightleftharpoons \text{complex} \longrightarrow \text{products}. \tag{7}$$

From the kinetic data a value of $2.6 \times 10^4\ M^{-1}$ was estimated for the complex formation quotient.

3.4. *Oxidation of alcohols by Ce(IV)*

Reports of the kinetics of ceric oxidation of a variety of different alcohols have been made. The substrate molecules include mono-, di- and trihydroxyalkanes, cycloalkanols (cyclic alkane alcohols) and phenols (table 3). Hydroxyacids have also been investigated and will be discussed in the section on carboxylic acids. Ceric oxidation of carbohydrates is discussed along with aldehydes and ketones. In those studies with excess substrate, the dominant products (where discussed) are the corresponding aldehydes and ketones. The most remarkable aspect of these investigations is that the resolved values for the stability constants of many of the precursor complexes exceed those observed in analogous Ce(IV)–carboxylic acid oxidations.

Dziegiec, Ignaczak and co-workers have reported the results of ceric oxidation of a variety of aliphatic alcohols in perchloric acid medium (Dziegiec 1981, Ignaczak et al. 1980, Ignaczak and Dziegiec 1978, Ignaczak and Markiewicz 1981). They observe that the reaction occurs through the formation of an intermediate complex. The stability constants derived from the kinetic analysis are 5.1 (± 2.7) M^{-1} for the mono alcohols, 15 (± 11) M^{-1} for the diols wherein the OH groups are not on adjacent carbons, and 52 (± 37) M^{-1} for the 1,2 (or 2,3) diols and glycerine. Although the authors reject the possibility of chelation in these precursor complexes, the large increase in K under the most favorable conditions appears to indicate otherwise. The rates of oxidation of the alcohols in the precursor complexes show little variation except that the C1 and C2 species are oxidized about an order of magnitude more slowly. Prakash et al. (1979) concur that chelation is significant in the ceric oxidation of a series of C4, C5, and C6 diols, although they observe lower rates for oxidation of the alcohol in the Ce(IV)–alcohol complex. The results of Ignaczak and Dziegiec are in good agreement with the prior literature as reviewed earlier by Benson (1976).

The rate of oxidation of hexane-1,6-diol (Behari et al. 1982a) by Ce(IV) has also been investigated in sulfuric acid media. Straightforward second-order kinetics (first order in both [Ce(IV)] and [diol]) appears to be appropriate in the molar sulfate/perchlorate solutions. Based on the observed acid and bisulfate ion dependence, the authors conclude that the two reactive Ce(IV) species are $CeSO_4^{2+}$ and $Ce(SO_4)_2$, which react with the diol in an outer-sphere electron transfer reaction

$$CeSO_4^{2+} + \text{diol} \xrightarrow{k} \text{Ce(III)} + \text{R}^{\bullet}, \tag{8}$$

$$Ce(SO_4)_2 + \text{diol} \xrightarrow{k'} \text{Ce(III)} + \text{R}^{\bullet}. \tag{9}$$

The second-order rate parameters determined (at 50°C) for the $CeSO_4^{2+}$ pathway compares favorably with the combined rate–equilibrium constants reported for the same reactions in perchloric acid.

Although the results of Balasubramanian and Venkatasubramanian (1975) indicate the existence of precursor complexes in the ceric oxidation of alicyclic alcohols (in 50% acetic acid), more recent results from Hanna and Fenton (1983) report that straightforward second-order kinetics apply in the oxidation of C5, C6, and C7 cyclic

TABLE 3
Ceric oxidation of alcohols.

Substrate	Conditions	Rate and equilibrium constants	Ref.[a]
Methanol	1.72 M $HClO_4$	$K=6.2(\pm 1.4)\ M^{-1}$ $k_1=6.4(\pm 0.4)\times 10^{-3}\ s^{-1}$ (20°C)	[1]
Ethanol		$K=4.6(\pm 0.3)\ M^{-1}$ $k_1=5.5(\pm 0.2)\times 10^{-3}\ s^{-1}$	
n-Propanol		$K=4.9(\pm 0.3)\ M^{-1}$ $k_1=5.5(\pm 0.2)\times 10^{-3}\ s^{-1}$ (20°C)	
Isopropanol		$K=0.65(\pm 0.01)\ M^{-1}$ $k_1=2.35(\pm 0.15)\times 10^{-1}\ s^{-1}$ (50°C)	
n-Butanol	2.15 M $HClO_4$ 20–55°C	$K=7.3(\pm 0.4)\ M^{-1}$ $k_1=2.3(\pm 0.1)\times 10^{-2}\ s^{-1}$ (30°C)	[2]
Isobutanol		$K=9.6(\pm 1.0)\ M^{-1}$ $k_1=1.1(\pm 0.1)\times 10^{-2}\ s^{-1}$ (25°C)	
sec-Butanol		$K=4.3(\pm 0.3)\ M^{-1}$ $k_1=3.3(\pm 0.3)\times 10^{-2}\ s^{-1}$ (40°C)	
t-Butanol		$K=3.6(\pm 0.4)\ M^{-1}$ $k_1=1.2(\pm 0.1)\times 10^{-2}\ s^{-1}$ (45°C)	
Ethylene glycol	1.72 M $HClO_4$	$K=19\ M^{-1}$ $k_1=2.6\times 10^{-3}\ s^{-1}$ (20°C)	[3]
Glycerine		$K=77\ M^{-1}$ $k_1=5.1\times 10^{-2}\ s^{-1}$ (20°C)	
1,2-Propanediol		$K=23(\pm 6)\ M^{-1}$ $k_1=2.7(\pm 0.1)\times 10^{-2}\ s^{-1}$ (20°C)	[1]
1,3-Propanediol		$K=8.5(\pm 1.4)\ M^{-1}$ $k_1=2.5(\pm 0.2)\times 10^{-2}\ s^{-1}$ (10°C)	
1,4-Butanediol		$K=12.8\ M^{-1}$ $k_1=1.0(\pm 0.1)\times 10^{-2}\ s^{-1}$ (25°C)	[2]
Butane-2,3-diol	1.72 M $HClO_4$, 0–50°C	$K=91(\pm 25)\ M^{-1}$ $k_1=1.19(\pm 0.23)\times 10^{-2}\ s^{-1}$ (0°C)	[4]
Butane-1,3-diol		$K=31(\pm 2)\ M^{-1}$ $k_1=4.2(\pm 0.2)\times 10^{-2}\ s^{-1}$ (45°C)	
cis-2-butene-1,4-diol		$K=7.9(\pm 0.4)\ M^{-1}$ $k_1=2.9(\pm 0.2)\times 10^{-2}\ s^{-1}$ (20°C)	

TABLE 3 (cont'd)

Substrate	Conditions	Rate and equilibrium constants	Ref.[a]
Propane-1,3-diol	0.5–2.0 M $HClO_4$, 12–20°C	$K=63(\pm 4)\ M^{-1}$ $k_1=1.0\times 10^{-3}\ s^{-1}$ (20°C)	[5]
Butane-1,3-diol		$K=72(\pm 6)\ M^{-1}$ $k_1=6.0\times 10^{-4}\ s^{-1}$	
Butane-1,4-diol		$K=58\ M^{-1}$ $k_1=1.25\times 10^{-3}\ s^{-1}$	
Pentane-1,5-diol		$K=8.0(\pm 1.8)\ M^{-1}$ $k_1=1.67\times 10^{-3}\ s^{-1}$	
Hexane-1,6-diol		$K=15(\pm 4)\ M^{-1}$ $k_1=2.38\times 10^{-3}\ s^{-1}$	
3-Methoxy-1-butanol		$K=18.7(\pm 2.7)\ M^{-1}$ $k_1=2.00\times 10^{-3}\ s^{-1}$	
Hexane-1,6-diol	0.5–4.0 M $HClO_4$/ H_2SO_4	$k=8.69\times 10^{-2}\ M^{-1}\ s^{-1}$ $k'=5.50\times 10^{-4}\ M^{-1}\ s^{-1}$	[6]
Cyclopentanol	2.0 M Na, $HClO_4$, 25°C	$k_2=2.10\ M^{-1}\ s^{-1}$	[7]
Cyclohexanol		$k_2=0.73\ M^{-1}\ s^{-1}$	
Cycloheptanol		$k_2=1.48\ M^{-1}\ s^{-1}$	
Cyclopentanol	50% HOAc, 0.5 M $HClO_4$, 30–40°C	$K=3.00\ M^{-1}$ (30°C) $k_1=4.0\times 10^{-2}\ s^{-1}$	[8]
Cyclohexanol	$I=0.7\ M$	$K=10.0\ M^{-1}$ $k_1=1.0\times 10^{-2}\ s^{-1}$	
Cycloheptanol		$K=4.32\ M^{-1}$ $k_1=4.2\times 10^{-3}\ s^{-1}$	
Cyclooctanol		$K=8.27\ M^{-1}$ $k_1=3.3\times 10^{-3}\ s^{-1}$	
trans,2-Chlorocyclohexanol		$k_2=0.0115\ M^{-1}\ s^{-1}$	
trans,2-Phenylcyclohexanol		$k_2=0.45\ M^{-1}\ s^{-1}$	
4,4'Biphenyldiol	4.0 M $HClO_4$, 6–16°C	$k_2=1.3(\pm 0.2)\times 10^6\ M^{-1}\ s^{-1}$ (16°C)	[9]
Hydroquinone phosphate	0.5 M H_2SO_4, 18.3–39.3°C	$k_2=9.5(\pm 0.2)\times 10^4\ M^{-1}\ s^{-1}$	[10]
Hydroquinone sulfate		$k_2=5.9(\pm 0.2)\times 10^4\ M^{-1}\ s^{-1}$	
Hydroquinone		$k_2=1.90(\pm 0.08)\times 10^4\ M^{-1}\ s^{-1}$	
Methylhydroquinone		$k_2=1.04(\pm 0.05)\times 10^4\ M^{-1}\ s^{-1}$	
Hydroquinonesulfonate		$k_2=1.42(\pm 0.08)\times 10^5\ M^{-1}\ s^{-1}$	
Methoxyhydroquinone		$k_2=1.80(\pm 0.07)\times 10^5\ M^{-1}\ s^{-1}$	
Hydroxyhydroquinone		$k_2=2.87(\pm 0.08)\times 10^5\ M^{-1}\ s^{-1}$	
Bromohydroquinone		$k_2=1.37(\pm 0.02)\times 10^5\ M^{-1}\ s^{-1}$	

[a] References: [1] Ignaczak et al. (1980); [2] Dziegiec (1981); [3] Ignaczak and Dziegiec (1978); [4] Ignaczak and Markiewicz (1981); [5] Prakash et al. (1979); [6] Behari et al. (1982a); [7] Hanna and Fenton (1983); [8] Balasubramanian and Venkatasubramanian (1975); [9] Pelizzetti et al. (1976); [10] Holwerda and Ettel (1982).

alcohols in aqueous $HClO_4$. The second-order rates are about an order of magnitude faster than the combined rate-stability constants reported by Balasubramanian and Venkatasubramanian. Hanna and Fenton also report the rate of oxidation of the corresponding α-hydroxy acids, which will be discussed in sect. 3.6.

Finally, there are two reports describing the oxidation of phenols. The first is from Pelizzetti et al. (1976) in which 4,4′-biphenyl diol is oxidized by Ce(IV) in perchlorate medium at reduced temperature. The product of oxidation, which proceeds at stopped-flow lifetimes, is the corresponding 4,4′-biphenoquinone. The oxidation reaction was studied under second-order conditions and indicates no acid concentration dependence with very rapid rates. The activation parameters are $\Delta H^* = 38$ (± 13) $kJ\,mol^{-1}$ and $\Delta S^* = -13$ (± 42) $J\,mol^{-1}\,K^{-1}$, consistent with an intramolecular rate-determining step.

Holwerda and Ettel (1982) report on the oxidation of substituted hydroquinones and hydroquinone esters 4-hydroxyphenylsulfuric and 4-hydroxyphenylphosphoric acids in acidic sulfate solutions. These reactions also occur on the stopped-flow time scale, and the rates appear to be directly proportional to the acidity. The rates for oxidation of the mono-substituted hydroquinones ($-CH_3$, $-SO_3H$, $-OCH_3$, $-OH$, $-Br$) are nearly constant. Within this series the thermodynamic driving force ΔG^0_{12} (as defined in the Marcus formalism) is small and correlates with ΔG^*_{12} (according to the Marcus relationship) for H_2Q–OH, H_2Q–OCH_3, H_2Q–Br and H_2Q–CH_3, implying the expected outer-sphere reaction pathway. The oxidation of H_2Q and H_2Q–SO_3H does not conform to this model, suggesting a contribution from an inner-sphere process. The oxidation of the phosphoric and sulfuric acid esters exhibits a nonlinear acid concentration dependence with a positive intercept, implying the existence of acid concentration-dependent and -independent rate processes. This is in contrast to the acid concentration independence reported earlier for the same reactions in perchlorate solutions (Wells and Kuritsyn 1969). The acid concentration dependence is rationalized in terms of weak protonation of coordinated sulfate in the ceric complex.

3.5. *Oxidation of aldehydes and ketones by Ce(IV)*

The next stage of oxidation for organic compounds is to aldehydes and ketones. In acid solutions, aldehydes and ketones are often hydrolyzed to produce ketohydrates (two OH groups attached to the same carbon atom). They also undergo keto–enol tautomerism to convert into more acidic species which are then more capable of forming ionic bonds with metal ions. The principal kinetic question is whether the keto, enol, or ketohydrate form is the one involved in the oxidation reaction. The results of the most complete investigations of ceric oxidation of ketones and aldehydes are contained in table 4.

The simplest of the aldehydes is formaldehyde, whose oxidation by Ce(IV) in 2.0 *M* perchlorate media has been studied by Husain (1977). It is presumed that formaldehyde exists as a ketohydrate in acid solution (the hydration constant is $\sim 10^4\ M^{-1}$). Michaelis–Menten kinetics describe the results, indicating the formation of a precursor complex. A detailed mechanism permits a calculation of the equilibrium quotients for formation of the Ce(IV)–formohydrate complex and the ionization of a

TABLE 4
Ceric oxidation of ketones and aldehydes.

Substrate	Conditions	Rate and equilibrium constants	Ref.[a]
Formaldehyde	Na, $HClO_4$, 2.0 *M*, 16.0–48.9°C	$K = 5.5(\pm 0.6)\,M^{-1}$ $k_1 = 0.185(\pm 0.02)\,s^{-1}$ $K' = 1.83(\pm 0.2)\,M$ $k' = 0.100(\pm 0.010)\,s^{-1}$ (28.5°C)	[1]
2-Furfural	25% acetic acid, 40–60°C	$k_3 = 0.0249\,M^{-2}\,s^{-1}$	[2]
Benzaldehyde	60% acetic acid, H_2SO_4, 55°C	$k_2 = 1.07 \times 10^{-1}\,M^{-1}\,s^{-1}$	[3]
p-Methylbenzaldehyde		$k_2 = 1.18 \times 10^{-1}\,M^{-1}\,s^{-1}$	
p-Chlorobenzaldehyde		$k_2 = 7.28 \times 10^{-2}\,M^{-1}\,s^{-1}$	
m-Chlorobenzaldehyde		$k_2 = 6.58 \times 10^{-2}\,M^{-1}\,s^{-1}$	
m-Nitrobenzaldehyde		$k_2 = 3.60 \times 10^{-2}\,M^{-1}\,s^{-1}$	
p-Nitrobenzaldehyde		$k_2 = 3.51 \times 10^{-2}\,M^{-1}\,s^{-1}$	
Methyl(isopropyl)ketone	$H_2SO_4/HClO_4$, 1.0–3.0 *M*, 30–50°C	$k_2 = 6.8 \times 10^{-3}\,M^{-1}\,s^{-1}$ (2.0 *M* H_2SO_4, 40°C)	[4]
Diethylketone	H_2SO_4 (2.0 *M*), $I = 3.0\,M$, 40°C	$k_2 = 5.2 \times 10^{-3}\,M^{-1}\,s^{-1}$	[5]
Cyclohexanone	$H_2SO_4/HClO_4$, 1.5–2.0 *M*, 30–40°C	$k_2 = 3.0 \times 10^{-2}\,M^{-1}\,s^{-1}$ (0.25 *M* H^+, 1.25 *M* HSO_4^-, 30°C)	[6]
2-methylcyclohexanone		$k_2 = 2.7 \times 10^{-2}\,M^{-1}\,s^{-1}$ (0.25 *M* H^+, 1.20 *M* HSO_4^-, 30°C)	
Phenacyl bromide	40–70% acetic acid/H_2SO_4	$k_2 = 1.25 \times 10^{-2}\,M^{-1}\,s^{-1}$ (3.0 *M* H_2SO_4, 45°C)	[7]
p-Methylphenacyl bromide		$k_2 = 4.27 \times 10^{-2}\,M^{-1}\,s^{-1}$	
p-Nitrophenacyl bromide		$k_2 = 3.75 \times 10^{-2}\,M^{-1}\,s^{-1}$	
m-Nitrophenacyl bromide		$k_2 = 2.36 \times 10^{-2}\,M^{-1}\,s^{-1}$	
p-Methoxyphenacyl bromide		$k_2 = 2.11 \times 10^{-2}\,M^{-1}\,s^{-1}$	
m-Bromophenacyl bromide		$k_2 = 2.02 \times 10^{-2}\,M^{-1}\,s^{-1}$	
p-Bromophenacyl bromide		$k_2 = 1.71 \times 10^{-2}\,M^{-1}\,s^{-1}$	
ℓ-Sorbose	0.8–2.05 *M*, $NaHClO_4$ 60°C	$k_2 = 1.15 \times 10^{-3}\,M^{-1}\,s^{-1}$	[8]
ℓ-Sorbose	$HClO_4$, 1.0 *M*	$K = 120\,M^{-1}$ $k_1 = 0.3\,s^{-1}$ $k_1' = (5.1–8.9) \times 10^{-4}\,s^{-1}$	[9]
d-Galactose		$K = 100\,M^{-1}$ $k_1 = 0.2\,s^{-1}$ $k_1' = (7.7–8.8) \times 10^{-5}\,s^{-1}$	
d-Glucose		$K = 34\,M^{-1}$ $k_1 = 0.06\,s^{-1}$ $k_1' = (6.4–7.8) \times 10^{-5}\,s^{-1}$	
d-Mannose		$K = 38\,M^{-1}$ $k_1 = 0.03\,s^{-1}$ $k_1' = (2.4–3.7) \times 10^{-5}\,s^{-1}$	
d-Fructose		$K = 140\,M^{-1}$ $k_1 = 0.3\,s^{-1}$ $k_1' = (1.8–4.5) \times 10^{-4}\,s^{-1}$	

TABLE 4 (cont'd)

Substrate	Conditions	Rate and equilibrium constants	Ref.[a]
1-Arabinose		$K = 67\,M^{-1}$	
		$k_1 = 0.4\,s^{-1}$	
		$k_1' = (1.7\text{–}2.3) \times 10^{-4}\,s^{-1}$	
d-Ribose		$K = 1000\,M^{-1}$	
		$k_1 = 0.1\,s^{-1}$	
		$k_1' = (5.5\text{–}6.6) \times 10^{-5}\,s^{-1}$	
d-Xylose		$K = 42\,M^{-1}$	
		$k_1 = 0.3\,s^{-1}$	
		$k_1' = (1.8\text{–}2.3) \times 10^{-4}\,s^{-1}$	

[a] References: [1] Husain (1977); [2] Gopalan and Kannamma (1984); [3] Saiprakash et al. (1975); [4] Singh et al. (1978); [5] Saxena et al. (1978); [6] Behari et al. (1982b); [7] Kishan and Sundaram (1985); [8] Kale and Nand (1982); [9] Virtanen et al. (1987).

proton from the complex

$$Ce^{4+} + CH_2(OH)_2 \xrightleftharpoons{K_1} Ce(CH_2(OH)_2)^{4+}, \tag{10}$$

and

$$Ce(CH_2(OH)_2)^{4+} \xrightleftharpoons{K_2} Ce(CH_2(OH)O)^{3+} + H^+. \tag{11}$$

The intramolecular electron transfer reaction proceeds through two parallel pathways involving the two complexes

$$\left[\mathrm{Ce(IV)}\ \mathrm{C(H)(OH)(H)(OH)}\right] \xrightarrow{k_1} \mathrm{Ce(III)} + {}^{\cdot}\mathrm{C(OH)(H)(OH)} \tag{12}$$

$$\left[\mathrm{Ce(IV)}\ \mathrm{C(H)(O^-)(H)(OH)}\right] \xrightarrow{k_2} \mathrm{Ce(III)} + {}^{\cdot}\mathrm{C(OH)(H)(OH)} \tag{13}$$

Although the reaction was previously reported to be outer-sphere, Hargreaves and Sutcliffe (1955) did suggest the existence of a precursor complex under reaction conditions which made kinetic measurements inaccessible to their conventional techniques.

The following results indicate that Ce(IV)–aldehyde precursor complexes are not important when aromatic ring systems are substituted for one of the hydrogens in formaldehyde. Saiprakash et al. (1975) report that the substituted benzaldehydes are oxidized without the formation of complexes in 60% acetic acid/H_2SO_4. Gopalan and Kannamma (1984) report that for the oxidation of 2-furfural in 25% acetic acid/sulfuric acid, the oxidation reaction is first-order in [Ce(IV)], [substrate] and $[H^+]$ from 0.25–1.25 M H_2SO_4.

Likewise, Kishan and Sundaram (1985, 1980) report that substituted phenacyl bromides (C_6H_5COBr) are oxidized (in 40–70% acetic acid/H_2SO_4) without complex formation. The reaction is catalyzed by acid, but the pseudo-first-order fits depend on the initial [Ce(IV)]. This observation is attributed to the presence of an unreactive Ce(IV) trimer, which has been reported to be present in acetic acid solutions. The rate of reaction of phenacyl bromides substituted with either electron-withdrawing or electron-donating substituents is faster than that of unsubstituted phenacyl bromide. The activation parameters for the *p*-methyl and *p*-methoxyl substituents suggest that a different mechanism operates for these systems. Where comparable substituents exist, the rates for oxidation of phenacyl bromides compare favorably with those for benzaldehyde cited above.

In explaining the ceric oxidation of mono-ketones in aqueous sulfuric acid solutions, Singh et al. (1978) and Saxena et al. (1978) have proposed the formation of a precursor complex in the oxidation of methyl(isopropyl)ketone and diethylketone, respectively. The existence of a complex is not supported by the data, as the reaction is first order in the substrate in both cases. The authors also argue for the enol as the reactive substrate species without offering substantial proof of its existence. We may safely note that in media of 3.0 M total ionic strength, 2.0 M H_2SO_4, $T = 40°C$, the second-order rate parameter indicates that methyl(isopropyl)ketone and diethylketone are oxidized at nearly identical rates.

Behari et al. (1982b) find that ceric oxidation of cyclohexanone and methylcyclohexanone in sulfuric acid solutions do conform to the Michaelis–Menten rate law. By comparing the rate of oxidation with that for enolization (determined by reaction rate for the substrate with iodine), the authors establish that the reaction must involve the ketonic form of the substrate. Earlier results reported by Benson (1976) agree with this assessment and suggest further that a C–H bond is broken in the rate-determining step. The authors are unable to resolve the rate and equilibrium parameters for electron transfer and precursor complex stability. The cyclic ketones are oxidized nearly an order of magnitude faster and under milder conditions than the aliphatic ketones.

Carbohydrates are aldehyde or ketone derivatives of polyhydric alcohols, of C-5 or C-6 carbon backbones. They exist in both straight chain and cyclic forms. The most important carbohydrates are starches, sugars, celluloses and gums. While a number of investigations of the kinetics of ceric oxidations of carbohydrates have been reported, only the following two are complete enough to discuss the kinetic results.

Kale and Nand (1982) investigated the oxidation of ℓ-sorbose in perchlorate medium. The rate is first-order in both Ce(IV) and substrate indicating that the reaction proceeds without formation of a precursor complex. From the observed acid concentration dependence, the authors determine a value for the first hydrolysis constant of Ce(IV) of 0.66 M. The authors conclude that the rate-determining step is probably a one-electron transfer from the keto group of sorbose which upon re-arrangement produces formaldehyde as the first oxidation product.

In contrast, Virtanen et al. (1987) report that Ce(IV) oxidation of eight mono-saccharides (including ℓ-sorbose) in perchloric acid occurs with the formation of two different precursor complexes of differing reactivity. The first one forms upon mixing in the stopped-flow instrument. This complex dissociates partly by oxidation of the substrate and partly by the formation of a second, more stable complex. The second complex is oxidized much more slowly than the first one. The authors rationale for the observed complexity of the system is to suggest that the initial complex is between the open chain form of the sugars and the metal ion, which is subsequently either oxidized or stabilized by forming a second complex, perhaps involving a ring form of the sugar.

3.6. *Oxidation of carboxylic acids by Ce(IV)*

The oxidation of carboxylic acids by ceric ion has been by far the greatest focus for study of ceric oxidations. It is generally expected that such oxidations should occur with the formation of precursor complexes, as these species form the thermo-dynamically strongest complexes of all non-hetero-atom organic compounds with the lanthanides. Only few thermodynamic data exist for ceric–carboxylate complexes, for obvious reasons. However, a moderately large thermodynamic data base for the stability of carboxylate complex exists for metal ions which could be considered reasonable analogues of Ce(IV).

The closest redox-stable analogue of Ce(IV) is thorium(IV), for which a large data base of thermodynamic parameters is available for the carboxylic acid complexes (Martell and Smith 1977). Using the ionic radii of Shannon (1976) and recalling that the stability of lanthanide and actinide complexes is derived almost exclusively from electrostatics, we can estimate that a 16% increase in the log of the stability quotients for thorium (since $\Delta G \propto Z^2/r \propto \log K_{eq}$) should provide a reasonable estimate for the corresponding complexes of cerium(IV) [$r_{Ce}(CN = 8) = 0.97$ Å, $r_{Th}(CN = 10) = 1.13$ Å, $(1/r_{Ce})/(1/r_{Th}) = 1.16$, CN = coordination number].

Since there is some uncertainty with respect to the identity of Ce(IV) species in acidic solutions, the pre-equilibria can be formulated to occur by at least three different reactions. If we assume that Ce^{4+} is the dominant cerium species and the precursor complex is deprotonated, the appropriate equilibrium and constant are described by:

$$Ce^{4+} + HL \rightleftharpoons CeL^{3+} + H^+, \quad K_{eq} = \beta_{CeL}/K_a, \tag{14}$$

where $\beta_{CeL} = [CeL]/([Ce][L])$ and $K_a = [HL]/([H][L])$. Alternatively, if the principal Ce(IV) species is $Ce(OH)^{3+}$, as suggested by Baes and Mesmer (1976), the appropriate thermodynamic relationship and equilibrium constant are given by

$$Ce(OH)^{3+} + HL \rightleftharpoons CeL^{3+} + H_2O, \quad K_{eq} = \beta_{CeL}/(K_a K_h), \tag{15}$$

where K_h is the first hydrolysis constant for Ce(IV) ($[CeOH][H]/[Ce] = 10^{1.0}$ *M*). The third possibility is that the precursor complexes are of the ion–dipole type,

TABLE 5
Estimated thermodynamic parameters for cerium(IV)–carboxylate complexes.

Carboxylic acid	pK_a	$\log \beta_{CeL}$	β_{CeL}/K_a	$\beta_{CeL}/(K_a K_h)$
Formic	3.53	3.15	0.4 (±0.2)	0.04 (±0.02)
Acetic	4.57	3.95	0.2 (±0.1)	0.02 (±0.01)
Propionic	4.67	4.00	0.2 (±0.1)	0.02 (±0.01)
Isobutyric	4.64	3.91	0.2 (±0.1)	0.02 (±0.01)
Cl-acetic	2.64	2.83	1.5 (±0.8)	0.2 (±0.1)
Glycolic	3.62	4.06	2.8 (±1.4)	0.3 (±0.1)
Lactic	3.64	4.27	4.3 (±2.1)	0.4 (±0.2)
α-Hydroxyisobutyric	3.77	4.49	5.2 (±2.5)	0.5 (±0.2)
Mandelic	3.17	3.94	5.9 (±3.0)	0.6 (±0.3)
Benzilic (diphenylhydroxyacetic)	2.80	3.92	13 (±5)	1.3 (±0.6)
Glyoxylic	2.91	4.45	35 (±15)	3.5 (±1.5)
Pyruvic	2.45	3.74	19 (±10)	2.0 (±1.0)
Malonic ($pK_1 + pK_2$)	7.67	8.6	8.5 (±4.0)	0.8 (±0.4)

formed without deprotonation of the carboxylic acid. In this case the available thermodynamic data can provide only inferred help.

Using the data of Martell and Smith (1977) and Baes and Mesmer (1976), we have compiled a table of estimated equilibrium quotients for the above reactions (table 5), wherein HL represents various carboxylic acids considered in this review. We will refer to this table periodically to consider the validity of the authors formulation of precursor complexes.

If precursor complexes are formed in sulfuric acid solutions, they will almost certainly be mixed sulfate–carboxylate complexes, and the thermodynamic data will be of only peripheral use. In nitrate and especially perchlorate media, the thermodynamic data should be directly comparable as Ce(IV)–NO_3^- complexes are weak and perchlorate complexes are unknown in aqueous solutions. It should be possible to compare the calculated constants with the kinetically derived data to gain further insight into the nature of the precursor complexes, and to assess the validity of authors claims with respect to complex stability and stoichiometry.

In the following discussion we will consider the oxidation of carboxylic acids in the general order: monocarboxylic acids, hydroxy carboxylic acids, keto carboxylic acids, polycarboxylic acids, esters, and finally ascorbic acid (see table 6). Oxidation of aminopolycarboxylic acids has been reviewed in sect. 3.3 of this report.

Benson (1976) indicates that ceric oxidation of simple carboxylic acids occurs through decarboxylation with co-generation of alkyl radicals. Vasudevan and Mathai (1978) have investigated the oxidation of a series of monocarboxylic acids by Ce(IV) in 1 *M* perchloric acid. The reactions conform to the Michaelis–Menten rate law, indicating the existence of the expected precursor complexes. The intramolecular electron transfer rates and equilibrium quotients for the formation of the precursor complex are given in table 6. Comparing the kinetically derived stability constants with the β_{CeL}/K_a values in table 5, good agreement is found for the strongest acid (Cl-acetic), but for the others the thermodynamic prediction is 5–15 times lower than the

TABLE 6
Ceric oxidation of carboxylic acids.

Substrate	Conditions	Rate and equilibrium constants	Ref.[a]
Formic acid	1.0 M $HClO_4$, 30–65°C	$K = 7.43\ M^{-1}$ $k_1 = 5.11 \times 10^{-5}\ s^{-1}$	[1]
Acetic acid		$K = 1.09\ M^{-1}$ $k_1 = 3.11 \times 10^{-5}\ s^{-1}$	
Chloroacetic acid		$K = 1.30\ M^{-1}$ $k_1 = 1.06 \times 10^{-5}\ s^{-1}$	
Propionic acid		$K = 2.34\ M^{-1}$ $k_1 = 1.68 \times 10^{-5}\ s^{-1}$	
Isobutyric acid		$K = 1.56\ M^{-1}$ $k_1 = 3.08 \times 10^{-5}\ s^{-1}$	
Pivalic acid		$K = 1.96\ M^{-1}$ $k_1 = 9.75 \times 10^{-4}\ s^{-1}$	
Phenylacetic acid	Acetonitrile	$K = 0.15\ M^{-1}$ $k_1 = 2.29 \times 10^{-4}\ s^{-1}$	
Propionic acid	1.50 M $HClO_4$, 45–60°C	$k_2 = 5.06 \times 10^{-4}\ M^{-1}\ s^{-1}$	[2]
Glycolic acid	Na, H_2SO_4, I = 1.45 M, 30–50°C	$k_2 = 1.26 \times 10^{-2}\ M^{-1}\ s^{-1}$ (30°C, 0.25 M H_2SO_4) $k_2 = 3.19 \times 10^{-2}\ M^{-1}\ s^{-1}$ (30°C, 0.05 M H_2SO_4)	[3]
Glycolic acid	H, $NaNO_3/HClO_4$	$K = 119(\pm 13)\ M^{-1}$ $k_1 = 0.37\ s^{-1}$	[4]
Lactic acid	$I = 1.5\ M$, 6.4–30°C	$K = 191(\pm 14)\ M^{-1}$ $k_1 = 2.25\ s^{-1}$	
α-Hydroxyisobutyric acid		$K = 372(\pm 46)\ M^{-1}$ $k_1 = 2.57\ s^{-1}$	
Mandelic acid		$K = 645(\pm 120)\ M^{-1}$ $k_1 = 92\ s^{-1}$	
Mandelic acid	$M_2SO_4/MClO_4$ (M = H, Li, Na, K), 0.79–3.0 M 35–45°C	$k_2 = 1.0(\pm 0.2) \times 10^{-2}\ M^{-1}\ s^{-1}$ ($Ce(SO_4)$ + Subst) $k'_2 = 1.22(\pm 0.04) \times 10^{-1}\ M^{-1}\ s^{-1}$ ($HCe(SO_4)_3$ + Subst)	[5]
p-Methylmandelic acid	$(H, Li)_2SO_4/(H, Li)ClO_4$, 25°C	$k_2 = 1.1(\pm 0.3) \times 10^{-2}\ M^{-1}\ s^{-1}$ $k'_2 = 1.44(\pm 0.05) \times 10^{-1}\ M^{-1}\ s^{-1}$	[6]
p-Chloromandelic acid	$(H, Na)_2SO_4/(H, Na)ClO_4$, 25–45°C	$k_2 = 1.2(\pm 0.3) \times 10^{-2}\ M^{-1}\ s^{-1}$ $k'_2 = 1.46(\pm 0.02) \times 10^{-1}\ M^{-1}\ s^{-1}$	[7]
p-Nitromandelic acid	$H_2SO_4/HClO_4$, 2.00 M	$k_2 = 1.3(\pm 0.4) \times 10^{-2}\ M^{-1}\ s^{-1}$ $k'_2 = 1.76(\pm 0.02) \times 10^{-1}\ M^{-1}\ s^{-1}$	[8]
p-Methoxymandelic acid	25–45°C	$k_2 = 6.82\ M^{-1}\ s^{-1}$	
Substituted benzilic acids	1.0 M H_2SO_4	$k_2 = 11\ M^{-1}\ s^{-1}$ (p-Z = H) $k_2 = 99\ M^{-1}\ s^{-1}$ (p-Z = OCH_3) $k_2 = 4.7\ M^{-1}\ s^{-1}$ (o-Z = Cl)	[9]

TABLE 6 (cont'd)

Substrate	Conditions	Rate and equilibrium constants	Ref.[a]
	Acetonitrile	$k_2 = 1.48 \times 10^4\ M^{-1}s^{-1}$ (p-Z = H)	
		$k_2 = 5.15 \times 10^3\ M^{-1}s^{-1}$ (o-Z = Cl)	
		$k_2 = 1.44 \times 10^4\ M^{-1}s^{-1}$ (p-Z = Cl)	
		$k_2 = 1.88 \times 10^4\ M^{-1}s^{-1}$ (p-Z = CH_3)	
		$k_2 = 2.66 \times 10^4\ M^{-1}s^{-1}$ (p-Z = OCH_3)	
		$k_2 = 3.78 \times 10^3\ M^{-1}s^{-1}$ (p-Z = NO_2)	
	50% Acetic acid, 0.2 M $HClO_4$	$k_2 = 2.98 \times 10^3\ M^{-1}s^{-1}$ (p-Z = H)	
		$k_2 = 2.32 \times 10^3\ M^{-1}s^{-1}$ (o-Z = Cl)	
		$k_2 = 3.04 \times 10^3\ M^{-1}s^{-1}$ (p-Z = Cl)	
		$k_2 = 4.20 \times 10^3\ M^{-1}s^{-1}$ (p-Z = CH_3)	
		$k_2 = 8.20 \times 10^3\ M^{-1}s^{-1}$ (p-Z = OCH_3)	
		$k_2 = 2.54 \times 10^3\ M^{-1}s^{-1}$ (p-Z = NO_2)	
Benzilic acid	1.45 M H_2SO_4, 25°C	$k_2 = 4.63(\pm 0.04)\ M^{-1}s^{-1}$	[10]
	0.5 M $HClO_4$	$K = 870$ (Ce(IV) + HLH = Ce(IV)LH + H) $k_1 = 95\ s^{-1}$ $K' = 7$ (Ce(IV)LH + HLH = Ce(IV)$(HL)_2$ + H) $k'_1 = 20\ s^{-1}$	
1-hydroxycyclopentanoic acid	Na/$HClO_4$, 2.00 M, 25°C	$K_{CeHLH} = 132\ M^{-1}$ $k_{CeHLH} = 0.225\ s^{-1}$ $K_{CeLH} = 126$ $k_{CeLH} = 2.59\ s^{-1}$	[11]
1-hydroxycyclohexanoic acid		$K_{CeHLH} = 105\ M^{-1}$ $k_{CeHLH} = 1.21\ s^{-1}$ $K_{CeLH} = 105$ $k_{CeLH} = 3.92\ s^{-1}$	
1-hydroxycycloheptanoic acid		$K_{CeHLH} = 147\ M^{-1}$ $k_{CeHLH} = 4.77\ s^{-1}$ $K_{CeLH} = 105$ $k_{CeLH} = 19.1\ s^{-1}$	
Glyoxylic acid	0.95 M H_2SO_4, 40–50°C	$k_2 = 18.54(\pm 0.35)\ M^{-1}s^{-1}$ (40°C, 0.95 M H_2SO_4)	[12]
Pentaamine glyoxylato cobalt (III)	35–45°C	$k_2 = 0.88(\pm 0.04)$ (40°C, 0.95 M H_2SO_4)	

TABLE 6 (cont'd)

Substrate	Conditions	Rate and equilibrium constants	Ref.[a]
Glyoxylic acid	$H_2SO_4/HClO_4$, 1.0 *M*, 25°C	$K = 30.77\ M^{-1}$ $k_1 = 0.68\ s^{-1}$	[13]
Phenylglyoxylic acid		$K = 8.33\ M^{-1}$ $k_1 = 1.31\ s^{-1}$	
Pyruvic acid		$K = 26.68\ M^{-1}$ $k_1 = 0.051\ s^{-1}$	
2-Oxo-butyric acid		$K = 10.25\ M^{-1}$ $k_1 = 0.52\ s^{-1}$	
Levulinic acid	H_2SO_4, 0.25–6.75 *M*	$K = 14.5\ M^{-1}$ $k_1 = 1.6\ s^{-1}$	[14]
Malic acid	H, $NaClO_4$, 1.50 *M*, 10.6–25.0°C	$K_{CeHL} = 13(\pm 5)\ M^{-1}$ $K_2K_H = 150$ $K_{CeL} = 0.52(\pm 0.06)\ s^{-1}$	[15]
Malonic acid	Na, $HClO_4$, 1.50 *M*, 2.4–35.0°C	$K_{CeH2L} = 224(\pm 60)$ $k_0 = 2.71(\pm 0.08) \times 10^{-1}$ (18.0°C) $k_2 = 2.75(\pm 0.08)\ M^{-1}\,s^{-1}$ (35.4°C)	[16]
Malonic acid	H_2SO_4, 1.0 *M*, 20°C	$k_2 = 0.24\ M^{-1}\,s^{-1}$	[17]
Malonic acid	25% Acetic acid, H_2SO_4, 1.0 *M*, 35°C	$k_2 = 4.16\ M^{-1}\,s^{-1}$ ($Ce(SO_4)_2$ + Mal) $k'_2 = 0.363\ M^{-1}\,s^{-1}$ ($Ce(SO_4)_3$ + Mal)	[18]
Malonic acid diethyl ester		$k_2 = 2.97 \times 10^{-1}\ M^{-1}\,s^{-1}$ $k_2^{\circ} = 2.90 \times 10^{-2}\ M^{-1}\,s^{-1}$	
Methylmalonic acid	Na/$HClO_4$, 5.0 *M*, 25.0°C	$k_2 = 2.04(\pm 0.02)\ M^{-1}\,s^{-1}$ ($[HClO_4] = 3.0\ M$)	[19]
Ethylmalonic acid		$k_2 = 1.76(\pm 0.08)\ M^{-1}\,s^{-1}$ ($[HClO_4] = 3.0\ M$)	
Ascorbic acid	H_2SO_4, 0–60% acetic acid, 30°C	$k_2 = 4.5 \times 10^{-4}\ M^{-1}\,s^{-1}$ ($[H_2SO_4] = 1.03\ M$)	[20]
	HNO_3, 30°C	$K = 1.93 \times 10^3\ M^{-1}$ $k = 1.7 \times 10^{-2}\,s^{-1}$	
	$HClO_4$	$K = 6.52 \times 10^2\ M^{-1}$ $k = 2.5 \times 10^{-2}\,s^{-1}$	

[a] References: [1] Vasudevan and Matthai (1977); [2] Tewari and Tripathy (1977); [3] Prasad and Choudhary (1979); [4] Amjad et al. (1977); [5] Arcoleo et al. (1979); [6] Arcoleo et al. (1977); [7] Calvaruso et al. (1981a); [8] Calvaruso et al. (1981b); [9] Hanna and Sarac (1977a); [10] Hanna and Sarac (1977b); [11] Hanna and Fenton (1983); [12] Mohanty and Nanda (1985); [13] Sarac (1985); [14] Prasad and Prasad (1979); [15] Amjad and McAuley (1974); [16] Amjad and McAuley (1977); [17] Foersterling et al. (1987); [18] Vaidya et al. (1987); [19] Tischler and Morrow (1983); [20] Rajanna et al. (1979).

kinetically derived value. This suggests that the latter precursor complexes may be of the ion–dipole type. Tewari and Tripathi (1977) confirm the Michaelis–Menten rate law and slow kinetics for the ceric oxidation of propionic acid, but are unable to resolve stability and rate constants.

As the following discussion indicates, substitution of more readily oxidizable moieties into the organic substrate increases the rate of oxidation of carboxylic acids without exception. The rate of oxidation of glycolic acid has been investigated in sulfate medium (Prasad and Choudhary 1979), nitrate medium (Amjad and McAuley 1977), and mixed perchlorate/sulfate medium (Calvaruso et al. 1983). Prasad and Choudhary (1979) find that in molar sulfate solutions, no precursor complex is formed. Variation of the acidity in the range 0.05 *M*–0.25 *M* indicates a shift from inverse acid concentration dependence at low acidity to direct acid concentration dependence at $[H^+] > 0.1\ M$. The authors interpret this shift as a result of a change in the reactive cerium species from $Ce(SO_4)_3OH^{3-}$ at low acid to $Ce(SO_4)_2$ and $Ce(SO_4)_3{}^{2-}$ at higher acidity. The outer-sphere mechanism agrees with earlier literature cited by Amjad and McAuley (1977).

As the sulfate is removed and replaced by perchlorate, probable participation of a precursor complex is again indicated (Calvaruso et al. 1983). These authors observe complex dependence of the reaction rate on sulfate concentration and acidity, but unfortunately arrive at a rate law too complex to resolve any useful constants. They suggest that the oxidation occurs simultaneously through three transition states involving one-, two-, and three-coordinated sulfate molecules ($CeSO_4L$, $Ce(SO_4)_2L$, and $HCe(SO_4)_3L$). These authors also conclude (Calvaruso et al. 1984) that the oxidation of lactic (2-hydroxypropionic) and atrolactic (2-hydroxy-2-phenylpropionic) acids follow this same complex pathway.

Complete elimination of sulfate from the medium promotes the formation of a moderately strong precursor complex and permits resolution of the constants describing the kinetic results. Amjad et al. (1977) investigated the ceric oxidation of glycolic, lactic, α-hydroxyisobutyric and phenylglycolic (mandelic) acids in perchloric/nitric acid solutions. The reactions occur at stopped-flow lifetimes and proceed through moderately strong precursor complexes. The presumed reaction for the formation of the principal reactive intermediate is

$$CeOH^{3+} + HL \rightleftharpoons CeL^{3+} + H_2O, \tag{16}$$

for which the equilibrium quotients are contained in table 6. If we correct their kinetically derived stability quotients for the water molecule generated in the precursor complex formation reaction, the constants are within the range of values expected from the thermodynamic calculations in table 5.

The rate of oxidation of the keto acid analogues of the above hydroxy acids is the subject of reports from Sarac (1985) and Mohanty and Nanda (1985). The results of Sarac are closely related to those reported by Amjad et al. (1977) (differing in the media: sulfate and mixed nitrate/perchlorate, respectively), while those of Mohanty and Nanda can be directly compared with those of Prasad and Choudhary (1979) (both sulfate).

Sarac reports that glyoxylic (oxoacetic), pyruvic (2-oxopropionic), phenylglyoxylic (2-oxo-2-phenylpropionic) and 2-oxobutyric acids are oxidized in sulfate media with formation of a precursor complex. The resolved rates of oxidation correlate most strongly with the percent concentration of $CeSO_4{}^{2+}$, which the author interprets to indicate an intermediate complex of the form $CeSO_4L^+$. The order for the resolved rate constants is phenylglyoxylic > glyoxylic > oxobutyric > pyruvic, which does not correlate with the order for the corresponding hydroxyacids from Amjad et al. (1977). Except for glyoxylic acid, the keto acids undergo electron transfer at a slower rate than the corresponding hydroxy acids, although this may be partly attributable to the presence of sulfate. The results of Prasad and Prasad (1979) on the oxidation of levulinic (4-oxopentanoic) acid support the conclusions of Sarac.

Mohanty and Nanda (1985) investigated oxidation of the glyoxylic acid and pentaamineglyoxylatocobalt(III) complex under second-order conditions and determined second-order rate parameters directly. Although they assume that an intermediate complex is involved, their data cannot be used to establish whether this is true. The data indicate that the free glyoxylic acid is oxidized at a 30-fold faster rate than the complexed glyoxylic acid. The second-order rate constant for oxidation of glyoxylic acid (in sulfate) is 1500 times faster than that of glycolic acid (Prasad and Choudhary 1979).

The kinetics of oxidation of substituted mandelic acids (α-hydroxyphenylacetic acid) in sulfate and mixed perchlorate/sulfate solutions are the subject of the four reports from Calvaruso et al. (1981a, b) and Arcoleo et al. (1977, 1979). The oxidation reaction yields (substituted) benzaldehydes upon decarboxylation of mandelic acid. The oxidation of the *p*-nitro and *p*-methoxy derivatives in sulfuric acid proceeds without significant complex formation. The authors conclude that the reaction proceeds through two reactive Ce(IV) species,

$$\mathrm{CeSO_4^{2+}} + \mathrm{Subst} \xrightarrow{k_2} \mathrm{Ce(III)} + \mathrm{Subst}^{\bullet}, \tag{17}$$

and the protonated species,

$$\mathrm{H(Ce(SO_4)_3)^-} + \mathrm{Subst} \xrightarrow{k_2'} \mathrm{Ce(III)} + \mathrm{Subst}^{\bullet}. \tag{18}$$

The oxidation of the *p*-chloro derivative in perchlorate medium does indicate the existence of a stable precursor complex. Both the slope and intercept of the $1/k_{obs}$ versus 1/[subst] plot exhibit a dependence on acidity, implying the involvement of a protonated species either in the formation of the precursor complex or in the electron transfer reaction. The relative rates for the oxidation of the substituted mandelic acids are $CH_3O– \gg –NO_2 > –Cl = –CH_3 > –H$. The oxidation of the *p*-methoxy derivative is characterized by 50% lower ΔH^* and negative ΔS^*, suggesting a different mechanism operating for the oxidation of this substrate.

The rate of ceric oxidation of substituted benzilic (2,2-diphenyl-2-hydroxyacetic) acid in sulfuric acid, aqueous perchloric/acetic acid and acetonitrile is the subject of two reports from Hanna and Sarac (1977a, b). The reaction proceeds like the other α-hydroxycarboxylic acids by oxidative decarboxylation, producing substituted benzophenones and CO_2. The primary interest in the first of these two reports (Hanna and Sarac 1977a) is in the organic chemical aspects of the reactions. However, it is observed that the relative rates vary with the media in the order $H_2SO_4 > HClO_4$/acetic acid > acetonitrile. Although little mechanistic information exists, it is apparent that the oxidation proceeds via an inner-sphere, electron transfer process.

In the companion paper (Hanna and Sarac 1977b) the oxidation of benzilic acid is reported with emphasis on learning the mechanistic implications with respect to Ce(IV). As is generally observed, the rates are much slower in sulfuric acid solutions and the existence of a stable precursor complex is doubtful. The authors calculate the equilibrium speciation of Ce among the first two hydrolysis products, free metal ion, and 1:1, 1:2, 1:3 Ce–SO_4 complexes and find that the rate of reaction is most strongly correlated with $CeSO_4^{2+}$. In the presence of sulfuric acid a precursor complex of the form $CeSO_4 \cdot HL$ is proposed. In perchlorate solutions there is some evidence for a reactive $CeHL^{3+}$ species and an unreactive $Ce(HL)_2^{2+}$ species. The precursor complex stability constant and first-order oxidation rate parameter for the former intermediate are in excellent agreement with those reported by Amjad et al. (1977) for mandelic acid.

Hanna and Fenton (1983) report that the oxidation of C5, C6, and C7 cycloalkane α-hydroxycarboxylic acids to the corresponding ketones in perchlorate solutions conform to the Michaelis–Menten rate law with both slope and intercept of the $1/k_{obs}$ versus 1/[subst] plot exhibiting a direct acid concentration dependence. For the acids, the relative rates are C7 > C6 > C5, whereas for the alcohols C5 > C7 > C6 is observed. The intermediate complexes between Ce(IV) and the acids are of the form $Ce(HLH)^{4+}$ and $Ce(LH)^{3+}$ with the equilibrium quotients resolved for both precursor complexes in all systems (the respective parameters are noted by the subscripts CeHLH and CeLH in table 6). No analogue thermodynamic data are available for comparison, but the kinetically derived stability constants are in substantial agreement with those of Amjad et al. (1977) for the noncyclic hydroxyacids. The deprotonated intermediate $Ce(LH)^{3+}$ is indicated as the most reactive species. The intramolecular electron transfer rate constants are in good agreement with other results reported for hydroxy and keto acids where comparable mechanisms are involved. A comparison of the reaction rate as a function of acidity with Ce(IV) speciation suggests $CeOH^{3+}$ as the kinetically most significant cerium species.

Amjad and McAuley (1974) have investigated the oxidation of malic (2-hydroxybutane-1,4-dicarboxylic) acid in perchloric acid. The reaction occurs at stopped-flow lifetimes and is unaffected by either Ce(III) or nitrate. The Michaelis–Menten plot is linear with finite positive intercept indicating precursor complex formation. The resolved values for the formation constants are comparable to those reported for the hydroxy and keto monocarboxylates, suggesting that the second carboxylate group is not bound in the activated complex. The intramolecular electron transfer rate parameters are less than those reported by Hanna and Fenton for the cyclic α-hydroxycarboxylic acids.

The rate of oxidation of citric acid by Ce(IV) in sulfuric acid has been reported by Tripathi et al. (1980). The reaction rate dependence on the concentration of the citric acid indicates very weak precursor complex formation, if any (i.e. the reaction is very nearly first-order in citric acid concentration). The authors propose that $Ce(SO_4)_2$ retards the rate of oxidation and $Ce(SO_4)_3^{2-}$ is the most reactive ceric species. It is interesting to note that citric acid, which might be considered the strongest complexant of the acids reported here, does not form a precursor complex in sulfuric acid while the keto acids apparently do.

The following five reports consider the rate of oxidation of malonic acid, its esters and related compounds. This species is somewhat unusual among organic acids as the α-hydrogens of the esters are moderately acidic, permitting this reactive center to be the focus of a fair amount of chemistry. This series also permits an assessment of the role of the carboxylic acid group in the formation and stability of the precursor complex, as the esters would ordinarily be considered much weaker complexants for cerium(IV) than the acid. In addition, this system is significant as malonic acid is the oxidizable substrate most often encountered in the Belousov–Zhabotinsky oscillating reaction (see Barkin et al. (1977) for a description of the B–Z oscillator), which will be briefly discussed in sect. 3.7.

Amjad and McAuley (1977) investigated the oxidation of malonic acid by Ce(IV) in perchlorate solutions. They observe that at low temperatures the reaction mechanism indicates the formation of a precursor complex, but as the temperature increases, the order of reaction with respect to the substrate changes to unity, implying that the intramolecular reaction is no longer dominant. Ce(III) has no effect on the rate of reaction. The authors propose two intermediate species, CeH_2L^{4+} and $CeHL^{3+}$, depending on the acidity of the reaction. In the high-temperature regime, the results are best explained by considering a weak CeH_2L^{4+} species as the reactive intermediate, accounting for the second-order overall rate law. The results of a flow-ESR experiment indicate that the initial radical species is $^{\bullet}CH(COOH)_2$, consistent with the known acidity of the α-hydrogen atom.

A recent ESR investigation of malonic acid oxidation by Ce(IV) in perchloric acid (Brusa et al. 1988) indicates that two radical species are significant: a malonic acid radical (Mal$^{\bullet}$) and a Ce(IV)–malonate radical complex (Ce(IV)–Mal$^{\bullet}$). The existence of the radical complex was confirmed by the loss of ESR signal for the reaction in H_2SO_4. The reaction of a second Ce(IV) ion with the radical complex produces 2-hydroxymalonic acid as the organic product. Analysis of the ESR data indicates that the Ce(IV)–Mal$^{\bullet}$ complex is 200 times stronger than the Ce–H_2Mal complex proposed by Amjad and Mc Auley (1977). Mechanistically, the similarity of the magnetic parameters for Mal$^{\bullet}$ and Ce(IV)–Mal$^{\bullet}$ requires that metal–carbon bonds are not involved in the complex.

The rate of ceric oxidation of malonic acid and its diethyl ester in acetic acid/sulfuric acid solutions has recently been reported by Vaidya et al. (1987). They find no evidence for precursor complex formation in either system. The reactive Ce(IV) species appear to be $Ce(SO_4)_2$ (k_2) and $Ce(SO_4)_3^{2-}$ (k'_2). The second-order rate parameter for the oxidation of malonic acid is 40 times greater than that for the ester. Oxidation of the ester is proposed to occur through the enol form yielding a malonyl radical analogous to that identified by Amjad and McAuley. Foersterling et al. (1987) find that the second-order rate constant for malonic acid oxidation by Ce(IV) in sulfuric acid is in excellent agreement with the value of Vaidya et al. They observe that Ce(III) does inhibit the reaction in sulfuric acid, which they attribute to a reversible Ce(IV)–malonic acid rate-controlling step.

Tischler and Morrow (1983) report on the oxidation of methyl and ethyl malonic acid (substitution at the α-carbon) to assess the role of the acidic α-hydrogen in malonic acid oxidation (in perchloric/nitric acid). The pseudo-first-order rate constants are linearly dependent on [malonic acid] with zero intercept, confirming the absence of a stable precursor complex. The observed rate constants exhibit a complex acid dependence, which can be related to a protonated intermediate and Ce(IV) hydrolysis. Substitution of a second methyl or ethyl group, eliminating both acidic α-hydrogens, slows the rate of oxidation dramatically, and even permits stopped-flow spectrophotometric evidence for the existence of a relatively stable, strongly colored complex. (The constants in table 6 are derived from Tischler and Morrow's data for the specific conditions listed in the table.)

The oxidation of ascorbic acid by Ce(IV) is the subject of a report by Rajanna et al. (1979). The initial oxidation product is dehydroascorbic acid. Contrary to most of the studies discussed in this review, the substrate is the limiting reagent in this study, which was conducted in sulfuric, perchloric and nitric acid solutions. In sulfuric acid the rate-determining step appears to require a bimolecular reaction, while in perchloric and nitric acids oxidation via an intramolecular process is indicated. In sulfuric acid, $Ce(SO_4)_2$ is presumed to be the active ceric species. A Michaelis–Menten plot of $1/k_{obs}$ versus 1/[Ce(IV)] is linear with finite positive intercept, from which the constants quoted in table 6 are determined.

3.7. *Ceric oxidation of nonmetals (including Belusov–Zhabotinsky oscillating reaction)*

In this section we will consider the rates for ceric oxidation of a variety of nonmetal-reducing reagents (table 7). These studies include two investigations of halide oxidation, three of Se(IV) and Te(IV), a collection of organosulfur compounds, and finally a brief summary of the voluminous literature available on the Belusov–Zhabotinsky cerium–bromate oscillating reactions.

Nazar and Wells have reported results of the oxidation of both iodide (Nazar and Wells 1985) and bromide (Wells and Nazar 1979) in perchloric acid solutions. The former system is well behaved, proceeding at stopped-flow lifetimes and first order in both [Ce(IV)] and $[I^-]$. The rate law also exhibits acid-dependent and -independent pathways according to the rate law given by

$$d[I_3^-]/dt = (k_0 + k_1[H^+])\,[Ce(IV)][I^-]. \tag{19}$$

The authors reject a simple mechanism which relates the acid dependent process to Ce(IV) hydrolysis in favor of a more complex mechanism involving multiple transition states. This mechanism requires high-order Ce(IV)–I^- complexes, for which there is no corroborating evidence (even in reasonable analogue systems). If the first hydrolysis constant recommended by Baes and Mesmer (1976) are taken, the simpler explanation of acid concentration dependence due to Ce(IV) hydrolysis becomes reasonable, and probably is the more plausible explanation for the observed results. The specifics of this proposed alternate mechanism are discussed below.

The bromide reaction also proceeds at stopped-flow lifetimes, much faster than was reported earlier in sulfuric acid solutions by King and Pandow (1953). They observed that bromide was oxidized through a rate law having both first- and second-order dependence on $[Br^-]$. This was explained by suggesting two pathways for the

TABLE 7
Ce(IV) oxidation of nonmetal species.

Substrate	Conditions	Rate and equilibrium constants	Ref.[a]
I^-	H, $NaClO_4$, 2.00 *M*, 25.5°C	$k_0 = 4.1\ (\pm 1.5) \times 10^3\ M^{-1}\,s^{-1}$ $k_1 = 3.7\ (\pm 0.1) \times 10^4\ M^{-2}\,s^{-1}$	[1]
Br^-	H, $NaClO_4$, 2.00 *M*, 17.2–44.8°C	$k_0 = 1.1 \times 10^2\ M^{-2}\,s^{-1}$ $k_1 = 3.6 \times 10^2\ M^{-4}\,s^{-1}$	[2]
Se(IV)	H, $NaClO_4$, 3.0–6.0 *M*	$K = 126\ M^{-1}$ (50°C) $k_1 = 3.3 \times 10^{-4}\ s^{-1}$	[3]
Dimethyl sulfoxide	H, $NaClO_4$, 1.5 *M*	$K = 19.0\ M^{-1}$ (25.0°C) $k_1 = 1.0 \times 10^{-3}\ s^{-1}$	[4]
Thiohydantoin[b]	H_2SO_4, 0.5–2.0 *M*, 30% acetic acid, 25–40°C	$k_2 = 0.095\ M^{-1}\,s^{-1}$ (25°C, $H_2SO_4 = 1.0\ M$)	[5]
Rhodanine[c]	30–45°C	$k_2 = 1.27 \times 10^{-1}\ M^{-1}\,s^{-1}$ (35°C, $H_2SO_4 = 1.0\ M$)	

[a] References: [1] Nazar and Wells (1985); [2] Wells and Nazar (1979); [3] Dikshitulu et al. (1981); [4] Pratihari et al. (1976); [5] Roy et al. (1978).
[b] structure: [c] structure:

H₂C—NH / O=C C=S / N / H

H₂C—NH / O=C C=O / S

destruction of the precursor complex $Ce(SO_4)_2Br^-$, a unimolecular electron transfer to produce $Br^{\bullet}$, and a bimolecular reaction with a second Br^- to produce Br_2^-. In perchloric acid, Wells and Nazar (1979) observe second-order dependence on both $[Br^-]$ and $[H^+]$ (rate law as in eq. (20)), which they rationalize in terms of a mechanism requiring complexes with Ce–Br stoichiometry of 1:3,

$$-\,d[Ce(IV)]/dt = (k_0 + k_1[H]^2)\,[Ce(IV)][Br^-]^2. \tag{20}$$

As was the case for the iodide system, there is no supporting evidence in the literature for Ce(IV)–Br^- complexes with so high a M:L stoichiometry, particularly at the low total bromide concentrations (0.02–0.1 *M*) used in this study. The proposed mechanism also requires that the Ce(IV)–Br^- complexes are hydrolyzed more readily than the noncomplexed Ce(IV), which is contrary to normal thermodynamic considerations.

An alternate mechanism which could explain the observed results can be derived if 1:1 and 1:2 hydrolysis products of Ce(IV) are considered relevant. The pre-equilibria

$$Ce^{4+} \xrightleftharpoons{K_{h1}} Ce(OH)^{3+} + H^+, \tag{21}$$

$$Ce(OH)^{3+} \xrightleftharpoons{K_{h2}} Ce(OH)_2^{2+} + H^+, \tag{22}$$

$$Ce^{4+} + X^- \xrightleftharpoons{K_c} CeX^{3+} \tag{23}$$

account for Ce(IV) species, where X^- is the halide ion and the dominant ceric species are hydrolysis products $Ce(OH)^{3+}$ and $Ce(OH)_2^{2+}$. For the reduction of the substrate, the following reactions may be considered,

$$Ce(OH)_2^{2+} + X^- \xrightarrow{k_{h2}} Ce(III) + X, \quad (24)$$

$$Ce(OH)^{3+} + X^- \xrightarrow{k_{h1}} Ce(III) + X, \quad (25)$$

$$CeX^{3+} + X^- \xrightarrow{k_c} Ce(III) + X_2^-. \quad (26)$$

The first two of these reactions describe, respectively, the acid-dependent and -independent pathways for the oxidation of I^-. The second-order dependence on $[Br^-]$ and $[H^+]$ can be explained by the latter reaction if the dominant reactive Ce(IV) hydrolyzed species (in 0.2–2.0 *M* $HClO_4$) is $Ce(OH)_2^{2+}$.

Dikshitulu and co-workers report the results of an investigation of ceric oxidation of Se(IV) (Dikshitulu et al. 1981b) and Te(IV) (Dikshitulu et al. 1981a, 1980). The reactions are first-order in Ce(IV) concentration and of fractional order in each of the substrate concentrations suggesting the appropriateness of Michaelis–Menten analysis. The rate law for Ce(IV) + Se(IV) (or Te(IV) in perchloric acid) in independent of acidity in the acid concentration range 3.5–6.0 *M* at 50°C. The Te(IV) reaction exhibits significant acid effects in nitric and sulfuric acid solutions, i.e. inverse dependence on $[HNO_3]$ and direct dependence on $[H_2SO_4]$ (although the latter may represent a combined effect of acidity and sulfate complexation). In the presence of Ce(III) the selenite reaction appears to achieve second-order dependence on [Ce(IV)]. The stability quotient determined kinetically for Ce(IV)–selenite, $K_{eq} = 127\ M^{-1}$, was confirmed by separate spectrophotometric measurements. The retardation effect of Ce(III) is attributed to oxidation of Ce(III) by the unstable Se(V) intermediate produced in the initial oxidation step. The ratio of the rate constants for Se(V) reaction with Ce(IV) versus Ce(III) is 2.0.

The speciation of Te(IV) is less straightforward than that of Se(IV) but, probably, is dominated by $TeO(OH)^+$ and $Te(OH)_3^+$ in acid solutions. In addition to the above-mentioned acidity effect, the Ce(IV)–Te(IV) reaction exhibits an inverse dependence on both $[NO_3^-]$ and $[HSO_4^-]^2$. The $1/k_{obs}$ versus 1/[Te(IV)] or 1/[Te(VI)] plots are linear, indicating complex formation with each. The reversible reaction through the pentavalent substrate species observed for Se(IV) is not observed in the tellurium system. In this system, the observed retardation effect of Ce(III) is attributed to complex formation with Te(IV) or Te (VI).

Although perhaps more properly considered among the organic substrates, the following are reports of the oxidation of a few organosulfur compounds. Dimethyl sulfoxide (DMSO) is a common polar nonaqueous solvent which solvates the lanthanides quite admirably. The rate of oxidation of DMSO by Ce(IV) in perchloric acid solutions is the subject of a report from Pratihari et al. (1976). The reaction conforms to the Michaelis–Menten rate law and the rate is enhanced by increasing acidity. The acid concentration dependence is attributed to hydrolysis of

Ce(IV). The Ce(IV)–DMSO stability quotient and intramolecular rate constant are $K = 19.0\ M^{-1}$, $k_1 = 1.0 \times 10^{-3}\ s^{-1}$.

Thioorganic compounds play a significant role as electron donors in biological systems. The following reports deal with the ceric oxidation of heterocyclic sulfur compounds which either directly or indirectly impact on this bioinorganic chemistry. Roy et al. (1978) investigated the ceric oxidation of thiohydantoin and rhodanine in mixed sulfuric/acetic acid media (structures in table 7). The oxidation products are the disulfides formed by the linkage of two substrate molecules through the thioketo groups. The reaction exhibits good first-order behavior in both Ce(IV) and substrate concentrations, without indication of precursor complex formation. Both the disulfide reaction products and the absence of precursor complexes are in agreement with earlier results for the oxidation of thiocarboxylic acids summarized by Benson (1976). Two reaction pathways are oxidation of TH^+ (protonated thioketone) and T to form radical sulfur compounds which then rapidly combine. Thiohydantoin reacts faster than rhodanine because the replacement of heterocycle S by imino (NH) inhibits thiol formation.

The Ce(IV)/Ce(III)–(Br^-/BrO_3^-) system is one of several redox systems in which oscillating oxidation–reduction cycles can be observed. This particular system, with organic substrates like malonic acid (among others), is known as the Belousov–Zhabotinsky oscillator. This system has been studied for more than 15 years and has been the subject of several reviews and a number of symposia. The Ce–Br oscillating reaction system has been described in terms of a seven-step rate process (all reversible reactions), the Field–Koros–Noyes (FKN) mechanism (Field et al. 1972).

The key species in the FKN mechanism is $BrO_2^{\bullet}$, which functions as both oxidant for Ce(III) and reductant for Ce(IV). This species is formed in the reaction between BrO_3^- (bromate) and $HBrO_2$ (bromous acid). Neither $BrO_2^{\bullet}$ nor $HBrO_2$ are stable in aqueous media and thus are produced and consumed in the FKN mechanism as intermediate species. A lucid description of the specifics of the oscillating reaction system has been provided by Barkin et al. (1977). A more recent contribution from Bar-Eli (1985) indicates that in a continuously stirred tank reactor (CSTR) only four or five of the total 14 reaction rate constants are required to describe the oscillations. Yoshida and Ushiki (1982) studied the oxidation of Ce(III) by BrO_3^- and found an induction period precedes a burst of Ce(IV) production, followed by continued slow generation of Ce(IV). The length of the induction period was highly dependent on the age of the reactant solution.

In an investigation of a specific step in the FKN mechanism, Sullivan and Thompson (1979) studied (by stopped-flow) the kinetics of Ce(IV) reduction by bromous acid ($HBrO_2$). The substrate was stabilized with respect to its rapid disproportionation in acid solution by preparing the reagent in slightly basic medium. This solution was then mixed in the stopped-flow instrument with Ce(IV) in sulfuric acid. Good second-order behavior is observed in dilute (0.3 M) acid media. However, in 1.5 M H_2SO_4/1.5 M $HClO_4$, closely approximating the conditions of the B–Z oscillating reaction, no reaction is observed. The second-order rate parameter for this reaction in the FKN mechanism is $2 \times 10^7\ M^{-1}\,s^{-1}$. These authors conclude that the

FKN estimates for the title reaction and for the disproportionation of $HBrO_2$ are too large. Their results suggest that the Ce(IV)–$HBrO_2$ reaction is unimportant in concentrated sulfuric acid with disproportionation being the principal pathway for bromous acid disappearance from solution. This discussion serves to demonstrate the folly of overemphasis of computer modeling to the exclusion of experimental measurement of key data.

3.8. *Cerium reactions with metal ions and their complexes*

In contrast to redox reactions involving organic species, reactions between metal ions and their complexes tend to be more straightforward and readily interpretable. One reason such reactions are easier to interpret is that the thermodynamic driving force for the reaction is usually well known, and multiple pathways to multiple products are much less common. In the following section we will discuss the results of a collection of studies of both Ce(III) oxidation and Ce(IV) reduction reactions by transition metals and their complexes.

3.9. *Oxidation of Ce(III)*

The estimated value for the reduction potential of the Ag(II)/Ag(I) couple is about 2 V, which provides an adequate thermodynamic driving force for the oxidation of Ce(III) to Ce(IV) in perchloric acid solutions. The oxidation of Ce(III) by Ag(II) in 1.0 M $HClO_4$ is characterized by a second-order rate constant of $3.2 \times 10^3\ M^{-1}\,s^{-1}$ at 25°C (Arselli et al. 1984). The comparison with the rates determined for Fe(II), Co(II), Mn(II) and V(V) as reductants led the authors to postulate that the mechanism proceeded via an inner-sphere activated complex.

The radiolysis of solutions of ethylenediaminetetraacetato–Ce(III) and triethylenetetraaminehexaacetato–Ce(III) produces Ce(IV) as well as degradation of the organic ligands (Hafez et al. 1978, 1979). The authors postulate that the oxidation results from the reaction of the radiolytically produced H_2O_2 with Ce(III).

These results may be compared with those reported by Rizkalla et al. (1986) wherein the reaction between H_2O_2 and ethylenediaminetetraacetato–Ce(IV) was studied. These authors report that, while there was catalytic decomposition of the H_2O_2, there was no evidence for decomposition of the ligand. They determined two second-order rate parameters (0.054 and 0.171 $M^{-1}\,s^{-1}$, which are attributed to the decomposition of H_2O_2 by the mono- and dihydroxo–Ce(IV)–EDTA complexes.

3.10. *Oxidation of peroxocompounds of Ti and Zr by Ce(IV)*

The reactions of Ce(IV) with peroxo complexes of Ti and Zr have been reported by Thompson (1984, 1985) in two seminal studies. The overall reaction in molar perchloric acid (25°C) for the peroxotitanium–Ce(IV) system is

$$TiO_2^{2+} + 2Ce(IV) + H_2O \rightleftharpoons TiO^{2+} + O_2 + 2Ce(III) + 2H^+. \qquad (27)$$

A mechanism consistent with the results obtained in a thorough kinetic study of this system identified reactions of Ce(IV) in the following steps:

$$TiO_2^{2+} + Ce(IV) \xrightarrow{k} TiO_2^{3+} + Ce(III), \tag{28}$$

$$TiO_2^{3+} + Ce(IV) + H_2O \xrightarrow{k'} O_2 + TiO^{2+} + Ce(III) + 2H^+. \tag{29}$$

The rate constants for these two reactions are $k = 1.1 \times 10^5\ M^{-1}\,s^{-1}$, and $k' = 400\ M^{-1}\,s^{-1}$.

A marked difference in the structure of the Zr(IV)–peroxo complex which Thompson formulated as $Zr_4(O_2)_2(OH)_4^{8+}$ is reflected in a mechanism wherein a rate-determining ring-opening step of the Zr(IV) complex is followed by a rapid reaction between the Ce(IV) and a pendant peroxo moiety. An earlier study of this system by Berdnikov et al. (1978) utilized ESR techniques. However, the interpretation of those results are suspect as the authors did not take into account the high degree of polymerization of the Zr–peroxo complex.

3.11. *Oxidation of Cr and Mo complexes by Ce(IV)*

The results of kinetic studies on the oxidation of a series of five Cr(III) tetradentate macrocycle complexes by Nair et al. (1987) were reportedly not complicated by secondary reactions of the Ce(IV) with the ligands since the overall stoichiometry was given as

$$Cr(III) + 3Ce(IV) \rightleftharpoons Cr(VI) + 3Ce(III). \tag{30}$$

The reported mechanism identifies the observed rate parameter as the product of the rapid equilibrium

$$Ce(OH)_2^{2+} + Cr(III) \rightleftharpoons Cr(IV) + Ce(III), \tag{31}$$

followed by the rate-determining step,

$$Cr(IV) + Ce(IV) \longrightarrow Cr(V) + Ce(III). \tag{32}$$

The observed variation in the rate with $[H^+]$ and sulfate ion concentration are presumed to be in equilibria prior to the redox equilibrium. The second-order rate parameters are reproduced in table 8. The observed variations were rationalized in terms of differences in the thermodynamic driving forces for the pre-equilibrium and changes in rearrangement energies of the ligands in the rate-determining step. The contribution of rearrangement energies is difficult to quantify since there is only a difference of a factor of three between the *cis*- and *trans*-cyclam complexes.

The measured value for the ratio $[Ce(IV)]/[Mo_2O_4] = 1.97\ (\pm 0.05)$ was the basis for Chappelle et al. (1984) to write the overall reaction as

$$Mo_2O_4^{2+} + 2Ce(IV) + 4H_2O \rightleftharpoons 2Mo(VI) + 2Ce(III) + 8H^+. \tag{33}$$

The empirical form of the rate law is

$$-d[Ce(IV)]/dt = k[Mo_2O_4^{2+}][Ce(IV)] \tag{34}$$

TABLE 8
Ceric oxidation of Cr(III) tetradentate macrocycle compounds (I = 1.00 M $LiClO_4$, T = 30°C).

Substrate[b]	Second-order rate constants	Ref.[a]
$[Cr(Me_6[14]ane)H_2O]^{3+}$	$1.37 \times 10^{-1}\ M^{-1}s^{-1}$	[1]
$[Cr(Me_6[14]ane)(CHCl_2)H_2O]^{2+}$	$1.28\ M^{-1}s^{-1}$	
$Cr(C_{10}H_{23}N_4)(H_2O)_2{}^{3+}$	$1.3 \times 10^{-2}\ M^{-1}s^{-1}$	
cis-$[Cr(cyclam)(H_2O)_2]^{3+}$	$1.05\ M^{-1}s^{-1}$	
trans-$[Cr(cyclam)(H_2O)_2]^{3+}$	$3.75 \times 10^{-1}\ M^{-1}s^{-1}$	

[a] Reference: [1] Nair et al. (1987).
[b] Abbreviations used:
$Me_6[14]$ane 5,7,7,12,14,14-hexamethyltetraazacyclotetradecane.
$C_{10}H_{23}N_4$ $H_2N(CH_2)_2NC(CH_3)CH_2C(CH_3)_2NH(CH_2)_2NH_2$.
cyclam 1,4,8,12-tetraazacyclotetradecane.

with the hydrolysis of Ce(IV) accommodated by the expression

$$k = k_h K_h/([H^+] + K_h) \tag{35}$$

where $k_h = 2.73 \times 10^4\ M^{-1}s^{-1}$ and $K_h = 0.46\ M$ (25°C, $I = 2.0\ M$ ($H/NaClO_4$)). An unusual aspect of the results was the interpretation of spectrophotometric evidence to postulate the formation of an ion pair between Ce(IV) and the aquomolybdenum dimer.

3.12. *Oxidation of Fe, Co and Ni compounds by Ce(IV)*

The use of Ce(IV) to probe the kinetics of outer-sphere electron transfer reactions has centered on compounds of the title metal ions whose primary coordination spheres (in the reduced state at least) are substitution inert on the time scale of electron transfer reactions. For example, Cyfert et al. (1980) investigated the effects of the variation of ionic strength on the rate of $Fe(phen)_3^{2+}$ oxidation by Ce(IV) in 0.125 M H_2SO_4 with different supporting electrolytes (table 9). Variation in the concentrations of NaCl or $NaClO_4$ in the range of 0.1–1.0 M changed the rate constants from 1.08 to $0.81 \times 10^5\ M^{-1}s^{-1}$ and 1.03 to $0.56 \times 10^5\ M^{-1}s^{-1}$, respectively (25°C). The variation of $[Na_2SO_4]$ from 0.05–0.35 M changed the rate parameter from 1.05 to $0.45 \times 10^5\ M^{-1}s^{-1}$. The authors suggest that the predominant reactive form of Ce(IV) is the tris(sulfato) complex.

The effect of the specific anion in Ce(IV) oxidations is demonstrated by a comparison of the above results with those presented by Vincenti et al. (1985a). The rate of oxidation of $Fe(Ebipy)_3^{2+}$ (Ebipy is 4,4-bis(ethoxycarbonyl)-2,2′-bipyridine) at 25°C in 0.05 M HNO_3 changes from $3.8 \times 10^4\ M^{-1}s^{-1}$ at ionic strength 2.0 M (Na/HNO_3) to $3.1 \times 10^2\ M^{-1}s^{-1}$ at ionic strength 0.05 M. At constant ionic strength of 2.0 M, the second-order rate constant increased from $3.8 \times 10^4\ M^{-1}s^{-1}$ at 0.05 M HNO_3 to $1.7 \times 10^6\ M^{-1}s^{-1}$ at 2.00 M HNO_3. The empirical form of the

TABLE 9
Ceric oxidation of iron coordination compounds.

Substrate[b]	Conditions	Second-order rate constants	Ref.[a]
$Fe(phen)_3^{2+}$	0.1 M $NaClO_4$, 0.125 M H_2SO_4	$1.03 \times 10^5\ M^{-1}s^{-1}$	[1]
	1.0 M $NaClO_4$, 0.125 M H_2SO_4	$0.56 \times 10^5\ M^{-1}s^{-1}$	
	0.1 M NaCl, 0.125 M H_2SO_4	$1.08 \times 10^5\ M^{-1}s^{-1}$	
	1.0 M NaCl, 0.125 M H_2SO_4	$0.81 \times 10^5\ M^{-1}s^{-1}$	
	0.05 M Na_2SO_4, 0.125 M H_2SO_4	$1.05 \times 10^5\ M^{-1}s^{-1}$	
	0.35 M Na_2SO_4, 0.125 M H_2SO_4	$0.45 \times 10^5\ M^{-1}s^{-1}$	
$Fe(Ebipy)_3^{2+}$	0.05 M HNO_3	$3.1 \times 10^2\ M^{-1}s^{-1}$	[2]
	0.05 M HNO_3, 2.0 M $NaNO_3$	$3.8 \times 10^4\ M^{-1}s^{-1}$	
$Fe(dmbpy)_3^{2+}$	0.05 M HNO_3	$3.2 \times 10^4\ M^{-1}s^{-1}$	
$Fe(bipy)_3^{2+}$	0.05 M HNO_3	$6.0 \times 10^3\ M^{-1}s^{-1}$	
$Fe(Clphen)_3^{2+}$	0.05 M HNO_3	$2.1 \times 10^3\ M^{-1}s^{-1}$	
$Fe(nphen)_3^{2+}$	0.05 M HNO_3	$5.1 \times 10^2\ M^{-1}s^{-1}$	
$Fe(bbipy)_3^{2+}$	0.05 M HNO_3	$1.3 \times 10^2\ M^{-1}s^{-1}$	[3]
$Fe(pbipy)_3^{2+}$	0.05 M HNO_3	$2.3 \times 10^2\ M^{-1}s^{-1}$	
$Fe(mbipy)_3^{2+}$	0.05 M HNO_3	$1.8 \times 10^2\ M^{-1}s^{-1}$	

[a] References: [1] Cyfert et al. (1980); [2] Vincenti et al. (1985a); [3] Vincenti et al. (1985b).
[b] Abbreviations used:

phen	1,10-phenanthroline.
Ebipy	4,4′-bis(ethoxycarbonyl)-2,2′-bipyridine.
dmbpy	4,4′-dimethyl-2,2′-bipyridine.
bipy	2,2′-bipyridine.
Clphen	5-chloro-1,10-phenanthroline.
nphen	5-nitro-1,10-phenanthroline.
bbipy	4,4′-bis(butoxycarbonyl)-2,2′-bipyridine.
pbipy	4,4′-bis(propoxycarbonyl)-2,2′-bipyridine.
mbipy	4,4′-bis(methoxycarbonyl)-2,2′-bipyridine.

rate law calculated from the data presented is

$$-\mathrm{d}[\mathrm{Ce(IV)}]/\mathrm{d}t = (k_0 + k_1[\mathrm{H}^+])[\mathrm{Ce(IV)}][\mathrm{FeL_3}], \qquad (36)$$

where $k_0 = 3.6(\pm 3.3) \times 10^4 M^{-1}s^{-1}$, $k_1 = 8.47(\pm 0.35) \times 10^5 M^{-2}s^{-1}$. The authors' interpretation of the predominant path first-order in $[H^+]$ as representing the reaction of Ce^{4+}_{aq} must be viewed with reservation. The variation of the observed second-order rate parameters for the reactions of Ce(IV) with a series of Fe(II) phenanthroline and bipyridine complexes was correlated satisfactorily with the reduction potentials. Based on the observed linearity, and a number of additional assumptions, an estimate of the self-exchange reaction for Ce^{4+}/Ce^{3+} was calculated.

The kinetics of the electron transfer reactions between Ce(IV) and a series of substituted alkoxycarbonylbipyridine complexes (table 9) were determined in homogeneous solutions and in the presence of varying concentrations of sodium dodecylsulfate (SDS) (Vincenti et al. 1985b). The results show a complex dependence of the rates of the electron transfer reaction on the concentration of the micelle. There is a surprising increase in rate at concentrations below the critical micelle concentration, which the authors attribute to the formation of Fe(II)–SDS complexes.

The results presented in the following three reports are on systems in which the electron transfer step to Ce(IV) is rapid and the subsequent chemistry is complex. In Holecek et al. (1979) the first step in the ceric oxidation of ferrocene produces unstable ferricenium cations which subsequently decompose with oxidation of the ligand. In Soria et al. (1980) and Chum and Helene (1980) the initial electron transfer process results in both metal-centered oxidation of the tris(2-pyridinial-α-methyl-(methylimine))–Fe(II) complex as well as oxidation of the α-methyl group to an aldehyde group, with no change in the oxidation state of iron.

The pioneering work of Taube (1970) on induced electron transfer reactions has been extended by Srinivasan and Gould (1981) to the reactions

$$RC(OH)HCOOCo(NH_3)_5^{2+} + Ce(IV) \rightleftharpoons RCO + CO_2 + Ce^{3+} + H^+ + Co^{2+} \quad (37)$$

where R is CH_3, C_6H_6 and $(C_6H_6)_2$ (table 10). The first step in the reaction sequence that can be identified is the formation of a 1:1 complex between Ce(IV) and the Co(III) compound followed by the first-order decay of the complex. The Co(III) metal ion is reduced by the radical produced in the Ce(IV)–complex reaction. The electron transfer rate for the lactate complex (R = CH_3) compares favorably with that reported by Amjad et al. (1977) for free lactic acid, but the mandelate (R = C_6H_6) complex reaction is 25 times slower than that for the free acid (see sect. 3.6).

TABLE 10
Ceric oxidation of pentaamine Co(III) carboxylate complexes.

Substrate[b]	Conditions	Rate and equilibrium constants	Ref.[a]
$LacCo(NH_3)_5{}^{2+}$	1.0 *M*	$k_1 = 2.7\ s^{-1}$ (Co(II) = 100%)	[1]
$ManCo(NH_3)_5{}^{2+}$	1.0 *M*	$k_1 = 3.6\ s^{-1}$ (Co(II) = 96%)	
$BenzCo(NH_3)_5{}^{2+}$	1.0 *M*	$k_1 = 3.7\ s^{-1}$ (Co(II) = 89%)	
	0.5 *M*	$k_1 = 1.84\ s^{-1}$	

[a] Reference: [1] Srinivasan and Gould (1981).
[b] Abbreviations used:
Lac lactate ($CH_3(OH)HCOO^-$).
Man mandelate ($C_6H_5C(OH)HCOO^-$).
Benz benzilate ($(C_6H_5)_2C(OH)COO^-$).

The reinvestigation of the oxidation of oxalatopentaamine Co(III) with Ce(IV) (in $HClO_4$) by Dash et al. (1984) confirmed the original mechanism for this reaction proposed by Saffir and Taube (1960). There was no evidence cited for the formation of a complex between Ce(IV) and the cobalt salt.

The reactions between Ce(IV) and a series of $(NH_3)_5$Co(III)R complexes where R is glycine, alanine, *N*-acetylglycine and *N*-benzoylglycine were reported by Subramani and Srinivasan (1985) to proceed by paths that involve the formation of a complex between Ce(IV) and the Co(III) complexes, followed by electron transfer reactions. In contrast to the earlier results, the product distribution studies report that only about 20% of the Co(III) is reduced to Co(II). The rates of disappearance of Ce(IV) are essentially the same as for the oxidation of the free ligands.

The study by Linn and Gould (1987) reported that Ce(IV) did not react with biphosphitopentaamineCo(III), whereas the hypophosphito complex reacts to yield decreasing amounts of Co(II) as the initial Ce(IV) concentration is increased. The proposed mechanism invokes competing reaction paths, internal electron transfer to the Co(III) center, and further oxidation by Ce(IV).

The results reported by Geiger et al. (1985) in a study of Ce(IV) oxidation of Co(II)-, Co(III)- and Ni(II)-tetrasulfophthalocyanine complexes provide evidence that the primary step is the formation of a ligand radical, rather than participation of the metal ion in the redox reaction. The combination of stopped-flow spectrophotometric and ESR data demonstrated that the Co(III)- and Ni(II)-phthalocyanine radicals disproportionate with regeneration of the original compound and ligand degradation products. In contrast, the Co(II) complex reaction proceeds via the formation of a radical and subsequent rapid reaction to produce the parent compound.

3.13. *Oxidation of Os and Rh coordination compounds*

The reaction between Ce(IV) and $[Os(NH_3)_5CO]^{2+}$ reported by Buhr and Taube (1979) yields a dinitrogen species $[(Os(NH_3)_4CO)_2N_2]^{4+}$. The electron transfer step with Ce(IV) is rapid and the kinetic studies reported are concerned with the subsequent disproportionation reactions.

The results of kinetic studies by Ali et al. (1983) of the reaction

$$\mathrm{Ce(IV)} + \mathrm{Rh_2(OAc)_4} \rightleftharpoons \mathrm{Ce(III)} + \mathrm{Rh_2(OAc)_4^+} \tag{38}$$

in perchloric and sulfuric acids demonstrated a reduction in the rate by a factor of three for the change of the medium. The rate constant in $HClO_4$, along with the values obtained for the Fe(II) and VO^{2+} reactions, were satisfactorily correlated in terms of the Marcus theory.

4. Oxidation/reduction reactions of Eu(II)/Eu(III) in solution

The standard potential of the Eu(II)/Eu(III) couple in acid solution is $E^0 = -0.35$ V, which implies that Eu(II) is nearly as powerful a reducing reagent as Cr(II). The electron transfer processes of Eu^{3+}/Eu^{2+} are in general not as facile as are those

TABLE 11
Europium redox reactions.

Reactant[c]	Conditions	Rate parameters	Ref.[a]
$CH_3COCOOH$	0.5–1.4 M	$k_1 = 2280\ M^{-1}s^{-1}$	
		$k_2/k_{-1} = 0.90$	[1]
$^{\bullet}C(CH_3)_2OH$	0.4 M	$k_2 = 3.7 \times 10^4\ M^{-1}s^{-1}$	
(Eu(III) reduced)		(H_2O)	[2]
		$k_2 = 8.8 \times 10^3\ M^{-1}s^{-1}$	
		(D_2O)	
$^{\bullet}C(CH_3)_2OH$	0.04 M	$k_2 = 1.5 \times 10^5\ M^{-1}s^{-1}$	
(Eu(II) oxidized)		(H_2O)	
		$k_2 = 3.1 \times 10^4\ M^{-1}s^{-1}$	
		(D_2O)	
N-methylphenothiazine[b]	–	$k_1 = 3.3 \times 10^6\ s^{-1}$	[3]
		$k_2 = 2.1 \times 10^9\ M^{-1}s^{-1}$	
H_2O_2	0.1–0.0005 M	$k_2 = 2.1\ M^{-1}s^{-1}$	[4]
$MoO(O_2)_2$		$k_2 = 1.9 \times 10^3\ M^{-1}s^{-1}$	
$MoO(OH)(O_2)^-$		$k_2 = 3.0 \times 10^3\ M^{-1}s^{-1}$	
$WO(OH)(O_2)_2^-$	0.0005 M	$k_2 = 2.8 \times 10^4\ M^{-1}s^{-1}$	
O_2	0.1–0.0005 M	$k_2 = 6.0 \times 10^4\ M^{-1}s^{-1}$	
Ferrichrome	pH = 4.1	$k_2 = 8.64 \times 10^2\ M^{-1}s^{-1}$	[5]
Ferrichrome A	pH = 2.4	$k_1 = 322\ M^{-1}s^{-1}$	
		$k_2K_a = 1.46 \times 10^{-1}\ s^{-1}$	
$Co(NH_3)_5PO_4$	pH = 0.3	$k_2 = 29\ M^{-1}s^{-1}$	[6]
$Co(NH_3)_5NO_2^{2+}$	pH = 0.3	$k_2 = 43\ M^{-1}s^{-1}$	
$Co(NH_3)_5ONO^{2+}$	pH = 0.3	$k_2 = 58\ M^{-1}s^{-1}$	
$Co(NH_3)_5PO_3H_2^{2+}$	0.1 M	$k_2 = 7.9\ M^{-1}s^{-1}$	[7]
$Co(NH_3)_5PO_2H_2^{2+}$		$k_2 = 14.5\ M^{-1}s^{-1}$	
t-$Ru(en)_2Cl_2^+$	0.2–0.4 M	$k_2 = 2.4 \times 10^3\ M^{-1}s^{-1}$	[8]
t-$Ru(en)_2Br_2^+$		$k_2 = 3.65 \times 10^3\ M^{-1}s^{-1}$	
t-$Ru(en)_2I_2^+$		$k_2 = 5.15 \times 10^3\ M^{-1}s^{-1}$	
t-Ru(2,3,2-tet)Cl_2^+		$k_2 = 3.5 \times 10^3\ M^{-1}s^{-1}$	
t-Ru(cyclam)Cl_2^+		$k_2 = 5.85 \times 10^3\ M^{-1}s^{-1}$	
t-Ru([15]aneN_4)Cl_2^+		$k_2 = 6.20 \times 10^3\ M^{-1}s^{-1}$	
t-Ru(teta)Cl_2^+		$k_2 = 8.55 \times 10^3\ M^{-1}s^{-1}$	
t-Ru(tetb)Cl_2^+		$k_2 = 7.35 \times 10^3\ M^{-1}s^{-1}$	

[a] References: [1] Konstantatos et al. (1982); [2] Muralidharan and Espenson (1984); [3] Moroi et al. (1979); [4] Schwane and Thompson (1989); [5] Kazmi et al. (1984); [6] Matusinovic and Smith (1981); [7] Linn and Gould (1987); [8] Poon and Tang (1984).

[b] $2.7 \times 10^{-4}\ M$ micellar concentration (decyl and lauryl sulfates).

[c] Abbreviations used:

t	*trans* isomer.
en	ethylenediamine.
2,3,2-tet	3,7-diazanonane-1,9-diamine.
cyclam	1,4,8,12-tetraazacyclotetradecane.
[15]aneN_4	1,4,8,12-tetraazacyclopentadecane.
teta	C-meso-5,5,7,12,12,14-hexamethyl-1,4,8,11-tetraazacyclotetradecane.
tetb	C-rac-5,5,7,12,12,14-hexamethyl-1,4,8,11-tetraazacyclotetradecane.

of the d-transition elements. The only other 4f transition elements that have divalent oxidation states accessible in normal aqueous solutions are Sm and Yb with respective reduction potentials of -1.55 V and -1.05 V. The rate parameters for the reactions discussed in sections 4.1 to 4.3 are reproduced in table 11.

4.1. *Reactions with organic compounds*

The reduction of pyruvic acid by Eu(II) is reported by Konstantatos et al. (1982) to occur through a complex rate law involving backward reaction of the Eu(III) product with the organic radical species formed in the initial reaction:

$$-\mathrm{d}[\mathrm{Eu(II)}]/\mathrm{d}t = k_1 k_2 [\mathrm{Eu(II)}]^2 [\mathrm{H}^+][\mathrm{L}]/(k_{-1}[\mathrm{Eu(III)}] + k_2[\mathrm{Eu(II)}]). \quad (39)$$

Since Eu(II) is constrained to react as a one-electron donor, oxidation with organic compounds leads to the formation of an organic radical in the first step of the reaction sequence. Therefore, it is remarkable that the sole product reported for the oxidation of Eu(II) by the keto form of pyruvic acid is lactic acid.

The results of Muralidharan and Espenson (1984) utilize an ingenious preparation of the 2-hydroxy-2-propyl radical to study the kinetics of both oxidation of Eu(II) and reduction of Eu(III):

$$\mathrm{Eu_{aq}^{3+}} + (\mathrm{CH_3})_2\mathrm{COH} \rightleftharpoons \mathrm{Eu_{aq}^{2+}} + (\mathrm{CH_3})_2\mathrm{CO} + \mathrm{H}^+, \quad (40)$$

$$\mathrm{Eu_{aq}^{2+}} + (\mathrm{CH_3})_2\mathrm{COH} \rightleftharpoons \mathrm{Eu_{aq}^{3+}} + (\mathrm{CH_3})_2\mathrm{CHOH}. \quad (41)$$

The comparison of the results obtained in H_2O and D_2O is the basis for the authors to postulate that the second reaction proceeds via a hydrogen atom abstraction mechanism.

The kinetics of the reduction of Eu^{3+} by the excited state of *N*-methylphenothiazine is completed within the time of the laser pulse and the backward reaction

$$\mathrm{Eu^{2+}} + \mathrm{MPTH^+} \longrightarrow \mathrm{Eu^{3+}} + \mathrm{MPTH} \quad (42)$$

is characterized by two stages. Moroi et al. (1979) report that the first stage is a rapid first-order process where $k_1 = 4 \times 10^6\ \mathrm{s}^{-1}$ followed by a second-order process with $k_2 = 3 \times 10^9\ M^{-1}\,\mathrm{s}^{-1}$. A kinetic model is developed which accounts for the observations in terms of the elementary processes that occur in the micellar environment.

4.2. *Reduction of hydrogen peroxide and peroxo-metal complexes*

The results obtained in a comprehensive study of the oxidation of Eu(II) by the title oxidants have been recently reported by Schwane and Thompson (1989). These investigators compared the rates of reactions of Eu(II) with those determined for Fe(II), SO_2^- and the methylviologen-radical cation. They concluded that the predominant factor that characterizes the reaction rates is not the free energy change of the reaction.

4.3. *Reduction of Fe and Co coordination compounds*

The results of kinetic studies on the reduction of the siderophores ferrichrome and ferrichrome A by Eu(II) were presented by Kazmi et al. (1984). The comparison of the kinetic data for the reduction of the same siderophores with V(II) and Cr(II) led the authors to postulate an outer-sphere electron transfer mechanism providing an adequate interpretation of the data. Apparent values of the activation free energy for the self-exchange reaction of Eu(II)/Eu(III) were calculated from the reduced form of the Marcus equation.

The report by Matusinovic and Smith (1981) presented the kinetic results on the reduction of a series of pentaamineCo(III) complexes by Eu(II) when the latter was generated in situ electrochemically. The value for the rate parameter was obtained from the variation of the limiting DC polarographic potential as a function of the substrate concentrations. The values of the rate parameters calculated by this technique compared well with those obtained by stopped-flow spectrophotometric methods.

The rates of Eu(II) reduction of a series of carboxylatopentaamineCo(III) (Srinivasan et al. 1981, 1982a) and (carboxylato)bis(hydroxo)bis(triamineCo(III)) compounds (Srinivasan et al. 1982b, c) have been reported. Comparisons of these rates with those obtained for reduction by a variety of other d-transition metals provide additional evidence for the nonfacile kinetic behavior of Eu(II).

4.4. *Reactions with Ru, Rh and lanthanide compounds*

The values for the rates of oxidation of Eu(II) by Ce(IV) were determined by Grzeszczuk and Smith (1986) by the use of an electrochemical technique that produced Eu(II) in situ. The values calculated for the second-order rate parameters (pH = 1.00, 25°C, $I = 1.0\ M$) in perchlorate, nitrate and chloride solutions were, respectively, 4.2, 4.0 and $3.5 \times 10^4\ M^{-1}\,s^{-1}$. The authors note that the decrease in rate parameters is correlated with the decrease in formal potentials of the Ce(IV)/Ce(III) couple and is qualitatively consistent with the Marcus theory.

The reduction of Yb(III) and Sm(III) by $Ru(bpy)_3^+$ was studied by Connolly et al. (1986) ($I = 1.0\ M$, $T = 23$°C). The respective values for the second-order rate parameters were $1.2 \times 10^5\ M^{-1}\,s^{-1}$ and $<2 \times 10^4\ M^{-1}\,s^{-1}$. The calculation of the Yb self-exchange reaction rate, based on the Marcus formalism and utilizing two different data sets, results in values that varied by $10^5\ M^{-1}\,s^{-1}$. The authors conclude that the self-exchange rate of Yb(III)/Yb(II) has not been experimentally defined.

One aspect of the problems encountered in the description of outer-sphere electron transfer reactions within the Marcus formalism is that the self-exchange rate of Eu(III)/Eu(II) has not been measured. Therefore, estimates have been calculated for this rate from measurements of the rates of reactions of systems that are well characterized, i.e. the equilibrium constant is known. In addition, there is substantiating evidence that the reaction is outer-sphere, and the value for self-exchange rate of the other reactant has been measured or calculated from an independent set of measurements. The usual simplifying assumption is that the probability of electron transfer in the activated complex is unity.

The analysis advanced by Balzani et al. (1981) was based on a comparison of literature data for the reactions of $Ru(NH_3)_6^{2+,3+}$, $Fe^{2+,3+}$, $Eu^{2+,3+}$ with a series of reactants that have well-characterized and nonlabile primary coordination spheres. The $\log k_{12} - \Delta G$ plots for the Fe and Ru reactions were consistent with the interpretation that the reactions were adiabatic over the entire range of ΔG. The data for $Eu^{2+,3+}$ was interpreted to indicate that the electron transfer reactions were strongly nonadiabatic for electron transfer between ground states with moderate driving force. When the thermodynamic driving force becomes too large, the authors postulate that new excited state channels become important. The important conclusion drawn was that because of the apparent nonadiabaticity of the electron transfer reactions of Eu ions, it is not possible to use the Marcus crossrelations to calculate self-exchange or electron transfer rates for reactions involving $Eu^{2+,3+}$.

The values for the rates of oxidation of Eu(II) by a series of Ru(II) amines have been reported by Poon and Tang (1984). These authors used the results that had been previously reported for the reduction of these same amines by Cr(II) and V(II) as the basis for the postulate that the Eu(II) exchange reactions were outer-sphere. They then calculated self-exchange reactions for the Ru(III) complexes from the data obtained with V(II), and used the data to calculate apparent self-exchange reactions for Eu(II)/Eu(III). The authors concluded that the values so calculated (within the range 2.3×10^{-4} to $1.7 \times 10^{-3}\ M^{-1}\,s^{-1}$) were in reasonable agreement with other estimates.

The importance of a nonadiabatic path for outer-sphere electron transfer reactions of Eu(III)/Eu(II) was again examined by Yee et al. (1983) via a study of a series of reactions with Eu(III)/(II) cryptates (table 12). The cryptate (polyoxadiazamacrobicyclic) ligands form thermodynamically stable and substitution inert complexes with both Eu(III) and Eu(II), markedly changing the primary coordination spheres of the Eu ions. The dramatic variation in the values for the Eu exchange reactions with such a change is demonstrated by the respective calculated values for $Eu_{aq}^{3+,2+}$, $Eu(2.2.2)^{3+,2+}$ and $Eu(2.2.1)^{3+,2+}$ of 5×10^{-6}, 4×10^{-2} and $10\ M^{-1}\,s^{-1}$. These values calculated from the cross reactions are consistent with the values of the Franck–Condon barriers estimated from structural data.

The nonadiabatic pathways for outer-sphere electron transfer would be characterized by small values of activation enthalpies and large negative entropies of activation. The entropy of activation for the $Eu_{aq}^{3+,2+}$ self-exchange reaction was calculated to be $-107\ J\,mol^{-1}\,K^{-1}$ similar in magnitude to that measured for d-transition metals. The authors concluded that there was no compelling reason to invoke a nonadiabatic path for electron transfer reactions involving $Eu_{aq}^{3+,2+}$ ions.

The results obtained in a study of the reactions of one of the same cryptands, $Eu(2.2.1)^{3+,2+}$, with the excited states of a series of poly(pyridine)Ru(II) complexes show a qualitative agreement with the previous study. The detailed analysis of the data by Sabbatini et al. (1986), however, invokes the availability of nonadiabatic reaction paths for the cross reactions. The fact that in this later study, the thermody-

TABLE 12
Redox reactions of Eu-cryptate complexes.

Eu-cryptate[b]	Co-reactant[b]	Second-order rate parameter	Ref.[a]
$Eu(2.2.1)^{3+}$	V^{2+}	$0.5\ M^{-1}s^{-1}$	[1]
$Eu(2.2.1)^{3+}$	Eu^{2+}	$\approx 0.2\ M^{-1}s^{-1}$	
$Eu(2.2.1)^{2+}$	$Co(NH_3)_6^{3+}$	$0.043\ M^{-1}s^{-1}$	
$Eu(2.2.2)^{3+}$	V^{2+}	$1.5\ M^{-1}s^{-1}$	
$Eu(2.2.2)^{3+}$	Eu^{2+}	$1.4\ M^{-1}s^{-1}$	
$Eu(2.2.1)^{3+}$	$Ru(bpy)_2(dcbpy)^{2+}$	$5.6 \times 10^6\ M^{-1}s^{-1}$	[2]
$Eu(2.2.1)^{3+}$	$Ru(bpy)_2(nbpy)^{2+}$	$2.1 \times 10^7\ M^{-1}s^{-1}$	
$Eu(2.2.1)^{3+}$	$Ru(bpy)_2(dmbpy)^{2+}$	$3.3 \times 10^7\ M^{-1}s^{-1}$	
$Eu(2.2.1)^{3+}$	$Ru(bpy)_3^{2+}$	$4.9 \times 10^7\ M^{-1}s^{-1}$	
$Eu(2.2.1)^{3+}$	$Ru(dmbpy)_2(bpy)^{2+}$	$7.1 \times 10^7\ M^{-1}s^{-1}$	
$Eu(2.2.1)^{3+}$	$Ru(dmbpy)_3^{2+}$	$5.6 \times 10^7\ M^{-1}s^{-1}$	
$Eu(2.2.1)^{3+}$	$Ru(dmphen)_3^{2+}$	$6.0 \times 10^7\ M^{-1}s^{-1}$	
$Eu(2.2.1)^{3+}$	$Ru(tmphen)_3^{2+}$	$3.4 \times 10^7\ M^{-1}s^{-1}$	
$Eu(2.2.1)^{3+}$	$Ru(bpy)_2(DMCH)^{+}$	$5.9 \times 10^7\ M^{-1}s^{-1}$	
$Eu(2.2.1)^{3+}$	$Ru(bpy)_3^{+}$	$7.0 \times 10^8\ M^{-1}s^{-1}$	
$Eu(2.2.1)^{2+}$	$Ru(dmbpy)_3^{2+}$	$1.3 \times 10^9\ M^{-1}s^{-1}$	
$Eu(2.2.1)^{2+}$	$Ru(bpy)(isobiq)_2^{2+}$	$1.3 \times 10^9\ M^{-1}s^{-1}$	
$Eu(2.2.1)^{2+}$	$Ru(bpy)_2(DMCH)^{2+}$	$1.0 \times 10^9\ M^{-1}s^{-1}$	
$Eu(2.2.1)^{2+}$	$Ru(bpy_2)(isobiq)^{2+}$	$1.2 \times 10^9\ M^{-1}s^{-1}$	
$Eu(2.2.1)^{2+}$	$Ru(bpy)_3^{2+}$	$1.3 \times 10^9\ M^{-1}s^{-1}$	
$Eu(2.2.1)^{2+}$	$Ru(bpy)(biq)_2^{2+}$	$1.7 \times 10^9\ M^{-1}s^{-1}$	
$Eu(2.2.1)^{2+}$	$Ru(bpy)_2(nbpy)^{2+}$	$2.3 \times 10^9\ M^{-1}s^{-1}$	
$Eu(2.2.1)^{2+}$	$Ru(bpy)_3^{3+}$	$1.3 \times 10^9\ M^{-1}s^{-1}$	

[a] References: [1] Yee et al. (1983); [2] Sabbatini et al. (1982).
[b] Abbreviations used:

(2.2.1)	4,7,13,16,21-pentaoxa-1,10-diazabicyclo[8.8.5]tricosene.
(2.2.2)	4,7,13,16,21,24-hexaoxa-1,10-diazabicyclo[8.8.8]hexacosane.
bpy	2,2′-bipyridine.
dcbpy	4,4′-dichloro-2,2′-bipyridine.
nbpy	4-nitro-2,2′-bipyridine.
dmbpy	4,4′-dimethyl-2,2′-bipyridine.
dmphen	5,6-dimethyl-1,10-phenanthroline.
tmphen	3,4,7,8-tetramethyl-1,10-phenanthroline.
DMCH	6,7-dihydro-5,8-dimethyldibenzo[b, j] [1,10]-phenanthroline.
biq	biquinoline.
isobiq	isobiquinoline.

namic driving forces for the reactions were much greater than in the previous study of Eu-cryptate redox reactions should be noted.

5. Summary and prospects

Because of their importance as magnetic resonance imaging agents, among many other significant commercial uses, the solution chemistry of the lanthanide elements will likely remain a hot topic for investigation for many years to come. Partly because of the commercial interests, some significant advances have been made in the last few years in the kinetics of complexation and electron transfer reactions of the lanthanides in aqueous solutions.

Perhaps most significant among recent achievements are the results of Merbach's investigation of the hydration dynamics of the lanthanides (Cossy et al. 1988, 1989b), and more specific experimental determination (and calculation) of lanthanide hydration numbers (Cossy et al. 1989a, Miyakawa et al. 1988). The development of high-field NMR spectrometers has led to this major advance, and promises to provide similar advances in resolution of questions in ligand exchange kinetics. In general, the most significant advances of recent years have been a direct result of the development and application of analytical methods capable of performing measurements in the millisecond to microsecond time regime.

Judging by the total number of literature citations which we considered, and by the relative proportions of this review, ceric oxidation of many different substrates is the most widely investigated aspect of lanthanide solution chemistry. In addition to the literature cited herein, over 100 reports dealing with cerium(IV) oxidation of organic compounds were found which relate to the use of cerium(IV) as a synthetic reagent, or to the chemistry of the substrate following ceric oxidation. No real consensus has developed on the probable reactive cerium(IV) species in the various media. This failing is partly a result of uncertainties related to the hydrolytic equilibria of Ce(IV) in non-complexing acidic solutions, and partly a result of many author's inability to recognize these difficulties and take precautions in preparation of Ce(IV) reagents.

Although outer-sphere electron transfer reactions are observed in cerium(IV) oxidations, particularly in sulfate media, this pathway is not the primary one. The dominant route for ceric oxidation of organic substrates is through the formation of relatively stable precursor complexes. The nature of the precursor complexes is not clear. In systems wherein these species are reported to exist, little difference is noted between the stability of analogous monofunctional Ce(IV)–alcohol, –aldehyde and –carboxylate complexes. This is contrary to what is typically observed in normal metal–ligand complexes. Polyfunctional organic substrates appear to form significantly stronger precursor complexes. The highest values reported for stability of precursor complexes are for the carbohydrates, which are not normally considered as strong complexants (Virtanen et al. 1987), and the α-hydroxycarboxylic acids, which are (Amjad et al. 1977, Hanna and Fenton 1983). Perhaps the similarity of precursor complex stability of these diverse species argues most strongly for metal ion solvation effects as the explanation for the large K's under conditions in which only weak complexes should be expected.

In contrast to cerium, outer-sphere electron transfer appears to be the dominant reaction route for europium redox reactions with both organic species and transition metal complexes. Interpretation of europium redox reactions in terms of the Marcus theory of outer-sphere electron transfer reactions is limited by the enduring controversy over the self-exchange rate for the Eu(II)/Eu(III) couple. Since the self-exchange rate for the Eu(II)/Eu(III) couple has so far proven inaccessible to direct measurement, many attempts have been made to infer a value for this process by application of Marcus theory to analogous systems wherein good Marcus behavior is observed. As the results summarized above indicate, this procedure has not provided an acceptable value for the self-exchange rate, although the relative order of magnitude appears to be known.

References

Ali, R.B., K. Sarawek, A. Wright and R.D. Cannon, 1983, Inorg. Chem. **22**, 351–353.

Amjad, Z., and A. McAuley, 1974, J. Chem. Soc. Dalton Trans., 2521–2526.

Amjad, Z., and A. McAuley, 1977, J. Chem. Soc. Dalton Trans., 304–308.

Amjad, Z., A. McAuley and U.D. Gomwalk, 1977, J. Chem. Soc. Dalton Trans., 82–88.

Arcoleo, G., G. Calvaruso, F.P. Cavasino and C. Sbriziolo, 1977, Inorg. Chim. Acta **23**, 227–233.

Arcoleo, G., G. Calvaruso, F.P. Cavasino and E. Di Dio, 1979, Int. J. Chem. Kinet. **11**, 415–425.

Arselli, P., C. Baiocchi, E. Mentasti and J.S. Coe, 1984, J. Chem. Soc. Dalton Trans., 475–477.

Baciocchi, E., L. Mandolini and C. Rol, 1980, J. Am. Chem. Soc. **102**, 7597–7598.

Baes Jr, C.F., and R.E. Mesmer, 1976, in: The Hydrolysis of Cations (Wiley, New York) pp. 138–146.

Balasubramanian, T.R., and N. Venkatasubramanian, 1975, Rev. Roum. Chim. **20**, 1301–1310.

Balzani, V., F. Scandola, G. Orlandi, N. Sabbatini and M.T. Indelli, 1981, J. Am. Chem. Soc. **103**, 3370–3378.

Bar-Eli, K., 1985, J. Phys. Chem. **89**, 2855–2860.

Barkin, S., M. Bixon, R.M. Noyes and K. Bar-Eli, 1977, Int. J. Chem. Kinet. **9**, 841–862.

Behari, K., N.N. Pandey, R.K. Khanna and R.S. Shukla, 1982a, Z. Phys. Chem. (Leipzig) **263**, 734–740.

Behari, K., H. Chatterjee and N.N. Pandey, 1982b, Z. Phys. Chem. (Leipzig) **262**, 273–281.

Benson, D., 1976, in: Mechanism of Oxidation by Metal Ions (Elsevier, Amsterdam) pp. 10–111.

Berdnikov, V.M., L.L. Makarshin and L.S. Ryvkina, 1978, React. Kinet. Catal. Lett. **9**, 211–216.

Breen, P.J., W.DeW. Horrocks Jr and K.A. Johnson, 1986, Inorg. Chem. **25**, 1968.

Brücher, E., I. Bányai and L. Krusper, 1984, Acta Chim. Hung. **116**, 39.

Brusa, M.A., L.J. Perissinotti and A.J. Colussi, 1988, Inorg. Chem. **27**, 4474.

Buhr, J.D., and H. Taube, 1979, Inorg. Chem. **18**, 2208–2212.

Calvaruso, G., F.P. Cavasino and C. Sbriziolo, 1981a, Int. J. Chem. Kinet. **13**, 135–148.

Calvaruso, G., F.P. Cavasino and C. Sbriziolo, 1981b, Int. J. Chem. Kinet. **13**, 1029–1040.

Calvaruso, G., F.P. Cavasino, C. Sbriziolo and R. Triolo, 1983, Int. J. Chem. Kinet. **15**, 417–432.

Calvaruso, G., F.P. Cavasino and C. Sbriziolo, 1984, Int. J. Chem. Kinet. **16**, 1201–1211.

Chang, C.A., P. H.-L. Chang, V.K. Manchanda and S. Kasprzyk, 1988, Inorg. Chem. **27**, 3786.

Chappelle, G.A., A. MacStay, S.T. Pittenger, K. Ohashi and K.W. Hicks, 1984, Inorg. Chem. **23**, 2768–2771.

Chum, H.L., and M.E.M. Helene, 1980, Inorg. Chem. **19**, 876–880.

Connolly, P., J.H. Espenson and A. Bakac, 1986, Inorg. Chem. **25**, 2169.

Cossy, C., L. Helm and A.E. Merbach, 1988, Inorg. Chem. **27**, 1973.

Cossy, C., A.C. Barnes, A.E. Merbach and J.E. Enderby, 1989a, J. Chem. Phys. **90**, 3254.

Cossy, C., L. Helm and A.E. Merbach, 1989b, Inorg. Chem. **28**, 2699–2703.

Cyfert, M., B. Latko and M. Wronska, 1980, Monatsh. Chem. **111**, 619–626.

Dash, A.C., R.K. Nanda and P. Mohanty, 1984, Indian J. Chem. Sect. A **23**A, 162.

Dikshitulu, L.S.A., V.H. Rao and S.N. Dindi, 1980, Indian J. Chem. Sect. A **19**A, 203–206.

Dikshitulu, L.S.A., V.H. Rao and S.N. Dindi, 1981a, Indian J. Chem. Sect. A **20**A, 784–787.

Dikshitulu, L.S.A., P. Vani and V.H. Rao, 1981b, J. Inorg. & Nucl. Chem. **43**, 1261–1265.

Dziegiec, J., 1981, Pol. J. Chem. **55**, 495–502.

Eberson, L., and E. Oberrauch, 1979, Acta Chem. Scand. Ser. B **833**, 343–351.

Fay, D.P., and N. Purdie, 1969, J. Phys. Chem. **74**, 1160.

Fay, D.P., and N. Purdie, 1970, J. Phys. Chem. **73**, 544.

Field, R.J., E. Koros and R.M. Noyes, 1972, J. Am.

Chem. Soc. **94**, 8649.

Foersterling, H.D., R. Pahcl and H. Schreiber, 1987, Z. Naturforsch. A **42**, 963–969.

Gasco, M.R., and M.E. Carlotti, 1979, Pharm. Acta Helv. **54**, 26–31.

Geiger, D.K., G. Ferraudi, K. Madden, J. Granifo and D.P. Rillema, 1985, J. Phys. Chem. **89**, 3890–3894.

Gopalan, R., and E. Kannamma, 1984, Indian J. Chem. Sect. A **23**A, 518–519.

Grzeszczuk, M., and D.E. Smith, 1986, J. Electroanal. Chem. Interfacial Electrochem. **198**, 401–407.

Hafez, M.B., H. Roushdy and N. Hafez, 1978, J. Radional. Chem. **43**, 121–129.

Hafez, M.B., W. Higazi and N. Hafez, 1979, J. Radional. Chem. **49**, 45–53.

Hanna, S., and M.E. Moehlenkamp, 1983, J. Org. Chem. **48**, 826–832.

Hanna, S., and S.A. Sarac, 1977a, J. Org. Chem. **42**, 2069–2073.

Hanna, S., and S.A. Sarac, 1977b, J. Org. Chem. **42**, 2063–2068.

Hanna, S., R.R. Kessler, A. Merbach and S. Ruzicka, 1976, J. Chem. Educ. **53**, 524–527.

Hanna, S.B., and J.T. Fenton, 1983, Int. J. Chem. Kinet. **15**, 925–944.

Hardwick, T.J., and E. Robertson, 1951, Can. J. Chem. **29**, 828.

Hargreaves, G., and L.H. Sutcliffe, 1955, Trans. Faraday Soc. **51**, 1105.

Holecek, J., K. Handlir and J. Klikorka, 1979, Collect. Czech. Chem. Commun. **44**, 1060–1069.

Holwerda, R.A., and M.L. Ettel, 1982, Inorg. Chem. **21**, 830–832.

Husain, M., 1977, J. Inorg. & Nucl. Chem. **39**, 2249–2252.

Ignaczak, M., and J. Dziegiec, 1978, Pol. J. Chem. **52**, 2467–2472.

Ignaczak, M., and M. Markiewicz, 1981, Pol. J. Chem. **55**, 671–677.

Ignaczak, M., J. Dziegiec and M. Markiewicz, 1980, Pol. J. Chem. **54**, 1121–1128.

Kale, A., and K.C. Nand, 1982, Gazz. Chim. Ital. **112**, 391–396.

Kazmi, S.A., A.L. Shorter and J.V. McArdle, 1984, Inorg. Chem. **23**, 4332–4341.

Kemp, T.J., P. Moore and G.R. Quick, 1980, J. Chem. Soc. Perkin Trans. **2**, 291–295.

King, E.L., and M.L. Pandow, 1953, J. Am. Chem. Soc. **75**, 3063.

Kishan, B.H., and E.V. Sundaram, 1980, J. Ind. Chem. Soc. **57**, 1079–1082.

Kishan, B.H., and E.V. Sundaram, 1985, J. Ind. Chem. Soc. **62**, 654–656.

Konstantantos, J., E. Vrachnou-Astra, N. Katsaros and D. Katakis, 1982, Inorg. Chem. **21**, 12–16.

Laurenczy, G., and E. Brücher, 1984, Inorg. Chim. Acta **95**, 5.

Lincoln, S.F., 1986, in: Advances in Inorganic and Bioinorganic Mechanisms, Vol. 4, ed. A.G. Sykes (Academic Press, London) pp. 217–287.

Linn Jr, D.E., and E.S. Gould, 1987, Inorg. Chem. **26**, 3442–3446.

Martell, A.E., and R.M. Smith, 1977, in: Critical Stability Constants, Vol. 3, Other Organic Ligands (Plenum, New York) .

Matusinovic, T., and D. Smith, 1981, Inorg. Chem. **20**, 3121–3122.

McAuley, A., 1981, in: Inorganic Reaction Mechanisms, Vol. 7, ed. A.G. Sykes (Royal Society of Chemistry, London) pp. 85–87.

Miyakawa, K., Y. Kaizu and H. Kobayashi, 1988, J. Chem. Soc. Faraday Trans. I, 1517–1529.

Mohanty, N.K., and R.K. Nanda, 1985, Bull. Chem. Soc. Jpn. **58**, 3597–3601.

Moroi, Y., P.P. Infelta and M. Graetzel, 1979, J. Am. Chem. Soc. **101**, 573–579.

Morss, L.R., 1985, Yttrium, lanthanum and the lanthanide elements, in: Standard Potentials in Aqueous Solution, eds A.J. Bard, R. Parsons and J. Jordan (Marcel Dekker, New York) pp. 587–629.

Muralidharan, S., and J.H. Espenson, 1984, Inorg. Chem. **23**, 636–639.

Nair, B., T. Ramasami and D. Ramaswamy, 1987, Int. J. Chem. Kinet. **19**, 277–298.

Nazar, A.F.M., and C.F. Wells, 1985, J. Chem. Soc. Faraday Trans. I, 801–812.

Nikitenko, S.I., L.I. Martynenko and N.I. Pechurova, 1984a, Zh. Neorg. Khim. **29**, 2801.

Nikitenko, S.I., L.I. Martynenko, E.G. Afonin and N.I. Pechurova, 1984b, Zh. Neorg. Khim. **29**, 933.

Nikitenko, S.I., L.I. Martynenko and N.I. Pechurova, 1985, Russ. J. Inorg. Chem. **30**, 788.

Nyssen, G.A., and D.W. Margerum, 1970, Inorg. Chem. **9**, 1314–1320.

Pelizzetti, E., E. Mentasti and C. Baiocchi, 1976, J. Inorg. & Nucl. Chem. **38**, 557–561.

Poon, C.K., and T.W. Tang, 1984, Inorg. Chem. **23**, 2130.

Prakash, A., R.N. Mehrotra and R.C. Kapoor, 1979, J. Chem. Soc. Dalton Trans., 205–210.

Prasad, S., and J. Choudhary, 1979, Indian J. Chem. Sect. A **17**A, 167–169.

Prasad, S., and R.K. Prasad, 1979, Indian J. Chem. Sect. A **17**A, 491–494.

Pratihari, H.K., A.K. Roy and P.L. Nayak, 1976, J. Indian Chem. Soc. **53**, 4–6.

Purdie, N., and M.M. Farrow, 1973, Coord. Chem. Rev. **11**, 189.

Rajanna, K.C., Y.R. Rao and P.K. Saiprakash, 1979, Indian J. Chem. Sect. A **17**A, 66–69.

Rizkalla, E.N., L.H. Lajunen and G.R. Choppin, 1986, Inorg. Chim. Acta **119**, 93–98.

Roy, A.R., B.C. Singh, P.L. Nayak and M.K. Rout, 1978, J. Inorg. & Nucl. Chem. **40**, 749–752.

Sabbatini, N., S. Dellonte, A. Bonazzi, M. Ciano and V. Balzani, 1986, Inorg. Chem. **25**, 1738–1742.

Saffir, P., and H. Taube, 1960, J. Am. Chem. Soc. **82**, 13.

Saiprakash, P.K., B. Sethuram and T.N. Rao, 1975, J. Indian Chem. Soc. **52**, 973–975.

Sarac, A.S., 1985, Int. J. Chem. Kinet. **17**, 1333–1345.

Saxena, P.K., B. Singh, R.K. Shukla and B. Krishna, 1978, J. Indian Chem. Soc. **55**, 56–59.

Schelenz, T., J. Stein and C.R. Kramer, 1982, Z. Chem. **22**, 143–144.

Schwane, L.M., and R.C. Thompson, 1989, Inorg. Chem. **28**, 3938.

Shannon, R.D., 1976, Acta. Cryst. A **32**, 751–767.

Singh, B., P.K. Saxena, K.M. Richards and B. Krishna, 1978, Z. Phys. Chem. (Leipzig) **259**, 801–807.

Soria, D., M.L. DeCastro and H.L. Cham, 1980, Inorg. Chim. Acta **48**, 121–128.

Srinivasan, V.S., and E.S. Gould, 1981, Inorg. Chem. **20**, 208–211.

Srinivasan, V.S., C.A. Radlowski and E.S. Gould, 1981, Inorg. Chem. **20**, 2094–2097.

Srinivasan, V.S., N. Rajasekar, A.N. Singh, C.A. Radlowski, J.C. Heh and E.S. Gould, 1982a, Inorg. Chem. **21**, 2924–2928.

Srinivasan, V.S., A.N. Singh, K. Wieghardt and E.S. Gould, 1982b, Inorg. Chem. **21**, 2531–2537.

Srinivasan, V.S., A.N. Singh, C.A. Radlowski and E.S. Gould, 1982c, Inorg. Chem. **21**, 1240–1242.

Subramani, K., and V.S. Srinivasan, 1985, Inorg. Chem. **24**, 235–238.

Sullivan, J.C., and R.C. Thompson, 1979, Inorg. Chem. **18**, 2375–2379.

Taube, H., 1970, Electron Transfer Reactions of Complex Ions in Solution (Academic Press, New York).

Tewari, I.C., and S.R. Tripathi, 1977, J. Inorg. & Nucl. Chem. **39**, 2095–2097.

Thompson, R.C., 1984, Inorg. Chem. **23**, 1794–1798.

Thompson, R.C., 1985, Inorg. Chem. **24**, 3542–3547.

Tischler, F., and J.I. Morrow, 1983, Inorg. Chem. **22**, 2286–2289.

Tripathy, S.N., and R.K. Prasad, 1980, Ind. J. Chem. Sect. A **19**A, 214–217.

Trubacheva, L.V., and N.I. Pechurova, 1981, Zh. Neorg. Khim. **26**, 76–79.

Vaidya, V.K., S.N. Joshi and G.V. Bakore, 1987, Z. Phys. Chem. (Leipzig) **268**, 364–368.

Vasudevan, R., and I.M. Mathai, 1978, Ind. J. Chem., Sect. A **16A**, 1097–1099.

Vincenti, M., C. Minero, E. Pramauro and E. Pelizzetti, 1985a, Inorg. Chim. Acta **110**, 51–53.

Vincenti, M., E. Pramauro, E. Pelizzetti, S. Diekmann and J. Frahm, 1985b, Inorg. Chem. **24**, 4533.

Virtanen, P.O.I., R. Lindroos, E. Oikarinen and J. Vaskuri, 1987, Carbohydr. Res. **167**, 29–37.

Wells, C.F., and L.V. Kuritsyn, 1969, J. Chem. Soc. A 2575.

Wells, C.F., and A.F.M. Nazar, 1979, J. Chem. Soc. Faraday Trans. I, **75**, 816–824.

Yee, E.L., J.T. Hupp and M.J. Weaver, 1983, Inorg. Chem. **22**, 3465–3470.

Yoshida, T., and Y. Ushiki, 1982, Bull. Chem. Soc. Jpn. **55**, 1772–1776.

Handbook on the Physics and Chemistry of Rare Earths, Vol. 15
edited by K.A. Gschneidner, Jr. and L. Eyring

Chapter 103

HYDRATION AND HYDROLYSIS OF LANTHANIDES

EMIL N. RIZKALLA AND GREGORY R. CHOPPIN
Department of Chemistry, Florida State University, Tallahassee, FL 32306, USA

Contents

Abbreviations

Ac	acetate
Acac	acetylacetonate
BEDTA	*N*-phenylethylenediaminetriacetate
ClAc	monochloroacetate
Cl_2Ac	dichloroacetate
DCTA	diaminocyclohexanetetraacetate
Dipic	2,6-dipicolinate
DMF	dimethylformamide
DMSO	dimethylsulfoxide
DTPA	diethylenetriaminepentaacetate
EDA	(1,2-ethanedioxy)diacetate
EDAP	ethylenediamine-*N*,*N*-diacetate-*N*′,*N*′-di-3-propionate
EDTA	ethylenediaminetetraacetate
EGTA	1,10-diaza-4,7-dioxadecane-1,1,10,10-tetraacetate
ENDADP	ethylenediamine-*N*,*N*′-diacetate-*N*,*N*′-di-3-propionate
ES	ethylsulfate
HEDTA	*N*-hydroxyethylethylenediamine-triacetate
Iso-nic	iso-nicotinate
MAE	poly(maleic acid-co-ethylene)
MEDTA	*N*-methylethylenediaminetriacetate
Nic	nicotinate
NTA	nitrilotriacetate
Ox	oxalate
PAA	polyacrylic acid

PAAm	poly(acrylic acid-co-acrylamide)
PDTA	1,2-propylenediaminetetraacetate
PMA	poly(methacrylic acid)
TCTP	tricapped trigonal prism
TDA	thiodiacetate
TMDTA	trimethylenediaminetetraacetate
TMEDTA	tetramethylenediaminetetraacetate
TP	terpyridyl

1. General

The number and the geometric distribution of solvent molecules around a metal ion in solution are often of major importance for the chemical behavior of the cation. For transition elements, the geometry of the coordination sphere is usually fixed by orbitals of the metal which form bonds with the solvent molecules and/or the ligands in complexation reactions. By contrast, the strongly ionic nature of the bonding of the f-block elements minimizes the importance of the metal orbitals and the geometry of the coordination sphere reflects an optimal balance of minimization of steric hindrance and maximum strength of electrostatic forces.

In the solid state, the coordination number of the f-element cations can differ significantly from that in solution. Factors such as the charge, size and number of the ligands are important in both solid and solution systems, but in solids the lack of shielding by the bulk solvent from the effects of other cations and ligands can result in different electrostatic and structural balance within a crystal compared to a solution system.

We define the hydration number as the average number of water molecules in the first sphere about the metal ion. The residence time of these molecules is determined generally by the nature of the bonding to the metal ion. For the f-element cations, ion–dipole interactions result in fast exchange between the hydration layer and the bulk solvent. The techniques for studying the nature (number and/or structure) of the hydration shell can be classified as either direct or indirect methods. The direct methods include X-ray and neutron diffraction, luminescence and NMR (nuclear magnetic resonance) relaxation measurements. The indirect methods involve compressibility, NMR exchange and absorption spectroscopy measurements.

This review is concerned with the hydration and hydrolysis of the lanthanide cations. The reported structures in solids and in solution are compared and the thermodynamics of hydration and the kinetics of exchange reviewed.

2. Lanthanides hydration

2.1. *Solid state hydrates*

The most reliable data on the structure of solid compounds are obtained from X-ray and neutron diffraction measurements. The number of hydrated lanthanide crystalline compounds investigated by these techniques is rather large and this review is limited to those solids in which the primary coordination shells are occupied predominantly by water molecules or in which the metal hydration is of interest for some relationship to hydration of the lanthanides in aqueous solution.

The crystal structures of lanthanide salts of bromate (Helmholz 1939, Singh 1944), trifluoromethylsulfonate (Lincoln 1986) and ethylsulfate (Fitzwater and Rundle 1959) are examples of systems in which the cations are strongly hydrated. In these solids, the lanthanide cation is surrounded by nine water molecules in a tricapped trigonal prismatic structure (Broach et al. 1979, Mackay et al. 1977) (fig. 1a). The anions are located in the mirror planes and linked by hydrogen bonding to the waters of the hydrated metal cation. In the bromate compound, the tricapped trigonal prism (TCTP) has a regular array of symmetry D_{3h}. The more bulky ethylsulfate anion reduces the symmetry to C_{3h} as a consequence of the rotation of the three axial oxygen atoms along the three-fold axis (Albertson and Elding 1977).

For a series of compounds in which the anion is constant but the cations are varied across the lanthanide family, variations in the hydration structures can be attributed to the differences in the cationic radii. For the three types of compounds described in the previous paragraph, the Ln–O prismatic bond distance is shorter than that of the Ln–O equatorial bond (Albertson and Elding 1977). For the same anion, the contraction in the Ln–O prismatic bond distance across the lanthanide series is twice as large as that of the equatorial Ln–O bond distance. The cell dimension was found to vary linearly with the ionic radii r of the various lanthanides; the plots of c versus r have slopes of 1.00 and of 0.63, respectively, for the bromate and ethylsulfate compounds compared to an expected slope value of 3.0. Albertson and Elding (1977) attributed the disagreement between the expected and the observed slopes to the rigidity of other parts of the structure in the c direction as well as to difference in the hydrogen bonding in the bromate and ethylsulfate systems. In a plot of the cell dimension a as a function

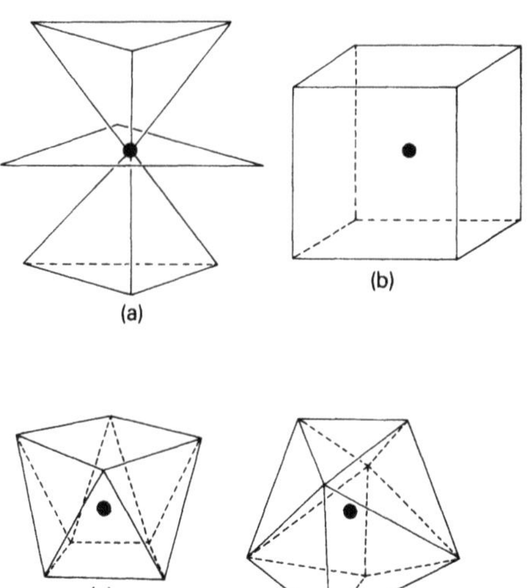

Fig. 1. Coordination polyhedra: (a) tricapped trigonal prism (CN = 9); (b) cube (CN = 8); (c) square antiprism (CN = 8); (d) dodecahedron with two nonequivalent sites (CN = 8).

of r, two lines were obtained which intersected at Tb^{3+}. The isostructural series of lanthanide oxydiacetates, $Na_3[Ln(C_4H_4O_5)_3]\cdot 2NaClO_4\cdot 6H_2O$ showed these same trends (Albertson 1970).

Albertson and Elding (1977) estimated a value of 1.1 Å as the best fit radius of the cavity for a hard-sphere model of the TCTP structure with equal prismatic and equatorial Ln–O distances. In this model, lanthanides of smaller radii than Gd^{3+} ($r = 1.107$ Å) for coordination number CN = 9 (Shannon 1976) are expected to destabilize the structure. These calculations were based on the assumption that the van der Waals radius of oxygen is 1.50 Å.

Structures of the hydrates $LnCl_3\cdot 7H_2O$ (Ln = La, Pr) (Habenschuss and Spedding 1978, 1979a) and $LnCl_3\cdot 6H_2O$ (Ln = Nd, Sm, Gd, Lu) (Kepert et al. 1983, Marezio et al. 1961) provide interesting comparisons. In the heptahydrates, the metal ions complete a nine coordination (ennea) sphere by forming dimers of the type $[(H_2O)_7Ln\text{-}\mu Cl, \mu' Cl\text{-}Ln(H_2O)_7]$. The hexahydrate structures have two of the chlorine atoms in the primary coordination sphere for a net coordination number of eight. The structure has antiprismatic geometry (fig. 1c) with the chlorine atoms *cis* to each other in one of the square faces. Hydrates of coordination number six were reported also for the lanthanide perchlorates (Glaser and Johansson 1981), for which the structures are alternating arrays of hydrated octahedral lanthanide ions, $Ln(H_2O)_6{}^{3+}$ and perchlorate anions along the four-fold symmetry axes.

Recently, an interesting series of mixed crown ether hydrate complexes of the lanthanides were isolated and identified (Rogers 1987a, b, Rogers and Kurihara 1986, 1987a). In many systems where there is a structural limitation due either to cavity size or to steric interferences, the crown ether forms hydrogen bonds with the water of the hydrated metal ion. This usually results in a coordination number of eight. In these structures, the geometry of the hydration about the cation is affected by the structure of the crown ether so as to optimise hydrogen bonding. Thus, for crown ethers with smaller cavities (e.g. 15-crown-5 or 15-C-5), the water molecules are distributed in a distorted dodecahedron geometry (fig. 1d), whereas for those with larger cavities (e.g. 18-C-6), a bicapped trigonal prismatic structure is favored with distortions towards dodecahedral symmetry (Rogers 1987b). For complexes in which the lanthanide ion is within the cavity of the crown ether, a coordination number of nine is found. Examples are $[Tb(H_2O)_5(12\text{-}C\text{-}4)]Cl_3\cdot 2H_2O$ (Rogers 1987a), $[Sm(H_2O)_4(15\text{-}C\text{-}5)]\text{-}(ClO_4)_3\cdot 15\text{-}C\text{-}5\cdot H_2O$ (Lee et al. 1983) and $[LnCl(H_2O)_2(18\text{-}C\text{-}6)]Cl_2\cdot 2H_2O$ (Ln = Sm, Gd, Tb) (Rogers and Kurihara 1987b, Rogers et al. 1987). These retain the distorted TCTP geometry except for the 12-C-4 complex, which has a capped square antiprismatic structure due to the ligand stereochemistry. It appears from these structures as well as from others (Guerriero et al. 1987) that the prismatic water molecules are more likely to be retained in the substitution reactions than are those which occupy capping positions.

Hydrated lanthanide sulfates have been studied for structure and for thermal and kinetic stability (Niinisto and Leskela 1987). Except for lanthanum and cerium, the sulfates tend to crystallize as octahydrates with all water molecules coordinated to the central ion (Dereigne et al. 1972). For Ce^{3+} and La^{3+}, the most stable sulfates are the nine coordination hydrates. An X-ray diffraction study of $Ce_2(SO_4)_3\cdot 9H_2O$

(Dereigne and Pannetier 1968) indicated the presence of two crystallographically independent cerium ions; one is 9-coordinated by six water molecules and three oxygen atoms of the sulfate anions in a trigonal prismatic structure, while the other is 12-coordinated by six water molecules and bidentate binding to the oxygen atoms of three sulfate anions. The remaining three water molecules are hydrogen bonded to the sulfate anions. Stable pentahydrate and tetrahydrates were also isolated for lighter lanthanides (La, Ce and Nd).

Studies on these hydrates by differential thermal analysis (DTA), thermogravimetric analysis (TGA) and differential scanning calorimetry (DSC) showed that the initial decomposition involved elimination of all water molecules via a multi-step process with the formation of the dihydrate and monohydrate intermediates. Further heating led to formation of the stable oxysulfates and, subsequently to the sesquioxides (Wendlandt 1958, Wendlandt and George 1961, Niinisto et al. 1982):

$$Ln_2(SO_4)_3{\cdot}8H_2O(s) \longrightarrow Ln_2(SO_4)_3{\cdot}nH_2O(s) + (8\text{-}n)H_2O(g),$$

$$Ln_2(SO_4)_3{\cdot}nH_2O(s) \longrightarrow Ln_2(SO_4)_3(s) + nH_2O(g),$$

$$Ln_2(SO_4)_3(s) \longrightarrow Ln_2O_2SO_4(s) + 2SO_3(g),$$

$$Ln_2O_2SO_4(s) \longrightarrow Ln_2O_3(s) + SO_3(g).$$

Luminescence measurements of the decomposition products of $Eu_2(SO_4)_3{\cdot}8H_2O$ (Brittain 1983) supported these conclusions. Figure 2 shows the spectra obtained for the $^5D_0 \rightarrow {}^7F_0$ and $^5D_0 \rightarrow {}^7F_1$ transition regions for europium sulfate before (a) and after (b) drying at 110°C. For the original compound the low intensity of the bands for the $^5D_0 \rightarrow {}^7F_0$ transition was attributed to the high symmetry of the structure. The intensification of these bands upon heating to 110°C is consistent with a phase transformation to a species of lower symmetry even though there was no dehydration. Within the temperature range of 175 to 350°C, complete dehydration takes place and the doublet assigned to the $^5D_0 \rightarrow {}^7F_0$ (580 nm) transition collapses into a broad, asymmetric band indicating the presence of a number of species with different amounts of hydration (Brittain 1983).

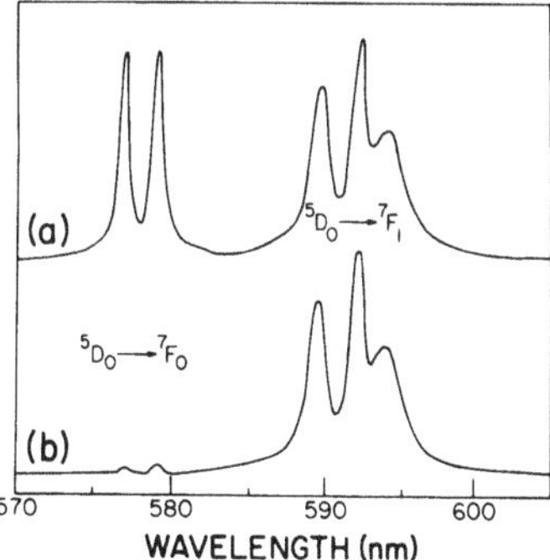

Fig. 2. Luminescence $^5D_0 \rightarrow {}^7F_0$ and $^5D_0 \rightarrow {}^7F_1$ transitions obtained for $Eu_2(SO_4)_3 \cdot 8H_2O$ at (a) 25 °C and (b) 110 °C.

The rates of the isothermal dehydration of these salts were studied at 130, 150 and 170°C using TGA (Saito 1988). Linear plots of $[1 - (1 - \alpha)^{1/3}]$ versus time (α is the degree of reaction) reflect a surface-controlled dehydration mechanism. The activation energies calculated from these data, which agreed with those reported earlier (Nathans and Wendlandt 1962), are inversely proportional to the cationic radius r_c. Such a relation is expected for the ionic interactions dominant in lanthanide complexes. The entropy of activation ΔS^* for these solid phase reactions varies from $-226\ \mathrm{J\,K^{-1}\,mol^{-1}}$ for La to $-123\ \mathrm{J\,K^{-1}\,mol^{-1}}$ for Tm. The variation in entropy is related to changes in the degree of rotational freedom of water molecules in the activated states as a result of lattice expansion. The linearity of the plot of ΔE^* versus ΔS^* was interpreted as evidence for a common dehydration mechanism independent of the particular lanthanide cation present.

Thermal studies of other lanthanide salts have been reviewed by Niinisto and Leskela (1987). For most salts the total dehydration occurs in a single-step reaction, although for a few systems (e.g. the selenate and iodate hydrates) dehydration occurs via a multi-step process.

2.2. *Hydrated lanthanides in solution*

The hydration of lanthanide ions and its effect on the physical and chemical properties of these ions in aqueous solutions have been discussed (usually, in disagreement) for almost three decades. Spedding's laboratory pioneered a systematic investigation of various properties of lanthanide salts in aqueous solution. Among the properties studied were apparent molal volumes (Spedding et al. 1966b, 1975b), relative viscosities (Spedding and Pikal 1966, Spedding et al. 1975a), apparent molal heat capacities (Spedding and Jones 1966), heats of dilution (Spedding et al. 1966a), and electrical conductances (Spedding et al. 1974a). Plots of these properties as a function of the lanthanide ionic radius showed a general pattern of an early minimum and a mid-series maximum (an "S-shape") proceeding from La through Lu (fig. 3). This shape was interpreted as reflecting a change in the number of waters in the primary coordination sphere of these ions. La^{3+} to Nd^{3+} were assigned to one isostructural series and Tb^{3+} to Lu^{3+} to another. The hydrated ions in the middle of the series (Nd^{3+}–Tb^{3+}) were assumed either to have structures that represented some transition between those of the two major groups or to exist in an equilibrium mixture between the other two structures. Thermodynamic properties of the chloride and perchlorate systems were consistent with a model in which the cations retain their hydration number up to the point of solubility saturation. However, with increasing concentration of lanthanide salt, the S-shaped pattern becomes less prominent for the chloride than for the perchlorate data (Spedding and Pikal 1966, Spedding et al. 1974a, b). Although the data did not allow an unambiguous assignment of the hydration numbers, the authors proposed that the difference in hydration number between light and heavy lanthanide ions seems not to exceed unity and suggested that the probable species is for the light cations $Ln(H_2O)_9^{3+}$ and for the heavy ones $Ln(H_2O)_8^{3+}$. The apparent molal volume data for both dilute and moderately concentrated solutions revealed an overall increase in the cation hydration as the

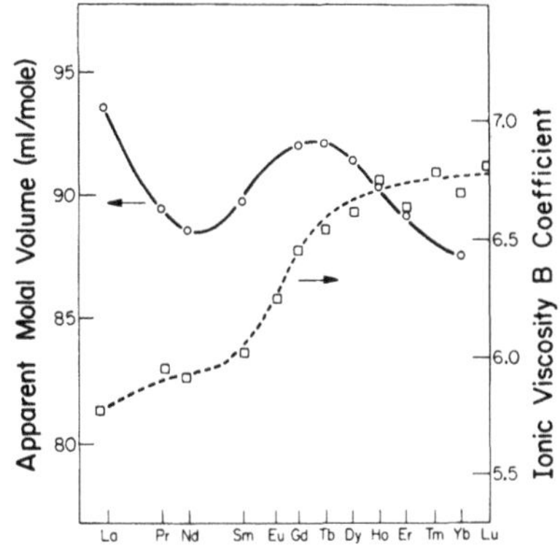

Fig. 3. Variation of the apparent molal volumes and the ionic viscosity coefficient B across the lanthanide series.

charge density Z/r_c increased across the series. This would seem to be explained by the model in which the inner-sphere hydration number is constant for the lighter, La^{3+}–Nd^{3+}, as well as the heavier, Tb^{3+}–Lu^{3+}, ions; being smaller for the latter. A steadily increasing second (outer)-sphere hydration occurs through the series from La^{3+} to Lu^{3+}.

In the next sections, a variety of experimental data are reviewed to ascertain the validity of this model.

2.2.1. *Experimental studies*

2.2.1.1. *Diffraction methods.* Bombardment of aqueous solutions of electrolytes by neutrons or X-rays causes scattering which is characteristic of the microscopic structure of the system. X-rays are preferentially scattered by heavy atoms whereas neutrons are scattered best by the lightest atoms. Direct determination of the number and geometry of the water molecules in the primary hydration sphere of the lanthanides have been attempted by both techniques.

For a metal ion M surrounded by a number n of oxygen atoms, the variation of the number dn with radial distance dr is expressed by

$$\mathrm{d}n = 4\pi\rho_0 g(r) r^2\,\mathrm{d}r, \tag{1}$$

where $\rho_0 = N_0/V$ (N_0 is the number of water molecules contained in a sample of volume V) and $g(r)$ is the probability of finding a water molecule at a distance r from a metal ion placed at the origin. High fluxes of neutrons or X-rays are normally required to obtain sufficient scattering for analysis. The equations relating the intensity of the coherently scattered radiation $I(k)$ to the scattering vectors k lead to the general equation (Enderby et al. 1987, Nielson and Enderby 1979)

$$g(r) = 1 + (2\pi^2 sr)^{-1}\int [S(k) - 1]k\sin(kr)\,\mathrm{d}k, \tag{2}$$

where ρ is the total number density, $S(k)$ is the partial structure average, $k = (4\rho\sin\theta)/\lambda_0$ (λ_0 is the incident wavelength) and θ has a value equal to one half of the scattering angle. The values of $S(k)$ can be obtained from the relationships

$$F(k) = \sum_{M}\sum_{O} C_M C_O B_M B_O [S(k) - 1], \tag{3}$$

$$I(k) = N\left[\sum_{M} C_M B_M{}^2 + F(k)\right], \tag{4}$$

where C_M and C_O are the atomic fractions of metal and oxygen, respectively, N is the total number of nuclei, B_M and B_O are the mean coherent scattering lengths. The Fourier transform of $F(k)$ for neutron diffraction is given by

$$G(r) = (2\pi^2 s r)^{-1} \int F(k) k \sin(kr)\,\mathrm{d}k.$$

$G(r)$ is the total radial distribution function (RDF) and is the linear combination of all the radial distribution functions, i.e.

$$G(r) = \sum_{M}\sum_{O} C_M C_O B_M B_O [g(r) - 1]. \tag{5}$$

In the X-ray technique, the manner of normalization of the intensities can lead to significant differences in the RDFs. Habenschuss and Spedding (1979b, c, 1980) reported a detailed X-ray diffraction study of concentrated lanthanide chloride solutions (3.2 to 3.8 M) in which the data were analyzed with different constraints imposed on the resolution of the peaks in the RDF. Figure 4 shows the normalized RDFs $G(r)$ for 10 concentrated solutions. The three peaks at ca. 2.5, 3.0 and 5.0 Å correspond to Ln^{3+}–H_2O, H_2O–H_2O and Cl–H_2O interactions, respectively. Resolution of the Ln^{3+}–H_2O peak from the net RDF was based on the assumption that all the peaks had Gaussian distributions which appeared acceptable for the first peak but was more questionable for the other peaks (Habenschuss and Spedding 1979c). In the least-squares fit, the Cl–H_2O distance was kept the same for all lanthanides while

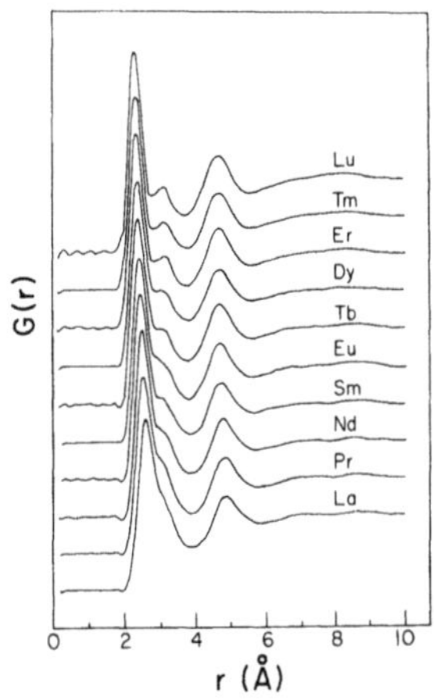

Fig. 4. Normalized radial distribution function $G(r)$ as a function of the distance r of $LnCl_3$ solutions.

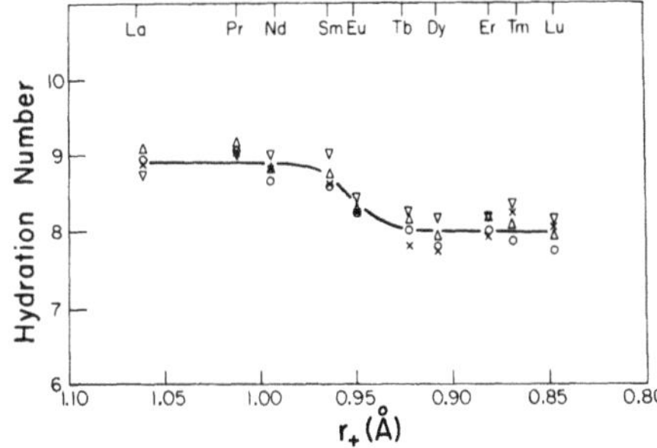

Fig. 5. Plot of the hydration number as a function of ionic radius of the lanthanide ions using different constraints: (i) △: $d_{H_2O-H_2O} = 4.5$ Å, $d_{Cl-H_2O} = 3.25$ Å, 6 for Cl–H_2O coordination; (ii) ○: 4.0 Å, 3.28 Å, 9; (iii) X: 4.0 Å, 3.23 Å, 7; (iv) ▽: no constraints.

the H_2O–H_2O distance was a floating variable since the latter is expected to reflect the variations in lanthanide ionic radii. Figure 5 shows the average Ln^{3+}–H_2O coordination calculated from the area of the first peaks in the RDFs analyzed with various constraints. The results indicate that the exact constraints imposed on the second and third peaks are not of critical importance provided the same constraints are used throughout the series of lanthanide cations and are within physically reasonable bounds (Habenschuss and Spedding 1979c). As expected, the position of the first peak (fig. 4) shifts in a regular fashion to lower values as the cation radius decreases (i.e. as Z of the Ln^{3+} increases).

In another study (Smith and Wertz 1975, 1977, Steele and Wertz 1976, 1977), a coordination number of eight was estimated for La, Nd and Gd ions in solutions of 1.7–2.6 M $LnCl_3$, taking into account the complexation of chloride ion at high concentrations. The structure of the aquo ions has a pseudo-cubic (D_{3h}) geometry. Johanssen and Wakita (1985) found a coordination number of eight as the best fit for the theoretical simulation of their X-ray data for aqueous perchlorate and selenate solutions. Corrections for inner-sphere complexation by selenate ion were included. Table 1 lists the estimates of the Ln–O distances and the respective hydration numbers obtained in these different studies using X-ray diffraction and rather concentrated solutions.

The ambiguity in the interpretation of a single diffraction pattern is avoided in neutron diffraction studies by measuring the diffraction patterns from solutions that are identical in all respects except for the isotopic nature of the metal ion. The algebraic difference between any two sets of measurements provides the lanthanide–water distance correlation (Narten and Hahn 1982). Measurements on solutions of $NdCl_3$ in D_2O (Narten and Hahn 1982) were consistent with a hydration number greater than eight for the light elements. The RDF has twin peaks assigned to an $\bar{r}_{Nd-O}$ of 2.48 Å [compared to 2.51 Å from Habenschuss and Spedding (1979c)] and r_{Nd-D} of 3.13 Å. The absence of broadening or of splitting in the two peaks indicated that the nearest-neighbor water molecules have the same average position and orientation around the metal ion. Such a symmetric distribution of the deuterium atoms suggest a restricted motion of the latter about the oxygen center. The analysis indicated that, on the average, Nd^{3+} ion had 8.5 water molecules in its primary hydration sphere. The

TABLE 1
Cation hydration as determined by diffraction methods.

Solute	Concentration (M)	Ln–O distance (Å)	Hydration number	Method	Ref.[a]
$LaCl_3$	1.74, 2.10, 2.67	2.48	8.0	X-ray	[1]
	3.81	2.58	9.1	X-ray	[2]
$LaBr_3$	2.66	2.48	8.0	X-ray	[3]
$La(ClO_4)_3$	4.62	2.57	8.0	X-ray	[4]
$La(SeO_4)_3$	0.72	2.56	8.0	X-ray	[4]
$PrCl_3$	3.80	2.54	9.2	X-ray	[2]
$NdCl_3$	1.73	2.41	8.0	X-ray	[5]
	3.37	2.51	8.9	X-ray	[2]
	2.85	2.48	8.5	Neutron diffraction	[6]
$SmCl_3$	3.23	2.47	8.8	X-ray	[7]
$Sm(ClO_4)_3$	3.66	2.46	8.0	X-ray	[4]
$EuCl_3$	3.23	2.45	8.3	X-ray	[7]
$GdCl_3$	2.66	2.37	8.0	X-ray	[8]
$TbCl_3$	3.49	2.41	8.2	X-ray	[9]
$Tb(ClO_4)_3$	1.40, 4.07	2.40	8.0	X-ray	[4]
$Tb(SeO_4)_3$	1.26	2.38	8.0	X-ray	[4]
$DyCl_3$	2.38	2.37	7.4	Neutron diffraction	[10]
	3.29	2.40	7.9	X-ray	[9]
$ErCl_3$	3.54	2.37	8.2	X-ray	[9]
$Er(ClO_4)_3$	4.59	2.36	8.0	X-ray	[4]
$Er(SeO_4)_3$	0.81, 1.05	2.34	8.0	X-ray	[4]
$TmCl_3$	3.63	2.36	8.1	X-ray	[9]
$LuCl_3$	3.61	2.34	8.0	X-ray	[9]
$Y(ClO_4)_3$	1.39, 4.48	2.37	8.0	X-ray	[4]
$Y(SeO_4)_3$	0.80, 0.95	2.33	8.0	X-ray	[4]

[a] References: [1] Smith and Wertz (1975); [2] Habenschuss and Spedding (1979c); [3] Smith and Wertz (1977); [4] Johanssen and Wakita (1985); [5] Steele and Wertz (1977); [6] Narten and Hahn (1982); [7] Habenschuss and Spedding (1980); [8] Steele and Wertz (1976); [9] Habenschuss and Spedding (1979b); [10] Annis et al. (1985).

tilt angle* between the Nd–O axis and the plane containing the water molecule is estimated to be 24°.

A similar study of $DyCl_3$ solutions gave a tilt angle of 17° and a primary hydration number of 7.4 (Annis et al. 1985). This difference between the Nd^{3+} and Dy^{3+} hydration numbers supports the conclusions of Spedding's studies that a difference of one water molecule exists between the hydration sphere of the heavy cations and that of the light ones (Habenschuss and Spedding 1980). The sharpness of the scattering

* A tilt angle of 0° corresponds to the dipole moment of the water molecule pointing away from the ionic core, while an angle of 55° implies that one of the "lone pair" orbitals points towards the core (Annis et al. 1985).

peaks for both cation systems indicated well-defined hydration boundaries around the metal ions.

In fig. 6 a clear discontinuity is seen in Ln^{3+}–OH_2 bond distances of curve A between Nd and Tb with an offset value of 0.045 Å. This value agrees with the analysis of Sinha (1976) who calculated a 0.047 Å decrease in the effective La^{3+} radius per unit decrease in the coordination number. In the solid heptahydrate (La–Pr) and hexahydrates (Nd–Lu), the offset value in the Ln^{3+}–OH_2 distances between the heavy and the light lanthanides is only 0.039 Å. The smaller value was attributed to the differences in the structures of the heptahydrate and hexahydrates (the former form dimers with chloride bridging).

The peak in the X-ray RDF at 5 Å (fig. 4) was assigned to Ln–O distances for water molecules in the second (outer) hydration sphere. Curve B in fig. 6 shows the dependency of the position of this peak on the lanthanide radius. Both ion pair interactions $[Ln(H_2O)_x]^{3+}$–Cl^- and secondary solvation $[Ln(H_2O)_x]^{3+}$–H_2O are expected to be responsive to differences in ionic radii of the lanthanide ions as well as to changes in the inner-sphere hydration. The decrease in the Ln^{3+}–H_2O distance between La^{3+} and Lu^{3+} (including the hydration change offset) is 0.24 Å (fig. 6, curve A), in good agreement with the difference of 0.22 Å for the peak at ca. 5 Å.

In summary, although the different X-ray studies agreed on the assignment of a hydration number of eight for heavy lanthanides, values of both eight and nine were found in the analysis of the lighter cations. This disagreement was considered by Cossy et al. (1988) to be due to the lack of consideration of inner-sphere complexation by the anion in some cases. The results of the neutron diffraction data agreed within 0.5 water molecules with the hydration number of Nd^{3+} and Dy^{3+} from X-ray diffraction data and also support a difference of one water molecule in the primary hydration spheres of the light and the heavy lanthanides.

2.2.1.2. *Fluorescence studies.* The fluorescence yield and the fluorescent lifetimes of lanthanide cations in solution are functions of the nature and arrangement of the species associated with the metal in its primary coordination sphere. Based on their fluorescent properties, these elements have been classified into three categories (Crosby 1966, Whan and Crosby 1962, Sioni and Lovgren 1987)

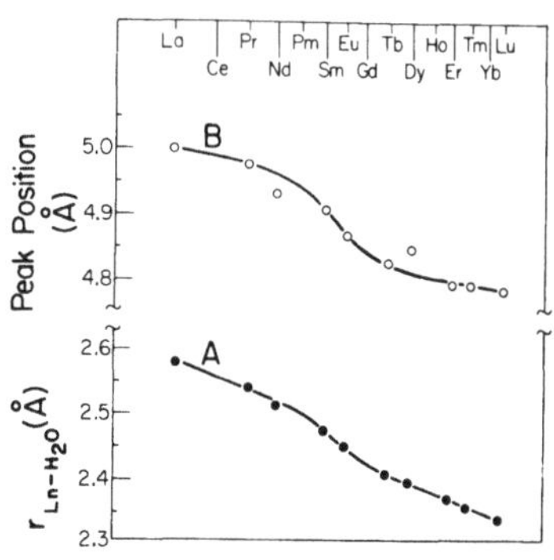

Fig. 6. Plots of the Ln^{3+}–H_2O bond length and the position of the peak at ca. 5 Å in the x-ray RDF curves of fig. 4.

(1) Those exhibiting strong fluorescence as aquated cations but with weak molecular fluorescence and phosphorescence when complexed; e.g. Eu^{3+} and Tb^{3+} and, to a much lesser extent, Dy^{3+} and Sm^{3+}. The strong fluorescence of the cations is the result of efficient intramolecular energy transfer and relatively weak decay.

(2) $La^{3+}(4f^0)$, $Gd^{3+}(4f^7)$ and $Lu^{3+}(4f^{14})$ which show no fluorescence as cations but whose complexes can exhibit strong molecular fluorescence and phosphorescence. The absence of fluorescence in the cations is due to the absence of suitable energy levels in the cations which can accept energy from the triplet state of the phosphor ligand.

(3) The remaining lanthanides for which both the fluorescence of the cations and the fluorescence and phosphorescence of the complexes are weak. This behavior is the result of efficient singlet-to-triplet energy transfer and closely spaced energy levels in the cations which enhance nonradiative transitions.

The first group has received particular attention as probes for hydration studies. Following excitation, these ions relax to sublevels of the ground electronic state; fig. 7 shows the most favorable transitions for Eu^{3+} and Tb^{3+} ions. In aqueous solutions, the dominant mode of quenching of these excited ions occurs via coupling of the electronic excitations to O–H vibrations of the water molecules in the solvation shell, although de-excitation by coupling with some ligands can also occur. If the latter is negligible, the rate of quenching is proportional to the number of O–H bonds present in the primary coordination sphere of the cation (Kropp and Windsor 1965, 1967). The reciprocal of the excited-state lifetime (the exponential decay constant k_{obs}) is the sum of the individual rate constants for all the de-excitation processes. In aqueous solution, this can be expressed as

$$k_{obs} = k_F + k_x + k_{H_2O}, \tag{6}$$

where k_F is the rate constant for the radiative transition ${}^5D_0 \rightarrow {}^7F_i$ (rate of photon emission); k_{H_2O} is the rate constant for the nonradiative energy transfer to the O–H oscillators in the first coordination sphere and k_x represents the rate constants for energy loss (nonradiative) by all other pathways. The value of k_{H_2O} generally domi-

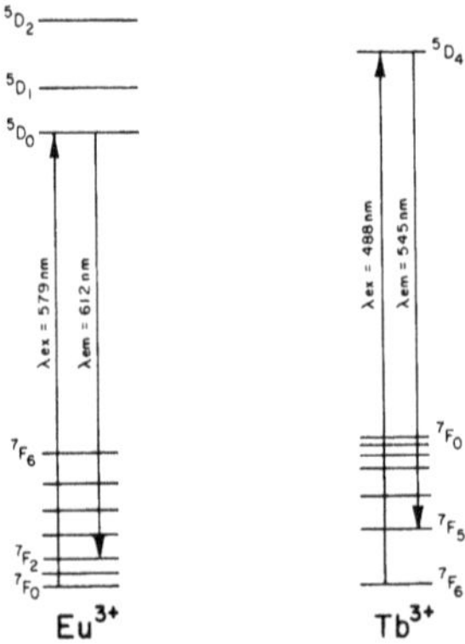

Fig. 7. Partial energy diagrams of Eu^{3+} and Tb^{3+} ions. The most probable excitation and decay path (luminescent emission) as well as the wavelengths for excitation and emission are indicated.

nates the other terms; e.g. the values of k_F, k_x and k_{H_2O} for Tb^{3+} ion in aqueous solution are 0.11, 0.19 and 2.15 ms^{-1}, while for Eu^{3+} ion, the corresponding values are 0.19, 0.25 and 9.5 ms^{-1}, respectively (Kropp and Windsor 1965, Dawson et al. 1966, Stein and Wurzberg 1975). While k_F and k_x for both ions are comparable, the greater value of k_{H_2O} for Eu^{3+} ion results from the small energy gap between the emitting excited state and the highest level of the ground manifold (Horrocks and Sudnick 1981). In deuterated media this energy difference is greater, so the vibronic coupling with the O–D oscillator is much less efficient. For D_2O solutions, eq. (6) reduces to

$$k_{obs} \approx k_F + k_x. \tag{7}$$

This difference for the cations in H_2O and in D_2O allows determination of the contribution to the rate of ion fluorescence of the coordinated water. Combination of eqs. (6) and (7) leads to

$$k_{obs}(H_2O) - k_{obs}(D_2O) = k_{H_2O}. \tag{8}$$

The value of $k_{obs}(D_2O)$ is generally obtained by extrapolating the values of k_{obs} for solutions with different mole fractions of H_2O in H_2O–D_2O mixtures to pure D_2O.

In hydration studies, the relation of k_{H_2O} to the number of water molecules in the primary coordination sphere of the lanthanide is usually calibrated with a series of solid compounds of known hydrate structure, e.g. crystalline Eu^{3+} and Tb^{3+} compounds in which n varies from zero to nine isolated from H_2O and from D_2O solutions. The results of lifetime studies for crystalline hydrates reported by Horrocks and Sudnick (1979) are listed in table 2*. For crystals with $n = 0$, all the measured k values were identical whether prepared from H_2O or from D_2O solution. This confirms that only O–H oscillators directly bound to the metal ion can significantly affect the luminescence decay constants.

The decay rates for Eu^{3+} and Tb^{3+} ions in pure crystals and host crystals doped with variable contents of Eu^{3+} or Tb^{3+} ions were the same; this shows that the rate constants are independent of the metal concentration. Figure 8 presents the correlation of Δk_{obs} $(= k_{(H_2O)} - k_{(D_2O)})$ with n, the number of water molecules in the first hydration sphere. The best fit for the linear relation between n and k_{H_2O} from eq. (8) is

$$n = A_{Ln} k_{H_2O}, \tag{9}$$

with A_{Ln} equal to 1.05 and 4.65 for Eu^{3+} and Tb^{3+} ions, respectively. The uncertainty in the hydration numbers estimated by this method is ± 0.5. The use of eq. (9) for solution data assumes that the solid and solution hydrates show the same spectral behavior. It has been shown that the value of n can be calculated satisfactorily from $k_{obs(H_2O)}$ by the relationship (Barthélémy and Choppin 1989a) $n = 1.05 k_{obs(H_2O)} - 0.70$.

Luminescence lifetimes for aqueous EuX_3 [$X = Cl^-$ (Kropp and Windsor 1965, Horrocks et al. 1977, Tanaka and Yamashita 1984, Breen and Horrocks 1983); NO_3^-

* The abbreviations used for various ligands are explained in the list of abbreviations at the beginning of this chapter.

TABLE 2
Luminescence decay constants for Eu^{3+} and Tb^{3+} crystalline hydrates and the number (n) of water molecules in the primary coordination sphere.

Solid[b]	n	Decay constant (ms^{-1})				Ref.[a]
		Eu^{3+}		Tb^{3+}		
		k_{H_2O}	k_{D_2O}	k_{H_2O}	k_{D_2O}	
$[Ln(H_2O)_9](ES)_3$	9	9.35	0.61	2.37	0.37	[1]
	9	9.01	1.33	2.40	0.66	[2]
$[Ln, Y(1:10)(H_2O)_9](ES)_3$	9	8.85		2.33		[2]
$[Ln, Y(1:30)(H_2O)_9](ES)_3$	9	8.62		2.42		[2]
$[Ln(H_2O)_9](BrO_3)_3$	9			2.38	0.40	[1]
	9	8.69		2.37		[3]
$[Ln(H_2O)_6Cl_2]Cl$	6	8.25	0.71	2.08	0.53	[1]
	6	8.33		2.07		[2]
	6	7.69	0.61	2.06	0.42	[4]
	6	8.19		2.07		[3]
$[Ln, Y(1:10)(H_2O)_6Cl_2]Cl$	6	8.47		2.07		[2]
$[Ln, Y (1:30)(H_2O)_6Cl_2]Cl$	6	8.54	2.63	2.07	0.91	[2]
$[Ln(H_2O)_5Cl(TP)]Cl_2 \cdot 3H_2O$	5	4.99	0.63	3.38	2.10	[1]
$[Ln_2(H_2O)_8(SO_4)_3]$	4	4.79	0.58	1.39	0.40	[1]
	4	4.90		1.39		[2]
	4	5.26		1.38		[3]
$[Ln(H_2O)_4(TDA)]Cl$	4	5.35	1.30	1.84	0.77	[1]
$Na[Ln(H_2O)_3(EDTA)] \cdot 5H_2O$	3	2.90	0.56	1.15	0.44	[1]
$[Ln_2(H_2O)_6(Ox)_3] \cdot 4H_2O$	3	3.53	0.76	1.27	0.59	[1]
$[(Ln, Gd)_2(H_2O)_6(Ox)_3] \cdot 4H_2O$	3	3.59		1.22		[1]
$[Ln(H_2O)_2(NTA)] \cdot H_2O$	2	2.72	0.53	0.94	0.49	[1]
$[Ln, Gd(H_2O)_2(NTA)] \cdot H_2O$[c]	2	2.65		0.87		[1]
$[Ln(H_2O)_2(Acac)_3] \cdot H_2O$	2	3.10	1.00	1.33	0.90	[1]
$[Ln(H_2O)_2(Nic)_3]$	2	2.49	1.02	1.14	0.70	[1]
$[Ln(H_2O)_2(Iso\text{-}Nic)_3]$	2	2.80	0.97	1.18	0.75	[1]
$Na_3[Ln(Dipic)_3] \cdot 15H_2O$	0	0.77	0.72	0.83	0.82	[1]

[a] References: [1] Horrocks and Sudnick (1979); [2] Heber (1967); [3] Heber and Hellwege (1967); [4] Freeman and Crosby (1964).
[b] The abbreviations of the ligands are explained in the list of abbreviations at the beginning of this chapter.
[c] Ln:Gd = 1:100.

(Kropp and Windsor 1965, Bunzli and Yersin 1979, 1984, Tanaka and Yamashita 1984, Breen and Horrocks 1983); ClO_4^- (Bunzli and Yersin 1979, Tanaka and Yamashita 1984); SCN^- (Breen and Horrocks 1983); Ac^- (Kropp and Windsor 1967, Barthélémy and Choppin 1989a) and $Cl_nH_{3-n}CCOO^-$ (Barthélémy and Choppin 1989a)], $Eu_2(SO_4)_3$ (Tanaka and Yamashita 1984) and TbX_3 [X = Cl^- (Kropp and Windsor 1965, Horrocks et al. 1977); NO_3^- (Kropp and Windsor 1965, Bunzli and Vukovic 1983); ClO_4^- (Bunzli and Vukovic 1983) and Ac^- (Kropp and Windsor

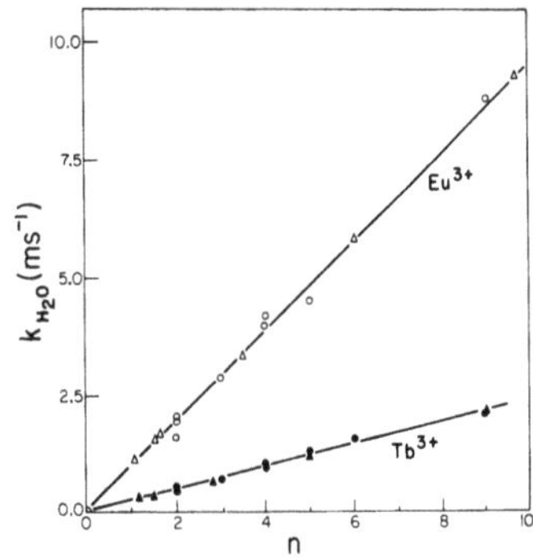

Fig. 8. Plot of Δk_{obs} versus n, the number of water molecules in the primary coordination sphere of the cation. (○, ●): data for crystalline solids; (△, ▲): data for aqueous solutions (Eu^{3+} and Tb^{3+}, respectively) (see Horrocks and Sudnick 1979, for the identity of the solids and solutions).

1967)] were studied as a function of the anion and lanthanide concentrations. For weakly complexing anions such as ClO_4^-, the decay lifetime is independent of both the salt and the total perchlorate concentrations. Measurements up to 6 M total perchlorate concentration were reported to show no change in the number of water molecules in the primary coordination sphere of the metal ion, indicating that any metal–perchlorate interaction must have an exclusively outer-sphere nature (Bunzli and Yersin 1979, Breen and Horrocks 1983). Tanaka and Yamashita (1984), however, observed a decrease in the luminescence lifetime of $Eu(ClO_4)_3$ with an increase in the perchlorate concentration, which suggests an increase in the hydration number of the Eu^{3+} ion. However, it would seem more likely that the decrease in lifetime is due to a perturbation of the hydrate equilibria of Eu^{3+} ion caused by outer-sphere complexation of the perchlorate anion (Breen and Horrocks 1983). The estimated hydration number for Eu^{3+} from the perchlorate studies varied from 9.6 ± 0.5 to 9.0 ± 0.5 (Horrocks and Sudnick 1979, Barthélémy and Choppin 1989a) which is somewhat larger than the estimated value of 8.3 from X-ray studies of 3.23 M chloride solution (Habenschuss and Spedding 1980). Similarly, for Tb^{3+} ion the hydration number 9.0 from luminescence data is greater than those determined by other methods. These results, in addition to the fact that the lanthanide bromate and ethylsulfate salts crystallize as the nonahydrates, led Horrocks and Sudnick (1979) to propose a coordination number of 10 for the cations La–Nd and 9 for Tb–Lu with a decrease for Sm to Gd from 10 to 9.

The interpretation of the fluorescence data from some solution systems have not always agreed with that from other techniques. Thermodynamic, X-ray and neutron diffraction, and absorption spectroscopic data support a dominant outer-sphere complexation by the Cl^-, NO_3^- and SCN^- ions at concentrations below 1 M (Choppin 1971, Habenschuss and Spedding 1980, Narten and Hahn 1982, Choppin and Bertha 1973, Bukietynska and Choppin 1970, Choppin and Ketels 1965). In the luminescence studies, the increase in decay lifetimes, the appearance of a new band at ca. 580 nm associated with $^5D_0 \rightarrow {}^7F_0$ transition and the band shape modification with increasing concentration of the anion have been interpreted as reflecting an increased dehydration, presumably due to inner-sphere complexation (Bunzli and Yersin 1979, Tanaka and Yamashita 1984, Breen and Horrocks 1983). These conclusions were based on the assumption that complexation in the inner coordination sphere is the same for Eu^{3+} in the excited $^5D_0$ state and in the ground $^7F_0$ state.

A spectroscopic study of Eu^{3+}–Br^- showed that the increase in the Br^- concentration causes a larger enhancement in the intensity and band area of the $^5D_1 \rightarrow {}^7F_J$ transitions than for the $^7F_J \rightarrow {}^5D_1$ transitions. These modifications in the spectra were attributed to changes in the structure and nature of the inner solvation sphere of Eu^{3+} in the excited state as compared to that of the ground state (Marcantonatos et al. 1984). The differences in intensity between absorption and emission bands would, therefore, reflect formation of inner-sphere complexes by Br^- in the excited state while outer-sphere complexation would dominate the ground state. It was proposed (Marcantonatos et al. 1981, 1982) that excitation of Eu^{3+} ion to the $^5D_1$ state would result in an expansion of the 4f and a shrinkage of the 5p orbitals with an overall decrease in the metal ion radius. The consequent contraction of the inner shell would be expected to produce more compact and less easily disrupted outer hydration spheres for both $(^5D_1)\,Eu(H_2O)_8{}^{3+}$ and $(^5D_1)Eu(H_2O)_9{}^{3+}$ with a possible increase in k_{obs}.

The differences in behavior of the Eu^{3+} ion in the ground and excited states may also account for the disagreement in the extent of outer-sphere complexation by chloroacetates as obtained from ^{139}La-NMR data (Choppin 1980), solvent extraction (Choppin 1980) and luminescence measurements (Barthélémy and Choppin 1989a). Although NMR shift data and solvent extraction studies gave estimates of 60% and 25% inner-sphere character in the $Eu(ClAc)^{2+}$ and $Eu(Cl_2Ac)^{2+}$ complexes, respectively, luminescence lifetimes data provided estimates of 100% and 50% inner-sphere character for the two complexes.

Marcantonatos et al. (1984) suggested that the water substitution reaction proceeds via an I_d mechanism, based on a comparison of the activation parameters for the excited state reaction

$$Eu(H_2O)_8^{3+} + Br^- \leftrightarrow EuBr(H_2O)_7^{2+} + H_2O$$

($\Delta H^* = 23\ kJ\,mol^{-1}$ and $\Delta S^* = 69\,J\,K^{-1}\,mol^{-1}$) with ground state reaction

$$M(NH_3)_5H_2O^{3+} + Cl^- \leftrightarrow M(NH_3)_5Cl^{2+} + H_2O$$

($\Delta H^* = 25$ and $14\ kJ\,mol^{-1}$ and $\Delta S^* = 67$ and $54\ J\,K^{-1}\,mol^{-1}$ for $M = Cr^{3+}$ and Co^{3+}, respectively) in which water substitution was shown to follow the I_d mechanism (Duffy and Early 1967, Taube 1960, Langford and Gray 1965).

2.2.1.3. *Nuclear magnetic resonance studies.* Measurements of cation hydration numbers by ^1H- and ^{17}O-NMR using peak areas, paramagnetic enhancement of relaxation rates induced in water molecules and contact chemical shifts have been applied to a variety of systems (Fratiello 1970). The peak area method is generally used when the distinction between coordinated and free water molecules is feasible due to slow exchange. In this situation, the hydration number n is determined from the integrated areas of the respective peaks and the knowledge of the total water content and the metal concentration. Even for slow exchange, this technique loses sensitivity in the presence of paramagnetic ions as a result of peak broadening. For the lanthanide cations, the exchange between hydrate and solvent water is too fast to allow the use of this technique. The exchange rates can be decreased by cooling the samples to temperatures between −50 to −100°C if the freezing point of the solvent is lowered, e.g. by using a mixture of water–acetone and Freon-12 (CCl_2F_2). In this solvent mixture, the participation of acetone in the primary solvation shell is negligible as long as the mole fraction of water exceeds 0.1 (Tanaka et al. 1988). However, little is known about how such a change in the solvent may affect other properties such as inner-sphere complexation by the anion.

Table 3 summarizes the results at low temperatures in this solvent mixture. These data show the hydration numbers *n* to be markedly sensitive to the type and the concentration of the anion. Low estimates of *n* were attributed to either replacement of water in the primary coordination sphere by the ClO_4^- or NO_3^- ions or to changes in water activity in solution as a result of hydrogen bonding between water and acetone (Brucher et al. 1985). Such hydrogen bonding has been detected by ^{1}H-NMR spectroscopy in water–acetone mixtures at very low water content (Takahashi and Li 1966) and could involve either A . . . HOH or A . . . HOH . . . A species, depending on the relative concentrations of water and acetone.

TABLE 3
Data for Ln^{3+} hydration in water–acetone–freon-12 solutions from ^{1}H-NMR measurements.

Salt	Hydration number	Ref.[a]
$Y(NO_3)_3$	2.4	[1]
$La(ClO_4)_3$	5.5–6.4	[2, 3]
$La(NO_3)_3$	3.1–2.9	[2]
$Ce(ClO_4)_3$	4.5–5.5	[3]
$Pr(ClO_4)_3$	6	[4]
$Nd(ClO_4)_3$	6	[4]
$Tb(ClO_4)_3$	2	[4]
$Dy(ClO_4)_3$	3	[4]
$Ho(ClO_4)_3$	3	[4]
$Er(ClO_4)_3$	4.2–6.4	[4]
$Er(NO_3)_3$	1–2	[4]
$Tm(NO_3)_3$	2	[4]
$Yb(ClO_4)_3$	8	[5]
$Lu(ClO_4)_3$	4.5–6.1	[4]
	6.3–8.7	[6]
$Lu(NO_3)_3$	2.1–2.2	[4]
	6	[7]

[a] References: [1] Fratiello et al. (1970); [2] Fratiello et al. (1989); [3] Fratiello et al. (1971); [4] Fratiello et al. (1973); [5] Kovun and Chernyshov (1979); [6] Brucher et al. (1985); [7] Scherbakov and Golubovskaya (1976).

Nuclear magnetic relaxation rates have been used to investigate the coordination number. In an investigation of the line-width broadening of ^{139}La in various perchlorate solutions, Nakamura and Kawamura (1971) attributed the decreases in the values of $\Delta\nu/\eta_r$ ($\Delta\nu$ is the relaxation rate and η_r is the relative viscosity) to a possible equilibrium between the nonahydrates and octahydrates for lanthanum ion. This conclusion was disputed by Reuben (1975) who proposed that this apparent anomaly reflected an erroneous estimate of the corrections of the linewidths for peaks due to the effect of the finite modulation amplitude and/or of partial saturation. Measurement of the transverse relaxation rates by the pulse method gave results consistent with a constant hydration number for lanthanum ion (Reuben 1975).

A recent ^{17}O-NMR kinetic study of water exchange of Gd^{3+} and Tb^{3+} ions over the temperature range 258–370 K showed the absence of inflection points in plots of the chemical shift versus T^{-1} for the individual elements. This was taken to indicate a constant hydration number over the entire temperature range (Cossy et al. 1988). By contrast, an inflection point was definitely observed in solvation studies with DMF medium (Pisaniello and Merbach 1982, Pisaniello et al. 1982). Reilley et al. (1976) showed that the contact and dipolar contributions to ^1H and ^{17}O shifts calculated for the heavy (Gd^{3+}–Yb^{3+}) ions in aqueous media are different in both sign and magnitude than those for the light (Ce^{3+}–Eu^{3+}) ions. This difference was taken as an evidence for a difference in hydration geometry for the metal cations of the two groups. Linear relationships were shown to exist between n in Gd^{3+}–aminopolycarboxylate complexes in solution and $1/T_1$ (Chang et al. 1990) and $1/T_2$ (Brittain et al. 1991), where T_1 is the spin–lattice relaxation time and T_2, the spin–spin relaxation time of the protons of the hydrate waters.

2.2.1.4. *Spectroscopic methods.* Hydration numbers have been inferred from comparisons of the characteristics of the absorption and the Raman spectra of lanthanide ions in aqueous solution and in crystalline hydrates of known structure (Freed 1942, Krumholz 1958). For the region of $^4I_{9/2} \rightarrow {}^2P_{1/2}$ and $^4I_{9/2} \rightarrow {}^4G_{7/2}$ transitions in Nd^{3+}, the bands obtained for dilute (ca. 0.1 M) solutions of NdX_3 ($X = Cl^-$, Br^-, BrO_3^-, ClO_4^-) were compared with those obtained for the solids $Nd(BrO_3)_3 \cdot 9H_2O$ (CN = 9), $NdCl_3 \cdot 6H_2O$ (CN = 8) and $Nd_2(SO_4)_3 \cdot 8H_2O$ (CN = 8) (Krumholz 1958, Davidenko and Lugina 1968, Karraker 1968). The spectrum of the Nd^{3+} in solution was very similar to that of crystalline $Nd(BrO_3)_3 \cdot 9H_2O$, which was interpreted as indicating a hydration number of 9 for the aqueous ion with symmetry D_{3h} similar to that about the cation in the solid hydrate. The selection rules for D_{3h} symmetry predict a maximum of five possible bands for the $^4I_{9/2} \rightarrow {}^2P_{1/2}$ transition. The absence of extra bands in the spectrum of the aqueous solution led to assignment of D_{3h} symmetry to the aquo ions (Davidenko and Lugina 1968). Cossy et al. (1988) interpreted the presence of a single peak at 427.1 nm which was insensitive to variations in temperature and concentration as showing the presence of only the nonahydrate ion.

For solutions in which the concentrations of HCl, $HClO_4$ or LiCl were increased to near saturation, the band shapes changed (fig. 9) and the oscillator strength increased.

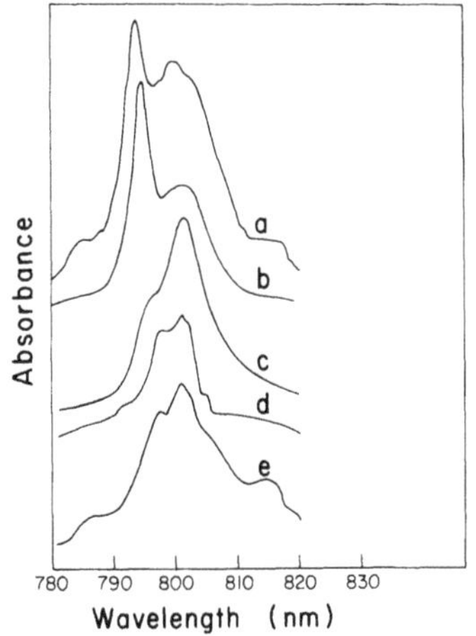

Fig. 9. Spectra of Nd^{3+} ${}^1I_{9/2} \rightarrow {}^2H_{9/2}$, $F_{5/2}$ transitions; (a) solid $Nd(BrO_3)_3 \cdot 9H_2O$; (b) 0.0535 *M* Nd^{3+} in water; (c) 0.0535 *M* Nd^{3+} in 11.4 *M* HCl; (d) solid $NdCl_3 \cdot 6H_2O$; (e) solid $Nd_2(SO_4)_3 \cdot 8H_2O$.

These effects were attributed to a decrease in CN from nine to eight as a result of a salting-out effect by the strongly hydrated H^+ and Li^+ cations.

The intensity changes in the hypersensitive bands of Nd^{3+}, Ho^{3+} and Er^{3+} associated with changes in solute concentration were explained as due to the action of an increased electromagnetic field in the medium surrounding the lanthanide (Karraker 1968, Jorgensen and Judd 1964). This is consistent with the theoretical model of hypersensitivity in which an enhancement in strength of the electric vector of the electromagnetic radiation across the ion increases the intensity of normally weak quadrupole transitions. The formation of inner-sphere complexes with the anions at higher concentrations could produce changes in symmetry and, thus in the spectrum. Polarization of the hydrate molecules by outer-sphere complexation and salting-out effects probably also play roles.

Unlike that of Nd^{3+}, the spectra of Er^{3+} and Ho^{3+} aquated ions did not resemble those of the corresponding $Ln(BrO_3)_3 \cdot 9H_2O$ crystals but showed marked similarity to the spectra of the $LnCl_3 \cdot 6H_2O$ solids. The latter crystals consist of the polyhedra $[LnCl_2(H_2O)_6]^+Cl^-$, so the agreement with the solution spectra suggests that the aquo ions may have a coordination number of eight for both Er^{3+} and Ho^{3+} ions. The similarity between the spectra of aqueous and solid states does not imply that in dilute solutions (0.10 *M*) the lanthanide ion exists as the dichlorohexaaquo ion, but rather that the point-charge potential of Cl^- ion has roughly the same effect as the dipolar potential of the water molecule, i.e. replacing Cl^- by H_2O in the coordination polyhedron is not accompanied by changes in the crystal field splittings or in "effective" symmetry (Svoronos et al. 1981). For Eu^{3+}, the solution spectra was significantly different from both the $EuCl_3 \cdot 6H_2O$ and the $Eu(BrO_3)_3 \cdot 9H_2O$ solids suggesting that the hydration number of Eu^{3+} is neither eight nor nine.

The Raman spectra of solid ethylsulfate salts, $[Ln(H_2O)_9](ES)_3$, showed a regular shift in ν(Ln–O) frequency with cationic radius (Yamauchi et al. 1988). Kanno and Hiraishi (1980, 1982, 1988) and Yamauchi et al. (1988) have reported Raman studies of lanthanide chlorides and nitrates at room and at liquid nitrogen (in the glassy state)

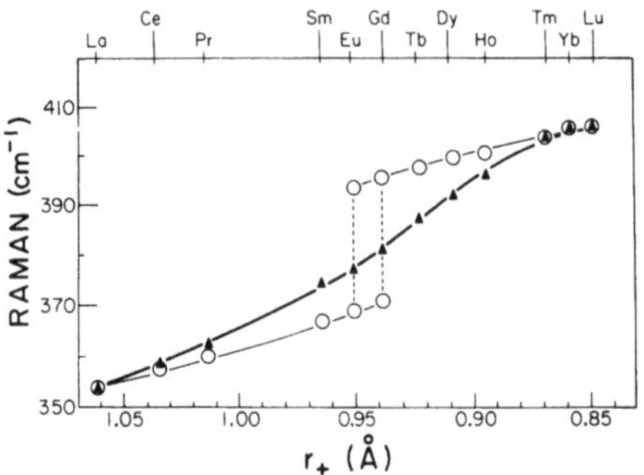

Fig. 10. Raman stretching frequencies for glassy-state solutions of lanthanide chlorides (○) and nitrates (△) in water.

temperatures and with variable ratios of water to salt. Plots of the frequency of the Raman stretching band for Ln^{3+}–OH_2 in glassy solutions of chloride and nitrate versus the ionic radius of the lanthanide ions showed a change in slope in the middle of the series (fig. 10). For glassy solutions of $EuCl_3$ and $GdCl_3$ the Raman band was split into a doublet, the peaks of which fitted the extrapolated Ln–O curves for the light and for the heavy lanthanides, respectively (fig. 10). This was interpreted as reflecting an equilibrium in $Eu^{3+}_{(aq)}$ and $Gd^{3+}_{(aq)}$ between two different inner-sphere hydration numbers. Increasing the Ln salt:water ratio for the Eu^{3+} and Gd^{3+} solutions was accompanied by changes in the relative peak intensity in accordance with a shift in equilibrium in favor of the formation of the higher hydrate. This apparently contradicting behavior was attributed in part to the lack of sufficient water molecules necessary to saturate and stabilize the second hydration sphere. As a result, the cation may compensate by having a higher hydration number. This could also explain the apparent discrepancy between the absorption spectroscopy and the luminescence lifetime for concentrated $Eu(ClO_4)_3$ solutions noted in the earlier section.

Since the shift in the Raman band arises from a change in the bond polarizability with variation in bond length, it was suggested that the Ln^{3+}–OH_2 interaction involves some degree of electron sharing. However, such sharing must be small as the band intensity is so low.

The evidence presented in this section from visible spectral data supports a hydration number of nine for the light lanthanides. Although there is evidence for a lower hydration number for the heavy lanthanides, alternate explanations for the observed differences in spectral characteristics between the lighter and the heavier members of the series cannot be ruled out of consideration.

2.2.1.5. *Compressibility method.* In a hard-sphere model, the solvated cation can be regarded as embedded in a spherical cavity within a continuous solvent (Padova 1964). The production by the high electrostatic field of the cation of dielectric saturation in the surrounding hydration sphere would result in an incompressible hydration shell (Passinsky 1938). This assumption leads to the equation

$$n = \frac{n_1}{n_2}\left(1 - \frac{V\beta}{n_1 V_1^0 \beta_0}\right), \tag{10}$$

where n is the hydration number, n_1 and n_2 are the number of solvent and solute molecules in volume V, V_1^0 is the molar volume of water, and β and β_0 are the compressibilities of the solution and of water, respectively. The adiabatic compressibility β is obtained from

$$\beta = (\rho U^2)^{-1}, \tag{11}$$

where U is the ultrasonic velocity and ρ is the density.

Table 4 lists the literature values of n obtained by this technique. Obviously, these hydration numbers are greater than the values determined by other techniques. This discrepancy was interpreted in terms of a continuum effect of the trivalent ion which results in a partial electrostriction of water molecules outside the first hydration sphere. The electric field strength E of an ion is expressed as (Conway 1981, Hahn 1988)

$$E = z_m/\varepsilon x^2, \tag{12}$$

where z_m is the ionic charge, ε is the dielectric constant and x is the distance from that ion. With this E value, the volume reduction per water molecule can be calculated

TABLE 4
Data on hydration of Ln^{3+} from adiabatic compressibility.

Cation	ClO_4^-		Cl^-		NO_3^-		Ref.*
	m	*n*	*m*	*n*	*m*	*n*	
Y^{3+}	–					15.6	[1]
La^{3+}	0.006–0.44	12.8 [2]				12.8	[1]
		11 [1]		0.37–0.57		10.4	[3]
Ce^{3+}						12.4	[1]
Pr^{3+}	0.006–0.40	13.1 [2]	0.37–0.70	15.7 [5]	0.63–0.96	10.1	[3, 5]
	0.55–0.75	12.0 [5]					
Nd^{3+}		12 [4]	0.45–0.72	17.1 [5]	0.40–0.65	10.9	[3, 5]
	0.006–0.39	12.5 [2]					
	0.36–0.61	12.8 [5]					
Sm^{3+}	0.006–0.42	11.1 [2]			0.29–0.38	12.0	[3]
Eu^{3+}	0.005–0.44	10.6 [2]	0.14–0.25	13–19 [5]	0.41–0.67	13–15	[5]
	0.27–0.41	13–17[5]					
Gd^{3+}	0.005–0.41	11.4 [2]			0.32–0.47	12.0	[3]
Tb^{3+}	0.006–0.39	11.7 [6]			0.28–0.42	13.0	[3]
Dy^{3+}	0.004–0.40	12.0 [6]			0.31–0.45	12.8	[3]
Ho^{3+}	0.005–0.43	12.7 [6]			0.31–0.56	12.8	[3]
Er^{3+}		12 [4]			0.44–0.60	12.2	[3]
	0.005–0.37	12.4 [6]					
Tm^{3+}	0.005–0.39	12.9 [6]					
Yb^{3+}		13 [4]					
	0.005–0.38	13.0 [6]					

* References: [1] Janenas (1978); [2] Jezowska-Trzebiatowska et al. (1978a); [3] Jezowska-Trzebiatowska et al. (1977); [4] Padova (1964); [5] Jezowska-Trzebiatowska et al. (1976); [6] Jezowska-Trzebiatowska et al. (1978b).

from the relationship

$$V = [1 - (KE^2 + 1)^{-G}]\, V^{f}_{(H_2O)}, \tag{13}$$

where $K = 1.1 \times 10^{-11}$, $G = 0.15$ and $V^{f}_{(H_2O)}$ is the molar volume of pure water (18.07 cm^3/mol). For lanthanide ions, the average volume reduction is 6.9 cm^3/mol in the primary hydration shell and 1.9 cm^3/mol in the secondary shell. These values are 38% and 10% of the molar volume of pure water and indicate that the electro-striction is much smaller in the secondary than in the primary shell. Apparently, electro-striction cannot account for the large hydration numbers obtained by compressibility measurements.

Compressibility data would reflect cumulative effects rather than the influence of cationic hydration alone. For example, comparison of the compressibility data for the chloride and perchlorate solutions shows significant differences which cannot be ascribed to differences in complexation alone and may also reflect anion hydration. For a particular anion, there appears to be a shallow minima in the middle region of the lanthanide series which has been interpreted in terms of a structural change in the first-hydration sphere (Jezowska-Trzebiatowska et al. 1976) associated with the formation of an iceberg (cluster) of highly structured water with enhanced hydrogen bonds.

2.2.1.6. *Other methods.* Glass formation of the aqueous solutions of lanthanide perchlorates, chlorides, nitrates and trifluoroacetates were studied by Kanno and Akama (1987). The transition temperature T_g decreased in the order $F_3Ac^- >$ $NO_3^- > Cl^- > ClO_4^-$. This was ascribed to changes in the solution viscosity and to steric factors associated with the configurational restrictions to anionic rotations which did not involve disruptions of the nearest-neighbor environment. Plots of T_g values versus radius of the lanthanide ions showed a discontinuity, with the transitional region depending on the type of anion in solution (fig. 11). This transition

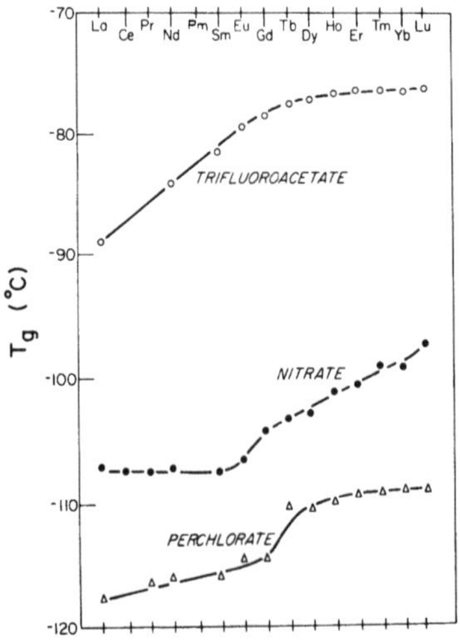

Fig. 11. Glass transition temperatures T_g across the lanthanide series in the presence of trifluoroacetate, nitrate and perchlorate anions.

extended over fewer lanthanides in the perchlorate system than in the chloride, nitrate or trifluoroacetate systems, presumably due to the differences in the complexing ability of the latter anions. Taking into account the results of other hydration studies such as Raman (Kanno and Hiraishi 1980, 1982) and transport and thermodynamic data (Spedding et al. 1974a, b, 1975a, Rard 1985, Choppin and Bertha 1973), the authors (Kanno and Akama 1987) assigned the extended S-shaped T_g behavior for the chloride and perchlorate solutions to a change in the hydration number (primary sphere) from nine to eight across the series. The rapid rise in T_g values in the middle of the series was taken to reflect an overall expansion in the secondary hydration sphere (Kanno and Akama 1987). Since glass transitions occur at a temperature for which there are relatively rapid translational motions of molecules or ions in a liquid, higher T_g's would be expected for a solution in which metal ions have a larger hydration sphere that dampens such motions.

The dependence of T_g on salt concentration was studied for $LaCl_3$, $EuCl_3$ and $LuCl_3$ solutions for molar ratios R of water:salt from 16 to 28. At the lower R values, $[T_g(Lu^{3+}) - T_g(Eu^{3+})]$ is greater than $[T_g(Eu^{3+}) - T_g(La^{3+})]$ whereas the reverse is true at the higher R values. This was explained as being due to a shift in the equilibrium between the nonaaquoeuropium(III) and octaaquoeuropium(III) ion in favor of the higher hydrate. This is consistent with the Raman spectra of the glassy solution which indicated an increase in higher hydration number for lower R values (Kanno and Hiraishi 1980, 1982).

A related study on the crystallization and desolvation energies of the perchlorate, chloride and nitrate salts was reported (Onstott 1985, Onstott et al. 1984). Data of solvent activity, heats of dilution and heats of solvation of crystalline hydrates of Spedding et al. (1975b, 1976, 1977a, b, c), Rard et al. (1977a, b), Rard and Miller (1979) and Rard and Spedding (1981) were used to predict the thermodynamic parameters for crystallization from the following relationships (Onstott et al. 1984)

$$\Delta G_c = -X^{-1} RT \ln A_e, \tag{14}$$

$$\Delta H_c = -\Delta H_{sat}, \tag{15}$$

$$\Delta S_c = \Delta H_c T^{-1} - \Delta G_c T^{-1}. \tag{16}$$

In these equations, ΔG_c is the free energy for formation of one mole of crystalline lanthanide hydrate from a saturated solution, X is the mole fraction of lanthanide precipitated for each mole of solvent water vaporized, A_e is the water activity of the electrolyte at saturation and ΔH_{sat} is the difference between the heat of solution of the crystal hydrate to form a solution of molality M and the heat of dilution of the saturated solution of molality M. Figure 12 is a plot of ΔS_c for the hexahydrates of $LnCl_3$ and octahydrates of $Ln(ClO_4)_3$. For $Dy(ClO_4)_3 \cdot 8H_2O$, the $T\Delta S_c$ is ≈ 0. This was interpreted as indicating a minimum in the geometric rearrangement needed for the phase transfer reaction which, presumably, means that the size of the Dy^{3+} cation is optimum for matching the geometry of the perchlorate anion and solvent water in crystallization in the saturated electrolyte solution. Based on this model, a negative entropy change in the perchlorate systems would reflect the effect of cation crowding from La to Dy while a positive entropy would be associated with the formation of a

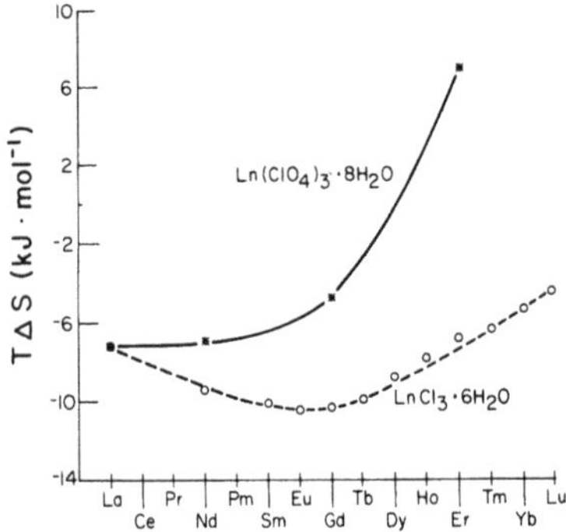

Fig. 12. Entropy change on crystallizing hydrates from saturated solutions of $Ln(ClO_4)_3{\cdot}8H_2O$ (∗) and $LnCl_3{\cdot}6H_2O$ (○) (Onstott et al. 1984).

more rigid anion structure in the electrolyte without compression of the inner-sphere water about the cations (Er–Lu).

2.2.1.7. *Preferential solvation in mixed solvents.* In a solvent S the solvation reaction can be written as

$$Ln^{3+}_{(g)} + nS_{(\ell)} \overset{K_s}{\longleftrightarrow} LnS^{3+}_{n(\ell)}, \tag{17}$$

where $Ln^{3+}_{(g)}$ is the free ion in vacuo, $LnS^{3+}_{n(\ell)}$ the solvated ion in the liquid solvent and K_s the solvation equilibrium constant. The free energy of transfer of the cation from solvent A to solvent B (ΔG_{tr}) is the difference in the standard free energy changes for the solvation equilibria in the two solvents. Thus, for

$$LnA^{3+}_n + mB \leftrightarrow LnB^{3+}_m + nA,$$

we have

$$\Delta G_{tr} = -RT\ln\beta, \tag{18}$$

$$\beta = K_B/K_A. \tag{19}$$

For mixed solvation

$$LnA^{3+}_n + iB \overset{K_i}{\longleftrightarrow} LnA_{n-i}B^{3+}_i + iA,$$

the free energy of transfer from A to any mixture of A and B is given by (Cox et al. 1974)

$$\Delta G_{tr} = -nRT\ln\varphi_A - RT\left\{1 + \sum_{i=1}^{n} K_i\left(\frac{\varphi_B}{\varphi_A}\right)^i\right\}, \tag{20}$$

where φ_A and φ_B are the volume fractions of A and B, respectively. In this treatment, the coordination number of the metal ion has been assumed to be constant, which is not necessarily correct.

The behavior of mixed solvents can be discussed in terms of two limiting cases in which one solvent acts as a better donor than the other. First, consider A to be the

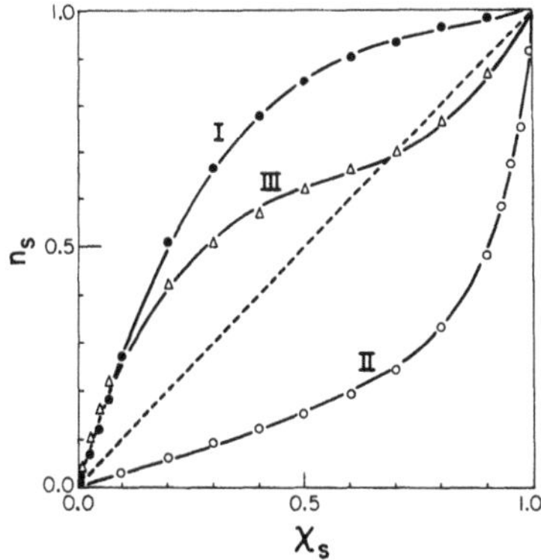

Fig. 13. Relationship between the mole fraction in the first hydration sphere (n_s) and the mole fraction in the bulk (x_s). (I) water + DMSO; (II) water + methanol; (III) water + DMF.

better donor; i.e. $\beta_i \ll 1$. In this case, addition of B to A results in the dilution of solvent A and the value of i/n increases more slowly than the increase in mole fraction of B in the solvent mixture. This is the case of Eu^{3+} solvation by mixtures of water + methanol, water + acetone, water + acetonitrile, and water + 1,4-dioxane (Tanaka et al. 1988, Lis and Choppin 1991), in which the metal ion is preferentially solvated by water molecules when the mole fraction of water is $\geqslant 0.1$. An example of this case is shown in curve II of fig. 13.

In the second case of $\beta_i \gg 1$, addition of small amounts of B results in a rapid increase in i/n until a constant maximum value is reached for which the concentration of B is sufficient to solvate the lanthanide ion completely. An example of this case is the solvation of Eu^{3+} in a water + DMSO mixture (curve I in fig. 13). Less regular behavior can be expected for systems in which strong interactions exist between solvent molecules A and B or between coordinated and bulk solvent molecules as well as for those in which the steric barrier increases markedly with the degree of solvation. Such an irregular pattern is observed in the solvation of Eu^{3+} in a water + DMF mixture. In this system, i/n is larger than the bulk solvent mole fraction of DMF below 0.7, but the reverse occurs at higher mole fractions of DMF (curve III in fig. 13).

Solvation studies in completely anhydrous solvents have been reported by Bunzli and co-workers (Bunzli and Mabillard 1986a, Bunzli and Vuckovic 1983, 1984, Bunzli and Yersin 1982, 1984, Bunzli et al. 1982a, b). The solvation numbers were obtained by spectrofluorimetry (life-time) measurements and vibrational spectroscopy. The affinity sequence was reported to be: DMSO > DMF ~ H_2O > acetone > acetonitrile (Bunzli and Vuckovic 1983), which is in accord with the expected basicity sequence. When ligand anions displaced solvents in the cationic coordination sphere, some systems, typically the poorer coordinating solvents, had a small expansion in coordination number. An example is the formation of $Ln(NO_3)_5^{2-}$ ($Ln = Eu^{3+}, Tb^{3+}$) with CN = 10 in CH_3CN but not in DMF, and the 9-coordinated systems $Ln(NO_3)_3(DMSO)_3$ and $Ln(NO_3)_3(H_2O)_3$ ($Ln = Eu^{3+}, Tb^{3+}, Er^{3+}$) in CH_3CN. In these complexes the nitrate ion acts as a bidentate ligand. Addition of water to the solutions caused reversion to coordination number eight for the heavy elements. Also, the CN was found to increase slightly as the concentration of lanthanide salt decreased (Bunzli and Mabillard 1986a, Aslanov et al. 1972a, b), an effect noted also in the Raman glassy solutions studies and the T_g studies (Kanno and Hiraishi 1980, 1982).

Relaxation measurements in mixed solvents indicated a change in solvation number upon inner-sphere complexation. Ultrasound data of Er^{3+} solutions were consistent with the structural change accompanying complexation with ligands such as Br^-, I^- (Silber and Bordano 1979), NO_3^- (Reidler and Silber 1974a, Silber and Kromer 1980), ClO_4^- (Silber 1974) but not with Cl^- (Reidler and Silber 1974b). By contrast, such a change with chloride ion complexation was reported for Nd^{3+} and Gd^{3+} (Silber et al. 1978). These authors attributed this structural change to steric crowding within the inner solvation shell. However, no explanation was offered for the lack of such change for the Er and Nd chloride systems. The Er^{3+} solvation would be expected to be more sensitive to crowding than the Nd^{3+} because of their relative cationic radii. Similar explanations were offered for complexation of Y^{3+} and La^{3+} ions in water–methanol solvent (Silber and Mioduski 1984).

In aqueous solutions, Cl^- and $ClO_4{}^-$ form outer-sphere complexes with the lanthanides (Choppin 1971, Rinaldi et al. 1979). Inner-sphere complexation of lanthanide ions by $ClO_4{}^-$ and Cl^- ligands in weakly and moderately coordinating solvents has been inferred from spectroscopic and conductometric measurements (Bunzli and Mabillard 1986b, Hamze et al. 1986, Kozachenko and Batyaev 1971, Zholdakov et al. 1971). The enhancement in the complexing ability of these anions in such media as compared to that in water was rationalized in terms of the difference in strength of the Ln–H_2O and Ln–solvent bond as well as the difference in the dielectric constants of water and the nonaqueous solvent.

2.2.1.8. *Hydration of lanthanide complexes.* X-ray diffraction studies of the solid complexes of KLn(EDTA)$(H_2O)_x$ showed the number of water molecules in the coordination sphere to be three for the lighter lanthanides and two for the heavier ones for total coordination numbers of nine and eight, respectively, since EDTA is hexadentate (Hoard et al. 1967). Ots (1973) measured a maximum at europium for the heat capacity change ΔC_p^0 for the formation of lanthanide–EDTA complexes. This maximum was taken as a strong evidence for hydration equilibrium between the complexed species,

$$\mathrm{Ln(EDTA)(H_2O)}_x \leftrightarrow \mathrm{Ln(EDTA)(H_2O)}_{x-1} + \mathrm{H_2O},$$

rather than between the free ions. This conclusion was strengthened by the nearly constant values of the partial molal heat capacities for various lanthanide perchlorates (Grenthe et al. 1973), although such constancy is in conflict with the results reported by Spedding et al. (1966b, 1975b). Variations in the visible spectra of $Eu(EDTA)^-$ (Geier et al. 1969) with temperature were also interpreted in terms of a shift in the equilibria between the hydrated complexed species. Kostromina et al. (1969) and Kostromina and Tananaeva (1971) proposed that changes in the absorption spectra of europium were due to a variation between four and three water molecules in the first coordination shell of the Eu(EDTA) complex, which coincides with a change of the denticity of EDTA between five and six to maintain a constant CN of nine.

Based on ^{1}H-NMR spectral analysis of the Ln(EDTA) complexes, Elgavish and Reuben (1976) proposed that in LiCl medium, the hexachelated Ln(EDTA)$(H_2O)_{x-1}$ complexes are shifted to pentachelation at low temperatures with the probable formation of LnLi(EDTA)$(H_2O)_x$. In solutions of $NaClO_4$, ^{1}H-NMR measurements at a higher magnetic field (270 MHz) (Baisden et al. 1977, Choppin et al. 1982) showed that across the whole lanthanide series, the four carboxylate groups are equivalent within the NMR-time domain for $Ln(EDTA)^-$, and EDTA behaves as a hexadentate ligand. The different results may be due to stronger competition by Li^+ than by Na^+ with Ln^{3+} for carboxylate bonding.

Brucher et al. (1984) attributed the discontinuity in the molal volumes of Ln(EDTA) across the series to structural changes in the complex, as the size of the metal ion is increased. The increase in the metal–ligand bond distances results in greater lability of the metal–functional-group interaction. The gradual decrease in bonding of the carboxylate groups could result in an increase in the hydration number (Brucher et al. 1984). Thus, for the elements Tb–Lu a hexadentate nature was assumed for EDTA with coordination of two water molecules in the inner sphere, whereas for the elements La–Gd the carboxylate groups spend less time coordinated to the cation, allowing some increase in its time-averaged hydration.

Fluorescence kinetic measurements were applied to the determination of hydration numbers in lanthanide–polyaminocarboxylate complexes; the results are given in table 5 (Brittain et al. 1991, Brittain and Jasinski 1988). Measurements of hydration numbers as a function of the solution pH for Eu^{3+} and Tb^{3+} complexes showed the presence of three buffer regions: the first one at low pH in which the hydration number of the cations is equivalent to that of the free ions; the second one in the pH range of 4–8 in which complexation decreases the hydration numbers; and the third one at high pH (ca. 9–12) being associated with the formation of ternary hydroxo complexes.

TABLE 5
Hydration numbers of lanthanide polyaminocarboxylate complexes[a].

Ligand	Number of donor groups	Average n for EuL, TbL
NTA (1:1)	4	4.5
(1:2)	8	1.0[b]
MEDTA	5	3.6
BEDTA	5	3.8
HEDTA	5 (6?)	3.1
EDTA	6	2.6
TMDTA	6	2.4
TMEDTA	6	2.0
PDTA	6	2.6
DCTA	6	2.3
EDAP	6	2.5
ENDADP	6	2.7
DTPA	7 (8?)	1.1
EGTA	8	1.0

[a] Brittain et al. (1989); [b] Brittain and Jasinski (1988).

At pH > 6, the ${}^5D_4 \rightarrow {}^7F_J$ ($J = 3$–6) transitions of the Tb(NTA) spectrum were enhanced in intensity by formation of the ternary hydroxo complex which caused self-association and formation of oligomers (Brittain and Jasinski 1988). In all these systems, the total coordination number, i.e. the sum of the number of ligand donor groups and the number of primary water molecules, was 8.8 ± 0.5 for Eu^{3+} and 8.5 ± 0.5 for Tb^{3+} complexes. Such nonintegral hydration numbers were interpreted as the equilibrium weighted average of mixtures of species with different hydration numbers (Brittain and Jasinsky 1988).

The hydration of Eu^{3+}–dicarboxylate [$^{-}OOC(CH_2)_nCOO^{-}$ ($n = 1$–4)] complexes was studied by luminescence lifetimes (Barthélémy and Choppin 1989b). The decrease in hydration number for the formation of EuL^+ [L^{2-} = malonate ($n = 1$), succinate ($n = 2$), glutarate ($n = 3$) and adipate ($n = 4$)] was 2.3, 1.5, 1.6 and 1.5 ± 0.5, respectively. For the ternary complexes $Eu(NTA)L^{2-}$, the corresponding change in the number of water molecules in the first coordination sphere was 2.2, 1.5, 1.1 and 1.0, respectively. These data indicate that all of the dicarboxylate ligands have a chelate structure in the binary complexes, whereas in the ternary complexes only malonate acts as a chelating ligand. These conclusions agreed with interpretation of the thermodynamic data of complexation of these systems (Choppin et al. 1986, Niu and Choppin 1987).

Whether or not the coordination number of the lanthanide ion in aqueous complexes remains constant as ML, ML_2, etc., the form of the species is an open question. Using the fluorescence technique, Albin et al. (1984) reported the hydration number of $Eu(EDA)^+$ to be seven and of $Eu(EDA)_2^-$ to be three. Since EDA is a tetradentate ligand, these values lead to CN = 11 in both complexes. Lifetime measurements on the crystalline hydrate $Na[Eu(EDA)_2(H_2O)_x]$ are consistent with the presence of two water molecules in the primary sphere (CN = 10), reflecting that the solid compounds may not be valid structural models for similar solution species.

Knowledge of the hydration state of metal ions bound to a functional group in a polymer can be helpful in determining the coordination environment of the metal (Kido et al. 1988, Vesala and Kappi 1985). For example, an increase in terbium dehydration was measured by the fluorescence method when complexation by monomeric carboxylate ligands was replaced by binding to polycarboxylate chains (PAA or PMA). This change was attributed to an increase in the strength of binding of Tb^{3+} to the polyelectrolyte. Some comparisons are listed in table 6. The greater dehydration of Tb^{3+} when bound to PMA as compared to binding to PAA and MAE (ca. 0.5 water molecule) is probably associated with changes in flexibility of the polymer chains imposed by the branched methyl groups which, presumably, interferes with the formation of multidentate metal ion binding.

Similar conclusions were proposed for site characterization of proteins and nucleic acids (table 6). Addition of one, two and four equivalents of Eu^{3+} to the Ca^{2+} ion binding sites of apocalmodulin (Mulqueen et al. 1985) indicated that the four sites are equivalent with respect to coordination since, in all cases, the metal ion retained two water molecules in its coordination sphere. The total coordination number of Ln^{3+} in these sites was estimated to be 6 to 8 (Horrocks 1982), which implies that the protein supplies at least four ligating groups with 3–4 carboxylate ligands in each binding loop (Kretsinger 1980).

TABLE 6
Hydration number of Ln(III)-carboxylates, polycarboxylates, proteins and nucleic acids.

Ligand	Ln^{3+}	n	Ref.[a]
Propionic acid	Tb	6.0	[1]
Dimethylmalonic acid	Tb	6.3	[1]
2,3-Dimethylsuccinic acid	Tb	5.8	[1]
Glutaric acid	Tb	5.6	[1]
PAA	Tb	3.5	[1]
MAE	Tb	3.5	[1]
PMA	Tb	3.9	[1]
PAAm	Tb	3.8	[1]
α-Lactalbumin	Eu	2.1 for site I	[2]
	Eu	4.0 for site III	[2]
Glutamine synthetase	Eu	4.1 for site I	[3]
	Gd	4.0 for site I	[3]
Calmodulin	Eu	2.1 for sites I–IV	[4]
Parvalbumin	Eu	1.0 for all sites	[5]
Thermolysine	Eu and Tb	2 for site I	[6]

[a] References: [1] Kido et al. (1988); [2] Bunzli et al. (1987); [3] Eads et al. (1985); [4] Mulqueen et al. (1985); [5] Rhee et al. (1981); [6] Horrocks et al. (1977).

2.2.1.9. *Rates of water exchange.* The rates of water exchange between the inner coordination sphere of lanthanide ions and bulk water for the reaction

$$Ln(H_2O)_n^{3+} + H_2O^{\#} \rightleftharpoons Ln(H_2O)_{n-1}(H_2O^{\#})^{3+} + H_2O,$$

where $H_2O^{\#}$ represents an isotopically labelled water, have been measured using both ^{17}O-NMR and ultrasound techniques. For highly labile systems, Cossy et al. (1987, 1988) and Southwood-Jones et al. (1980) showed that the transverse relaxation rates (T_2^{-1}), the chemical shift ($\Delta\omega$), the longitudinal relaxation rates (T_1^{-1}) and the partial molar fraction of solvent bound to the metal ion (P_m) can be related by the expression (Swift and Connick 1962, Zimmermann and Brittain 1957)

$$T_2^{-1} - T_1^{-1} = P_m \Delta\omega_m^2/k, \tag{21}$$

where $\Delta\omega_m$, given by

$$\Delta\omega = P_m \Delta\omega_m, \tag{22}$$

is the chemical-shift difference between the free and the bound solvent, and k is the rate of exchange. The results for the ions from Gd^{3+} to Yb^{3+} are summarized in table 7. Attempts to estimate exchange rates for the lighter elements were unsuccessful, probably due to high labilities (within the limits of diffusion control). The rate constants k were calculated using a value of P_m which corresponds to a constant hydration number of nine. The assumption of constancy of n was based on the observation that ^{17}O-NMR chemical shift data for aqueous lanthanide solutions varies linearly with temperature rather than with the S-shaped dependence seen for DMF solutions.

TABLE 7
Kinetic parameters for water exchange of $[Ln(H_2O)_9]^{3+}$ ions[a].

Ln^{3+}	k^b ($10^8\ s^{-1}$)	ΔH^* ($kJ\ mol^{-1}$)	ΔS^* ($J\ K^{-1}\ mol^{-1}$)	ΔV^* ($cm^3\ mol^{-1}$)
Gd^{3+}	10.60	12.0	−32	–
Tb^{3+}	4.96	12.1	−38	−5.7
Dy^{3+}	3.86	16.6	−25	−6.0
Ho^{3+}	1.91	16.4	−32	−6.6
Er^{3+}	1.18	18.4	−29	−6.9
Tm^{3+}	0.81	22.7	−17	−6.0
Yb^{3+}	0.41	23.3	−22	–

[a] Cossy et al. (1987); Cossy et al. (1988); Southwood-Jones et al. (1980).
[b] At 298 K.

Enthalpies and entropies of activation for the H_2O exchange reactions calculated from Arrhenius plots are also listed in table 7. A decrease in hydration number from nine to eight results in an increase in the rate constant by a factor of $\frac{9}{8}$ and an increase in ΔS^* by one unit with no change in ΔH^*. The k values decrease regularly with decreasing ionic radius, as expected from the electrostatic nature of the water–metal bond. In contrast, in DMF solvent the rate of DMF exchange with $Ln(DMF)_8{}^{3+}$ does not vary regularly across the cationic series. The steric crowding of the large DMF molecules apparently negates any effect of cation radius (Pisaniello et al. 1983).

The differences in solvent exchange in H_2O and DMF are also reflected in the exchange mechanisms. The activation volumes for water exchange are negative and, within experimental error, essentially constant for the different lanthanides; this suggests a concerted associative (I_a) mechanism. By contrast, the data for DMF exchange are more consistent with a gradual transition from an I_d to a D mechanism between Tb^{3+} and Yb^{3+}.

The interpretation of the ultrasound relaxation data is based on the assumption that the rate-controlling step is the loss of water from the cation solvation shell via the Eigen mechanism of complex formation (Eigen and Tamm 1962)

$$Ln^{3+}_{(aq)} + X^{n-}_{(aq)} \underset{k_{-1}}{\overset{k_1}{\rightleftarrows}} [Ln^{3+}(H_2O)X^{n-}]_{(aq)} \underset{k_{-2}}{\overset{k_2}{\rightleftarrows}} LnX^{3-n}_{(aq)}.$$

The relaxation time τ for the rate-determining step is given by the relation

$$\tau^{-1} = k_{-2} + k_2, \tag{23}$$

where k_1 and k_2 are determined by the following relations:

$$\varphi = \Theta/(\Theta + K_{1n}^{-1}), \tag{24}$$

$$\Theta \approx [Ln^{3+}_{(aq)}] + [L^{n-}_{(aq)}], \tag{25}$$

$$K_n = k_n/k_{-n}. \tag{26}$$

$K_c = K_1/(1 + K_2)$ is called the stability constant of the complex. A summary of the

TABLE 8
Ultrasound data for the rate of water exchange (X_{H_2O} = the mole fraction of H_2O).

Ion	k_2 ($10^8\ s^{-1}$)	Medium	Ref.[a]
La^{3+}	2.1	Aqueous, SO_4^{2-}	[1]
Ce^{3+}	3.3	Aqueous, SO_4^{2-}	[1]
Pr^{3+}	4.4	Aqueous, SO_4^{2-}	[1]
Nd^{3+}	5.2	Aqueous, SO_4^{2-}	[1]
	1.8	Aqueous, NO_3^-	[2]
	6.0	Aqueous, NO_3^-	[3]
	1.5	Aqueous, NO_3^-	[4]
	4.5	Aqueous methanol (X_{H_2O} = 0–1.0), Cl^-	[5]
Sm^{3+}	7.4	Aqueous, SO_4^{2-}	[1]
Eu^{3+}	6.6	Aqueous, SO_4^{2-}	[1]
Gd^{3+}	6.7	Aqueous, SO_4^{2-}	[1]
	5.0	Aqueous, NO_3^-	[3]
	0.24	Aqueous methanol (X_{H_2O} = 0–1.0), Cl^-	[5]
Tb^{3+}	5.2	Aqueous, SO_4^{2-}	[1]
Dy^{3+}	4.2	Aqueous, SO_4^{2-}	[1]
Ho^{3+}	2.8	Aqueous, SO_4^{2-}	[1]
Er^{3+}	1.9	Aqueous, SO_4^{2-}	[1]
	0.6	Aqueous, NO_3^-	[2]
	1.2	Aqueous methanol (X_{H_2O} = 0.69), ClO_4^-	[6]
	0.094	Aqueous methanol (X_{H_2O} = 0.23), Br^-	[7]
	1.0, 0.77	Aqueous methanol (X_{H_2O} = 0.026, 0.05), Cl^-	[8]
	0.56	Aqueous DMSO (X_{H_2O} = 0.163), NO_3^-	[9]
	0.18	Aqueous DMSO (X_{H_2O} = 0.3), Cl^-	[10]
Tm^{3+}	1.4	Aqueous, SO_4^{2-}	[1]
Yb^{3+}	~0.8	Aqueous, SO_4^{2-}	[1]
Lu^{3+}	~0.6	Aqueous, SO_4^{2-}	[1]

[a] References: [1] Fay et al. (1969); [2] Silber et al. (1972); [3] Garnsey and Ebdon (1969); [4] Silber and Fowler (1976); [5] Silber et al. (1978); [6] Silber (1974); [7] Silber and Bordano (1979); [8] Reidler and Silber (1974b); [9] Silber and Kromer (1980); [10] Silber (1978).

estimated k_2 values obtained for aqueous and mixed solvents media is given in table 8. The only available complete set of data in aqueous media is that for the sulfate salts (Fay et al. 1969), for which k_2 as a function of ionic radius was found to increase from La^{3+} to a maximum at Sm^{3+} and decrease almost linearly from Gd^{3+} to Lu^{3+}.

The similarities of k_2 values for water and water + methanol mixtures led Silber and his co-workers (Silber 1974, Silber and Fowler 1976, Silber and Bordano 1979, Silber et al. 1972, 1978, Reidler and Silber 1974b, Garnsey and Ebdon 1969) to suggest that the slow step in the exchange is the cation–ligand bond formation rather than the cation–solvent exchange. The low rates of exchange observed for water + dimethyl-sulfoxide (DMSO) mixtures were attributed to a structural reorganization (Silber 1978, Silber and Kromer 1980). Alternatively, the preferential solvation by water in water + methanol and by DMSO in water + DMSO mixtures (Tanaka et al. 1988) plus the possibility of changes in the solvation number can explain the similarities and differences in the rate values of the three systems.

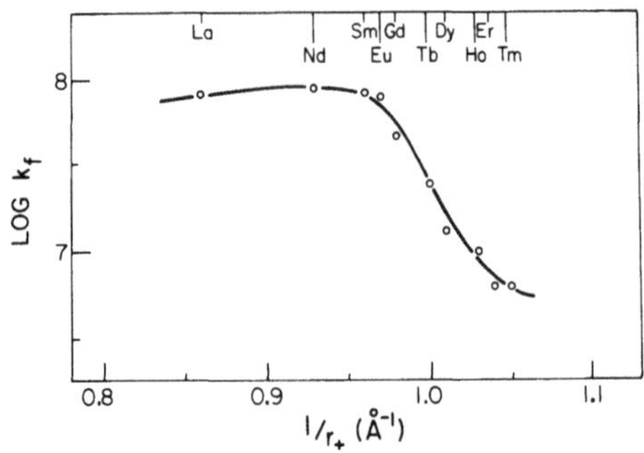

Fig. 14. Plot of the log of the formation rate constant, $\log k_f$, of the lanthanide oxalate complex as a function of the inverse cation radius.

Support for the conclusion that the rate of water release is the controlling step in these reactions is obtained from the similarity in the values of the rate of formation of oxalate (Graffeo and Bear 1968) and murexide (Geier 1965, 1968) complexes. These were measured by pressure jump and temperature jump techniques, respectively. Figure 14 shows the variation in $\log k_f$ of lanthanide oxalates with ionic radii. If the structure and number of water molecules in a complex remains the same throughout the lanthanide series, a gradual decrease in k_f should be observed due to the decreasing ionic size. Alternatively, a decrease in the primary hydration number at some point in the series would result in an accelerated increase in the strength of the metal ion–water bond and, therefore, in a slower substitution rate. We can see that the rate of substitution of oxalate for solvation water remains essentially constant from lanthanum to europium followed for the subsequent cations by a regular decrease in k_2 values. The decrease in the rate of exchange between europium and dysprosium was attributed to an increase in the net size of the hydrated ion in that region since a decrease in the primary hydration number would result in a large secondary solvation sphere.

In conclusion, many questions remain about the mechanism of water exchange. While activation volume data support an association mechanism (I_a), the ultrasound and kinetic data are interpreted in terms of an overall dissociative mechanism. Kinetic results show a discontinuity in plots of $\log k$ values versus lanthanide radius in the region of Sm^{3+}–Tb^{3+} which would seem to be good evidence for a change in hydration number in this region.

2.2.2. *Theoretical and thermodynamic studies*

CNDO quantum chemical studies on the hydrated ions of lanthanides were reported by Jia (1987). The calculations were performed assuming the following:

(1) The structures of the primary hydration sphere in solution are similar to those reported for the hydrated crystals (Albertson and Elding 1977), i.e. for the lighter cations $n = 9$ with tricapped trigonal prism (TCTP) structure and D_{3h} symmetry, and for the heavier cations $n = 8$ with square prismatic structure and D_{2h} symmetry. The geometric parameters (Ln–H_2O and H_2O–H_2O distances) were taken from the X-ray data on solutions from Habenschuss and Spedding (1979b, c, 1980).

(2) The interaction between the metal ions and the water molecules in the primary coordination sphere is not purely ion–dipole but has a small covalent contribution

involving the 6s, 6p and 5d orbitals of the lanthanide ions, the 2s and 2p orbitals of the oxygen atom and the 1s orbital of the hydrogen atom. This is consistent with the interpretation of Lewis et al. (1962) that the ^{17}O-NMR shift in aqueous lanthanide perchlorate solutions was associated with the isotropic part of the indirect contact hyperfine interaction of the unpaired lanthanide electrons with the oxygen nucleus. The magnitude and variation of the shift with the number of the 4f electrons were taken as an evidence for a very weak covalent bond between the 2s orbital of oxygen and the 6s orbital of the lanthanide. The 4f, 5d and 6p orbitals of the lanthanide apparently play a very minor, if any, role in the bonding.

(3) Perturbations of the hydrogen-bonded interactions between water molecules (coordinated and uncoordinated) were neglected.

The simulated curves for the hydration energy for the aquo ions with hydration numbers eight and nine indicated the presence of breaks at Nd^{3+} and Tb^{3+}. The hydration numbers for Pm^{3+}, Sm^{3+}, Eu^{3+} and Gd^{3+} were obtained by interpolating the data between the two standard curves and were estimated to be 8.8, 8.5, 8.3 and 8.1, respectively. These values seems to be in reasonable agreement with the estimates from X-ray studies for Sm^{3+} (8.6–8.7) and Eu^{3+} (8.4) (Habenschuss and Spedding 1979b, c, 1980). These ions were proposed as existing in a transitional configuration between TCTP and square prismatic which contrasts to the earlier assumption that they existed in an equilibrium mixture of the individual nonaaquo and octaaquo species (Spedding et al. 1977a).

Mioduski and Siekierski (1975) proposed that the structural changes occurred by a gradual displacement of a water molecule from the primary hydration sphere of the ions as the series proceeded from neodymium through terbium. This displacement leads to configurations with fractional hydration numbers for the Nd–Tb cations as the displaced water molecule belongs partially to the inner hydration sphere and partially to the outer hydration sphere. This explanation is consistent with the optical spectral observations of Geier and Karlen (1971) and the NMR data of Reuben and Fiat (1969) which gave no indication for an equilibrium between different hydration forms.

Another semi-empirical approach, based on an electrostatic hydration model for the primary sphere and a Born continuum treatment for the secondary solvation effects, was used by Goldman and Morss (1975) and by Tremaine and Goldman (1978). The value of the hydration number was taken as an adjustable parameter to fit the experimental free energy (ΔG^0_h) and enthalpy (ΔH^0_h) of hydration. This gave an estimate of n across the lanthanide series of 5.6 which seems to be too small in view of the experimental data reviewed in earlier sections.

Kobayashi et al. (1987) and Okada et al. (1985) used discrete vibrational X_α (DV-X_α) calculations with a self-consistent charge approach to estimate the charge distribution on $[Ce(H_2O)_9]^{3+}$. The net charges were the central metal ion, +0.35; the equatorial waters, +0.27 (O = −0.10; H = +0.19); the vertex waters, +0.31 (O = −0.11; H = +0.21). The model did not take into account electron delocalization between the metal and the ligands, and the charge transferred from the ligands to the metal was attributed to the dipole moments induced in the ligands by the electrostatic potential of the cation. The agreement of the calculated transition energies in the 5d ← 4f excitations and of the ligand field splitting of the $^2D(5d^1)$ multiplet in the D_{3h} ligand field with the experimental values provide support for this model.

The large Stokes shift observed between the absorption and emission maxima of $[Ce(H_2O)_9]^{3+}$ doped in La(ethylsulfate)$_3 \cdot 9H_2O$ crystal was attributed to a deformation in the TCTP structure in the excited state. In the rapidly fluctuating solution environment, this deformation leads to dissociation of one of the equatorial waters followed by a shift of one of the vertex waters to an equatorial position to form the excited state $[Ce(H_2O)_8]^{3+}$. The appearance of a new band was the evidence for the formation of this excited state species of lower hydration (Jorgensen and Brinen 1963, Kaizu et al. 1985, Miyakawa et al. 1988).

Figure 15 shows plots of the equilibrium distances r_{M-O} calculated for the TCTP $Ln(H_2O)_9^{3+}$ species and for the cubic, square antiprism and dodecahedron $Ln(H_2O)_8^{3+}$ species. The results agree well with the X-ray data of Habenschuss and Spedding (1979b, c, 1980) which indicated a transition from the TCTP ($n = 9$) structure of La–Nd group to the cubic structure of the Gd–Lu group.

Regardless of the atomic number of the metal ion, the calculated bond energies U of different structures favor formation of the TCTP structure. This is not true for the hydration enthalpies, ΔH_h^0, calculated from the relationship

$$\Delta H_h^0 = U + \Delta H_B - \tfrac{5}{2} nRT + n\Delta H_v. \tag{27}$$

In eq. (27), ΔH_B is the Born hydration enthalpy, n is the difference in the total number of moles between the gaseous and solution species and ΔH_v is the enthalpy of vaporization of water. According to these calculations, the hydration stabilizes the TCTP structure of $La(H_2O)_9^{3+}$ by 15 kJ mol^{-1} relative to the square antiprism structure of $La(H_2O)_8^{3+}$. By contrast, $[Lu(H_2O)_8]^{3+}$ is stabilized in the square antiprism structure by 1 kJ mol^{-1} relative to $[Lu(H_2O)_9]^{3+}$ in the TCTP structure. The difference between the hydration free energies $[\Delta(\Delta G_h^0)]$ of the octahydrate and

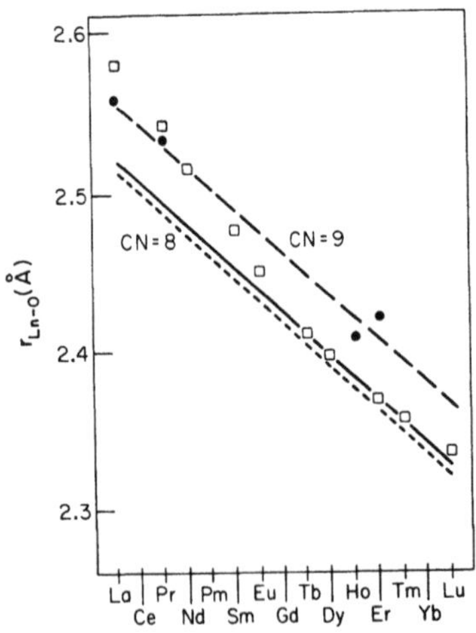

Fig. 15. Comparison of the experimental values of Ln–O distances for $Ln(H_2O)_n^{3+}$ in solution (□) and crystals (●). The crystals all have CN = 9. Also shown as lines are the Ln–O calculated distances for TCTP (CN = 9) and the cubic (solid) square antiprism and dodecahedral (CN = 8) structures.

nonahydrate ions across the lanthanide series indicate a preference for nonahydration for the earlier members and octahydration for the later ones (Miyakawa et al. 1988).

Direct measurements of the hydration values ΔG^0_h, ΔH^0_h and ΔS^0_h are not feasible and they are usually calculated indirectly using semi-empirical methods (Born–Haber thermochemical cycles). These cycles lead to the general equations (Rosseinsky 1965, Morss 1976, Bratsch and Lagowski 1985b)

$$\Delta H^0_h(Ln^{z+}) = \Delta H^0_f(Ln^{z+}_{(aq)}) - \Delta H^0_f(Ln^{z+}_{(g)}) + z[\Delta H^0_f(H^+_{(g)}) + \Delta H^0_h(H^+_{(aq)})], \quad (28)$$

$$\Delta S^0_h(Ln^{z+}) = S^0(Ln^{z+}_{(aq)}) - S^0(Ln^{z+}_{(g)}) + z[S^0(H^+_{(g)}) + \Delta S^0_h(H^+_{(aq)})], \quad (29)$$

$$\Delta G^0_h(Ln^{z+}) = \Delta G^0_f(Ln^{z+}_{(aq)}) - \Delta G^0_f(Ln^{z+}_{(g)}) + z[\Delta G^0_f(H^+_{(g)}) + \Delta G^0_h(H^+_{(aq)})]. \quad (30)$$

In these equations, the reference states of $H^+_{(aq)}$ are, by convention, equal to zero as are the functions $\Delta H^0_f(e^-_{(g)})$ and $\Delta G^0_f(e^-_{(g)})$. The absolute entropies for the gaseous ions are calculated from statistical mechanics (Bratsch and Lagowski 1985a) and agree fairly well with the experimental values reported by Bertha and Choppin (1969), who interpreted the S-shaped dependence of standard state entropies on ionic radius in terms of a change in the overall hydration of the cation across the lanthanide series. Hinchey and Cobble (1970) proposed that this S-shaped relationship was an artifact of the method of data treatment and calculated a set of entropies from lanthanide

TABLE 9

Gibbs free energies[a,b] of formation (ΔG^0_f, in kJ mol^{-1}) and hydration (ΔG^0_h, in kJ mol^{-1}) of the lanthanide ions; $T = 298$ K.

	$\Delta G^0_{(f)}$, $Ln^{n+}_{(g)}$			ΔG^0_f, $Ln^{n+}_{(a)}$			ΔG^0_h		
	+2	+3	+4	+2	+3	+4	+2	+3	+4
La	1999	3853	–	−322[c]	−693	–	−1419	−3193	–
Ce	1967	3916	7467	−257	−677	−508	−1322	−3240	−6171
Pr	1867	3954	7716	−377	−677	−319	−1342	−3278	−6231
Nd	1860	3992	7890	−402	−672	−197	−1360	−3311	−6283
Pm	1866	4027	7973	−414	−667	−160	−1378	−3341	−6329
Sm	1786	4058	8041	−510	−665	−134	−1394	−3370	−6371
Eu	1776	4183	8282	−535	−567	67	−1409	−3397	−6411
Gd	2121	4112	8349	−330[c]	−662	99	−1549	−3421	−6446
Tb	2030	4144	7944	−319	−659	−338	−1447	−3450	−6478
Dy	1953	4162	8129	−412	−672	−198	−1463	−3481	−6523
Ho	1984	4188	8259	−398	−675	−109	−1480	−3510	−6564
Er	2021	4214	8305	−376	−676	−100	−1495	−3537	−6601
Tm	1958	4241	8342	−453	−675	−97	−1509	−3563	−6635
Yb	1901	4313	8514	−524	−627	41	−1523	−3587	−6669
Lu	2273	4297	8658	−222[c]	−665	155	−1593	−3609	−6699
Y		4161			−689			−3497	

[a] Calculated for the lowest energy ground states.
[b] Johnson (1982); Bratsch and Lagowski (1985b).
[c] Corrected for LFSE by −120 kJ mol^{-1}.

chloride solutions. They concluded that the value of entropies of hydration varied linearly with the inverse square of the ionic radius of the cations, within the uncertainties of the calculations. A reinvestigation of the chloride system by Spedding et al. (1977a) confirmed the two-series pattern reported by Bertha and Choppin. The differences in the reported thermodynamic functions of hydration (Goldman and Morss 1975, Bratsch and Lagowski 1985a, b, Johnson 1982, Spedding et al. 1977a, Bettonville et al. 1987, David 1986, Mikheev et al. 1983, Morss 1971) reflects primarily the uncertainties in the reference-states values of hydration for $H^{+}_{(aq)}$. For example, the accepted value for $\Delta H^0_h(H^{+}_{(aq)})$ was taken by some authors as -1091 ± 10 kJ mol^{-1} (Goldman and Morss 1975, Morss 1971), while others preferred -1105 ± 17 kJ mol^{-1} (Bratsch and Lagowski 1985b). The propagated standard deviations of all terms involved in the calculation of the enthalpy of hydration of La^{3+} ion was estimated to be ± 46 kJ mol^{-1} (Morss 1971). Errors in ΔG^0_h are probably of the same order of magnitude. Although the relative error in ΔH_h or ΔG_h seems to be small (approximately 1–2%), that of ΔS_h is quite large (ca. 40–50%) because of the intrinsically small values of hydration entropies.

Tables 9–11 list the predicted thermodynamic functions for the hydration of divalent, trivalent and tetravalent lanthanides as calculated by Bratsch and Lagowski (1985b). Values for yttrium hydration are also included when available. The formation values refer to the reaction $Ln_{(m)} \rightarrow Ln^{n+}_{(a)}$ while the hydration values relate to the use of eqs. (28)–(30). The standard state ionic entropies given in table 10 are corrected for

TABLE 10

Enthalpies[a,b] of formation (ΔH^0_f, in kJ mol^{-1}) and of hydration (ΔH^0_h, in kJ mol^{-1}) of the lanthanide ions; $T = 298$ K.

	ΔH^0_f, $Ln^{n+}_{(g)}$			ΔH^0_f, $Ln^{n+}_{(a)}$			ΔH^0_h		
	+2	+3	+4	+2	+3	+4	+2	+3	+4
La	2049	3905	–	−324	−755	–	−1488	−3326	–
Ce	2014	3969	7522	−261	−743	−636	−1395	−3380	−6376
Pr	1914	4007	7774	−382	−744	−448	−1417	−3421	−6442
Nd	1908	4046	7950	−409	−739	−327	−1436	−3454	−6496
Pm	1913	4081	8033	−421	−732	−291	−1454	−3482	−6543
Sm	1832	4111	8102	−518	−731	−265	−1469	−3512	−6585
Eu	1822	4232	8339	−543	−634	−65	−1486	−3538	−6625
Gd	2170	4166	8408	−341	−732	−37	−1630	−3567	−6662
Tb	2078	4199	8003	−328	−732	−471	−1527	−3600	−6694
Dy	2002	4217	8189	−421	−747	−332	−1544	−3634	−6741
Ho	2033	4242	8320	−407	−750	−243	−1561	−3663	−6783
Er	2069	4269	8367	−386	−753	−235	−1575	−3692	−6821
Tm	2005	4296	8403	−463	−752	−233	−1589	−3717	−6855
Yb	1947	4370	8579	−534	−704	−95	−1598	−3740	−6889
Lu	2324	4352	8725	–	−744	−19	–	−3759	−6918
Y	–	4216	–	–	−763	–	–	−3640	–

[a] Calculated for the lowest energy ground states.
[b] Johnson (1982).

TABLE 11
Standard entropies[a,b] of formation (S_f^0, in $J K^{-1} mol^{-1}$) and hydration (ΔS_h, in $J K^{-1} mol^{-1}$) of lanthanide ions; $T = 298$ K.

	S_f^0, $Ln^{n+}_{(g)}$			S_f^0, $Ln^{n+}_{(a)}$			ΔS_h		
	+2	+3	+4	+2	+3	+4	+2	+3	+4
La	182	170	–	−5	−209	–	−231	−445	–
Ce	189	185	170	−12	−220	−431	−245	−471	−689
Pr	190	189	185	−17	−225	−434	−251	−480	−707
Nd	189	190	189	−22	−225	−437	−255	−481	−714
Pm	186	189	190	−25	−218	−439	−255	−473	−717
Sm	181	187	190	−27	−222	−441	−252	−475	−719
Eu	189	181	187	−27	−226	−443	−260	−473	−718
Gd	192	189	182	−36	−236	−455	−272	−491	−725
Tb	195	193	189	−29	−245	−447	−268	−504	−724
Dy	196	195	194	−31	−252	−448	−271	−513	−730
Ho	196	196	196	−31	−253	−450	−271	−515	−734
Er	194	196	196	−32	−258	−454	−270	−520	−738
Tm	190	194	196	−33	−257	−455	−267	−517	−739
Yb	173	190	194	−34	−258	−456	−251	−514	−738
Lu	179	173	191	–	−264	−457	–	−503	−736
Y	–	165	–	–	−249	–	–	−480	–

[a] Calculated for the lowest energy ground states and corrected for "electronic" entropy contribution.
[b] Johnson (1982).

the "electronic" entropy contribution

$$S_e^0 = R \ln(2J + 1) \tag{31}$$

where J is the quantum number for the total electronic angular momentum of the ground state. This correction assumes no population of higher states which is not completely valid for Sm^{3+} and Eu^{3+} at room temperature. In the case of divalent lanthanides, the lowest energy ground state is that of the configuration $4f^n$, except for La^{2+}, Gd^{2+} and Lu^{2+} which have ground state configurations of $5d^1$, $4f^7 5d^1$ and $4f^{14} 6s^1$, respectively.

The free energies and the enthalpies of hydration of the trivalent lanthanides have an approximately linear dependence on the cationic radii. The slight variations in the Pr–Nd and Er–Tm regions have been attributed to a small ligand field stabilization caused by the nonspherical symmetry of the f-electron subshell (Morss 1971). In general, the solvation enthalpy change determines the magnitude of the free energy change while the entropy term ($T\Delta S_h$ is $\approx$ 3–5% of ΔH_h^0 term) shows more distinct variation with changing ionic radii. The hydration entropy changes for the heavy lanthanides (Tb–Lu) for all oxidation states are more negative than those for the lighter elements, indicating more net ordering in hydration of the heavier cations. Also, the incremental increase between adjacent elements seems to be greater for the heavy lanthanides. Spedding et al. (1977a) interpreted these changes as a result of an increase in the ion–dipole interactions with decreasing ionic radius. The loss of one water molecule from the hydration sphere of the heavy elements (Spedding et al. 1977a, Bertha and Choppin 1969) would be accompanied by an increase in the surface charge density and, hence, an increase in the dipole–dipole interactions within the secondary hydration sphere.

3. Lanthanide hydrolysis

Hydrated cations can act as Brönsted acids by releasing protons from the bound water molecules to form hydroxo complexes:

$$q\,\mathrm{Ln(H_2O)}_n^{z+} \leftrightarrow \mathrm{Ln}_q\mathrm{(OH)}_{p\,\mathrm{(aq)}}^{qz-p} + p\mathrm{H}^+, \quad {}^*\beta_{pq} = \frac{[\mathrm{Ln}_q\mathrm{(OH)}_p^{zq-p}][\mathrm{H}^+]^p}{[\mathrm{Ln}^{z+}]^q}. \tag{32}$$

As the trivalent lanthanide cations have relatively large ionic radii, the polarization of the hydrate molecules is relatively weak and the acidic properties of these ions are less pronounced than those of other trivalent cations such as Fe^{3+}, Al^{3+}, etc.

Equilibrium constants of the hydrolysis reaction [eq. (32)] have been reported for the trivalent lanthanides and tetravalent cerium but not for the divalent lanthanide cations. The latter would be even weaker acids due to a lower charge density than the Ln^{3+} cations. A variety of mononuclear and polynuclear hydrolytic species have been reported and the values of the hydrolysis equilibrium constants in the literature are listed in tables 12 (mononuclear) and 13 (polynuclear). Baes and Mesmer (1976) have reviewed the hydrolysis reactions of the lanthanides; so the scope of this review is limited to the studies which have appeared since the publication of their book. The species are discussed in terms of mononuclear species $\mathrm{Ln(OH)}_q^{3-q}$, $q = 1$ to 6) and polynuclear species $\mathrm{Ln}_p(OH)_q^{3p-q}$.

TABLE 12
Hydrolysis constants of mononuclear lanthanide complexes.

Species	T (°C)	Medium		Method[a]	$\log {}^*\beta_{pq}$	Ref.[b]
$La(OH)^{2+}$	25	3.0	($LiClO_4$)	GE	−10.1 ± 0.1	[1]
	25	3.0	($LiClO_4$)	GE	−10.08 ± 0.01	[2]
	25	3.0	($NaClO_4$)	GE	−10.36 ± 0.06	[3]
	60	3.0	($LiClO_4$)	GE	−9.29 ± 0.03	[4]
	21.5	1.0	($NaClO_4$)	Sol.	−8.6	[5]
	25	0.3	($NaClO_4$)	GE	−9.08	[6]
	25	0.1	($LiClO_4$)	SE	−7.4	[7]
	20	0.1	($La(NO_3)_3$)	GE	−8.4	[8]
	25	0.05	($NaClO_4$)	GE	−9.06 ± 0.09	[9]
	25	0.001	(sulfate)	GE	−10.0 (aged)	[10]
	25	0			−8.5	[11]
$La(OH)_2^+$	21.5	1.0	($NaClO_4$)	Sol.	−17.9	[5]

TABLE 12 (cont'd)

Species	T (°C)	Medium		Method[a]	log $^*\beta_{pq}$	Ref.[b]
$La(OH)_3$	21.5	1.0	$(NaClO_4)$	Sol.	−27.3	[5]
$Ce(OH)^{2+}$	50	3.0	$(LiClO_4)$	GE	−9.11 ± 0.05	[12]
	RT[a]	1.0	$(NaClO_4)$	Sol.	−8.1	[13]
	25	0.001	(sulfate)	GE	−9.0	[10]
$Ce(OH)_2^+$	RT[a]	1.0	$(NaClO_4)$	Sol.	−16.3	[13]
$Ce(OH)_3$	RT[a]	1.0	$(NaClO_4)$	Sol.	−26.0	[13]
$Ce(OH)^{3+}$	1.6	2.0	(ClO_4)		0.9	[14]
	5	2.0	(ClO_4)		−0.06	[15]
	15	2.0	(ClO_4)		0.32	[15]
	25	2.0	(ClO_4)		0.72	[15]
	35	2.0	(ClO_4)		1.18	[15]
$Pr(OH)^{2+}$	25	3.0	$(NaClO_4)$	GE	−9.56 ± 0.03	[3]
	25	3.0	$(NaClO_4)$	Sol.	−8.5 ± 0.4	[16]
	25	0.3	$(NaClO_4)$	GE	−8.57	[6]
	25	0.001	(sulfate)	GE	−9.0	[10]
$Nd(OH)^{2+}$	25	3.0	$(NaClO_4)$	GE	−9.4 ± 0.4	[17]
	25	3.0	$(NaClO_4)$	Sol.	−8.5 ± 0.4	[16]
	25	3.0	$(NaClO_4)$	GE	−9.57	[6]
	21.5	1.0	$(NaClO_4)$	Sol.	−8.1	[18]
	25	0.3	$(NaClO_4)$	GE	−8.45	[6]
	25	0.1	$(LiClO_4)$	SE	−7.0	[7]
	25	0.001	(sulfate)	GE	−9.0	[10]
	25	0			−8.0	[11]
$Nd(OH)_2^+$	21.5	1.0	$(NaClO_4)$	Sol.	−16.2	[18]
$Nd(OH)_3$	21.5	1.0	$(NaClO_4)$	Sol.	−24.3	[18]
$Nd(OH)_4^-$	25	0			−37.4	[11]
$Sm(OH)^{2+}$	RT	1.0	$(NaClO_4)$	Sol.	−7.5	[19]
	25	0.3	$(NaClO_4)$	GE	−8.36	[6]
	25	0.1	$(LiClO_4)$	SE	−4.4	[7]
	25	0.001	(sulfate)	GE	−8.9	[10]
$Sm(OH)_2^+$	RT	1.0	$(NaClO_4)$	Sol.	−15.0	[19]
$Sm(OH)_3$	RT	1.0	$(NaClO_4)$	Sol.	−22.7	[19]
$Sm(OH)_4^-$	RT	1.0	$(NaClO_4)$	Sol.	−36.7	[20]
$Eu(OH)^{2+}$	25	1.0			−8.12 ± 0.02	[21]
	25	0.7		SE	−7.3 ± 0.2	[22]
	25	0.3	$(NaClO_4)$	GE	−8.33	[6]
	25	0.05		GE	−8.03 ± 0.03	[9]
	25	0.003			−8.7 ± 0.3	[23]
	25	0.001	(sulfate)	GE	−8.9	[10]
$Gd(OH)^{2+}$	25	3.0	$(LiClO_4)$	GE	−8.20 ± 0.01	[2]
	25	3.0	$(NaClO_4)$	GE	−9.2 ± 0.4	[24]
	RT	1.0	$(NaClO_4)$	Sol.	−7.3	[25]
	25	0.5	$(LiClO_4)$	SE	−7.5	[7]
	25	0.3	$(NaClO_4)$	GE	−8.37	[6]
	25	0.1	$(LiClO_4)$	SE	−7.1	[7]
	25	0.05		GE	−8.27 ± 0.03	[9]
$Tb(OH)^{2+}$	25	0.3	$(NaClO_4)$	GE	−8.18	[6]
$Dy(OH)^{2+}$	25	0.3	$(NaClO_4)$	GE	−8.12	[6]
$Ho(OH)^{2+}$	25	0.3	$(NaClO_4)$	GE	−8.06	[6]

TABLE 12 (cont'd)

Species	T (°C)	Medium		Method[a]	log $^*\beta_{pq}$	Ref.[b]
$Er(OH)^{2+}$	25	0.3	($NaClO_4$)	GE	−9.0	[26]
	21.5	1.0	($NaClO_4$)	Sol.	−6.3	[27]
		0.3	($NaClO_4$)	GE	−8.01	[6]
	25	0.1	($LiClO_4$)	SE	−5.5	[7]
$Er(OH)_2^+$	25	3.0	($LiClO_4$)	GE	−17.4 ± 0.1	[2]
	21.5	1.0	($NaClO_4$)	Sol.	−14.5	[11]
$Er(OH)_3$	21.5	1.0	($NaClO_4$)	Sol.	−23.1	[11]
$Er(OH)_4^-$	21.5	1.0	($NaClO_4$)	Sol.	−36.8	[11]
$Tm(OH)^{2+}$	25	0.3	($NaClO_4$)	GE	−7.97	[6]
$Yb(OH)^{2+}$	25	3.0	($NaClO_4$)	GE	−8.6 ± 0.2	[3]
	21.5	1.0	($NaClO_4$)	Sol.	−7.7	[28]
	25	0.3	($NaClO_4$)	GE	−7.94	[6]
	25	0.1	($LiClO_4$)	SE	−4.3	[7]
	25	0.05		GE	−8.01 ± 0.03	[9]
	25	0.001	(sulfate)	GE	−8.4	[10]
	25			Sol.	−8.4	[29]
$Yb(OH)_2^+$	21.5	1.0	($NaClO_4$)	Sol.	−15.5	[28]
$Yb(OH)_3$	21.5	1.0	($NaClO_4$)	Sol.	−23.2	[28]
$Yb(OH)_4^-$	21.5	1.0	($NaClO_4$)	Sol.	−37.5	[28]
$Yb(OH)_5^-$	21.5	1.0	($NaClO_4$)	Sol.	−51.9	[28]
$Yb(OH)_6^-$	21.5	1.0	($NaClO_4$)	Sol.	−66.2	[28]
$Lu(OH)^{2+}$	25	0.3	($NaClO_4$)	GE	−7.92	[6]
	25	0.05		GE	−7.66 ± 0.03	[9]
$Y(OH)^{2+}$	25	3.0			−9.08	[11]
		0.5		IE	−7.70 ± 0.02	[30]
	25	0.3	($NaClO_4$)	GE	−8.36	[6]
	25	0.05		GE	−8.04 ± 0.04	[9]
	25	0.001	(sulfate)	GE	−8.7	[10]
	25	0			−7.7	[11]
$Y(OH)_2^+$	25	3.0	($LiClO_4$)	GE	−17.0 ± 0.1	[2]

[a] Abbreviations used: GE = glass electrode; SE = solvent extraction; Sol. = solubility; IE = ion-exchange; RT = room temperature.

[b] References: [1] Biederman and Ciavatta (1961); [2] Amaya et al. (1973); [3] Burkov et al. (1982); [4] Ciavatta et al. (1987); [5] Kragten and Decnop-Weever (1987); [6] Frolova et al. (1966); [7] Guillaumont et al. (1971); [8] Wheelwright et al. (1953); [9] Usherenko and Skorik (1972); [10] Moeller (1946); [11] Smith and Martell (1976); [12] Ciavatta et al. (1988); [13] Kragten and Decnop-Weever (1978); [14] Baker et al. (1960); [15] Hardwick and Robertson (1951); [16] Tobias and Garrett (1958); [17] Burkov et al. (1973); [18] Kragten and Decnop-Weever (1984); [19] Kragten and Decnop-Weever (1979); [20] Kragten and Decnop-Weever (1983b); [21] Nair et al. (1982); [22] Caceci and Choppin (1983); [23] Schmidt et al. (1978); [24] Din Ngo and Burkov (1974b); [25] Kragten and Decnop-Weever (1980); [26] Burkov et al. (1975); [27] Kragten and Decnop-Weever (1983a); [28] Kragten and Decnop-Weever (1982); [29] Ivanov-Emin et al. (1970); [30] Davydov and Voronik (1983).

3.1. *Mononuclear complexes*

The most complete set of data for the formation of the monohydroxo complexes is that reported by Frolova et al. (1966) for 25 °C and an ionic strength of 0.3 M. These

TABLE 13
Hydrolysis constants of polynuclear complexes.

Species	T (°C)	Medium	Method[a]	$\log {}^*\beta_{pq}$	Ref.[b]
$La_2(OH)^{5+}$	25	3.0 ($LiClO_4$)	GE	-9.98 ± 0.10	[1]
	60	3.0 ($LiClO_4$)	GE	-8.93 ± 0.05	[2]
$La_2(OH)_2^{4+}$	25	3.0 ($NaClO_4$)	GE	-17.74 ± 0.07	[3]
$La_2(OH)_3^{3+}$	60	3.0 ($LiClO_4$)	GE	-22.47 ± 0.05	[2]
$La_3(OH)_3^{6+}$	60	3.0 ($LiClO_4$)	GE	-22.9 ± 0.2	[2]
$La_5(OH)_9^{6+}$	25	3.0 ($LiClO_4$)	GE	-71.4 ± 0.1	[1]
	25	0		-71.2	[4]
$Ce_2(OH)^{5+}$	50	3.0 ($LiClO_4$)	GE	-9.46 ± 0.2	[5]
$Ce_3(OH)_5^{4+}$	25	3.0 ($LiClO_4$)	GE	-35.7 ± 0.1	[6]
	RT[a]	1.0 ($NaClO_4$)	Sol.	-32.8	[7]
	25	0		-33.5	[4]
	50	3.0 ($LiClO_4$)	GE	-35.53 ± 0.07	[5]
$Ce_2(OH)_3^{5+}$	25	3.0 (NO_3)		-1.63	[8]
$Ce_2(OH)_4^{4+}$	25	3.0 (NO_3)		-2.29	[8]
$Ce_6(OH)_{12}^{12+}$	25	3.0 (NO_3)		-1.98	[8]
$Pr_2(OH)_2^{4+}$	25	3.0 ($NaClO_4$)	GE	-16.31 ± 0.05	[3]
$Nd_2(OH)_2^{4+}$	25	3.0 ($NaClO_4$)	GE	-13.93 ± 0.03	[9]
	21.5	1.0 ($NaClO_4$)	Sol.	-11.6	[10]
	25	0		-13.9	[4]
$Sm_3(OH)_4^{5+}$	RT[a]	1.0 ($NaClO_4$)	Sol.	-19.5	[11]
$Gd_2(OH)_2^{4+}$	25	3.0 ($NaClO_4$)	GE	-14.23 ± 0.03	[12]
$Gd_3(OH)_4{}^{5+}$	RT[a]	1.0 ($NaClO_4$)	Sol.	-19.0	[13]
$Er_2(OH)^{5+}$	25	3.0 ($NaClO_4$)	GE	-17.2	[14]
$Er_2(OH)_2^{4+}$	25	3.0 ($NaClO_4$)	GE	-13.72 ± 0.01	[14, 15]
$Yb_2(OH)_2^{4+}$	25	3.0 ($NaClO_4$)	GE	-13.32 ± 0.04	[12]
	25	3.0		-14.3	[4]
$Y_2(OH)_2^{4+}$	25	3.0 ($LiClO_4$)	GE	-14.04 ± 0.01	[14]
	25	0		-14.2	[4]
$Y_3(OH)_5^{4+}$	25	3.0		-33.80	[4]
	25	0		-31.6	[4]

[a] Abbreviations used: GE = glass-electrode; Sol. = solubility; RT = room temperature.
[b] References: [1] Biederman and Ciavatta (1961); [2] Ciavatta et al. (1987); [3] Burkov et al. (1982); [4] Smith and Martell (1976); [5] Ciavatta et al. (1988); [6] Biederman and Newmann (1964); [7] Kragten and Decnop-Weever (1978); [8] Danesi (1967); [9] Burkov et al. (1973); [10] Kragten and Decnop-Weever (1984); [11] Kragten and Decnop-Weever (1979); [12] Din Ngo and Burkov (1974b); [13] Kragten and Decnop-Weever (1980); [14] Amaya et al. (1973); [15] Burkov et al. (1975).

authors did not obtain a value of ${}^*\beta_{11}$ for Ce(III). This was attributed to the dominance of the formation of the trinuclear species, $Ce_3(OH)_5^{4+}$ by Ciavatta et al. (1987, 1988) who investigated this system at elevated temperatures. The latter authors also concluded that increasing the temperature enhances the formation of intermediate hydroxo species of low p and q values [e.g. $Ce(OH)^{2+}$ and $Ce_2(OH)^{5+}$]. They agreed with Arnek (1970) that ΔH_{pq} generally is less than zero and for a given cation $\Delta H_{pq}/p$ is constant to within $\pm 1.7\ \mathrm{kJ\,mol^{-1}}$.

Kragten and Decnop-Weever (1978, 1979, 1980, 1982, 1983a, b, 1984, 1987) conducted a series of solubility measurements over a wide range of pH (ca. 6–15). The dependence of the precipitation on pH allowed calculation of formation constants of mononuclear and polynuclear species which were internally consistent. The accuracy of the calculated constants was estimated to yield uncertainties of ca. ± 0.1–0.2 log units. The accuracy of successive stability constants (as p and/or q increased) decreased as a result of error accumulation in the fitting procedure. Nevertheless, the results, in general, indicate that the hydrolysis constants for stepwise formation of higher hydroxo species are of the same order of magnitude as those for the formation of $Ln(OH)^{2+}$.

3.2. *Polynuclear complexes*

Most of the evidence cited in the literature for the formation of polynuclear hydroxo lanthanide complexes is obtained from potentiometric measurements with computer fitting of rather smooth trends. Unfortunately, direct identification of the species formed has, thus far, been unobtainable due to the low solubility of the $Ln(OH)_3$ compounds and to the simultaneous presence of a number of hydrolyzed species in solution.

Biedermann and Ciavatta (1961) and Ciavatta et al. (1987, 1988) reported detailed studies of polynuclear complexation of La^{3+} and Ce^{3+} ions. Values of $^*\beta_{pq}/^*\beta_{11}^p$ were calculated for the following reactions:

$$\text{(i) } Ln(OH)^{2+} + Ln^{3+} \longleftrightarrow Ln_2(OH)^{5+} \qquad [\log(^*\beta_{12}/^*\beta_{11}) = 0.4 \pm 0.1],$$

$$\text{(ii) } 3Ln(OH)^{2+} \longleftrightarrow Ln_2(OH)_3^{3+} + La^{3+} \qquad [\log(^*\beta_{32}/^*\beta_{11}^3) = 5.4 \pm 0.2],$$

$$\text{(iii) } 3Ln(OH)^{2+} \longleftrightarrow Ln_3(OH)_3^{6+} \qquad [\log(^*\beta_{33}/^*\beta_{11}^3) = 4.9 \pm 0.3].$$

No difference was found between the values for the La^{3+} and for the Ce^{3+} reactions, indicating that the polymerization process is primarily a property of the hydroxy anion but relatively independent of the identity of the cation. This cationic independence is unlikely to be valid over a wider range of the lanthanide series. Although there have been some attempts to account for the polymeric species formed, at present it would seem more important to seek firmer evidence for the species which are formed than to propose explanations for those which have been reported but whose existence is open to question.

Estimates for the enthalpies of formation of $La(OH)^{2+}$ and $La_2(OH)^{5+}$ were obtained from temperature data at 25 °C and 60 °C and found to be 41 ± 7 and 56 ± 8 kJ mol^{-1}, respectively (Ciavatta et al. 1987). Such values derived from data at only two temperatures should be used with caution.

3.3. *Solubility of lanthanide hydroxides*

Moeller and Kremers (1945) and Aksel'rud (1963) have published extensive reviews on the solubility of lanthanide hydroxides and on the trend in basicity within these compounds. Baes and Mesmer (1976) used these reports to predict "smoothed" values for the solubility products. Table 14 presents recent experimental results at 1.0 *M* ionic strength together with values recommended by Smith and Martell (1976) at zero ionic strength. The values $^*K_{so}$ are for the reaction

$$Ln(OH)_{3(c)} + 3H^+_{(aq)} \longleftrightarrow Ln^{3+}_{(aq)} + 3H_2O_{(1)}.$$

Suzuki et al. (1986) measured the onset of precipitation of $Ln(OH)_3$ by optical detection. To avoid local concentration effects, the hydroxide ion was generated electrolytically. The pH values associated with initial precipitation in this study were substantially lower than those reported earlier (Moeller and Kremer 1945), presumably, due to differences in the sensitivity of detection. A plot of the pH of initial precipitation as a function of the radius of the lanthanide cations was an S-shaped

TABLE 14
Solubility products of lanthanide hydroxides.

Ln	T(°C)	Medium	log $^*K_{so}$	Method and remarks	Ref.[a]
La	21.5	1.0 ($NaClO_4$)	22.8	Solubility, corrected for $La(OH)_n^{3-n}$	[1]
		0.38 ($NaNO_3$)	22.8	Fresh precipitation (ppt)	[2]
	20	0.1	23.8	Solubility	[3]
	25	0.05 (sulfate)	23.2	pH, aged precipitation (ppt)	[4]
		0.01	20	pH, aged precipitation (ppt)	[5]
	25	0	22.7	Oscillopolarography	[5]
	25	0	20.3		[6]
			23.1	Fresh precipitation (ppt)	[7]
			19.4		[8]
Ce	25	1.0 ($NaClO_4$)	20.1	Solubility, corrected for $Ce(OH)_n^{3-n}$	[9]
	25	0	19.0	Oscillopolarography	[5]
	25	0	20.8		[6]
Pr	20	0.1	22.5	Solubility	[3]
	25	0	20.5		[6]
Nd	21.5	1.0 ($NaClO_4$)	19.4	Solubility, corrected for $Nd(OH)_n^{3-n}$	[10]
	20	0.1	22.3	Solubility	[3]
	25	0	20.3	Oscillopolarography	[11]
	25	0	18.9		[6]
Sm	RT	1.0 ($NaClO_4$)	17.5	Solubility, corrected for $Sm(OH)_n^{3-n}$	[12]
	20	0.1	21.3	Solubility	[3]
	25	0	16.8	Oscillopolarography	[13]
	25	0	16.6		[6]
Eu	25	0	16.4		[6]
Gd	25	3.0 ($NaClO_4$)	15.2	Aged precipitation (ppt)	[14]
	20	0.1	21.7	Solubility	[3]
	25	0	17.6	pH	[15]
	25	0	16.3		[6]
	25	0	16.5		[6]

TABLE 14 (cont'd)

Ln	T (°C)	Medium	log *K_{so}	Method and remarks	Ref.[a]
Dy	25	0	17.0	Oscillopolarography	[13]
	25	0	16.4		[6]
Ho	25	0	16.1		[6]
Er	21.5	1.0 ($NaClO_4$)	18.0	Solubility	[16]
	25	0	17.1		[6]
Yb	25	3.0 ($NaClO_4$)	15.4	Aged ppt.	[14]
	21.5	1.0 ($NaClO_4$)	18.0	Solubility	[17]
	25	0	16.9	Oscillopolarography	[11]
	25	0	17.0		[6]
	25		18.5	Solubility	[18]
Lu	25	0	15.9		[6]
Y	20	0.1	21.5	Solubility	[3]
	25	0	18.8		[6]

[a] References: [1] Kragten and Decnop-Weever (1987); [2] Ziv and Shestakova (1965); [3] Meloche and Vratny (1959); [4] Moeller and Vogel (1951); [5] Buchenko et al. (1970); [6] Smith and Martell (1976); [7] Sadolin (1927); [8] Mironov and Odnosevtev (1957); [9] Kragten and Decnop-Weever (1978); [10] Kragten and Decnop-Weever (1984); [11] Azhipa et al. (1967); [12] Kragten and Decnop-Weever (1979); [13] Kovalenko et al. (1966); [14] Aksel'rud (1963); [15] Din Ngo and Burkov (1974a); [16] Kragten and Decnop-Weever (1983a), [17] Kragten and Decnop-Weever (1982); [18] Ivanov-Emin et al. (1970).

curve, which is, probably associated with the change in hydration number in the lanthanide series.

Two mechanisms were proposed for the hydrolysis reaction. Since the hydroxide ions were generated electrolytically, the lanthanide ion may have reacted at the cathode surface with replacement of coordinated water molecules to form $Ln(OH)^{2+}$ or $Ln(OH)_2^+$. Polynuclear complex formation could also have been formed by this mechanism. Alternatively, the aquo complexes could have been electrolyzed at the cathode surface to form the hydroxo complexes. The authors (Suzuki et al. 1986) suggested, without explanation, that hydrolysis of the elements with large ionic radii and higher hydration number occurs more likely by the first mechanism. However, the second mechanism would seem more likely for the formation of all hydrolyzed species.

4. Summary and conclusions

Based on the data reviewed in this chapter, we draw the following general conclusions about lanthanide hydration and hydrolysis.

For solid lanthanide salts in which the cation coordination sphere is saturated with nine water molecules, the oxygen atoms form a tricapped trigonal prismatic structure surrounding the cation. Within an isostructural series, the changes in bond length are directly proportional to the changes in ionic radius. No evidence has been found in solids for a change in the hydration number when the primary coordination sphere is fully occupied by water molecules. However, changes in coordination number have been observed when the primary coordination sphere of the metal ion is shared by ligands and water.

In solution, a majority of the data supports a change in hydration number across the lanthanide series. The light lanthanides (La^{3+}–Nd^{3+}) form a series with a hydration number of nine (primary sphere) and TCTP geometry, whereas the heavier elements (Tb^{3+}–Lu^{3+}) apparently form octahydrates with square antiprismatic structure. For the hydrated ions in the middle of the series (Nd^{3+}–Tb^{3+}), either there is some type of transitional structure between these two geometries or both hydrate structures exist in rapid equilibrium. These conclusions are supported by data from X-ray and neutron diffraction, fluorescence, Raman and visible spectroscopy, thermodynamic and exchange kinetics measurements. They are also in agreement with data on apparent molal volumes, relative viscosities, molal heat capacities, heats of dilution, electrical conductances and entropies of hydration. Theoretical predictions of the enthalpies of hydration also favor a change in hydration number across the series. By contrast, NMR relaxation measurements would seem to provide evidence for a constant hydration number for all lanthanides.

Data on aquation rates are limited and in poor agreement with each other. Both dissociative and associative exchange mechanisms have been proposed for water exchange. Another area of uncertainty is the extent to which, if any, the chemistry of lanthanides in excited states differs significantly from that of ions in the ground state. It has been proposed that the coordination number and the relative strength of metal–water and metal–ligand bonds are changed when the lanthanide ion is raised to an excited state.

Finally, little systematic study has been devoted to lanthanide hydrolysis. Speciation in these systems is uncertain, reflecting difficulties in analysis of the types of measurements normally used. The development of techniques with better detection limits, such as fluorescence and photoacoustic spectroscopy, as well as studies at tracer levels (below concentrations for precipitation and polymerization) should result in more definitive data on the nature of the hydrolytic species present in solution.

Acknowledgments

The preparation of this research was aided by a grant from the Division of Chemical Sciences U.S.D.O.E. One of us (E.N.R.) acknowledges also the support from the C.N.R.S. and Institut Curie, Section de Physique et Chemie, Paris, France, during part of the time of preparation of this chapter.

References

Aksel'rud, N.V., 1963, Russ. Chem. Revs. **32**, 353.
Albertson, J., 1970, Acta Chem. Scand. **24**, 3527.
Albertson, J., and I. Elding, 1977, Acta Crystallogr. B **33**, 1460.
Albin, M., G.K. Farber and W.DeW. Horrocks Jr, 1984, Inorg. Chem. **23**, 1648.
Amaya, T., H. Kakihana and M. Maeda, 1973, Bull. Chem. Soc. Jpn. **46**, 1720.

Annis, B.K., R.L. Hahn and A.H. Narten, 1985, J. Chem. Phys. **82**, 2086.

Arnek, R., 1970, Ark. Kemi **32**, 55.

Aslanov, L.A., L.I. Soleva and M.A. Porari-Koshits, 1972a, Russ. J. Struct. Chem. **23**, 1021.

Aslanov, L.A., L.I. Soleva, M.A. Porari-Koshits and S.S. Goukhberg, 1972b, Russ. J. Struct. Chem. **13**, 610.

Azhipa, L.T., P.N. Kovalenko and M.M. Evstifeev, 1967, Russ. J. Inorg. Chem. **12**, 601.

Baes Jr, C.F., and R.E. Mesmer, 1976, The Hydrolysis of Cations (Wiley-Interscience, New York).

Baisden, P.A., G.R. Choppin and B.B. Garrett, 1977, Inorg. Chem. **16**, 1367.

Baker, F.B., T.W. Newton and M. Khan, 1960, J. Phys. Chem. **64**, 109.

Barthélémy, P.P., and G.R. Choppin, 1989a, Inorg. Chem. **28**, 3354.

Barthélémy, P.P., and G.R. Choppin, 1989b, private communication.

Bertha, S.L., and G.R. Choppin, 1969, Inorg. Chem. **8**, 613.

Bettonville, S., J. Goudiakas and J. Fujer, 1987, J. Chem. Thermodyn. **19**, 595.

Biedermann, G., and L. Ciavatta, 1961, Acta Chem. Scand. **15**, 1347.

Biedermann, G., and L. Newman, 1964, Ark. Kemi **22**, 203.

Bratsch, S.G., and J.J. Lagowski, 1985a, J. Phys. Chem. **89**, 3310.

Bratsch, S.G., and J.J. Lagowski, 1985b, J. Phys. Chem. **89**, 3317.

Breen, P.J., and W.DeW. Horrocks Jr, 1983, Inorg. Chem. **22**, 536.

Brittain, H.G., 1983, J. Less-Common Met. **93**, 97.

Brittain, H.G., and J.P. Jasinski, 1988, J. Coord. Chem. **18**, 279.

Brittain, H.G., G.R. Choppin and P.P. Barthélémy, 1991, to be submitted for publication.

Broach, R.W., J.M. Williams, G. Felcher and D.G. Hinks, 1979, Acta Crystallogr. B **35**, 2317.

Brücher, E.J., Cs.E. Kukri and R. Kiraly, 1984, Inorg. Chim. Acta **95**, 135.

Brücher, E.J., J. Glaser, I. Grenthe and I. Puigdomench, 1985, Inorg. Chim. Acta **109**, 111.

Buchenko, L.I., P.N. Kovalenko and M.M. Evstifeev, 1970, Russ. J. Inorg. Chem. **15**, 1666.

Bukietynska, K., and G.R. Choppin, 1970, J. Chem. Phys. **2**, 2875.

Bunzli, J.-C.G., and C. Mabillard, 1986a, J. Less-Common Met. **126**, 379.

Bunzli, J.-C.G., and C. Mabillard, 1986b, Inorg. Chem. **25**, 2750.

Bunzli, J.-C.G., and M.V. Vukovic, 1983, Inorg. Chim. Acta **73**, 53.

Bunzli, J.-C.G., and M.V. Vukovic, 1984, Inorg. Chim. Acta **95**, 105.

Bunzli, J.-C.G., and J.-R. Yersin, 1979, Inorg. Chem. **18**, 605.

Bunzli, J.-C.G., and J.-R. Yersin, 1982, Helv. Chim. Acta **65**, 2498.

Bunzli, J.-C.G., and J.-R. Yersin, 1984, Inorg. Chim. Acta **94**, 301.

Bunzli, J.-C.G., C. Mabillard and J.-R. Yersin, 1982a, Inorg. Chem. **21**, 4214.

Bunzli, J.-C.G., J.-R. Yersin and C. Mabillard, 1982b, Inorg. Chem. **21**, 1471.

Bunzli, J.-C.G., P.M. Morand and J.M. Pfefferle, 1987, J. Phys. (Paris) **48**, C7-625.

Burkov, K.A., L.S. Lilich, N. Din Ngo and A.Yus. Smirnov, 1973, Russ. J. Inorg. Chem. **18**, 797.

Burkov, K.A., L.S. Lilich and N. Din Ngo, 1975, Izv. Vyssh. Uchebn. Zaved. SSSR, Khim. i Khim. Tekhnol. **18**, 181.

Burkov, K.A., E.A. Bus'ko and I.V. Pichugina, 1982, Russ. J. Inorg. Chem. **27**, 643.

Caceci, M.S., and G.R. Choppin, 1983, Radiochim. Acta **33**, 101.

Chang, C.A., H.G. Brittain, J. Telser and M.F. Tweedle, 1990, Inorg. Chem. **29**, 4468.

Choppin, G.R., 1971, Pure & Appl. Chem. **27**, 23.

Choppin, G.R., 1980, in: Lanthanide and Actinide Chemistry and Spectroscopy, Am. Chem. Soc. Symp. Ser. No. 131, ed. N.M. Edelstein (Am. Chem. Soc., Washington, DC) ch. 8, p. 173.

Choppin, G.R., and S.L. Bertha, 1973, J. Inorg. & Nucl. Chem. **35**, 1309.

Choppin, G.R., and J. Ketels, 1965, J. Inorg. & Nucl. Chem. **27**, 1335.

Choppin, G.R., P.A. Baisden and E.N. Rizkalla, 1982, in: Rare Earths in Modern Science and Technology, Vol. 3, eds G.J. McCarthy, H.B. Silber and J.J. Rhyne (Plenum, New York) p. 187.

Choppin, G.R., A. Dadgar and E.N. Rizkalla, 1986, Inorg. Chem. **25**, 3581.

Ciavatta, L., M. Iuliano and R. Porto, 1987, Polyhedron **6**, 1283.

Ciavatta, L., R. Porto and E. Vasca, 1988, Polyhedron **7**, 1355.

Conway, B.E., 1981, Ionic Hydration in Chemistry and Biophysics (Elsevier, Amsterdam).

Cossy, C., L. Helm and A.E. Merbach, 1987, Inorg. Chim. Acta **139**, 147.

Cossy, C., L. Helm and A.E. Merbach, 1988, Inorg. Chem. **27**, 1973.

Cox, B.G., A.J. Parker and W.E. Waghorne, 1974, J. Phys. Chem. **78**, 1731.

Crosby, G.A., 1966, Mol. Cryst. **1**, 37.

Danesi, R.R., 1967, Acta Chem. Scand. **21**, 143.

David, F., 1986, J. Less-Common Met. **121**, 27.

Davidenko, N.K., and L.N. Lugina, 1968, Russ. J. Inorg. Chem. **13**, 512.

Davydov, Yu.P., and N.I. Voronik, 1983, Russ. J. Inorg. Chem. **28**, 1270.

Dawson, W.R., J.L. Kropp and M.W. Windsor, 1966, J. Chem. Phys. **45**, 2410.

Dereigne, A., and G. Pannetier, 1968, Bull. Chim. Soc. Fr. 174.

Dereigne, A., J.-M. Manoli, G. Pannetier and P. Herpin, 1972, Bull. Fr. Miner. Crystallogr. **95**, 269.

Din Ngo, N., and K.A. Burkov, 1974a, Russ. J. Inorg. Chem. **19**, 680.

Din Ngo, N., and K.A. Burkov, 1974b, Russ. J. Inorg. Chem. **19**, 1249.

Duffy, N.V., and J.-E. Early, 1967, J. Am. Chem. Soc. **89**, 272.

Eads, C.D., P. Mulqueen, W.DeW. Horrocks Jr and J.J. Villafranca, 1985, Biochem. **24**, 1221.

Eigen, M., and K. Tamm, 1962, Z. Elektrochem. **66**, 93, 107.

Elgavish, G.A., and J. Reuben, 1976, J. Am. Chem. Soc. **98**, 4755.

Enderby, J.E., S. Cummings, G.S. Hardman, G.W. Nielson, P.S. Salmon and N. Skipper, 1987, J. Am. Chem. Soc. **91**, 5851.

Fay, D.P., D. Litchinsky and N. Purdie, 1969, J. Phys. Chem. **73**, 544.

Fitzwater, D.R., and R.E. Rundle, 1959, Z. Kristallogr. **112**, 362.

Fratiello, A., 1970, in: Inorganic Reaction Mechanisms, Part II, ed. J.O. Edwards (Interscience, New York).

Fratiello, A., R.E. Lee and R.E. Schuster, 1970, Inorg. Chem. **9**, 391.

Fratiello, A., V. Kubo, S. Peak, B. Sanchez and R.E. Schuster, 1971, Inorg. Chem. **10**, 2552.

Fratiello, A., V. Kubo and A. Vidulich, 1973, Inorg. Chem. **12**, 2066.

Fratiello, A., V. Kubo-Anderson, T. Bollinger, C. Cordero, B. De Merit, T. Flores and R.D. Perrigan, 1989, private communication.

Freed, S., 1942, Rev. Mod. Phys. **14**, 105.

Freeman, J., and G.A. Crosby, 1964, J. Mol. Spectrosc. **13**, 399.

Frolova, U.K., V.N. Kumok and V.V. Serebrennikov, 1966, Izv. Vyssh. Uchebn. Zaved. SSSR, Khim. i Khim. Tekhnol **9**, 176.

Garnsey, R., and D.W. Ebdon, 1969, J. Am. Chem. Soc. **19**, 50.

Geier, G., 1965, Ber. Bunsenges. Physik. Chem. **69**, 617.

Geier, G., 1968, Helv. Chim. Acta **51**, 94.

Geier, G., and U. Karlen, 1971, Helv. Chim. Acta **54**, 135.

Geier, G., U. Karlen and A.V. Zelewsky, 1969, Helv. Chim. Acta **52**, 1967.

Glaser, J., and G. Johansson, 1981, Acta Chem. Scand. A **35**, 639.

Goldman, S., and L.R. Morse, 1975, Can. J. Chem. **53**, 2695.

Graffeo, A.J., and J.L. Bear, 1968, J. Inorg. & Nucl. Chem. **30**, 1577.

Grenthe, I., G. Hessler and H. Ots, 1973, Acta Chem. Scand. **27**, 2543.

Guerriero, P., U. Casellato, S. Sitran, P.A. Vigato and R. Graziani, 1987, Inorg. Chim. Acta **133**, 337.

Guillaumont, R., B. De Sire and M. Galin, 1971, Radiochem. & Radioanal. Lett. **8**, 189.

Habenschuss, A., and F.H. Spedding, 1978, Cryst. Struct. Commun. **7**, 535.

Habenschuss, A., and F.H. Spedding, 1979a, Cryst. Struct. Commun. **8**, 515.

Habenschuss, A., and F.H. Spedding, 1979b, J. Chem. Phys. **70**, 2797.

Habenschuss, A., and F.H. Spedding, 1979c, J. Chem. Phys. **70**, 3758.

Habenschuss, A., and F.H. Spedding, 1980, J. Chem. Phys. **73**, 442.

Hahn, R.L., 1988, J. Phys. Chem. **92**, 1688.

Hamze, M., J. Meullemeestre, M.J. Schwing and F. Vierling, 1986, J. Less-Common Met. **118**, 153.

Hardwick, T.J., and E. Robertson, 1951, Can. J. Chem. **29**, 818.

Heber, J., 1967, Phys. Kondens. Mater. **6**, 381.

Heber, J., and K.H. Hellwege, 1967, Optical Properties of Ions in Crystals, eds H.M. Crosswhite and H.W. Morse (Interscience, New York).

Helmholz, L., 1939, J. Am. Chem. Soc. **61**, 1544.

Hinchey, R.J., and J.W. Cobble, 1970, J. Am. Chem. Soc. **61**, 1544.

Hoard, J.L., B. Lee and M.D. Lind, 1967, J. Am. Chem. Soc. **87**, 1612.

Horrocks Jr, W.DeW., 1982, Adv. Inorg. Biochem. **4**, 201.

Horrocks Jr, W.DeW., and D.R. Sudnick, 1979, J. Am. Chem. Soc. **101**, 334.

Horrocks Jr, W.DeW., and D.R. Sudnick, 1981, Acc. Chem. Res. **14**, 384.
Horrocks Jr, W.DeW., G.F. Schmidt, D.R. Sudnick, C. Kittrell and R.A. Bernheim, 1977, J. Am. Chem. Soc. **99**, 2378.
Ivanov-Emin, B.N., E.M. Egorov, V.I. Romanynk and E.N. Siforova, 1970, Russ. J. Inorg. Chem. **15**, 628.
Janenas, V., 1978, Russ. J. Phys. Chem. **52**, 839.
Jezowska-Trzebiatowska, B., S. Ernst, J. Legendziewicz and G. Oczko, 1976, Bull. Accad. Pol. Sci. Ser. Sci. Chim. **24**, 997.
Jezowska-Trzebiatowska, B., S. Ernst, J. Legendziewicz and G. Oczko, 1977, Bull. Accad. Pol. Sci. Ser. Sci. Chim. **25**, 649.
Jezowska-Trzebiatowska, B., S. Ernst, J. Legendziewicz and G. Oczko, 1978a, Bull. Accad. Pol. Sci. Ser. Sci. Chim. **26**, 805.
Jezowska-Trzebiatowska, B., S. Ernst, J. Legendziewicz and G. Oczko, 1978b, Chem. Phys. Lett. **60**, 19.
Jia, Y.-Q., 1987, Inorg. Chim. Acta **133**, 331.
Johansson, G., and H. Wakita, 1985, Inorg. Chem. **24**, 3047.
Johnson, D.A., 1974, J. Chem. Soc. Dalton Trans., 1671.
Johnson, D.A., 1982, Some Thermodynamic Aspects of Inorganic Chemistry, 2nd Ed. (Cambridge University Press).
Jørgensen, C.K., and J.S. Brinen, 1963, Mol. Phys. **6**, 629.
Jørgensen, C.K., and B.R. Judd, 1964, Mol. Phys. **8**, 281.
Kaizu, Y., K. Miyakawa, K. Okada, H. Kobayashi, M. Sumitani and K. Yoshihara, 1985, J. Am. Chem. Soc. **107**, 2622.
Kanno, H., and Y. Akama, 1987, J. Phys. Chem. **91**, 1263.
Kanno, H., and J. Hiraishi, 1980, Chem. Phys. Lett. **75**, 553.
Kanno, H., and J. Hiraishi, 1982, J. Phys. Chem. **86**, 1488.
Kanno, H., and J. Hiraishi, 1988, J. Phys. Chem. **92**, 2787.
Karraker, D.G., 1968, Inorg. Chem. **7**, 473.
Kepert, D.L., J.M. Patrick and A.H. White, 1983, Aust. J. Chem. **36**, 477.
Kido, J., H.G. Brittain and Y. Okamoto, 1988, Macromolecules **21**, 1872.
Kobayashi, H., K. Okada, Y. Kaizu, N. Hamada and H. Adachi, 1987, Mol. Phys. **60**, 561.
Kostromina, N.A., and N.M. Tananaeva, 1971, Russ. J. Inorg. Chem. **16**, 1256.
Kostromina, N.A., T.V. Ternovaya and K.B. Yatsimirski, 1969, Russ. J. Inorg. Chem. **14**, 154.
Kovalenko, P.N., L.T. Azhipa and M.M. Evstifeev, 1966, Russ. J. Inorg. Chem. **11**, 1443.
Kovun, V.Ya., and B.N. Chernyshov, 1979, Koord. Khim. **5**, 53.
Kozachenko, N.N., and I.M. Batyaev, 1971, Russ. J. Inorg. Chem. **16**, 66.
Kragten, L., and L.G. Decnop-Weever, 1978, Talanta **25**, 147.
Kragten, L., and L.G. Decnop-Weever, 1979, Talanta **26**, 1105.
Kragten, L., and L.G. Decnop-Weever, 1980, Talanta **27**, 1047.
Kragten, L., and L.G. Decnop-Weever, 1982, Talanta **29**, 219.
Kragten, L., and L.G. Decnop-Weever, 1983a, Talanta **30**, 131.
Kragten, L., and L.G. Decnop-Weever, 1983b, Talanta **30**, 134.
Kragten, L., and L.G. Decnop-Weever, 1984, Talanta **31**, 731.
Kragten, L., and L.G. Decnop-Weever, 1987, Talanta **34**, 861.
Krestinger, R.H., 1980, CRC Crit. Rev. Biochem. **8**, 119.
Kropp, J.L., and M.W. Windsor, 1965, J. Chem. Phys. **42**, 1599.
Kropp, J.L., and M.W. Windsor, 1967, J. Phys. Chem. **71**, 477.
Krumholz, P., 1958, Spectrochim. Acta **10**, 274.
Langford, C.H., and H.B. Gray, 1965, Ligand Substitution Processes (Benjamin, New York).
Lee, T.J., H.R. Sheu, T.I. Chiu and C.T. Chang, 1983, Acta Crystallogr. C **39**, 1357.
Lewis, W.B., J.A. Jackson, J.F. Lemons and H. Taube, 1962, J. Chem. Phys. **36**, 694.
Lincoln, S.F., 1986, Adv. Inorg. & Bioinorg. Mechanisms **4**, 217.
Lis, S., and G.R. Choppin, 1991, Manuscript in preparation.
Mackay, A.I., J.L. Finney and R. Gotoh, 1977, Acta Crystallogr. A **33**, 98.
Marcantonatos, M.D., M. Deschaux and J.-J. Vuilleumier, 1981, Chem. Phys. Lett. **82**, 36.
Marcantonatos, M.D., M. Deschaux and J.-J. Vuilleumier, 1982, Chem. Phys. Lett. **91**, 149.
Marcantonatos, M.D., M. Deschaux and J.-J. Vuilleumier, 1984, J. Chem. Soc. Faraday Trans. II, **80**, 1569.

Marezio, M., H.A. Plettinger and W.H. Zachariasen, 1961, Acta Crystallogr. **14**, 234.
Meloche, C.C., and F. Vratny, 1959, Anal. Chim. Acta **20**, 415.
Mikheev, N.B., I.A. Rumer and L.N. Auerman, 1983, Radiochim. & Radional. Lett. **59**, 317.
Mioduski, T., and S. Siekierski, 1975, J. Inorg. & Nucl. Chem. **37**, 1647.
Mironov, N.N., and A.I. Odnosevtev, 1957, Russ. J. Inorg. Chem. **2**, 2022.
Miyakawa, K., Y. Kaizu and H. Kobayashi, 1988, J. Chem. Soc. Faraday Trans. I, **84**, 1517.
Moeller, T., 1946, J. Phys. Chem. **50**, 242.
Moeller, T., and H.E. Kremers, 1945, Chem. Rev. **37**, 97.
Moeller, T., and N. Vogel, 1951, J. Am. Chem. Soc. **73**, 4481.
Morss, L.R., 1971, J. Phys. Chem. **75**, 392.
Morss, L.R., 1976, Chem. Rev. **76**, 827.
Mulqueen, P., J.M. Tingey and W.DeW. Horrocks Jr, 1985, Biochemistry **24**, 6639.
Nair, G.M., K. Chander and J.K. Joshi, 1982, Radiochim. Acta **30**, 37.
Nakamura, K., and K. Kawamura, 1971, Bull. Chem. Soc. Jpn. **44**, 330.
Narten, A.H., and R.L. Hahn, 1982, Science **217**, 1249.
Nathans, M.W., and W.W. Wendlandt, 1962, J. Inorg. & Nucl. Chem. **24**, 869.
Nielson, G.W., and J.E. Enderby, 1979, Ann. Rep. Progr. Chem., Sect. C, R. Soc. Chem. **76**, 185.
Niinisto, L., and M. Leskela, 1987, in: Handbook on the Physics and Chemistry of Rare Earths, Vol. 9, eds K.A. Gschneidner Jr and L. Eyring (North-Holland, Amsterdam) ch. 59.
Niinisto, L., P. Saikkonen and R. Sonninen, 1982, The Rare Earths in Modern Science and Technology, Vol. 3, eds G.J. McCarthy, H.B. Silber and J.J. Rhyne (Plenum, New York).
Niu, C., and G.R. Choppin, 1987, Inorg. Chim. Acta **131**, 277.
Okada, K., Y. Kaizu, H. Kobayashi, K. Tanaka and F. Marumo, 1985, Mol. Phys. **54**, 1293.
Onstott, E.I., 1985, Inorg. Chem. **24**, 3884.
Onstott, E.I., L.B. Brown and E.J. Peterson, 1984, Inorg. Chem. **23**, 2430.
Ots, H., 1973, Acta Chem. Scand. **27**, 2344.
Padova, J., 1964, J. Chem. Phys. **40**, 691.
Passinsky, A., 1938, Acta Phys. Chim. URSS **8**, 835.
Pisaniello, D.L., and A.E. Merbach, 1982, Helv. Chim. Acta **65**, 573.
Pisaniello, D.L., P.J. Nichols, Y. Ducommun and A.E. Merbach, 1982, Helv. Chim. Acta **65**, 1025.
Pisaniello, D.L., L. Helm, P. Meier and A.E. Merbach, 1983, J. Am. Chem. Soc. **105**, 4528.
Rard, J.A., 1985, Chem. Revs. **85**, 555.
Rard, J.A., and D.G. Miller, 1979, J. Chem. Eng. Data **24**, 348.
Rard, J.A., and F.H. Spedding, 1981, J. Chem. Eng. Data **26**, 391.
Rard, J.A., H.O. Weber and F.H. Spedding, 1977a, J. Chem. Eng. Data **22**, 187.
Rard, J.A., L.E. Shiers, D.J. Heiser and F.H. Spedding, 1977b, J. Chem. Eng. Data **22**, 337.
Reidler, J., and H.B. Silber, 1974a, J. Inorg. & Nucl. Chem. **36**, 175.
Reidler, J., and H.B. Silber, 1974b, J. Phys. Chem. **78**, 424.
Reilley, C.N., B.W. Good and R.A. Allendoerfer, 1976, Anal. Chem. **48**, 1446.
Reuben, J., 1975, J. Phys. Chem. **79**, 2154.
Reuben, J., and D. Fiat, 1969, J. Chem. Phys. **31**, 4909.
Rhee, M.J., D.R. Sudnick, V.K. Arkle and W.DeW. Horrocks Jr, 1981, Biochem. **20**, 3328.
Rinaldi, P., G.C. Levy and G.R. Choppin, 1979, J. Am. Chem. Soc. **101**, 1350.
Rogers, R.D., 1987a, Inorg. Chim. Acta **133**, 175.
Rogers, R.D., 1987b, Inorg. Chim. Acta **133**, 347.
Rogers, R.D., and L.K. Kurihara, 1986, Inorg. Chim. Acta **116**, 171.
Rogers, R.D., and L.K. Kurihara, 1987a, Inorg. Chim. Acta **130**, 131.
Rogers, R.D., and L.K. Kurihara, 1987b, Inorg. Chem. **26**, 1498.
Rogers, R.D., L.K. Kurihara and E.J. Voss, 1987, Inorg. Chem. **26**, 2360.
Rosseinsky, D.R., 1965, Chem. Rev. **65**, 467.
Sadolin, E., 1927, Z. Anorg. Chem. **160**, 133.
Saito, A., 1988, Thermochim. Acta **124**, 217.
Schmidt, K.H., J.C. Sullivan, S. Gordon and R.C. Thompson, 1978, Inorg. & Nucl. Chem. Lett. **14**, 429.
Shannon, R.D., 1976, Acta Crystallogr. A **32**, 751.
Shcherbakov, V.A., and O.G. Golubovskaya, 1976, Russ. J. Inorg. Chem. **21**, 314.
Silber, H.B., 1974, J. Phys. Chem. **78**, 1940.
Silber, H.B., 1978, in: The Rare Earths in Modern Science and Technology, Vol. 1, eds G.J. McCarthy and J.J. Rhyne (Plenum, New York) p. 149.
Silber, H.B., and G. Bordano, 1979, J. Inorg. & Nucl. Chem. **41**, 1169.

Silber, H.B., and J. Fowler, 1976, J. Phys. Chem. **80**, 1451.

Silber, H.B., and L.U. Kromer, 1980, J. Inorg. & Nucl. Chem. **42**, 103.

Silber, H.B., and T. Mioduski, 1984, Inorg. Chem. **23**, 1577.

Silber, H.B., N. Scheinin, G. Atkinson and J.J. Grecsek, 1972, J. Chem. Soc. Faraday Trans., Ser. I, **68**, 1200.

Silber, H.B., D. Bouler and T. White, 1978, J. Phys. Chem. **82**, 775.

Singh, S., 1944, Z. Kristallogr. **105**, 384.

Sinha, S.P., 1976, Struct. & Bonding **25**, 69.

Smith Jr, L.S., and D.L. Wertz, 1975, J. Am. Chem. Soc. **97**, 2365.

Smith Jr, L.S., and D.L. Wertz, 1977, J. Inorg. & Nucl. Chem. **39**, 95.

Smith, R.M., and A.E. Martell, 1976, Critical Stability Constants, Vol. 4 (Plenum, New York).

Soini, E., and T. Lovgren, 1987, CRC Crit. Rev. Anal. Chem. **18**, 105.

Southwood-Jones, R.V., W.L. Earl, K.E. Newman and A.E. Merbach, 1980, J. Chem. Phys. **73**, 5909.

Spedding, F.H., and K.C. Jones, 1966, J. Phys. Chem. **70**, 2450.

Spedding, F.H., and M.J. Pikal, 1966, J. Phys. Chem. **70**, 2430.

Spedding, F.H., D.A. Csejka and C.W. DeKock, 1966a, J. Phys. Chem. **70**, 2423.

Spedding, F.H., M.J. Pikal and B.O. Ayers, 1966b, J. Phys. Chem. **70**, 2440.

Spedding, F.H., J.A. Rard and V.W. Seager, 1974a, J. Chem. Eng. Data **19**, 373.

Spedding, F.H., D.L. White, L.E. Shiers and J.A. Rard, 1974b, J. Chem. Eng. Data **19**, 369.

Spedding, F.H., L.E. Shiers and J.A. Rard, 1975a, J. Chem. Eng. Data **20**, 66.

Spedding, F.H., L.E. Shiers, M.A. Brown, J.L. Derer, D.L. Swanson and A. Habenschuss, 1975b, J. Chem. Eng. Data **20**, 81.

Spedding, F.H., J.L. Derer, M.A. Mohs and J.A. Rard, 1976, J. Chem. Data **21**, 474.

Spedding, F.H., J.A. Rard and A. Habenschuss, 1977a, J. Phys. Chem. **81**, 1069.

Spedding, F.H., C.W. DeKock, C.W. Pepple and A. Habenschuss, 1977b, J. Chem. Eng. Data **22**, 58.

Spedding, F.H., M.A. Mohs, J.L. Derer and A. Habenschuss, 1977c, J. Chem. Eng. Data **22**, 142.

Steele, M.L., and D.L. Wertz, 1976, J. Am. Chem. Soc. **98**, 4424.

Steele, M.L., and D.L. Wertz, 1977, Inorg. Chem. **16**, 1225.

Stein, G., and E. Wurzberg, 1975, J. Chem. Phys. **62**, 208.

Suzuki, Y., T. Nagayama, M. Sekine, A. Mizuno and K. Yamaguchi, 1986, J. Less-Common Met. **126**, 351.

Svoronos, D.R., E. Antic-Fidancev, M. Lemaitre-Blaise and P. Caro, 1981, Nouv. J. Chim. **5**, 547.

Swift, T.J., and R.E. Connick, 1962, J. Chem. Phys. **37**, 307.

Takahashi, F., and N.C. Li, 1966, J. Am. Chem. Soc. **88**, 1117.

Tanaka, F., and S. Yamashita, 1984, Inorg. Chem. **23**, 2044.

Tanaka, F., Y. Kawasaki and S. Yamashita, 1988, J. Chem. Soc. Faraday Trans. I, **84**, 1083.

Taube, H., 1960, J. Am. Chem. Soc. **82**, 524.

Tobias, R.S., and A.B. Garrett, 1958, J. Am. Chem. Soc. **80**, 3532.

Tremaine, P.R., and S. Goldman, 1978, J. Phys. Chem. **82**, 2317.

Usherenko, L.N., and N.A. Skorik, 1972, Russ. J. Inorg. Chem. **17**, 1533.

Vesala, A., and R. Kappi, 1985, Acta Chem. Scand. A **39**, 287.

Wendlandt, W.W., 1958, J. Inorg. & Nucl. Chem. **7**, 51.

Wendlandt, W.W., and T.D. George, 1961, J. Inorg. & Nucl. Chem. **19**, 245.

Whan, R.E., and G.A. Crosby, 1962, J. Mol. Spectrosc. **8**, 315.

Wheelwright, E.J., F.H. Spedding and G. Schwarzenbach, 1953, J. Am. Chem. Soc. **75**, 4196.

Yamauchi, S., H. Kanno and Y. Akama, 1988, Chem. Phys. Lett. **151**, 315.

Zholdakov, A.A., L.N. Lugina and N.K. Davidenko, 1971, Russ. J. Inorg. Chem. **16**, 1265.

Zimmermann, R.J., and W.E. Brittin, 1957, J. Phys. Chem. **61**, 1328.

Ziv, D.M., and A.I. Shestakova, 1965, Radiokhim. **7**, 175.

Handbook on the Physics and Chemistry of Rare Earths, Vol. 15
edited by K.A. Gschneidner, Jr. and L. Eyring

Chapter 104

MACROCYCLIC COMPLEXES OF THE LANTHANIDE(III), YTTRIUM(III) AND DIOXOURANIUM(VI) IONS FROM METAL-TEMPLATED SYNTHESES

LIDIA M. VALLARINO
Department of Chemistry, Virginia Commonwealth University, Richmond, VA 23284, USA

Contents

Introduction and history

The first indication that the elements of the 4f and 5f blocks can act as effective templates for the synthesis of otherwise inaccessible ligands came from the discovery by Day et al. (1975) that the bluish black uranyl complex of stoichiometry $UO_2(C_8H_4N_2)_5$ contains a dioxouranium(VI) ion bound into the cavity of a highly conjugated five-nitrogen macrocyclic ligand. This complex, formed by the condensation of 1,2-dicyanobenzene in the presence of a uranyl salt, had been known for over 20 years but was considered to be an ordinary phthalocyanine complex including an extra molecule of 1,2-dicyanobenzene (Frigerio 1962, Bloor et al. 1964, Kirin et al. 1967); a patent for the synthesis of this "phthalocyanine" complex had actually been obtained (Frigerio 1962).

Four years after the report of the structure of this uranyl superphthalocyanine, Hart and co-workers (Backer-Dirks et al. 1979) showed that the nitrates of the two larger lanthanide(III) ions could promote the cyclic (2 + 2) condensation of 2,6-diacetylpyridine and 1,2-diaminoethane, yielding complexes of a six-nitrogen ligand $C_{22}H_{26}N_6$. The macrocyclic structure of the La(III) complex of this ligand was clearly established by X-ray crystallography. Even though these authors encountered difficulties in extending the metal-templated synthesis to the lanthanides smaller than Ce(III), their work represents a landmark as it reports the first deliberate synthesis of an otherwise inaccessible macrocyclic complex of the f-block elements.

It might be interesting to speculate why it took over 25 years for chemists to extend to the f-block elements the rather simple synthetic approach that had been so extensively and so successfully employed for the transition elements of the 3d-block since the initial report of Thompson and Bush (1964). Perhaps, research in this direction was initially discouraged by consideration of the well-known chemical features of the lanthanide(III) ions. These ions have crystal field stabilization energies that are negligible compared to those of the transition elements; they form extremely labile complexes; they have large ionic radii that favor high coordination numbers of 8 to 12; they exhibit a variety of easily interchangeable geometries – all features that would not be expected to favor the role of these metals as templates in macrocyclic syntheses. Most likely, however, sustained efforts to develop this new area of lanthanide and actinide chemistry were stimulated in recent years by the interest in the unique properties of certain f-block elements. For example, two of the lanthanides, Eu(III) and Tb(III), exhibit narrow-band luminescence in solution at room temperature. Certain Eu(III) complexes have been reported to give laser action in solution; the Nd(III) ion also provides laser action when present in aprotic solvents. The compounds of Tb(III), Dy(III), Er(III) and Ho(III) have the highest known magnetic moments, in the range of $9\,\mu_B$–$11\,\mu_B$ at room temperature. Several lanthanide(III) ions are useful as shift reagents in NMR (nuclear magnetic resonance) spectroscopy and one, Gd(III), has been shown to be an effective contrast agent for in vivo NMR imaging. The effective sequestering of the radioactive actinides continues to be a topic of vital interest and a radioactive isotope of the lanthanide-like yttrium(III) ion, ^{90}Y, has been suggested for use as an auto-imaging biological marker.

This survey covers the literature on the macrocyclic complexes of metal-template origin for the lanthanide elements, the yttrium(III) ion and uranium; certain aspects of this area have been recently reviewed (Bombieri 1987, Fenton and Vigato 1988)*. The survey is divided into two main sections, one dealing with the complexes of the lanthanides and yttrium, the other with the complexes of uranium; each section is further subdivided according to the type of ligand. A comprehensive listing of the complexes and their chief properties is given in tables. Schematic structural formulas of the macrocyclic ligands, I through XXXVII, are shown at the end of this chapter**.

* The lanthanide complexes of phthalocyanines, which have attracted intense interest owing to their electrical properties, are not included in this survey because they do not possess a true metal-macrocyclic structure. Even the smallest lanthanide, Lu(III), cannot be accommodated inside the four-N-donor cavity of the phthalocyanine system. In these complexes, the metal ion is always considerably displaced from the plane of the ligand and sandwich-type structures are common.

** Macrocycles that represent variations of the same structural type are numbered sequentially; this occasionally results in an inversion of the numbering sequence within the text.

1. Complexes of the lanthanide(III) and yttrium(III) ions

1.1. *Syntheses and structures*

1.1.1. (2 + 2) *ligands with pyridine head-units and aliphatic diimine side-chains*

1.1.1.1. *Compounds obtained from* 2,6-*diacetylpyridine.* The compounds of this class are listed in table 1, together with selected physical properties.

The formation of the La(III) and Ce(III) trinitrato complexes of ligand I from the Schiff-base condensation of 2,6-diacetylpyridine and 1,2-diaminoethane in the presence of the metal nitrates (2:2:1 mole ratio in refluxing methanol) was reported by Backer-Dirks et al. (1979), who also established the structure of the resulting La(III) complex, $[La(NO_3)_3(I)]$. In this compound the La(III) ion is 12-coordinate, being linked to the six N-donor atoms of a symmetric 18-member macrocyclic ligand and to three bidentate chelating nitrates (fig. 1). The macrocyclic ligand itself departs considerably from planarity and consists of two approximately flat sections hinged at the flexible $=N-CH_2-CH_2-N=$ side chains to give a "folded butterfly" configuration. As in the case of the $[La(NO_3)_3(18\text{-crown-}6)]$ complex (Backer-Dirks et al. 1980), two of the nitrates are situated on the convex side of the folded macrocycle, while the other is on the opposite side. Selected structural data for $[La(NO_3)_3(I)]$ are given in table 2. The interest of this new class of lanthanide complexes was immediately recognized, but at first their metal-templated synthesis appeared to have limited applicability, as it could not be successfully extended beyond cerium (Backer-Dirks et al. 1979). However, the use of lanthanide salts other than nitrates has since then allowed the metal-templated synthesis of macrocyclic complexes of ligand I with all the lanthanide(III) ions (except synthetically obtained radioactive promethium) as well as with yttrium(III).

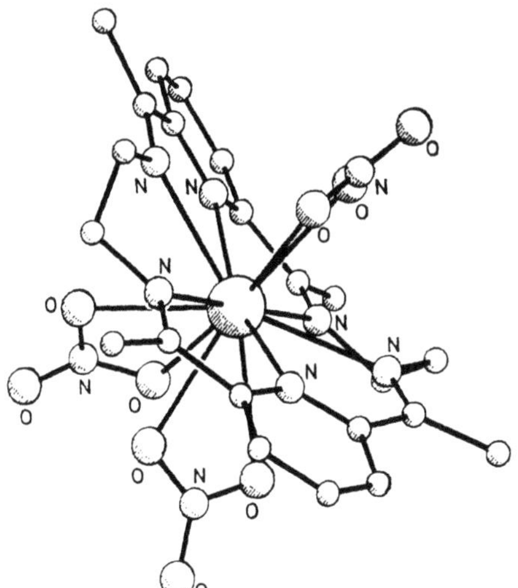

Fig. 1. The molecular structure of $[La(NO_3)_3(I)]$ (Backer-Dirks et al. 1979, Arif et al. 1987).

TABLE 1

(2 + 2) Macrocyclic complexes of the lanthanide(III) and yttrium(III) ions with pyridine head-units and aliphatic side-chains

Compound[a]	Color	Other data[j]	Ref.[b]
Ligand I = $C_{22}H_{26}N_6$			
*$[La(NO_3)_3(I)]$	Pale pinkish brown	dec. 240–245°C	[1, 2]
		X-ray structure: fig. 1, table 2	[1, 2]
		^1H NMR: table 20; ^{13}C NMR: table 19	[2]
		UV $\lambda_{max}(H_2O)$: 297 nm	[2]
		Λ_M: 328 $\Omega^{-1}\,cm^2\,mol^{-1}$ (10^{-3} M, H_2O); IR[c]	[2]
$[Ce(NO_3)_3(I)]$	Deep yellow	dec. 240–245°C	[2]
		^1H NMR: table 20; ^{13}C NMR: table 19	[1, 2]
		UV $\lambda_{max}(H_2O)$: 242 nm; IR[c]	[2]
		Λ_M: 341 $\Omega^{-1}\,cm^2\,mol^{-1}$ (10^{-3} M, H_2O)	
$[La(NO_3)_2(H_2O)(I)](NO_3)\cdot H_2O$	Colorless	–	[2]
*$[Ce(NO_3)_2(H_2O)(I)](NO_3)\cdot H_2O$	Orange	X-ray structure: fig. 3, table 2; IR[c]	[2]
$[Pr(NO_3)_2(H_2O)(I)](NO_3)\cdot H_2O$	Pale green	dec. 300°C; TGA; IR[c]; ^1H NMR: table 20; ^{13}C NMR: table 19	[2]
*$[\{Nd(NO_3)(H_2O)_2(I)\}_2](NO_3)(ClO_4)_3\cdot 4H_2O$	Lilac	X-ray structure: fig. 4, table 2; IR[c]; ^1H NMR: table 20	[2]
$[\{Sm(NO_3)(H_2O)_2(I)\}_2](NO_3)(ClO_4)_3\cdot 4H_2O$	Cream	IR[c]; ^1H NMR	[2]
$[\{Eu(NO_3)(H_2O)_2(I)\}_2](NO_3)(ClO_4)_3\cdot 4H_2O$	Colorless	dec. 300°C; TGA; IR[c]; ^1H NMR	[2]
$[\{Gd(NO_3)(H_2O)_2(I)\}_2](NO_3)(ClO_4)_3\cdot 4H_2O$	Colorless	IR[c]; ^1H NMR	[2]
$[\{Tb(NO_3)(H_2O)_2(I)\}_2](NO_3)(ClO_4)_3\cdot 4H_2O$	Colorless	IR[c]; ^1H NMR	[2]
$[\{Dy(NO_3)(H_2O)_2(I)\}_2](NO_3)(ClO_4)_3\cdot 4H_2O$	Colorless	IR[c]; ^1H NMR	[2]
$[\{Ho(NO_3)(H_2O)_2(I)\}_2](NO_3)(ClO_4)_3\cdot 4H_2O$	Pale yellow	IR[c]; ^1H NMR	[2]
$[\{Er(NO_3)(H_2O)_2(I)\}_2](NO_3)(ClO_4)_3\cdot 4H_2O$	Pale pink	IR[c]; ^1H NMR	[2]
$Y(I)Br_3\cdot 5H_2O$	Colorless	dec. above 270°C; TGA	[3]
		IR(cm^{-1}): 3500–3000 (vs, br) $\nu(OH_2)$, 1660 (vs) ν(C=N, imine), 1605 (vs) ν(C=N, py), 1430 (m), 1385 (vs), 1330 (vw), 1303 (w), 1268 (s), 1230 (sh), 1190 (s), 1155 (vw), 1125 (vw), 1065 (vw), 1003 (m), 975 (vw), 930 (vw), 880 (vw), 808 (s), 735 (m), 710 (vs), 640 (vw), 600–400 (w, vbr, with peak at 580) $\rho(OH_2)$	[3]
$Y(I)(BF_4)_3\cdot 2H_2O$	Colorless	dec. above 250°C; TGA	[3]
		IR(cm^{-1}): 1050 (vs, br) $\nu_3(BF_4^-)$, 510 (m) $\nu_4(BF_4^-)$; other absorptions as in tribromide analog	[3]

$Y(I)(CH_3COO)_2Cl \cdot 4H_2O \cdot CH_3OH$	Colorless	dec. above 250°C; TGA; ^1H NMR: table 20	[3]
		IR(cm^{-1}): 1550 (vs) $\nu_{as}(COO)$, 1440 (vs) $\nu_s(COO)$, 655 (s) $\delta(COO)$; other absorptions as in tribromide analog	[3]
$Y(I)(CH_3COO)_{0.5}Cl_{2.5} \cdot 5H_2O$	Colorless	dec. above 250°C; TGA; IR: similar to diacetate chloride except for less intense –COO absorptions.	[3]
*$[Y(CH_3COO)_2(I)][Y(CH_3COO)(H_2O)(I)]$-$(CH_3COO)(ClO_4)_2$	Cream	dec. *ca.* 250°C	[3]
		X-ray structure: fig. 6, table 3	[3]
		IR(cm^{-1}): 1060–1100 (vs, br) $\nu_3(ClO_4^-)$, 630 (m) $\nu_4(ClO_4^-)$; other absorptions as in diacetate chloride analog	[3]
$Y(I)(HOC_6H_4COO)_2Cl \cdot 4H_2O$	Colorless	dec. above 250°C; TGA	[3]
$Y(I)(H_2NC_6H_4COO)_2Cl \cdot 4H_2O$	Colorless	dec. above 250°C; TGA	[3]
$Y(I)(OC_4H_3COO)_2Cl \cdot 4H_2O$	Colorless	dec. above 250°C; TGA	[3]
$La(I)(CH_3COO)_2Cl \cdot nH_2O$	Cream	dec. 220–240°C; IR[d]	[4]
		^1H NMR: table 20; ^{13}C NMR, table 19	[4]
$Ce(I)(CH_3COO)_2Cl \cdot nH_2O$	Light rose	dec. *ca.* 240°C; IR[d]	[4]
$Pr(I)(CH_3COO)_2Cl \cdot nH_2O$	Cream	dec. *ca.* 240°C; IR[d]; ^1H NMR: table 20; ^{13}C NMR: table 19	[4]
*$[Nd(CH_3COO)_2(I)]Cl \cdot 4H_2O$	Pale lavender	dec. *ca.* 240°C; TGA; IR[d]	[4]
		^1H NMR: table 20; ^{13}C NMR: table 19	[5a]
		X-ray structure: table 3	[5a]
$Sm(I)(CH_3COO)_2Cl \cdot nH_2O$	Cream	IR[d]	[4]
*$[Eu(CH_3COO)_2(I)]Cl \cdot 4H_2O$	Cream	dec. *ca.* 240°C; TGA; IR[d]	[4]
		^1H NMR: table 20; ^{13}C NMR: table 19	[5a]
		X-ray structure: fig. 2a, table 3	[5a]
		UV $\lambda_{max}(H_2O)$; luminescence spectrum (emission, excitation, lifetime)	[6]
		Luminescence spectrum	[7]
*$[Eu(CH_3COO)_2(I)](CH_3COO) \cdot 9H_2O$		X-ray structure: fig. 2c, table 3	[5b]
*$[Gd(CH_3COO)_2(I)]Cl \cdot 4H_2O$	Cream	IR[d]; TGA	[4]
		X-ray structure: fig. 2b, table 3; NMR relaxivity	[7]
$Tb(I)(CH_3COO)_2Cl \cdot nH_2O$	Cream	IR[d]	[4]
$Dy(I)(CH_3COO)_2Cl \cdot nH_2O$	Cream	IR[d]	[4]
$Ho(I)(CH_3COO)_2Cl \cdot nH_2O$	Cream	IR[d]	[4]
$Er(I)(CH_3COO)_2Cl \cdot nH_2O$	Light peach	IR[d]	[4]
$Tm(I)(CH_3COO)_2Cl \cdot nH_2O$	Cream	IR[d]	[4]
$Yb(I)(CH_3COO)_2Cl \cdot nH_2O$	Cream	IR[d]	[4]

TABLE 1 (cont'd)

Compound[a]	Color	Other data[j]	Ref.[b]
$Lu(I)(CH_3COO)_2Cl \cdot nH_2O$	Cream	IR[d]; ^1H NMR; ^{13}C NMR	[4]
*$[Lu(CH_3COO)(H_2O)(I)][Lu(CH_3COO)(CH_3OH)(I)](OH)_2(ClO_4)_2 \cdot CH_3OH$	Light amber	dec. above 210°C; TGA	[8]
		X-ray structure: fig. 7, table 3	[8]
		^1H NMR: table 20; ^{13}C NMR: table 19	[8]
		IR(cm^{-1}): 3370–3000 (m, br) $\nu(OH)$, 1090 (s, br) $\nu_3(ClO_4^-)$, 620 (m) $\nu_4(ClO_4^-)$; other absorptions similar to La diacetate chloride complex	[8]
$La(I)(OH)(ClO_4)_2$	Cream	dec. *ca.* 250°C; IR(cm^{-1}): *ca.* 3600 (s, sharp) $\nu(OH)$, 1120, 1100, 1045 (partly resolved, all very strong) $\nu_3(ClO_4^-)$, 620 (m, split) $\nu_4(ClO_4^-)$, 445 (s) $\delta(LaOH)$, 365 (w) $\nu(LaOH)$; other absorptions similar to Y tribromide complex	[4]
$Ce(I)(OH)(ClO_4)_2$	Rose red	dec. *ca.* 250°C; μ_{eff}: 2.3μ_B; IR[e]	[4]
$Pr(I)(OH)(ClO_4)_2$	Cream	dec. *ca.* 250°C; μ_{eff}: 3.3μ_B; IR[e]	[4]
$Nd(I)(OH)(ClO_4)_2$	Cream	dec. *ca.* 250°C; μ_{eff}: 3.2μ_B; IR[e]	[4]
$Sm(I)(OH)(ClO_4)_2$	Cream	dec. *ca.* 250°C; μ_{eff}: 1.7μ_B; IR[e]	[4]
$Eu(I)(OH)(ClO_4)_2 \cdot H_2O$	Cream	dec. *ca.* 250°C; IR[e]; TGA	[4]
$Gd(I)(OH)(ClO_4)_2 \cdot 2H_2O$	Cream	dec. *ca.* 250°C; μ_{eff}: 7.2μ_B; IR[e]	[4]
$Tb(I)(OH)(ClO_4)_2 \cdot 2H_2O$	Cream	dec. *ca.* 250°C; IR[e]	[4]
$Ho(I)(OH)(ClO_4)_2$	Cream	dec. *ca.* 250°C; IR[e]	[4]
$Er(I)(OH)(ClO_4)_2$	Cream	dec. *ca.* 250°C; μ_{eff} = 9.5μ_B; IR[e]	[4]
$Yb(I)(OH)(ClO_4)_2 \cdot nH_2O$	Cream	dec. *ca.* 250°C; IR[e]	[4]
*$[Y(NCS)_3(I)]$	Colorless	dec. above 250°C; TGA; X-ray structure: table 4	[9]
		IR spectrum: nearly identical to that of $Y(I)Br_3 \cdot 5H_2O$, except H_2O absorptions are absent and NCS^- absorptions appear at 2062 cm^{-1} [$\nu(C{\equiv}N)$] and 813 cm^{-1} [$\nu(C{-}S)$]	[9]
*$[Eu(NCS)_3(I)]$	Colorless	dec. above 250°C; TGA; X-ray structure: fig. 5a, b, table 4	[9]
		Luminescence emission spectrum; IR spectrum identical to that of the Y analog	
$[La(H_2O)_2(I)](ClO_4)_3$[f]	Yellowish orange	TGA; UV–VIS; IR; ^1H NMR	[10]
$Ln(I)(ClO_4)_3 \cdot 2(NH_2CH_2CH_2NH_2) \cdot 2H_2O$ (Ln = La–Tb)		TGA; DTA; IR; conductance	[11]

Ligand II = $\mathbf{C_{18}H_{18}N_6}$			
La(II)$(NO_3)_3$	Pale pink	IR(cm^{-1}): 3450 (br) ν(OH), 1625 (s) ν(C=N, imine), 1590 (s) ν(C=N, py); 1380 (m), 1035 (s) and 812 (m): NO_3^- absorptions; others: 1440, 1320 and 1160	[12]
		^{1}H NMR: table 20; ^{13}C NMR: table 19	[12]
Ce(II)$(NO_3)_3$	Yellow	IR[g]	[12]
Pr(II)$(NO_3)_3$	White	IR[g]	[12]
Eu(II)$(NO_3)_3$	Milky	IR[g]	[12]
*[Sm(NO_3)(OH)(H_2O)(II)]$(NO_3)(CH_3OH)_2$	Off-white	IR[g]; X-ray structure: fig. 8, table 5	[13]
Y(II)$(CH_3COO)_2Cl\cdot 4H_2O$	Off-white	IR(cm^{-1}): 3480–3200 (vs, vbr) $\nu(OH_2)$, 3068 (vw), 2922 (vw), 2862 (vw), 1662 (vs) and 1594 (vs) ν(C=N), 1540 (vs) ν_{as}(COO), 1456 (vs) ν_s(COO), 1424 (sh), 1385 (w), 1338 (w), 1277 (m), 1253 (sh), 1162 (m), 1103 (w), 1077 (w), 1045 (m–s), 1006 (m–s), 983 (vw), 956 (vw), 936 (vw), 824 (m), 810 (m), 748 (w), 680 (s) δ(COO), 645 (w), 617 (vw), 602 (vw)	[5]
Ligand III = $\mathbf{C_{18}H_{20}N_6O}$			
Nd(III)$(NO_3)_3\cdot H_2O$	White	IR(cm^{-1}): 3450 (br) ν(OH), 3200 (sh) ν(NH), 1660 (s) ν(C=N, imine), 1598 (s) ν(C=N, py); 1380 (m), 1078 (s) and 820 (m): NO_3^- absorptions; others: 1495, 1270 and 1160	[12]
Sm(III)$(NO_3)_3\cdot H_2O$	White	IR[h]	[12]
Gd(III)$(NO_3)_3\cdot H_2O$	White	IR[h]	[12]
Tb(III)$(NO_3)_3\cdot 2H_2O$	White	IR[h]	[12]
Dy(III)$(NO_3)_3\cdot H_2O$	White	IR[h]	[12]
Ho(III)$(NO_3)_3\cdot 2H_2O$	White	IR[h]	[12]
Tm(III)$(NO_3)_3\cdot 2H_2O$	White	IR[h]	[12]
Lu(III)$(NO_3)_3\cdot 4H_2O$	White	IR[h]; ^{1}H NMR; ^{13}C NMR (see text)	[12]
Ligand IV = $\mathbf{C_{20}H_{22}N_6}$			
La(IV)$(NO_3)_3\cdot 2H_2O$	Pale pink	MS: highest peak 364 (metal-free tetraimine)	[12]
		^{1}H NMR, δ(ppm/TMS), DMSO-d_6: 8.90 (d, 4H, H–C=N); 8.41 (t, 2H, γH-py); 8.18 (d, 4H, βH-py); 4.28 (br, m, 4H, CH_2); 3,80 [br, t, 2H, CH(CH_3)]; 1.36 (br, s, 6H, CH_3)	[12]
		IR(cm^{-1}): 3450 (br) $\nu(OH_2)$, 1650 (s) ν(C=N, imine), 1592 (s) ν(C=N, py); 1383 (m), 1042 (m) and 815 (s): NO_3^- absorptions; others: 1460, 1320, 1165 and 736	[12]
Ce(IV)$(NO_3)_3\cdot 3H_2O$	Yellow	IR: similar to La analog	[12]

TABLE 1 (cont'd)

Compound[a]	Color	Other data[j]	Ref.[b]
$Pr(IV)(NO_3)_3 \cdot 2H_2O$	White	IR: similar to La analog	[12]
$Nd(IV)(NO_3)_3 \cdot 3H_2O$[i]	Pale pink	MS: highest peak 364 (metal-free tetraimine)	[12]
		IR(cm^{-1}): 3450 (br) $\nu(OH_2)$, 3240 (br) ν(NH), 1665 (s) ν(C=N, imine), 1590 (s) ν(C=N, py); 1384 (m), 1050 (s) and 820 (m): NO_3^- absorptions; others: 1490, 1278, 1162 and 738	[12]
$Sm(IV)(NO_3)_3 \cdot 2H_2O$[i]	Milky	IR[h]	[12]
$Eu(IV)(NO_3)_3 \cdot H_2O$[i]	White	IR[h]	[12]
$Gd(IV)(NO_3)_3 \cdot 2H_2O$[i]	White	IR[h]	[12]
$Tb(IV)(NO_3)_3 \cdot 3H_2O$[i]	Milky	IR[h]	[12]
$Dy(IV)(NO_3)_3 \cdot 2H_2O$[i]	White	IR[h]	[12]
$Ho(IV)(NO_3)_3 \cdot 2H_2O$[i]	Pale gray	IR[h]	[12]
$Er(IV)(NO_3)_3 \cdot 2H_2O$[i]	Milky	IR[h]	[12]
$Tm(IV)(NO_3)_3 \cdot 3H_2O$[i]	Milky	IR[h]	[12]
$Yb(IV)(NO_3)_3 \cdot 2H_2O$[i]	White	IR[h]	[12]
$Lu(IV)(NO_3)_3 \cdot 2H_2O$[i]	White	IR[h]	[12]
Ligand V = $C_{20}H_{22}N_6$			
$La(V)(NO_3)_3$	White	MS: highest mass 346 (metal-free tetraimine)	[12]
		^1H NMR, δ(ppm/TMS), DMSO-d_6: 8.92 [s, 4H, H–C=N]; 8.45 [t, 2H, γH-py]; 8.18 [d, 4H, βH-py]; 3.96 [m, 8H, N–CH_2]; 3.78 [m, 4H, –CH_2–]	[12]
		IR(cm^{-1}): 3450 (br) $\nu(OH_2)$, 1650 (s) ν(C= N, imine), 1590 (s) ν(C=N, py); 1384 (m), 1098 (s) and 820 (s): NO_3^- absorptions; others: 1430, 1320, 1170 and 732	[12]
$Ce(V)(NO_3)_3 \cdot 2H_2O$	Orange	IR[g]	[12]
$Pr(V)(NO_3)_3 \cdot 2H_2O$	Milky	IR[g]	[12]
$Nd(V)(NO_3)_3 \cdot H_2O$	White	IR[g]	[12]
$Sm(V)(NO_3)_3 \cdot 2H_2O$	White	IR[g]	[12]
$Eu(V)(NO_3)_3 \cdot 2H_2O$	White	IR[g]	[12]
$Gd(V)(NO_3)_3 \cdot 2H_2O$	White	IR[g]	[12]
$Tb(V)(NO_3)_3 \cdot 4H_2O$	Milky	IR[g]	[12]
$Dy(V)(NO_3)_3 \cdot 2H_2O$	White	IR[g]	[12]
$Ho(V)(NO_3)_3 \cdot H_2O$	Pale gray	IR[g]	[12]

$Er(V)(NO_3)_3 \cdot H_2O$	White	IR[g]	[12]
$Tm(V)(NO_3)_3 \cdot 3H_2O$	Pale gray	IR[g]	[12]
$Yb(V)(NO_3)_3 \cdot 3H_2O$	White	IR[g]	[12]
$Lu(V)(NO_3)_3 \cdot H_2O$	White	IR[g]	[12]

[a] Coordination formulas (in square brackets, with negative ligands immediately following the metal symbol) are used when given in the reference. However, in the absence of evidence from X-ray analysis or other unambiguous data, coordination formulas for macrocyclic metal complexes of template-synthesis origin should be considered as tentative. Compounds for which the structure has been determined by X-ray analysis are marked with an asterisk. Empirical formulas are written with the macrocyclic ligand immediately following the metal symbol.

[b] References: [1] Backer-Dirks et al. (1979); [2] Arif et al. (1987); [3] Bombieri et al. (1989b); [4] De Cola et al. (1986); [5a] Benetollo et al. (1990a); [5b] Bombieri et al. (1989c); [6] Sabbatini et al. (1987); [7] Smith et al. (1989); [8] Bombieri et al. (1986); [9] Bombieri et al. (1989a); [10] Radecka-Paryzek (1980); [11] Wang and Miao (1984); [12] Abid and Fenton (1984b); [13] Abid et al. (1984).

[c] Absorptions (cm^{-1}) present in all complexes of this class are 1668–1630 (s) ν(C=N, imine) 1588–1585 (s) ν(C=N, py); 1450–1411 and 1340–1293: coordinated NO_3^-. Additional absorptions present in hydrated nitrate complexes are 1383, 1371 and 1352 (La), and 1378 (Sm): ionic NO_3^-; 3530, 3500: lattice water, and 3350–3260: coordinated water (in dihydrates). Additional absorptions present in perchlorate complexes are 1080 (vs) $\nu_3(ClO_4^-)$ and 630 (m) $\nu_4(ClO_4^-)$.

[d] IR spectrum very similar to that of $Y(I)(CH_3COO)_2Cl \cdot 4H_2O \cdot CH_3OH$.

[e] IR spectrum very similar to that of $La(I)(OH)(ClO_4)_2$.

[f] Synthesis could not be reproduced (De Cola et al. 1986).

[g] Spectrum very similar to that of La analog.

[h] Spectrum very similar to that of Nd analog.

[i] It is suggested that the ligand in this complex may have a carbinolamine structure similar to that of III (Abid and Fenton 1984b).

[j] Abbreviations used:

br	broad	TMS	tetramethylsilane
d	doublet	UV	ultraviolet
dec.	decomposes	vbr	very broad
DTA	differential thermal analysis	VIS	visible
FAB	fast-ion-bombardment mode	vs	very strong
IR	infrared	vvs	very very strong
m	medium	vvw	very very weak
MS	mass spectrometry	vw	very weak
NMR	nuclear magnetic resonance	w	weak
py	pyridine	δ	in NMR: chemical shift; in IR: deformation mode
s	strong	ν	in IR: stretching mode
sh	shoulder	ν_{as}	in IR: antisymmetric stretching mode
t	triplet	ν_s	in IR: symmetric stretching mode
TGA	thermogravimetric analysis	ρ	in IR: rocking mode

TABLE 2

Selected crystallographic data, bond lengths and fold angles for $[La(NO_3)_3(I)]$, $[Ce(NO_3)_2(H_2O)(I)](NO_3)\cdot H_2O$ and $[\{Nd(NO_3)(H_2O)_2(I)\}_2](NO_3)(ClO_4)_3\cdot 4H_2O$.[a]

Compound	$[La(NO_3)_3(I)]$		$[Ce(NO_3)_2(H_2O)(I)](NO_3)\cdot H_2O$		$[\{Nd(NO_3)(H_2O)_2(I)\}_2](NO_3)(ClO_4)_3\cdot 4H_2O$	
Formula	$C_{22}H_{26}N_9O_9La$		$C_{22}H_{30}N_9O_{11}Ce$		$C_{44}H_{68}N_{15}O_{29}Cl_3Nd_2$	
Space group	$P2_1/c$		$P\bar{1}$		$P\bar{1}$	
Mol. wt.	699.40		736.65		1521.69	
a (Å)	8.623 (1)		8.961 (4)		11.970 (1)	
b (Å)	18.221 (2)		9.079 (2)		14.880 (2)	
c (Å)	17.039 (2)		18.247 (3)		19.678 (3)	
α (deg)			86.12 (1)		72.11 (2)	
β (deg)	99.01 (2)		103.66 (2)		76.58 (3)	
γ (deg)			100.63 (3)		86.55 (2)	
V (Å^3)	2644.2		1417.3		3244.15	
Z	4		2		2	
D_c (g cm^{-3})	1.75		1.82		1.64	
Coordinate bond lengths (Å)	La–N(1)	2.746 (5) py	Ce–N(1)	2.720 (9) imine	N(11)–Nd(1)	2.584 (15) imine
	La–N(2)	2.672 (6) imine	Ce–N(2)	2.726 (5) py	N(13)–Nd(1)	2.614 (12) py
	La–N(3)	2.729 (6) imine	Ce–N(3)	2.719 (6) imine	N(15)–Nd(1)	2.671 (16) imine
	La–N(4)	2.764 (5) py	Ce–O(11)	2.707 (8)	N(12)–Nd(1)	2.660 (15) imine
	La–N(5)	2.704 (6) imine	Ce–O(21)	2.668 (7)	N(14)–Nd(1)	2.573 (14) py
	La–N(6)	2.727 (7) imine	Ce–O(1)	2.569 (5)	N(16)–Nd(1)	2.628 (12) imine
	La–O(71)	2.767 (5)			O(11)–Nd(1)	2.598 (14)
	La–O(72)	2.730 (5)			O(5)–Nd(1)	2.516 (13)
	La–O(81)	2.705 (5)			O(12)–Nd(1)	2.617 (13)
	La–O(82)	2.749 (5)			O(6)–Nd(1)	2.493 (11)
	La–O(91)	2.718 (5)				
	La–O(92)	2.689 (5)				
Fold angle (deg)[b]	153.3		121.0		116.1	

[a] Arif et al. (1987).

[b] Angle formed by the two quasi-planar sections of the macrocycle that contain the pyridine rings.

Mixed diacetate chloride complexes*, $M(I)(CH_3COO)_2Cl \cdot nH_2O$ (M = La, Ce, Pr, Nd, Sm, Eu, Gd, Tb, Dy, Ho, Er, Tm, Yb, Lu; $n = 3$–6) were described by De Cola et al. (1986). The analogous Y(III) derivative, $Y(I)(CH_3COO)_2Cl \cdot 4H_2O \cdot CH_3OH$, was described by Bombieri et al. (1989b). The syntheses proceeded with ease in refluxing anhydrous methanol; after work-up and crystallization, the complexes were isolated in 70–85% yields. The structures of the Nd and Eu diacetate chloride derivatives (Benetollo et al. 1990a) and of the Gd analog (Smith et al. 1989) have been established by X-ray crystallography. These complexes form 10-coordinate species of formula $[M(CH_3COO)_2(I)]Cl \cdot 4H_2O$, in which the two bidentate chelating acetates occupy opposite sides of the macrocycle and are staggered by approximately 90° with respect to one another; the acetate on the concave side is oriented along the macrocycle foldline (fig. 2a, b). Selected structural data for these complexes are given in table 3. A comparison of the data shows that the M–O(acetate) distances decrease from Nd(III) to Gd(III) as expected, following the decrease in metal ionic radii. In contrast, the M–N(pyridine) and M–N(imine) distances for Nd(III) are shorter than the corresponding distances for Eu(III), a fact also reflected in the more compact unit cell of the Nd(III) complex. An analysis of the nonbonded contacts in the crystal lattices of the Nd and Eu analogs does not reveal any features that could account for this anomalous variation in metal–ligand bond lengths. An almost identical 10-coordinate complex cation is also present in the triacetate species, $[Eu(CH_3COO)_2(I)](CH_3COO) \cdot 9H_2O$. The most distinctive feature of this latter complex (Bombieri et al. 1989c) is the presence of a large number of water molecules, arranged to form parallel channels within the crystal lattice (fig. 2c).

The use of a hydrated metal perchlorate as the template was reported to produce a complex formulated as the 8-coordinate $[La(H_2O)_2(I)](ClO_4)_3$ (Radecka-Paryzek 1980), as well as a series of ethylenediamine (en) adducts, $M(I)(ClO_4)_3 \cdot 2en \cdot 2H_2O$ (M = La, Ce, Pr, Nd, Sm, Eu, Gd, Tb) (Wang and Miao 1984). These syntheses could not be reproduced; instead, diperchlorate hydroxide compounds were obtained, $M(I)(OH)(ClO_4)_2 \cdot nH_2O$ (M = La, Ce, Pr, Nd, Sm, Eu, Gd, Tb, Dy, Ho, Er, Yb; $n = 0$–2) (De Cola et al. 1986).

From the diacetate chloride complexes, compounds containing other counterions, e.g. SCN^- (Bombieri et al. 1989a), Cl^-, Br^- (Bombieri et al. 1989b), or ClO_4^- (Bombieri et al. 1986) have been obtained by metathesis in methanol solution. Trinitrate or dinitrate perchlorate complexes of all lanthanide(III) ions except the three smallest ones, Tm(III), Yb(III) and Lu(III), have also been obtained by transmetallation from the barium perchlorate complex $Ba(I)(ClO_4)_2$ and the lanthanide nitrate in aqueous solution at room temperature (Arif et al. 1987). Structures have been determined for several of these compounds.

The Ce(III) trinitrate complex has the coordination formula $[Ce(NO_3)_2(H_2O)(I)](NO_3) \cdot H_2O$ (Arif et al. 1987). In this species the two nitrates, both bidentate and chelating, are situated on opposite sides of the macrocycle, and the

* Here and throughout this chapter, the symbol/formula of the macrocyclic ligand is placed immediately after the symbol of the metal in the empirical formulas of complexes. When the coordination formulas are known, they are written in the conventional manner, with the negative ligands immediately following the metal symbol.

OW2
OW4
CL1
C22
C26
C19
C21
C1
O3
C25
C4
N1
C18
C20
C2
N2
O4
C6
OW3
C3
C5
C7
N6
OW1
C17
EU1
C9
N3
C16
C8
C14
O1
N5
C10
N4
C11
O2
C15
C23
C12
C13
C24
(a)
C10
C9
C11
C7
C8
C12
N4
C14
C6
C13
O1
N3
N5
C24
C15
C5
C23
Gd
O4
O3
C16
C4
O2
N6
N2
C17
C2
N1
C18
C3
C1
C19
C22
C20
C21
(b)

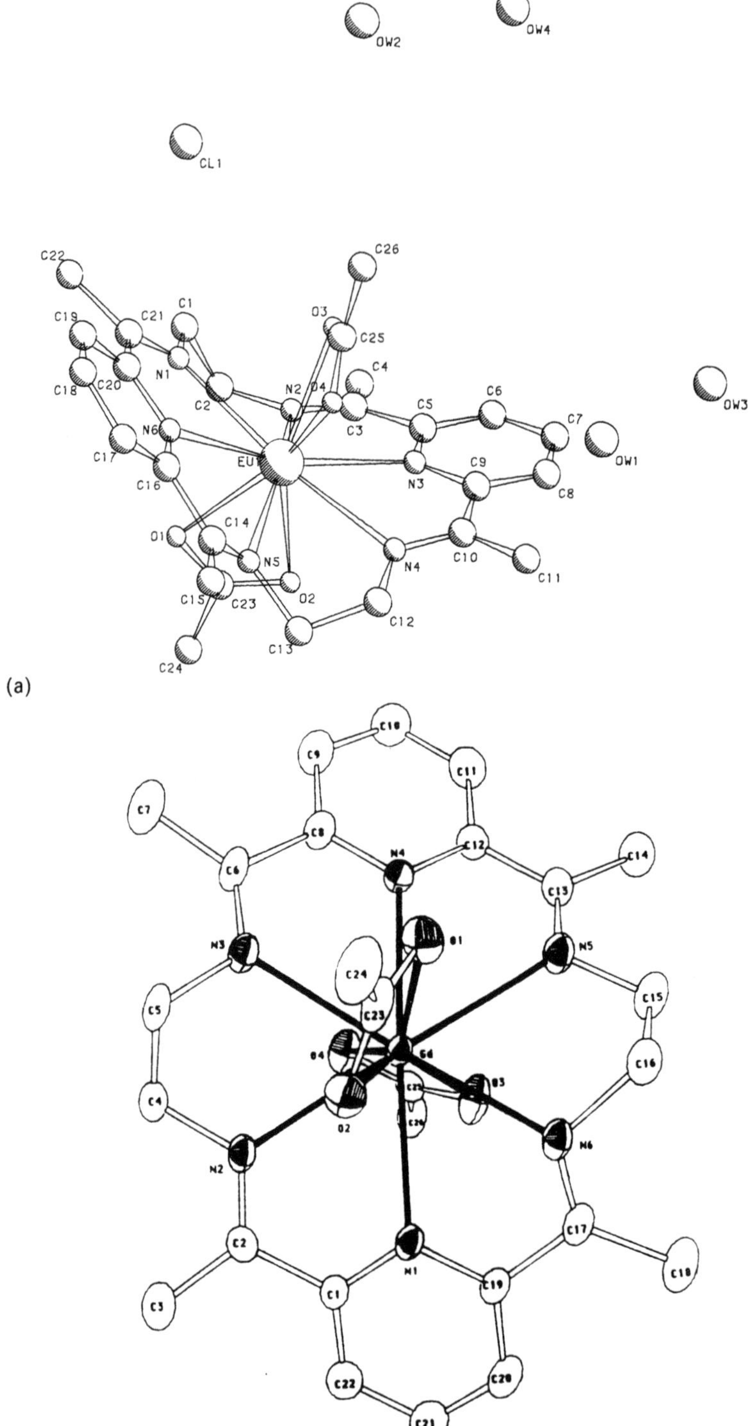

Fig. 2a, b.

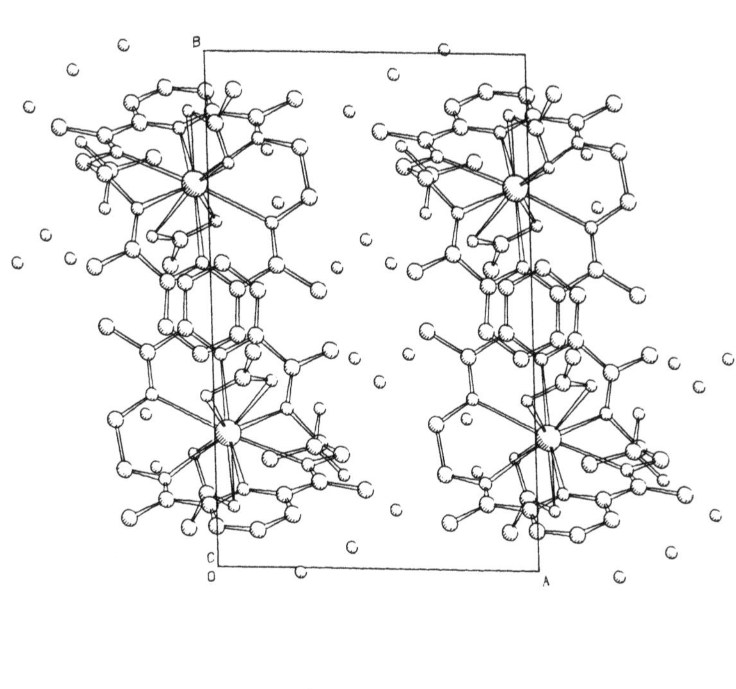

Fig. 2. (a) Structure of $[Eu(CH_3COO)_2(I)]Cl \cdot 4H_2O$ (Benetollo et al. 1990a). (b) Structure of $[Gd(CH_3COO)_2(I)]Cl \cdot 4H_2O$ (Smith et al. 1989). (c) View of the unit cell of $[Eu(CH_3COO)_2(I)]$-$(CH_3COO) \cdot 9H_2O$ (Bombieri et al. 1989c).

coordinated water molecule flanks the nitrate on the convex side (fig. 3). The folding of the macrocycle is much more pronounced than in the La(III) trinitrato complex (the fold angle is 121.0° instead of 153.3°) and there is a considerable difference between the two "wings" of the macrocycle. One wing has an average Ce–N distance of 2.722 Å, while the other is more strongly bonded, with an average Ce–N distance of 2.625 Å. Complexes of the same formula have also been obtained with La(III) and Pr(III), and reportedly are isostructural with the Ce(III) complex, but no individual X-ray data are available. Selected structural data are listed in table 2.

With the nitrate–ligand I combination, but in the presence of perchlorate, the smaller Nd(III) ion gives a 10-coordinate complex, $[\{Nd(NO_3)(H_2O)_2(I)\}]_2$-$(NO_3)(ClO_4)_3 \cdot 4H_2O$ (Arif et al. 1987). In the complex cation, the folding of the macrocycle (fold angle 116.1°) is more pronounced than in the previously cited La(III) and Ce(III) trinitrate derivatives; the two water molecules occupy the convex side of the folded macrocycle, while the bidentate chelating nitrate is situated on the concave side (fig. 4). It was observed that this is a direct demonstration of the smaller steric

TABLE 3
Selected crystallographic data, bond lengths and fold

$[Lu(O_2CCH_3)(H_2O)(I)][Lu(O_2CCH_3)(CH_3OH)(I)](OH)_2(ClO_4)_2(CH_3OH)$[a]

Formula	$C_{25}H_{36}N_6O_{8.5}ClLu$	
Mol. wt.	766.0	
Crystal system	Triclinic	
Space group	$P\bar{1}$	
a (Å)	13.051 (3)	
b (Å)	11.351 (2)	
c (Å)	11.400 (2)	
α (deg)	108.95 (3)	
β (deg)	104.63 (2)	
γ (deg)	100.40 (3)	
V (Å^3)	1481	
Z	2	
D_c (g cm^{-3})	1.72	
Coordinate bond lengths (Å)		
Lu–N(1)	2.555 (5)	py
Lu–N(2)	2.501 (6)	imine
Lu–N(3)	2.608 (7)	imine
Lu–N(4)	2.552 (4)	py
Lu–N(5)	2.471 (6)	imine
Lu–N(6)	2.581 (7)	imine
Lu–O(31)	2.325 (6)	H_2O or CH_3OH
Lu–O(32)	2.368 (5)	bidentate acetate
Lu–O(51)	2.232 (6)	bidentate acetate
Fold angle (deg)	114.4 (1)	

requirements of a bidentate chelating nitrate compared to two (sterically unhindered) monodentate O-donor ligands. Complexes having the same stoichiometry as the Nd(III) complex were also obtained with smaller lanthanides, Sm(III) to Er(III) (Arif et al. 1987), and were considered to have the same structure. Selected structural data for the Nd(III) complex are listed in table 2.

Structures have been determined for the isostructural complexes of formula $[M(NCS)_3(I)]$, in which M is Eu(III) or Y(III) (Bombieri et al. 1989a). The crystals of these complexes contain two distinct molecules, illustrated in fig. 5a, b for the Eu(III) species and referred to as a and b molecules; the packing of these molecules in the unit cell is shown in fig. 5c. Selected structural data for both the Eu(III) and the Y(III) complexes are given in table 4. In both the a and b molecules the Eu(III) ion is 9-coordinate, being linked to the six nitrogen atoms of the macrocycle and to the three nitrogen atoms of the isothiocyanato ligands, two on the convex side of the macrocycle and the third on the opposite side. However, the macrocyclic ligands in the a and b molecules have different fold angles as well as different torsion angles about the $=N-CH_2-CH_2-N=$ side chains. Also, each of the Eu–N (macrocycle) bond distances in molecule a, which has the larger fold angle (111.3° compared to 102.1° for angles for lanthanide-acetate complexes of ligand I.

$[Y(O_2CCH_3)(OCOCH_3)(I)][Y(O_2CCH_3)(H_2O)(I)](CH_3COO)(ClO_4)_2$[b]		
Formula	$C_{26}H_{33}N_6O_{8.5}ClY$	
Mol. wt.	689.94	
Crystal system	Triclinic	
Space group	$P\bar{1}$	
a (Å)	13.052 (2)	
b (Å)	11.362 (2)	
c (Å)	11.426 (2)	
α (deg)	108.87 (4)	
β (deg)	104.73 (4)	
γ (deg)	100.26 (4)	
V (Å^3)	1486.60	
Z	2	
D_c (g cm^{-3})	1.54	
Coordinate bond lengths (Å)		
Y–N(1)	2.579 (8)	py
Y–N(2)	2.551 (8)	imine
Y–N(3)	2.623 (8)	imine
Y–N(4)	2.584 (8)	py
Y–N(5)	2.498 (8)	imine
Y–N(6)	2.597 (8)	imine
Y–O(31)	2.355 (6)	bidentate acetate
Y–O(32)	2.415 (6)	bidentate acetate
Y–O(41)	2.274 (7)	H_2O or monodentate acetate
Fold angle (deg)	115.7 (1)	

molecule b), is shorter than the corresponding distance in molecule b; an opposite trend is observed for the average Eu–NCS bond distances. The NCS^- ligands, although different in detail, show the same structural pattern in both the a and b molecules. In each case the NCS^- group is linear and the Eu–NCS linkage is bent at the N atom, with an average M–N–C angle of 142° (individual M–N–C angles range from 127.0 to 151.7°); the C–N distances range from 1.08 to 1.26 Å, while the C–S distances range from 1.55 to 1.69 Å. Considering that the two main resonance forms for a coordinated isothiocyanato ligand are M–N≡C–S (M–N–C angle 180°; Pauling bond lengths 1.15 Å (N≡C), 1.81 Å (C–S)) and M–N=C=S (M–N–C angle 120°; Pauling bond lengths 1.29 Å (N=C), 1.61 Å (C=S)), the observed bond angles and distances suggest that both forms must contribute to the Eu–NCS bonding. The situation in $[Eu(NCS)_3(I)]$ thus differs appreciably from that in $\{(C_4H_9)_4N\}_3[Er(NCS)_6]$, the only other lanthanide(III) isothiocyanato complex for which the crystal structure has been reported (Martin et al. 1968). In the octahedral $[Er(NCS)_6]^{3-}$ ion, the Er–N–C–S groups were found to be nearly linear, with the N–C bond (1.10 Å) and the S–C bond (1.61 Å) only slightly shorter than expected for the M–N≡C–S resonance form.

TABLE 3

$[Eu(O_2CCH_3)_2(I)](Cl)(H_2O)_4$[c]			$[Nd(O_2CCH_3)_2(I)](Cl)(H_2O)_4$[c]		
Formula	$C_{26}H_{40}N_6O_8ClEu$		Formula	$C_{26}H_{40}N_6O_8ClNd$	
Mol. wt.	752.05		Mol. wt.	744.33	
Crystal system	Triclinic		Crystal system	Triclinic	
Space group	$P\bar{1}$		Space group	$P\bar{1}$	
a (Å)	13.692 (2)		*a* (Å)	13.271 (2)	
b (Å)	12.758 (2)		*b* (Å)	12.788 (2)	
c (Å)	10.058 (2)		*c* (Å)	10.089 (2)	
α (deg)	73.94 (2)		α (deg)	73.75 (2)	
β (deg)	107.73 (3)		β (deg)	107.72 (3)	
γ (deg)	110.83 (3)		γ (deg)	111.15 (3)	
V (Å^3)	1535.88		V (Å^3)	1492.99	
Z	2		Z	2	
D_c (g cm^{-3})	1.63		D_c (g cm^{-3})	1.66	
Coordinate bond lengths (Å)[f]			Coordinate bond lengths (Å)[f]		
Eu–N(1)	2.651 (3)	imine	Nd–N(1)	2.619 (3)	imine
Eu–N(2)	2.590 (3)	imine	Nd–N(2)	2.609 (3)	imine
Eu–N(3)	2.638 (3)		Nd–N(3)	2.587 (4)	py
Eu–N(4)	2.665 (3)	imine	Nd–N(4)	2.628 (4)	imine
Eu–N(5)	2.624 (3)	imine	Nd–N(5)	2.641 (3)	imine
Eu–N(6)	2.639 (4)	py	Nd–N(6)	2.591 (4)	py
Eu–O(1)	2.505 (3)	bidentate acetate	Nd–O(1)	2.511 (3)	bidentate acetate
Eu–O(2)	2.461 (3)	bidentate acetate	Nd–O(2)	2.500 (4)	bidentate acetate
Eu–O(3)	2.500 (3)	bidentate acetate	Nd–O(3)	2.545 (3)	bidentate acetate
Eu–O(4)	2.458 (3)	bidentate acetate	Nd–O(4)	2.498 (4)	bidentate acetate
Fold angle (deg)	132.37 (6)		Fold angle (deg)	132.82 (6)	

[a] Bombieri et al. (1986). [b] Bombieri et al. (1989b). [c] Benetollo et al. (1990a). [d] Smith et al. (1989). [e] Bombieri et al. (1989c). [f] Numbering schemes: for Eu and Nd, as in fig. 2a; for Gd, as in fig. 2b.

The yttrium(III) diacetate perchlorate complex having stoichiometry $Y(I)(CH_3COO)_2(ClO_4)\cdot 0.5H_2O$ was found to contain two complex entities of slightly different formulas, $[Y(CH_3COO)_2(I)]^+$ and $[Y(CH_3COO)(H_2O)(I)]^{2+}$, present in a 1:1 ratio in the crystal lattice (Bombieri et al. 1989b). The charge of the +1 cation is balanced by a disordered ionic perchlorate, while that of the +2 cation is balanced by a disordered ionic perchlorate and an ionic acetate. In each case, the Y(III) attains 9-coordination by bonding to six nitrogen atoms of the macrocycle, to the two oxygen atoms of a bidentate chelating acetate on the convex side of the macrocycle, and to either a monodentate acetate (fig. 6a) or a water molecule (fig. 6b) on the opposite side. The coordination geometry is the same in both complex cations and is best described, despite the distortion imposed by the presence of a short-span bidentate acetate, as a monocapped square antiprism. Selected structural data are given in table 3.

A closely related structure was found in the lutetium(III) complex of stoichiometry $Lu(I)(CH_3COO)(OH)(ClO_4)\cdot 0.5H_2O\cdot CH_3OH$ (Bombieri et al. 1986). Again,

(cont'd)

$[Gd(O_2CCH_3)_2(I)]Cl(H_2O)_4$[d]			$[Eu(O_2CCH_3)_2(I)](CH_3COO)(H_2O)_9$[e]		
Formula	$C_{26}H_{40}N_6O_8ClGd$		Formula	$EuO_{15}N_6C_{28}H_{53}$	
Mol. wt.	757.34		Mol. wt.	865.72	
Crystal system	Triclinic		Crystal system	Triclinic	
Space group	$P\bar{1}$ (no. 2)		Space group	$P\bar{1}$	
a (Å)	10.032 (2)		a (Å)	11.006 (2)	
b (Å)	12.765 (2)		b (Å)	15.318 (3)	
c (Å)	13.668 (3)		c (Å)	12.663 (2)	
α (deg)	69.190 (9)		α (deg)	84.56 (2)	
β (deg)	72.405 (9)		β (deg)	115.14 (3)	
γ (deg)	74.07 (1)		γ (deg)	94.20 (2)	
V (Å^3)	1532 (1)		V (Å^3)	1922.68	
Z	2		Z	2	
D_c (g cm^{-3})	1.64		D_c (g cm^{-3})	1.49	
Coordinate bond lengths (Å)[f]			Coordinate bond lengths (Å)[f]		
Gd–N(3)	2.565 (6)	imine	Eu–N(1)	2.611 (3)	py
Gd–N(6)	2.602 (6)	imine	Eu–N(2)	2.572 (5)	imine
Gd–N(1)	2.627 (6)	py	Eu–N(3)	2.649 (4)	imine
Gd–N(4)	2.642 (6)	py	Eu–N(4)	2.599 (4)	py
Gd–N(2)	2.643 (6)	imine	Eu–N(5)	2.566 (4)	imine
Gd–N(5)	2.654 (6)	imine	Eu–N(6)	2.633 (3)	imine
Gd–O(3)	2.442 (5)	bidentate acetate	Eu–O(1)	2.496 (4)	bidentate acetate
Gd–O(1)	2.446 (5)	bidentate acetate	Eu–O(2)	2.453 (3)	bidentate acetate
Gd–O(2)	2.484 (5)	bidentate acetate	Eu–O(3)	2.490 (4)	bidentate acetate
Gd–O(4)	2.495 (5)	bidentate acetate	Eu–O(4)	2.519 (5)	bidentate acetate
Fold angle (deg)	132.82 (6)		Fold angle (deg)	122.34 (7)	

two slightly different coordination entities, $[Lu(CH_3COO)(H_2O)(I)]^{2+}$ and $[Lu(CH_3COO)(CH_3OH)(I)]^{2+}$, are present in a 1:1 ratio in the crystal lattice. In both species the coordination number of Lu(III) is 9, and the bidentate chelating acetate is situated on the convex side of the macrocycle (fig. 7). On the opposite side, either a water molecule (in a) or a methanol molecule (in b) completes the coordination sphere. A hydroxide ion, occupying different positions in a and b, and a disordered perchlorate ion balance the charges of each complex cation; clathrated methanol (one molecule for every two complex cations) is also present. Selected structural data are given in table 3.

1.1.1.2. *Compounds obtained from* 2,6-*diformylpyridine.* The compounds of this class are listed in table 1. When the lanthanide-templated synthesis was carried out using 2,6-diformylpyridine, 1,2-diaminoethane and the metal nitrate as precursors, two series of macrocyclic complexes were obtained (Abid and Fenton 1984b). With the larger members of the lanthanide series, La(III)–Pr(III) and also with Eu(III), the trinitrate complexes of the tetraimine macrocycle II were formed; those of La(III),

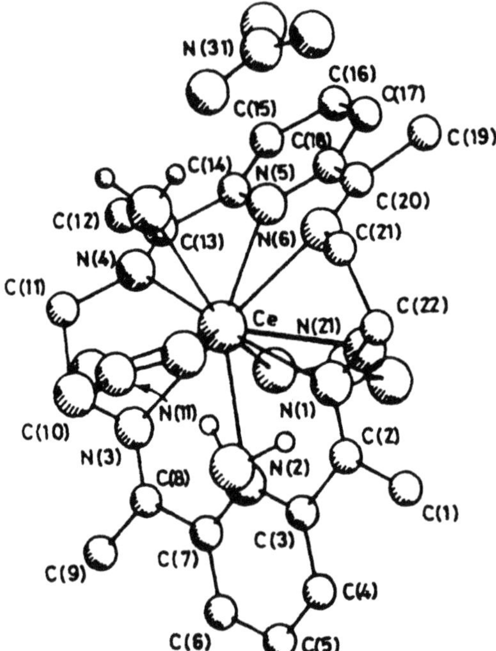

Fig. 3. Structure of the complex cation in $[Ce(NO_3)_2(H_2O)(I)](NO_3)\cdot H_2O$ (Arif et al. 1987).

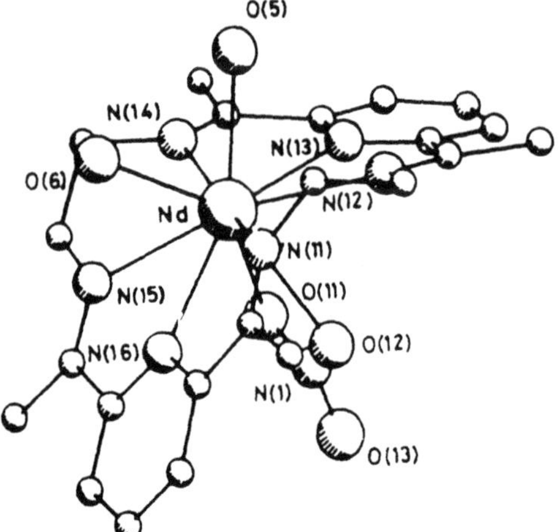

Fig. 4. Structure of the complex cation in $[\{Nd(NO_3)(H_2O)_2(I)\}_2](NO_3)(ClO_4)_3\cdot 4H_2O$ (Arif et al. 1987).

Ce(III), and Pr(III) were anhydrous; the Eu(III) complex crystallized with one molecule of water. With the smaller lanthanides, Nd(III)–Lu(III), except Eu(III), complexes of the carbinolamine ligand III, in which a molecule of water is added across one of the C=N bonds, were formed instead.

When first isolated from the template reaction, the complexes of the carbinolamine had formula $M(III)(NO_3)_3\cdot nH_2O$, with $n = 0$–3; upon recrystallization, however, they changed to mixed nitrate–hydroxide complexes of the tetraimine ligand II. A compound of this series for which an X-ray structure is available is the dinitrate hydroxide $[Sm(NO_3)(OH)(H_2O)(II)](NO_3)\cdot CH_3OH$ (Abid et al. 1984). In this complex the Sm(III) ion is 10-coordinate, being bonded to the six N-donor atoms of the macrocyclic ligand, to one bidentate chelating NO_3^- group on one side of the macrocycle, and to the hydroxide group and a water molecule on the opposite side (fig. 8). The coordination polyhedron can be described as an irregular antiprism capped on its "square" faces by the two N-atoms of the pyridine rings. The Sm–N(pyridine) bond lengths are shorter than the Sm–N(imine) bond lengths, and all Sm–N distances are significantly shorter than those found in the Sm(III) complex of cryptand 222 (Burns 1979). A remarkable difference is found between the Sm–O

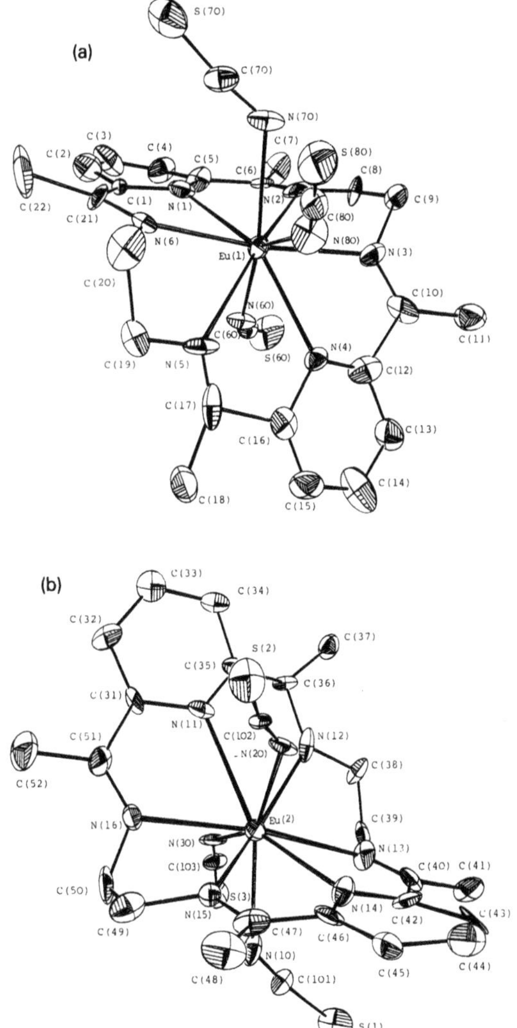

Fig. 5a, b.

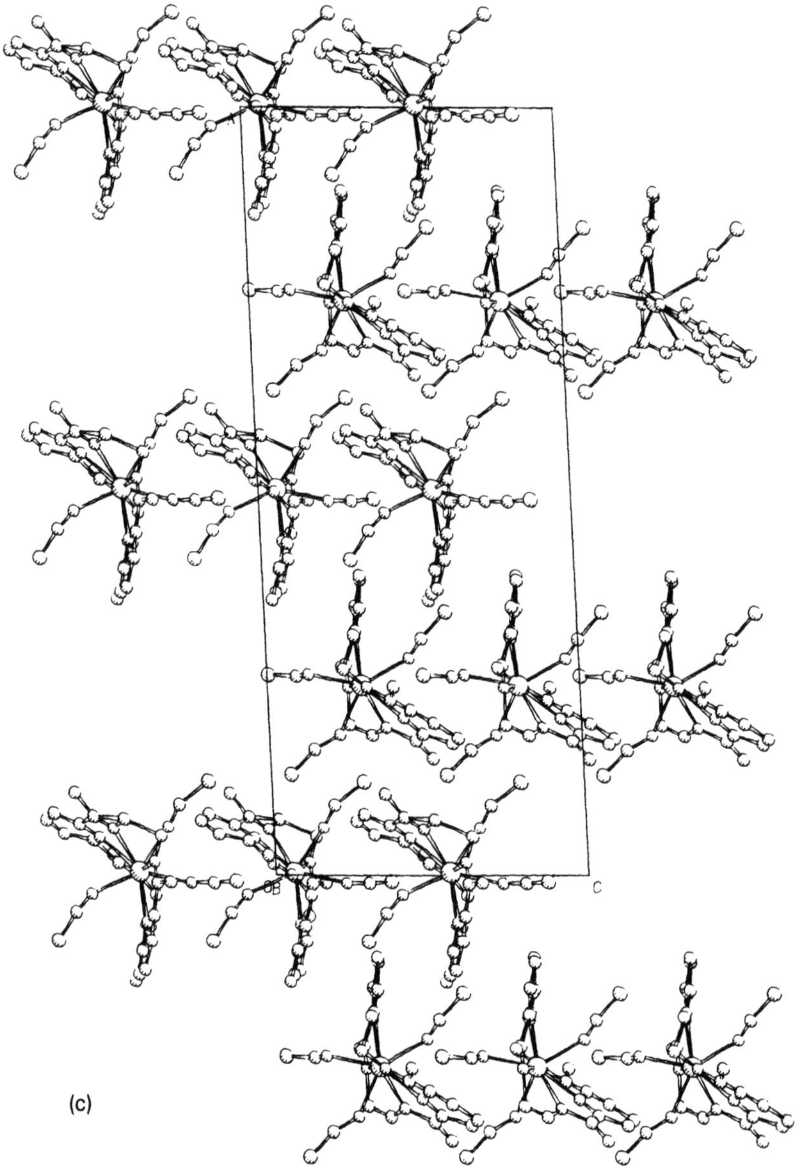

Fig. 5. (a), (b) Two slightly different [Eu(NCS)$_3$(I)] molecules present in the unit cell of the complex. (c) Packing of the molecules in the unit cell of [Eu(NCS)$_3$(I)] (Bombieri et al. 1989a).

distance of the OH^- group (2.46 Å) and the Sm–O distance of the water (2.52 Å) and bidentate nitrate groups (2.51 and 2.52 Å). In the description of this structure, there is no direct mention of the macrocycle conformation. The two Sm–N (pyridine) bonds appear to be aligned, and the planes of the two halves of the macrocycle containing

TABLE 4
Selected crystallographic data, bond lengths and fold angles for [Eu(NCS)$_3$(I)] and [Y(NCS)$_3$(I)].[a]

	Eu complex			Y complex		
Formula	$C_{25}H_{26}N_9S_3Eu$			$C_{25}H_{26}N_9S_3Y$		
Mol. wt.	700.70			637.64		
Crystal system	Monoclinic			Monoclinic		
Space group	Cc			Cc		
a (Å)	33.244 (3)			33.226 (3)		
b (Å)	11.976 (2)			11.986 (2)		
c (Å)	14.164 (2)			14.102 (2)		
β (deg)	92.72 (5)			92.84 (5)		
V (Å^3)	5632.8			5609.2		
Z	8			8		
D_c (g cm^{-3})	1.65			1.51		
Coordinate bond lengths (Å)	Eu(1)–N(1)	2.656 (8)	py	Y(1)–N(1)	2.58 (4)	py
	Eu(1)–N(2)	2.668 (9)	imine	Y(1)–N(2)	2.57 (3)	imine
	Eu(1)–N(3)	2.538 (8)		Y(1)–N(3)	2.50 (4)	
	Eu(1)–N(4)	2.542 (7)	py	Y(1)–N(4)	2.57 (3)	py
	Eu(1)–N(5)	2.592 (8)	imine	Y(1)–N(5)	2.63 (3)	imine
	Eu(1)–N(6)	2.489 (8)		Y(1)–N(6)	2.56 (3)	
	Eu(1)–N(60)	2.475 (8)	NCS$^-$	Y(1)–N(60)	2.40 (3)	NCS$^-$
	Eu(1)–N(70)	2.527 (8)		Y(1)–N(70)	2.39 (3)	
	Eu(1)–N(80)	2.514 (7)		Y(1)–N(80)	2.49 (4)	
	Eu(2)–N(11)	2.719 (8)	py	Y(2)–N(11)	2.65 (4)	py
	Eu(2)–N(12)	2.625 (7)	imine	Y(2)–N(12)	2.54 (3)	imine
	Eu(2)–N(13)	2.578 (8)		Y(2)–N(13)	2.46 (2)	
	Eu(2)–N(14)	2.569 (8)	py	Y(2)–N(14)	2.55 (3)	py
	Eu(2)–N(15)	2.639 (7)	imine	Y(2)–N(15)	2.72 (3)	imine
	Eu(2)–N(16)	2.560 (8)		Y(2)–N(16)	2.51 (4)	
	Eu(2)–N(10)	2.429 (9)	NCS$^-$	Y(2)–N(10)	2.53 (4)	NCS$^-$
	Eu(2)–N(20)	2.439 (8)		Y(2)–N(20)	2.40 (3)	
	Eu(2)–N(30)	2.411 (9)		Y(2)–N(30)	2.35 (3)	
Fold angles (deg)	111.3, 102.1			111.1, 103.2		

[a] Bombieri et al. (1989a).

the pyridine rings appear to be slightly "twisted" with respect to one another, rather than "folded" as in all derivatives of macrocycle I. Selected structural data are given in table 5.

The easy isolation of complexes of the carbinolamine macrocycle III appears to be a feature unique to the combination of 2,6-diformylpyridine, 1,2-diaminoethane, and certain lanthanide(III) nitrates. No carbinolamine complexes were isolated from the same diamine and dicarbonyl precursors when the condensation was carried out in the presence of yttrium(III) acetate (Bombieri et al. 1989b). Similarly, the use of 1,2- or 1,3-diaminopropane in place of 1,2-diaminoethane yielded only trinitrate complexes of the tetraimine macrocycles IV and V (Abid and Fenton 1984b). Complexes of ligand IV are expected to exist in various isomeric forms, since a (2 + 2) cyclic condensation involving an unsymmetrically substituted diamine precursor may yield

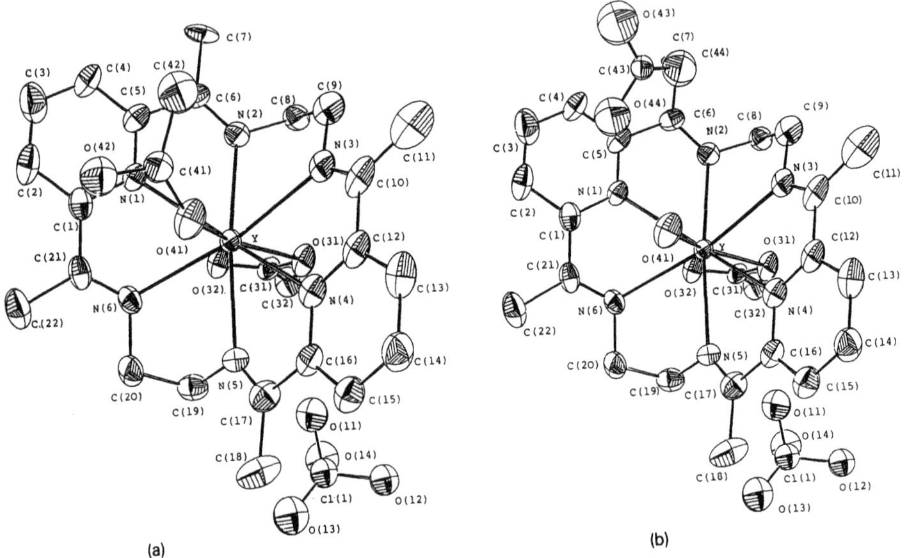

Fig. 6. Structure of the two complex cations present in the crystal of $Y(I)(CH_3COO)_2(ClO_4)\cdot 0.5H_2O$: (a) $[Y(CH_3C\underline{O}\underline{O})(CH_3CO\underline{O})(I)]ClO_4$, (b) $[Y(CH_3C\underline{O}\underline{O})(H_2\underline{O})(I)](CH_3COO)(ClO_4)$ (Bombieri et al. 1989b).

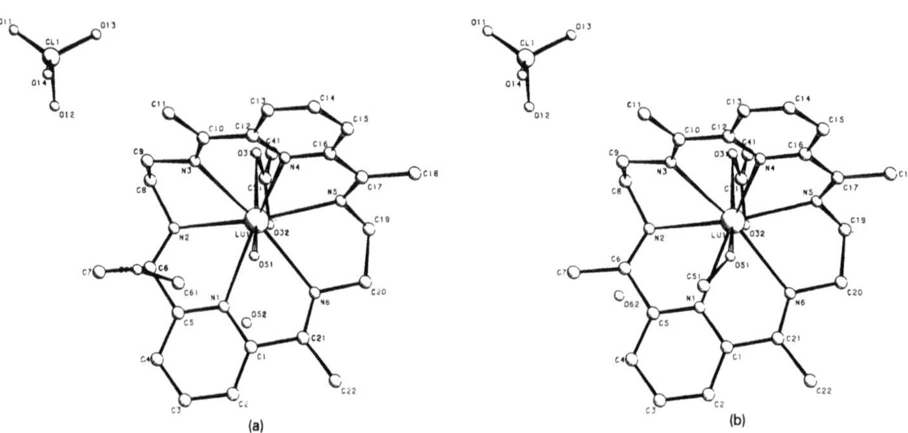

Fig. 7. Structures of the two complex cations present in $[Lu(CH_3COO)(CH_3OH)(I)][Lu(CH_3COO)(H_2O)(I)]$ $(OH)_2(ClO_4)_2\cdot CH_3OH$ (Bombieri et al. 1986). (a) Species with coordinated H_2O. (b) Species with coordinated CH_3OH.

two constitutional isomers, depending on whether the two substituents are adjacent to the same (*cis*) or different (*trans*) pyridine rings. In addition, the substituted diamine has a chiral carbon atom, so stereoisomers are possible when the precursor is racemic. The multiplicity observed for certain signals in the ^1H NMR spectra of the La(III)

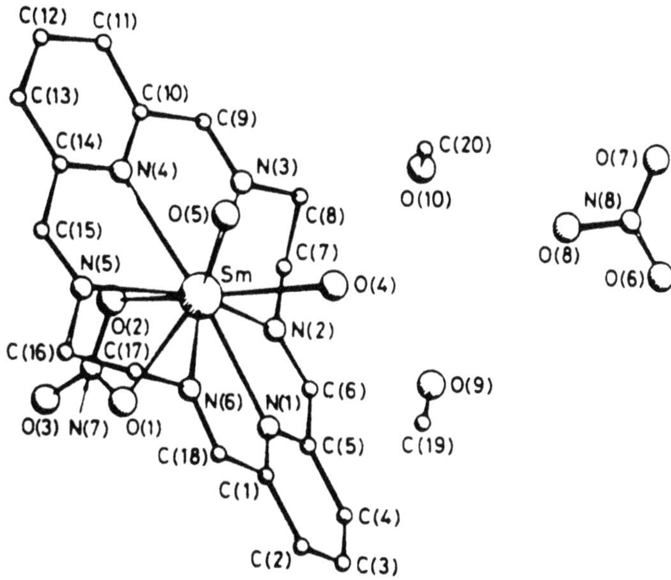

Fig. 8. Structure of the complex cation in $[Sm(NO_3)(OH)(H_2O)(II)](NO_3)\cdot CH_3OH$ (Abid and Fenton 1984b).

complex of ligand IV were indeed considered to indicate the existence of isomers, but no individual forms were identified (Abid and Fenton 1984b).

1.1.2. (2 + 2) *ligands with pyridine head-units and hydrazine side-chains*

The known compounds of this class are listed in table 6. The (2 + 2) Schiff-base condensation of hydrazine and 2,6-diacetylpyridine in the presence of the perchlorates of the smaller lanthanides, Tb(III)–Lu(III), has been reported to yield complexes of the 14-member, six-nitrogen macrocyclic ligand VI (Radecka-Paryzek 1981a), formulated as $[M(H_2O)_2(VI)](ClO_4)\cdot 4H_2O$. With the larger lanthanides, complexes of the "three-quarter cycle" diketone VII were instead isolated; complexes of 2,6-diacetylpyridine were also occasionally obtained.

On the basis of IR, UV–VIS, and proton NMR spectra, the complexes of ligand VI were assigned an octahedral coordination geometry, with the 6-coordinate metal ion linked to the two pyridine nitrogens, two opposite imine nitrogens, and two water molecules situated on opposite sides of the macrocycle. The macrocycle itself was considered to be planar, with full conjugation throughout the cavity (Radecka-Paryzek 1981a). In the absence of crystallographic or other definitive evidence, this suggested structure should be considered as highly tentative.

1.1.3. (2 + 2) *Ligands with pyridine head-units and aromatic diimine side-chains*

The compounds reported to contain ligands of this class are listed in table 7. Considerable controversy has surrounded the metal complexes of ligand VIII since the acid catalyzed condensation of 2,6-diacetylpyridine with 1,2-diaminobenzene was first reported to yield a yellow crystalline product, $C_{30}H_{26}N_6$, formulated as an

TABLE 5
Selected crystallographic data, bond lengths, and fold angle for $[Sm(NO_3)(OH)(H_2O)(II)](NO_3)(CH_3OH)_2$[a].

Formula	$C_{20}H_{29}N_8O_{10}Sm$		
Mol. wt.	691.4		
Crystal system	Triclinic		
Space group	$P\bar{1}$		
a (Å)	16.02 (1)		
b (Å)	9.64 (1)		
c (Å)	8.29 (1)		
α (deg)	91.06 (3)		
β (deg)	95.26 (3)		
γ (deg)	92.18 (3)		
V ($Å^3$)	1273		
Z	2		
D_c (g cm^{-3})	1.67		
Coordinate bond lengths (Å)	Sm–N(1)	2.66 (1)	py
	Sm–N(2)	2.64 (1)	imine
	Sm–N(3)	2.62 (1)	imine
	Sm–N(4)	2.65 (1)	py
	Sm–N(5)	2.62 (1)	imine
	Sm–N(6)	2.60 (1)	imine
	Sm–O(1)	2.51 (1)	NO_3^-
	Sm–O(2)	2.52 (1)	NO_3^-
	Sm–O(4)	2.52 (1)	H_2O
	Sm–O(5)	2.46 (1)	OH
Fold angle (deg)	173.3		

[a] Abid and Fenton (1984b).

18-member macrocyclic ligand with a six-nitrogen-donor cavity (Stotz and Stoufer 1970). A dark-brown bimetallic copper(II) complex, $[Cu_2(C_{30}H_{26}N_6)](NO_3)_4$, was also described and considered to involve Cu–Cu interactions (Stotz and Stoufer 1970). Re-investigation of these species by X-ray crystallography showed the yellow compound, $C_{30}H_{26}N_6$, to have the tricyclic structure IX, consisting of a central 12-member ring flanked by two identical 7-member rings (Cabral et al. 1982a). Also, the true formula of the copper(II) complex was established to be $[Cu(C_{30}H_{26}N_6)(H_2O)](ClO_4)_2 \cdot H_2O$; in this complex, the organic moiety had re-arranged to give a cyclic ligand with an asymmetric 4-nitrogen cavity encapsulating a single Cu(II) ion (Cabral et al. 1982b).

The existence of metal complexes of ligand VIII was brought back into question by reports that trinitrato macrocycles $M(VIII)(NO_3)_3 \cdot 2H_2O$ could be obtained for the larger lanthanides, La(III)–Nd(III), from condensation of the precursors in the presence of the hydrated metal nitrates (Radecka-Paryzek 1981a, b, 1985, 1986). Hydrated triisothiocyanates of Eu(III), and Lu(III), M(VIII) $(NCS_3) \cdot 2H_2O$ (Zhang and Tong 1986), and a lanthanum(III) perchlorate derivative (Wang and Miao 1984)

TABLE 6
(2 + 2) Macrocyclic complexes of the lanthanide(III) ions with pyridine head-units and hydrazine side-chains

Compound[a]	Color	Other data[d]	Ref.[b]
Ligand VI = $C_{18}H_{18}N_6$			
$[Tb(H_2O)_2(VI)](ClO_4)_3 \cdot 4H_2O$	Yellow	UV λ_{max}, nm (diffuse reflectance and CH_3CN solution): 225, 255, 293, 315, 370 IR(cm^{-1}): 1570: ν(C=N, imine); 1593, 1458, 630 and 423: py absorptions; 3390, 865 and 560: OH absorptions; 1095 and 625: ClO_4^- absorptions; other absorptions: 370, 315, 290, 260 and 220	[1]
$[Dy(H_2O)_2(VI)](ClO_4)_3 \cdot 4H_2O$	Yellow	UV, IR[c]	[1]
$[Ho(H_2O)_2(VI)](ClO_4)_3 \cdot 4H_2O$	Yellow	UV, IR[c]	[1]
$[Er(H_2O)_2(VI)](ClO_4)_3 \cdot 4H_2O$	Yellow	UV, IR[c]	[1]
$[Tm(H_2O)_2(VI)](ClO_4)_3 \cdot 4H_2O$	Yellow	UV, IR[c]	[1]
$[Yb(H_2O)_2(VI)](ClO_4)_3 \cdot 4H_2O$	Yellow	UV, IR[c]	[1]
$[Lu(H_2O)_2(VI)](ClO_4)_3 \cdot 4H_2O$	Yellow	UV, IR[c]	[1]
		^{1}H NMR, δ(ppm/TMS), CD_3CN: 8.3–7.7 (6H, py); 2.6 (6H) and 2.28 (6H) both CH_3	[1]

[a] See footnote a in table 1.
[b] Reference: [1] Radecka-Paryzek (1981a).
[c] Spectrum similar to Tb analog.
[d] See footnote j in table 1.

were also reported. Attempts to repeat these works yielded primarily the tricyclic organic compounds IX, either as the free base or as nonstoichiometric adducts of the partly protonated form with metal salts (Benetollo et al. 1989). However, true macrocyclic complexes of La(III), Pr(III) and Nd(III) with ligand VIII were eventually isolated in minute yields (1–5% relative to the precursors) and their structure was unambiguously established by X-ray analysis of a Pr(III) dinitrato perchlorate species of formula $[Pr(NO_3)_2(CH_3OH)(VIII)](ClO_4) \cdot 0.5H_2O \cdot 0.5CH_3OH$ (Benetollo et al. 1989). In the complex cation of this species, the 11-coordinate Pr(III) ion is linked to the six nitrogen atoms of the macrocyclic ligand, to a bidentate chelating nitrate and to a methanol molecule on one side of the macrocycle, and to another bidentate chelating nitrate on the opposite side. The macrocycle itself is doubly bent into a saddle shape, with the two benzene rings directed toward the hemisphere containing the NO_3^- and CH_3OH ligands and the two pyridine rings directed toward the hemisphere containing the single NO_3^- (fig. 9). The bending occurs chiefly at the bonds adjacent to the *o*-phenylene groups and results in angles of 113.0° (py–py) and 112.1° (bz–bz). Selected structural data for this compound are given in table 8.

The condensation of freshly sublimed 1,2-diaminobenzene with 2,6-diformylpyridine in the presence of a lanthanide nitrate in refluxing methanol yielded macrocyclic complexes of the general type $M(X)(NO_3)_{3-n}(OH)_n \cdot mH_2O$ (Benetollo et al. 1991). With the larger lanthanides, La–Gd, n was zero and the complexes were isolated in

TABLE 7
(2 + 2) Macrocyclic complexes of the lanthanide(III) and yttrium(III) ions with pyridine head-units and aromatic side chains.

Compound[a]	Color	Other data[j]	Ref.[b]
Ligand VIII = $\mathbf{C_{30}H_{26}N_6}$			
$[La(NO_3)_3(VIII)]\cdot 0.5H_2O$	Yellow	IR(cm^{-1}): 3630 (s) $\nu_{as}(OH_2)$, 3550 (s) $\nu_s(OH_2)$, 3115 (sh), 3100 (m), 3030 (vw), 2985 (vw), 2940 (vw), 1635 (s) ν(C=N), 1595 (s) ν(C=N), 1490 (sh), 1455 (vs, vbr) ν(N=O) of bidentate chelating NO_3^-, 1425 (sh), 1375 (m), 1325 (vs, br) $\nu(NO_2)$ of bidentate chelating NO_3^-, 1305 (sh), 1255 (s), 1235 (m), 1195 (w), 1160 (vw), 1145 (vw), 1115 (w), 1035 (m), 995 (m), 980 (w), 955 (vw), 815 (w), 804 (vs), 795 (w), 765 (s), 740 (m–s), 725 (m–s), 705 (sh), 650 (vw), 555 (sh), 545 (m), 525 (vw), 505(vw), 495 (vw), 410 (vs, vbr) $\rho(OH_2)$	[1]
$[(Pr(NO_3)_3(VIII)]\cdot 2H_2O$	Yellow	IR: identical to La analog	[1]
$[Nd(NO_3)_3(VIII)]\cdot 3H_2O$	Yellow	IR: identical to La analog	[1]
*$[Pr(NO_3)_2(CH_3OH)(VIII)]$-$(ClO_4)\cdot 0.5(CH_3OH)\cdot 0.5H_2O$	Yellow	X-ray structure: fig. 9, table 8	[1]
$Nd(VIII)(OH)(NO_3)_{0.5}$-$(ClO_4)_{1.5}\cdot 2H_2O$	Yellow	IR(cm^{-1}): 3600 (sh), 3540–3350 (m, vbr) ν(OH), 3120 (sh), 3110 (m), 3020 (sh), 2980 (vw), 2940 (w), 1630 (s) ν(C=N), 1595 (vs) ν(C=N), 1480 (vs) ν(N=O) of bidentate chelating NO_3^-, 1380 (s), 1310 (s) $\nu(NO_2)$ of bidentate chelating NO_3^-, 1250 (s), 1195 (m), 1160 (vw), 1130 (w), 1115 (vw), 1085 (vs) ν_3 of ionic ClO_4^-, 1035 (w), 1000 (m), 985 (m), 975 (w), 900 (m), 825 (w), 815 (m), 805 (m), 785 (w), 765 (m), 740 (m), 710 (sh), 705 (sh), 630 (w), 605 (m) ν_4 of ionic ClO_4^-, 555 (w), 505 (w), and 495 (vw)	[1]
$[La(H_2O)_2(VIII)](NO_3)_3$[c]	Yellowish brown	UV λ_{max}(nm, CH_3CN): 225, 245, 278, 358. MS: highest peak $m/e = 470$, corresponding to metal-free $C_{30}H_{26}N_6$.	[2]
		IR(cm^{-1}): 1609 (s) ν(C=N); 1590, 1572, 1460, 1415, 620 and 420: pyridine absorptions; 1360 and 840: NO_3^- absorptions	[2]
$[Ce(NO_3)_3(VIII)]\cdot 2H_2O$[c]		UV λ_{max}(nm, CH_3CN): 225, 240, 280 and 355 assigned to $\pi \rightarrow \pi^*$ transitions of ligand	[3]
		IR(cm^{-1}): 3400 [ν(OH)]; 1610 [ν(C=N, imine)]; 1590 and 1570: pyridine absorptions; 1465 and 1288: NO_3^- absorptions; 300 [ν(Ce–N)]	[3]
$[Pr(NO_3)_3(VIII)]\cdot 2H_2O$[c]	–	UV, IR: similar to Ce analog	[3]
$[Nd(NO_3)_3(VIII)]\cdot 2H_2O$[c]	–	UV, IR: similar to Ce analog	[3]
$Eu(VIII)(NCS)_3(H_2O)_2$	–	IR, conductance, $\mu_{eff} = 3.05\mu_B$	[4]
$[Lu(NCS)_2(H_2O)_2(VIII)]NCS$	–	IR, conductance	[4]

Compound	Colour	IR	Assignment	Ref.
Ligand X = $C_{26}H_{18}N_6$				
Trinitrates				
$La(X)(NO_3)_3 \cdot 1.5H_2O$	Yellow	IR(cm^{-1}):[d] 1443 (vs, br) ν_5; 1335, 1310 (vs, br) ν_1[e]; 1008 (m) ν_2	N–O stretching modes of coordinated NO_3^-	[5]
$Ce(X)(NO_3)_3 \cdot H_2O$	Yellow	IR(cm^{-1}):[d] 1435 (vs, br) ν_5; 1340 (vs, br), 1310 (sh) ν_1[e]; 1010 (m) ν_2		[5]
$Pr(X)(NO_3)_3 \cdot 1.5H_2O$	Yellow	IR(cm^{-1}):[d] 1443 (vs, br) ν_5; 1331, 1300 (vs, br) ν_1[e]; 1002 (m) ν_2		[5]
$Nd(X)(NO_3)_3 \cdot 4.5H_2O$	Yellow	IR(cm^{-1}):[d] 1445 (vs, br) ν_5; 1340, 1310 (vs, br) ν_1[e]; 1005 (m) ν_2		[5]
$Sm(X)(NO_3)_3 \cdot H_2O$	Yellow	IR(cm^{-1}):[d] 1443 (vs, br) ν_5; 1337, 1302 (vs, br) ν_1[e]; 1002 (m) ν_2		[5]
$Eu(X)(NO_3)_3$	Yellow	IR(cm^{-1}):[d] 1445 (vs, br) ν_5; 1340, 1315, 1290 (vs, br) ν_1[e]; 1002 (m) ν_2		[5]
$Gd(X)(NO_3)_3$	Yellow	IR(cm^{-1}):[d] 1445 (vs, br) ν_5; 1345, 1320, 1290 (vs, br) ν_1[e]; 1000 (m) ν_2		[5]
Others				
$La(X)(CF_3SO_3)_3 \cdot H_2O$		IR(cm^{-1}):[d] 1301–1166 (vvs, vbr, with maxima at 1301, 1275, 1235, 1166) $\nu(SO_3)(E)$, $\nu(CF_3)(E)$; 1029 (vs) $\nu(SO_3)(A_1)$; 637 (vs) $\delta(SO_3)(A_1)$; 512 (m) $\delta(CF_3)(E)$, $\delta(SO_3)(E)$		[5]
$Eu(X)(CF_3SO_3)_3 \cdot 2H_2O$		IR(cm^{-1}):[d] 1316–1168 (vvs, vbr, with maxima at 1316, 1279, 1240, 1168) $\nu(SO_3)(E)$, $\nu(CF_3)(E)$; 1029 (vs) $\nu(SO_3)(A_1)$; 635 (vs) $\delta(SO_3)(A_1)$; 516 (m) $\delta(CF_3)(E)$, $\delta(SO_3)(E)$		[5]
$La(X)(ClO_4)_2(OH) \cdot 1.5H_2O$		IR(cm^{-1}):[d] 1080 (vs, br) ν_3; 625 (m) ν_4: (Cl–O) antisymm. stretching and bending of modes of ionic ClO_4^-		[5]
$La(X)(CH_3COO)_3 \cdot 2.5CH_3OH \cdot 3.5H_2O$		IR(cm^{-1}):[d] 1567 (vvs, br); 1415 (vs, br)	stretching modes of $-COO^-$ group	[5]
$Ce(X)(CH_3COO)_3 \cdot 5H_2O$		IR(cm^{-1}):[d] 1565 (vs, br); 1410 (vs, br)		[5]
$Ce(X)(CH_3COO)_2(ClO_4) \cdot H_2O$		IR(cm^{-1}):[d] 1531 (vs), 1451 (vs), 1418 (s) $\nu(COO^-)$; 1104 (vs, br) ν_3, 623 (m) ν_4 (ionic ClO_4^-)		[5]
$Tb(X)(NO_3)_{1.5}(OH)_{1.5} \cdot 0.5H_2O$		IR(cm^{-1}):[d] 1467 (s, br) ν_5; 1380, 1345 (vs, br), 1279 (s) ν_3, ν_1	(N–O) stretching modes of coordinated (ν_5, ν_1) and ionic (ν_3) NO_3^-	[5]
$Dy(X)(NO_3)_{1.5}(OH)_{1.5} \cdot 1.5H_2O$		IR(cm^{-1}):[d] 1467 (s, br) ν_5; 1380, 1345 (vs, br[e]), 1280 (s) ν_3, ν_1		[5]
$Er(X)(NO_3)_{1.5}(OH)_{1.5}$		IR(cm^{-1}):[d] 1470 (s, br) ν_5; 1375, 1340 (vs, br[e]), 1300 (sh) 1280 (s) ν_3, ν_1		[5]
$Yb(X)(NO_3)_2(OH) \cdot H_2O$		IR(cm^{-1}):[d] 1500 (vs) ν_5; 1390, 1345 (vs, br[e]), 1295 (s) ν_3, ν_1)		[5]
$La(X)(NO_3)_{1.75}(Cl)_{1.25} \cdot 1.5H_2O$		IR(cm^{-1}):[d] 1443 (vs, br) ν_5; 1314 (vs, br) ν_1, 1002 (m) ν_2	(N–O) stretching modes of coordinated NO_3^-	[5]
$Ho(X)(NO_3)_{1.75}(Cl)_{1.25} \cdot H_2O$		IR(cm^{-1}):[d] 1519 (vs) ν_5; 1277 (s) ν_1, 1007 (m–w) ν_2		[5]
$Ho(X)(NO_3)_2(ClO_4) \cdot CH_3OH$		IR(cm^{-1}):[d] 1517 (vs) ν_5; 1281 (s) ν_1 (coord. NO_3^-); 1102 (vs, br) ν_3, 623 (m) ν_4 (ionic ClO_4^-)		[5]
$Tm(X)(SCN)_2(ClO_4) \cdot 2H_2O$		IR(cm^{-1}):[d] 2050 (vs) $\nu(C{\equiv}N)$ of NCS^-; 1090 (s) ν_3, 620 (m) ν_4 (ionic ClO_4^-)		[5]

[a] See footnote a in table 1.

[b] References: [1] Benetollo et al. (1989); [2] Radecka-Paryzek (1981b); [3] Radecka-Paryzek (1985a, b); [4] Zhang and Tong (1986); [5] Benetollo et al. (1991).

[c] Synthesis could not be reproduced, see text (Benetollo et al. (1989)).

[d] Notations used in the assignments follow Nakamoto (1986) for all anions except $CF_3SO_3^-$, where Batchelor et al. (1977) are followed.

[e] Broad absorption with partly resolved maxima.

[f] See footnote j in table 1.

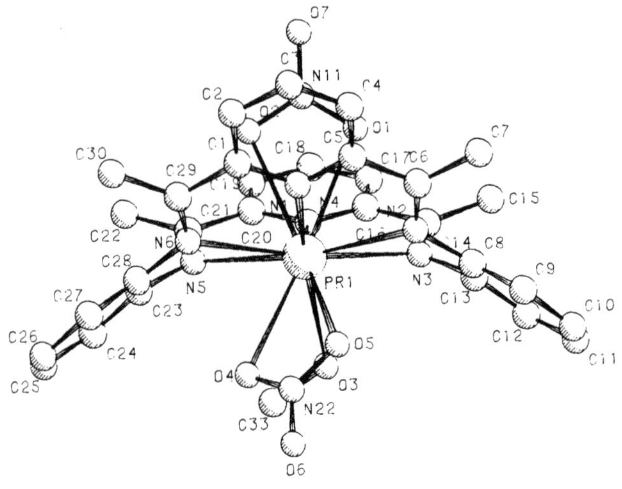

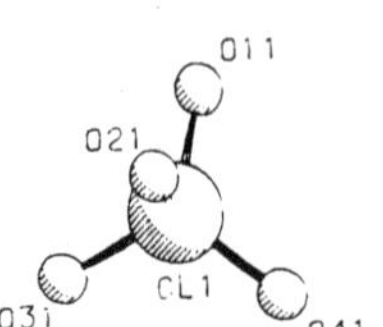

Fig. 9. Structure of $[Pr(NO_3)_2(CH_3OH)(VIII)](ClO_4)\cdot 0.5H_2O\cdot 0.5CH_3OH$ (Benetollo et al. 1989). The clathrated solvent molecules are omitted for clarity.

good yields (60% or better) as the trinitrates. With the smaller lanthanides, Gd(III)–Lu(III), partial substitution of nitrate by hydroxide occurred; also, the yields progressively decreased and became minimal (less than 2%) for Lu(III). The metal-templated synthesis was also successful with the lanthanide trifluoromethylsulfonates and, limited to the larger lanthanides, with the acetates. The major organic byproduct of these reactions was 2,6-bis(2-benzimidazyl)pyridine. The metal–macrocycle entities of ligand X were slowly decomposed by strong bases but were inert to acids (e.g. they were not decomposed by contact with methanol–HCl over a period of days). In contrast, the exocyclic anionic ligands were labile and readily underwent metathesis. Complexes of ligand X with the nitrates of La–Nd were also obtained by transmetallation from the bimacrocyclic $Ba(X)_2(ClO_4)_2$ complex (Gray and Hart 1987); the compounds thus obtained are identical with those of the template synthesis.

TABLE 8
Selected crystallographic data, bond lengths, and fold angles in $[Pr(NO_3)_2(CH_3OH)(VIII)](ClO_4)\cdot 0.5(CH_3OH)\cdot 0.5(H_2O)$[a].

Formula	$C_{31.5}H_{33}N_8O_{12}ClPr$		
Mol. wt.	892.01		
Crystal system	Monoclinic		
Space group	$P2_{1/n}$		
a (Å)	20.198 (3)		
b (Å)	14.208 (2)		
c (Å)	12.727 (2)		
β (deg)	104.25 (4)		
V (Å^3)	3540		
Z	4		
D_c (g cm^{-3})	1.67		
Coordinate bond lengths (Å)	Pr–N(1)	2.651 (4)	py
	Pr–N(4)	2.661 (4)	
	Pr–N(2)	2.681 (4)	imine
	Pr–N(3)	2.660 (4)	
	Pr–N(5)	2.664 (4)	
	Pr–N(6)	2.657 (4)	
	Pr–O(1)	2.705 (4)	NO_3^-
	Pr–O(2)	2.675 (4)	
	Pr–O(4)	2.690 (4)	NO_3^-
	Pr–O(5)	2.557 (4)	
	Pr–O(2)	2.592 (4)	CH_3OH
Fold angles (deg)	113.0 (py–py); 112.1 (bz–bz)		

[a] Benetollo et al. (1989).

1.1.4. (2 + 2) *ligands with furan head-units and aliphatic or aromatic diimine side-chains*

The Schiff-base condensation of 2,5-diformylfuran with 1,2-diaminoethane, 1,2-diaminopropane, or 1,3-diaminopropane, in the presence of the nitrates of the larger lanthanides, produced three series of closely related complexes: $M(XI)(NO_3)_3\cdot nH_2O$ (M = La–Eu, except Pr; $n = 0$–2) (Abid and Fenton 1984a, Zang et al. 1987), $M(XII)(NO_3)_3\cdot nH_2O$ (M = La, Ce, Pr; $n = 0$–2) (Abid and Fenton 1984a), and $M(XIII)(NO_3)_3\cdot nH_2O$ (M = La, Ce, Pr; $n = 0, 1$) (Abid and Fenton 1984a). With the nitrates of the smaller lanthanides, Gd–Lu, and with 1,2-diaminoethane as the precursor, species of the unusual stoichiometry $3M(NO_3)_3 2(XI)\cdot 4H_2O$ were obtained. Mass spectrometry suggested that these trimetallic species still contained discrete macrocyclic units (Abid and Fenton 1984a). Lanthanide(III) trinitrate complexes were also reported for a macrocycle with *o*-phenylenediimine side-chains (Zhang et al. 1987). The compounds of this class are listed in table 9.

1.1.5. (1 + 1) *ligands with a single pyridine head-unit*

The metal-templated condensation of 2,6-diacetylpyridine with several aliphatic α,ω-diamines was reported to produce macrocyclic complexes of ligands XIV–XVII,

TABLE 9
(2 + 2) Macrocyclic complexes of the lanthanide(III) ions with furan head-units.

Compound	Color	Other data[d]	Ref.[a]
Ligand XI = $C_{16}H_{16}N_4O_2$			
La(XI)$(NO_3)_3$		IR(cm^{-1}): 1614 (s) ν(C=N); 1382 (sh), 1018 (s) and 818 (m): NO_3^- absorptions; others: 1445, 1320, 1248 and 1040	[1]
		MS: highest peak m/e = 296, corresponding to metal-free ligand	[1]
		^{1}H NMR[b] δ(ppm/TMS, DMSO-d_6): two sets of resonances (1) 8.58 (s, 1H, H–C=N); 7.37 (s, 1H, furan-H); 4.00 (s, 2H, CH_2); (2) 8.12 (s, 1H, H–C=N); 6.96 (s, 1H, furan-H); 3.86 (s, 2H, CH_2)	[1]
Ce(XI)$(NO_3)_3$		IR: similar to La analog, except no absorption at 1040 cm^{-1}	[1]
		MS[c]	[1]
Pr(XI)$(NO_3)_3$		IR: similar to Ce analog; MS[c]	[1]
Nd(XI)$(NO_3)_3$		IR: similar to Ce analog; MS[c]	[1]
Sm(XI)$(NO_3)_3 \cdot H_2O$		IR(cm^{-1}): 1620 (s) ν(C=N); 1385 (m), 1030 (s) and 819 (m): NO_3^- absorptions; others: 3300 (br); 1450, 1310 and 1255	[1]
		MS[c]	[1]
Eu(XI)$(NO_3)_3 \cdot 2H_2O$		IR: similar to Eu analog	[1]
Gd_3(XI)$(NO_3)_9 \cdot 4H_2O$		MS[c]: highest peak m/e = 296 (metal-free ligand)	[1]
Er_3(XI)$(NO_3)_9 \cdot 4H_2O$		MS: similar to Gd analog	[1]
Ligand XII = $C_{18}H_{20}N_4O_2$			
La(XII)$(NO_3)_3$		IR(cm^{-1}): 1625 (s) ν(C=N); 1386 (sh), 1035, and 838, NO_3^- absorptions; others: 1450, 1310, 1065 and 980.	[1]
		MS: highest peak m/e = 324, corresponding to metal-free ligand	[1]
Ce(XII)$(NO_3)_3 \cdot H_2O$		MS[c]; IR: similar to La analog, except for additional broad absorption at 3450 cm^{-1}	[1]
Pr(XII)$(NO_3)_2 \cdot H_2O$		MS[c]; IR: similar to Ce analog	[1]
Ligand XIII = $C_{18}H_{20}N_4O_2$			
La(XIII)$(NO_3)_3$		IR(cm^{-1}): 1618 (s) ν(C=N); 1386 (m), 1030 (s) and 818 (m): NO_3^- absorptions; others: 1453, 1315 and 1245.	[1]
		MS: highest peak m/e = 324, corresponding to metal-free ligand	[1]
Ce(XIII)$(NO_3)_3$		MS[c]; IR: similar to La analog	[1]
Pr(XIII)$(NO_3)_3$		MS[c]; IR: similar to La analog	[1]

[a] Reference: [1] Abid and Fenton (1984a).
[b] Spectrum taken immediately after preparing solution. Set (1) assigned to La-macrocyclic complex; set (2) assigned to "free ligand" (unknown as a free species). After 1 h, only set (2) remained.
[c] Results similar to those of La analog.
[d] see footnote j in table 1.

in which one section of the ligand is flexible while the other is rigidly planar. Trinitrate complexes of ligand XIV, having general formulas M(XIV)$(NO_3)_3$ for M = La(III) or Ce(III) and M(XIV)$(NO_3)_2(ClO_4) \cdot H_2O$ for M = Nd(III) or Sm(III), were reported by Arif et al. (1985) and are listed in table 10. The true macrocyclic nature of these complexes was established by crystallographic analysis of the $[La(NO_3)_3(XIV)]$ species (Arif et al. 1985). This complex contains a 12-coordinated La(III) linked to three ether oxygen atoms, three nitrogen atoms (one of the pyridine, two of the imine

TABLE 10
(1 + 1) Macrocyclic complexes with a single pyridine head-unit

Compound[a]	Color	Other data[c]	Ref.[b]
Ligand XIV = $C_{17}H_{25}N_3O_3$			
*[La(NO_3)$_3$(XIV)]	Pale yellow	X-ray structure: fig. 10, table 11	[1]
		IR(cm^{-1}): 1648 [ν(C=N), imine)]; 1590 [ν(C=N, py)]; 1475 and 1448: B_1 of coordinated NO_3^-; 1321 and 1300: A_1 of coordinated NO_3^-; 1095, 1089, 1071, 1064, 1040 and 1005: O–C–O stretching region; ionic NO_3^- absorptions absent	
		^{1}H NMR δ(ppm/TMS, CD_3CN): 8.30 (γH-py); 8.17(βH-py); 4.12 (NC$\underline{H}_2$CH$_2$O); 3.90 (NCH$_2$C$\underline{H}_2$O); 3.86 (OCH$_2$CH$_2$O, unresolved at 400 MHz); 2.53 (CH_3)	[1]
[Ce(NO_3)$_3$(XIV)]	Golden yellow	IR, similar to La analog	[1]
		^{1}H NMR δ(ppm/TMS, CD_3CN): 9.75 [(γ + β)H-py]; 2.83 [NC$\underline{H}_2$CH$_2$O]; 1.78 (s, CH_3); 0.56 (NCH$_2$C$\underline{H}_2$O); 0.04 (OCH$_2$CH$_2$O)	[1]
Nd(XIV)(NO_3)$_2$$ClO_4$·$H_2O$	Lilac	IR(cm^{-1}): ligand absorptions similar to La analog; also: 3550 (w, br) uncoordinated H_2O; 1505 and 1290: coordinated NO_3^- (ionic NO_3^- absorptions absent); 1085 and 620: ionic ClO_4^-	[1]
Sm(XIV)(NO_3)$_2$(ClO_4)·H_2O	White	IR: similar to Nd analog	[1]

[a] See footnote a in table 1.
[b] Reference: [1] Arif et al. (1985).
[c] See footnote j in table 1.

groups) and three bidentate chelating nitrato groups, two on one side of the macrocycle and one on the other (fig. 10). The structure of this compound thus closely resembles that of the previously described [La(NO_3)$_3$(I)]. The three-nitrogen-donor portion of the macrocycle is approximately coplanar, and the triether portion has a conformation similar to that found in crown ether complexes. The ligand as a whole is considerably folded, the angle between the best planes defined by N(11), N(12), N(13), La and O(1), O(2), O(3), La being 135.38°. Thus, the bending of the macrocycle is considerably more pronounced than in the La–trinitrato complex of ligand I (fig. 1 and table 2). Selected structural data for [La(NO_3)$_3$(XIV)] are given in table 11.

Complexes of a triether ligand related to XIV, but with partly aromatic side-chains, have been reported (Zhang and Wang 1986). The compounds were formulated as M(ligand)(NCS)$_3$·2H_2O (M = La–Gd) and M (ligand) (NO_3)$_3$·4H_2O (M = La–Eu). A La(III) perchlorate complex of ligand XV, formulated as La(XV)(ClO_4)$_3$·4H_2O has been mentioned (Wang and Miao 1984).

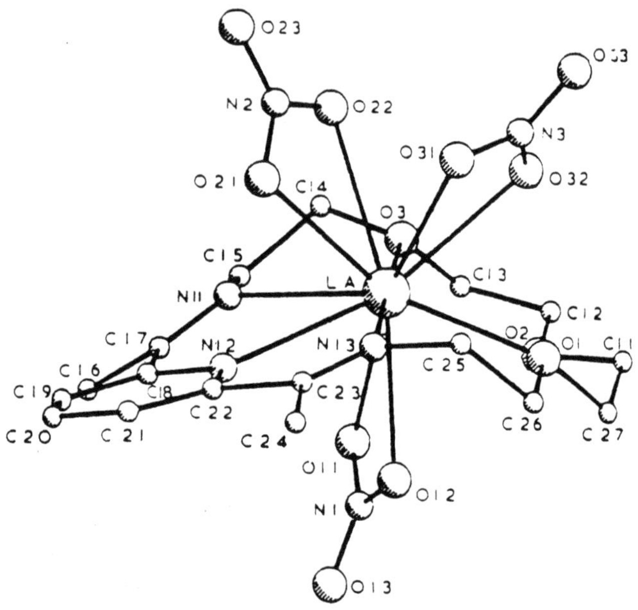

Fig. 10. Structure of [La$(NO_3)_3$(XIV)] (Arif et al. 1985).

TABLE 11
Selected crystallographic data and bond lengths for [La$(NO_3)_3$(XIV)].[a]

Formula	$C_{17}H_{25}N_6O_{12}La$		
Mol. wt.	644.33		
Crystal system	Orthorhombic		
Space group	Pbca		
a (Å)	21.657		
b (Å)	17.291		
c (Å)	12.792		
V (Å^3)	4790.5		
Z	8		
D_c (g cm^{-3})	1.70		
Coordinate bond lengths (Å)	La–O(1)	2.746 (9)	ether
	La–O(2)	2.781 (9)	
	La–O(3)	2.634 (9)	
	La–O(11)	2.713 (9)	NO_3^-
	La–O(12)	2.643 (9)	
	La–O(21)	2.655 (9)	NO_3^-
	La–O(22)	2.752 (9)	
	La–O(31)	2.695 (8)	NO_3^-
	La–O(32)	2.653 (8)	
	La–N(11)	2.755 (10)	imine
	La–N(12)	2.734 (9)	py
	La–N(13)	2.680 (10)	imine

[a] Arif et al. (1985).

1.1.6. *Potentially anionic compartmental ligands*

The compounds of this class are listed in table 12. The two-step (2 + 2) condensation of 2,6-diformyl-4-chlorophenol with an α,ω-primary diamine incorporating an additional donor group (S or NH) produced the 8-donor ligands H_2-XVIII and H_2-XIX (Guerriero et al. 1987). The first step of the synthesis involved the (nontemplated) condensation of one molecule of the diamine with two molecules of the dicarbonylphenol, to give an open "three-quarter cycle" diketone. This was treated with a lanthanide nitrate and the resulting complex then acted as a template in the subsequent condensation with a second molecule of the same diamine. The reaction was carried out in the presence of a strong base (LiOH, NaOH) and complexes of the doubly deprotonated macrocycle (dinegative ion form) were obtained, having the general formula M(ligand^{2-})(NO_3)·nH_2O, where M = La, Gd, Dy, and n = 2–4. Direct complexation of the metal salt with the pre-synthesized, protonated macrocyclic ligand, in the absence of added base, resulted instead in complexes of the general formula M(H_2-ligand)$(NO_3)_3$·nH_2O, where M = La, Nd, Gd, Tb, Dy, Ho, Er, Yb, and n = 0–5. For ligand XVIII, complexation was accompanied by ring contraction to an imidazolidine-bearing macrocycle of smaller cavity size and decreased denticity, H_2-XX. The structure of the ring-contracted species was established by X-ray analysis of the Tb(III) (Guerriero et al. 1987) and Eu(III) (Bünzli et al. 1988) complexes. In these complexes, of formula [M$(NO_3)_2$(H_2-XX)]NO_3 (M = Tb(III) or Eu(III)), the metal ion is 9-coordinate, being bound to five donor atoms from the macrocyclic ligand (two oxygens of the phenol groups, two imine nitrogens, one secondary amine nitrogen) and to two bidentate chelating NO_3^- groups (fig. 11). The coordination polyhedron around the metal ion is best described as a distorted tricapped trigonal prism. The two diimine sections containing the phenol rings are planar, but the organic macrocycle as a whole is considerably puckered. Selected structural data are given in table 13.

A series of bimetallic macrocyclic complexes of the lanthanides has been obtained by the lanthanide-templated condensation of 2,6-diformyl-4-methylphenol and diethylenetetramine (Kahwa et al. 1986). A three-fold excess of the amine precursor promoted the formation of bimetallic complexes of the 10-donor, doubly deprotonated ligand H_2-XXI, having stoichiometry M_2(XXI)$(NO_3)_4$·nH_2O, M_2(XXI)$(NO_3)_3$(OH), M_2(XXI)$(NO_3)_2(OH)_2$, or M(XXI)$(ClO_4)_2(OH)_2$. The anhydrous tetranitrato species, M_2(XXI)$(NO_3)_4$ (M = La, Ce) were obtained in both a yellowish orange and an off-white form. On the basis of mass spectral data, these complexes were assigned a homobinuclear structure, with both lanthanide ions accommodated within the ligand cavity.

An extensive series of homo-bimetallic and hetero-bimetallic complexes of the 10-donor, doubly deprotonated ligand H_2-XXXI has been recently obtained (Guerriero et al. 1990). Both the direct complexation of the lanthanide salt(s) with the pre-synthesized ligand, in the presence of a base, and the metal-templated synthesis from 2,6-diformyl-4-methylphenol and 1,8-diamino-3,5-dioxaoctane were found to be suitable for the preparation of the complexes. The homobinuclear complexes were obtained as homogeneous microcrystals (5–10 μm thick platelets). The heterobinuclear complexes were microcrystalline powders for which all evidence indicated,

TABLE 12
Macrocyclic complexes of the lanthanide(III) ions with potentially anionic ligands.

Compound[a]	Color	Other data[h]	Ref.[b]
Ligand H_2-XVIII = $C_{24}H_{28}Cl_2N_6O_2$			
Gd(XVIII)$(NO_3)\cdot 4H_2O$	Yellowish orange	IR(cm^{-1}): 1660 (s) ν(C=N, imine), 1638 (s) ν(C=C, bz), 1543 (s) $\nu(C\text{–}O^-)$, 3277 (br) ν(NH), 1386 (vs) $\nu(NO_3^-$ ionic)	[1]
Dy(XVIII)$(NO_3)\cdot 3H_2O$	Yellowish orange	IR: similar to Gd analog	[1]
Ligand H_2-XIX = $C_{24}H_{26}Cl_2N_4O_2S_2$			
La(XIX)$(NO_3)\cdot 4H_2O$	Yellowish orange	IR(cm^{-1}): 1655 (s) ν(C=N, imine), 1634 (s) ν(C=C, bz), 1638 (s) $\nu(C\text{–}O^-)$, 1386 (vs) $\nu(NO_3^-$ ionic)	[1]
Dy(XIX)$(NO_3)\cdot 2H_2O$	Yellowish orange	IR: similar to La analog	[1]
Ligand H_2-XX = $C_{24}H_{28}Cl_2N_6O_2$ = ring-contracted H_2-XVIII			
La(H_2-XX)$(NO_3)_3\cdot 3H_2O$	Yellowish orange	IR[c]	[1]
Nd(H_2-XX)$(NO_3)_3\cdot 3H_2O$	Yellowish orange	IR[c]	[1]
Gd(H_2-XX)$(NO_3)_3$	Yellowish orange	IR[c]	[1]
*[Tb$(NO_3)_2$(H_2-XX)]NO_3	Yellowish orange	X-ray structure; fig. 11, table 13	[1]
		IR(cm^{-1}): 3224 (br) ν(NH), 1662 (s) ν(C=N, imine), 1637 (s) ν(C=C, bz), 1542 (s) $\nu(C\text{–}O^-)$, 1386 (vs) $\nu(NO_3^-$ ionic)	[1]
*[Eu$(NO_3)_2$(H_2-XX)]NO_3	(a) Yellow crystals (unstable)		
	(b) Red crystals (stable)	X-ray structure of red form, table 13; IR: identical to Tb analog	[2]
		Luminescence spectrum (emission, excitation, lifetime)	[2]
Ho(H_2-XX)$(NO_3)_3\cdot 3H_2O$	Yellowish orange	IR[c]	[1]
Er(H_2-XX)$(NO_3)_3\cdot 5H_2O$	Yellowish orange	IR[c]	[1]
Yb(H_2-XX)$(NO_3)_3$	Yellowish orange	IR[c]	[1]
Ligand H_2-XXI = $C_{30}H_{44}N_8O_2$			
La_2(XXI)$(NO_3)_4$	Off-white	UV–VIS λ_{max} (nm, solid reflectance): *ca.* 435 (sh), *ca.* 370 (dominant peak)	[3]
		IR(cm^{-1}): 1640–1630 (vs) ν(C=N), 1550–1545 (s) $\nu(C\text{–}O^-)$, 1500–1200: absorptions of NO_3^-; dec. 250° C	[3]

$Ce_2(XXI)(NO_3)_4$	Off-white	UV–VIS[d], IR[d], dec.[d]	[3]
$Pr_2(XXI)(NO_3)_4 \cdot 2H_2O$	Off-white	UV–VIS[d], IR[d], dec.[d]	[3]
$Nd_2(XXI)(NO_3)_4 \cdot 2H_2O$	Off-white	UV–VIS[d], IR[d], dec.[d]	[3]
$Sm_2(XXI)(NO_3)_4 \cdot 2H_2O$	Off-white	UV–VIS[d], IR[d], dec.[d]	[3]
$Eu_2(XXI)(NO_3)_4 \cdot 2H_2O$	Off-white	UV–VIS[d], IR[d], dec.[d]; FAB-MS: 760, 762, corresponding to $\{Eu(NO_3)(XXI) + H^+\}$ (151 Eu, 153 Eu)	[3]
$Gd_2(XXI)(NO_3)_4 \cdot 2H_2O$	Off-white	UV–VIS[d], IR[d], dec.[d]; FAB-MS: 763–769, corresponding to $\{Gd(NO_3)(XXI) + H^+\}$ with appropriate isotopic distribution	[3]
$La_2(XXI)(NO_3)_4$	Yellow	IR[d], dec.[d]; UV–VIS λ_{max}(nm, solid reflectance): 435 (sh), *ca.* 370 (dominant peak)	[3]
$Ce_2(XXI)(NO_3)_4$	Yellowish orange	IR[d], dec.[d], UV–VIS[e]	[3]
$Pr_2(XXI)(NO_3)_4$	Yellowish orange	IR[d], dec.[d], UV–VIS[e]	[3]
$Nd_2(XXI)(NO_3)_4$	Yellowish orange	IR[d], dec.[d], UV–VIS[e]	[3]
$Eu_2(XXI)(NO_3)_4$	Yellowish orange	IR[d], dec.[d], UV–VIS[e]; FAB-MS: highest peak 549, corresponding to H_3-XXI^+	[3]
$Ho_2(XXI)(NO_3)_4$	Yellowish orange	IR[d], dec.[d], UV–VIS[e]	[3]
$Tm_2(XXI)(NO_3)_4$	Yellowish orange	IR[d], dec.[d], UV–VIS[e]	[3]
$Yb_2(XXI)(NO_3)_4$	Yellowish orange	IR[d], dec.[d], UV–VIS[e]; FAB-MS: highest peak 549, corresponding to H_3-XXI^+	[3]
$Lu_2(XXI)(NO_3)_4$	Yellowish orange	IR[d], dec.[d], UV–VIS[e]	[3]
$La_2(XXI)(NO_3)_2(OH_2)$	Deep orange	IR[d], dec.[d], UV–VIS[e]; λ_{max}(nm, solid reflectance): 457 (dominant peak), 435	[3]
$Nd_2(XXI)(NO_3)_2(OH)_2$	Orange	IR[d], dec.[d], UV–VIS[e]	[3]
$Eu_2(XXI)(NO_3)_3(OH)$	Orange	IR[d], dec.[d], UV–VIS[e]	[3]
$Gd_2(XXI)(NO_3)_3(OH)$	Orange	IR[d], dec.[d], UV–VIS[e]	[3]
$Gd_2(XXI)(NO_3)_2(OH)_2$	Orange	IR[d], dec.[d], UV–VIS[e]	[3]
$Tb_2(XXI)(NO_3)_3(OH)$	Orange	IR[d], dec.[d], UV–VIS[e]	[3]
$Dy_2(XXI)(NO_3)_3(OH)$	Orange	IR[d], dec.[d], UV–VIS[e]	[3]
$Ho_2(XXI)(NO_3)_3(OH)$	Orange	IR[d], dec.[d], UV–VIS[e]	[3]
$Er_2(XXI)(NO_3)_3(OH)$	Orange	IR[d], dec.[d], FAB-MS: highest peak 549, corresponding to H_3-XXI^+	[3]
$Y_2(XXI)(NO_3)_3(OH)$	Orange	IR[d], dec.[d], VIS–UV[e]	[3]
$Nd_2(XXI)(ClO_4)_3(OH)$	Orange	IR: similar to La(III)-nitrate complex, except for absorptions at *ca.* 1100 and 630 cm^{-1}, both split, assigned to coordinated ClO_4^-; UV–VIS[e]	[3]

TABLE 12 (cont'd)

Compound	Color	Other data[h]	Ref.[a]
$Sm_2(XXI)(ClO_4)_3(OH)$	Orange	IR: similar to Nd analog; VIS–UV[e]	[3]
$Gd_2(XXI)(ClO_4)_3(OH)$	Orange	IR: similar to Nd analog; VIS–UV[e]	[3]
Ligand H_2-XXII = $C_{24}H_{30}N_6O_2$			
$La_2(XXII)(NO_3)_4 \cdot 5H_2O$		Diamagn.; IR(cm^{-1}): 1645–1630 [ν(C=N)]; 1450, 1300, 1035–1030 and 815: absorptions of bidentate NO_3^-; ^{13}C NMR: table 14	[4]
$Pr_2(XXII)(NO_3)_4 \cdot 5H_2O$		$\mu_{eff} = 3.43\mu_B$; IR[f]	[4]
$La_2(XXII)(NCS)_4 \cdot 4H_2O$		Diamagn.; IR: similar to nitrato complex, except for anion absorptions; coordinated NCS^-: 2050 cm^{-1}; ^{13}C NMR: table 14	[4]
$Pr_2(XXII)(NCS)_4 \cdot 4H_2O$		$\mu_{eff} = 3.43\mu_B$; IR[f]	[4]
$Nd_2(XXII)(NCS)_4 \cdot 4H_2O$		$\mu_{eff} = 3.49\mu_B$; IR[f]	[4]
$Sm_2(XXII)(NCS)_4 \cdot 4H_2O$		$\mu_{eff} = 1.60\mu_B$; IR[f]	[4]
$Eu_2(XXII)(NCS)_4 \cdot 4H_2O$		$\mu_{eff} = 3.41\mu_B$; IR[f]	[4]
Ligand H_2-XXXI = $C_{28}H_{34}N_4O_6Cl_2$			
$La_2(XXXI)(NO_3)_4 \cdot H_2O$	Pale yellow	Diamagn.; IR[g]; FAB-MS, 1054 ($M^+ - NO_3$)	[5]
$Pr_2(XXXI)(NO_3)_4 \cdot 2H_2O$	Pale yellow	$\mu_{eff} = 4.9\mu_B$; IR[g]	[5]
$Sm_2(XXXI)(NO_3)_4 \cdot H_2O$	Pale yellow	$\mu_{eff} = 2.19\mu_B$; IR[g]	[5]
$Eu_2(XXXI)(NO_3)_4 \cdot H_2O$	Pale yellow	$\mu_{eff} = 4.90\mu_B$; IR[g]; detailed luminescence spectrum	[5]
$Gd_2(XXXI)(NO_3)_4 \cdot 2H_2O$	Pale yellow	$\mu_{eff} = 11.17\mu_B$; IR[g]	[5]
$Tb_2(XXXI)(NO_3)_4 \cdot 2H_2O$	Pale yellow	$\mu_{eff} = 13.43\mu_B$; IR[g]; FAB-MS, 1094 ($M^+ - NO_3$)	[5]

$Dy_2(XXXI)(NO_3)_4 \cdot H_2O$	Pale yellow	$\mu_{eff} = 14.78\mu_B$; IR[g]	[5]
$LaSm(XXXI)(NO_3)_4 \cdot 2H_2O$	Whitish yellow	$\mu_{eff} = 1.55\mu_B$; IR[g]	[5]
$LaEu(XXXI)(NO_3)_4 \cdot 2H_2O$	Whitish yellow	$\mu_{eff} = 3.5\mu_B$; IR[g]; luminescence spectrum	[5]
$LaGd(XXXI)(NO_3)_4 \cdot 2H_2O$	Whitish yellow	$\mu_{eff} = 7.85\mu_B$; IR[g]	[5]
$LaTb(XXXI)(NO_3)_4 \cdot 2H_2O$	Whitish yellow	$\mu_{eff} = 9.5\mu_B$; IR[g]	[5]
$LaDy(XXXI)(NO_3)_4 \cdot H_2O$	Whitish yellow	$\mu_{eff} = 10.45\mu_B$; IR[g]	[5]
$DyEu(XXXI)(NO_3)_4 \cdot H_2O$	Whitish yellow	$\mu_{eff} = 11.02\mu_B$; IR[g]; luminescence spectrum	[5]
$DyGd(XXXI)(NO_3)_4 \cdot 2H_2O$	Whitish yellow	$\mu_{eff} = 13.2\mu_B$; IR[g]	[5]
$GdEu(XXXI)(NO_3)_4 \cdot 2H_2O$	Whitish yellow	$\mu_{eff} = 8.64\mu_B$; IR[g]; luminescence spectrum	[5]
$GdTb(XXXI)(NO_3)_4 \cdot H_2O$	Whitish yellow	$\mu_{eff} = 12.32\mu_B$; IR[g]	[5]
$EuTb(XXXI)(NO_3)_4 \cdot 2H_2O$	Whitish yellow	$\mu_{eff} = 10.12\mu_B$; IR[g]; luminescence spectrum	[5]
Ligand H_3-XXIII = $C_{36}H_{45}N_9O_3$			
$La_3(XXIII)(NO_3)_6 \cdot 3CH_3OH$	Yellow	^{1}H NMR, table 15; FAB-MS: highest peak 1375, corresponding to $La_3(NO_3)_5$ (XXIII)	[6, 7]

[a] See footnote a in table 1.

[b] References: [1] Guerriero et al. (1987); [2] Bünzli et al. (1988); [3] Kahwa et al. (1986); [4] Sakamoto (1987); [5] Guerriero et al. (1990); [6] Fenton et al. (1987); [7] Fenton et al. (1988).

[c] Spectrum very similar to that of Tb analog.

[d] Similar to "off-white" La analog.

[e] Similar to yellowish orange La analog. For additional data, see text.

[f] Spectrum similar to La analog.

[g] Spectra were nearly identical for all members of this series; representative absorptions (cm^{-1}) of La_2 complex: 1651 (s), 1559 (s) and 1542 (s): ν(C=N) and $\nu(C–O^-)$; 1460 (br), 1295 (s), 1036 (m), 817 (m): coordinated NO_3^-.

[h] See footnote j in table 1.

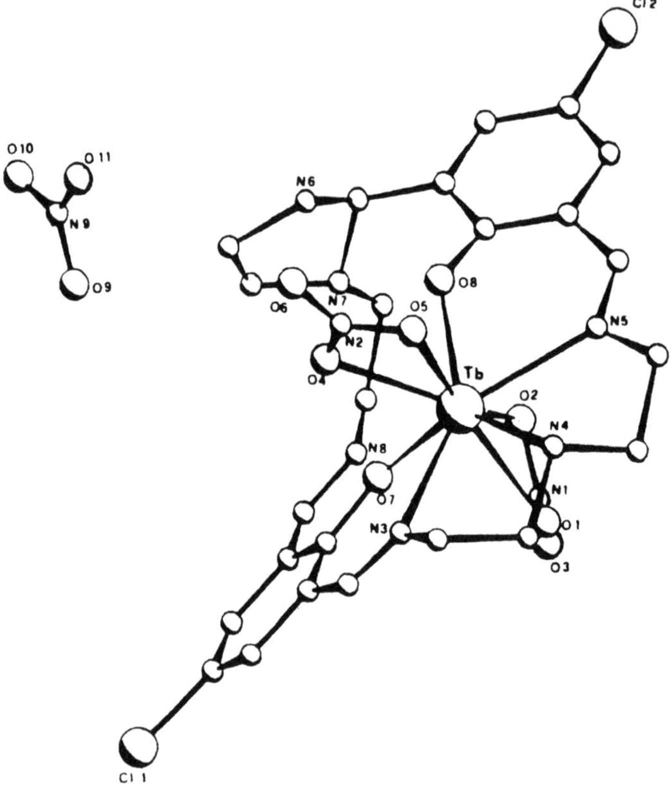

Fig. 11. Structure of $[Tb(NO_3)_2(H_2\text{-XX})](NO_3)$ (Guerriero et al. 1987).

although did not unambiguously prove, a true hetero-occupancy of the macrocycle. A detailed luminescence study was carried out to investigate the possibility of metal–metal energy transfer within the macrocyclic cavity.

The condensation of 2,6-diacetylpyridine with 1,3-diamino-2-hydroxypropane in the presence of a lanthanide nitrate or thiocyanate was reported (Sakamoto 1987) to produce bimetallic complexes of the doubly deprotonated (2 + 2) macrocyclic ligand H_2-XXII, having formula $M_2(\text{XXII})(NO_3)_4 \cdot 5H_2O$ (M = La, Pr) and $M_2(\text{XXII})(NCS)_4 \cdot 4H_2O$ (M = La, Pr, Nd, Sm, Eu). The carbon-13 NMR spectra of the La(III) complexes are given in table 14. The same combination of diamine and diketone precursors, in the presence of lanthanum(III) nitrate, was also reported to give a trimetallic species, $La_3(\text{ligand}^{3-})(NO_3)_6 \cdot 3MeOH$, postulated to contain the triply deprotonated (3 + 3) macrocyclic ligand H_3-XXIII (Fenton et al. 1987). The structural assignment of this complex was supported by a detailed study of the proton NMR spectrum in D_2O (table 15). The only apparent difference in the method of synthesis of these (2 + 2) and (3 + 3) species is the initial order of addition of the precursors: in Sakamoto's work, 1,3-diamino-2-hydroxypropane was added to a

TABLE 13
Selected crystallographic data and bond lengths for $[Tb(NO_3)_2(H_2\text{-XX})]NO_3$[a] and $[Eu(NO_3)_2(H_2\text{-XX})]NO_3$[b].

Formula	$C_{24}H_{26}N_9O_{11}Cl_2Tb$	$C_{24}H_{26}N_9O_{11}Cl_2Eu$
Mol. wt.	846	839.4
Crystal system	Monoclinic	Monoclinic
Space group	C_2/c	C_2/c
a (Å)	23.668 (5)	23.783 (3)
b (Å)	14.328 (7)	14.377 (3)
c (Å)	19.482 (5)	19.346 (3)
β (deg)	91.82 (5)	91.76 (4)
V (Å^3)	6603	6611
Z	8	8
D_c (g cm^{-3})	1.70	1.69
Coordinate bond lengths (Å)	Tb–O(1) 2.47 (1) } nitrate	Eu–O(1) 2.48 (1) } nitrate
	Tb–O(2) 2.51 (1) } nitrate	Eu–O(2) 2.53 (1) } nitrate
	Tb–O(4) 2.42 (2) } nitrate	Eu–O(4) 2.48 (2) } nitrate
	Tb–O(5) 2.43 (2) } nitrate	Eu–O(5) 2.48 (2) } nitrate
	Tb–O(7) 2.26 (1) } phenol	Eu–O(7) 2.28 (1) } phenol
	Tb–O(8) 2.24 (1) } phenol	Eu-O(8) 2.28 (1) } phenol
	Tb–N(3) 2.55 (1) py	Eu–N(3) 2.57 (1) py
	Tb–N(4) 2.52 (2) NH	Eu–N(4) 2.54 (2) NH
	Tb–N(5) 2.55(2) py	Eu–N(5) 2.58 (2) py

[a] Guerriero et al. (1987).
[b] Bünzli et al. (1988).

methanol solution of metal nitrate and 2,6-diacetylpyridine, whereas in the work of Fenton et al., 2,6-diacetylpyridine was added to a methanol solution of the metal nitrate and 1,3-diamino-2-hydroxypropane.

1.1.7. *Mono- and bi-cyclic poly-amine ligands without Schiff-base linkages*

The compounds of this class are listed in table 16. Lanthanide complexes of the “dibridged” monocyclic amine XXXII were obtained by the condensation of tris(2-aminomethyl)amine (tren) with an excess of the formaldehyde derivative, bis(dimethylamino)methane, in the presence of a lanthanide trifluoromethylsulfonate (triflate) as template (Smith and Raymond 1985, Smith et al. 1988). The reaction was carried out with a 2:1 mole ratio of tren to triflate, and the mixture was refluxed in anhydrous acetonitrile under dry nitrogen or argon. The complexes had the general formula M(XXXII)(triflate)$_3\cdot n$CH$_3$CN and their structures were established by X-ray analysis of the $[La(CF_3SO_3)_2(XXXII)](CF_3SO_3)\cdot CH_3CN$ and $[Yb(CF_3SO_3)(XXXII)](CF_3SO_3)_2\cdot CH_3CN$ species (Smith and Raymond 1985), as illustrated in fig. 12. Selected structural data are given in table 17.

A fully encapsulated Yb(III) complex of the “tribridged” cage-ligand XXXIII, having formula Yb(XXXIII)(triflate)$_3\cdot$CH$_3$CN, was obtained (Smith et al. 1988,

TABLE 14

^{13}C NMR chemical shifts of complexes of ligand H_2-XXII[a].

Complex	Chemical shift (ppm)					
	C=N	Pyridine	NCS^-	CH	CH_2	CH_3
$La_2(XXII)(NO_3)_4(H_2O)_5$	168.55, 166.67	154.87, 125.76	–	75.68	60.00, 58.42	16.50, 16.03
	167.56	141.67, 125.23	–	73.49	59.24	16.26, 15.85
$La_2(XXII)(NCS)_4(H_2O)_4$	169.67, 168.44	154.76, 141.49	131.10	75.85	59.83, 58.95	16.97, 16.38
	168.79, 167.32	154.64, 125.99		74.50	59.24, 58.77	16.50, 16.09
		154.35, 125.46				
		141.90				
$Ba_2(H_2\text{-XXII})(ClO_4)_2(H_2O)_2$[b]	165.86	155.05, 123.88	–	68.69	57.47	16.09
		139.67				

[a] Sakamoto (1987).

[b] Included for comparison.

TABLE 15
Chemical shifts and coupling constants for the major product of the $La(NO_3)_3$-template synthesis of ligand XXIII[a].

Chemical shifts (ppm)	Integral	Assignment	Coupling constants (Hz)
8.450	2H	$2H^{11}$	$J(H^1–H^2) = 8$
8.445	1H	H^1	$J(H^{10}–H^{11}) = 8$
8.348	2H	$2H^{10}$	$J(H^{11}–H^{12}) = 8$
8.324	2H	$2H^{12}$	$J(H^6–H^5) = 2$
8.260	2H	$2H^2$	$J(H^6–H^4) = 12$
4.97	2H	$2H^6$	$J(H^6–H^8) = 5.5$
4.75	1H	H^{16}	$J(H^4–H^5) = 12$
4.32	2H	$2H^7$	$J(H^7–H^8) = 14$
4.25	2H	$2H^8$	$J(H^{16}–H^{15}) = 3$
4.16	2H	$2H^{15}$ (*cis*)	$J(H^{16}–H^{14}) = 11$
3.96	2H	$2H^5$	$J(H^{15}–H^{14}) = 13$
3.62	2H	$2H^{14}$ (*trans*)	
3.49	2H	$2H^4$	
2.74	6H	$2CH_3{}^9$	
2.68	6H	$2CH_3{}^{13}$	
2.57	6H	$2CH_3{}^3$	

[a] Fenton et al. (1988).

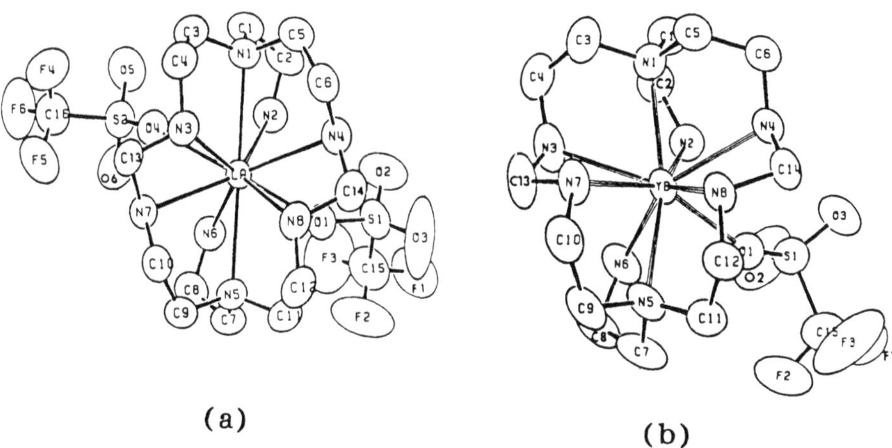

Fig. 12. Structures of the complex cations in the complexes $M(XXXII)(CF_3SO_3)_3 \cdot CH_3CN$: (a) $[La(CF_3SO_3)_2(XXXII)]^+$; (b) $[Yb(CF_3SO_3)(XXXII)]^{++}$ (Smith and Raymond 1985).

Raymond and Smith 1988) from the same precursors used for the synthesis of the dibridged analog, but using a larger excess of the formaldehyde reagent and longer reflux time. The isolation and purification of the product were laborious and the yields quite low (3–5%). The structure of this compound was established primarily on the basis of fast-atom-bombardment (FAB) mass spectrometry.

TABLE 16
Complexes of the lanthanide(III) and yttrium(III) ions with cage-like polycyclic amines.

Compound[a]	Other data[b]	Ref.[c]
Ligand XXXII = $C_{14}H_{36}N_8$		
$[La(CF_3SO_3)_2(XXXII)](CF_3SO_3)\cdot CH_3CN$	X-ray structure: fig. 12a, table 17; IR[d]	[1, 2]
	NMR: ^1H, δ 4.1 (mult), 3.8 (br), 3.0 (mult), 2.75 (mult), 2.5 (mult); ^{13}C, δ 63.0, 59.4, 58.8, 55.9, 47.3, 44.8 and 40.6.	[1, 2]
	Positive ion FAB-MS: molecular ion = $[La(trif)_2(L)]^+$ at $m/z = 753$ (with appropriate isotope distribution)	[1, 2]
$Ce(XXXII)(CF_3SO_3)_3$	IR[d]; ^1H NMR: δ 19.05 (area = 1), 18.73 (area = 1), 11.67 (area = 1), 11.21 (area = 1), 10.71 (area = 2), 9.76 (area = 1), 6.81 (area = 2), 3.08 (area = 1), 2.23 (area = 1), 1.32 (area = 1), 0.80 (area = 1), −2.08 (area = 1), −4.10 (area = 1), −5.87 (area = 1), −8.65 (area = 1) and −12.52 (area = 1)	[1, 2]
	Positive ion FAB-MS: molecular ion = $[Ce(trif)_2(L)]^+$ at m/z (with appropriate isotope distribution; no value given)	[1, 2]
$Pr(XXXII)(CF_3SO_3)_3$	IR[d]; ^1H NMR: δ 32.84, 30.79, 23.06, 21.95, 21.22, 16.90, 13.43, 9.99, 8.63, 4.52, 3.93, 2.05, −0.27, −3.22, −9.57, −9.86, −18.94 and −28.21.	[1, 2]
	Positive ion FAB-MS: molecular ion = $[Pr(trif)_2(L)]^+$ at m/z 755 (with appropriate isotope distribution)	[1, 2]
$Eu(XXXII)(CF_3SO_3)_3$	IR[d]; ^1H NMR: δ 26.19 (mult), 25.70, 17.01, 15.58, 14.77, 12.44 (mult), 3.00, −0.93, −0.99, −1.05, −2.38, −6.37, −6.39, −8.36, −8.43, −9.34, −9.36, −9.69, −9.75, −10.09, −10.15, −14.90 (mult), −15.99, −16.05, −17.50, −17.58, −19.41, −19.47, −21.25 (br), −25.52 and −25.58.	[1, 2]
	Positive ion FAB-MS: molecular ion = $[Eu(trif)_2(L)]^+$ at m/z 767 (with appropriate isotope distribution)	[1, 2]

Compound	Data	Ref.[c]
$[Yb(CF_3SO_3)(XXXII)](CF_3SO_3)_2 \cdot CH_3CN$	X-ray structure: fig. 12b, table 17; IR[d]	[1, 2]
	^1H NMR: δ 95 (vbr), 65 (vbr), 45 (vbr), 37, 29, 20, 12, 7 (mult), −6, −12, −21, −32, −43, −58, −78, −100 (vbr), −144 (vbr) and −198 (vbr)	
	Positive ion FAB-MS: molecular ion = $[Yb(trif)_2(L)]^+$ at m/z 788 (with appropriate isotope distribution)	[1, 2]
$Y(XXXII)(CF_3SO_3)_3$	IR[d]; NMR: ^1H, δ 4.1 (mult), 3.9 (br), 3.25 (mult), 3.1 (mult), 2.9 (br), 2.8 (mult), 2.6, 2.55; ^{13}C, δ 59.54, 59.14, 55.85, 49.92, 46.35, 46.29, 41.79, 41.13	[1, 2]
	Positive ion FAB-MS: molecular ion = $[Y(trif)_2(L)]^+$ at m/z 703	[1, 2]
Ligand XXXIII = $C_{15}H_{36}N_8$		
$Yb(XXXIII)(CF_3SO_3)_3 \cdot CH_3CN$	IR[d]; ^1H NMR in CD_3CN: δ 97.6, 95.2, 80.6, 79.0, 75.6, 66.2, 57.5 (br), 44.0, 43.3, 41.6, 38.5, 36.1, 34.8, 32.4, 30.1, 28.2, 27.2, 25.4, 24.3, 21.4, 19.5, 17.9, 15.8, 14.4, 11.8, 9.3, 7.9, 6.9, 6.2, −6.1, −7.5, −8.3, −10.8, −14.5, −16.9, −17.8, −22.6, −28.9, −30.1, −36.1, −40.7, −41.8, −44.5, −49.4, −49.7, −52.5, −56.6, −58.6, −66.2, −68.9, −79.4, −81.6 and −110. ^1H NMR in D_2O: δ 42.3, 33.1, 28.4, 25.5, 23.8, 20.5, 18.7, 10.1, 5.7, −1.4, −4.9, −13.4, −18.9, −21.1, −24.9, −25.5, −41.4, −42.3, −53.4, −61.6, −64.1, −68.7 and −73.7	[1, 2]
	Positive ion FAB-MS: m/z calculated for $[Yb(XXXIII)(CF_3SO_3)_2CH_3CN]^+$: 841 (with appropriate isotope distribution pattern); Polarogram	[3]

[a] See footnote a in table 1.
[b] See footnote j in table 1.
[c] References: [1] Smith and Raymond (1985); [2] Smith et al. (1988); [3] Raymond and Smith (1988).
[d] Complete list of IR absorptions given in Ref. [2] without assignments.

TABLE 17
Selected crystallographic data and bond lengths for complexes of a two-strand cyclic amine, $[La(CF_3SO_3)_2(XXXII)](CF_3SO_3)\cdot CH_3CN$ and $[Yb(CF_3SO_3)(XXXII)](CF_3SO_3)_2\cdot CH_3CN$.[a]

Formula	$C_{19}H_{39}N_9O_9S_3F_9La$		$C_{19}H_{39}N_9O_9S_3F_9Yb$	
Mol. wt.	943.68		977.81	
Space group	$P\bar{1}$		$P\bar{1}$	
a (Å)	10.716 (2)		9.594 (2)	
b (Å)	13.733 (2)		10.940 (2)	
c (Å)	14.211 (2)		18.072 (2)	
α (deg)	98.87 (2)		74.068 (10)	
β (deg)	97.54 (2)		94.698 (13)	
γ (deg)	107.54 (2)		81.901 (13)	
V (Å^3)	1936 (1)		1754.4 (5)	
Z	2		2	
D_c (g cm^{-3})	1.62		1.85	
Coordinate bond lengths (Å)	La–O(1)	2.585 (3) triflate	Yb–O(1)	2.390 (3) triflate
	La–O(4)	2.632 (3) triflate	Yb–N(1)	2.559 (3) ≡N
	La–N(1)	2.814 (3) ≡N	Yb–N(2)	2.442 (3) $-NH_2$
	La–N(2)	2.669 (4) $-NH_2$	Yb–N(3)	2.611 (4) =NH
	La–N(3)	2.756 (3) =NH	Yb–N(4)	2.568 (3) =NH
	La–N(4)	2.701 (3) =NH	Yb–N(5)	2.611 (3) ≡N
	La–N(5)	2.816 (3) ≡N	Yb–N(6)	2.453 (3) $-NH_2$
	La–N(6)	2.677 (3) $-NH_2$	Yb–N(7)	2.468 (3) =NH
	La–N(7)	2.685 (3) =NH	Yb–N(8)	2.475 (3) =NH
	La–N(8)	2.751 (3) =NH		

[a] Smith and Raymond (1985).

1.1.8. (2 + 2) *ligands with pyridine head-units, aliphatic side-chains and peripheral functionalities*

Several lanthanide complexes of macrocyclic ligands with pendant peripheral substituents, XXXIV–XXXVI, have recently been obtained by the metal-templated condensation of 2,6-diacetylpyridine with an appropriately carbon-substituted 1,2-diaminoethane in the presence of a lanthanide acetate or triflate (Gootee et al. 1989; Gribi et al. 1989). The complexes have the general formula M(macrocycle)X_3·$n$solvent (M = La, Eu, Gd, Tb, Y; X = CH_3COO^-, $CF_3SO_3^-$; n = 2–4; solvent = H_2O, CH_3OH). The synthesis and characterization of these products follow the same pattern described for their nonsubstituted analogs; the presence of the pendant –OH or $-NH_2$ on the carbon chain of the 1,2-diaminoethane precursor does not affect the macrocycle formation, except for requiring somewhat longer reaction times. Aside from any additional reactivity due to the pendant functionalities, the properties of these lanthanide macrocycles closely resemble those of the nonsubstituted analogs. However, the substituted complexes tend to form glassy or powdery solids rather than well-formed crystals. This is not surprising, since these species may be obtained as a mixture of isomers, as indicated for the complexes of ligand IV. A study of the isomers of these functionalized macrocycles is in progress in the author's laboratory, in an effort to establish whether the metal template effect offers some measure of structural control.

1.2. *General properties*

1.2.1. *Colors, thermal stability and solubility*

The macrocyclic complexes of the lanthanide(III) and yttrium(III) ions are high-melting solids that can occasionally be obtained as well-formed crystals. In the absence of colored counterions, the complexes of the (2 + 2) ligands with pyridine head-units and aliphatic diimine side-chains are colorless or show the light colors (table 1) expected from the f–f transitions of the metal ion. For example, the diacetate chloride complex of ligand I with Nd(III) has a very light lavender color and that of Er(III) has a light peach color. The fairly intense yellow or rose colors of the Ce(III) derivatives of ligands I and II have been attributed to essentially $f \rightarrow \pi^*$ transitions, but no detailed spectral studies of these systems have been reported. The complexes of (2 + 2) ligands with pyridine head-units and aromatic diimine side-chains (table 7) or hydrazine side-chains (table 6) have colors ranging from bright yellow to orange for all lanthanides, and so do the complexes of potentially ionizable, phenol type ligands (table 12). The UV–VIS absorptions of a number of these complexes have been reported but not assigned; the values are listed in the tables under "Other Data".

The metal–macrocycle entities considered in this section are generally stable to atmospheric conditions; an exception are the complexes of the monocyclic polyamine ligand XXXII, which had to be handled in an inert (dry N_2 gas) atmosphere (Smith and Raymond 1985). Most metal–macrocycle entities are also extremely stable to heat, while solvent molecules present in the crystal lattice are often lost at very low temperatures. For example, the hydrated diacetate chloride complexes of ligand I begin to lose water below 60°C and lose one acetate between 180 and 220°C, but the metal–macrocycle entities themselves remain unchanged up to 240–260°C (De Cola et al. 1986). Thermogravimetric analyses of various complexes have been reported and are listed in the tables under "Other Data". The thermal behavior of representative complexes of ligand I is illustrated by the thermograms in fig. 13.

The solubility of the complexes in various solvents depends on the ligand–counterion combination. In general, the (2 + 2) complexes of macrocycles with aliphatic diimine side-chains are more soluble than their aromatic counterparts; for each series, the triacetates and diacetate chlorides are generally more soluble in both water and organic solvents than are the triisothiocyanates, trinitrates and mixed-anion salts containing carboxylate–chloride, carboxylate–perchlorate, or nitrate–perchlorate combinations. The (1 + 1) complexes are generally soluble in various organic solvents.

1.2.2. *Inertness in solution*

One of the most striking and potentially useful properties of certain (2 + 2) complexes is the inertness of the metal–macrocycle entities with respect to ligand hydrolysis or metal release in solution, associated with a high lability of the exocyclic ligands (anions or solvent molecules). For example, the metal–macrocycle entities of

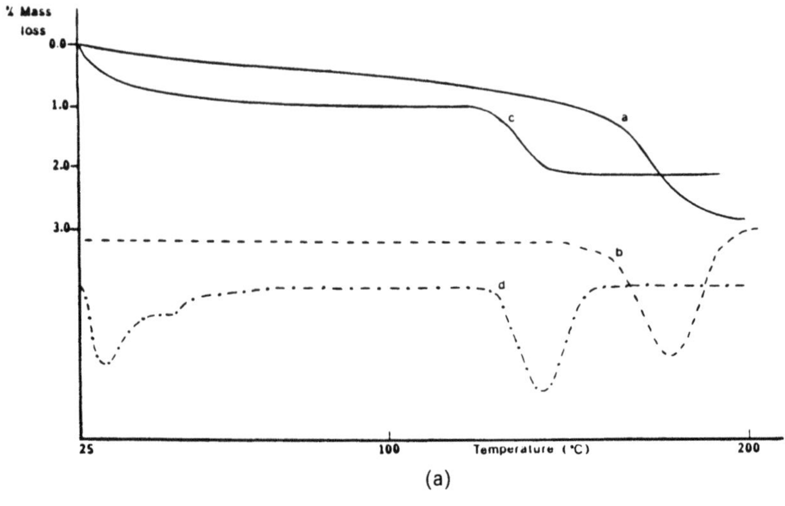

(a)

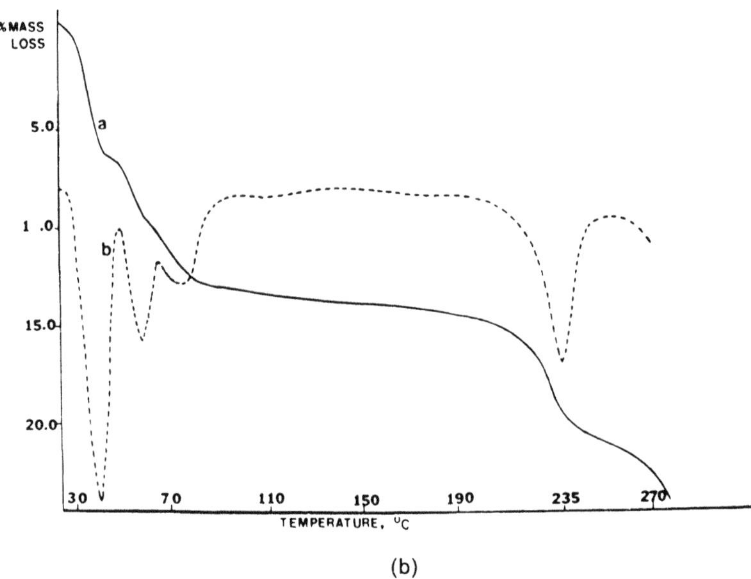

(b)

Fig. 13. Thermogravimetric behavior of (a) $[Lu(CH_3COO)(0.5CH_3OH–0.5H_2O)(I)](OH)$-$(ClO_4)\cdot 0.5CH_3OH$ (curves a and b represent mass loss and its first derivative; and curves c and d represent the same after rehydration by exposure to the atmosphere (Bombieri et al. 1986)); (b) Gd(I)-$(CH_3COO)_2Cl\cdot 6H_2O$ (curve a shows percentage mass loss versus temperature; curve b shows its first derivative (De Cola et al. 1986)).

ligand I do not show any sign of decomposition in water, even over a period of days, as proven by their unchanged NMR spectra. At room temperature, the complexes of ligand I are also resistant to strong acids and bases. For example, treatment of the diacetate chloride salts with gaseous hydrogen chloride or concentrated hydrobromic acid in methanol–diethylether solution results simply in anion metathesis and constitutes a convenient method for the preparation of the corresponding trihalides. Addition of anionic ligands that form highly insoluble precipitates with the "free" lanthanides (e.g. hydroxide, oxalate) or have high binding constants (e.g. 2,6-pyridinedicarboxylic acid) does not cause removal of the metal ion from the macrocycle entity; rather, precipitation of insoluble salts of the intact metal–macrocycle cation may occur.

This inertness, which contrasts dramatically with the ease of decomposition observed for lanthanide(III) complexes of nonmacrocyclic ligands containing similar donor systems, is specific of the complexes of ligands I and II (18-member cavity, six nitrogen donors with almost planar geometry) and must arise from an especially favorable combination of metal ionic radius, ligand cavity size, and bonding features of the donor atoms. This suggestion is supported by the observation that the complexes of ligand XI, which has an 18-member tetraimine cavity but contains furan instead of pyridine head-units, decompose rapidly when dissolved in water. Also, the lanthanum(III) nitrate complex of ligand V, which contains pyridine head-units but has a 20-member cavity, undergoes facile transmetallation with copper(II) (Abid and Fenton 1984b), while the complexes of the smaller lanthanides (Tb–Lu) with ligand VI (pyridine head-units, 14-member cavity) are easily hydrolyzed to derivatives of an open-cycle diketone (Radecka-Paryzek 1981a).

The flexibility of the $-CH_2-CH_2-$ diimine side chains must be an additional favorable factor in the stabilization of the lanthanide complexes of ligands I and II, as the corresponding complexes of ligands VIII and X, which also have 18-member cavities with pyridine head-units but include rigid *o*-phenylenediamine side-chains, are considerably less inert to metal release or rearrangement. For example, the proton and carbon-13 NMR spectra of $[La(NO_3)_3(X)]$ in freshly prepared DMSO-d_6 solutions show the resonances expected for ligand X. Over a period of days, however, the spectrum changes and new resonances appear, indicating the presence of a new organic species, either free or complexed with the La(III) ion. Also, the addition of small quantities of strong acids or bases to a methanol solution of the complexes of ligand VIII or X causes fleeting color changes (red for H^+ addition, bluish green for OH^- addition). Since the metal-free organic molecule related to macrocyclic ligand VIII behaves somewhat as an acid–base indicator with the same range of colors, the color changes observed for the complexes may be considered as a sign of minor decomposition; upon neutralization the original trinitrato complexes are recovered without detectable degradation (Benetollo et al. 1989).

1.2.3. *Crystal structures*

Single-crystal X-ray studies have been reported for a number of complexes of the lanthanide(III) and yttrium(III) ions with various macrocycles and are listed under "Other Data". Selected crystallographic data for each compound are reported in individual tables.

The majority of the structures are for complexes of macrocycle I and show several distinctive features. First, in each case all six N-donor atoms of the macrocyclic cavity are bound to the central metal ion and the bond distances are in good agreement with

TABLE 18
Fold angles for complexes of the lanthanides and yttrium with ligand I.

Complex	Coordination number of metal ion	Ionic radius (Å)	Macrocycle fold angle (deg)	Ref.[a]
$[La(NO_3)_3(I)]$	12	1.216	153.3	[1]
$[Ce(NO_3)_2(H_2O)(I)]^+$	11	1.196	121.0	[1]
$[Nd(NO_3)(H_2O)_2(I)]^{++}$	10	1.163	116.1	[1]
$[Nd(CH_3COO)_2(I)]^+$	10	1.163	132.8	[2a]
$[Eu(CH_3COO)_2(I)]^{+b}$	10	1.131	137.4	[2a]
$[Eu(CH_3COO)_2(I)]^{+c}$	10	1.131	122.5	[2b]
$[Gd(CH_3COO)_2(I)]^+$	10	1.131	132	[3]
$[Eu(NCS)_3(I)]$	9	1.120	111.3, 102.1	[4]
$[Y(NCS)_3(I)]$	9	1.075	111.1, 103.2	[4]
$[Y(CH_3COO)(H_2O)(I)]^{++}$	9	1.075	115.7	[5]
$[Y(CH_3C\underline{O}\underline{O})(CH_3CO\underline{O})(I)]^+$	9	1.075	115.7	[5]
$[Lu(CH_3COO)(H_2O)(I)]^{++}$	9	1.032	114.4	[6]
$[Lu(CH_3COO)(CH_3OH)(I)]^{++}$	9	1.032	114.4	[6]
$[Pr(NO_3)_2(CH_3OH)(VIII)]^+$	11	1.141	113.0 (py–py) 112.1 (bz–bz)	[7]

[a] References: [1] Arif et al. (1987); [2a] Benetollo et al. (1990a); [2b] Bombieri et al. (1989c); [3] Smith et al. (1989); [4] Bombieri et al. (1989a); [5] Bombieri et al. (1989b); [6] Bombieri et al. (1986); [7] Benetollo et al. (1989).

[b] Chloride counterions.

[c] Acetate counterions.

the values expected on the basis of the metal ionic radius. Second, the coordination sphere of the metal ion always includes two or three exocyclic ligands (anions and solvent molecules), at least one of which is anionic. Third, the macrocycle always departs from planarity, and displays a "folded butterfly" configuration consisting of two approximately planar "wings" containing the pyridine rings and hinged at the $-CH_2-CH_2-$ side-chains. The angle formed by the two "wings" varies from complex to complex, generally decreasing as the metal ionic radius and coordination number decrease. The bulk of the exocyclic ligands and the presence of hydrogen-bonding species external to the coordination sphere also affect the bending of the macrocycle, as illustrated by the values of the angles summarized in table 18. A marked departure from planarity is also observed for the organic moieties present in other lanthanide and yttrium macrocycles. It is significant, however, that the six-donor-atom rings of the few systems that have been subjected to ring-puckering analysis were found to be nearly planar.

In general, attempts to classify the coordination geometry of the metal ion in terms of a regular polyhedron have not been successful; in those examples where a geometry was identified, major distortions were observed.

1.3. *Spectral studies*

The structural identification of macrocyclic complexes obtained by metal-templated condensation of precursors rather than by complexation of metal salts with well-identified, pre-synthesized ligands must rely primarily on crystallographic data. However, once the structure of at least one member of a series has been established by single-crystal X-ray diffraction, the identification and characterization of related complexes may be conveniently obtained from chemical analysis, including thermogravimetric studies, and from spectroscopic measurements. Infrared (IR) absorption spectra, mass spectra (MS), especially in the fast-atom-bombardment (FAB) mode, proton and carbon-13 nuclear magnetic resonance (NMR) spectra (chiefly for the diamagnetic species), and excitation–emission (EX–EM) spectra (for some of the Eu(III) and Tb(III) complexes) have been widely used, either singly or in conjunction with one another.

1.3.1. *Infrared spectra*

Infrared spectra have been routinely measured for most macrocyclic complexes of the lanthanide(III) and yttrium(III) ions. Reported frequencies and assignments are given in the tables for the individual complexes. In general, each spectrum consists of a multitude of intense absorptions arising from the organic moiety; the pattern of these absorptions has excellent fingerprint value and often includes at least some features that are diagnostic of the macrocyclic structure. For example, the two strong and sharp bands at approximately 1650 and 1600 cm^{-1}, arising from the C=N stretching modes of the imine and pyridine systems, respectively, are characteristic of Schiff-base macrocycles with pyridine head-units (tables 1, 7, 10). The absorptions of solvent molecules and polyatomic anions (coordinated or ionic) are also clearly recognizable and diagnostic. It should be kept in mind, however, that minor structural variations

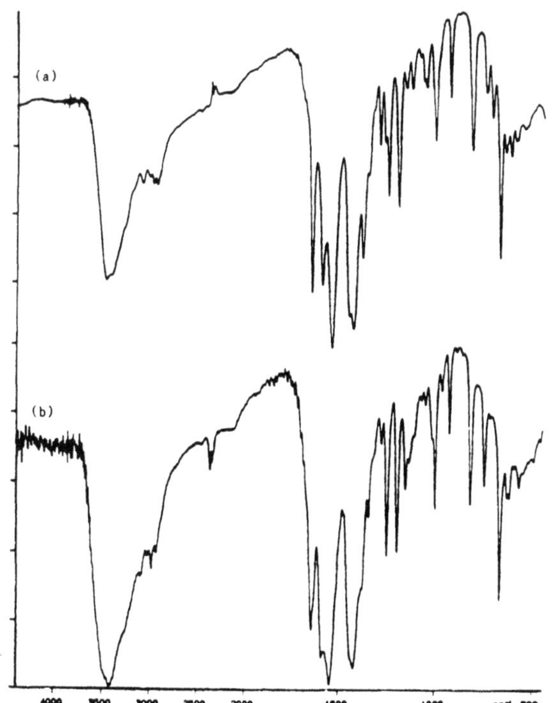

Fig. 14. Infrared spectra of (a) La(I)$(CH_3COO)_2Cl\cdot nH_2O$ (De Cola et al. 1986) and (b) Y(IV)$(CH_3COO)_2Cl\cdot nH_2O$ (Gootee et al. 1989).

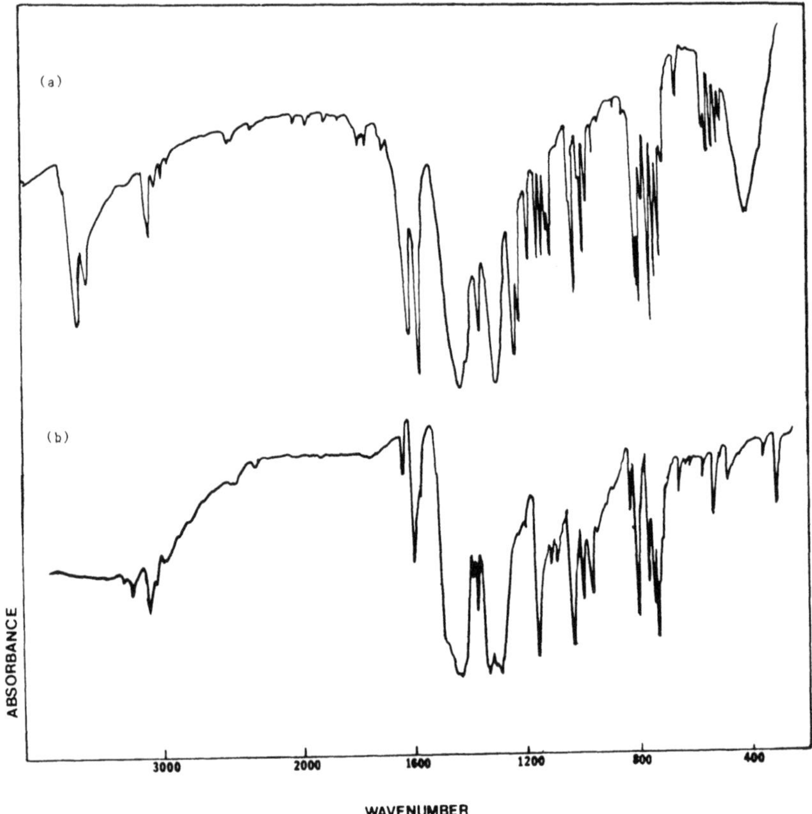

Fig. 15. Infrared spectra of (a) $[Pr(NO_3)_3(VIII)]\cdot H_2O$ (Benetollo et al. 1989), (b) $[Eu(NO_3)_3(X)]$ (Benetollo et al. 1991).

may not be detectable by IR spectroscopy; for example, the two spectra in fig. 14 are nearly identical yet belong to different macrocycles.

The spectra shown in fig. 15 illustrate the usefulness of IR spectroscopy in the identification of complexes for which direct X-ray data are not available. Spectrum (a) belongs to the trinitrato complex $[Pr(NO_3)_3(VIII)]\cdot H_2O$, which is a true macrocyclic complex of ligand VIII since it produces the X-ray-analyzed $[Pr(NO_3)_2(H_2O)(VIII)]$-$ClO_4\cdot 0.5H_2O\cdot 0.5CH_3OH$ by partial anion metathesis. Spectrum (b) belongs to the compound of stoichiometry $Eu(X)(NO_3)_3$, obtained from the condensation of 2,6-diformylpyridine with freshly sublimed 1,2-diaminobenzene in the presence of Eu(III) nitrate. A comparison of the two spectra provides reliable evidence that the latter complex is also a true macrocyclic species and may be formulated as $[M(NO_3)_3(X)]$.

1.3.2. *Nuclear magnetic resonance spectra*

Proton NMR spectra have been measured for many lanthanide macrocycles; carbon-13 spectra have been reported for selected species. Information on all reported NMR spectra is included in the tables under the individual complexes.

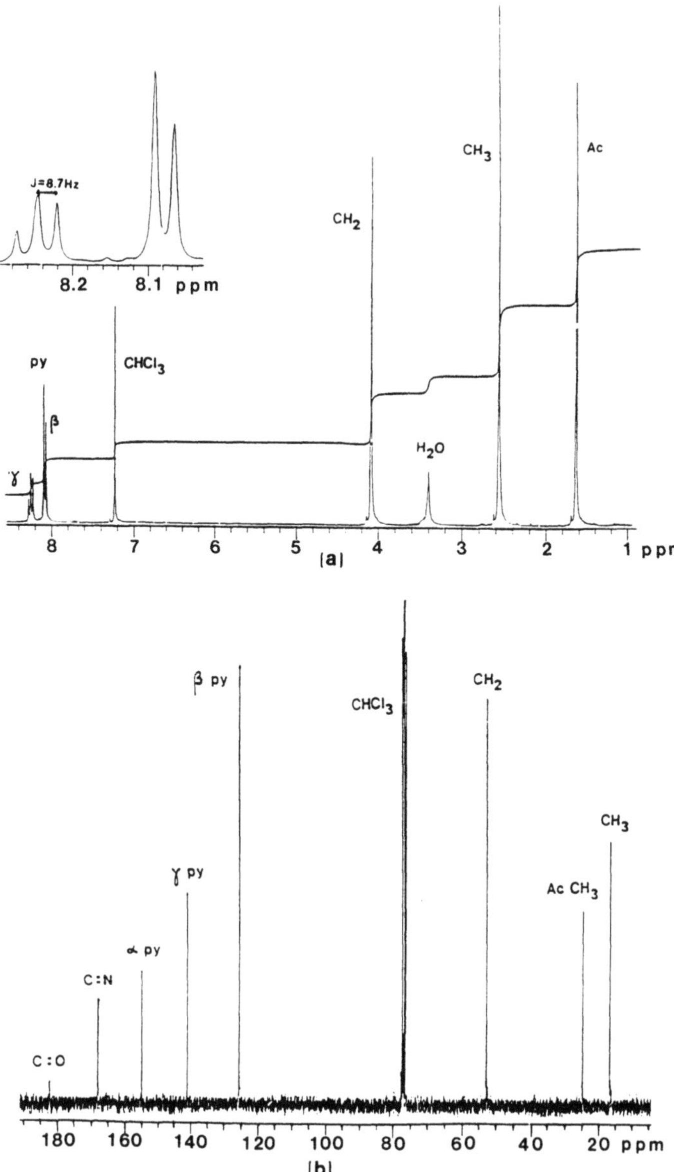

Fig. 16. NMR spectra of La(I) $(CH_3COO)_2Cl \cdot nH_2O$ in $CHCl_3$: (a) proton; (b) carbon-13 (Fonda and Vallarino 1989).

The spectra of the symmetric (2 + 2) macrocycles are rather simple; representative examples are shown in fig. 16. For the diamagnetic complexes of ligands I and II, the carbon-13 spectra (table 19) are readily assigned by analogy with related organic molecules; proton spectra (table 20) are unambiguously assigned on the bases of coupling and integration. The simplicity of these spectra and the fact that only one

TABLE 19
Representative carbon-13 nuclear magnetic resonance spectra of complexes of ligands I, II and VIII.

Complex[a]	Solvent	C=O (acetate)	C=N	αC–py	γC–py	βC–py	CH_2 (side chain)	CH_3 (ring)	CH_3 (acetate)	Benzene (side chain)	Ref.[b]
Ligand I = $C_{22}H_{26}N_6$											
Diamagnetic species											
*[La$(NO_3)_3$(I)]	D_2O	–	171.8	154.5	142.0	126.7	52.0	15.9	–	–	[1, 2]
La(I)$(CH_3COO)_2Cl \cdot nH_2O$	DMSO-d_6	182.6	169.0	154.4	141.3	126.4	51.8	16.1	23.8	–	[3]
Lu(I)$(CH_3COO)(ClO_4)(OH)\cdot(CH_3OH)\ 0.5(H_2O)$	DMSO-d_6	183.3	168.4	152.4	141.7	125.4	51.6	15.6	23.3	–	[4]
$[UO_2(I)]X_2$ ($X = ClO_4^-, NO_3^-, I^-$)	DMSO-d_6	–	177.2	153.5	143.8	128.2	52.8	18.1	–	–	[5]
Paramagnetic species											
*[Ce$(NO_3)_2(H_2O)$(I)]$(NO_3)\cdot H_2O$	D_2O	–	175.3	175.3	148.3	136.6	46.1	23.8	–	–	[2]
*[Pr$(NO_3)(H_2O)$(I)]$(NO_3)\cdot H_2O$	D_2O	–	193.6	193.6	152.6	148.7	46.7	37.5	–	–	[2]
Ce(I)$(CH_3COO)_2Cl\cdot nH_2O$	$CDCl_3$	173.5	165.5	158.7	142.6	129.6	42.8	21.3	33.6	–	[6]
Pr(I)$(CH_3COO)_2Cl\cdot nH_2O$	$CDCl_3$	164.3	157.2	155.2	139.8	132.9	26.2	27.9	54.6	–	[6]
*[Nd$(CH_3COO)_2$(I)]$Cl\cdot 4H_2O$	$CDCl_3$	–	171.3	163.6	139.6	137.7	46.9	37.6	–	–	[6, 9]
Sm(I)$(CH_3COO)_2Cl\cdot nH_2O$	$CDCl_3$	192.4	170.1	155.3	141.9	125.0	53.5	15.0	21.6	–	[6]
*[Eu$(CH_3COO)_2$(I)]$Cl\cdot 4H_2O$	$CDCl_3$	–	161.7	136.3	148.4	104.7	57.4	–17.6	–	–	[6, 9]
Ligand II = $C_{18}H_{18}N_6$											
Diamagnetic species											
La(II)$(NO_3)_3$	DMSO-d_6	–	165.1	150.1	142.3	129.1	59.4	–	–	–	[7]
Ligand VIII = $C_{30}H_{26}N_6$											
$[UO_2(VIII)](ClO_4)_2$	DMSO-d_6	–	180.0	155.4	141.3	129.6	–	21.13	–	178.7 130.6 123.8	[8]

[a] See footnote a in table 1.

[b] References: [1] Backer-Dirks et al. (1979); [2] Arif et al. (1987); [3] De Cola et al. (1986); [4] Bombieri et al. (1986); [5] De Cola et al. (1985c); [6] Fonda and Vallarino (1989). (Assignments of proton-bearing carbons have been confirmed by heteronuclear correlation experiments. Assignments of quaternary carbons are tentative); [7] Abid and Fenton (1984b); [8] Benetollo et al. (1989); [9] Benetollo et al. (1990a).

TABLE 20
Representative proton nuclear magnetic resonance spectra of complexes of ligands I, II and VIII.

Complex[a]	Solvent	H–C=	βH–py	γH–py	CH_2	CH_3 (ring)	CH_3 (acetate)	H-benzene	Ref.[b]
Ligand I = $C_{22}H_{26}N_6$									
Diamagnetic species									
*[La(NO_3)$_3$(I)]	D_2O	–	8.21	8.21	4.04	2.59	–	–	[1]
La(I)(CH_3COO)$_2$Cl·$n$$H_2O$	DMSO-d_6	–	8.34	8.34	3.97	2.50	1.50	–	[2]
Lu(I)(CH_3COO)(ClO_4)(OH)·(CH_3OH)·0.5(H_2O)	DMSO-d_6	–	8.32	8.32	3.97	2.56	1.34	–	[3]
[UO_2(I)]X_2 (X = ClO_4^-, NO_3^-, I^-)	DMSO-d_6	–	8.75	8.75	4.62	2.84	–	–	[4]
Y(I)(CH_3COO)$_2$Cl·4H_2O·CH_3OH	DMSO-d_6	–	8.31	8.31	3.95	2.53	1.35	–	
Paramagnetic species									
*[Ce(NO_3)$_2$(H_2O)(I)](NO_3)·H_2O	D_2O	–	13.04	12.18	0.74	3.79	–	–	[1]
[Pr(NO_3)$_2$(H_2O)(I)](NO_3)·H_2O	D_2O	–	18.13	15.63	3.92	6.60	–	–	[1]
*[{Nd(NO_3)(H_2O)$_2$(I)}]$_2$(NO_3)(ClO_4)$_3$·4H_2O	D_2O	–	13.81	11.70	11.88	5.35	–	–	[1]
Ligand II = $C_{18}H_{18}N_6$									
[La(II)(NO_3)$_3$]	DMSO-d_6	8.93	8.17 d	8.45 t	4.07	–	–	–	[5]
Ligand VIII = $C_{30}H_{26}N_6$									
[UO_2(VIII)](ClO_4)$_2$	DMSO-d_6	–	9.16–8.8 (complex pattern)		–	3.37	–	7.95–7.61 (complex pattern)	[6]

[a] See footnote a in table 1.
[b] References: [1] Backer-Dirks et al. (1979); [2] De Cola et al. (1986); [3] Bombieri et al. (1986); [4] De Cola et al. (1985c); [5] Abid and Fenton (1984b); [6] Benetollo et al. (1989).

resonance is observed for each type of carbon atom proves that in solution the macrocycle retains a symmetric structure, at least on the NMR time scale. When the macrocyclic ligands are unsymmetric or include fairly long and flexible aliphatic side-chains, the NMR spectra become more complex. For most diamagnetic species, however, reliable assignments can still be made.

The assignment of the spectra is less straightforward for the complexes of para-magnetic lanthanides, since proton chemical shifts vary widely from lanthanide to lanthanide and inversions in the order of the resonances relative to the diamagnetic species often occur. Furthermore, severe broadening of the signal occurs for some metal centers, so that coupling information is lost. For some macrocycles, most proton signals may still be confidently assigned on the basis of integration; assignment of the carbon resonances requires the assistance of heteronuclear correlation (Fonda and Vallarino 1989). In addition, the NMR spectrum of each paramagnetic lanthanide complex depends strongly on the solvent, although not on the concentration within the range usually employed. Unlike the macrocycle resonances of the diamagnetic species, which are not affected by changes of counterions, those of the paramagnetic species may also shift with changes in the counterion combination. These effects, which are illustrated by the selected data in tables 19 and 20, should be kept in mind when attempting to compare NMR spectra of paramagnetic macrocycles reported by different authors.

1.3.3. *Mass spectra*

Mass spectrometry has been used with varied success in the characterization of lanthanide macrocyclic complexes, as their high formula mass and low volatility, together with the ionic character of many complexes, render them somewhat unsuited to this technique. When the parent metal–macrocycle, with or without the associated anions, is present in the spectrum (usually obtained in the FAB mode), this may be considered as a meaningful evidence of the complex existence. When, however, only a mass corresponding to the metal-free macrocycle is observed, great caution should be exercised in the interpretation of the results, as the species detected may actually be quite a different compound. For example, the MS spectra of nonstoichiometric metal adducts of compound IX ($C_{30}H_{26}N_6$ = 470.22 amu) show a peak at an m/e ratio of 470, which could be erroneously considered to prove the existence of metal–macro-cycle complexes of ligand VIII (also $C_{30}H_{26}N_6$). Mass spectral values for individual compounds are listed in the tables under "Other Data".

1.3.4. *Luminescence of the europium(III) and terbium(III) macrocycles*

Relatively few luminescence studies have been reported for lanthanide macrocycles of metal-template origin; they are indicated in the tables under the individual compounds.

The luminescence of the Eu(III) diacetate chloride complex of ligand I has been investigated in detail by Sabbatini et al. (1987), both in the solid state and in aqueous solution. Luminescence decay measurements as a function of temperature (300 and 77 K) and solvent (H_2O and D_2O) indicate the absence of low-lying charge transfer levels and show that under the conditions of the measurements (10^{-3} M complex,

TABLE 21
Luminescence lifetimes and quantum yields of Eu(I)$(CH_3COO)_2$Cl in aqueous solution[a].

Solvent	T (K)	τ (ms)[b]	Φ_{cm}
H_2O	300	0.70	–
	77	1.4	–
D_2O	300	2.0	$\sim 10^{-1}$,[c] $\sim 6\times 10^{-3}$ [d]
	77	2.1	–

[a] Sabbatini et al. (1987). Emission from the $^5D_0$ excited state of Eu^{3+}.
[b] 1.0×10^{-3} M complex in the presence of 1.0×10^{-1} M CH_3COONa. $\lambda_{exc} = 300$, 337 or 394 nm.
[c] Excitation in the $^5L_6$ Eu^{3+} absorption ($\lambda = 394$ nm).
[d] Excitation in the intraligand absorption ($\lambda = 300$ nm).

0.1 M added acetate) only one water molecule is, on the average, coordinated to the Eu(III) ion. The Eu(III) luminescence in this complex is quite efficient for f–f excitation but very inefficient for excitation into the ligand (table 21); the latter results chiefly in nonradiative decay within the ligand itself. Taking into consideration the detailed features of the emission spectrum, Sabbatini et al. concluded that the most likely site symmetry for the $\{EuN_6\}$ environment in aqueous solution is C_{6v}. This symmetry implies an essentially planar six-nitrogen ring and is in remarkably good agreement with the results of the puckering analysis of this ring obtained from X-ray analysis of the crystalline compound. A luminescence study under different conditions has been reported by Smith et al. (1989), who suggested the presence of three to four coordinated water molecules.

In an effort to find conditions resulting in a more intense ligand-mediated luminescence of the {Eu–(I)} system, a systematic search for exocyclic emission enhancers was carried out (De Cola et al. 1985a,b). The following criteria were used in the selection of potential enhancers: (1) The enhancer should be a good ligand for Eu(III), i.e. it should contain hard donor atoms, preferably polar oxygen or aromatic nitrogen atoms, in order to favor stable coordination. (2) The enhancer should have a rigid, π-bonded, preferably chelating structure, in order to promote absorption of ultraviolet light and efficient ligand-to-metal energy transfer. (3) The enhancer should not contain H atoms directly attached to the donor atom(s) in order to prevent vibrational quenching of the metal ion luminescence.

Emission–titration experiments for the Eu–(I) diacetate chloride in methanol solution were carried out with the following compounds in neutral or anionic form: 1,10-phenanthroline; 2,2′-bipyridine; tribenzylphosphine oxide; benzoic, 2-pyridinecarboxylic, 3-pyridinecarboxylic, 2-furancarboxylic, 2-thiophenecarboxylic, and 2-pyrrolecarboxylic acids; various pyridinedicarboxylic acids (2,3-; 2,4-; 2,5-; 2,6- and 3,5-). These experiments showed that neutral ligands of the phenanthroline and bipyridine type do not enhance the Eu emission even though they produce intensely luminescent complexes with the "free" Eu(III) ion. Aromatic monocarboxylato ligands without heteroatoms readily replace the acetate(s) in the original substrate, as shown by the isolation of crystalline substitution products (e.g. Eu(I)(benzoate)$_2$Cl·4H$_2$O), but cause little or no luminescence enhancement; in contrast, aromatic monocarboxylato ligands containing N, O or S heteroatoms are effective enhancers provided the heteroatoms are not directly linked to H atoms. Finally, pyridinedicarboxylato ligands cause emission enhancement but also result in the slow precipitation of insoluble salts of the Eu–macrocycle cation. The luminescence enhancing effect of these aromatic carboxylates may arise from several cooperative factors, such as higher binding constant relative to the original acetate, better protection from quenching by the solvent, and most importantly, increased light absorption and more effective energy transfer to the Eu(III) ion.

2. Uranium complexes

The known uranium–macrocycle complexes of metal-templated origin are listed in table 22; all contain the *trans*-dioxouranium(VI) ion, UO_2^{2+} (uranyl), as the metal center. It is evident that the UO_2^{2+} ion, with its intermediate size (1.00 Å for 8-coordination) and its preference for equatorially directed bonds, is a very effective template for the synthesis of macrocyclic complexes having appropriate cavity size and a planar or quasi-planar arrangement of five or six donor atoms.

2.1. *Five-nitrogen "super-phthalocyanine" complex*

The condensation of 1,2-dicyanobenzene in the presence of anhydrous uranyl chloride, in dimethylformamide solution, produced a highly insoluble substance of stoichiometry $UO_2(C_8H_4N_2)_5$, originally believed to be a uranyl phthalocyanine complex (Frigerio 1962, Bloor et al. 1964, Kirin et al. 1967). Bluish black single crystals were obtained by Day et al. (1975) by slow evaporation of a hot 1,2,4-trichlorobenzene solution and their X-ray analysis showed the compound to be the uranyl complex of the highly conjugated five-nitrogen macrocycle XXXVII. The coordination geometry in this complex may be best described as an axially compressed pentagonal bipyramid (fig. 17). The U(VI) ion is 7-coordinate, being linked to the five N-donor atoms of the macrocyclic cavity and to the two *trans*-oxo ligands. The five-nitrogen girdle is essentially planar; its plane is perpendicular to the O–U–O axis and practically coincides with that of the entire 70-atom macrocycle, making an angle of only 2.4° with it. Selected structural data for this complex are listed in table 23.

2.2. (2 + 2) *macrocyclic complexes with pyridine head-units*

Complexes of the macrocyclic ligands I and II were obtained in nearly quantitative yields by the Schiff-base condensation of 1,2-diaminoethane with 2,6-diacetyl- or 2,6-diformylpyridine in the presence of uranyl acetate, followed by anion metathesis (De

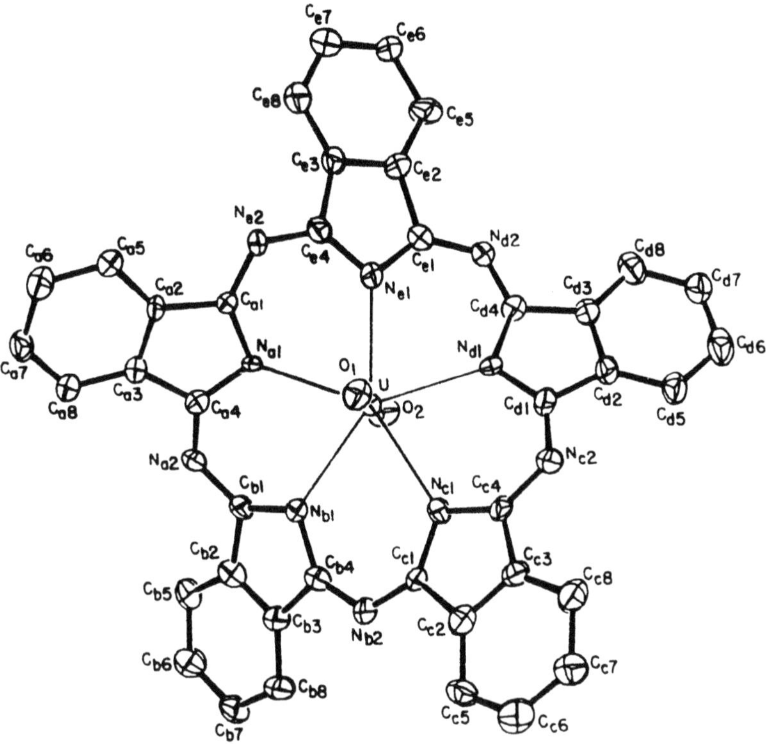

Fig. 17. Structure of the "uranyl-super-phthalocyanine" complex [UO_2(XXXVII)] (Day et al. 1975).

Cola et al. 1985c). The [UO_2L]X_2 salts, where L = I or II and X = I^-, SCN^-, NO_3^- or ClO_4^-, were obtained as crystalline solids of colors ranging from golden brown to almost black. The complexes are stable toward air and moisture; they are fairly soluble in dimethylsulfoxide and dimethylformamide, sparingly soluble or insoluble in other common solvents. Their IR spectra show the following features: (1) The absence of OH, NH, and C=O absorptions. (2) The typical patterns of the macrocyclic ligands I or II, identified by comparison with complexes of known crystal structures. (3) A sharp absorption for the uranyl ion (ν_{as}(U=O): 930 cm^{-1}). (4) The absorptions characteristic of the uncoordinated anions (NO_3^-: 1350 s, 835 m, 720 w; ClO_4^-: 1100 s, 620 s; NCS^-: 2050 s). The carbon-13 NMR spectra in DMSO-d_6 solution show the macrocyclic moieties to have a symmetrical structure, as only one resonance appears for each type of carbon atom; the proton spectra confirm the symmetry of the organic ligands. Taken together, the IR and NMR spectra of the [UO_2L]$^{2+}$ complexes conclusively show that the U(VI) is coordinated to all six ring N-atoms and suggest an effective D_{6h} site symmetry for the central metal ion in solution. A magnetic circular dichroism study (De Cola et al. 1984, 1985a) based on crystal field analysis (Gorller-Warland and Colen 1984) supports this assignment.

It would be interesting to know whether the bending of the macrocycle observed for the lanthanide complexes of ligand I also exists in the crystalline derivatives of the

TABLE 22
Macrocyclic complexes of the dioxouranium(VI) ion.

Compound[a]	Color	Other data[f]	Ref.[b]
$[UO_2(I)](X)_2$ ($X = NO_3^-$, ClO_4^-, I^-, NCS^-)	Golden brownish black	IR(cm^{-1}): absorptions characteristic of ligand I as given for the Y(III) bromide complex in table 1; $\nu_{as}(UO_2)$: 930 (s), sharp; absorptions of ionic counterions: 1350 (s), 835 (m) and 720 (w) for NO_3^-; 1100 (s, br), 620 (m) for ClO_4^-; 2060 (s) for SCN^- ^1H NMR: table 20; ^{13}C NMR: table 19	[1]
$[UO_2(II)](X)_2$	Golden brownish black	IR(cm^{-1}): absorptions of ligand II as given for the Y(III) diacetate chloride in table 1; absorptions of $\{UO_2\}$ moiety and counterions as in corresponding complexes of ligand I	[1]
$[UO_2(VIII)](ClO_4)_2$	Olive green	IR(cm^{-1}): 3095 (w), 3025 (vvw), 2980 (vvw), 2940 (vw), 1615 (m) and 1595 (s): ν(C=N), 1490 (m), 1450 (m), 1375 (s), 1340 (vw), 1270 (s), 1195 (vw), 1155 (vw), 1100–1085 (vs, br) $\nu_3(ClO_4^-)$, 1015 (vw), 1011 (m), 990 (sh), 945 (s) $\nu_{as}(UO_2)$, 830 (sh), 820 (m–s), 800 (vw), 770–765 (m), 750–745 (m), 725 (w), 620 (s) $\nu_4(ClO_4)$, 570 (w), 530 (w), 500 (vw), 440 (vw) ^1H NMR: table 20; ^{13}C NMR: table 19	[2]
*$[UO_2(XIX)]$	Yellowish orange	X-ray structure: fig. 18, table 24; IR(cm^{-1}): 1632 [ν(C=N, imine)]; 896 [$\nu_{as}(UO_2)$]	[3]
$[UO_2(XXV)]$	Orange powder	UV–VIS[c]; IR(cm^{-1}): 3247 [ν(NH)]; 1633 [ν(C=N)]; 1308 [$\nu(C–O^-)$]; 896 [$\nu_{as}(UO_2)$]; 256 [bend (UO_2)]	[4, 6]
$UO_2(XXV)Ni(ClO_4)_2 \cdot 2H_2O$	Yellow powder	$\mu_{eff} = 3.03\mu_B$; UV–VIS[c]; IR(cm^{-1}): similar to simple UO^{2+} analog except for additional $\nu(C–O^-)$ band at 1674; also: 1124–1097 [$\nu_3(ClO_4^-)$]; 627 [$\nu_4(ClO_4^-)$].	[4]
$UO_2(XXV)Cu(ClO_4)_2$	Green powder	$\mu_{eff} = 1.95\mu_B$; UV–VIS[c]; IR[e]	[4]
$UO_2(XXV)Co(ClO_4)_2$	Yellow powder	$\mu_{eff} = 3.88\mu_B$; UV–VIS[c]; IR[e]	[4]
$UO_2(XXVI)$	Orange red powder	UV–VIS[c]; IR[d]	[4]
$UO_2(XXVI)Ni(ClO_4)_2 \cdot 3H_2O$	Yellow powder	UV–VIS[c]; IR[e]	[4]

$UO_2(XXVI)Cu(ClO_4)_2 \cdot 2EtOH$	Green powder	UV–VIS[c]; IR[e]	[4]
$UO_2(XXVI)Co(ClO_4)_2 \cdot 2H_2O$	Yellow powder	$\mu_{eff} = 3.94\mu_B$; UV–VIS[c]; IR[e]	[4]
$UO_2(XXVII)$	Orange red powder	UV–VIS[c], IR[d]	[4]
$UO_2(XXVII)Ni(ClO_4)_2 \cdot 3H_2O$	Yellow powder	$\mu_{eff} = 3.23\mu_B$; UV–VIS[c]; IR[e]	[4]
$UO_2(XXVII)Cu(ClO_4)_2 \cdot EtOH$	Green powder	UV–VIS[c]; IR[e]	[4]
$UO_2(XXVII)Co(ClO_4)_2 \cdot 2H_2O$	Yellow powder	UV–VIS[c]; IR[e]	[4]
$UO_2(XVIII)$ or			
$UO_2(XVIII) \cdot H_2O$	Orange red powder	UV–VIS[c]; IR[d]	[4]
$UO_2(XVIII)Ni(ClO_4)_2$	Yellow powder	$\mu_{eff} = 1.67\mu_B$; UV–VIS[c] $\lambda_{max}(cm^{-1})$: 24700, 11000; IR[e]	[4, 6]
$UO_2(XIX) \cdot H_2O$	–	UV–VIS λ_{max}(nm): 680, 440 (s), 375 (DMSO) and 470 (s), 375 ($CHCl_3$);	[5]
		IR(cm^{-1}): 1632, 1552 and 1523 [ν(C–N, C–O, C–C)]; 896 [$\nu_{as}(UO_2)$]	[6]
$UO_2(XXVIII) \cdot H_2O$	–	UV–VIS λ_{max}(nm): 375 (DMSO), 385 ($CHCl_3$)	[5]
		IR(cm^{-1}): 1631 and 1549 [ν(C–N, C–O, C–C)]; 901 [$\nu_{as}(UO_2)$]	[5]
$UO_2(XXIV) \cdot MeOH$	–	UV–VIS λ_{max}(nm): 570 (s), 480 (s), 415 (DMSO); 520 (s), 420 (s), 385 ($CHCl_3$)	[5]
		IR(cm^{-1}): 3219 [ν(N–H)]; 1635 and 1553 [ν(C–N, C–O, C–C)]; 892 [$\nu_{as}(UO_2)$]	[5]
$UO_2(XIX)Cu(ClO_4)_2 \cdot 3EtOH$	–	IR(cm^{-1}): 1636 (s) and 1554 (m): ν(C–O, C–N, C–C), 896 $\nu_{as}(UO_2)$, 1146 (s), 1120 (s), 1091 (s), 1112 (s) $\nu(ClO_4^-)$	[5]
$UO_2(XXIX)$	Yellowish orange	IR(cm^{-1}): 3245 [ν(N–H)]; 1630 [ν(C=N)]; 889 [$\nu_{as}(UO_2)$]	[7]
$UO_2(XXX)$	Yellowish orange	IR(cm^{-1}): 1630 [ν(C=N)]; 903 [$\nu_{as}(UO_2)$]	[7]
*[$UO_2(XXXVII)$]	Bluish black	X-ray structure: fig. 17, table 23	[8]

[a] See footnote a in table 1. Complexes of ligands XVIII–XXVIII are listed in an order that follows their presentation in the original papers.

[b] References: [1] De Cola et al. (1985c); [2] Benetollo et al. (1989); [3] Casellato et al. (1986b); [4] Vidali et al. (1975); [5] Casellato et al. (1985); [6] Casellato et al. (1986a); [7] Bullita et al. (1989); [8] Day et al. (1975).

[c] General features: λ_{max}: 26 700 cm^{-1}, very intense, tentatively assigned to UO_2 macrocycle C.T.; 22 000 cm^{-1}, uranyl moiety; <7000 cm^{-1}, vibrations of macrocycle. Also: UO_2–Co(II) complex; 16 000 sh, 10 500 cm^{-1}; UO_2–Ni(II) complex: 18 000, 14 000, 10 000 cm^{-1}; UO_2–Cu(II) complex: 16 600 cm^{-1}.

[d] Similar to complex of XXV, with minor frequency variations.

[e] Similar to UO_2–Ni complex of XXV, with minor variations.

[f] See footnote j in table 1.

TABLE 23
Selected crystallographic data and bond lengths for $[UO_2(XXXVII)]$[a].

Formula	$C_{40}H_{20}N_{10}O_2U$		
Mol. wt.	–		
Crystal system	Monoclinic		
Space group	$P2_{1/c}$		
a (Å)	8.210 (3)		
b (Å)	21.667 (7)		
c (Å)	18.462 (5)		
β (deg)	103.16 (2)		
Z	4		
D_c ($g\,cm^{-3}$)	1.891		
Coordinate bond lengths (Å)	$U–O_1$	1.745 (9)	dioxouranium
	$U–O_2$	1.743 (7)	dioxouranium
	$U–N_{al}$	2.523 (8)	macrocycle
	$U–N_{bl}$	2.548 (9)	macrocycle
	$U–N_{cl}$	2.526 (9)	macrocycle
	$U–N_{dl}$	2.533 (9)	macrocycle
	$U–N_{el}$	2.490 (9)	macrocycle
	$O_1–M–O_2$ angle 179.2		macrocycle

The U atom is displaced toward O_1 by 0.02 Å from the mean plane of the macrocycle.

[a] Day et al. (1975).

trans-dioxouranium ion, where no additional coordination of counterions or solvent occurs. Definitive information on this aspect can only come from X-ray analysis; unfortunately, none of the crystals obtained so far (thin needles or leaflets) have proved suitable for X-ray studies.

A complex of formula $[UO_2(VIII)](ClO_4)_2$ has been obtained, as a mixture with the purely organic tricyclic compound IX, from the condensation of 1,2-diaminobenzene with 2,6-diacetylpyridine in the presence of uranyl perchlorate in methanol (Benetollo et al. 1989). Repeated extraction of the mixture with boiling tetrachloroethylene yielded the uranyl–macrocycle as an analytically pure olive green powder, stable toward the atmosphere and insoluble or sparingly soluble in water and organic solvents. The macrocycle portion of the IR spectrum of the solid $[UO_2(VIII)](ClO_4)_2$ complex is identical to that of the X-ray analyzed praseodymium(III) complex of the same ligand, indicating a very similar structure. The proton and carbon-13 spectra of the complex in DMSO-d_6 solution are remarkably simple and consistent with a symmetric structure involving a $\{trans\text{-}UO_2(N_6)\}$ coordination sphere; the resonances arising from the C=N and pyridine carbon atoms match the NMR spectra of the aliphatic macrocycles. The NMR spectra did not change with time, suggesting that in this solvent the $\{UO_2–(VIII)\}$ entity is more inert to metal release and/or structural changes than its lanthanum(III) analog.

2.3. *Macrocyclic complexes of potentially anionic compartmental ligands*

The syntheses of the uranyl complexes of the macrocyclic ligands H_2-XVIII through H_2-XXX represent examples of stepwise metal-templated condensation reactions. For ligands H_2-XVIII, H_2-XXIV, H_2-XXVII, H_2-XXIX, and H_2-XXX (Vidali et al. 1975, Casellato et al. 1985, Casellato et al. 1986a, b, Zanello et al. 1986, Bullita et al. 1989) the first step of the synthesis involves the reaction of 2,6-diformyl-4-chlorophenol with diethylenetriamine in the presence of a uranyl salt, to give uranyl complexes of an open "three-quarter cycle" diketone ligand. This further condenses with a second molecule of an appropriate diamine, yielding a uranyl macrocycle which still has free donor sites and can coordinate the relatively small transition metal ions of the 3d series, such as Co(II), Cu(II) and Ni(II). A similar synthetic procedure has been used to produce uranyl complexes of the symmetric and unsymmetric bicompartmental macrocyclic ligands, H_2-XIX and H_2-XXVIII. The crystal structure of the complex containing ligand H_2-XIX in dianion form has shown (Casellato et al. 1986b) that the symmetric $[N_4O_2S_2]$ macrocycle has the usual "butterfly fold". The UO_2^{2+} ion is linked in its equatorial plane to the $[N_2O_2S]$ donor set of one compartment, leaving the other $[N_2S]$ compartment free to accommodate a smaller metal ion (fig. 18). Selected structural data for this complex are given in table 24. A number of heterobinuclear complexes of these compartmental ligands have also been synthesized and their magnetic and UV–VIS spectral properties have been reported.

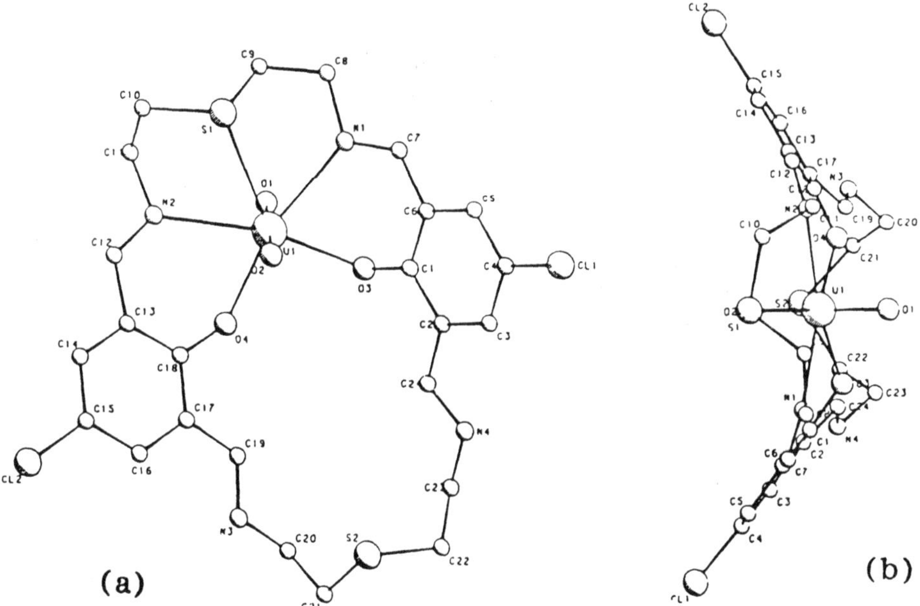

Fig. 18. Structure of $[UO_2(XIX)]$: (a) Front view. (b) Side view showing the wings of the ligand inclined with respect to the coordination plane (Casellato et al. 1986b).

TABLE 24
Selected crystallographic data, bond lengths, and angles for [UO_2(XIX)][a].

Formula	$C_{24}H_{24}Cl_2N_4O_4S_2U$		
Mol. wt.	795		
Crystal system	Orthorhombic		
Space group	Pbca		
a (Å)	26.654 (3)		
b (Å)	22.871 (3)		
c (Å)	8.875 (5)		
V (Å^3)	5410 (3)		
Coordinate bond lengths (Å)	U–O(1)	1.79 (1)	dioxouranium
	U–O(2)	1.78 (1)	
	U–O(3)	2.22 (1)	phenol
	U–O(4)	2.25 (1)	
	U–N(1)	2.60 (1)	imine
	U–N(2)	2.54 (1)	
	U–S(1)	3.018 (4)	thioether

[a] Casellato et al. (1986b).

3. Overview

It is by now well established that all lanthanide(III) ions, as well as the yttrium(III) and uranyl ions, are effective templates for the synthesis of macrocyclic complexes; the scarcity of information on similar systems involving 5f-block elements other than uranium appears to be due to lack of investigation rather than to any intrinsic failure of these elements to function as templates in macrocyclic syntheses.

The vast majority of the macrocyclic complexes of the lanthanide(III), yttrium(III) and uranyl ions obtained so far by metal-templated synthesis are of the Schiff-base type; even the few known examples of "simple polyamine" complexes actually result from the cyclic condensation of a diamine with a (modified) carbonyl precursor. In general, the metal-templated synthesis is facilitated by the presence of oxygen-donor anions, such as nitrate, acetate, or trifluoromethylsulfonate; lanthanide(III) thiocyanates have also been successfully used. With a few exceptions, the outcome of the synthesis appears to be independent of the order of addition of the reactants. No deliberate attempts have been made to investigate the detailed mechanism of these metal-templated cyclic condensation reactions.

The size of the metal ion has an evident control on the synthesis of the monometallic macrocyclic complexes. For example, all lanthanide(III) ions promote the formation of the six-nitrogen-donor macrocycles with an 18-atom cavity and flexible aliphatic diimine side-chains, even though the yields become progressively lower as the metal atom size decreases. However, when the six-nitrogen, 18-atom macrocycles contain rigid *o*-phenylene side chains, the smaller lanthanides actually fail to promote the formation of true macrocyclic complexes. Parallel effects are observed for other systems: the six-nitrogen, 20-atom macrocycles are obtained only with the larger lanthanides, whereas the six-nitrogen, 14-atom macrocycles are obtained only with the smaller lanthanides.

The metal ion control appears to be less stringent for the complexes of the larger macrocyclic ligands, many of which may be independently synthesized without the assistance of a metal ion. When the ligand is sufficiently large and suitably shaped, bimetallic complexes may be obtained in which two metal centers (identical or different) occupy adjacent sites within the macrocyclic cavity. A number of these complexes have recently attracted interest as models for bimetallic enzymes; those that contain lanthanide pairs including at least one luminescent center offer the opportunity to study metal–metal energy transfer at a controlled distance in solution. The structures and properties of these bimetallic systems have been recently investigated (Guerriero et al. 1990).

Both the synthetic trends and the reactivity of the metal–macrocycle complexes considered in this review find some rationalization in the structural data obtained from X-ray analysis. In the series of complexes containing the six-nitrogen, 18-atom macrocycles with aliphatic side-chains, the macrocyclic moiety can flex to a different extent in each metal complex, thus allowing the most favorable metal–donor distance to be achieved for all donor atoms. These favorable metal–donor distances, together with appropriate geometries of the donor atoms and low steric strain, must account for the unique inertness displayed by these metal–macrocycle entities under conditions (dilute aqueous solution, presence of acids, bases and competing ligands) that would cause most other lanthanide complexes to decompose instantly.

The inertness of these metal–macrocycle entities provides the key to new ways of utilizing the characteristic properties of the individual metal ions (Vallarino 1989). As early as 1976, for example, it was suggested by Leif et al. that the nearly monochromatic luminescence of Eu(III) and Tb(III) chelates might provide a new approach to the multiparameter analysis of cells. The long decay times of the europium and terbium emissions also make possible detection in a time-gated mode, thus effectively removing interference from the intrinsic background fluorescence of the analyte (Leif et al. 1977). Targeting to biological systems has been attempted with radioactive ^{90}Y (for auto-imaging or therapy, Moi et al. 1985) and with gadolinium(III) (as contrast enhancer in NMR imaging, Lauffer 1987). These and other uses requiring an aqueous environment would obviously take advantage of the attachment of an inert metal–macrocycle to the desired substrate via a peripheral coupling functionality. The synthesis of several functionalized macrocycles has recently been reported (Gootee et al. 1989, Gribi et al. 1989) and represents an important new avenue in the macrocyclic chemistry of these lanthanide systems.

Acknowledgement

I express my deep appreciation to Diane Holmes for her invaluable secretarial assistance. I also thank Dr. K. K. Fonda for her help in the final stages of the manuscript.

Appendix: Schematic structural formulas of the macrocyclic ligands I–XXXVII

I R=CH_3 $C_{22}H_{26}N_6$
II R=H $C_{18}H_{18}N_6$

III $C_{18}H_{20}N_6O$

IV $C_{20}H_{22}N_6$

V $C_{20}H_{22}N_6$

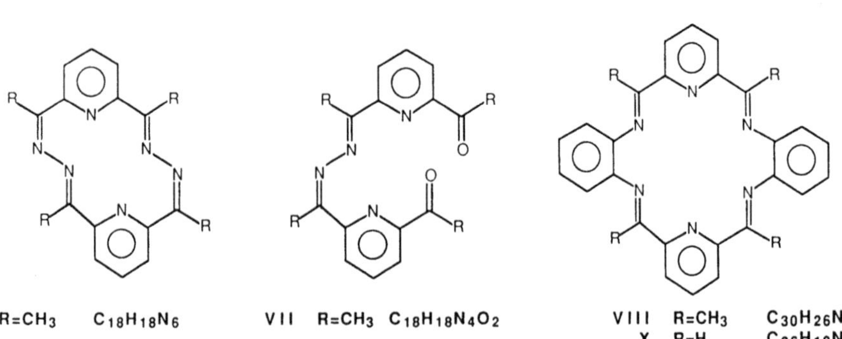

VI R=CH_3 $C_{18}H_{18}N_6$

VII R=CH_3 $C_{18}H_{18}N_4O_2$

VIII R=CH_3 $C_{30}H_{26}N_6$
X R=H $C_{26}H_{18}N_6$

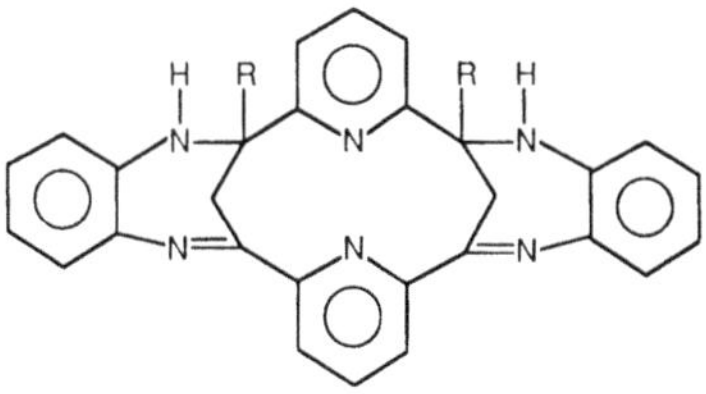

H R R H
N N N N N N
IX R=CH_3 $C_{30}H_{26}N_6$

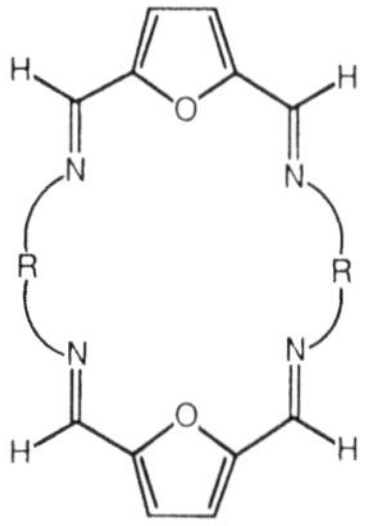

H H O N N R R N N O H H

XI R= $-CH_2CH_2-$ $C_{16}H_{16}N_4O_2$
XII R= $-CH_2CH_2CH_2-$ $C_{18}H_{20}N_4O_2$
XIII R= $-CH_2CH(Me)-$ $C_{18}H_{20}N_4O_2$

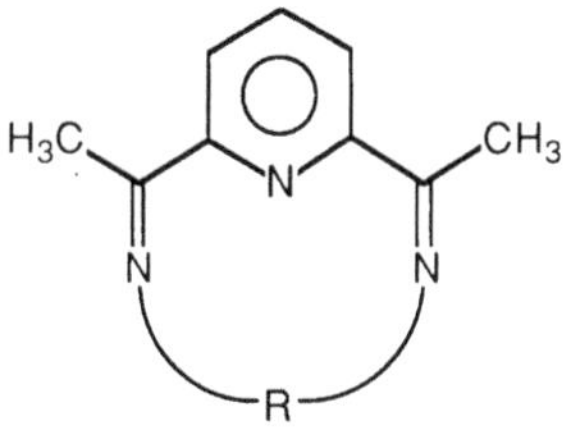

H_3C N CH_3 N N R

XIV R= O O O $C_{17}H_{25}N_3O_3$

XV R= NH NH NH $C_{17}H_{28}N_6$

XVI R= NH NH $C_{15}H_{23}N_5$

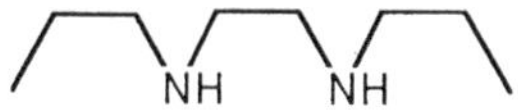

XVII R= N-H N-H $C_{16}H_{25}N_5$

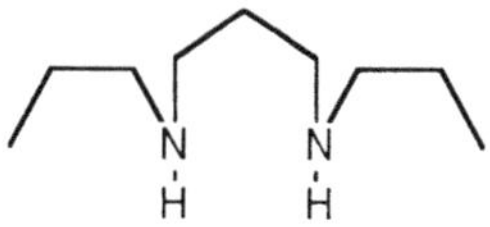

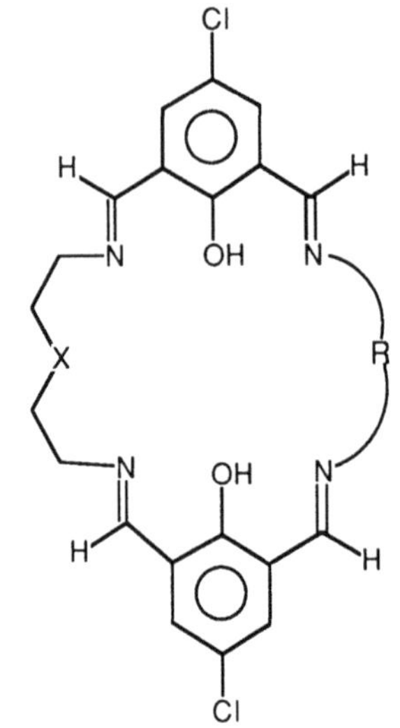

X	R	Symbol	Formula
NH	$CH_2CH_2NHCH_2CH_2$	H_2-XVIII	$C_{24}H_8N_6O_2Cl_2$
S	$CH_2CH_2SCH_2CH_2$	H_2-XIX	$C_{24}H_{26}N_4O_2S_2Cl_2$
NH	$CH_2CH_2NHCH_2CH_2$ (ring-contracted)	H_2-XX	$C_{24}H_{28}N_6O_2Cl_2$
NH	$CH_2CH_2SCH_2CH_2$	H_2-XXIV	$C_{24}H_{27}N_5O_2SCl_2$
NH	CH_2CH_2	H_2-XXV	$C_{22}H_{23}N_5O_2Cl_2$
NH	$CH_2(CH_3)CH$	H_2-XXVI	$C_{23}H_{25}N_5O_2Cl_2$
NH	$CH_2CH_2CH_2$	H_2-XXVII	$C_{23}H_{25}N_5O_2Cl_2$
S	CH_2CH_2	H_2-XXVIII	$C_{22}H_{22}N_4SO_2Cl_2$
NH	$CH_2CH_2N(CH_2CH_2)$ with N-substituent $(CH_2)_{11}CH_3$	H_2-XXIX	$C_{36}H_{52}N_6O_2Cl_2$
$N(CH_2)_{11}CH_3$	$CH_2CH_2N(CH_2CH_2)$ with N-substituent $(CH_2)_{11}CH_3$	H_2-XXX	$C_{48}H_{76}N_6O_2Cl_2$

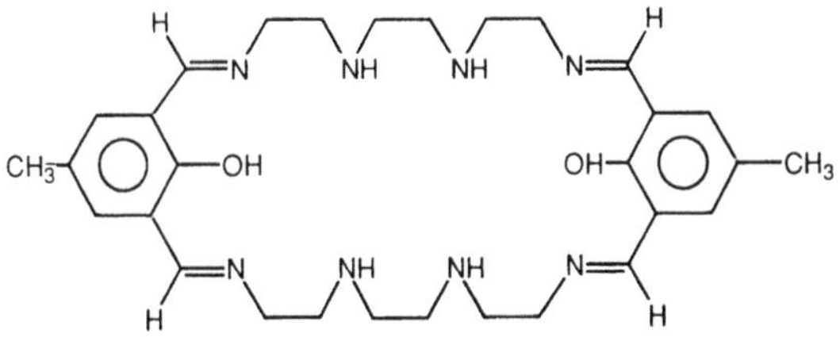

H_2-XXI $C_{30}H_{44}N_8O_2$

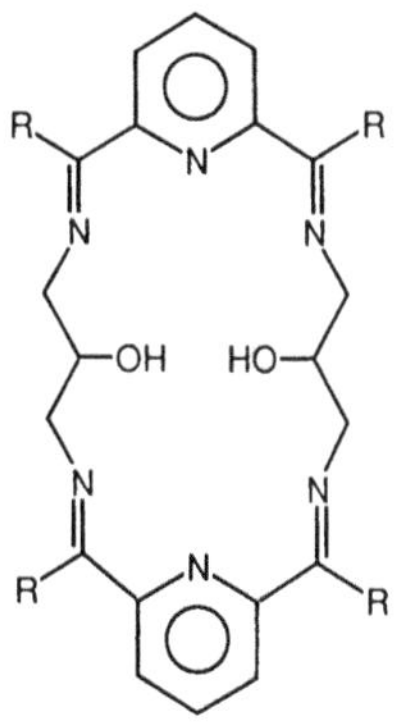

H_2-XXII $C_{24}H_{30}N_6O_2$

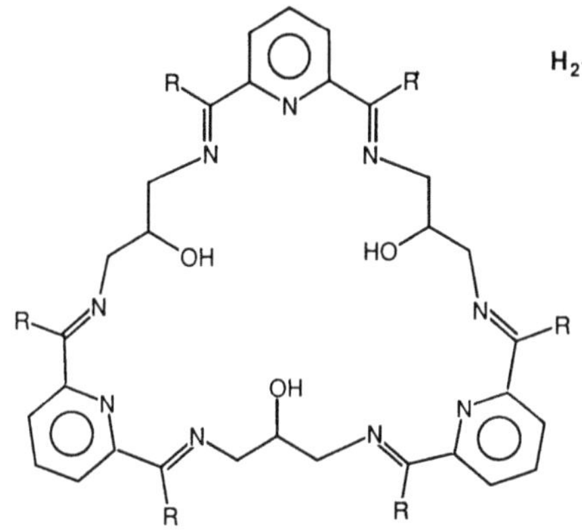

H_2-XXIII R = CH_3 $C_{36}H_{45}N_9O_3$

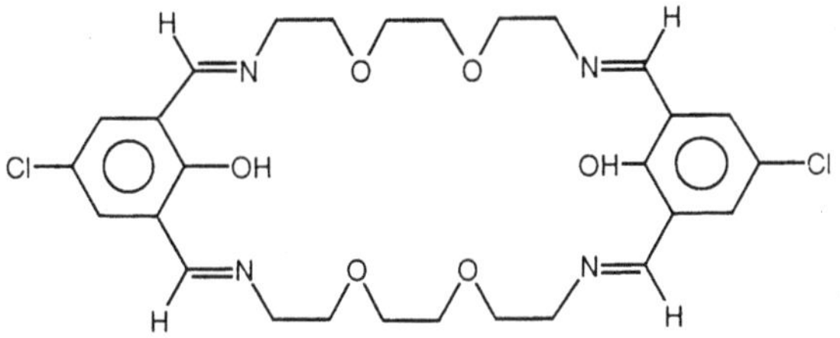

H_2-XXXI $C_{30}H_{34}N_4O_6Cl_2$

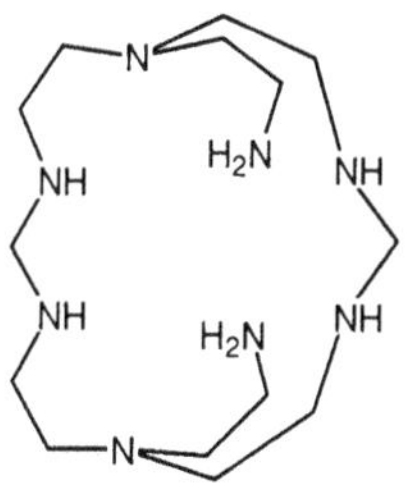

XXXII $C_{14}H_{36}N_8$

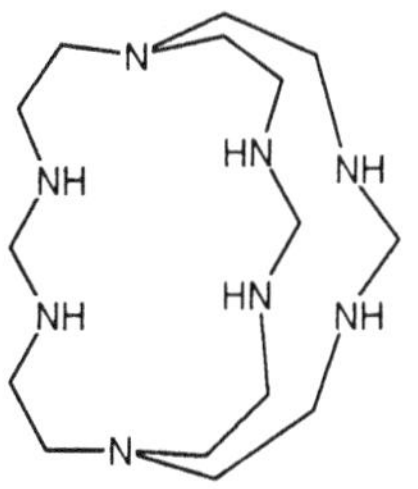

XXXIII $C_{15}H_{36}N_8$

XXXVII $C_{40}H_{20}N_{10}$

XXXIV R= $-CH_2-OH$ $C_{24}H_{30}N_6O_2$
XXXV R= $-CH_2-C_6H_4-OH$ $C_{36}H_{38}N_6O_2$
XXXVI R= $-CH_2-C_6H_4-NH_2$ $C_{36}H_{38}N_8$

References

Abid, K.K., and D.E. Fenton, 1984a, Inorg. Chim. Acta **82**, 223.

Abid, K.K., and D.E. Fenton, 1984b, Inorg. Chim. Acta **95**, 119.

Abid, K.K., D.E. Fenton, U. Casellato, P.A. Vigato and R. Graziani, 1984, J. Chem. Soc. Dalton Trans., 351.

Arif, A.M., C.J. Gray, F.A. Hart and M.B. Hursthouse, 1985, Inorg. Chim. Acta **109**, 179.

Arif, A.M., J.D.J. Backer-Dirks, C.J. Gray, F.A. Hart and M.B. Hursthouse, 1987, J. Chem. Soc. Dalton Trans., 1665.

Backer-Dirks, J.D.J., C.J. Gray, F.A. Hart, M.B. Hursthouse and B.C. Schoop, 1979, J. Chem. Soc. Chem. Commun. 774.

Backer-Dirks, J.D.J., J.E. Cooke, A.M.R. Galas, J.S. Ghotra, C.J. Gray, F.A. Hart and M.B. Hursthouse, 1980, J. Chem. Soc. Dalton Trans., 2191.

Batchelor, R.J., J.N.R. Ruddick, J.R. Sams and F. Aubke, 1977, Inorg. Chem. **16**, 1414.

Benetollo, F., G. Bombieri, L. De Cola, A. Polo, D.L. Smailes and L.M. Vallarino, 1989, Inorg. Chem. **28**, 3447.

Benetollo, F., A. Polo, G. Bombieri, K.K. Fonda and L.M. Vallarino, 1990a, Polyhedron **9**, 1411.

Benetollo, F., G. Bombieri, K.K. Fonda, A. Polo and L.M. Vallarino, 1990b, Am. Chem. Soc., 200th National Meeting, Washington, DC, August 1990, Abstr. No. 458.

Benetollo, F., G. Bombieri, K.K. Fonda, A. Polo, J.R. Quagliano and L.M. Vallarino, 1991, Inorg. Chem. **30**, 1345.

Bloor, J.E., J. Schlabitz, C.C. Walden and A. Demerdache, 1964, Can. J. Chem. **42**, 2201.

Bombieri, G., 1987, Inorg. Chim. Acta **139**, 21.

Bombieri, G., F. Benetollo, A. Polo, L. De Cola, D.L. Smailes and L.M. Vallarino, 1986, Inorg. Chem. **25**, 1127.

Bombieri, G., F. Benetollo, A. Polo, L. De Cola, W.T. Hawkins and L.M. Vallarino, 1989a, Polyhedron **8**(17), 2157.

Bombieri, G., F. Benetollo, W.T. Hawkins, A. Polo and L.M. Vallarino, 1989b, Polyhedron **8**(15), 1923.

Bombieri, G., F. Benetollo, A. Polo and L.M. Vallarino, 1989c, Am. Chem. Soc., 41st Southeastern Regional Meeting, Winston-Salem, NC, Nov. 1989, Abstr. No. 479.

Bullita, E., U. Casellato, P. Guerriero and P.A. Vigato, 1987, Inorg. Chim. Acta **139**, 59.

Bullita, E., P. Guerriero, S. Tamburini and P.A. Vigato, 1989, J. Less-Common Met. **153**(2), 211.

Bünzli, J.-C.G., E. Moret, U. Casellato, P. Guerriero and P.A. Vigato, 1988, Inorg. Chim. Acta **150**, 133.

Burns, J.H., 1979, Inorg. Chem. **18**, 3044.

Cabral, J. de O., M.F. Cabral, M.G.B. Drew, F.S. Esho, O. Haas and S.M. Nelson, 1982a, J. Chem. Soc. Chem. Commun. 1066.

Cabral, J. de O., M.F. Cabral, M.G.B. Drew, F.S. Esho and S.M. Nelson, 1982b, J. Chem. Soc. Chem. Commun. 1068.

Casellato, U., D. Fregonà, S. Sitran, S. Tamburini, P.A. Vigato and D.E. Fenton, 1985, Inorg. Chim. Acta **110**, 181.

Casellato, U., P. Guerriero, S. Tamburini, P.A. Vigato and R. Graziani, 1986a, Inorg. Chim. Acta **119**, 215.

Casellato, U., S. Sitran, S. Tamburini, P.A. Vigato and R. Graziani, 1986b, Inorg. Chim. Acta **114**, 111.

Day, V.W., T.J. Marks and W.A. Wachter, 1975, J. Am. Chem. Soc. **97**, 4519.

De Cola, L., D.D. Shillady and L.M. Vallarino, 1984, Am. Chem. Soc., 36st Southeastern Regional Meeting, Raleigh, NC, Abstr. No. 252.

De Cola, L., D.L. Smailes and L.M. Vallarino, 1985a, Am. Chem. Soc., Annual Meeting, Miami, FL, 1985, Abstr. No. 149.

De Cola, L., D.L. Smailes and L.M. Vallarino, 1985b, 10th National Conference of Photochemistry, Ravenna, Italy, 1985, Abstr. No. 14.

De Cola, L., D.L. Smailes and L.M. Vallarino, 1985c, Inorg. Chim. Acta **110**, L1.

De Cola, L., D.L. Smailes and L.M. Vallarino, 1986, Inorg. Chem. **25**, 1729.

Fenton, D.E., and P.A. Vigato, 1988, Chem. Soc. Rev. **17**, 69.

Fenton, D.E., S.J. Kitchen, C.H. Spencer, P.A. Vigato and S. Tamburini, 1987, Inorg. Chim. Acta **139**, 55.

Fenton, D.E., S.J. Kitchen, C.M. Spencer, S. Tamburini and P.A. Vigato, 1988, J. Chem. Soc. Dalton Trans., 685.

Fonda, K.K., and L.M. Vallarino, 1989, Am. Chem. Soc., 41st Southeastern Regional Meeting, Winston-Salem, NC, Nov., 1989, Abstr. No. 311.

Frigerio, N.A., 1962, U.S. Patent 3 027 391.

Gootee, W.A., K.T. Pham and L.M. Vallarino, 1989, Am. Chem. Soc., 41st Southeastern Regional Meeting, Winston-Salem, NC, Nov. 1989, Abstr. No. 313.

Gorller-Walrand, C., and W. Colen, 1984, Inorg. Chim. Acta **84**, 183.

Gray, C.J., and F.A. Hart, 1987, J. Chem. Soc. Dalton Trans. 2289.

Gribi, C., D.L. Smailes, A.K. Twiford and L.M. Vallarino, 1989, Am. Chem. Soc., 41st Southeastern Regional Meeting, Winston-Salem, NC, Nov. 1989, Abstr. No. 312.

Guerriero, P., U. Casellato, S. Tamburini, P.A. Vigato and R. Graziani, 1987, Inorg. Chim. Acta **129**, 127.

Guerriero, P., P.A. Vigato, J.-C.G. Bünzli and E. Moret, 1990, J. Chem. Soc. Dalton Trans., 647.

Kahwa, I.A., J. Selbin, T.C.-Y. Hsieh and R.A. Lane, 1986, Inorg. Chim. Acta **118**, 179.

Kirin, J.S., P.N. Moskalev and V.Ya. Mishin, 1967, J. Gen. Chem. USSR **37**, 265.

Lauffer, R.B., 1987, Chem. Rev. **87**, 902 and references therein.

Leif, R.C., S.P. Clay, H.G. Gratzner, H.G. Haines and L.M. Vallarino, 1976, The Automation of Uterine Cancer Cytology, in: Tutorials of Cytology, Chicago, IL, 1976, eds G.L. Wied, G.F. Bahr and P.H. Bartels, p. 313.

Leif, R.C., R.A. Thomas, T.A. Yopp, B.D. Watson, V.R. Guarino, D.H.K. Hindman, N. Lefkove and L.M. Vallarino, 1977, Clin. Chem. **23**, 1492.

Martin, J.L., L.C. Thompson, L.J. Rodanovich and M.D. Glick, 1968, J. Am. Chem. Soc. **90**, 4493.

Moi, M.K., C.F. Meares, W.C. Cole and S.J. DeNardo, 1985, Am. Chem. Soc., 189th National Meeting, Miami, Fl., 1985, Abstr. No. 30.

Nakamoto, K., 1986, in: Infrared and Raman Spectra of Inorganic Coordination Compounds (Wiley, New York) pp. 231–233.

Radecka-Paryzek, W., 1980, Inorg. Chim. Acta **45**, L147.

Radecka-Paryzek, W., 1981a, Inorg. Chim. Acta **52**, 261.

Radecka-Paryzek, W., 1981b, Inorg. Chim. Acta **54**, L251.

Radecka-Paryzek, W., 1985, Inorg. Chim. Acta **109**, L21.

Radecka-Paryzek, W., 1986, in: Rare Earths Spectroscopy, Proc. Int. Symp. 1984, p. 186; Chem. Abstr. No. **104**, 140937b.

Raymond, K.N., and P.H. Smith, 1988, Pure & Appl. Chem. **60** (8), 1141.

Sabbatini, N., L. De Cola, L.M. Vallarino and G. Blasse, 1987, J. Phys. Chem. **18**, 4681.

Sakamoto, M., 1987, Bull. Chem. Soc. Jpn. **60**(4), 1546.

Smith, P.H., and K.N. Raymond, 1985, Inorg. Chem. **24**, 3469.

Smith, P.H., Z.E. Reyes, C.-W. Lee and K.N. Raymond, 1988, Inorg. Chem. **27**, 4154.

Smith, P.H., J.R. Brainard, D.E. Morris, G.D. Jarvinen and R.R. Ryan, 1989, J. Am. Chem. Soc. **111**, 7437.

Stotz, R.W., and R.C. Stoufer, 1970, J. Chem. Soc. Chem. Commun. 1682.

Thompson, M.C., and D.H. Bush, 1964, J. Am. Chem. Soc. **86**, 3651.

Vallarino, L.M., 1989, J. Less-Common Met. **149**, 121.

Vidali, M., P.A. Vigato, U. Casellato, E. Tondello and O. Traverso, 1975, J. Inorg. & Nucl. Chem. **37**, 1715.

Wang, G., and L. Miao, 1984, Gaodeng Xuexiao Huaxue Xuebao **5**(3), 281; Chem. Abstr. No. **101**, 182591c.

Zanello, P., A. Cinquantini, P. Guerriero, S. Tamburini and P.A. Vigato, 1986, Inorg. Chim. Acta **117**, 91.

Zang, Y., D. Wang, G. Wang and C. Zeng, 1987, Wuji Huaxue **3**(1), 66; Chem. Abstr. No. **107**, 227850d.

Zhang, R., and B. Tong, 1986, Huaxue Tongbao, **2**, 35; Chem. Abstr. **104**, 236176d.

Zhang, W., and G. Wang, 1986, Huaxue Xuebao **44** (12), 1261; Chem. Abstr. No. **106**, 148322j.

ERRATA

Vol. 7, ch. 52, Buschow

1. Page 425, (table A2), second column, the entry for $Er_{1-x}Ni_x$ for $x = 0.31$: delete decimal point, and align the value (620) in the column.

Vol. 12, Contents 1–11

1. Page xi, ch. 66, the second author's name is J.H. Weaver instead of "J.H. Waver".

Vol. 13, Contents of Volumes 1–12

1. vii: Page number for Errata "459" should be "453".
2. vii: Page number for Subject index "461" should be "455"
3. Page xi, ch. 66, the second author's name is J.H. Weaver instead of "J.H. Waver".

SUBJECT INDEX

www.ingramcontent.com/pod-product-compliance
Ingram Content Group UK Ltd.
Pitfield, Milton Keynes, MK11 3LW, UK
UKHW021436280726
14060UKWH00001BA/115